ECOSYSTEMS OF THE WORLD 5

TEMPERATE DESERTS AND SEMI-DESERTS

Edited by

Neil E. West

Department of Range Science and the Ecology Center
Utah State University
Logan, Utah (U.S.A.)

ELSEVIER SCIENTIFIC PUBLISHING COMPANY

Amsterdam — Oxford — New York 1983

ELSEVIER SCIENTIFIC PUBLISHING COMPANY
1 Molenwerf
P.O. Box 211, 1000 AE Amsterdam, The Netherlands

Distributors for the United States and Canada:

ELSEVIER SCIENCE PUBLISHING COMPANY INC.
52 Vanderbilt Avenue
New York, N.Y. 10017, U.S.A.

Library of Congress Cataloging in Publication Data
Main entry under title:

Temperate deserts and semi-deserts.

(Ecosystems of the world ; 5)
Includes bibliographical references and indexes.
1. Deserts. 2. Arid regions. 3. Desert ecclogy.
I. West, Neil E. II. Series.
GB611.T42 574.5'2652 80-28760
ISBN 0-444-41931-4

ISBN 0-444-41931-4 (Vol.5)
ISBN 0-444-41702-8 (Series)

Printed in The Netherlands

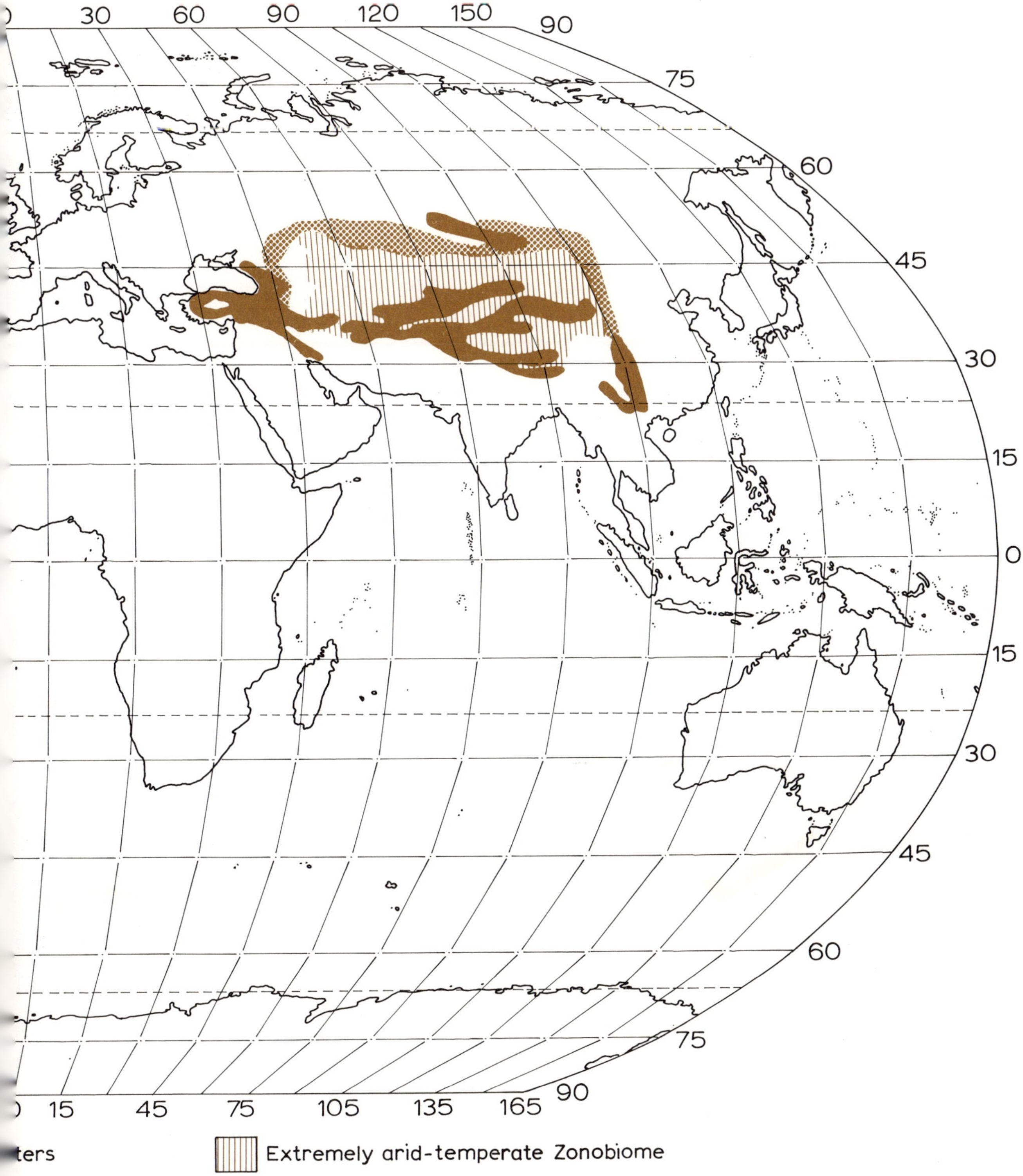

30
60
90
120
150
90
75
60
45
30
15
0
15
30
45
60
75
90
15
45
75
105
135
165
ters
Extremely arid-temperate Zonobiome

ECOSYSTEMS OF THE WORLD 5

TEMPERATE DESERTS AND SEMI-DESERTS

ECOSYSTEMS OF THE WORLD

Editor in Chief:

David W. Goodall

CSIRO Division of Land Resources Management, Wembley, W.A. (Australia)

I. TERRESTRIAL ECOSYSTEMS

A. Natural Terrestrial Ecosystems

1. Wet Coastal Ecosystems
2. Dry Coastal Ecosystems
3. Polar and Alpine Tundra
4. Mires: Swamp, Bog, Fen and Moor
5. Temperate Deserts and Semi-Deserts
6. Coniferous Forest
7. Temperate Deciduous Forest
8. Natural Grassland
9. Heathlands and Related Shrublands
10. Temperate Broad-Leaved Evergreen Forest
11. Mediterranean-Type Shrublands
12. Hot Desert and Arid Shrubland
13. Tropical Savannas
14. Tropical Rain Forest Ecosystems
15. Wetland Forests
16. Ecosystems of Disturbed Ground

B. Managed Terrestrial Ecosystems

17. Managed Grassland
18. Field Crop Ecosystems
19. Tree Crop Ecosystems
20. Greenhouse Ecosystems
21. Bioindustrial Ecosystems

II. AQUATIC ECOSYSTEMS

A. Inland Aquatic Ecosystems

22. River and Stream Ecosystems
23. Lakes and Reservoirs

B. Marine Ecosystems

24. Intertidal and Littoral Ecosystems
25. Coral Reefs
26. Estuaries and Enclosed Seas
27. Ecosystems of the Continental Shelves
28. Ecosystems of the Deep Ocean

C. Managed Aquatic Ecosystems

29. Managed Aquatic Ecosystems

PREFACE

This volume reviews current understanding of the most xeric ecosystems at mid-latitudes where freezing temperatures are consistently encountered each winter. To some these are the "cold deserts" and "shrub steppes". However, to many cold deserts is the label used for the drier areas near the poles, and shrub steppes are not necessarily restricted to mid-latitudes or temperate climates. Accordingly, we have chosen the less traditional but more precise definitions implied in the title.

After a short introduction by the editor, coverage of the huge Eurasian area is reviewed by Professor Heinrich Walter of Hohenheim University, Stuttgart (Federal Republic of Germany). His knowledge of the Russian literature was enhanced by cooperation from Professor L.E. Rodin of Leningrad. Professor E.O. Hilbig of Martin Luther University, Halle–Wittenberg (German Democratic Republic) gave assistance on the chapter concerning Mongolia. Professor E.O. Box of the University of Georgia (U.S.A.) translated Walter's German manuscript into English. Professor S.-W. Breckle of the University of Bielefeld (Federal Republic of Germany) contributed a chapter on the temperate deserts and semi-deserts of Afghanistan and Iran. This chapter was greatly strengthened by his recent travels and research there.

The section dealing with North America was synthesized by the editor, although numerous colleagues gave assistance in obtaining written reports and providing figures and counsel on interpretation of his conclusions.

Professor Alberto Soriano coordinated a team of scientists who summarized the understanding of the ecosystems of the more arid portions of Patagonia.

Finally, the editor has taken the liberty in the concluding chapter to compare the three portions of the globe having these kinds of environments. He offers speculations on why they are alike in general, but different in details. He also points out some topics where information is lacking and thus research opportunities are available.

We hope that our efforts at synthesis here will help current and future generations of students, researchers and land managers become more quickly familiar with these ecosystems and interest at least a few of them toward contributing their ideas toward improving our basic understanding and management of these areas.

The editor wishes to extend thanks to the writers of the various chapters for their willingness to contribute manuscripts and acceptance of some suggestions that were offered. He also wishes to thank Dr D.W. Goodall, Editor-in-Chief of the Series, for his constant encouragement and help. The editor also owes a considerable debt of gratitude to Ms Patricia Johnson for her careful reading of the final typescript and preparation of the indices. Finally, Mrs Janet Alderman should be congratulated for her considerable patience and care in typing the several drafts of the typescript.

NEIL E. WEST
Professor of Range Science
Utah State University

LIST OF CONTRIBUTORS

ELGENE O. BOX
Department of Geography
University of Georgia
Athens, Ga, 30601 (U.S.A.)

S.-W. BRECKLE
Faculty of Biology
University of Bielefeld
Bielefeld (Federal Republic of Germany)

PABLO CANEVARI
Argentinian Museum of Natural
Sciences "Bernardino Rivadavia"
470 Angel Gallardo Avenue
Buenos Aires (Argentina)

MARCELO CANEVARI
Argentinian Museum of Natural
Sciences "Bernardino Rivadavia"
470 Angel Gallardo Avenue
Buenos Aires (Argentina)

JOSE M. GALLARDO
Argentinian Museum of Natural
Sciences "Bernardino Rivadavia"
470 Angel Gallardo Avenue
Buenos Aires (Argentina)

W. HILBIG
Martin-Luther University
Halle–Wittenberg (German
Democratic Republic)

ROLANDO J.C. LEÓN
Department of Ecology
Faculty of Agronomy
University of Buenos Aires
Buenos Aires (Argentina)

ARRIGO A. MARCOLIN
National Institute of Agricultural
Technology
San Carlos de Bariloche,
Río Negro (Argentina)

CHIARA P. MOVIA
Department of Ecology
Faculty of Agronomy
University of Buenos Aires
Buenos Aires (Argentina)

MAURICIO RUMBOLL
Argentinian Museum of Natural
Sciences "Bernardino Rivadavia"
470 Angel Gallardo Avenue
Buenos Aires (Argentina)

ALBERTO SORIANO
Department of Ecology
Faculty of Agronomy
University of Buenos Aires
Buenos Aires (Argentina)

JORGE A. VALLERINI
Fuera de Radio
San Carlos de Bariloche,
Río Negro (Argentina)

WENCESLAO G. VASINA
Argentinian Museum of Natural
Sciences "Bernardino Rivadavia"
470 Angel Gallardo Avenue
Buenos Aires (Argentina)

WOLFGANG VOLKHEIMER
Argentinian Museum of Natural
Sciences "Bernardino Rivadavia" and
CONICET
470 Angel Gallardo Avenue
Buenos Aires (Argentina)

HEINRICH WALTER
Botanisches Institut der Universität
7000 Stuttgart–Hohenheim 70
(Federal Republic of Germany)

CONTENTS

Section IV. South American deserts and semi-deserts

Section V. Comparison and contrast between continents

Chapter 1

APPROACH

NEIL E. WEST

DEFINITIONS

The purview of this volume is the ecosystems that dominate the arid and semi-arid regions of the world's temperate zones. "Arid" pertains to those landscapes that receive less than 200 mm average total precipitation per year. "Semi-arid" includes those areas that receive between 200 and 500 mm average annual precipitation. The temperate zone is characterized by considerable seasonal variation in temperature. A period of cold temperatures dependably occurs, when some precipitation occurs in the solid state and soil moisture is frozen. The majority of landscapes where such climates occur are in the interior of continents where, without the moderating influences of oceans, extreme diurnal temperature fluctuations are found. A location in the interior of the continents is not a requirement for producing the desert shrublands and semi-desert shrub steppes. South America, however, has the only exception covered in this volume.

The above definitions follow the worldwide classification of ecosystems proposed by Walter and Box (1976). Our focus is on Zonobiome VIIa, the desert subzonobiome of arid temperate regions. This comprises about 4% of the world's land surface. We will also mention related orobiomes and pedobiomes that occur both within and adjacent to this zonobiome.

The concept of desert as being any area devoid of vegetation, irrespective of climate, is not used here. All of the ecosystems which are dealt with have shrub dominance. The dominance of shrubs declines as effective soil moisture increases in the semi-arid zone. The relative importance of grasses and thus the influences of fire and grazing increase along this same gradient. We therefore have to interpret the successional status of the shrub–grass mixtures and decide where grasses would prevail if fires were frequent and livestock were removed. Making a hard and fast demarcation between steppe (low grasslands) and shrub steppe is probably impossible, except on maps of the very smallest scale. It is inevitable that there will thus be some overlap in coverage between this and Volume 8 in this Series, dealing with grasslands. Hopefully, this discussion from two points of view will help the reader understand better what favors shrub dominance versus that of grass. This topic is of such practical importance that the small amount of overlap appears well justified.

DISTRIBUTION

A map of the worldwide distribution of the kinds of ecosystems under scrutiny here is given on the inside covers. Three areas of the world are involved: Central and Middle Asia, the interior western United States in North America, and part of the Patagonian region of Argentina (Table 1.1).

The Eurasian–Middle Eastern portion has, by far, the greatest extent (about 84% of the total area) (4.9×10^6 km^2) and diversity, and is thus accorded a high proportion of the pages here. The understanding of the ecosystems in this vast area with a poor transportation network and few scientists has rarely been advanced beyond the descriptive stages. An exception is the Karakum region, of which more can be said (Ch. 6).

The temperate arid and semi-arid portions of North America constitute about 0.8×10^6 km^2. This is about 8% of the United States, about 3% of the continent and 13% of the world's temperate

TABLE 1.1

Area of temperate deserts and semi-deserts on each continent and for the world with percentages of total world temperate desert and continental areas

Continent, subdivision	Total area ($\times 10^6$ km^2)	Percentage of world's temperate deserts and semi-deserts	Percentage of the continent
EURASIA			
Kazakho-Dzungarian deserts and semi-deserts	3.59		
Central Asian deserts and semi-deserts	2.46		
High, cold mountain deserts and semi-deserts	1.11		
Irano-Turanian deserts and semi-deserts	0.53		
Total desert and semi-desert for continent	4.92[1]	84	15
NORTH AMERICA[2]			
Great Basin–Colorado Plateau sagebrush	0.18		
Intermountain sagebrush steppes	0.45		
Intermountain salt-desert shrublands	0.17		
Colorado Plateau–Mohavian blackbrush	0.01		
Southeast Utah Grama–Galleta–Threeawn shrub steppe	0.005		
Total for continent	0.78	13	3
SOUTH AMERICA			
Patagonian semi-deserts	0.15		
Total for continent	0.15	3	1
Total	5.85	100	

[1]Note totals for subdivision will not add to totals given for continents since different map bases were used. The climatic classifications of Walter and Box (1976) were used for the totals for continents. The subdivisions of Eurasia depicted in Fig. 2.1 were used. Küchler's (1970) potential natural vegetation map units were the basis of the North American subdivisions.
[2]Sagebrush: *Artemisia* spp.; blackbrush: *Coleogyne ramosissima*; grama: *Bouteloua* spp.; galleta: *Hilaria jamesii*; threeawn: *Aristida* spp.

arid and semi-arid landscape. Probably because of the higher relative density of scientists, especially in the last decade, much more can be said about the functional processes of these ecosystems than their counterparts elsewhere in the world. They thus receive more attention on a pages per hectare basis.

The temperate semi-arid portion of South America is smallest both on absolute (about 149 000 km^2) and relative (3%) bases. Only a rudimentary, descriptive understanding of these ecosystems is available at this time. Very few scientists have worked in these areas. The treatment to follow reflects the paucity of ecological research in South America. Hopefully, the comparisons with other continents afforded by this and other volumes in this series will help frame further research and encourage administrative support for accomplishing it.

ORGANIZATION OF THE VOLUME

We will first review what is known about the ecosystems of temperate arid and semi-arid environments of each continent. The progression will be ordered by areal extent, with Eurasia first, North America second, and South America third. I will then attempt to summarize by first drawing out the similarities between the ecosystems of each continent. This will be followed by a highlighting of differences, and a rationale for their existence. The volume will end with my assessment of the deficiencies in the understanding of these ecosystems and thus opportunities for further research.

REFERENCES

Küchler, A.W., 1970. Potential natural vegetation (map at scale of 1:7 500 000). In: U.S. Geol. Survey, *The National Atlas of the U.S.A.* U.S. Government Printing Office, Washington, D.C., pp. 90–91 (revision of 1964 map at scale of 1:3 168 000).

Walter, H. and Box, E.O., 1976. Global classification of natural terrestrial ecosystems. *Vegetatio*, 32: 75–81.

Chapter 2

OVERVIEW OF EURASIAN CONTINENTAL DESERTS AND SEMI-DESERTS[1]

H. WALTER and E.O. BOX

INTRODUCTION

The vegetation of the deserts and semi-deserts of Eastern Europe, Middle and Central Asia has been described in a German version by the first author (Walter, 1974). The present contribution emphasizes those aspects which are most important for ecosystems research and includes material from the recent Russian literature. L.E. Rodin and S.V. Sveshnikova of the Komarov Botanical Institute, U.S.S.R. Academy of Science at Leningrad, along with their colleagues, have looked through the first manuscript. They provided very valuable suggestions and sent additional recent materials, which have been included in this presentation. We would like to express to them our most special thanks for this support. Their assistance was especially valuable since the regions treated are not known to us from personal observation.

We are most grateful to S.-W. Breckle for compiling the chapter on Afghanistan, a country in which he was active scientifically for several years.

E.O. Box accepted the task of translating the German manuscript into English and overseeing the corrections during the printing.

The treatment of the individual regions of the vast Eurasion desert area is not uniform because the available source materials are of a quite varied nature. Regions which have been studied in detail, such as the Karakum Desert and the eastern Pamir, appear alongside regions which have been studied very little ecologically, such as Central Asia and Afghanistan. In these latter cases one must be content with general descriptions of environmental conditions and of the vegetation.

Most of the material summarized here appears in this volume for the first time in English, having remained unknown in the West until now.

OVERVIEW

The extension of the Eurasian land mass over almost 200° of longitude and the screening off of its northern part from the south by the immense Himalayan mountain system and other ranges causes the formation of a vast, continental arid and temperate zonobiome, which stretches from the Danube delta in the west across 100° of longitude to Manchuria in the east. This zonobiome (ZB VII) can be divided into two subzonobiomes (Walter, 1976; Walter and Box, 1976; Walter, 1979):

(1) the temperate semi-arid subzonobiome represented by the semi-arid Eurasian steppes,

(2) the temperate arid subzonobiome represented by the Asian deserts.

The transition between the two is formed by the semi-desert subzono-ecotone, in Europe occurring especially in the Caspian Lowland.

In the present volume only the semi-deserts and deserts will be emphasized. They cover the entire southern part of the zonobiome, beginning in the west in the Aral–Caspian region and extending to the east to the Great Wall of China, along the great bend of the Huang Ho in the Ordos region. The Asian deserts are distinguished from subtropical deserts by their cold winters with often very low temperatures produced by the East Siberian winter high-pressure system.

The western part of these deserts is called Middle Asia in the Russian literature, with only the eastern

[1] Manuscript completed November, 1977.

part called Central Asia. Indeed, the two parts are so different from each other, both climatically and floristically, that it is appropriate to follow this division and to discuss them separately. The precipitation regime, in particular, is quite different in the deserts of Middle Asia from that in the deserts of Central Asia. The boundary between the two regions is formed by Dzungaria, which is inclined more toward the west, and the great mountain systems of the Tien Shan (Tyan Shan) and the Pamiro-Alay (Fig. 2.1).

In Middle Asia the remnants of eastward-moving storms from the Atlantic are still noticeable. Precipitation falls more in winter, with a maximum in March. The soil is thoroughly wetted in the springtime, and the spring vegetation is well developed. Precipitation amounts decrease in this region, without exception, as one goes from west to east.

In Central Asia, on the other hand, frontal and cyclonic storms from East Asia move to the west in summer. The winter and spring months are extremely dry. Springtime vegetation is completely lacking, and plant growth does not begin until after the first summer rains. Precipitation amounts decrease in this region from east to west and reach a minimum in the almost completely rainless desert of the Takla Makan (Tarim Basin), on the eastern slopes of the Tien Shan.

While the deserts of Middle Asia extend to the south toward Afghanistan and Iran, the entire southern part of Central Asia is taken up by the high plateau of Tibet and Pamir, the average altitude of which is around 4000 m. It differs from the lower-lying northern deserts by having a very short summer, with frost occurring throughout the year. The transition is formed by the Tsaidam Desert, with an altitude of around 3000 m.

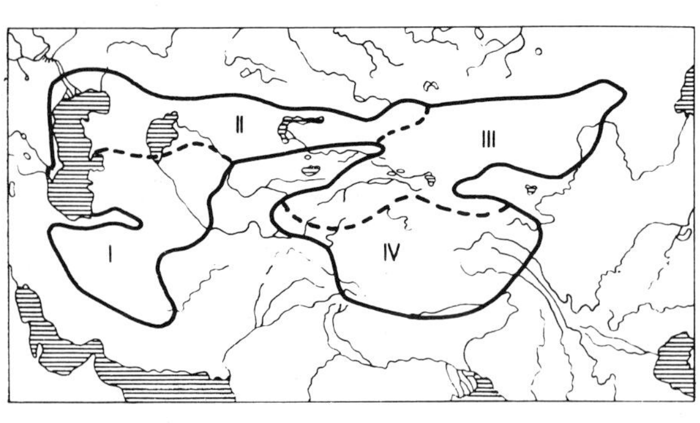

Fig. 2.1. Eurasian desert regions of the temperate climatic zone (from Petrov, 1966–67). *I* = Irano-Turanian; *II* = Kazakho-Dzungarian; *III* = Central Asian deserts; *IV* = cold high-mountain deserts. High-mountain ranges lie between the various desert regions (see map inside covers).

Of the various Asian deserts, those belonging to the U.S.S.R. in Middle Asia are well studied, those in Iran and Afghanistan less so. The last blank spaces on the maps of Central Asia disappeared only in the present century.

The first reports of Inner Asia came to Europe by way of the silk trade carried on in the thirteenth and fourteenth centuries over the land routes between Europe and the East. The reports of the Venetian Marco Polo (Agakhanyantz, 1972) are especially well known. Actual geographical exploration, however, began only in the second half of the previous century, first by Russian explorers from the north, including Semenov in 1857, Severzov in 1864 and 1878, Fedchenko between 1868 and 1871, Przevalski between 1870 and 1878, and Potanin between 1876 and 1886. These were followed by explorers from other nations, coming mostly from the south and west, including Richthofen (1868–72), Carey (1885–86), Younghusband (1887–90), Bonvalet (1887), Dutreuil de Rhins (1891–94), Littledale (1895), and Sven Hedin (1894–97, 1899–1902). Expeditions became more numerous in the first three decades of this century, and will not be mentioned individually. As a result of all this, the climate, geomorphology, geological structure, and also the ethnic groups of Central Asia are largely known today.

For the understanding of ecological systems, however, knowledge is not yet sufficient. Only recently have functional aspects of the vegetation been considered more closely, whereas earlier the focus was mainly on the flora. The most important of these recent vegetation studies is the two-volume work by Petrov (1966–67).

Two regions in which research stations for continuing studies are available, the Karakum Desert in Middle Asia and the eastern Pamir region in Central Asia, have been so thoroughly studied that one can make rather precise statements about the ecosystems. Similar studies have recently been started in Kazakhstan. In the other regions one must limit oneself to abiotic factors and descriptive reports of the vegetation.

The most important biotic components of ecosystems are the plants, that is, the producers through which the material cycles and energy flows are initiated. The consumers play, quantitatively, a very unimportant role in ecosystems. Two material cycles in ecosystems must always be distinguished, a short and a long cycle (Fig. 2.2). The above-ground consumers are completely excluded from the short cycle, which carries 90 to 99% of the energy flow. Plant litter enters directly into the soil and is decomposed there by the decomposers, including soil fauna at the beginning of the process. Only in the long cycle are materials passed through many food chains with very numerous and different consumers, whereby different regulatory circuits, in the cybernetic sense, with negative feedback, are set up by the selective eating habits of both herbivores and carnivores. Competition between the plant species which make up a community is influenced by these regulatory circuits, and it is through these interrelationships that the maintenance of the dynamic steady-state of the ecosystem is guaranteed. The significance of the long cycle is, thus, regulatory and not a result of quantitative relationships. It should be an urgent task of ecosystem research to clarify these various feedback patterns, whereas the very difficult quantitative description of the individual food chains and of secondary production is not so urgent.

Ecosystem equilibrium is easily destroyed by the intervention of man, for example by the extermination of carnivores or hunting of large ungulates. Still more disruptive is the interference of man through the introduction of his domesticated animals, which are quite foreign to most ecosystems. Precisely this factor is especially important in deserts with low primary production, since the desert is almost universally used as pasture land. For this reason, the Asian deserts, which are only thinly settled but which have served for millennia as pasture land for nomads, are seldom found in their original, natural condition.

The assessment of primary production in deserts is especially difficult. For ecosystems which are in equilibrium with their environments, the mean primary production over many years (i.e., the annual newly formed organic matter, which we call "growth") must theoretically equal the annual litter (i.e., the amount of biomass which dies annually). The remaining phytomass must be constant over many years.

Because the climatic elements vary so much from

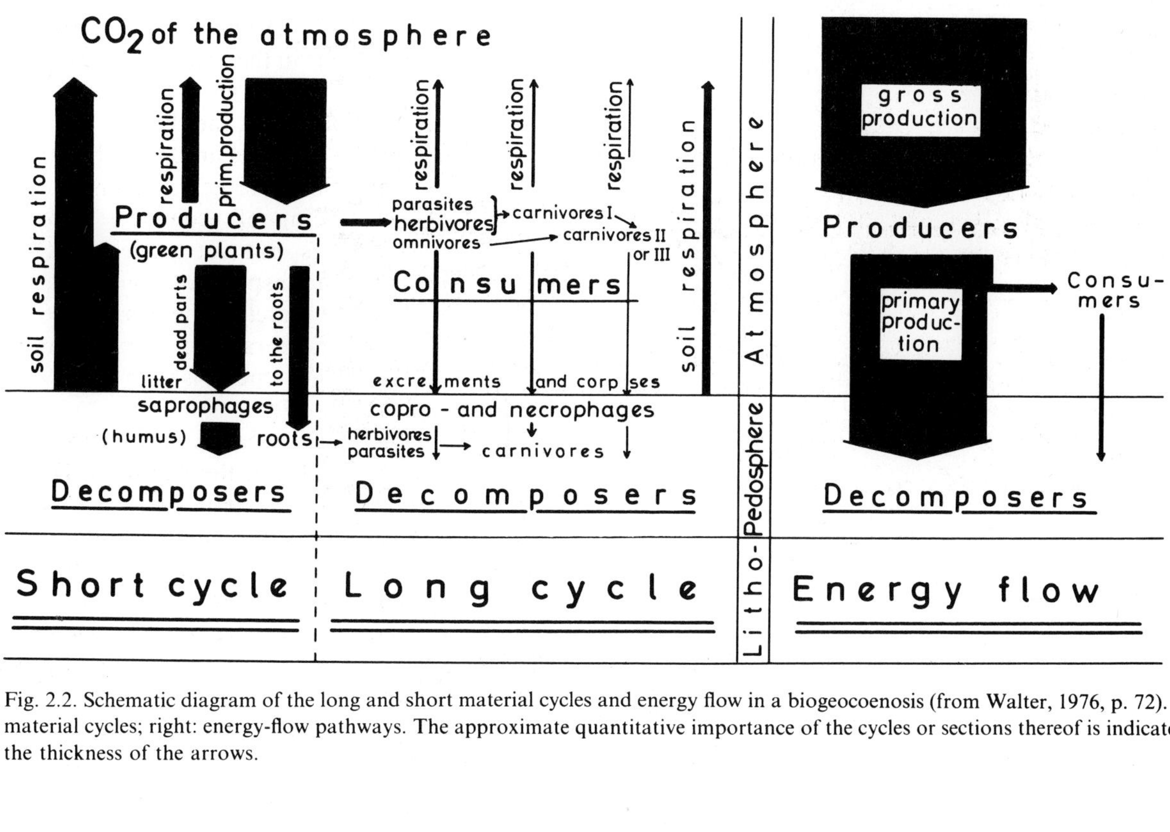

Fig. 2.2. Schematic diagram of the long and short material cycles and energy flow in a biogeocoenosis (from Walter, 1976, p. 72). Left: material cycles; right: energy-flow pathways. The approximate quantitative importance of the cycles or sections thereof is indicated by the thickness of the arrows.

year to year in dry areas, however, the situation is much more complicated. For example, rainy years often immediately follow dry years. In a good rainy year the growth will be quite large and the annual litter relatively small. In a dry year, though, growth will be slight and the litter much greater. One can best estimate primary production, therefore, by considering it to be half of the annual growth plus litter. For this reason (as Rodin explained to us on March 29, 1977), wherever possible both annual growth (Russian *prirost*) and annual litter (Russian *opad*) are given in the tables, rather than annual primary production.

Over an experimental area of 1 ha all shrubs were closely examined over the course of a year for new growth and for newly dead parts. The litter for the year was carefully collected, though it was not always easy to distinguish it from older litter.

The portion of the annual growth consumed by herbivory could, in general, not be determined. All data represent actually determined values. One must naturally keep in mind that the dying parts begin decomposing immediately after separation from the mother plant, so that the parts which were collected later have already lost a portion of their dry weight.

In addition, data on the weight of all dead parts (i.e., the *mortmasse*) are given in most tables. The above-ground dead biomass consists of the annual litterfall plus the already partially decomposed soil surface litter, including dead branches and stems, plus the dead parts within the crowns and the standing dead stems (Russian *otpad* or *sukhostoi*). The below-ground dead biomass includes all dead roots, as well as dead rhizomes, tubers, and bulbs.

In both the Middle Asian and the Central Asian deserts one can distinguish individual natural geological–geographical units, which we can also take, for scientific purposes, as ecological units. These are usually composed of many, but at least one, biogeocoenosis complex and should simply be denoted as biomes, or true biomes (eu-biomes), of

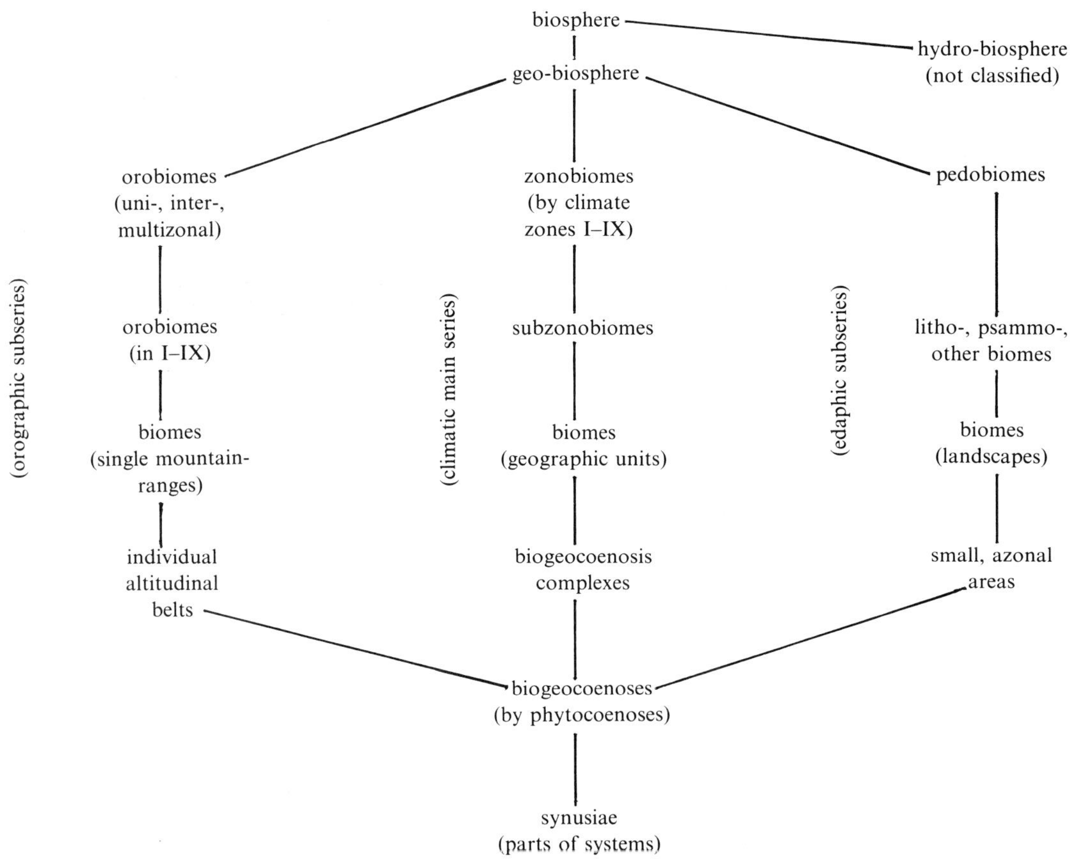

Fig. 2.3. Outline of a classification scheme for the world's ecosystems (Walter and Box, 1976).

the larger subzonobiome. These biomes are the actual object of this study. Beginning in the west, the following classification system arises:

A. Middle Asian biome complex

1. Caspian Lowland Biome (with northern and southern subbiomes) in the region of the retreating Caspian Sea.
2. Kazakh Desert Biome, primarily composed of the remnant plateau with many insular mountains, plus the Betpak-Dala Desert and the northern sand deserts Taukum and Muyunkum.
3. Ustyurt Plateau Desert Biome, with rocky soil.
4. Southern Sand-Desert Biome, with the subbiomes Kyzylkum and Karakum, of which the last, as best-studied, will be treated in detail.
5. Loess-Desert Biome (northern and southern), which reaches over into Afghanistan.
6. Biomes of Afghanistan.

B. Biome Complexes of the Iranian region

C. Central Asian biome complexes of lower-lying deserts

1. Dzungarian Biome, as transition from Middle Asia.
2. Gobi Desert Biome.
3. Pei Shan Desert Biome.
4. Tarim Basin Biome, with the Takla Makan.
5. Kansu (Hessi) Corridor Biome.
6. Ala Shan Desert Biome.
7. Ordos Desert Biome, in the bend of the Huang Ho.

D. Biomes of the cold Central Asian deserts above 3000 m

1. Tsaidam Desert Biome, as transition from C to D.
2. Tibetan Plateau Biome.
3. Eastern Pamir Biome, which is better studied and will be treated in more detail than will western Pamir.

Within these biomes, biogeocoenosis complexes with azonal vegetation, for which the soil is of greater importance than the climate, can also be distinguished. Such units are called pedobiomes (Walter, 1976; Walter and Box, 1976). Within the subzonobiome the vertical dimension must also be considered, that is, the biogeocoenosis complexes of mountains with different altitudinal belts, for which we use the term orobiome (Walter, 1976, 1979; Walter and Box, 1976). Our classification of ecological units is based on the scheme in Fig. 2.3.

For explanation of climate diagrams, see Walter (1979, pp. 25–31) or Walter et al. (1975).

REFERENCES

Agakhanyantz, O.E., 1958. Short survey of the Badakhahan vegetation. *Sb. Statei Tadzhik. Fil. Geogr. Obshsh. S.S.R., Stalinabad*, 99: 47–67 (in Russian).

Agakhanyantz, O.E., 1972. Der Pamir im Buch des Marco Polo. *Peterm. Geogr. Mitt.*, 116: 53–57.

Petrov, M.P., 1966–67. *Deserts of Central Asia*. Nauka, Leningrad, Vol. 1: 274 pp.; Vol. 2: 288 pp. (in Russian).

Walter, H., 1974. *Die Vegetation Osteuropas, Nord- und Zentralasiens. Band VII, Vegetationsmonographien der einzelnen Großräume*. Fischer-Verlag, Stuttgart, 452 pp.

Walter, H., 1976. *Die ökologischen Systeme der Kontinente (Biogeosphäre)*. Fischer-Verlag, Stuttgart, 131 pp.

Walter, H., 1979. *Vegetation of the Earth and Ecological Systems of the Geo-Biosphere*. Springer-Verlag, New York, N.Y., 2nd ed., 274 pp.

Walter, H. and Box, E., 1976. Global classification of natural terrestrial ecosystems. *Vegetatio*, 32: 75–81.

Walter, H., Harnickell, E. and Mueller-Dumbois, D., 1975. *Climate-Diagram Maps*. Springer-Verlag, Berlin.

Chapter 3

CASPIAN LOWLAND BIOME[1]

H. WALTER and E.O. BOX

CLIMATIC CONDITIONS

The climate of the Caspian Lowland represents the transition from the steppe climate of the southern part of Eastern Europe to the desert climate of Middle Asia. It is therefore, an ecotone, that is, a region of ecological tension in which one vegetation type, the steppe, is replaced by another type, the desert. We can speak of the ecotone as a semidesert. In such an ecotone both vegetation types occur side-by-side under the same macro-climatic regime. Micro-climatic conditions and/or differing edaphic conditions on different sites give the advantage to one vegetation type or the other.

In his classic works, of which we cite those appearing in German periodicals (1925/26, 1927, 1930) and the large Russian summary (1936), B. Keller[2] emphasized the peculiarity of the semidesert and regarded it as a special zone. On the other hand, even in the Russian literature (Levina, 1961), one sees it today described more often as a transition zone.

The climate of the Caspian Lowland becomes increasingly arid from north to south and from west to east. This is seen from the climate diagrams of stations in the region (Fig. 3.1). Malyy Uzensk, Ural'sk and Aktyubinsk lie just outside the lowland in the dry steppe with 300 mm annual precipitation and a mean annual temperature of about 5°C. Akhtubinsk and Verkhni Baskunchak on the bend in the Volga, as well as the more eastern and northern Uil'skoye, are typical for the northern border of the lowland: the precipitation amounts to only about 250 to 260 mm, and the summer drought is quite pronounced.

Astrakhan at the mouth of the Volga, Gur'yev at the mouth of the Ural River, and Fort Shevchenko (Fort Aleksandrovsk) on the Mangyshlak Peninsula exhibit a desert climate, which is ameliorated slightly by their locations on the sea coast. The annual precipitation is 156 to 177 mm, and the drought period quite outweighs the barely noticeable humid season.

In winter the Siberian high-pressure system extends either into the Caspian Lowland or close enough to it, so that northeast winds bring cold Siberian air into the region. The winters are very cold, with minima from −30°C on the coast to −40°C inland. Winter thaw periods are rare. Summers, on the other hand, are hot, with July means around 25° to 26°C and maxima up to 40°C. Dust storms are especially bad in summer, being brought by southeast winds which pick up the heated surface dust. These *sukhovey* (Russian: dry winds) can cause heavy damage to cultivated land. The potential evapotranspiration, estimated as 1400 mm annually, exceeds the annual precipitation almost tenfold (see Walter, 1974). The climate is thus typically continental and arid.

GEOLOGY AND GEOMORPHOLOGY

The vast plain of the Caspian Lowland includes the entire region north of the Caspian Sea from the Terek Delta on the northwest side to the Mangyshlak Peninsula on the northeast side. The lowland is delimited in the north by the steep descent of the Obshchi Syrt (*syrt*: plateau) north of Ural'sk, in the west by the Volga bend at Volgograd and the Yergeni Hills, which fall off sharply by

[1] Manuscript completed November, 1977.

[2] The first work of Keller and Dimo appeared as early as 1907.

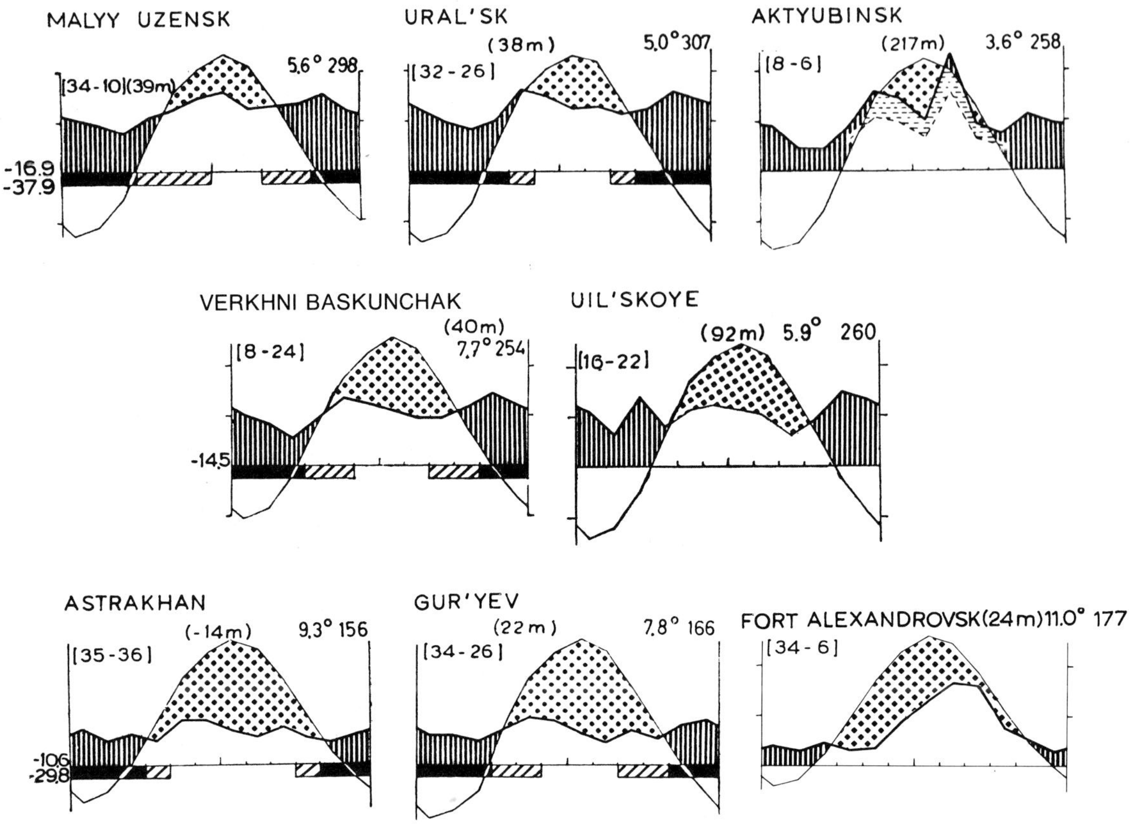

Fig. 3.1. Climate diagrams from stations in the Caspian Lowland, the locations of which are shown in Fig. 3.2 (from the *Klimadiagramm Weltatlas* of Walter and Lieth): Malyy Uzensk, Ural'sk, Aktyubinsk, Verkhni Baskunchak, Uil'skoye, Astrakhan, Gur'yev, and Fort Shevchenko (= Fort Alexandrovsk).

200 m toward the lowland, and in the east by spurs of the Ural Mountains and the Ustyurt Plateau (Fig. 3.2). At the end of the Tertiary the lowland was subjected to repeated flooding caused by tectonic movement or by greater water flow into the Caspian Sea, as occurred three times during the Pleistocene, caused by the melting of the ice cap at the end of each glacial period. The lowland was under water as recently as 10 000 years ago. At the time of the last major flooding the water level was 78 m higher than it is today, so that there was a connection with the Azov Sea through the Manych Basin. The water level then fell to 50 m below sea level, so that, at the time of the last flooding, it approached the current situation.

The surface of the lowland slopes so slightly from the north to the Caspian Sea that the descent can scarcely be noticed. The surface is above global sea level in the north and sinks slowly below sea level to the surface of the Caspian Sea, which in 1929 was 26 m below sea level. In the following ten to fifteen years the water level dropped 2 to 2.5 m more, and continues to sink. The shoreline has, as a result, advanced 25 to 60 m seaward. The water surface of the Caspian Sea was 420 000 km^2 in 1869 and has since decreased by about 10% to 370 000 km^2. This sinking has accelerated recently as a result of the regulation of the Volga and the construction of seven large reservoirs. The Kuybyshevskoye Reservoir is 300 km long and has a surface area of 6450 km^2, with a total volume of 53.3 km^3 and a mean depth of 8 m (Fig. 3.3A).

As a result of this shallow geometry, evaporation was increased. In addition, a part of this water is used for the extension of irrigated areas, so that the runoff of the Volga continually decreases. Serious problems have arisen both for shipping and the harbors on the Caspian Sea, and also for fishing

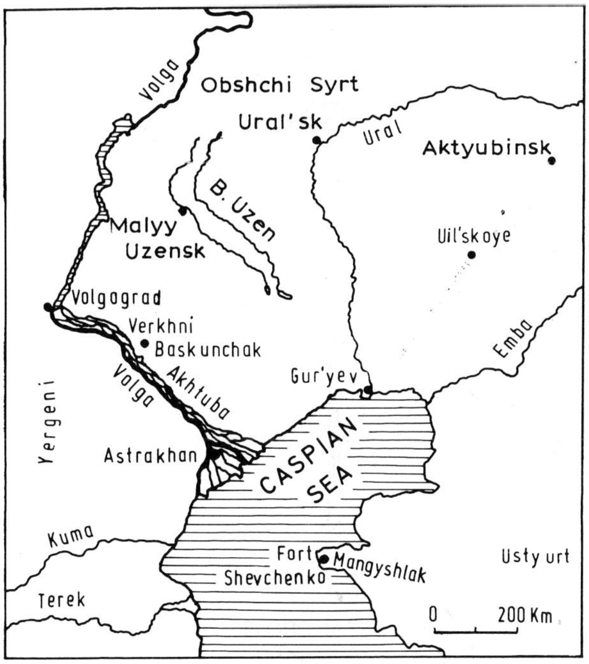

Fig. 3.2. Overview of the various regions of the Caspian Lowland, with locations of the climate stations from Fig. 3.1.

and the spawning grounds of many fish. This is especially true in the northern part of the Caspian Sea, which is only 4 m deep.

The Volga Delta, whose shoreline is 250 km long, is also expanding more and more into the Caspian Sea (Fig. 3.3A, B and C). A sharp coastline does not exist, since winds from the sea can cause water buildups which raise the level of the sea by as much as 2 m.

The salt content of the sea water varies greatly in the various coastal sections. In the east the concentration is 6 to 13‰, while in the vicinity of the large deltas it is only 0.57‰ — almost freshwater. Primary ecological succession differs correspondingly. For the coastal section between the Kuma and Terek deltas, to be discussed later, the salt content was 2‰.

The newly emerging land is colonized by *Phragmites communis*, which is cut and used for the production of cellulose. The reeds are replaced landwards by *Puccinellia gigantea*. Since the ground is constantly wet, a light salinization results from evaporation from the ground surface. As the groundwater table sinks deeper and capillary water transport is reduced, ephemerals appear. The vegetation of the older terraces has a semi-desert character. For the region between the Volga and the Terek rivers, the following zonation is characteristic (Miroshnichenko, 1966; Miroshnichenko and Miroshnichenko, 1972):

(1) 0 to 300 m from the coast: reed marsh composed of *Phragmites communis*, *Scirpus lacustris*, *S. tabernaemontani*, *S.* (*Bolboschoenus*) *maritimus*, *Typha angustifolia*, *T. laxmannii*, and some *Eleocharis parvula*[1]. The annual production in the outer 50-m zone is 0.83 t ha^{-1}, elsewhere only 0.35 t ha^{-1}.

(2) 300–800 m from the coast: marsh vegetation composed of *Alisma lanceolata*, *Najas marina*, *N. minor*, *Potamogeton filiformis*, *Ruppia maritima*, with *Butomus umbellatus*, *Sagittaria sagittifolia*, and others. The annual production in this zone, which is still shallowly covered by water, is 0.07 t ha^{-1}.

(3) First dry terrace (dry since 1955), still annually flooded by freshwater in springtime: *Phragmites* grows especially luxuriantly on salt-free marsh soils in a band of 2 to 4 km. The annual production of the above-ground phytomass reaches 12.0 t ha^{-1} of dry matter. Toward the outer extremity of this zone the groundwater table sinks to about 1 m, the production decreases to 7.3 t ha^{-1}, and *Salicornia herbacea*, *Suaeda maritima* ssp. *salsa*, and *Limonium gmelini* appear as undergrowth. There is evidence of readily soluble salts in the soil, SO_4^{2-} more than Cl^-.

(4) The next roughly 2.3 km wide zone is occupied by *Puccinellia gigantea*. There is more Cl^- in the soil here than SO_4^{2-}. This section emerged from the sea about 1950. The annual production decreases landwards from 2.8 to 1.25 t ha^{-1}, the height of the *Puccinellia* decreases from 50 to 22 cm, and the number of the accompanying halophytes (*Salsola soda* and *Spergularia salina*) increases. *Phragmites* is only 40 cm tall.

(5) Another 0.5 to 5 km wide belt has emerged since 1963 and is covered only with ephemerals: *Bromus japonicus*, *Descurainia sophia*, *Vulpia ciliata*, *Eremopyrum triticeum*, *Lepidium perfoliatum*. The

[1] Most species lists in Chapters 3 to 9 are arranged as follows: in communities with several strata, those of the upper stratum are listed first, followed by successively lower strata. In each stratum, the species are in taxonomic order, according to the main floras for the regions in question.

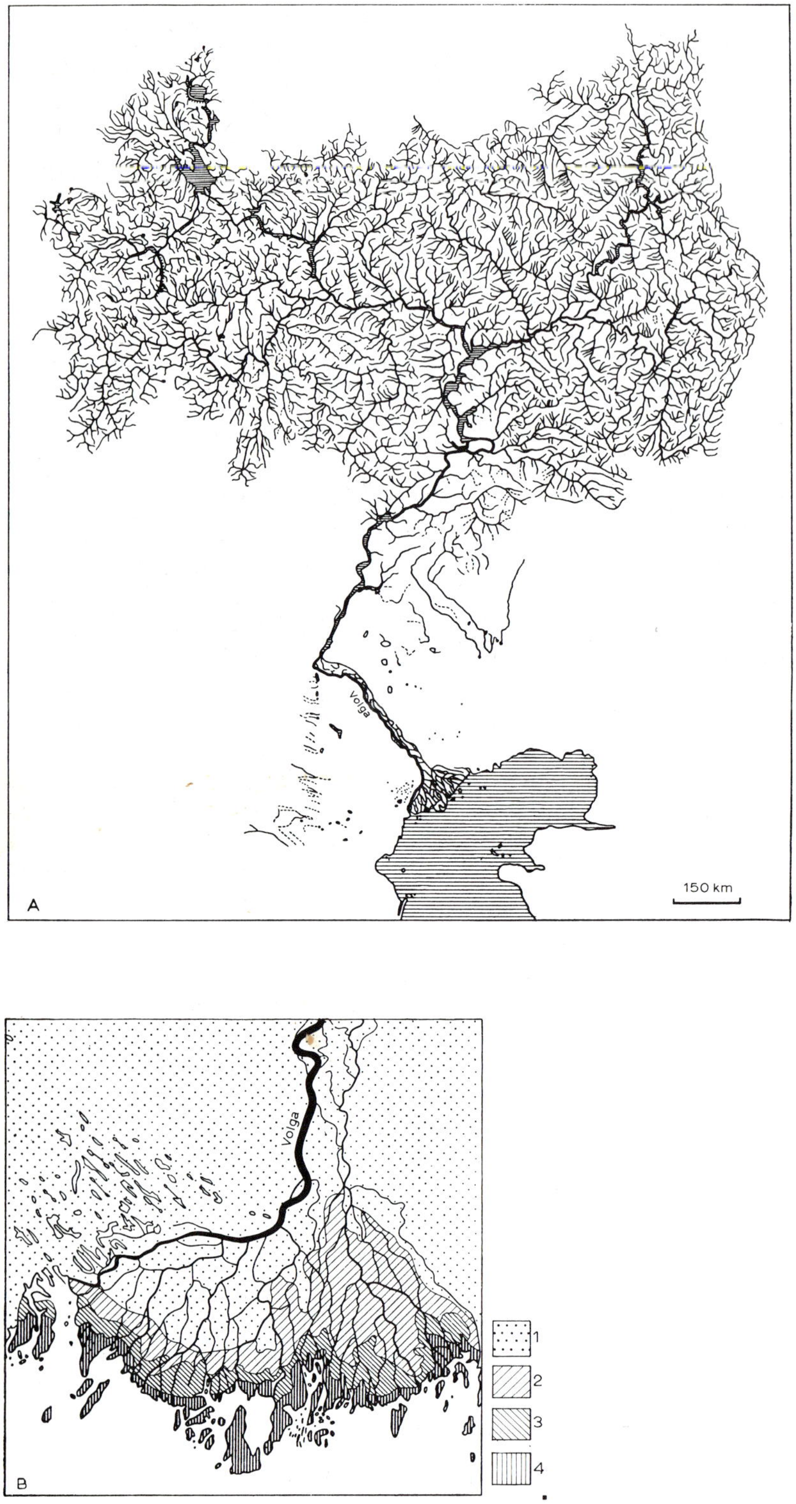

Fig. 3.3A, B.

Fig. 3.3.A. Catchment area of the Volga (with tributaries over 15 km in length), which flows into the Caspian Sea. The arid region of the Caspian Lowland is recognizable by the lack of tributaries flowing into the Volga (from Anonymous, 1968). B. The delta region of the Volga, which is continually extending itself into the Caspian Sea. New land areas: *1* = up to 1810; *2* = 1811 to 1873; *3* = 1874 to 1927; *4* = 1928 to 1950 (from Anonymous, 1968). C. Aerial photograph of the western main Volga arm near the outlet in the Caspian Sea (received from W.K. Leimbach). The photograph is oriented with north to the right. On the west shore the settlement Oranzherey, farther west the parallel "Yemeni" hills.

annual production is 0.05 to 0.2 t ha^{-1}.

(6) Second terrace, with groundwater table at a depth of 3 m: semi-desert vegetation with *Artemisia monogyna*, *Kochia prostrata*, *Poa bulbosa*, *Atriplex verrucifera*, and *Aeluropus litoralis*. The annual production is 1.46 t ha^{-1} or less. On the third terrace, with *Agropyron sibiricum* and *Kochia prostrata* on brown sandy loam, the annual production is only 0.42 to 0.63 t ha^{-1}.

The age of the plant cover over the entire Caspian Lowland also decreases from north to south and generally exhibits corresponding successional stages. The particulars of the successional stages, however, are rather complicated, since the recession of the sea has been interrupted intermittently by partial floodings. In addition, the Volga and the Ural rivers continually extend their deltas further to the south.

The Volga is the largest river in Europe, with a length of 3700 km and a catchment area of 1380 km^2 (Fig. 3.3A). Below the bend at Volgograd, the river divides into several arms and forms the 350 km long and 15 to 35 km wide Volga–Akhtuba floodplain,[1] which with the delta represents an oasis of 13 500 km^2.

The earlier channels and backwaters of both the Volga and the Ural can still be seen in the region. This is also true for the river valleys to the north (the Greater and Lesser Uzen', among others) which, between Kamyshin on the Volga and Ural'sk on the Ural, flow down from the Obshchi Syrt (100 to 200 m above sea level) toward the south and either disappear or run off into endorheic, salty lakes, as in the Kamysh–Samara district.

Permian and Jurassic loams were involved in the formation of the Obshchi Syrt, and these show a

[1]The Akhtuba is the main (leftmost) arm of the lower Volga.

certain salt content. The same is true of the heavy loams deposited from the Caspian Sea, which, especially in the south, alternate with sands deposited by the rivers (see Larin et al., 1954b). The dominance of the heavy loams in the northern half of the Caspian Lowland and of sands in the southern half creates very different situations and makes it necessary to distinguish two subbiomes, which will be handled separately (Fig. 3.4): (1) the northern semi-desert subbiome; and (2) the southern semi-desert or desert subbiome.

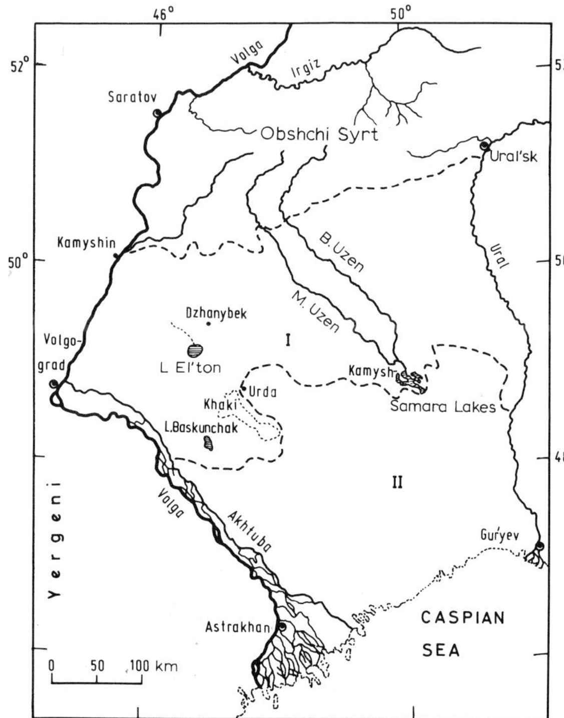

Fig. 3.4. Physiographic subdivisions of the Caspian Lowland between the rivers Volga and Ural (adapted from Levina, 1964): *I* = northern semi-desert; *II* = southern semi-desert or desert. Dotted area is limany; striped area is salt marshes or salt swamps.

NORTHERN SEMI-DESERT SUBBIOME

As a result of the almost perfect flatness of the northern semi-desert, the groundwater table is generally not deep, so that in the shallow basins numerous shallow, endorheic salt lakes are formed. One must distinguish two such types of lakes: (1) the three large lakes El'ton and Baskunchak in the west, plus Inder to the east of the Ural, next to a 60 m high salt dome (Nikitin, 1954b; see Fig. 3.25, p. 35); and (2) the numerous small salt lakes, called *limany* (Russian: *liman*, plural: *limany*), plus the salt swamps and marshes (Keller, 1936).

Lake El'ton is 20.5 km long and 14.8 km wide, while Lake Baskunchak is 19.2 km long and 10.2 km wide. Rock salt is deposited in both lakes, and in Lake El'ton this forms a layer 9 m thick, a total deposit of 10^9 tons. The salt deposits of Lake Baskunchak are being mined on a large scale. These salt deposits are derived from Permian underground salt beds or from the salt domes found in the region (Bogdo Hills and others). Saline waters are transported by springs to the salt lakes, where they become enriched through the evaporation of the lake water. In the smaller salt lakes and mud-filled marshes, on the other hand, the salt has been derived from earlier marine sediments.

Except for the Volga and the Ural, which do not originate in the region, the entire region is endorheic — that is it has no outlet for surface runoff. Thus, the marine salts locked into the soil can only be redistributed by the rainfall, primarily from higher levels of the meso- and micro-relief to lower levels. The higher levels lose their saltiness with time, while the salinity in the lower levels increases. This leads to the formation of very interesting, regular, macro- and micro-complexes of soil and vegetation.

Experimental area at Dzhanybek

The situation has been best studied in the years 1950 to 1970 in the region around the experimental station at Dzhanybek, 30 km north of Lake El'ton (Gordeyeva and Larin, 1965; Rode, 1974). We can go into this in more detail, using the summary of Olovyannikova (1976), since it serves as a typical example.

This endorheic region is almost ideally flat and level. The basic elements of the meso-relief are the limany and larger basins and the scarcely higher plains. The size of individual basins varies between 100 and 200 ha, and the relative depth is only 1 to 1.5 m. The limany are a bit deeper (2–3 m) and larger. The two forms together make up about 12 to 15% of the total area.

The vast plains exhibit another typical micro-relief with three basic elements: (1) the micro-depressions, (2) the micro-summits, and (3) the micro-slopes between them (Figs. 3.5 and 3.6). The three together form the basis for a biogeocoenosis complex. The difference in elevation in this micro-relief varies from a few centimeters to a few decimeters, while the horizontal extent of the individual elements is measured in meters or at most a few decameters. The micro-depressions are rounded and 5 to 30 cm deep (diameter 5 to 40 m), and they make up about 25% of the total area. A similar percentage of the total area is taken up by the micro-slopes, which form belts 1 to 3 m wide around the depressions. The slightly raised micro-plains make up about 50% of the total area. This micro-relief is a determining factor for soil and vegetation relationships.

The climate follows the general pattern for the Caspian Lowland. The following, more precise, data were gained as a result of 20 years of observations. Average annual precipitation is 274 mm, with 146 mm in the summer half-year and 128 mm in the winter half-year. Variations from year to year are large. The minimum for the 20-year period was 136 mm (67 mm in summer, 69 mm in winter), and the maximum was 484 mm (286 mm in summer, 198 mm in winter).

The average potential evapotranspiration is estimated as 950 mm (835–1281 mm) per year. The relative humidity of the air falls below 30% on an average of 64 (35–119) days out of the year. Hot duststorms (*sukhovey*) occur on an average of eighty (54–120) days out of the year.

The mean annual temperature is 6.9°C (January: −11°C, July: 24°C). The lowest mean annual temperature was 4.3°C, and the highest was 9.1°C. The variation in absolute temperature reaches almost 80°, with an absolute maximum of 41.1°C and absolute minimum of −36.1°C. The number of days with mean temperature above 5°C was 198, with mean temperature above 10°C was 168. A snow cover is present from November to March or,

Fig. 3.5. View across the northern semi-desert showing the micro-mosaic vegetation structure (photograph by B. Keller; from Walter, 1976, fig. 19).

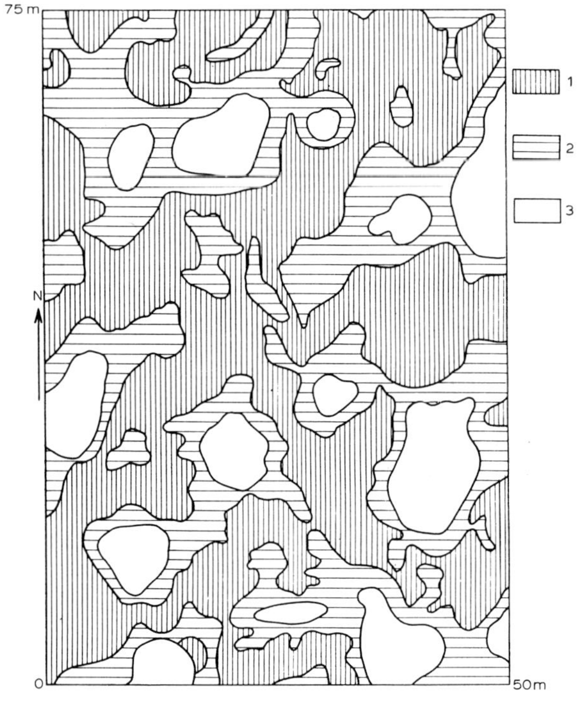

Fig. 3.6. Horizontal projection of the mosaic complex in the northern Caspian semi-desert near Dzhanybek (from Olovyannikova, 1976, p. 6). Legend: *1* = desert biogeocoenosis (*Artemisia pauciflora* + *Kochia prostrata*) on micro-rises; *2* = semi-desert biogeocoenosis (*Tanacetum achilleifolium*), on light chestnut earths on micro-slopes; *3* = steppe biogeocoenosis on dark soils of the micro-depressions.

at most, until the middle of April. Snow cover is seldom more than 10 cm deep on the flat areas but may be 40 to 50 cm deep in the depressions, so that the flat areas receive only about 60% of the mean annual winter precipitation, the slopes about 100%, and the depressions up to 250%. This means, of course, that one finds the zonal soil and zonal vegetation on the barely inclined slopes. Compared with the macroclimate, the flat areas are excessively dry and the depressions comparatively wet.

Snowmelt in the spring completely fills the micro-depressions roughly once every two years and, in snowy winters (roughly two in ten years), also fills the limany and the large basins of the meso-relief.

The basic substrate material is heavy clayey-loam deposits left by the Khvalyn transgression (Kovda, 1950). This material contains about 0.1 to 0.2% readily soluble salts and about 4.5% carbonates. The groundwater table in this region is about 6 to 7 m deep and is slightly salty. It receives water periodically from the depressions, under which freshwater lenses form but fluctuate greatly in size. After snowy winters the groundwater table rises in the spring, after snowmelt, almost to the surface.

Desert biogeocoenosis (Fig. 3.7)

About 50 to 60% of the broad plains of the semi-desert are occupied by the drier, desert-like, biogeocoenosis dominated by *Artemisia pauciflora* and *Kochia prostrata* or *Camphorosma monspeliacum* (Fig. 3.8). Various synusiae of ephemeroids (perennial geophytes) associate themselves with these dominants, in particular *Poa bulbosa* var. *vivipara* with *Tulipa biflora* and *T. schrenkii* (Fig. 3.9), *Ornithogalum fischerianum*, *Ferula nuda* and *F. caspica*. Synusiae of annual ephemerals are also present, primarily *Alyssum desertorum*, *Ceratocephalus falcatus*, and *Lepidium perfoliatum*. Less common are *Leymus* (*Elymus*) *ramosus* and *Salsola laricina*. The total number of species is 26, of which only about 7 to 14 are found in a single stand. The total vegetal cover ranges from 20 to 45%, and the height of the vegetation from about 15 to 20 cm.

The average annual above-ground production is about 890 kg ha^{-1} (390–1550 kg ha^{-1}). The soil type is a solonetz with a 5 to 10 cm thick bleached, light-gray layer (humus content 1.7 to 2.8% by weight), under which lies a dense solonetz (alkaline) horizon which is tough and chocolate-colored when wetted. Its salt content (primarily Na_2SO_4) increases downward and reaches 2.5 to 3.0% at a depth of 30 cm. Underneath lies a continuously dry, reddish-brown, sandy horizon. At a depth of 120 to 150 cm gypsum deposits appear. The maximum salt content is found at a depth of 2 m, below which it decreases and corresponds roughly to the salt content of the groundwater. The salt distribution in the soil profile is shown in Fig. 3.10.

In general, the soils are wetted, in the spring, only to a depth of about 20 to 30 cm. The salts are not leached out but are only redistributed temporarily within the profile. The actual evaporation from these soils amounts to about 210 mm, only a third of which is covered by the water supplies in spring. The roots of *Artemisia* are almost entirely limited (80–90%) to the top 0 to 35 cm of the

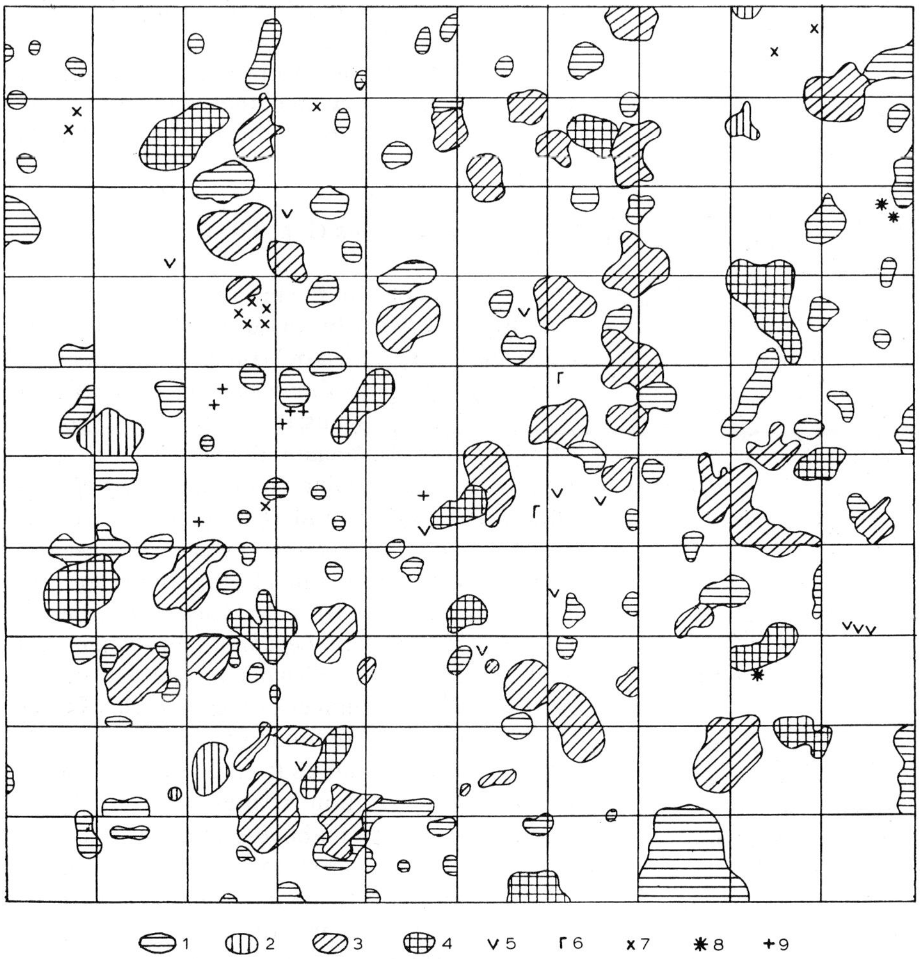

Fig. 3.7. Horizontal projection (1 m^2) of the desert biogeocoenosis near Dzhanybek (from Gordeyeva and Larin, 1965, p. 15). Legend: *1 = Poa bulbosa; 2 = Agropyron desertorum; 3 = Artemisia pauciflora; 4 = Kochia prostrata; 5 = Lepidium perfoliatum; 6 = Leymus ramosus; 7 = Echinopsilon sedoides; 8 = Ceratocarpus arenarius; 9 = Ceratocephalus orthoceras.*

profile. Those of *Kochia*, however, penetrate as far as 3 m down, that is, past the solonetz horizon (Fig. 3.11). The small amount of moisture at this depth comes up from even greater depths as the soil dries out above. The formation of salt crystals in these horizons comes about from the removal of water by the roots, primarily those of *Kochia* and *Salsola*.

Semi-desert biogeocoenosis (Fig. 3.12)

On the micro-slopes, which make up about 25% of the total area, one finds the zonal, slightly alkalized, chestnut earths, with a plant cover composed primarily of *Tanacetum* (*Pyrethrum*) *achilleifolium* and *Agropyron desertorum* with considerable *Poa bulbosa* var. *vivipara*. Single individuals of *Leymus ramosus*, *Festuca sulcata*, *Artemisia austriaca*, *A. pauciflora*, *Phlomis tuberosa*, *Kochia prostrata*, *Ornithogalum fischerianum*, *Tulipa biebersteiniana*, and the ephemerals *Alyssum desertorum* and *Lepidium perfoliatum* (Fig. 3.13) join the primary species.

The humus content of the uppermost soil horizon reaches 3%. In the spring the upper 120 cm are thoroughly wetted, but the water content scarcely changes at all below this layer.

The upper 50 to 60 cm are well permeated by roots. Individual roots of *Tanacetum* reach 120 cm deep and those of *Agropyron* 80 to 95 cm deep.

Fig. 3.8. Desert biogeocoenosis on crusted solonetz soil: *Artemisia pauciflora* + *Camphorosma monspeliacum* (the latter appearing seldom around Dzhanybek) (photograph by B. Keller; from Walter, 1976, fig. 22).

Readily soluble salts and gypsum are found only below 120 cm. The most common salt is Na_2SO_4, with a chloride maximum occurring slightly above a depth of 2 m. The groundwater contains 4 to 5 (sometimes as high as 10) g l^{-1} of salts.

Steppe biogeocoenosis (Fig. 3.14)

The conditions for growth are much more favorable in the micro-depressions, which make up about 25% of the total area. These accumulate more snow in winter, and in spring the soil receives about 200 mm of snowmelt water. The upper 100 to 150 cm of soil are continually wet. Water even filters down periodically to deeper levels, so that the groundwater table rises dome-like under the depressions. At such sites the groundwater contains only bicarbonates (0.3–5 g l^{-1}).

The chernozem-like profile is salt-free, and the humus horizon is 35 cm thick with 5 to 6% humus in the upper layers.

In the plant cover the synusiae of steppe bunch grasses are typically: *Festuca sulcata*, *Stipa lessingiana*, *S. capillata*, *Koeleria cristata* with *Agropyron pectiniforme*.[1] Forbs are less well represented: *Medicago romanica*, *Galium ruthenicum*, *Galatella villosa*, *Phlomis pungens*, and many others.

The total vegetal cover reaches 75 to 95%, and the average height of the plants is about 27 to 37 cm. The number of species per stand varies from 27 to 47. The annual production in this steppe biogeocoenosis is high, on the average about 2870 kg ha^{-1} (1370–4940 kg ha^{-1}).

[1]Terenozhkin (1954) explained the taxonomic positions and the distributions of the three *Agropyron* species (*A. desertorum*, *A. sibiricum* and *A. pectiniforme*).

Fig. 3.9. Desert biogeocoenosis in springtime, with the ephemeroid (geophyte) *Tulipa biflora* (center) next to a lighter-colored *Artemisia pauciflora* with fresh growth (photograph by B. Keller; from Walter, 1976, fig. 24).

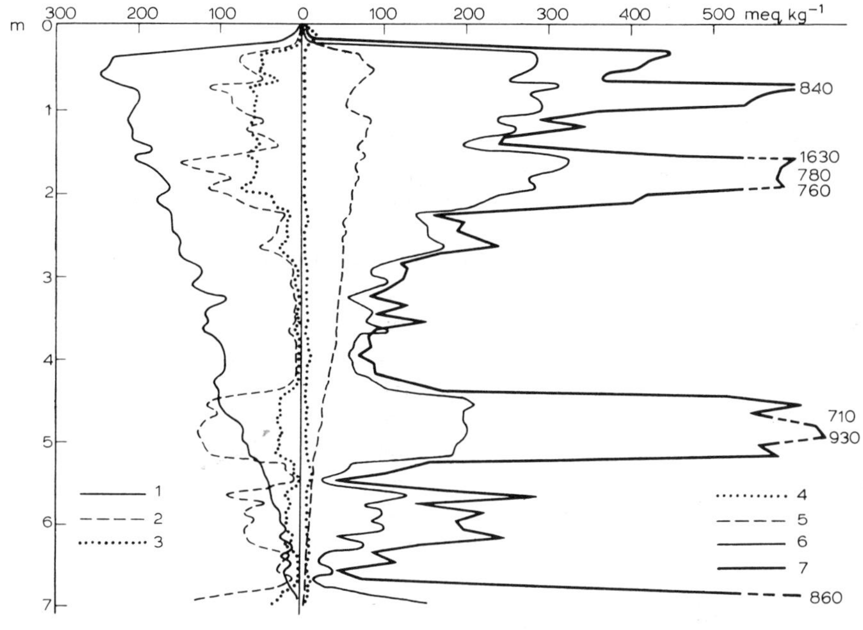

Fig. 3.10. Soil profile to a depth of 7 m, with data on the salt content (numbers above in meq kg^{-1} dry weight) of a solonchak-like solonetz. Cations: *1* = Na^+; *2* = Ca^{2+}; *3* = Mg^{2+}. Anions: *4* = HCO_3^-; *5* = Cl^-; *6* = SO_4^{2-} (soluble); *7* = SO_4^{2-} (with gypsum). From Olovyannikova (1976, p. 8).

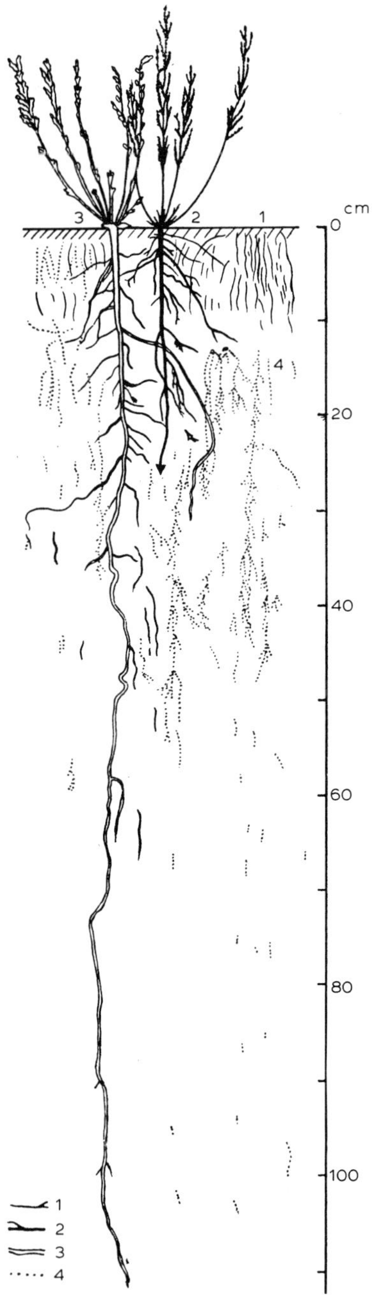

Fig. 3.11. Root systems of the desert biogeocoenosis (from Gordeyeva and Larin, 1965, p. 62): *1* = *Poa bulbosa*; *2* = *Artemisia pauciflora*; *3* = *Kochia prostrata*; *4* = *Leymus ramosus*. Depths are in centimeters.

The roots of the grasses end at a depth of about 2 m and those of *Medicago* and *Galium* at about 3 m (Fig. 3.15). Water uptake can still be seen at a depth of 3.5 m. The upper 2 m dry out to the wilting point in summer. The annual evapotranspiration reaches 421 mm.

In contrast to the case of the solonetz soils, it is the water supply in the deeper layers (50–300 cm) that is most important for the vegetation here. The larger and deeper the depression, the more hygrophytic is the vegetation and the more leached is the soil. In the shallow depressions with dark chestnut soil are found primarily *Festuca sulcata* and *Stipa lessingiana*, often with *Tanacetum achilleifolium* and the xerophytic steppe species *Jurinea multiflora* and *Onosma tinctorium*. *Glycyrrhiza glabra* is not common (Fig. 3.16). In the deep depressions, on the other hand, it is typical to find *Stipa rubens*, *Thymus marxchallianus*, *Arenaria longifolia*, *Veronica spicata*, and *Gypsophila paniculata*. In the deepest depressions one can even find the shrub *Spiraea hypericifolia*.

The more extensive depressions are today cultivated land or are covered by ruderal vegetation, as is often found on fallow ground, often with *Amygdalus nana*. Earlier these were true fragments of steppe with thick humus horizons and freshwater lenses underneath.

In deep erosional gullies (*balki*), a considerable amount of snow accumulates in winter, so that woodlands composed of *Rhamnus cathartica*, *Prunus spinosa*, *Rosa canina*, *Lonicera tatarica*, *Euonymus verrucosa* and individual small trees of *Malus sylvestris* are able to develop. Less frequently, stunted *Ulmus glabra* or, on the Yergeni slope, *Quercus robur* appear (Fig. 3.17).

The region in general

The studies of Keller and Novikov (in Keller, 1936) covered the entire north-western part of the Caspian Lowland. They were begun at a time when the region was still mostly unsettled.

The micro-mosaic structure just described is characterized as consisting of two members: plains and depressions on the one hand, and the transitional slopes on the other. Numerous complete stand samplings were conducted so as to include the lower plants, which in desert biogeocoenoses cover the surface between the seed plants and appear clearly only upon wetting.

Cyanophytes are present as ground covers: *Microcoleus vaginatus* and *Scytonema ocellatum*, with numerous *Stratonostoc commune*. In addition

Fig. 3.12. Horizontal projection (1 m^2) of the semi-desert biogeocoenosis (from Gordeyeva and Larin, 1965, p. 25): *1 = Agropyron desertorum*; *2 = Tanacetum achilleifolium*; *3 = Stipa lessingiana*; *4 = Artemisia pauciflora*; *5 = Kochia prostrata*; *6 = Jurinea multiflora*; *7 = Koeleria gracilis*; *8 = Leymus ramosus*; *9 = Lepidium perfoliatum*; *10 = Serratula xeranthemoides*; *11 = Ceratocephalus orthoceras*; *12 = Alyssum desertorum*; *13 = Erysimum leucanthemum*; *14 = Poa bulbosa*.

to these there appear individual lichens such as *Collema* sp., *Parmelia vagans*, *P. ryssolea* (Fig. 3.18), more seldom *Acarospora schleicheri* and others, as well as small bryophytes such as *Bryum argenteum*, *B. caespitosum*, *Pterygoneurum subsessile*, or *Riccia* spp. *Tortula desertorum* is lacking in the actual desert biogeocoenoses but is present in the others.

Actually, not only two-member but also three-member micro-mosaics occur. In these it is the raised portions, from which the salts are leached out by rain water, that have the light chestnut earths. The shallowest depressions, to which the salts are brought and from which the water evaporates rapidly without sinking into the soil, are characterized by solonetz soils. Chernozem-like steppe soils occur in the deeper depressions. On the solonetz soils, which are flooded often but only for a short time, the lower plants, especially the cyanophytes, develop quite luxuriantly.

Finally, there are also four-member mosaic structures which consist of: (1) crusted solonetz soils with *Artemisia pauciflora*, *Camphorosma monspeliacum*, and *Anabasis salsa*; (2) deep, columnar solonetz soils with *Kochia prostrata* and *Artemisia lercheana*; (3) burozems (brown earths) with *Tanacetum achilleifolium*, *Agropyron desertorum*, and *Artemisia lercheana*; and (4) leached

Fig. 3.13. An area with semi-desert vegetation: *Festuca sulcata*, *Tanacetum achilleifolium*, and lighter-colored *Artemisia*, identified by Keller as *A. maritima* var. *incana* (photograph by B. Keller; from Walter, 1976, fig. 21).

humus soils with steppe species, but also *Tanacetum*. Nikitin (1954a) even mentioned six-membered complexes.

The question of how these mosaic structures came about arises. One can imagine that the soil laid bare by the receding sea was very level and slightly salty, but that after rainfall many pools of water remained. These pools were very shallow, but the soil under them was much wetter than elsewhere and dissolved the salts. The result was a slight settling of the soil under the pools, so that after the next rain the pool was a little deeper and the water infiltrated a bit deeper into the soil. Plant growth was more abundant on such sites, the roots loosened the soil, and infiltration of the water was improved. Slight depressions arose, which caught more snow and caused increased wetting and increased sinking, until the micro-mosaic was finally clearly apparent.

The necessary conditions for the formation of such micro-mosaics are, therefore, a young, very level, maritime sedimentary deposit with a certain salt content. Only under such conditions will this type of semi-desert develop. On sandy deposits of earlier delta formations no such micro-mosaic will arise. The relief can be exaggerated by the earth mounds cast up by burrowing rodents, but the rodents are not the primary cause of the mosaic.

The most comprehensive overview of the entire northern Caspian Lowland was given by Levina (1964). She performed an analysis of the entire flora and of the spread of the most important constituent species, discussed the different synusiae and complexes, and evaluated the pertinent literature. She characterized the semi-desert mosaic elements as xerophilic desert elements in order to differentiate them from the halophilic desert elements found on solonetz soils. Levina has distin-

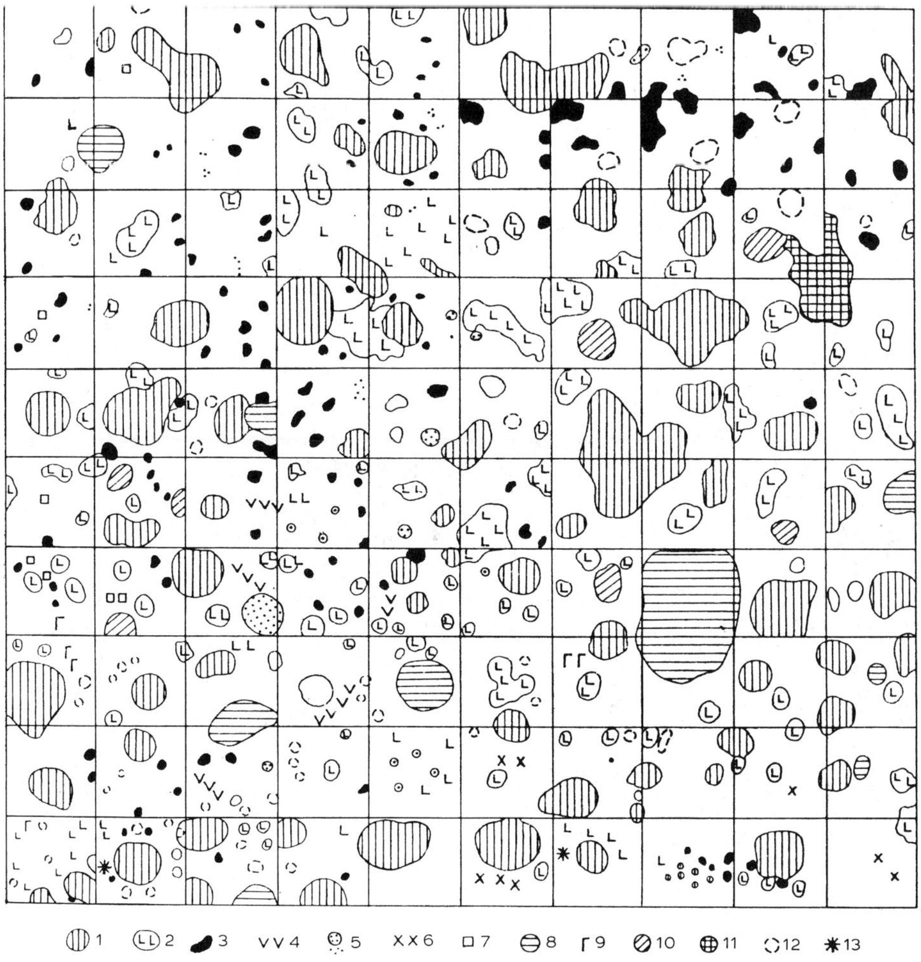

Fig. 3.14. Horizontal projection (1 m^2) of the steppe biogeocoenosis (from Gordeyeva and Larin, 1965, p. 33): *1* = *Agropyron pectiniforme*; *2* = *Festuca sulcata*; *3* = *Artemisia austriaca*; *4* = *Calium ruthenicum*; *5* = *Koeleria gracilis*; *6* = *Ceratocephalus orthoceras*; *7* = *Arenaria longifolia*; *8* = *Stipa capillata*; *9* = *Alyssum desertorum*; *10* = *Medicago romanica*; *11* = *Artemisia lercheana*; *12* = *Stipa lessingiana*; *13* = *Erysimum leucanthemum*.

guished a very large number of associations based on the dominance of the most important species.

The limany

The biogeocoenosis complexes around the limany are more complicated (see Leisle, in Keller, 1936). The larger of these mostly brackish lakes are remnants of earlier river courses. Their size varies from barely a hectare to many thousand hectares. After snowy winters they are filled in spring with the snowmelt, and into them are carried the salts leached from the higher areas. The salts accumulate more or less, depending upon whether the water in the lakes primarily evaporates or soaks into the soil, since the limany have no surface outlets.

The biogeocoenosis complexes around the individual limany were treated in detail by Levina (1956). According to their origin, the following types of limany can be distinguished: (1) limany into which rivers empty: (2) limany which form in riverine floodplains; and (3) limany on flat watersheds with small catchment areas.

As far as the vegetation around the limany is concerned, Levina identified the following types:

(1) Non-brackish or only slightly brackish meadows with *Agropyron repens* and either *Glycyrrhiza glabra* or *Calamagrostis epigeios* as dominants.

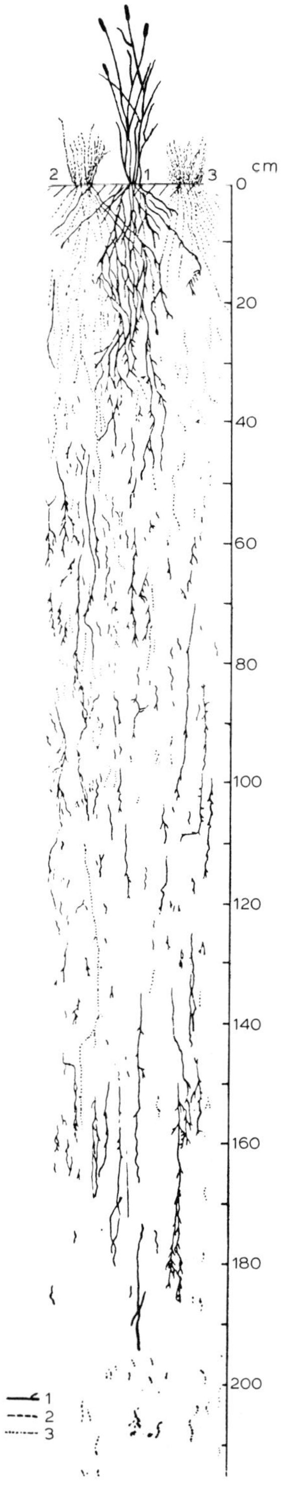

Fig. 3.15. Root systems of the steppe biogeocoenosis (from Gordeyeva and Larin, 1965, p. 67): *1* = *Agropyron pectiniforme*; *2* = *Festuca sulcata*; *3* = *Koeleria gracilis*.

Accompanying species are *Bromus inermis*, *Hierochloë odorata*, and *Beckmannia eruciformis*. Average harvests of the hay are 2.2 t ha^{-1} (1.9–2.5 t ha^{-1}).

(2) Brackish wet meadows with dominant *Puccinellia fomini*, *Juncus gerardi*, *Alopecurus ventricosus*, *Aeluropus litoralis*, *Eleocharis oxylepis*, and *Agropyron repens*. Associated species are *Artemisia monogyna*, *Petrosimonia triandra*, *P. oppositifolia*, *Suaeda altissima*, and *Atriplex litoralis*.

(3) Brackish dry meadows with dominant *Puccinellia scleroides*, *Artemisia monogyna*, and *Atriplex litoralis*. Accompanying are the species of type 2 above, plus *Salicornia europaea* and *Halocnemum strobilaceum*.

(4) Dry, slightly brackish meadows with *Agropyron pectiniforme*, *Bromus inermis*, *Poa angustifolia*, *Leymus ramosus*, *Carex praecoccus*, and *C. uralensis* as dominants. Associated are *Festuca sulcata*, *Agropyron repens*, *Artemisia austriaca*, *A. monogyna*, *Limonium gmelini*, and *Alopecurus ventricosus*.

(5) Slightly brackish marshes with dominant *Phragmites communis*, *Scirpus* (*Bolboschoenus*) *maritimus*, *Eleocharis eupalustris*, *Carex melanostachya*, and *Agropyron repens*. Associated are *Hierochloë odorata*, *Butomus umbellatus*, *Scirpus lacustris*, *Cirsium incanum*, *Euphorbia palustris*, *Lythrum virgatum*, *Rhaponticum salinum*, and *Alopecurus ventricosus*.

These biogeocoenoses are associated, in the individual limany, with quite different biogeocoenosis complexes, which are represented on numerous profiles. Exact stand surveys are also included by Levina (1956).

Solonchak soils with halophyte vegetation

These halobiomes occur around salt lakes, especially Lake El'ton and Lake Baskunchak, where the groundwater table is so high that the water can rise to the surface by capillarity and evaporate. The salts dissolved in the water accumulate on the surface and form a white crust. There arises a solonchak soil, which has a completely different salt and water distribution from that of the solonetz (Table 3.1).

The differences in the water content and the salt distribution between the solonetz and solonchak soils stand out clearly. Depending on these two

TABLE 3.1

Water content and salt distribution in solonetz and solonchak soil types (Keller, 1925/26)
(All numbers are percent of dry soil; the maximum value in each column is in italics)

Depth of soil samples (cm)	Solonetz soil with *Artemisia pauciflora* and *Camphorosma monspeliaca*				Transition soil type with *Halimione verrucifera*				Solonchak soil with *Salicornia herbacea*			
	water content	soluble salts	Cl	"SO_3"	water content	soluble salts	Cl	"SO_3"	water content	soluble salts	Cl	"SO_3"
Surface	3.3	0.111	0.005	0.025	2.5	0.054	0.005	0.015	9.1	*18.854*	*3.005*	*7.982*
10	13.9				15.5	0.437	0.052	0.102	23.1	4.580	0.593	2.096
15		0.184	0.040	0.007								
20		0.317	0.128	0.006	18.7	*2.612*	0.164	*1.254*	32.1	3.862	0.361	1.609
30	10.3	0.466	0.187	0.030								
40					18.5	1.683	0.273	0.671	35.8	3.059	0.334	1.154
45		*1.094*	*0.260*	*0.388*								
60		0.816	0.214	0.166	21.1	2.009	0.393	0.716	31.8	2.423	0.482	1.018
90		0.496	0.202	0.051								
100	9.0				23.0	1.693	*0.403*	0.505		2.398	0.472	1.013

Fig. 3.16. A deep depression in the steppe biogeocoenosis, with numerous awns of *Stipa capillata* visible, *Festuca sulcata* covered by *Glycyrrhiza glabra* (dark), and *Artemisia austriaca* (light). In addition, there occur: *Koeleria gracilis*, *Galium verum* s.l., *Veronica spicata*, *Medicago romana*, *Phlomis pungens* and others (photograph by B. Keller; from Walter, 1976, fig. 20).

Fig. 3.17. Last remnants of the extra-zonal oak forests: *Quercus robur* grove at the open end of a ravine in the Yergeni Hills (photograph by B. Keller; from Walter, 1976, fig. 23).

factors, there occurs also a regular change in the composition of the mostly very species-poor vegetation. According to Keller (1925/1926), the halophytes can be arranged along the following ecological gradient, beginning with the wettest and saltiest soils:

1. *Salicornia herbacea*
2. *Halocnemum strobilaceum* (Fig. 3.19) } solonchak soils
3. *Petrosimonia crassifolia*
4. *Atriplex verruciferum*
5. *Atriplex canum* (Fig. 3.20) } transitional soils
6. *Anabasis salsa* (Fig. 3.21)
7. *Nanophytum erinaceum*
8. *Artemisia pauciflora* with *Camphorosma monspeliacum* } mostly solonetz soils
9. *Kochia prostrata* with *Camphorosma monspeliacum*

The most extreme sites around the salt lakes, as for example around Lake Baskunchak, consist of hard table salt (NaCl) and carry no vegetation. The first pioneer species are generally *Salicornia europaea*, together with *Halocnemum strobilaceum* (Fig. 3.19). The soil is often still covered by a salt crust. These species are leafless, succulent salt plants, which store in their tissues considerable amounts of sodium chloride and whose assimilating stems are clearly jointed. However, while *Salicornia* grows upright and the plants stand alone, the shoots of *Halocnemum* lie along the ground and form rounded, flat cushions of often over a meter in diameter.

Only a few accompanying species occur in this community. Worthy of mention are *Frankenia pulverulenta*, *Petrosimonia crassifolia*, *Ofaiston monandrum*, *Bassia sedoides*, and *Puccinellia convoluta*. *Salicornia* also grows occasionally, together with *Suaeda maritima*, *Halimione pedunculata*, *Limonium caspium*, and others. In general, *Petrosimonia crassifolia* and *Atriplex verruciferum* occupy less extreme but still wet and very salty sites. Both have succulent leaves. Completely different is the situation on sites occupied by *Atriplex canum*, which, alone or with *Anabasis salsa*, often can form a desert-like community with only 20% vegetal cover over distances of many kilometers.

Fig. 3.18. Nomadic lichens in the semi-desert biogeocoenosis: *Parmelia vagans* (light) and *P. ryssolea* (dark) (photography by B. Keller; from Walter, 1976, fig. 25).

Fig. 3.19. *Halocnemum strobilaceum* (Kirghiz: *sarsan*) on the shore of a salt lake (from Keller, 1928).

The neighboring regions

Both the semi-desert and the halophyte communities can be found in quite similar form further to the west, particularly on the narrow isthmus of the Crimean Peninsula (Wulff, 1926; see Walter, 1974, pp. 349–353). This semi-desert forms a narrow strip, as transition to the feather-grass steppe, on both sides of the Sivash (the salt marshes between the Crimea and the continent). The halophyte communities occupy the lowest, middle part of the isthmus.

The Trans-Caucasian semi-deserts can also be mentioned. Only the Apsheron Peninsula near the city of Baku can really be called desert. The climate diagram, with 187 mm of rainfall annually, is only slightly different from that of Krasnovodsk (see Fig. 5.2) on the opposite Middle Asian coast of the Caspian Sea. The other arid parts of the Trans-Caucasus can all be considered to be semi-desert. They cover the entire Kura–Araks lowland and extend up the Kura River almost to Tbilisi, also along the west coast of the Caspian Sea from Baku almost to Derbent, and up the foothills to about 400 to 600 m. A larger area of montane semi-desert is found in southern Armeniya and in the region (Azerbaydzhan) around Nakhichevan', at a height of 1000 to 1500 m.

The Trans-Caucasian semi-deserts are distinguished from the north Caucasian semi-deserts by their more southern, almost mediterranean character. Along gradients of increasing rainfall they do not change directly into steppes but rather into open woodlands with single-standing, low trees of *Pistacia atlantica* and *Juniperus* species. These woodlands have only been preserved in very few places because elsewhere the woody plants have been totally removed. One cannot consider these

Fig. 3.20. *Atriplex canum* along Lake Baskunchak (from Keller, 1928).

semi-deserts to be subtropical, however, because the winter temperatures, at −16° to −26°C, are much too low.

The Mugan Steppe southeast of the lower course of the Araks River, the Mil Steppe between the Araks and the Kura, and the Shirvan Steppe north of the Kura are semi-deserts with extensive salty soils. These were originally a bay of the Caspian Sea, which was filled by deposits from the Kura and Araks Rivers (Beydeman et al., 1962); otherwise the soils of the semi-deserts are sierozems (grey desert soils) and, in the marginal areas, also chestnut earths (see Walter, 1974, pp. 774–779).

Large parts of the semi-deserts in the lowland have been converted into irrigated agricultural land, on which cotton, alfalfa, and wine grapes are grown. In the following summary, only the natural plant communities will be briefly treated.

Bolshakov and Rode (1972) provide more detailed data about the phytomass of the micromosaic complex. In Table 3.2, the various elements are numbered as follows:

I desert element on slight rises
II semi-desert element on the slopes
III steppe element in the depressions

TABLE 3.2

Phytomass of the micro-complexes in t ha^{-1}

Phytomass	Above ground	Below ground	Total	Depths of roots (cm)
I *Artemisia* Assoc.	0.9(0.5–1.9)	7.8	8.7	150–300
II *Pyrethrum-Agropyron* Assoc.	1.3(0.6–2.3)	10.9	12.2	150
III Grass steppe	2.8(1.8–4.4)	17.5	20.3	300

Fig. 3.21. *Anabasis salsa* (Kirghiz: *byurgun*) atop a *bogdo* (salt dome) along Lake Baskunchak. The soil is a rocky burozem. (From Keller, 1928.)

The same authors also provide data on the soil fertility (Table 3.3). The increase of phytomass and soil fertility from I to III is clearly seen from the data in the tables.

Finally one should mention the descriptions with vegetation map and profile by Nikitin (1954a, b) of the eastern Caspian Lowland, in the region of the rivers Ural and Emba. The region extends in the east up to the Mugodzhary Hills and in the south up to the Ustyurt Plateau. The mosaic structure in the northern part of the lowland (patchiness and stripe patterns in the vegetation) is stressed also in this work. These patterns can be intensified through the activity of the rodents. Nikitin presents two figures with horizontal projections of small sections of the plant cover. We refer the reader to the overview map (in the original work) of the vegetation of the whole region. More detailed data about species composition are also presented in the original. According to the composition of the natural vegetation, the region is divided into 22 natural geobotanical units with definite dominant biogeocoenosis complexes (map in Larin et al., 1954b).

TABLE 3.3

Yield of summer wheat on the different soils in t ha^{-1} for the years 1956–1958

Soil type	1956		1957		1958	
	dry-weight	grains	dry-weight	grains	dry-weight	grains
I Solonetz	2.92	0.83	2.26	0.71	3.05	0.95
II Chestnut earths	4.41	2.16	1.85	0.79	4.02	1.14
III Dark soils	5.37	2.09	3.10	1.16	7.53	2.87
Entire area	3.85	1.48	2.34	0.84	4.29	1.48

SOUTHERN SAND-DESERT SUBBIOME (Yakubov, 1955)

The southern, more desert-like, zone lies below sea level and is drier. The zonal soil type is burozem (brown earth), but in this region sands and vast areas of shifting dunes dominate. The raw material for these sands was provided by post-Tertiary deposits of the Caspian Sea, the alluvial sediments of the rivers,[1] to some extent also the Tertiary sandstones of the Yergeni Hills, and in particular the old, already colonized sand dunes of the Caspian Lowland. The sands are underlain by heavy loams. Thus, the groundwater table is mostly quite high.

The sands are shifted around by the wind, and in the process the 0.25 to 0.05 mm particle size increases its proportion of the total substrate. This size class makes up over 90% (up to 99%) of the entire weight of the sand. The sands are composed overwhelmingly of silicon dioxide.

The natural plant cover on eu-climatopes[2] is dominated by *Artemisia lercheana*. Grasses such as *Agropyron sibiricum*, *A. ramosum*, *A. cristatum*, *Festuca sulcata* (*beckeri*), and *Koeleria glauca*, among others, associate themselves sporadically with *Artemisia*. In the spring there are numerous ephemerals. We can regard this community as typical for the psammobiome. The water losses on eu-climatopes are balanced by the amount of rainfall reaching the soil.

The relief in the sand areas exhibits the following forms: (1) level sandy areas (*ashiki*); (2) slightly rolling sands with 2 to 3 m (up to 5 m) rises; (3) sand areas with hills up to 15 m and valleys between; (4) shifting sands (barkhan chains) with flat windward slopes (13–15° declination) and steep lee slopes (30–33° declination). High dunes form where the substrate is hilly and the sand covers the rises, often with a layer of only a few meters (maximum 7 to 8 m).

The sand areas were partly covered by plants originally, and the sand was anchored. On a map of 1840 only a few shifting-dune regions were indicated, but by 1864 they already occupied a large area. The cause of this change was overgrazing and attempts to cultivate the sand areas. These actions destroyed the plant cover and set the sand in motion.

The degree of deflation (wind erosion) of the sand depends on the relief. Steeper slopes which stand oblique to the wind direction and are thus hit with the full force of the wind are especially vulnerable. The sand which is removed is redeposited at places which are somewhat more protected from the wind. Plants which are rooted into firm substrate at such places are covered by the sand and die. In their place appears *Aristida pennata*, which always grows out of the sand again when it is covered. Totally vegetationless shifting dunes form only where sand is continuously blown away in large quantities. These are mostly separated from the deflation areas by a strip of *Aristida*. The greater the sand layer which is blown away, the larger become the shifting dunes. They can become as high as 20 m but then usually combine to form dune embankments.

One can distinguish several stages in the formation of the dunes, as Dubyanski (1911) showed quite early. We illustrate this by way of an example. The loose rock substrate is made up as follows: Underneath a 15 cm thick, loose sand layer, lightly crusted on top, there lies a thicker, yellow-brown sand layer of 20 to 25 cm thickness; under this lies still another sand layer, which is well cemented by calcium carbonate.

Stage 1. The surface crust is destroyed by the trampling of livestock, and the sand in the surface layer is blown away. The wind carries the dust far away, the coarser particles remain, and only the intermediate fine sand is set in motion along the ground. The sand is caught by bushes and every unevenness of the relief, and at these places primary dunes form. The next sand layer is laid bare but offers some resistance against the wind. The grasses which are blown out (*Agropyron*, *Eragrostis minor*) are killed; *Artemisia lercheana* and *Kochia prostrata* dry up. *Artemisia scoparia*, *Salsola*, *Ceratocarpus*, *Corispermum*, and *Alhagi* colonize the small primary dunes. If grazing ceases, the

[1]The alluvial sands are the remains of the delta deposits of the Volga and the Ural, which along their lower courses repeatedly changed course and grew together. The sands deposited by the Volga came from the Russian Plain, while those deposited by the Ural came from the Ural Mountains. The latter are characterized by a higher content of hornblende and are thereby easily recognizable (Yakubov, 1955).

[2]Sites where the unmitigated effect of overall climate is seen, that is, flat areas where the soil is neither too light or too heavy, so that precipitation does not run off but penetrates the soil and is retained as interstitial water (Walter, 1979, p. 13).

vegetation cover becomes complete within three to five years, and the relief which was set in motion by the wind slowly levels out somewhat.

Stage 2. If grazing is continued, the deflation area increases. The second sand layer is loosened and carried away, until the deflation assumes very large proportions. Whirlwinds attack the sandy substrate material in the depressions. Two types of landscapes are formed, one consisting of unevenly distributed blow-outs and the other of drifted dunes. Psammophytes, which can survive the blow-outs, remain on the blown-out landscape; the drifted dunes are colonized by additional species which can survive the repeated covering by sand. Totally vegetationless dunes arise only if the drifting is quite extreme. In such a case, the sand particles are continuously blown up the windward sides of the dunes until they reach the crest and roll down the steeper backside. When this occurs, the dune begins to wander and simply covers every obstacle which stands in its way.

The supply of sand for the dunes is cut off when sand wetted by capillary rise of groundwater is reached or when a clayey layer, which cannot be blown away, is laid bare. When this happens, a closed plant cover again forms, sooner or later, on both the blown-out sand areas and on the dunes. Several stages in this process can be recognized.

The first pioneer to take hold on the shifting sand in the Caspian region is *Leymus racemosus*, sometimes also the chenopod *Agriophyllum arenarium*. These species are soon followed by *Salsola ruthenica*, *S. kali*, and *Corispermum aralo-caspicum*. They, like the pioneer species, have large root systems which extend 5 to 7 m horizontally and only about 1 m deep. Individual plants of *Heliotropium micranthos*, *Tournefortia sibirica*, *Chondrilla ambigua*, *Aristida pennata*, *Linaria odorata*, *Melilotus polonicus*, *Artemisia scoparia*, and others appear in the dune valleys. Along the edge of the dune chains one finds individual bushes of *Calligonum* and *Tamarix*.

This colonization causes a settling and compression of the sand. In the surface layer organic materials accumulate, and the water content of the soil decreases as a result of plant transpiration. *Artemisia arenaria* expands its area, and the pioneer species can hold their own only along the crests of the dunes. Representatives of the zonal vegetation now soon appear; *Agropyron sibiricum*, *Ephedra*, *Bromus tectorum*, and *Poa bulbosa*. Here and there a moss cover of *Tortula desertorum* forms. This stage leads over to the final stage. *Artemisia* becomes ever more dominant and invades *Agropyron desertorum*, *Ephedra* becomes more numerous, *Eragrostis minor* appears, and the mosses (*Tortula* with *Bryum caespitosum*) begin to form a crust. The relief levels out continuously during this colonization, and finally the hills are only half a meter high.

If sand-drifting occurs on a wet, brackish area, it is colonized by salt-resistant halophytes. Sand is repeatedly dropped on top of these plants and their shoots continually grow out, so that each of the bushes, which once spread itself over the clayey surface, finally sits alone on its own miniature dune (*nebka*, plural: *nebkas*). The bushes involved are usually *Tamarix* species, *Nitraria schoberi*, or *Halocnemum strobilaceum*. The area between the dunelets is colonized by small halophytes and ephemerals. If the sand-drifting goes on further, it becomes more difficult for the bushes to get water, and they are replaced by Chenopodiaceae such as *Kalidium*, *Atriplex*, *Anabasis aphylla*, *Alhagi*, *Corispermum*, and *Peganum harmala* (Zygophyllaceae).

The groundwater table is usually not very deep under the very level relief of the lowland, and the groundwater is, in general, very salty. In sandy areas, though, one can also find layers which carry freshwater (Fig. 3.22). The vegetationless sand dunes serve as water reservoirs, since the water penetrates easily into the sand and is, except for the small amounts which remain in the upper layers, protected from evaporation. In these dune areas it is even possible for small freshwater lakes to form, around which, as in wet dune valleys, grow *Populus hybrida*, *P. alba*, *Elaeagnus angustifolia*, *Salix caspica*, *S. rosmarinifolia*, and, less often, *Alnus glutinosa*. The understory characteristic of such a swamp forest is also present. In most cases, though, the only woody plants are *Calligonum aphyllum*, *Tamarix* species, and, in the southeast, the arborescent endemic *Halimodendron* (*Astragalus*) *ammodendron*.

One often finds freshwater in the form of lenses of quite different sizes floating on top of the salty groundwater. An especially detailed investigation of the groundwater situation was made for the narrow spit on the east coast of the Caspian Sea between the sea and the Kara-Bogaz Bay (Fig. 3.23).

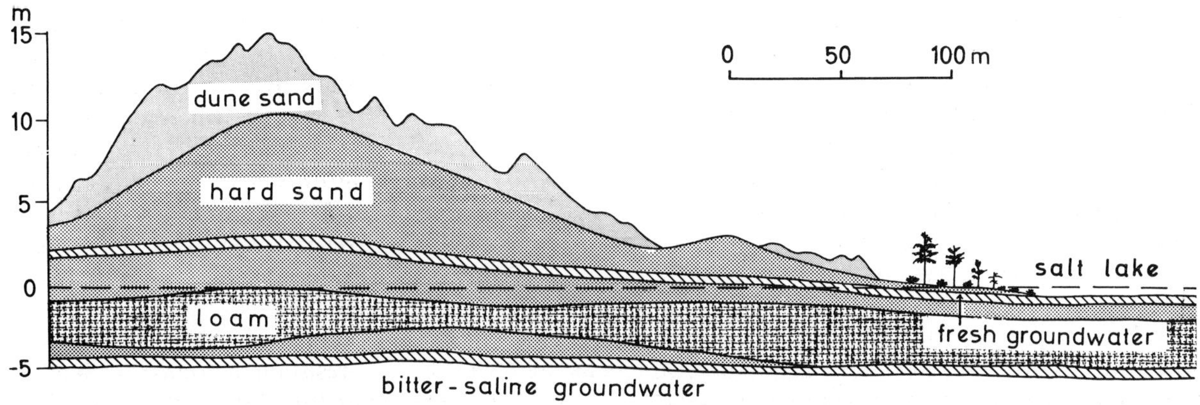

Fig. 3.22. Groundwater conditions in the sandy region of the Caspian Lowland (from Yakubov; in Walter, 1968, fig. 521). Groundwater is indicated by diagonal striping. The upper, fresh groundwater accumulates under the barkhan above an impervious layer in the stabilized sand; saline groundwater is found deeper, under a loamy layer. The vegetation grows where the fresh groundwater flows out into the frequently dried-out salt lake.

All wells in the area are sunk into this freshwater lens. The water table fluctuates considerably, depending on the precipitation of individual years. It sinks deeper in the summer, and sometimes the wells dry out completely.

As already mentioned, the water supply is greatest under vegetationless sand layers. As soon as plants become established, a large part of the rainfall which penetrates into the sand is taken up and transpired by the plants. This can be seen quite clearly in Fig. 3.24. Whereas the water content in sands under a plant cover is only slightly above the wilting point, even to a depth of 4 m, the water content in the vegetationless sands is around 7 to 9% and increases to 25% at 4.8 m, which amounts to saturation conditions. The large amount of water required by plants must be taken into consideration when one thinks about afforestation of bare sands. The stands of *Populus nigra* or *Salix acuminata* survive the first years in good condition because water from previous years is still available in the soil. With time, however, these water supplies are used up, since the transpiration of the woody plants is higher than the annual precipitation. Table 3.4 shows the increasing dryness of the soil quite clearly.

One sees how, over these four years, the water content of the deeper layers of the soil continually decreases in autumn: at 1.5 m from 7.6 to 2.7%, at 2 m from 8.9 to 2.9%. During the same period the

TABLE 3.4

Water content (% of dry weight) of sandy soil under a *Populus nigra* plantation (from Tomashevski in Yabukov, 1955)

Time of sampling	Sample depth (cm)						
	20	40	60	80	100	150	200
1914, April 3	9.6	10.5	8.3	8.6	9.0	9.4	9.6
Sept. 17	2.8	4.6	5.8	6.9	7.8	7.6	8.9
1915, April 12	5.2	5.7	6.4	7.9	8.6	7.8	9.1
Oct. 26	7.2	4.3	3.6	3.9	3.8	4.3	7.8
1916, March 26	9.8	10.1	5.8	6.0	5.0	7.0	9.0
Aug. 10	2.3	3.2	3.1	3.8	4.1	5.3	6.7
1917, April 7	6.5	7.3	7.2	7.8	5.6	6.1	7.6
Sept. 3	1.6	1.8	2.2	2.5	2.6	3.8	4.0
1918, April 11	5.7	6.5	6.4	5.2	4.0	2.6	3.5
Oct. 23	1.9	1.9	2.2	2.4	3.5	2.7	2.9

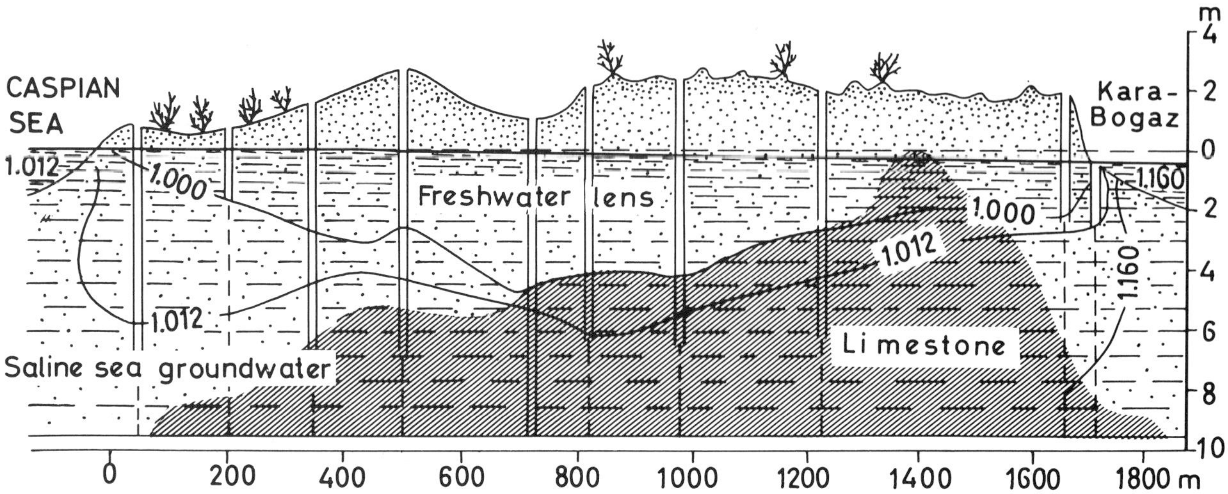

Fig. 3.23. Freshwater lens under sand dunes on the east coast of the Caspian Sea (from Yelovskaya; in Walter, 1968, fig. 522). Under the sand region there forms a freshwater lens, which floats on top of the saline groundwater. The numbers indicate the specific weight of the groundwater, as determined by borings. The specific weight is highest in the Kara-Bogaz Bay (1160 $cm^3\ cm^{-3}$). The high evaporation rate causes the water level in the bay to lie lower, so water from the Caspian Sea continuously flows through a narrow connecting channel into the bay.

groundwater table fell from 2.9 m in 1914 to 3.4 m in 1918, and the soluble salts in the groundwater increased from 2.6 $g\ l^{-1}$ in 1914 to 5.3 $g\ l^{-1}$ in 1918. Thus, it is not possible to afforest the sandy soils in such dry areas. At best one can lay out narrow windbreak strips.

In some areas, larger or smaller natural woodlands occur (Fig. 3.25). In region I, on the Naryn sands with especially good water relations, stands of the following occur: *Populus hybrida*, *P. alba*, *Elaeagnus angustifolia*, *Salix caspica*, and *S. rosmarinifolia*. These remind one of carr vegetation on other continents. *Clematis orientalis* climbs up the trees; *Rosa cinnamomea* forms thickets in the underbrush; reed thickets of *Holoschoenus vulgaris*, *Glycyrrhiza glabra*, *Equisetum ramosissimum*, and *Phragmites communis* are found in wet depressions; *Rhamnus cathartica*, with the Middle Asian elements *Calligonum aphyllum* and *Eremosparton aphyllum*, grows on slopes; and *Tamarix* species are found around salt pans.

Region II is more brackish. Between the sandhills one can find *Calligonum* and *Eremosparton*, and only individually *Populus* or *Elaeagnus*. On flat areas overlying brackish groundwater, *Nitraria schoberi* dunes form a nebka landscape, while *Halocnemum strobilaceum* indicates solonchak soils.

Woody plants are completely absent in region III. *Ceratoides papposa* (*Eurotia ceratoides*) grows in clumps. *Anabasis aphylla* grows on slightly loamy soils and *Halocnemum* on solonchak.

In region IV only *Tamarix ramosissima* and *T. gracilis* are found in the depressions between sandhills, whereas *Calligonum* and *Eremospartum* are found more as individuals.

In the coastal region V, woody plants are found only along small rivers and their dried-up arms, where the meadow-solonchak soils occur. In addition to *Tamarix*, *Kalidium caspicum* and *Halostachys caspica* can be found. Groups of *Salix* species appear in the reed stands on the sea coast.

With the exception of region I, the flora is rather poor and primarily Turanian. Colonization occurred in post-glacial time, proceeding from the east. The Yergeni Hills in the west form a sharp floristic boundary.

Afforestation experiments have been carried out with some success in region I. In the other regions plantations are more local, around settlements or along railroad lines.

Larin et al. (1954b) divided the southern part of the Caspian Lowland into 17 natural geobotanical units. These authors also stress the Naryn sand region southeast of Urda as being especially typical of the region. The soil water content at various sites is shown in Fig. 3.26.

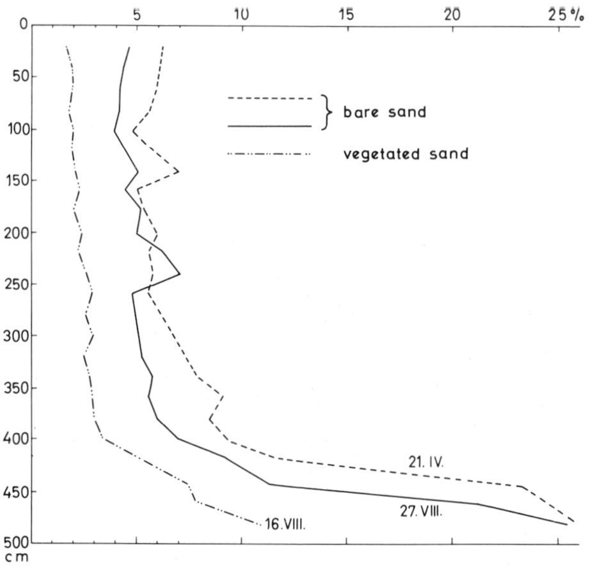

Fig. 3.24. Water content of bare and vegetation-covered sandy soil (from Yakubov; in Walter, 1968, fig. 523). Ordinate: depth of sampling. The water consumption by the plants causes the water content of the vegetated sand to be lower at all depths.

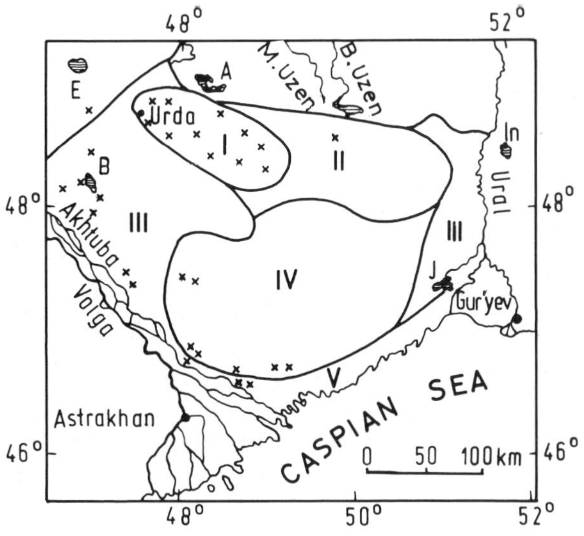

Fig. 3.25. Map of the sandy semi-desert (southern Caspian Lowland), with artificial forests marked by crosses (adapted from Yakubov, 1955). More details are given in the text. *In* = Lake Inder; *E* = Lake El'ton; *B* = Lake Baskunchak; *A* = Lake Aralsor; *J* = Lake Zhaltyr.

INTERFERENCE OF MAN IN THE CASPIAN LOWLAND

Interference by man in the Caspian Lowland goes back a very long way. Stone tools from the period 3000 to 2000 B.C. have been found, as well as Bronze Age tools from the Scythian period (700 to 500 B.C.). Various nomadic tribes (Sarmats, Khosars, Huns, Bulgars, Mongols) successively came through the area up into the thirteenth century, and the Golden Horde had its seat of power at Akhtuba on the lower Volga. At that time, however, the vegetation cover was only temporarily affected and was always able to recover. In the sixteenth century the Nogay inhabited the region between the Volga and the Ural but were driven out in the seventeenth century by the Kalmyks. It was not until 1801 that the present inhabitants, the Kazakhs with their several million sheep, came into the region. At this point the destruction of the plant cover and the blowing away of the sand, especially around the watering places, began. The heavy use of what little wood was available contributed to vegetation and soil destruction.

The result was that the good grazing areas became fewer and fewer. In drought years there was considerable loss of livestock. In the nineteenth century the total size of the herds of the Inner Bukeyev Horde, for example, fluctuated between 1.2 million and only 0.13 million.

The area of barkhans increased steadily, the result also of the uncontrolled agricultural use of sand areas by the Astrakhan Cossacks (Russian: *kazaki*). Even the shipping on the Volga was affected, since an average of 1.4 million m^3 of sand was blown annually into the Volga.

Questions of improved grazing in the steppe, semi-desert, and desert zones of the U.S.S.R. are discussed, for the individual regions, including also parts of the Caspian Lowlands, in 28 contributions to a collective volume (Larin, 1954a). We must content ourselves here with a general overview of the pertinent measures which can be taken.

The northern Caspian Lowlands remained relatively unchanged by man until well into the present century. The number of good watering places was small, so the plant cover was only lightly grazed by the nomads' herds in a manner similar to the original grazing by the wild animals. Today the wild horses (*tarpans*) and the wild asses (*kulans*) have been completely exterminated, while the wild antelopes (*Saiga tatarica*) are only to be found in a

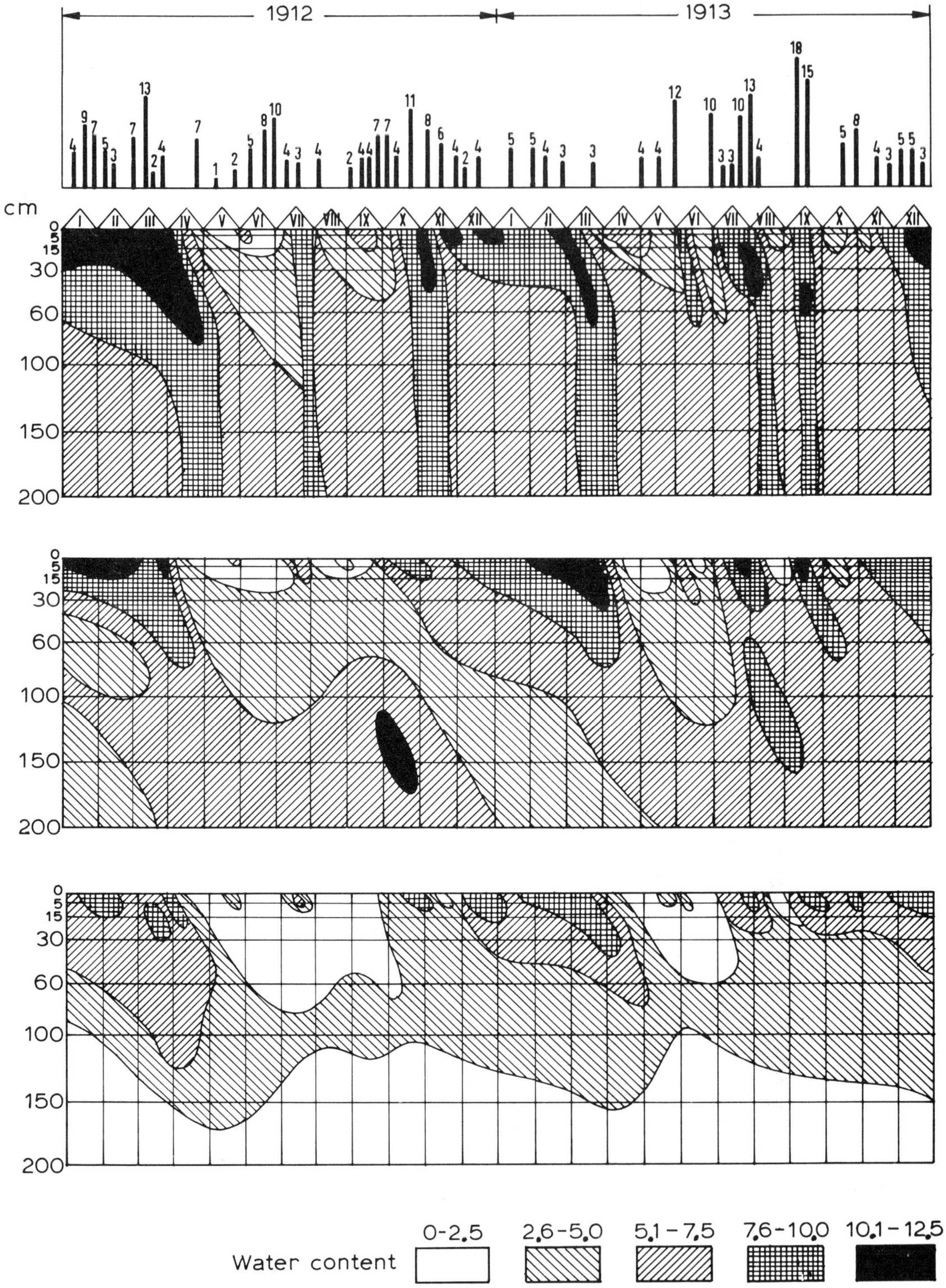

Fig. 3.26. Precipitation amounts (mm) during 1912–1913 and corresponding water content of the sandy soils: in a dune valley between barkhans (top), on a level dune area (middle), and in vegetation-covered sand in which no seepage reaches the groundwater level (by Tomashevski, from Yakubov, 1955, p. 399).

few protected areas. On the other hand, the hare (*Lepus europaeus*) and various rodents, such as species of the genera *Microtus*, *Citellus*, *Lagurus*, and others, are common everywhere.

Measures for the more intensive development of the region have been taken only in the last decades (Tsatsenkin, 1954; Levina, 1960, 1964).

Only the larger depressions with steppe vegetation are suited to agriculture (summer wheat). Grazing of the natural vegetation by sheep, goats, and camels occurs everywhere else. The herbage production of the natural pastures is given as 0.7 to 0.9 t ha^{-1} in rainy years and 0.4 to 0.5 t ha^{-1} otherwise. For grazing by horses or cattle, only the flat areas, in which the steppe vegetation makes up 30 to 40% of the total, and the meadow-like areas around the limany, are suitable (Levina, 1956).

When grazing is too heavy or irrational, the plant cover suffers. First the perennial steppe grasses disappear; *Artemisia* and *Tanacetum* are more resistant. The numbers of annuals and biennials, especially *Poa bulbosa* var. *vivipara*, increases. Finally, only grayish-white, primarily annual weeds remain, such as *Ceratocarpus arenarius*, *Bassia sedoides*, and occasionally *Atriplex lacinatum*. After these disappear, the ground is completely bare.

Since 1950 studies on the most rational use of the natural pastures have been underway at the experimental station at Dzhanybek under the direction of Larin (1954a, 1954b; Gordeyeva and Larin, 1965). The pasture has its best grazing quality in the spring, when the perennials come out and the ephemerals develop, but it is at this time that the plant cover suffers most from grazing. Rotation thus becomes necessary. A five-year rotation on 30 individual paddocks has proved to be sufficient in this region. Other rotation patterns must be tried out in other regions, especially for the larger livestock. A significant improvement in production can only be gained, however, if the solonetz soils and the solonetz-containing chestnut earths can be desalinized and the water supplies can be improved (Rode, 1974). The first condition requires a deep plowing of the soil in order to bring to the surface the gypsum deposits at a depth of 35 to 50 cm. The second condition requires the accumulation of snow on the ameliorated areas.

As a result of these measures the soils will become much more uniform, even though the micro-mosaic structure will not disappear completely. The permeability of the soils would increase three- to six-fold, the sodium content would decrease rapidly, and the wetting of the solonetz soils would increase by two- to five-fold.

The leaching out of the various ions occurs in the following sequence:

$$Cl^- > Na^+ > SO_4^{2-} > Mg^{2+}$$

Fifteen woody species were tried out as woodland strips. The best proved to be *Ulmus pinnato-ramosus*, a garden variety which is quite close to the Mongolian *Ulmus pumila* (Erpert, 1960). After this species came *Lonicera tatarica*, *Ribes aureum*, and *Elaeagnus angustifolia*, but the latter was later dropped. After numerous experiments the best manner of planting, the most favorable density, and the most rational distance between strips was determined [see the very detailed description in the contributions of Rode (1974) and of Olovyannikova (1976)]. If the plantings are too close together, the plants use up more water than can be supplied by snow accumulation.

On the areas roughly 40 m wide between strips are planted either feed crops (*Agropyron desertorum*, *Bromus inermis*, *Onobrychis*, *Medicago*) or summer wheat or silo-sorghum.

The improved area can make up 25% of the total area. The snow which is blown off the other 75% accumulates on this area.

Yields can be increased greatly, however, only if the soils are also irrigated. New problems are presented, however, by the irrigation of these heavy soils in the almost perfectly level terrain with no possibility for the drainage necessary to prevent salinization resulting from the rising groundwater table. These problems are not easy to solve.

In the sandy southern part of the Caspian Lowlands the problems are completely different. Freshwater lenses are well distributed throughout the regions of open sands. The groundwater table is very high in the dune valleys, so that natural wooded areas, as already mentioned, occur frequently. The main problem here is the shifting sand masses set in motion by excessive grazing and wood use.

Afforestation experiments in the region of the Naryn sands near Urda have been going on since the end of the previous century. Of the 4.5 million hectares of sand in the region between the Volga and the Ural, about 0.5 million hectares are found here. The prohibition of grazing here quickly led to a two- to three-fold increase in the wooded areas, as the wood-covered areas quickly spread into the dune valleys. During afforestation it is very important that the young plants are able to reach the groundwater with their roots in a short period of time. This is only possible in the dune valleys, since the maximum height of a capillary water column in sand is about 50 cm, of which 40 cm are attained within six hours. Rain water sinks very rapidly into the sand. The field capacity is only about 3 to 4.5% and the wilting point is about 0.5%, so that only up to 4% of the total pore volume can be filled with available water. This is too little for trees. Even grass needs so much water that there is no downward infiltration of water in sands under grass covers (Fig. 3.26).

The bare sands are, however, not completely sterile. As soon as the sand surface becomes stationary for a while, there forms a green film of cyanophytes (*Phormidium* and *Oscillatoria*) and *Chlamydomonas*, as well as diatoms (*Pinnularia* spp.). Consequently, decomposers (*Amoeba*, *Bodo*, *Oxytrichia*) and bacteria become established. Even nematodes have been observed (Zasukhin, 1930). This leads to the formation of a light crust, which holds more water in the top layer and permits more water to be lost by evaporation.

The most suitable species for afforestation in sandy areas have proven to be *Pinus sylvestris*, *Robinia pseudacacia*, *Alnus glutinosa*, poplar hybrids, and others. Cherries, pears, apples and wine grapes can be cultivated around settlements. Fresh groundwater which can be reached by the root systems is always a precondition. Gayel (1952, 1954) discusses all these questions in detail.

The blowing away of readily mobile sand can be prevented through spraying with a bituminous emulsion (Sakharov and Revut, 1954).

If it is possible to stabilize the sands through afforestation, which will lead to the recolonization of the dunes, then these areas can again be grazed if certain precautions are taken.

In contrast to the light-colored sands, the "black earths" have a particular importance in the Astrakhan region as winter pasture. The question of improving the quality of the plant cover by annual burning has also been studied here (Chizhikov, 1954). In general the pastures are not burned because *Artemisia*, which is especially important for feed in winter, is heavily damaged by fire. But range fires always occur as a result of lightning or carelessness, and the plants are still dense enough for the fire to spread. For example, in 1947 about 80 000 ha, or 7% of the total area, burned (Chizhikov, 1954).

A comparison of the composition of the pastures one (Plot 1) and four (Plot 4) years after fire, relative to an unburned area (Plot 0), can be made from Table 3.5.

A year after a fire the total cover is less. Those species which are favored by fire include the grasses *Festuca*, *Agropyron sibiricum*, and *Koeleria*, *Carex*, and the forbs *Kochia*, *Ephedra*, and *Pyrethrum*. *Stipa* is first damaged but recovers after four years. *Artemisia lercheana* is depressed considerably, and the lichens disappear completely for a longer period.

The generative phase of the grasses is favored by fire. For each 100 vegetative shoots, the number of flowering shoots increased after the fire by the following percentages:

Stipa capillata	121%
Festuca sulcata	133%
Agropyron sibiricum	339%
Koeleria gracilis	110%

The grasses rejuvenate well on the burned areas, since older tussocks are annihilated by the fire. The number of seedlings up to two years old increased by 658% for *Festuca sulcata* and by 530% for *Stipa sareptana*.

The geophytes *Allium paniculatum* and *Iris pumila* were not damaged by the fire. The bulblets of *Poa bulbosa*, on the other hand, lie so near the surface that they are burned. *Poa bulbosa* never completely disappeared from the burned areas, however. If the root crown is protected by drifted sand or dust, then the plants survive better.

In general, regular burning is not to be recommended because of the reduction in above-ground phytomass in the first year after the fire. This is especially true for winter pastures. When too much dead plant matter has accumulated on old pastures, however, then fire can be used to rejuvenate the stand. One should do the burning in the spring, when the fire will not be as hot, and the surviving plants will be less damaged. Of course, wind erosion

TABLE 3.5

Species percentages on unburned (Plot 0) areas and on neighboring areas, one (Plot 1) and four (Plot 4) years after fire (Chizhikov, 1954)

	Plots			
	0	1	0	4
Number of site samples:	23	47	6	125
Total cover (%):	19.5	15.0	22.0	23.0
Stipa capillata	24	19	58	67
Stipa sareptana	24	8.5	–	–
Stipa lessingiana	7	5	0.5	0.5
Festuca sulcata	14	20	0.5	0.5
Agropyron sibiricum	27	31	9	3
Agropyron desertorum	8	7	–	–
Poa bulbosa	2	2	0.5	1
Koeleria gracilis	6.5	12.5	2	3
Kochia prostrata	11	15	4.5	2
Artemisia lercheana	11	4	2	0.5
Artemisia scoparia	–	–	0.5	0.5
Ephedra distachya	1	2	0.5	2
Pyrethrum achilleifolium	0.5	1	–	–
Carex uralensis	1	2	–	2
Other species	4	3	18	17
Lichens (*Parmelia vagans*)	–	–	4	–

can cause further damage on burned areas if the soil is light.

The main problem for livestock-raising in the semi-deserts is the irregular sequence of drought years. These cause great losses of livestock resulting from lack of feed.

THE VOLGA–AKHTUBA FLOODPLAIN AND DELTA

This wetland region is an oasis in the middle of the semi-desert and desert region. As a result of the varying water level, though, the original vegetation development was very peculiar (Fursayev, 1954; see also Walter, 1974).

The high water does not come here at the beginning of the growing period, as elsewhere, but rather divides the growing period into two phases. The highest level (+4 m) is attained in mid-June. Spring begins in the south in March, long before thawing has begun on the upper courses. The ice cover on the middle Volga remains, on the average, not over 110 days, but above Kazan' it remains 140 to 150 days. For this reason, spring annuals on the flood plain are found which complete their life cycles before the coming of high water at the end of April or beginning of May. Included are, for example, *Ranunculus pedatus*, *Stenophragma thalianum*, *Androsace elongata*, *Draba nemorosa*, and *Cerastium anomalum*, and in the Volga–Akhtuba region, *Alyssum desertorum*, *Chorispora tenella*, *Agropyron orientale*, *A. prostratum*, and others.

When the high water falls at the end of July, other quick-growing species appear, which we can call summer ephemerals (e.g., *Scirpus hamulosus*, *S. supinus*, *S. michelianus*, *Marsilea aegyptiaca*, *Vandellia pyxidaria*, and others). These require only 10 to 15 days to develop, bloom, and produce ripe fruit. The perennial plants, which are covered by the high water in the middle of their development and remain under water for up to three months, lose parts of their above-ground shoots (*Agropyron repens*, *Tournefortia sibirica*) or lose their leaves and keep only the stem with the terminal buds (*Euphorbia uralensis*, *Cynanchum acutum*). Others possess the ability to develop aquatic forms under water (*Butomus umbellatus*, *Lysimachia nummularia*, *Nasturtium brachycarpum*, and others). The swampy regions on the lower Volga are known for the Indian lotus flower. It has not yet been finally decided whether this is the true *Nelumbo nucifera*, which was perhaps accidentally brought in from India, or a relict species, known as *Nelumbo caspica*, which has a smaller number of seeds on the placenta and a few other minor differences. This relict grows normally in the Talysh of the southeast Caucasus. An overview of the most important plant communities and of the successional sequences in the years 1935 to 1936 is given by Fursayev (1954).

Also interesting here is the occurrence of the pelican, a southern bird, and a number of herons. In order to protect these uncommon species, a nature-protection area was set up in 1919 which consists of three parts: a western, a central, and an eastern coastal strip of the delta. This is a paradise for birds. Pelicans nest here on floating islands which they make out of dry reeds, twigs from the pastures, etc. There can be up to 700 nests densely packed together next to each other. The rich fishing grounds provide their sustenance. A pelican eats about 1.5 to 2 kg of fish per day.

In addition to the countless bird species, wild pigs (*Sus scrofa*), ermine (*Mustela erminea*), foxes (*Vulpes vulpes*), and less often wild cats (*Felis syl-*

vestris) and fish otters (*Lutra lutra*) can be found in the reed thickets. In the winter, wolves (*Canis lupus*) from the north occasionally wander in.

The situation in the Volga–Akhtuba marsh area is fundamentally changed today as a result of the regulation of the Volga. The high-water peaks are caught by the reservoirs, so that the flooded areas have decreased. Land for cultivation can be obtained by building high-water dams, and these fields can be irrigated rationally. The entire area of the marshes amounts to 1 350 000 ha, of which 7% was cultivated in 1968. Vegetables, especially tomatoes, rice, pumpkins, and melons, as well as fruit and wine grapes, are cultivated. Areas which are flooded for a short period are used as meadows and pastures. The floodplain meadows which are under water for a long time in the spring are also spawning grounds for fish, and the fish are quite abundant in the many arms of the river.

Settlement by man in the marsh region is still quite low: two to three people per square kilometer.

REFERENCES

Anonymous, 1968. *Soviet Union, Russian Federation. The Southeast of Europe.* Nauka, Moscow, 795 pp. (in Russian).

Beydeman, I.N., Bespalova, S.G. and Rakhmanin, A.T., 1962. *Ecological–Geobotanical Investigations in the Kura–Araks Plain of Transcaucasia.* Bot. Inst. Akad. Nauk S.S.S.R., Moscow, Leningrad, 464 pp. (in Russian).

Bolshakov, A.F. and Rode, A.A., 1972. Soils of the solonetzic complex in the northern part of the Pre-Caspian Lowland and their biological productivity. In: L.E. Rodin (Editor), *International Symposium on Eco-physiological Foundations of Ecosystem Productivity in Arid Zones.* Nauka, Leningrad, pp. 122–124 (in Russian).

Chizhikov, O.I., 1954. The influence of burning on the composition and productivity of the winter grazing pastures in the semi-desert of the Pre-Caspian Lowland. *Pustyn. S.S.S.R. Ikh Osvoyeniye*, 2: 291–302 (in Russian).

Dubyanski, V., 1911. *Vegetation of the Russian Sand Deserts.* Nauka, Moscow, Leningrad, 414 pp. (in Russian).

Erpert, S.D., 1960. Peculiarities of the growth of annual shoots of the microphyllous elm (*Ulmus pinnato-ramosa* Dieck). *Soobch. Lab. Lesoved., Akad. Nauk S.S.S.R.*, 2: 88–196 (in Russian).

Fursayev, A.D., 1954. The vegetation of the Volga–Akhtuba floodplains and their improvement, in perspective. In: A.A. Rode (Editor), *Biogeocoenologic Basis for Cultivation of the Northern Pre-Caspian Lowlands.* Akad. Nauk S.S.S.R., Laboratoriya Lesovedeniya, Moscow, pp. 100–107 (in Russian).

Gayel, A.G., 1952. *Afforestation of Dune Sands in Arid Regions.* Nauka, Moscow, 218 pp. (in Russian).

Gayel, A.G., 1954. The sands of the region between the rivers Volga and Ural, and the possibility of their fixation and afforestation. *Pustyn. S.S.S.R. Ikh Ozvoyeniye*, 2: 430–448 (in Russian).

Gordeyeva, T.K. and Larin, I.B., 1965. *Natural Vegetation of the Pre-Caspian Semi-Desert As a Grazing Basis for Livestock.* Bot. Inst. Akad. Nauk S.S.S.R. Nauka, Moscow, Leningrad, 160 pp. (in Russian).

Keller, B.A., 1925/26. Die Vegetation auf der Salzböden der russischen Halbwüsten und Wüsten. *Z. Bot.*, 18: 113–117.

Keller, B.A., 1927. Die Halbwüste bei Krasnoarmeysk (Sarepta). In: *Vegetationsbilder, Series 18, Heft IV.* Fischer, Jena.

Keller, B.A., 1928. Die Vegetation der Salzböden in der grossen Halbwüste des Bundes d.S.S.R. In: *Vegetationsbilder, Series 18, Heft VIII.* Fischer, Jena.

Keller, B.A., 1930. Die Erdflechten und Cyanophyceen am unteren Lauf der Wolga und des Ural. In: *Vegetationsbilder, Series 20, Heft VIII.* Fischer, Jena.

Keller, B.A. (Editor), 1936. *The Vegetation of the Pre-Caspian Lowland Between the River Volga and Ural, I.* Akad. Nauk S.S.S.R., Moscow, Leningrad, 295 pp. (With contributions by B.A. Keller, G.N. Novikov, and F.F. Leisle; in Russian).

Kovda, V.A., 1950. *The Soils of the Pre-Caspian Lowland.* Nauka, Moscow, Leningrad, 353 pp. (in Russian).

Larin, I.V. (Editor), 1954a. *Problems of Improvement of the Quality of Fodder Resources in the Steppe, Semi-Desert, and Desert Zones of the U.S.S.R.* Akad. Nauk S.S.S.R. Bot. Inst., Leningrad, 386 pp. (in Russian).

Larin, I.V. et al., 1954b. The main regularities in the distribution of vegetation and the divisions of the Pre-Caspian Lowland between the rivers Volga and Ural. In: I.V. Larin (Editor), *Problems of Improvement of the Quality of Fodder Resources in the Steppe, Semi-desert, and Desert Zones of the U.S.S.R.* Akad. Nauk S.S.S.R., Bot. Inst. Leningrad, pp. 3–30 (in Russian).

Levina, F.Ya., 1956. The vegetation of the meadows and limany in the northern part of the Volga–Ural inter-riverine region. In: *Priroda i kormovye osobennosti rastitelnosti limanov Volga-Uralskovo mezhduryechya.* Nauka, Moscow, pp. 9–257 (in Russian).

Levina, F.Ya., 1960. The basic eco-biological and phytocoenological peculiarities of the semi-desert flora in the Volga–Ural inter-riverine region of the northern Pre-Caspian Lowland. In: *Voprosy senokosno-pastbishchenovo khozyaystva.* Nauka, Moscow, pp. 386–410 (in Russian).

Levina, F.Ya., 1961. Once more concerning the semi-desert as a zone. *Bot. Zh.*, 46: 728–731 (in Russian).

Levina, F.Ya., 1964. *The Semi-Desert Vegetation of the Northern Pre-Caspian Lowland and the Quality of Its Fodder.* Akad. Nauk S.S.S.R., Moscow, 336 pp. (in Russian).

Miroshnichenko, Yu.M., 1966. Succession of the soil-vegetational cover of the maritime zone of the northwestern Cis-Caspian area. *Bot. Zh.*, 51: 520–541 (in Russian).

Miroshnichenko, Yu.M. and Miroshnichenko, N.V., 1972. The influence of the sinking level of the Caspian Sea on succession and vegetational productivity. In: L.E. Rodin (Editor), *International Symposium on the Eco-physiological Foundations of Ecosystem Productivity in Arid Zones.* Nauka, Leningrad, pp. 124–128 (in Russian).

Nikitin, S.A., 1954a. The vegetation of the eastern part of the Pre-Caspian Lowland. *Pustyn. S.S.S.R. Ikh Osvoyeniye*, 2: 216–264 (with vegetation map 1: 1 250 000; in Russian).

Nikitin, S.A., 1954b. Vegetation between the rivers Kushum, Ural, and Emba and its fodder resources. In A.A. Rode (Editor), *Biogeocoenologic Basis for Cultivation of the Northern Pre-Caspian Lowlands.* Laboratoriya Lesovedeniya, Moscow, pp. 58–86 (in Russian).

Olovyannikova, I.N., 1976. *The Influence of Woodland on the Solonchak-like Solonetz Soils.* Akad. Nauk S.S.S.R., Laboratoriya Lesovedeniya. Moscow, 126 pp. (in Russian).

Rode, A.A. (Editor), 1974. *Biogeocoenologic Basis for Cultivation of the Northern Pre-Caspian Lowlands.* Akad. Nauk S.S.S.R., Laboratoriya Lesovedeniya, Moscow, 360 pp. (Contributions of G.S. Basykina, A.F. Bolshakov, S.D. Erpert, G.P. Maksimyuk, and A.A. Rode; in Russian).

Sakharov, N.G. and Revut, I.G., 1954. Stabilization of mobile sands with bituminous emulsions. *Pustyn. S.S.S.R. Ikh Osvoyeniye*, 2: 449–464 (in Russian).

Terenozhkin, I.I., 1954. The ecology and distribution of *Agropyron* species of the Southeast. *Pustyn. S.S.S.R. Ikh Osvoyeniye*, 2: 279–290 (in Russian).

Tsatsenkin, I.A., 1954. Measures for improvement and utilization of pastures and meadows in the western part of the Pre-Caspian Lowland and the Yergeni. *Pustyn. S.S.S.R. Ikh Osvoyeniye*, 2: 202–215 (in Russian).

Walter, H., 1968. *Die Vegetation der Erde, II. Die gemässigten und arktischen Zonen.* Fischer-Verlag, Jena–Stuttgart, 1001 pp.

Walter, H., 1974. *Die Vegetation Osteuropas, Nord- und Zentralasiens VII. Vegetationsmonographien der einzelnen Großräume.* Fischer-Verlag, Stuttgart, 452 pp.

Walter, H., 1976. *Die ökologischen Systeme der Kontinente (Biogeosphäre).* Fischer-Verlag, Stuttgart, 131 pp.

Walter, H., 1979. *Vegetation of the Earth and Ecological Systems of the Geo-Biosphere.* Springer-Verlag, New York, N.Y., 2nd ed., 274 pp.

Walter, H. and Lieth, H., 1960–1967. *Klimadiagramm-Weltatlas.* Fischer-Verlag, Jena.

Wulff, E., 1926. Vegetationsbilder der Krim. In: *Vegetationsbilder, Series 17, Band 1.* Fischer, Jena.

Yakubov, T.F., 1955. *Sand Deserts and Semi-deserts in the Northern Pre-Caspian Lowland.* Akad. Nauk, S.S.S.R., Pochvennyi Inst., Moscow, 530 pp. (in Russian).

Zasukhin, A.N., 1930. Materials to the problem of micro-organisms living in the mobile sands of the Kirghiz steppe. *Gidrobiolog. Zh.*, 9(4–6) (in Russian).

Chapter 4

SEMI-DESERTS AND DESERTS OF CENTRAL KAZAKHSTAN[1]

H. WALTER and E.O. BOX

GENERAL SITUATION

The entire region of Kazakhstan covers an area of 2.7 million km^2, stretching from 55.5°N to 41°N (over 1500 km) and from 46.5°E to 87.3°E (over 3000 km) (Fig. 4.1). The northern part belongs to the steppe zone, the western part (regions *4–6*) is a part of the Caspian Lowland, which was discussed in Chapter 3, and the eastern part (regions *33–46*) is mountainous. To be considered here are the regions *18* to *21* of semi-desert or desert-steppe, and the regions *26* to *27* of desert — all in Central Kazakhstan. The regions *24* to *25* and *28* to *32* will be treated under Middle Asia (Chapter 5).

In Central Kazakhstan only the region outlined by dots in Fig. 4.1, with the regions *19*, *20*, and *27*, has been ecologically studied in any detail (Fig. 4.2). (Anonymous, *Biocomplex Investigations in Kazakhstan*, 1969a, b, Volumes I and II).[2] No corresponding record exists for the other areas.

The region which has been studied represents 30 000 km^2 and stretches from north to south over 700 to 750 km and from west to east over 400 to 450 km. It is bounded in the south by the River Chu which separates the Betpak-Dala Desert from the sand desert of Muyunkum (Fig. 4.1). Within the region studied, continuing "stationary" (permanent plot) studies were carried out on three experimental areas (see Fig. 4.2, *IX*), of which the northern area occurs in the dry steppe, the middle in the semi-desert, and the southern in the desert.

CLIMATIC CONDITIONS

Most of the climatic data available are for the semi-desert or "desert steppe", which lies between roughly 48° and 49°N.

The climate of this zone is transitional between steppe and desert. The net energy income is 108 to 117 kJ cm^{-2} annually (in the steppe only 91 to 108 kJ cm^{-2}). The annual potential evaporation exceeds the annual precipitation by seven-fold in the northern part of the desert-steppe and by twenty-fold in the south. The annual precipitation ranges from 155 to 279 mm, of which 57 to 73% falls in the warm season.

The mean annual temperature ranges from 3° to 5°C. The winter is cold, lasting five months, during which calm, clear, high-pressure weather dominates. Advancing thrusts of warm air-masses with cloudiness and high winds, though, are not unknown during this period.

The mean January temperature is –15° to –20°C, with a mean minimum of −26°C and absolute minima around −40°C or colder. Snow falls from October to April. A snow cover remains from the middle of November until spring and is generally 20 to 30 cm thick. The mean length of the period of snow cover is 120 to 140 days, but in some years a snow cover may be completely lacking.

The mean July temperature is 22° to 25°C, but maxima of 40° to 45°C are not uncommon. The number of days with mean temperature about 25°C ranges from 30 to 45, and the number of days with temperature above 35°C is 10 to 20.

The growing season (mean temperature above 10°C) lasts 150 to 180 days (in the steppe only 140 to 150 days). Frosts until May 10 to 20 or even June 1 and early frosts from September 20 to October 10 are typical.

[1] Manuscript completed November, 1977.

[2] Volume III (1976) was received after the manuscript was completed. It is discussed on pp. 68–71.

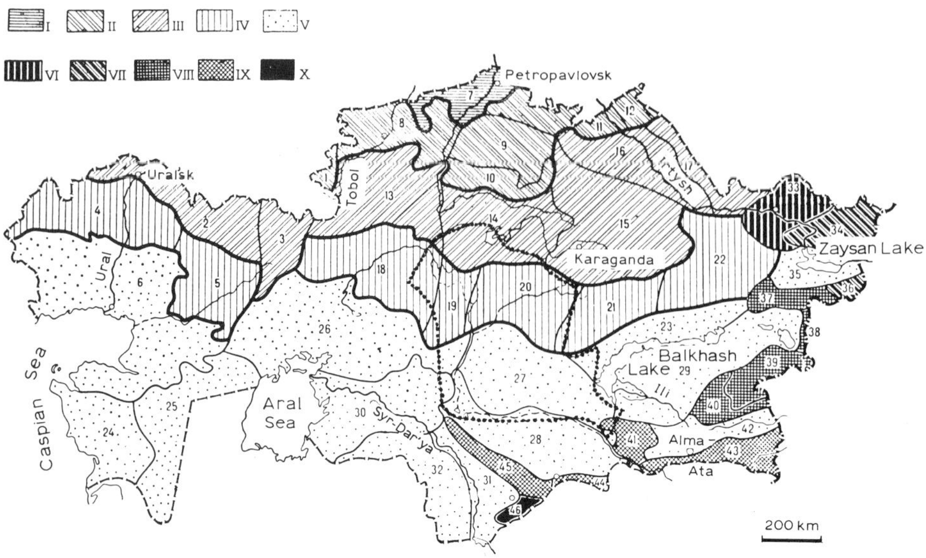

Fig. 4.1. Vegetation zones of the northern Kazakhstan region (from Gvozdezkiy and Nikolayev, 1971): *I* = forest-steppe: *II* = steppe; *III* = dry steppe; *IV* = desert-steppe (semi-desert); *V* = desert (southern semi-desert), *VI–X* = montane deserts and steppes. Subdivisions are as follows:

Northwest Kazakhstan: 1 = Trans-Ural steppe (west of the Tobol); *2* = Ilek dry steppe; *3* = Mugodzhary dry steppe; *4* = Volga–Ural semi-desert; *5* = Pre-Ural semi-desert; and *6* = Caspian Desert.

North and Central Kazakhstan: 7 = Ishim forest-steppe; *8* = Kustanay steppe zone; *9* = Kokchetau steppe zone; *10* = Atbasar steppe zone; *11* = Irtysh steppe; *12* = Barabin steppe; *13* = Turgay dry steppe; *14* = Tengiz dry steppe (around Lake Tengiz); *15* = Bayan-Aul dry steppe: *16* = dry steppe along the Irtysh; *17* = Kulundin dry steppe; *18* = Turgay semi-desert; *19* = Ulutau semi-dessert; *20* = Sarysu semi-desert; *21* = Kyzyltau semi-desert; 22 = Chinghiz semi-desert; *23* = northern Balkhash Desert.

Southwest and South Kazakhstan: 24 = Mangyshlak Desert; *25* = Ustyurt Desert; *26* = northern Pre-Ural Desert; *27* = Betpak-Dala Desert; *28* = Muyunkum Desert; *29* = southern Balkhash Desert; *30* = lower Syr-Dar'ya Desert; *31* = Syr-Dar'ya Desert; and *32* = northern Kyzylkum Desert.

East and southeast Kazakhstan: 33 = western Altay; *34* = southern Altay; *35* = Zaysan region; *36* = Saur region; *37* = Tarbagatay region; *38* = Barlyk–Maylin region; *39* = Dzungarian Alatau; *40* = west Dzungarian region; *41* = Chu–Ili region; *42* = Ili region; *43* = Trans-Ili region; *44* = Kirgiz Mountain region; *45* = Karatau region; *46* = Ugam region.

The region circumscribed by the dotted line is the region shown in Fig. 4.2.

Much more extreme are the conditions in the Betpak-Dala Desert. The amplitude of the temperature curve is very great. While the air temperature in summer rises to 42° to 44°C (65° to 70°C on the ground), it also sinks in winter to −40°C. Precipitation is only slightly above 100 mm. Its distribution over the year is irregular, but most of the rain falls in the summer as heavy showers. This rainfall is of little significance to the vegetation. The snow cover is present for $4\frac{1}{2}$ months but is not over 15 to 20 cm thick. The east and northeast winds, primarily in spring and autumn, are an important factor. The average annual wind speed is 6.2 m s^{-1}, reaching a maximum at about 13 h.

The only climate diagram from this entire region is that of Kokchetau, with eight years of observation (Fig. 4.3). It gives the conditions for the semi-desert zone, while the diagram from Semipalatinsk is valid for the steppe zone. Fig. 4.4 shows the corresponding climatogram of the dry steppe at the top and underneath one from the desert-steppe zone.

GEOLOGY AND GEOMORPHOLOGY

The largest part of central Kazakhstan is the remnant of an earlier mountain range, in particular a remnant plateau (Russian: *Kazakhskiy Melkosopochnik*, or Kazakh Hills) with low inselberge (Fig. 4.5). An old, folded mountain range has been almost completely eroded, and its remains are buried today under the loose rubble of Mesozoic and Cenozoic layers. Only in the region of the Betpak-Dala Desert is the topography that of an

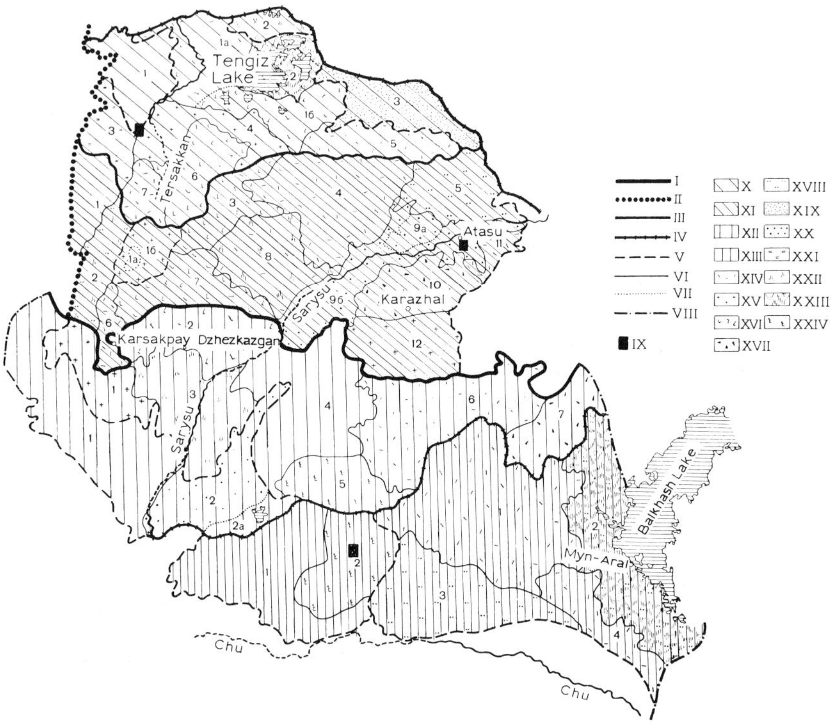

Fig. 4.2. Subdivision of the study region in the western part of the eroded hills of Central Kazakhstan and the Betpak-Dala Desert, with the three permanent study plots indicated by black rectangles (*IX*). *I* = Border between the Eurasian steppes and the Saharo-Gobi deserts; *II–VII* = further subdivisions of the landscape. Zonal patterns: *X* = forb-rich, xerophilic *Festuca–Stipa* steppes; *XI* = *Artemisia–Stipa* desert-steppe; *XII* = grass-rich deserts; *XIII* = grass-poor deserts; *XIV* = series on igneous rocks; *XV* = on granites; *XVI* = on limestone; *XVII* = on quartzites; *XVIII* = on salt-containing sandstone; *XIX* = on slightly sandy soils; *XX* = on sands; *XXI* = on Tertiary, gypsum-containing clays; *XXII* = on partially salty soils; *XXIII* = on solonetz; *XXIV* = bushland with *Spiraeanthus*. Lakes are indicated by horizontal striping (see Walter, 1974, fig. 177). The arabic numerals represent regions in the geobotanical subdivision of the U.S.S.R. (see Anonymous, 1969a).

arid, denuded high plateau. Paleozoic rocks, primarily of Silurian, Devonian, and Carboniferous age (sandstones, conglomerates, clay shales, and limestones) are quite widespread in this remnant landscape. These layers are interrupted by granite intrusions of various ages. Permian rocks (conglomerates, clays, and sandstones) and chalk deposits (clays, gravels, and sandstones) are less common.

The Silurian layers consist of sedimentary rocks, while the Devonian, on the other hand, are igneous formations (porphyry, lava, volcanic tuffs). Huge, colorful sandstone layers were deposited on the eroded surface of these layers.

Sediments from a shallow Carboniferous sea are widespread in the lower southwestern part of the remnant region. These are primarily limestones, marls, and quartzitic shales.

The climate was arid during the Mesozoic Era, so the erosion products covered the lower-lying formations.

As the result of the transgression of a Tertiary sea, red, green, and often multi-colored, calcareous muds filled up the old valleys.

The Pleistocene deposits were: (1) sandy-loamy inclined layers or rubble cells of different sizes, on which the recent soils developed; (2) alluvial clayey or sandy marine deposits in earlier river valleys; and (3) Pleistocene sandy loams.

The geomorphology of the remnant landscape is just as complicated as the geological structure. The climate remained arid throughout the Pleistocene, but the rivers arising in the mountains during the glacial periods brought more water, and the lower temperatures of the glacial periods reduced the evaporation. A large lake filled the entire Balkhash–Alakol' Lowland and even had an outlet

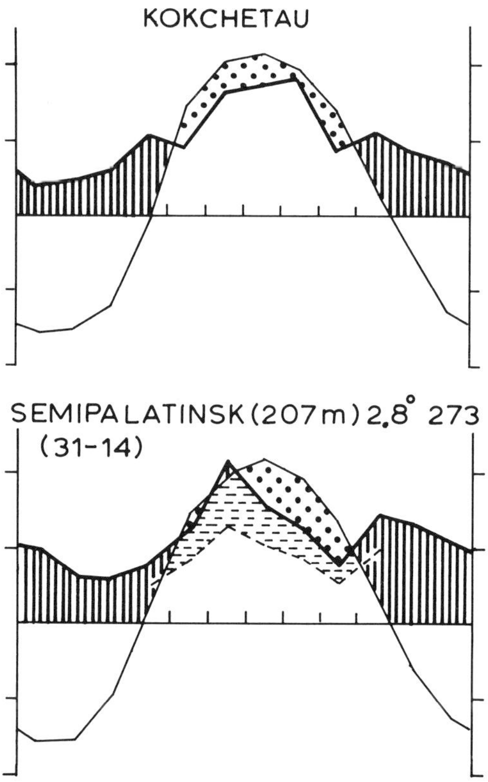

Fig. 4.3. Climate diagram of Kokchetau, on the border of the Kazakhstan semi-desert, and the diagram of Semipalatinsk (more to the north, in the steppe zone) (from Walter, 1974, fig. 176).

to the east, through the Dzungarian Gate, into the Ebi Nor Basin (Ai-pi Hu).

The relative height today of the peaks in the remnant landscape, of which only the summits stick out of the rubble which covers them, varies between 10 and 100 m. For marls and limestones the differences in height are about 30 to 35 m, with limestone ridges of 10 to 15 km length being fairly common. The tops are rounded or flat. The furrowed sides slope about 20 to 23° into salty valleys, in which the salts contained in the sediments and set free by erosion are able to accumulate. Small, rifted plates of 5 to 10 m in height, oriented along the direction of stratification, protrude from sandstone outcroppings. If quartzite shale is present, the slopes are covered with quartzite rocks which are resistant to weathering. Individual cone-shaped peaks may rise only 10 m out of the level, salty plains.

The remnant landscape is an endorheic region, in the middle of which is a large depression and a flat area covered with lakes. The largest lake is Lake Tengiz with an area of 1520 km^2. It has no outlet and is saline. The dense river network consists, in large part, of small watercourses which carry water only in the spring, after the melting of the snow. Only a few large rivers carry water the entire year, as for example the Nura River, which empties into Lake Tengiz. Even in this river 88% of the water flows in the spring, 3% in the summer, 2% in the autumn, and 7% in the winter. The continuously flowing rivers have formed wide valleys. The water in the watercourses becomes increasingly brackish in the summer as the water volume decreases. For this reason, a 490 km long canal from the River Irtysh was built in order to provide drinking water for the mining center of Karaganda, which otherwise must be provided from wells or springs. Groundwater, mostly in fissures, can often be found at a depth of 10 m.

The Betpak-Dala Desert is completely dry. It has no outlet either to Lake Balkhash or to the Chu River. Even the snowmelt in the spring runs off in little furrows and quickly soaks into the ground.

COMPONENTS OF THE ECOSYSTEMS

Soils

The zonal soil type in the semi-desert is a light chestnut earth, which represents the transition between the true chestnut earths of the dry steppe and the burozems (brown earths) of the desert.

Most light chestnut earths are slightly alkaline. The humus horizon is 35 to 40 cm thick. A prismatic or flaky alkaline layer is often found at a depth of 10 to 25 cm. Deeper, at 25 to 35 cm, a calcareous horizon is recognizable, and still deeper, at 80 to 100 cm, gypsum deposits. Readily soluble salts appear at 30 cm. These light chestnut earths, however, cover only small areas within the region, primarily where calcareous loams and clays are present.

The most widespread soil types are the rubbly, poorly developed soils of the slopes of the numerous hills. Their characteristics depend entirely on their lithological substrates. Solonetz and solonchak soils also cover large areas.

In the wetter lowlands and in the river valleys

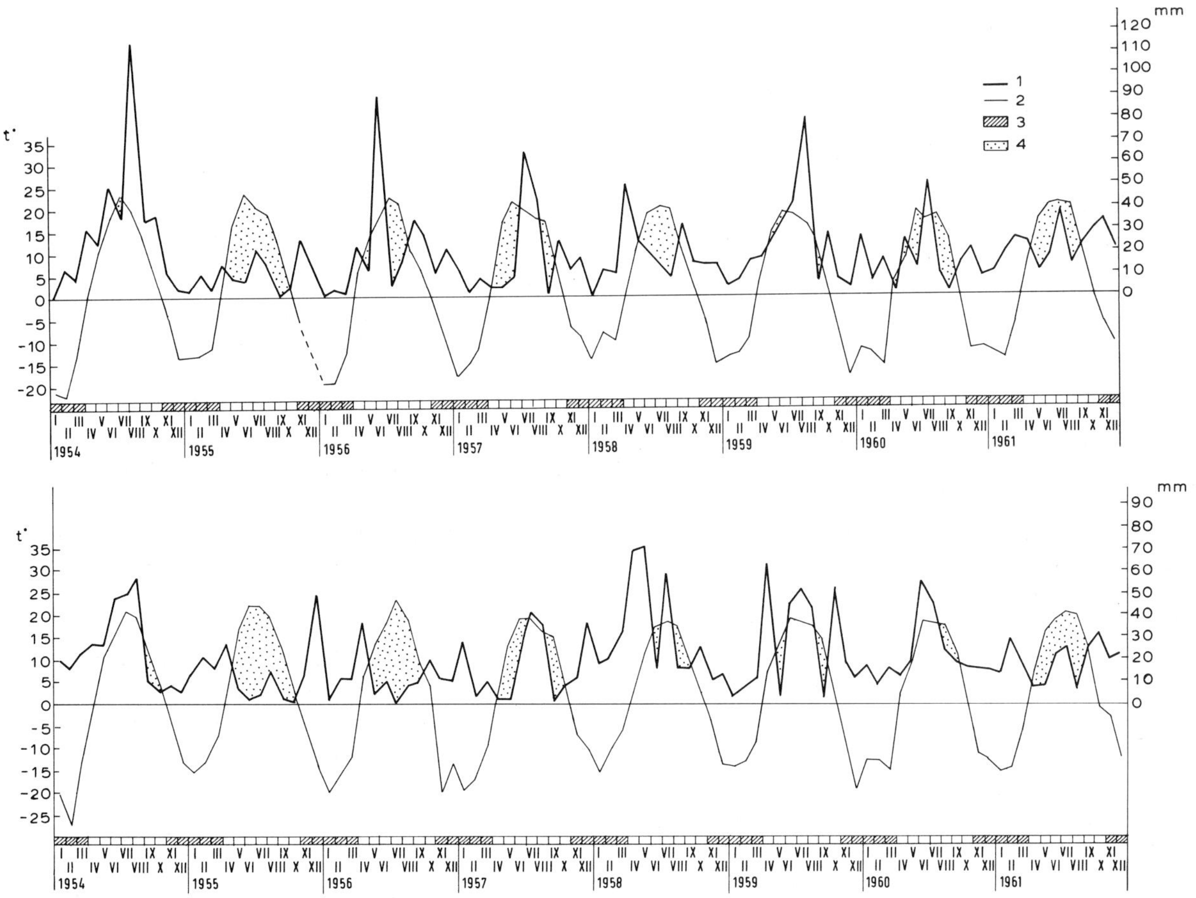

Fig. 4.4. Climatogram (1954–1961) of stations in Central Kazakhstan: from the dry steppe (above) and from the desert-steppe or semi-desert (below). Legend: *1* = precipitation curve; *2* = temperature curve; *3* = months with mean temperature below 0°C (November–March); *4* = drought periods, dotted (from Walter, 1974, fig. 178).

one finds meadow chestnut earths, alluvial meadow soils, and marsh soils.

The zonal burozems or siero-burozems are typical for the drier desert, the former in the northern part of the Betpak-Dala Desert.

The sparse vegetation and the rapid decomposition of organic remains do not permit the humus content of the soils to exceed 2%. Among the humic compounds, fulvic acids predominate. Effervescence with hydrochloric acid occurs right on the surface, although the actual calcareous horizon is found at a depth of 20 to 25 cm. Gypsum deposits appear at 70 to 100 cm. In the western part of the Betpak-Dala one often finds solonetz-like burozems in complexes with solonetz soils. In these, the upper half of the humus horizon is laminated and loose, while the lower half is brown and dense,

Fig. 4.5. Landscape of eroded mountains (Russian: *melkosopochnik*) in the granite region of Central Kazakhstan (Begazy Mountains). In the depression in the foreground is a small meadow with steppe elements (photograph by Bespalova; from Walter, 1974, fig. 175).

with a lumpy structure. The gypsum horizon is at about 40 to 45 cm.

Otherwise the soils on the slopes, even here, are mostly rubbly types of burozems.

The subdivision siero-burozem, in the desert, is the most xeromorphic type of burozem. In the Betpak-Dala region these are shallow, very dry soils, in which the dryness is accentuated by the type of substrate material. The humus content is only 0.8 to 1.6%. The calcareous accumulation occurs right on the surface. Sometimes a second carbonate maximum at a depth of about 50 cm can be found. The easily soluble salts are hardly present at all. Alkalinity is high. Considerable soluble gypsum and sulphates are found at a depth of 60 to 100 cm.

Flora

The plant species of the investigated area have been studied in particular detail, not only the higher plants but also the lower. Precise lists for the individual zones, with data on sites and affinities to particular flora elements, have been published (Anonymous, 1969a, pp. 230–356).

Vascular plants. The richness of the flora decreases with increasing dryness. The numbers of species identified were 487 for the dry steppe, 423 for the desert-steppe (semi-desert), and only 304 in the desert.

As can be seen in Table 4.1, the number of Holarctic species decreases from north to south, while the number of Mediterranean species, which are especially typical for the Turano-Iranian deserts, increases from north to south.

Lichens. Lichens were also carefully collected in this region for the first time. In all, 128 species were found, 94 in the steppe, 54 in the desert-steppe, and 58 in the desert. In the desert-steppe most of the lichens are found on solonetz soils but also on outcroppings of sandstone plates and quartzite. The same is true of the desert zone.

TABLE 4.1

Numbers of Holarctic, Mediterranean, and endemic species in the different zones (Anonymous, 1969a, b)

	Holarctic	Mediterranean	Endemics
Dry steppe zone	50%	48%	30 species
Desert-steppe zone	37.3%	60%	35 species
Desert zone	6.3%	91.4%	53 species

Bryophytes. A total of 31 species of bryophytes was identified, most of which are found only on wet sites. One can identify the following as xerophytic species: *Grimaldia gragans*, species of the genera *Pterygoneurum* and *Tortula*, *Bryum argenteum*, *B. caespiticium*, *Thuidium abietinum*, and of the Hepaticae, *Riccia sorocarpa* and *R. lamellata*.

Algae. Soil algae are very important in dry regions. Numbers of species identified include 99 Cyanophyta, 38 Chlorophyta, 18 Xanthophyta, and 15 Bacillariophyta. Important differences between the various zones could not be identified.

The algae form membranous synusiae on the soil surface but also appear, though more diffusely, to a depth of 3 to 5 cm. The Cyanophyta are most important. One can distinguish several synusiae: (1) algae films composed of *Stratonostoc commune* (primarily in the steppe); (2) similar films which also include *Scytonema ocellatum* and the gelatinous lichen *Collema*; (3) films primarily of *Microcoleus vaginatus*; and (4) films of *Phormidium* (primarily on *takyr* soils; see pp. 91–94). So-called "window-algae" can also be found, i.e., algae which form a green coating on the underside of translucent stones or on their side walls, where moisture remains longer. These are primarily Chlorococcales, but also Ulotrichales or Xanthophyta.

Bacteria. Even the synusiae of microbes were studied — a particularly difficult task. Bacteria were not only isolated and identified from soil samples but were also microscopically observed directly or studied as *Aufwuchs* on objects buried in the soil. The number of isolated species was very large.

Only a few important discoveries will be mentioned. *Clostridium pasteurianum* was very widespread, especially in the horizon of the tap-roots. *Azotobacter* could be isolated from soil under several different plant communities, even in salty soils. It seems to be less salt-sensitive than *Clostridium*. In all, sixteen different species were identified which can assimilate nitrogen from the air, including eight *Bacillus* species and two Actinomycetes.

One group of bacteria can dissolve largely insoluble phosphorus compounds by means of acids which they produce themselves. These compounds can then be drawn into the various material cycles. The Actinomycetes showed themselves to be capable of ammonification.

Fungi. Microscopic fungi were isolated from soil samples and from the rhizosphere of several plant species. In addition, numerous parasitic fungi of the Ustilaginales and Uredinales could be identified.

Vegetation of the desert-steppe

As a result of the very complicated geomorphological situation in this highly eroded area, the vegetation typical for this ecotone is found only on very small eu-climatopic areas on the gentle slopes at the foot of the hills, the soils with the least salt content. The dominants are the petrophyte communities (lithobiomes), the composition of which depends entirely on the type of rock substrate. These cover the slopes completely. There are many halophytic variations. Psammophyte communities are found on sandy alluvial deposits. In addition, meadow communities occur around springs and in floodplains; on sand with high groundwater even a stand of birches and aspens can appear.

Typical vegetation of the desert-steppe zone (semidesert zone). This is composed of two associations dominated by *Stipa*, *Festuca sulcata*, and *Artemisia gracilescens* (Figs. 4.6 and 4.7). In the northern part the *Stipa* species is *S. lessingiana*, while in the south it is *S. sareptana*, without feathery awns. The northern association appears on slightly alkalized light chestnut earths, the southern association on more solonetz-like soils. This is the most southerly type of desert-steppe and, on light, sandy soils, it even penetrates into the desert zone.

The vegetation of the desert-steppe is distinguished from the zonal vegetation of the dry-steppe zone by the large number of the eu-xerophytic semishrub *Artemisia gracilescens*, a typically Kazakh floral element. Additional characteristic species for the desert-steppe are: *Kochia prostrata*, *Ferula caspica*, *Tulipa patens*, *Poa bulbosa*, and the ephemerals

Fig. 4.6. Vegetation dominated by *Stipa lessingiana* in Central Kazakhstan (photograph by Bespalova; from Walter, 1974, fig. 179).

Fig. 4.7. The same *Stipa lessingiana*-dominated vegetation as in Fig. 4.6, but with *Stipa* in the fruiting stage (photograph by Bespalova; from Walter, 1974, fig. 180).

Alyssum desertorum and *Ceratocephalus orthoceras*. Distinguishing species for the southern association, as opposed to the northern, are: *Ferula ferulaeoides*, *Hymenolyma trichophyllum* (Apiaceae), *Rheum tataricum*, and *Rindera tetraspis* (Boraginaceae). It is very striking that *Festuca sulcata* continuously decreases to the south and is completely lacking at the southern boundary of the desert-steppe. The transitional desert-steppe vegetation (ecotone) occurs mostly in complexes with other communities on solonetz soils.

Vegetation on calcareous soils. This vegetation grows both on calcareous light chestnut earths and on calcareous solonetz soils. These soils are found either on old marine deposits or on rounded hills composed of Tertiary, gypsum-containing shales. These communities are not widespread.

The dominance of *Anabasis salsa*, sometimes with *Suaeda physophora* and *Atraphaxis decipiens*, is characteristic.

One can distinguish two series: (1) with *Artemisia pauciflora* ssp. *majcara*; and (2) with *Artemisia terrae-albae* var. *semiarida*. These are transitional to communities in which *Festuca sulcata* and *Stipa lessingiana* appear also. Character-species for these calcareous soils are *Linosyris tatarica*, *Serratula dissecta*, and *S. cardunculus*.

Psammophyte vegetation. This is found on the sand terraces of the Sarysu River. The vegetation consists of *Stipetum pennatae* and *Festucetum beckerii*, the latter on mobile sands. Most widespread, however, are hemi-psammophytic variants of the typical vegetation, which contain *Artemisia marschalliana*, as well as the psammophytes *Gypsophila paniculata* and *Syrenia sessiliflora* and psammo-petrophytes *Ephedra distachya* and *Hulthemia berberidiflora*, as well as some steppe species (*Astragalus testiculatus*, *Galium ruthenicum*, *Verbascum phoeniceum*). On wetter sites the following appear also: *Sophora alopecuroides*, *Glycyrrhiza aspera*, *Aneurolepidium angustum*, and *Calamagrostis epigeios*.

Halophyte vegetation. This occupies a very large area. An especially characteristic species is *Atriplex cana* (*kokpeg*), which occurs in association with various *Artemisia* species (*A. gracilescens*, *A. schrenkiana*, *A. terrae-albae*, *A. nitrosa* and others). Also important are the communities with *Artemisia pauciflora* together with *Psathyrostachys*, *Aneurolepidium*, *Anabasis salsa*, *Nanophyton erinaceum*, and *Halocnemum strobilaceum*. Stands of the very tall grass *Lasiagrostis splendens* are especially typical on slightly salty soils with high groundwater level (Fig. 4.8). *Atriplex cana* with *Camphorosma lessingii* are often found around springs or along the shores of salt lakes.

Of the species named there also occur various combinations and mosaic-like complexes.

Petrophyte vegetation. In place of the feathery *Stipa* species of the steppe zone, there appear in the desert-steppe, on shallow soils, *Stipa sareptana* and *Artemisia*, either *A. gracilescens* or *A. sublessingiana*, mostly in complexes with communities of the solonetz soils (*Atriplex cana*, *Anabasis salsa*, *Nanophyton*). More to the south and not far from the edge of the desert, *Ceratoides papposa* (*Eurotia ceratoides*) is important on the flat rocky tops of the hills. On rubble it is replaced by *Artemisia frigida* or other *Artemisia* species. *Allium galanthum*, *Anabasis truncata* and *Sedum albertii* are also characteristic on these sites.

Shrubs appear on rocky soils: *Caragana balchaschensis* and *C. frutex*, and on the metamorphic rocks in the northern part, also *Spiraea hypericifolia*, with *Festuca sulcata*, *Stipa sareptana*, and *Artemisia sublessingiana*. The number of different combinations of species is very large.

Meadow vegetation. The plant communities on non-saline, moist to wet sites correspond, in general, to those of the steppe, but they become impoverished with increasing aridity of the climate. The more sensitive species disappear: *Vicia*, *Lathyrus*, *Filipendula*, *Sanguisorba*, and others. The grasses *Agropyron repens*, *Bromus inermis*, and *Alopecurus* become more important. The more resistant species (*Aneurolepidium angustum*, *Calamagrostis epigeios*, *Medicago falcata*, *Glycyrrhiza uralensis*) and a number of steppe species increase in number. *Phragmites communis* grows in the water.

Vegetation of the desert zone

The zonal vegetation is distinguished from that of the desert-steppe by the absence of grasses, and by the presence of the more xerophilic *Artemisia* species and the various desert dwarf shrubs. Of course, there is no clearly contrasted zonal vegetation on eu-climatopes, but rather the vegetation merges smoothly with petrophytic variants. *Salsola*

Fig. 4.8. *Lasiagrostis splendens*, commonly called "*chiy*" (photograph by B. Keller, 1928).

arbusculiformis is a very important species, occurring with *Artemisia terrae-albae* and *A. turanica* but also with *Ferula ferulaeoides*, *Caragana balchaschensis* or others.

Ephemerals are quite numerous: *Poa bulbosa*, *Ranunculus polyrhizus*, *Rheum tataricum*, *Megacarpea megalocarpa*, *Rindera tetraspis*, *Euclidium syriacum*, *Tauscheria lasiocarpa*, and in southern variants also *Euphorbia rapulum*, *Koelpinia linearis*, and others.

In the more open, petrophytic variants of the *Salsola* community there appear also semi-shrubs and some forbs and grasses, such as *Thalictrum isopyroides*, *Stipa caucasica*, *S. orientalis*, *Pseudosedum lievenii*, *Rhamatophyllum frutescens*, and *Nanophyton erinaceum*. *Atraphaxis decipiens*, *Artemisia sublessingiana*, and *Silene viscosa* are found on southern slopes. In ravines, besides *Caragana balchaschensis*, are found also *Achnatherum caragana*, *Psathyrostachys juncea*, and a little *Stipa kirghisorum*, with *Ceratoides papposa* more around the edges. The ephemerals grow especially well under the shrubs.

Another type is the *Artemisia* communities on somewhat lighter, only slightly saline soils. *Artemisia terrae-albae* and *A. turanica* dominate in level places, *A. sublessingiana*, *A. juncea*, *A. obtusifolia*, and *A. frigida* in hilly terrain. On rocky and rubble soils there appear *Silene balchaschensis*, *Zygophyllum iliense*, *Tanacetum santolina*, and others.

For the most part, however, the various communities occur in complexes: the complex *Anabasis salsa–Artemisia* spp.–*Salsola arbusculiformis* or the complex *Atriplex cana–Artemisia pauciflora–Anabasis salsa* and many others.

The desert zone of the Kazakh Hills continues to the south in the form of the Betpak-Dala Desert Plateau (see Figs. 4.1 and 4.2), which forms a special landscape and geological–geomorphological unit. It lies south of the parallel 48°N, between the Sarysu River in the west and Lake Balkhash in the east, and continues southward to the Chu River.

The Betpak-Dala Desert, along with the Ustyurt Plateau, from which it is separated by the sandy region south of the Aral Sea, belongs to the northern group of deserts of the Aralo-Caspian region and reminds one, by its rocky soils and sparse vegetation, of the deserts around the Dzungarian Gate.

The most important dominant plant species are (1) *Salsola laricifolia*, (2) *Artemisia terrae-albae*, and (3) *Anabasis salsa*. Next come *Caragana grandiflora* and *Arthrophytum*. These species, in different combinations, make up the open-desert biogeocoenoses. A floristic peculiarity is the rare, endemic relict species *Spiraeanthus schrenkiana*, a shrubby member of the Rosaceae which occurs, together with the grass *Achnatherum*, and *Caragana balchaschensis*, in karst depressions of the central Betpak-Dala.

The almost complete absence of spring ephemerals in this desert is especially striking. This is caused both by the rocky or crusted soils and by the climate. As in Central Asia, spring rains are absent, and while the winter moisture is present, the temperatures are too low for the ephemerals (see *Results of the Betpak-Dala Desert Expedition 1935*).

Because there are no good watering places, this desert remains unsettled by man. It is only crossed, using the shortest path, by livestock herds which winter in the Chu Valley and graze in the summer in northern Kazakhstan. This intermediate region was earlier called the "Golodnaya Step'" (Hunger Steppe).

In the salty depressions grow various halophyte communities with *Anabasis salsa*, *A. ramosissima*, *Atriplex cana*, *Artemisia pauciflora*, and *Halocnemum strobilaceum*. *Anabasis salsa* often forms pure stands. On rubble *Nanophyton erinaceum* also appears, and in river valleys *Artemisia* species.

Associations such as *Anabasis salsa–Suaeda physophora* and *Anabasis salsa–Halocnemum strobilaceum* are characteristic on solonchak soils. *Halocnemum* grows on blistered solonchak, with *Suaeda corniculata*, *Saussurea turgaiensis*, and sometimes also *Lasiagrostis splendens* or *Aeluropus litoralis*. Halophytic shrub communities are formed by *Tamarix ramosissima*, *Nitraria schoberi*, *Kalidium caspicum*, and *Halostachys caspica*.

Widespread in the entire desert zone are the *takyry* (singular: *takyr*, see pp. 91–94), i.e., clayey plains which are slightly flooded in the spring but dry out and crack into polygon sections in the summer. The takyry have either no seed plants or very few, individual plants of *Anabasis salsa*, *A. truncata*, or *Crypsis aculeata*.

Small areas of dunes occur in the northwest Betpak-Dala. Part of the psammophyte vegetation is similar to that of Middle Asia. The character species there are, for example, *Haloxylon persicum* and *Carex physodes*.

The meadow vegetation is even more impoverished in the deserts than in the desert-steppe zone. *Agropyron repens*, *Calamagrostis epigeios*, and *Phragmites communis* still occur here. In the water with *Phragmites communis* can also be found *Typha latifolia*, *T. angustifolia*, *Scirpus tabernaemontani*, *S.* (*Bulboschoenus*) *maritimus*, *S.* (*B.*) *compactus*, *Juncus gerardii*, *J. jaxarticus*, *Carex diluta*, and *C. songarica*.

As the meadows become drier on the lower terraces, the soil quickly becomes saline, as can be seen by the occurrence of *Aeluropus litoralis* and *Lasiagrostis splendens*.

The reader is also referred to the short communication of Rustamov (1972), which contains a table of phytomass and litter production by six biogeocoenoses. These are primarily *Artemisia badhysi*–ephemeral communities (with *Salsola* spp.) on the plateau at 200 to 300 m elevation near Krasnovodsk.

The results of the I.B.P. (International Biological Program) studies in the northern Aral Lowland (Malyye Barsuki, Station Tereskent) and the sand region south of Lake Balkhash (Taukum Station) have been summarized in a recently published book (Rodin, 1977). We will come back to this at the end of this chapter.

Important animals

Most of the faunistic data for Central Kazakhstan are of the zoogeographic type, with general data for the particular species, their distribution and frequency, etc. The significance of the individual species in food chains, which are extremely complicated, is hardly mentioned. Quantitative data, which are seldom available for well-studied ecosystems, are completely lacking for this region.

The entomo-fauna has been studied most, with over 1500 species having been identified. Data have also been given for mammals, but birds, reptiles, and amphibians have not been studied.

Of the mammals, 29 species occur in the desert zone and 22 are rodents. They are genetically related to the fauna of the Mongolian and Turanian deserts. Some are quite widespread. The desert-steppes are a transitional zone faunistically as well as floristically: 20 species (14 rodents) are desert elements and 8 species (6 rodents) are steppe elements. Carnivores, such as the wolf and the fox, are quite widespread.

The main species in the desert are as follows:

Rodents: *Ochotona pallasi*, *Spermophilopsis leptodactylus*, *Citellus fulvus*, *C. intermedius*, *Allactaga severtzowi*, *Alactagulus acontion*, *Pygerethmus zhitkovi*, *Scirtopoda telum*, *Dipus sagitta*, *Cricetus eversmanni*, *Ellobius talpinus*, *Microtus socialis*.

Carnivores: *Vormela peregusna*, *Vulpes corsak*, *Felis ocreata*.

Ungulates: *Saiga tatarica*, *Gazella subgutturosa*.

Insectivores: *Erinaceus auritus*.

The following species occur in the desert steppe:

Rodents: *Ochotona pusilla*, *Lepus timidus*, *Sicista subtilis*, *Cricetus cricetus*, *Lagurus lagurus*, *Clethrionomys retilus*, *Arvicola terrestris*, *Microtus arvalis*, *Apodemus sylvaticus*.

Carnivores: *Mustela eversmanni*, *M. erminea*.

Insectivores: *Sorex macropygmaeus*, *S. minutus*, *S. arcticus*, *Neomys fodiens*.

Bats: *Nyctalus noctula*.

The most important group ecologically is the rodents. These exercise the greatest influence on the plants and represent the main food of the carnivores.

The rodents exhibit definite preference for certain biotopes, although the boundaries of their ranges do not coincide with those of phytocoenoses, but are rather somewhat larger. Rodents inhabit ten biotopes in the desert and seventeen in the desert-steppe.

The significance of individual biotopes is not the same in the different zones. For example, the biotopes with saline soils cover large areas in the desert but quite small areas in the steppe. Floodplains are important in the steppe as settlement areas but are essentially meaningless in the desert. Dunes are present as biotopes in the Betpak-Dala but are absent in the desert-steppe. The biotopes with shrubby *Haloxylon persicum* in the sand desert are replaced, in the desert-steppe, by biotopes with shrubby *Caragana* and *Spiraea* species. Least colonized are the takyry, on which one finds only narrowly specialized, *Allactaga* species.

The rodents are up to 50% stenotopic in the desert and desert-steppe.

Of the insects the following groups have been studied in some detail:

Homoptera: Euchenorryncha: These are plant-sucking cicadas, which are often limited to particular plant species or families but one-fifth of which are, in Kazakhstan, polyphagous. Halophytes are also attacked by sucking insects. These insects are especially active during the hot drought period. There are 365 species which have been identified.

Homoptera: Psylloidea: Leaf lice are very narrowly specialized and often multiply very rapidly on their host plants, which is very detrimental to the production of the plants. They are especially numerous in May and June. In all, 91 species have been found, of which 43 are restricted to woody plants and 48 to herbs but often especially to *Artemisia* as well.

Coleoptera: Among the beetles can be found phytophages, saprophages, and carnivores. The Tenebrionidae are especially well adapted to drought and do not need any water. In all, 357 beetle species have been identified.

Hymenoptera: Braconidae: The ichneumon-flies are parasites on other insects and thereby regulate the numbers of the plant damagers. In all, 268 species have been identified.

Diptera: Chloropidae: Most of the insects in this group are phytophagous in the larval stage. Their numbers have been estimated as 75 species.

Lepidoptera: Of the butterfly groups studied, 338 species have been identified.

INVESTIGATION OF THE ECOSYSTEMS

General description

In volume II of the large ecological work on Central Kazakhstan (Anonymous, 1969b) there are available interesting preliminary studies relating to the investigation of the various ecosystems. For the most important dominant steppe and desert species, the morphological–ecological characteristics during the entire period of development were studied, in particular the various means of environmental adaptation. The purpose was to provide for each species a bio-complex characterization or, in a single word, to characterize its "ecobiomorphs". In order to do this the position of each individual plant in the consortium was determined. A consortium is understood to be a community which is dependent on a single autotrophic plant. The plant is placed, for study purposes, in the middle, and the directly dependent organisms are arranged around it in circles: The parasites, herbivores, and also the micro-organisms of the rhizosphere (symbiotic, parasitic, or saprophytic). In another larger but concentric circle are found those organisms which are tied to those in the first circle, i.e., the primary carnivores or zooparasites. In a third circle are the secondary carnivores. Such consortia correspond to a certain degree, then, to the different food chains.

The species studied belong to various life forms. The spectrum of life forms shown in Table 4.2 is characteristic for the dry steppe and for the desert-steppe.

As examples with additional subdivisions, the following species can be named:

Shrubs
with runners: *Spiraea, Caragana*
with root shoots (suckers): *Hulthemia berberidifolia*

Dwarf shrubs
Ephedra distachya

Semi-shrubs
woody: *Ceratoides papposa*
true: *Atriplex cana, Suaeda physophora*
slightly lignified: *Artemisia gracilescens, Kochia prostrata*

Polycarpous forbs
multi-branched with tap-root: *Medicago romana*
with branched roots: *Nepeta panonica*

TABLE 4.2

Life-form fractions as percentages of the total number of species found in the dry steppe and desert (Anonymous, 1969a, b)

	No. of species	Shrubs	Dwarf shrubs	Semi-shrubs	Polycarpous herbs	Monocarpous herbs	
						perennials	annuals
Dry steppe	484	2.9	0.4	11.3	67.2	4.4	13.8
Desert	423	2.6	0.5	10.6	58.9	4.9	22.5

single-stemmed with tap-root: *Palimbia redeviva*
with branched roots: *Eremus altaicus*
with long rhizomes: *Phlomis puberula*
with short rhizomes: *Tanacetum achilleifolium*
with bunched growth: *Stipa* spp., *Festuca sulcata*

Monocarpous herbs
perennial with tap-roots: *Ferula ferulaeoides*
annual with tap-root: *Alyssum desertorum*
with branched roots: *Eremopyrum triticeum*

For the six most important communities in the Betpak-Dala Desert, with a total of 170 species, the following spectrum is given. A distinction is made between perennial ephemeroids (geophytes) and annual ephemerals (therophytes): trees 1%, shrubs 6%, semi-shrubs 21%, perennial herbs 23%, ephemeroids 23%, ephemerals 27%.

The following seventeen dominant species were studied:

Desert-steppe species (steppe elements): *Stipa lessingiana, S. sareptana, Festuca sulcata, Artemisia frigida, A. gracilescens, A. pauciflora, Atriplex cana, Tanacetum achilleifolium, Linosyris tatarica, Ferula songorica, Spiraea hypericifolia.*

Desert species, also appearing in the desert-steppe: *Spiraeanthus schrenkianus, Salsola arbuscula, Anabasis salsa, Nanophyton erinaceum, Artemisia terrae-albae, A. turanica.*

Of these species only a few will be discussed, as examples.

Bio-ecology of *Festuca sulcata*

Festuca sulcata, a species from the section Ovinae, has an extremely large range, which includes the entirety of the steppes from Hungary to the Altay Mountains and also includes the semi-deserts. This species was studied in great detail in Kazakhstan (Anonymous, 1969b, pp. 90–116).

Development

The phenological development during 1959 to 1961 is shown in Fig. 4.9. The phenology is almost exactly the same in the dry steppe and the desert-steppe. The leaves, which have overwintered with dried-out tips, begin to grow at the end of April but die after two or three weeks. During this time the spring leaves form. Depending upon rainfall, their growth lasts until the middle of June or into July. At this point the summer dormancy, which is almost absent in wet years, begins. In August these leaves are replaced by the autumn leaves, whose winter dormancy is forced by snowfall. Heading begins at the beginning of May, followed by blooming, on May 20 in 1961, but not until the middle of June in 1960. Anthesis lasts five to ten days and the formation of fruit about three weeks. Only a few plants bloom in dry years. The anemophilous flowers open about 16 h (4 p.m.) and close again around 19 h (7 p.m.). Ripening of the caryopses lasts fifteen to twenty days, seed-setting about six to seven (but as many as ten to twenty-five) days. Propagation is hemi-anemochorous (partly wind-dispersed) over a distance of 1 to 2 m. In 1958, a favorable year, 28% of the plants fruited and formed 1925 seeds m^{-2}; in 1961 it was only 78 seeds m^{-2}.

In the laboratory the germination potential for the next year was 40% and in the following year

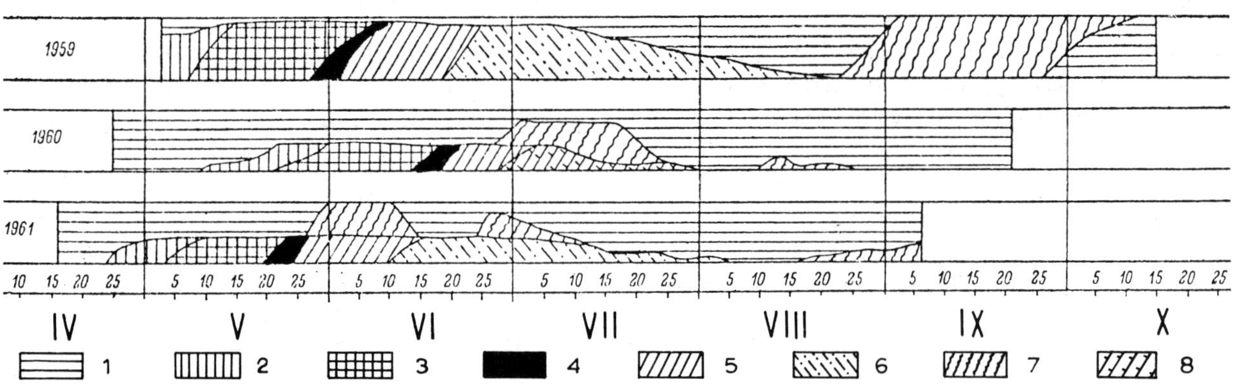

Fig. 4.9. Phenological spectra of *Festuca sulcata* in Central Kazakhstan during 1959–1961 (from Anonymous, 1969b, fig. 11). Legend: *1* = vegetative; *2* = formation of inflorescences; *3* = formation of panicles; *4* = anthesis; *5* = young fruit; *6* = ripe fruit; *7* = seed propagation; and *8* = dormancy.

97%; it remained at this level for ten years. Germination takes place both in autumn and in spring. Three small leaves and three rootlets were formed before summer dormancy. In 1960, 23% of the seedlings survived the summer drought. At the beginning of the winter the seedlings had roughly six leaflets and six rootlets. The danger was great that they would not survive the winter. The production of stems and the formation of the clumped form began in the second year (Fig. 4.10).

The generative shoots consist of the short rootstock, which increased in length about 0.12 to 0.15 cm annually, and the flowering stem, which was 12 to 22 cm long in dry years and reached 35 to 55 cm in wetter years. The rootstock grows for a few years and then begins to decompose, so the connection with the lateral shoots is broken, and several new parts form (particularization), each with three to four loci of shoot generation. These are held together in a clump, however, by the interlacing of the adventitous roots.

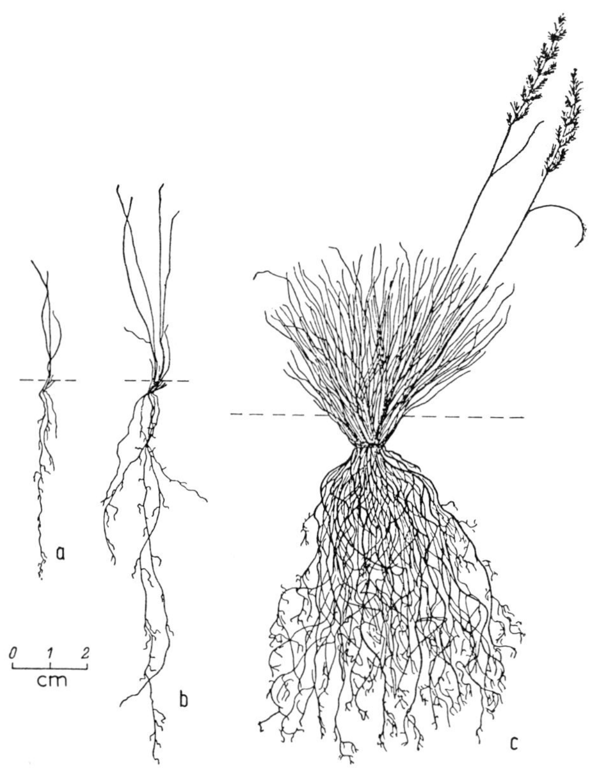

Fig. 4.10. Development stages of *Festuca sulcata* in the desert-steppe of Kazakhstan: a, seedling; b, two-year stage; c, bunch blooming for the first time (from Popova, in Anonymous, 1969b, fig. 12).

For *Festuca* one can distinguish three life-cycle phases:

(1) The juvenile phase (Fig. 4.10), which lasts a few years, during which the clumps reach a diameter of 3 to 4 cm. The number of individual shoots in the clump averages 60 to 70, with 13 to 15 generative shoots. At the end of this phase a dead section forms in the middle of the clump.

(2) The mature phase, which lasts about twenty years; the clumps attain a diameter of 5 to 6 cm. The dead middle part is, at the beginning, only 3 cm in diameter and is surrounded by 95 to 100 vegetative and 20 to 25 generative, living shoots. Toward the end of this phase, when the dead section has reached a diameter of 4 cm, the living ring is broken and the number of living shoots decreases.

(3) The senile phase, in which the dead center of the clump dominates. Out of the clump rise about 40 vegetative and 5 to 6 (as many as 20) generative shoots. This stage lasts three to five years, so a total age of about thirty years can be attained.

All numbers given are approximate averages; variation depends on the weather and is extremely large.

Ecophysiology

In the spring, photosynthesis on the natural site reaches maxima of 12 mg CO_2 (g dry wt.)$^{-1}$h^{-1}. The daily course depends only on the irradiation and the temperature. By the end of May or beginning of June, the assimilation capacity drops, after a few drought days, to about half of this value and later drops even more. Activity increases somewhat, however, after rainfall.

The water content of the leaves drops from 60% in May to about 40% in July. The extreme values of osmotic potential in the desert-steppe, during the vegetative period, were -22.7 to -59.9 bar in 1959, and -23.6 to -41.3 bar in 1960. The fluctuations are smaller in the dry steppe, for example, only -17.8 to -38.5 bar in 1962. The daily fluctuations are enormous: 3.4 to 13.5 bar in the desert-steppe and 6.0 to 15.2 bar in the dry steppe.

Transpiration, computed per gram of fresh weight, is intense. The maximum values were, in the desert-steppe, 1.7 g (g fresh wt.)$^{-1}$ h^{-1} and, in the dry steppe, 0.8 g (g fresh wt.)$^{-1}$ h^{-1}. During the drought periods the transpiration drops to one-fifth of these amounts. The springtime values are attained again after early autumn rains, but then there is a rapid decline after the middle of August.

Only 30% of the stomata are open fully even under favorable conditions, the others less open. In autumn only about 60% are open at all.

Productivity

The main period of growth for *Festuca* is in the spring. The greatest amount of above-ground biomass is found at the beginning of June, about 88 g m^{-2}. The mean weight of a single plant is 4 g. The proportion of green parts in a clump varies between 20 and 30% of the total mass (dead plus living).

The herbage yield (stems cut 3 to 4 cm above the ground) was 52 g m^{-2} at the time of panicle formation. The productivity of the dry steppe seems to be about 16% higher, but the numbers are not strictly comparable.

Chemical composition

Tables of the cation and anion content of the ash are available (Anonymous, 1969b). One cannot conclude very much from such numbers, however, because the cations have an effect on the function of the various organelles in the living cells and are, in some cases, separated in insoluble form within the cells (e.g., calcium oxalate).

The ash content varies, depending upon the degree of silicification, between about 7 and 8.5%. Of the total ash, 70 to 76% is SiO_2. The sodium and chlorine contents are not elevated, even though most of the soils are solonetz-like. The ash content was studied in spring, during flowering, and during the fruiting period. Certain variations could be identified. The same is true of the organic compounds. The cellulose content increases with age, while the protein content decreases correspondingly.

Festuca sulcata is the best livestock forage plant of the dry steppe and desert-steppe, and its hay is especially valued.

Consortional relations

The micro-organisms isolated from the rhizosphere of *Festuca* are cited (Anonymous, 1969b). There were some small differences between these and the microflora from the rhizosphere of the *Stipa* species.

Worthy of mention is that the germination of *Festuca* seems to benefit from the metabolic products from the cultures of its rhizosphere bacteria. Germination was increased by up to 40%, and growth was also greater. Conversely, root extracts from *Festuca* improve the growth of the rhizosphere microbes in pure cultures.

The most important rhizosphere fungi are *Penicillium* species (a total of 36 in the dry steppe and 28 in the desert-steppe). There were 17 *Aspergillus* species in each zone. Many Mucorales and Fungi Imperfecti were also present.

The most important parasitic fungus was *Puccinia festucae*.

Festuca is eaten by all grasshoppers which are present, also by various cicadas and bugs (Heteroptera). Of the beetles, only the soil larvae are important. Caterpillars of two pyralids and one satyrid are also found on *Festuca*.

In summary, one can say that *Festuca sulcata* is a very adaptive species, which is perhaps an explanation for its unusually wide distribution.

Bio-ecology of the *Stipa* species

Similarly detailed studies have been made on the two most important *Stipa* species of the dry steppe and desert-steppe, *Stipa lessingiana* (series Pennatae) and *Stipa sareptana* (series Capillatae).

The productivity of these two species is not much different from that of *Festuca sulcata*, as far as can be told from the poorly comparable values available. *Stipa lessingiana* has the greatest forage value of all the *Stipa* species which occur in the area. It is preferred in particular by horses.

Stipa sareptana serves as a good livestock forage only before fruiting, since the sharp caryopses bore into the skin of the grazing animals and injure them. The development of this species is briefly discussed.

This perennial grass reaches a height of 60 to 90 cm when blooming (30 to 40 cm for the vegetative shoots). The stem bases lie 2.5 to 4 cm deep in the ground. The roots go down to about 1 m and spread out laterally about 40 cm.

S. sareptana occurs from the Volga and the Don in the west to Outer Mongolia in the east. In Kazakhstan, the center of its range, it covers large areas on solonetz-like soils. In the desert zone it is found only on sandy or on rocky rubble soils (law of relative site-constancy with biotope change: see

Walter and Straka, 1970, p. 53; also Walter, 1974, p. 241).

The vegetative shoots have no intrinsic dormancy in the winter. The leaves from the previous year die in May and are replaced by younger leaves, which appear in April and are filled out in 1.5 to 2 months. In June the leaf tips begin to dry out. The potential osmotic pressure climbs from 29 bar in May to 40 bar in June, 44 bar in July, 47 bar in August, and 57 bar in September. Daily fluctuations also increase over this period, from 3 bar to 16 bar. The photosynthetic capacity, under optimal conditions, decreases in parallel from 7 mg CO_2 (g dry wt.)$^{-1}$ h^{-1} in June to only 1 mg CO_2 (g dry wt.)$^{-1}$ h^{-1} in July and August.

The generative shoots form in May, blooming follows in the middle of June, and the ripening of the fruit is completed by the end of July. From germination to the development of a dense clump and flowering maturity, four to six years have passed. At an age of 25 years the clumps begin to decompose into individual parts. The total age is generally about 50 to 60 years.

The maximum above-ground phytomass is attained in June (18 to 20 g m^{-2}).

The leaves are very strongly silicified (SiO_2 content of the ash: 60–80%). Sodium chloride content, on the other hand, is low.

Forty species of bacteria and numerous fungi, including 34 *Penicillium* species and 17 *Aspergillus* species, were isolated from the rhizosphere. In wetter years, the inflorescences are attacked by *Ustilago*.

The reader is also referred to the comparative-ecological investigations on two *Stipa* species (*S. kirghisorum* and *S. lessingiana*) in the years 1957 to 1960 carried out in Kazakhstan by Sveshnikova (1962) and by Lavrenko and Sveshnikova (1965). The results are summarized by Walter (1968, pp. 649–653). Also given are a phenological spectrum for the two species, data on the regulation of transpiration by these species through closing of the stomata and rolling up of the leaves at particular seasons, and data on the potential osmotic pressure from May to September. *Stipa kirghisorum* occupies the more favorable sites.

Bio-ecology of *Artemisia pauciflora*

The genus *Artemisia* is extremely important in Kazakhstan. It is represented by many species belonging to the subgenus *Seriphidium*. The following species were studied in the desert-steppe: *Artemisia frigida*, *A. gracilescens*, *A. pauciflora*, *A. terrae-albae*, and *A. turanica*. Of these, *Artemisia pauciflora* is chosen for detailed description here because it is the most salt-resistant of those studied. The *Artemisia* species are not true halophytes, but some resistant species can grow in saline soils.

Artemisia pauciflora is found from the Don River to Lake Zaysan in the semi-desert of the Altay Mountains. Its northern limit runs along 52° to 53°N and its southern limit along 44° to 45°N. The species is a Kazakho-northern Turanian element. It occurs as the dominant species in saline lowlands which receive supplementary water or have a high groundwater level. Its occurrence in semi-desert micro-complexes (see pp. 70–71) is especially characteristic. The 15 to 20 cm (up to 25 cm) tall plants are fairly regularly spaced, as a result of their root systems, about 20 to 30 cm apart. Their cover is about 40 to 50%. Of the eight to ten accompanying species, *Kochia prostrata*, *Anabasis salsa*, *Atriplex cana*, and also *Artemisia gracilescens* are the most common. The bare soil between the plants is filled in spring with ephemerals and ephemeroids. In addition there are the synusiae of soil algae and lichens, which cover 70 to 80% of the area. The algae cover the temporarily flooded sites, while the lichens prefer the higher ground.

Development

Artemisia overwinters with small green leaves which begin assimilation of carbon dioxide in the spring after the first warm days. In years of prolonged summer drought, only the terminal buds and leaf primordia remain in autumn, while the leaves themselves dry up. The new shoots in the spring come out at the end of April or beginning of May. Growth stops at the end of May, after the formation of up to fifteen leaves. The overwintering leaves turn yellow in the first half of May. In July even the young leaves have dried up, and only the terminal buds survive the summer drought.

The flowering shoots form in the middle of May and grow until the end of June. The development of the buds lasts fifteen to twenty days, even on into the drought period. When the plant blooms, all the leaves are already dead and the plant appears black and nearly lifeless. Ripening of the fruit lasts until the middle of September or even the beginning of

October. Dispersal of the achenes begins at the end of September.

In the autumn, three to five new, overwintering leaves grow out of the axillary buds of the vegetative shoots. The phenological course of development for the years 1958 to 1961 is shown schematically in Fig. 4.11. There are an average of three flowers in each head of the inflorescence. The flowers open around 8 to 9 h at temperatures of 17 to 22°C and a relative humidity of 48 to 65%. Flowers remain closed on rainy days. Pollen dispersal begins after about an hour and lasts for 1 to 2.5 h. Around 16 to 17 h the pistil begins to grow, whereas the stigma does not open before the next morning, in order to be pollinated over the course of the next day or two. The pollen grains of the anemophilous flowers are carried about 1.5 to 2.5 m; self-pollination does not occur.

In the drought year 1961, 25% of the flowers remained sterile. The number of achenes formed per square meter was:

1958: 32 640	1960: 1272
1959: 940	1961: 0

On the average, only one achene forms per head. The ripening of the achenes lasts thirty to forty days. One achene weighs 0.2 mg.

The fruit are barochorous (gravity-dispersed) and fall usually only 15 to 25 cm away from the mother plant, seldom as far as 40 to 50 cm.

Germination occurs in the spring. In pure stands of *A. pauciflora*, from 700 to 2240 seedlings have been counted per square meter, but only a few (none in 1961) survive until autumn, and these only between bushes, never underneath other bushes. The seedlings are also easily killed in the spring when the ground is flooded for a time by snowmelt.

The germination potential of the achenes is 68 to 90% in the laboratory, even after two years.

On the natural growing site, the seedling forms five leaves by the end of May, but these dry up during the summer. One or two more leaves develop from the terminal bud in the autumn. Branching does not occur until the second year. The plant is 5 to 6 cm tall by the third year. In the fourth and fifth years the first flowering stems are formed, by branching from the base. Development can proceed more quickly under good climatic conditions. The morphology of the stem system is very complicated in older plants.

The juvenile phase lasts two to four or up to six years. During the mature phase some branches die, but these are more than replaced by the growth of younger branches. Up to seven particularized parts (cf. Fig. 9.7) form at the base of the plant. During the senescent phase the shrub separates into individual, particularized, independent shrubs, which still bloom normally but already show a decrease in vegetative growth. Senile plants bloom only in good years.

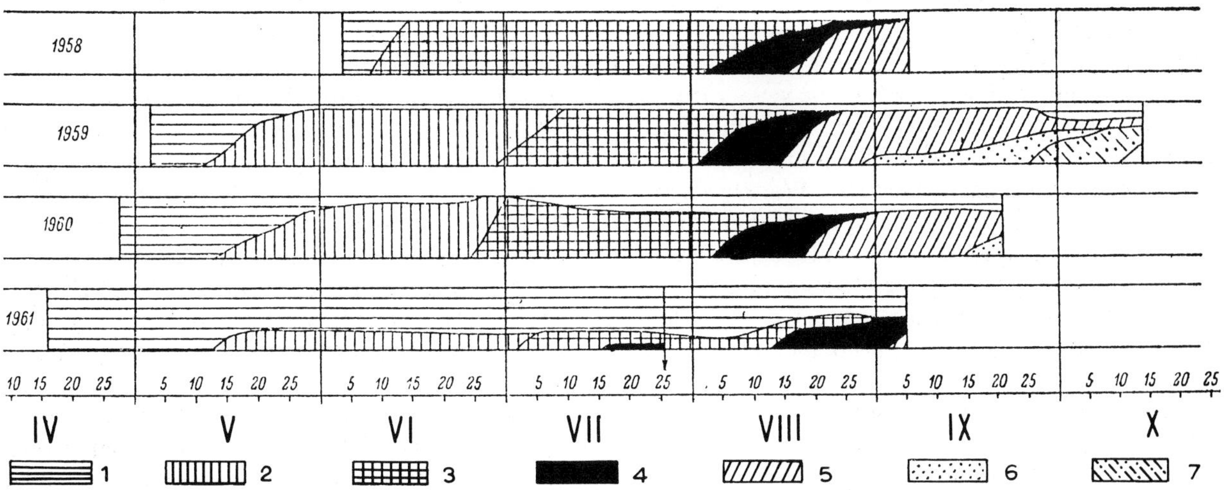

Fig. 4.11. Phenological spectra of *Artemisia pauciflora* (1958–1961) in the desert-steppe of Central Kazakhstan (from Bespalova, in Anonymous, 1969b, fig. 24). Legend: *1* = vegetative stage; *2* = flowering shoots; *3* = flower buds; *4* = anthesis; *5* = fruit; *6* = seed-propagation; *7* = dry branches. The much shorter developmental period during 1958 is especially noteworthy.

The total age is very difficult to estimate, but it must be thirty years or more. Lignification of the stems proceeds with age, from the base of the plant upwards.

Ecophysiology

Photosynthesis under natural conditions, with good illumination and water availability, amounts to 31 to 45 mg CO_2 (g dry wt.)$^{-1}$ h^{-1}. On cloudy days values of only 11 mg are attained. The assimilation efficiency is somewhat lower than for the other *Artemisia* species, but *A. pauciflora* also recovers more quickly after drought. Photosynthesis depends heavily on the water content of the soil.

The maximum water content of the leaves is 70 to 78%; the minimum, at permanent wilting, is 48 to 58%.

The osmotic potential varies from -35 bar in May to -49.4 bar at the end of the growing season. (For *A. gracilescens* these values are -27.7 bar and -56.4 bar.)

Transpiration, as one would expect from this malacophyllous species, is very high: 1 to 3 g (g fresh wt.)$^{-1}$ h^{-1}, even in June and July. The plant hardly reduces its transpiration at all, which leads to the early withering of the leaves during the drought period.

Productivity

The above-ground phytomass in an almost pure *A. pauciflora* stand fluctuated between 90 and 170 g m^{-2} over the course of the growing season of the most unfavorable year, 1961. The phytomass of the annual shoots fluctuated between 14 and 24 g m^{-2}.

A. pauciflora is eaten by sheep, goats, and horses, but not by cattle. Its nutritional value is roughly that of good hay.

Chemical composition

The ash content of *A. pauciflora* is higher than that of the other *Artemisia* species. The high content of sodium and, in late summer, of chlorine is striking in both cases. The plant evidently stores sodium chloride in the cellular fluid of its leaves — a result of the saline soil.

One finds considerable amounts of maltose and dextrins but little monosaccharide in the leaves. The cellulose content increases from 11.3 to 21.6%, but that of total nitrogen decreases from 3.6% to 2.5% of the total dry weight, since nitrogen-free dry weight increases (by 44%).

Consortional relations

Forty species of microbes were isolated from the rhizosphere of *A. pauciflora*, among which *Pseudomonas* species were especially well represented. *Clostridium pasteurianum* occurred also. The improved germination and seedling growth resulting from chemical exchanges with the rhizosphere bacteria, as mentioned earlier for *Festuca sulcata* (see p. 57), occur also for *A. pauciflora*. It was determined for *Stipa sareptana* that the active compounds were Vitamin B_6 and nicotinic acid. Leaf and root extracts of *A. pauciflora* promoted the growth of *Pseudomonas longa* considerably but inhibited *P. melochlora* and *P. rubiginosa*, which are in the rhizosphere of *Artemisia pauciflora*.

Sixty species of fungi were isolated from the rhizosphere of *A. pauciflora*, of which only one species, *Penicillium restrictum*, occurs widely.

Of the leafhoppers found on *A. pauciflora*, two are strongly specialized for attacking the genus *Artemisia*, as were also five leaf-louse species. On the other hand, of the numerous cicadas, most attack composites in general. Some are specialized for the genus *Artemisia* and some even for the subgenus *Seriphidium*.

The other insects found on *A. pauciflora*, such as bugs, beetles, butterflies, and caterpillars, are generally also *Artemisia* specialists.

In summary, it can be said that the different *Artemisia* species have, in general, similar ecological behavior, but the summer dormancy is distinct only in *A. pauciflora*. In *A. frigida*, which grows more to the north and is found exclusively on rocky biotopes in Kazakhstan, the summer dormancy is completely absent, probably because water is available on such sites even in the arid regions. *A. pauciflora* has been outcompeted on more favorable sites and dispersed to those biotopes which are, for *Artemisia* species, the most unfavorable, the saline soils with especially poor moisture conditions in the summer.

Bio-ecology of *Ferula songorica* (Anonymous, 1969b, pp. 220–238)

Ferula songorica, a very large umbellifer, is 110 to 185 cm tall when blooming and 50 to 80 cm tall otherwise. A single, erect stalk is formed, which branches only in the vicinity of the inflorescence. At the base are compound-pinnate rosette leaves 22 to 125 cm long and up to 50 cm wide, found at the ends of long petioles. The root is very thick, cylindrical, considerably branched, and penetrates 150 cm into the ground. The range of this species includes the central and eastern parts of Kazakhstan and the Altay Mountains. The biotopes it prefers are the deep valleys in the more hilly areas, the depressions in the flatter areas, and, in the east, rocky slopes (screes) and rubble piles, all of which are biotopes with relatively good water availability. As a result, shrubs (*Spiraea*, *Caragana*, etc.) are often found associated with *Ferula*. *Ferula* dominates in these communities. It is a hemi-ephemeroid[1].

Development

The reddish buds come out of the ground a week after the snow cover disappears. The buds consist of two scaly lower leaves and the rolled-up foliage leaf, which quickly opens (by the middle or end of May). The leaves begin to turn yellow by the middle or end of June, and from July on the plant observes summer dormancy. The only growth in autumn is that of the lower leaves, which takes place under the ground surface. These leaves overwinter at a depth of 3 to 6 cm.

The flowering shoot begins to grow in May, and flower buds form in June. Flowering lasts ten to fifteen days, from the end of June to the beginning of July. Ripening of the fruit requires a month and is completed after the stem has already dried up.

Flowering shoots can develop every two years, but then only in wetter years. In drought years, only those plants in the vicinity of springs or other underground water are able to bloom.

An inflorescence contains over 5000 flowers, some of which are unisexual. They open, beginning at 8 to 10 h, throughout the day. Single flowers are insignificant, but the entire inflorescence is found by insects of all different groups. *Ferula* is entomophilous, but self-pollination can occur if pollination is not performed by insects.

Seed production is low. Only 9 to 20% of the flowers form fruits, and, of these 50 to 75% are defective. Fruit propagation is performed autochorously by means of a slinging mechanism. Fruits are generally propelled about 30 cm, seldom as far as 3 m.

The germination potential in the laboratory was only 10% for 1959 but 44% for 1961.

The juvenile phase lasts 15 to 20 years (Fig. 4.12). The caudex at the base of 3 to 10 year old plants increases in length to 1.4 to 2 cm, and dormant buds are formed, which will not open until the mature phase, after the flowering stalk has died.

The younger mature phase lasts 16 to 22 years. The caudex is not yet large, but the previously dormant axillary buds form two or three shoots so the plant has several leaves.

The older mature phase extends over about thirty years. The caudex becomes 12 to 17 cm long and 8 to 13 cm thick. The main root and the lateral roots become thicker also. The plant develops four to twelve leaves, and the lateral branches become particularized. With time some of the particularized parts begin to die. It was not possible to determine how old a plant can actually become. It could be one hundred years or more. *Spiraea hypericifolia* and *Atriplex cana* in the desert-steppe reach a similar old age.

Ecophysiology

The temperature optimum for net photosynthesis is around 35°C and the maximum is around 50°C. The maximum potential rate of carbon dioxide assimilation is quite high in early summer [143 mg CO_2 (g dry wt.)$^{-1}$ h^{-1}] but lower (50 mg) at the end of the vegetation period (as a result of water stress) and lower still (39 mg) if light or temperature approaches limiting levels.

The water content of the leaves is very high (70 to 80%), and the diurnal fluctuation is comparatively small. Even the osmotic potential fluctuates only 12 bar, between −20 bar and −32 bar.

Transpiration is intense: 1.4 to 1.8 g (g fresh wt.)$^{-1}$ h^{-1}. It increases sharply in the morning up to 13 h. The stomata are wide open in spring but somewhat less open later on.

Productivity

The dry weight of the above-ground parts was about 133 g on the average. The plant is not eaten

[1]The leaves of hemi-ephemeroids develop underground and overwinter. Hemi-ephemeroids are, thus, somewhere between geophytes and hemicryptophytes.

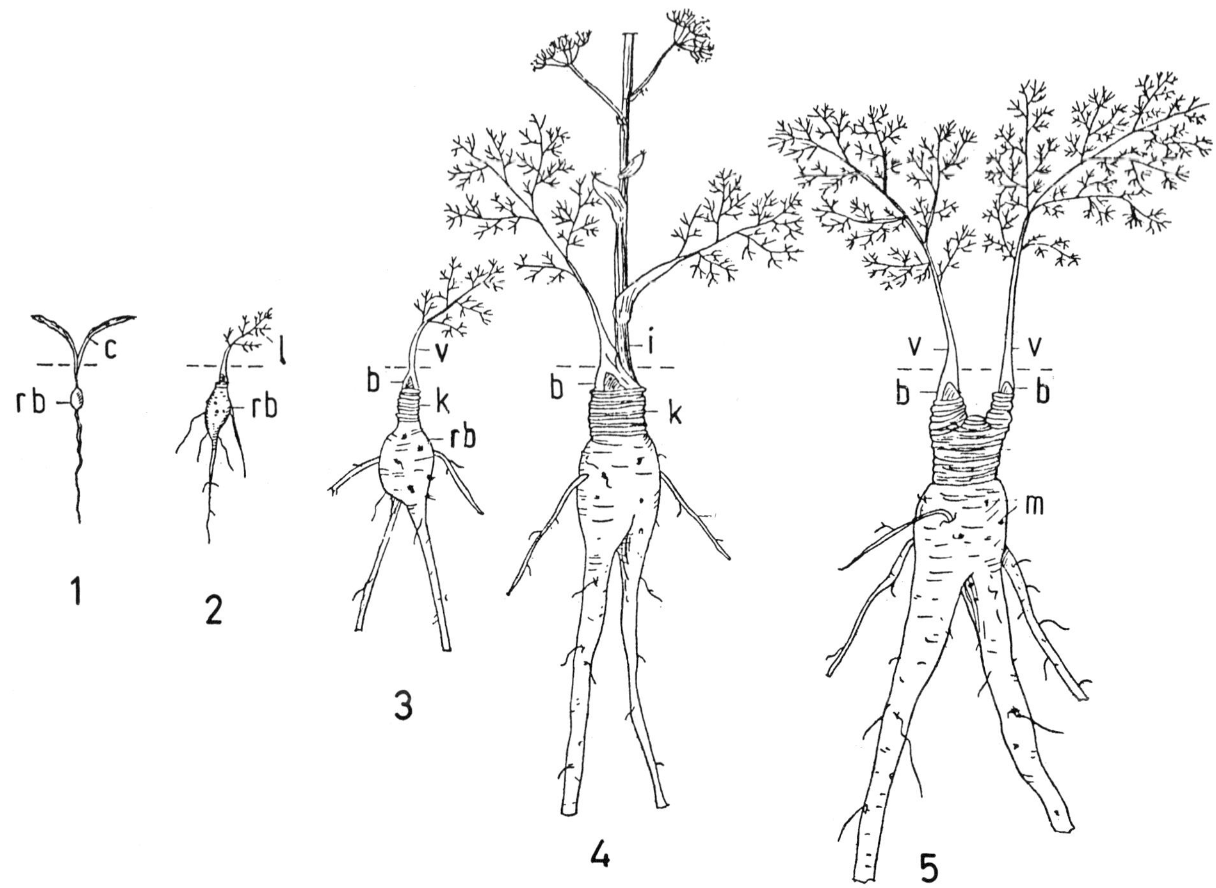

Fig. 4.12. Development of *Ferula songorica* in Kazakhstan (from Borisova, in Anonymous, 1969b, fig. 41): 1. Annual plant with only two cotyledons and a main root. 2. Two-year-old plant with only one foliage leaf but a well-developed root system. 3. Ten-year-old plant with a large foliage leaf (leaf scars of previous years and regenerative buds for the next year are recognizable on the caudex). 4. Sixteen-year-old plant blooming (only the lower part of the generative shoot is shown). 5. Twenty-five-year-old plant, which has separated after blooming roughly seven years earlier into two caudices (with the scar from the flowering shoot visible between them). Legend: *c* = cotyledon; *rb* = root bulb; *l* = leaf; *v* = vegetative shoot; *b* = apical bud; *k* = caudex; *m* = mammilae; *i* = inflorescence.

by livestock in its fresh state but probably is eaten as silage.

Chemical composition

The ash content of the stalk was 6 to 10% and of the leaves 9 to 11%. Other details were not given.

Consortional relations

Uromyces graminis parasitizes the leaves. Neither cicadas nor leaf-miner larvae occur on *Ferula* or on Apiaceae at all. Three species of bugs eat the fruits. Many beetles eat anthers and pollen, and some proboscis beetles carry out their development in seed buds. *Coccinella septempunctata* and others eat both pollen and the many leaf lice on *Ferula*. The caterpillars of some Microlepidoptera feed on the fruits. A large number of different insects, such as braconids, syrphids, and others, collect the nectar.

Of the 159 steppe elements, the most important bio-ecological characteristics were summarized in a large table (Anonymous, 1969b, pp. 264–285).

Bio-ecology of the desert elements (Anonymous, 1969b, pp. 286–322)

These were studied in less detail. Ash analyses and rhizosphere studies are lacking completely.

Production estimates are especially difficult in the desert because few years are really representative. Reproduction can also occur only in especially favorable years. We limit ourselves to three examples: *Salsola arbusculiformis*, *Anabasis salsa*, and *Nanophyton erinaceum*.

Salsola arbusculiformis

This chenopod is widespread on rubbly sites both in the Betpak-Dala Desert and on the Ustyurt Plateau. The soil is wetted 50 to 60 cm deep (up to 100 cm) on such sites.

This dwarf-shrub, which stands 20–80 cm tall and has somewhat spreading lateral branches, comes out at the end of March. The shoots are fully grown out by the end of May or beginning of June, at which time blooming begins in favorable years. The vegetative shoots are long but stop growing after a few years in order to form shorter shoots (brachyblasts) from the lateral buds. Up to sixteen leaflets can be held on these lateral shoots. The leaves are 0.7 to 1.3 cm long and succulent. At the beginning of the drought they turn yellow and are dropped.

Seedlings can continue to develop only in favorable years, but rejuvenation of the stands proceeds normally since many age-classes are present in a stand. The growth habit of the plants varies from tall, well-foliated plants, where water is available, to low, almost leafless plants on dry sites.

Salsola probably is a halophyte rather than a xerophyte. This is suggested by the succulence of the leaves and the high ash content (up to 32%). The Na^+, Cl^-, and SO_4^{2-} content of the transpiring organs, however, was not determined. The transpiration rate is low, about 0.2 to 0.3 g (g fresh wt.)$^{-1}$ h^{-1}. As is the case with most Chenopodiaceae with woody stem bases, segmentation into several individual stems occurs with age.

Salsola arbusculiformis is a livestock forage plant in spring and early summer, as long as the juicy leaflets have not yet fallen off. The raw protein content in spring is 10.7 to 16.8%.

Anabasis salsa

Anabasis salsa, a 25 cm-tall chenopodiaceous semi-shrub, is typical on saline soils in the desert-steppe. It occurs from the Caspian Lowlands into Dzungaria, but also in the southern Kazakh and northern Turanian deserts. It is native to the alkalized siero-burozems as well as the solonetz and *takyr* soils. The most common associated species are *Nanophyton erinaceum*, *Artemisia terrae-albae*, *Atriplex cana*, and *Suaeda physophora*.

The unpretentious flowering is completed within fifteen to twenty days, beginning in the middle of May or by the beginning of June. *Anabasis* is an anemophilous species. The number of seedlings depends on the weather conditions in the previous year. After the favorable year 1958 one could find in 1959 about 38 seedlings m^{-2}, of which 27 m^{-2} were still living on September 25. In 1960 only one seedling per square meter could be found, and in 1961 none. The development of the young plants proceeds slowly. Lateral branches appear only after five to seven years, by which time the plant is 3 to 5 cm tall. The primary shoot dies after about seven to ten years. Lateral shoots are plagiotropic. The annual growth of the shoots is about 0.2 to 1.5 cm.

The main root penetrates down 15 to 20 cm until it reaches the dense, brown, saline horizon, where it branches. The lateral roots go down further to 60 to 70 cm (Fig. 4.13). In addition to the permanent roots, ephemeral rootlets are formed after each rainfall, permitting efficient uptake of the available water. The cylindrical leaves and stalk have a two-layered epidermis, and the stalk also has a waxy layer. The number of stomata is small. Calcium oxalate crystals are found in the hypodermis. Underneath are two cell layers with chloroplasts, and in the center is a vascular bundle with sclerenchyma. The leaves are kept for the entire year. Transpiration is low: 0.2 to 0.85 g (g fresh wt.)$^{-1}$ h^{-1} (maximum: 1.44–1.68 g). The water content of the annual shoots drops from 78% in May to 51% in August, a result of the increasing salinity of the soil solution.

For pure stands of *Anabasis salsa* the cover percentage was 15 to 20%. The average annual dry mass was 460 kg ha^{-1} (= 46 g m^{-2}) in the favorable year 1958. In the subsequent dry years it decreased continually to 200 kg ha^{-1} (= 20 g m^{-2}) in 1961.

Anabasis can be grazed, but it contains the poisonous alkaloid anabasine. The anabasine content is lowest in autumn, when the shoots die back. *Anabasis* areas should thus be grazed by livestock only at this time.

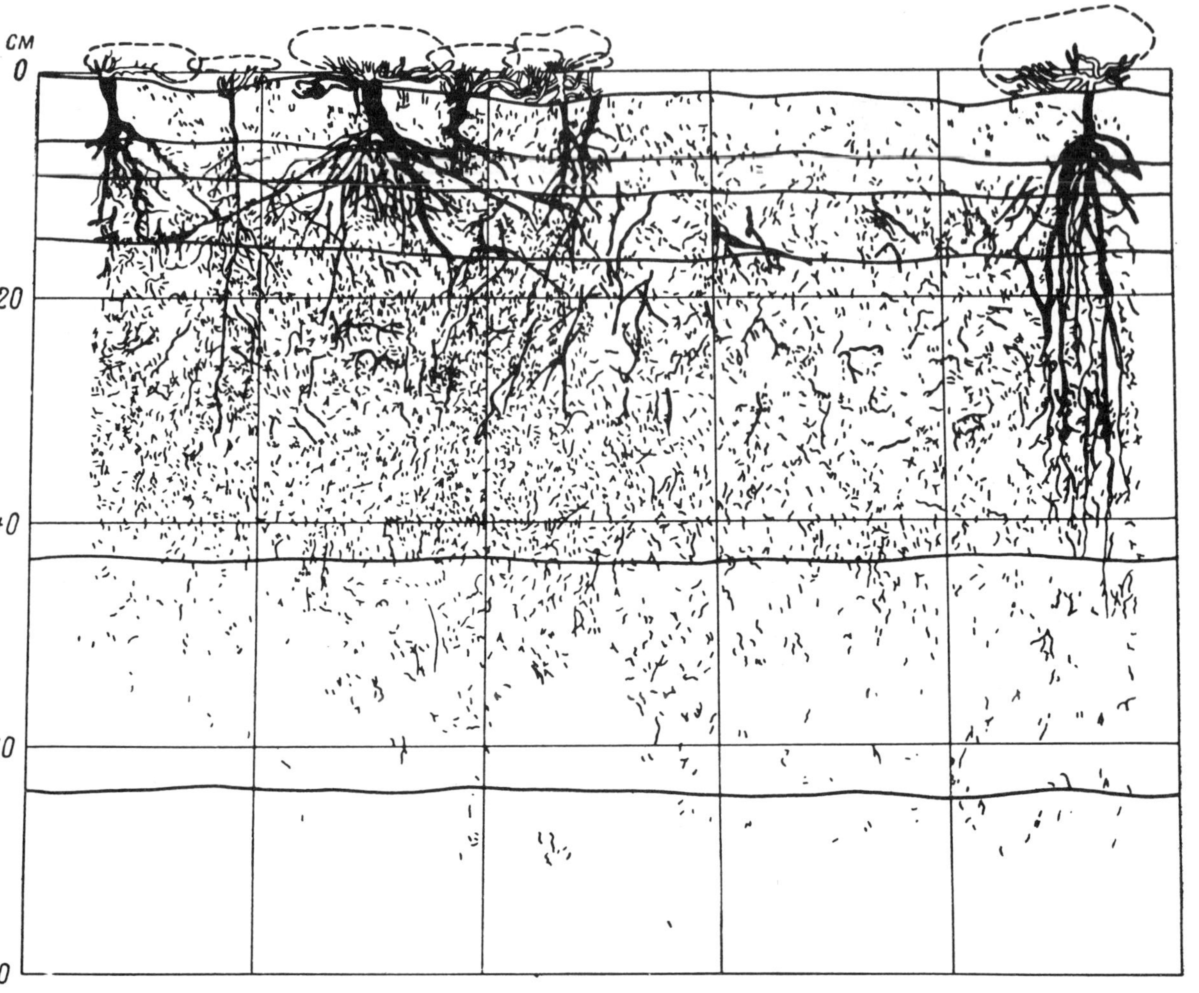

Fig. 4.13. Profile through the *Anabasis salsa* biogeocoenosis: open vegetation above ground; closed, interweaving root systems underground, very dense to 40 cm and reaching to a depth of 70 cm. The soil is a solonchak-like solonetz in the floodplain of the Kara-Kinghin in Kazakhstan (from Shalyt, 1952).

On the dead shoots three saprophytic Fungi Imperfecti were observed.

No studies were made on the characteristics of *Anabasis* as a halophyte. From the ash analyses of plants from Middle Asia, however, one knows that *Anabasis salsa* has a high sodium content but a relatively low chlorine content. The cations are equilibrated by the anions of the organic acids. Very often it is the oxalic-acid anion, but in *Anabasis* it is not only oxalic acid since it would then be as poisonous as *Halogeton glomeratus*. Like this species, *Anabasis salsa* is an alkali halophyte (Walter, 1968, p. 785).

Nanophyton erinaceum

This chenopod is a Kazakh–north Turanian element which also occurs in Dzungaria. It is actually a halo-petrophyte which is not so widespread as *Anabasis salsa*. It grows on rocky rubble soils but also on takyry.

The 5 to 10 cm tall semi-shrub forms loose cushions with diameters of 10 to 20 cm (Fig. 4.14). Reproduction is only by seeds.

The seedlings grow very slowly. Even for mature plants annual shoot growth of 0.6 to 3.5 cm is very good. The terminal shoot dies after three to five years. The lateral branches on the upper part of the

Fig. 4.14. *Nanophyton erinaceum* (Mongolian: *tar*) in Dzungaria (photograph by P. Hanelt; from Walter, 1974, fig. 250).

annual shoot are short and densely covered with small leaflets. A cushion with a diameter of 20 cm has about 850 shoots, all standing perpendicular to the ground surface. The roots penetrate 60 to 70 cm into the soil but also have many lateral roots at a depth of 7 to 10 cm. After rainstorms these take up the water from the only slightly saline soil.

The leaves are very xeromorphic. Palisade-like cells with chloroplasts are found only on the ventral side. The epidermis is heavily reinforced, and the few stomata are recessed. The entire basal part is covered with a wax layer which is as thick as the epidermal cells. The dorsal side is also protected against water loss: it has no waxy layer but also no stomata. The central vascular bundle of the leaf consists primarily of sclerenchyma. All of this points to a low rate of water conduction. Transpiration is, on the average, 0.4 to 0.45 g (g fresh wt.)$^{-1}$ h^{-1}, with a maximum of 1.56 g. The water content of the leaves of *Nanophyton* is 52% in May and 36% in August.

Nanophyton has value for livestock grazing and is palatable over the entire year.

On the dead leaves, two saprophytic Fungi Imperfecti were found.

Artemisia species

Two *Artemisia* species also occur in the desert, *A. terrae-albae* and *A. turanica.* These grow under conditions which are less favorable than in the desert-steppe and, thus, must restrict their transpiration considerably at the beginning of the drought period. This is done by opening the stomata only during the morning hours and by reducing the transpiring surface through loss of leaves. Otherwise, their ecological behavior is roughly that of the desert-steppe species of *Artemisia.*

A. terrae-albae and *A. turanica* are also important livestock forage plants. They should be grazed only in the spring and autumn, however, since the leaves contain too much volatile oil in the summer.

The ecological–biological method (behavioral research on plants)

In the previous sections we looked at a number of examples which showed how one can clarify the life cycles of the most important structural components of an ecosystem through detailed observation. This is a type of ecological research which we can call ecological–biological, and which represents a definite contrast to modern ecophysiology, which is so often carried to extremes.

Using the ecological–biological method, the researcher is an intense observer of the entire life cycle of the plant, not only observing certain phenological phases but also following the morphological changes and adaptations during the various growth processes. This latter task must include the collection of as many quantitative data as possible on the number of flowers formed, the fraction pollinated, the number of ripening fruit and the number of seeds they contain, the manner and distance of their propagation, the germination potential (percentage), the number of seedlings per unit area and their fate, the development of juvenile plants in years with different weather conditions, and so on. Simple measurements to clarify the water budget and assimilate budget, as well as the age structure of the population, must also be made. Finally, data on the negative influences of parasites and other pests and the positive influences of symbiotic relations are also necessary. Only in this manner can one determine the really critical life-cycle phases of particular species, which decide their presence or absence in a given ecosystem. Mostly it is not the mature phase of the life cycle which is endangered but rather the germination stage or the juvenile phase, i.e., the first stage or at least one in the early years of the plant.

It is clear that such a study presupposes the continuous presence of the researcher at the observation site, a condition which cannot be fulfilled when research is carried on incidentally and teaching or other academic duties have priority. Such research leads to a very intimate relationship with the research object, as in the case of the behavioral scientist in zoology with his animal subjects. We can and should, therefore, consider behavioral research on plants as well.

This sort of research does not require large expenditures for equipment. Certain simple instruments are sufficient to follow the environmental conditions continuously and to clarify the most important functions of the plants. For the most part, one can take samples, preserve them, and study them in the laboratory during the plant's dormant period. What is necessary is time at the site during the entire growing season, and this is usually lacking. It is often suggested, then, that, in place of a living observer, self-registering instruments should be called on. These are supposed to carry out "objective" measurements, instead of "subjective" observations, and thus lead to incontestable conclusions. This is the goal of modern ecophysiology.

This sort of research, which was first called "experimental ecology", developed as a reaction to the "speculative ecology" of the first part of this century. The adaptations of the plants had previously only been studied morphologically and anatomically, and scientists tried to interpret the characteristic features more through ingenuity than using ecological reason. As an example, we mention the long-discussed "physiological dryness" of the cold, peaty or salty soils, on which plants with xerophytic characteristics grow.

When scientists tried experimentally to demonstrate the accepted physiological functions, they often found that the interpretations were not correct and that relations were much more complicated than teleological ecology had assumed. With this began the period of transpiration and, later, photosynthesis measurements on the site, with very simple, rather insensitive, but transportable equipment. It was soon found, however, that the procedures necessary for conducting the measurements (e.g., cutting off of branches for transpiration measurements, cobalt-paper methods, the half-leaf method for determining assimilates, enclosure in chambers for photosynthesis measurements, etc.) so greatly destroyed the natural conditions that one could no longer accept the results.

Then began the period of construction of more complicated and precise instruments. Many ecologists became almost exclusively technicians, for which the construction of equipment was the end in itself. Instead of carrying out series of measurements with a new piece of equipment, they proceeded to construct an even better instrument. The costs of the equipment rose accordingly. The task of the ecologist at the site was no longer observation but rather the maintenance of equipment, so that it would function faultlessly. The direct relationship with the object of study was lost. Further advances with self-registering instruments provided so many numbers that computers had to be called in. The computer's assertions were then published, after statistical scrutiny. The subjectivity of the researcher was excluded, but at the same time also his ability to relate one idea to another and his intuitive understanding.

The great precision of the instruments and other equipment is not compatible, however, with the variation in values for the objects at the site or with the constantly changing environmental conditions. As a result, there exists the tendency to move the investigations into the laboratory and perform them under controlled conditions and with uniformly pre-cultivated plants. Ecology thus becomes pure physiology, with a certain amount of ecological thinking only in the posing of questions.

As useful as such information is, one should not overlook the fact that the results of laboratory studies cannot necessarily be extrapolated to the field, if for no other reason than because competition is not present in the laboratory. The synergism of the ecosystem is missing, and the clarification of this is in fact the main task of field ecology. Ecophysiology by instrumentation cannot replace observational ecobiology. The goal should be the cooperation of both approaches.

Actual ecology must remain close to nature. It requires a good physiological schooling combined with a developed talent for observation. Every partiality or over-estimation of a given method or manner of working should be avoided. Only then will we obtain correct insight into ecosystems.

Interference by man

Kazakhstan is a region which has only recently been opened up for settlement by man. The introduction of modern agricultural machines on a large scale, however, has led to the plowing up of most of the steppe zone, which still offers to some extent favorable possibilities for agriculture. The natural vegetation remains only in small patches.

The semi-desert zone is much less suited for agriculture because of the frequent drought years. This is the region of extensive cattle-raising on natural pastures. The same is true for this sort of grazing economy as was said about the Caspian Lowland. The small valleys with a denser plant cover are the most productive segments of the landscape.

The problems in raising cattle in the desert are considerable. Methods are being tried at the Experimental Station at Chelkar in the northern Aral Lowland. Favorable results could make possible an agricultural strategy based on rainfall in the northern part and on irrigation using the scant geologic water reserves available in the southern part (Maljugin, 1954). Such agriculture, even to a limited extent, can be very important in an arid region.

The most suitable soils for rainfall-based agriculture are the sandy soils of the light chestnut earth or burozem variety, which carry a natural vegetation of *Agropyron sibiricum* and *A. fragile*, with *Stipa joannis*, *S. sareptana*, and *Artemisia incana* (=*A. maritima* var. *incana*). Rainfall alone, however, is seldom enough, and the accumulation of snow in winter on the cultivated strips is usually necessary. To this purpose, the stalks of planted strips of millet are allowed to stand over the winter, and 10 m wide strips of watermelons are planted between them.

By seeding very lightly, yields of wheat and barley of suitable types can reach 0.5 to 0.6 t ha^{-1}. Seeding must occur as early as possible. A delay of seven to ten days means a 30% reduction in yield. Feed crops can also be grown in this manner. The major species are yellow alfalfa (*Medicago falcata*) or the blue variety (*M. coerulea* subsp. *trautvetteri*), as well as *Agropyron sibiricum*.

Greater possibilities are offered by irrigated agriculture using the water from slightly saline lakes or by using groundwater. It has been shown that the best method is frequent watering with small amounts, by which the water content of the soil is kept above 50 to 60% of field capacity. In this way one can grow watermelons and vegetables, as well as grains. Satisfactory results can be obtained from water subsidies of 4500 to 5000 m^3 ha^{-1} over the entire growing season. Furrow irrigation has been shown to be the most practical method.

Planting in raised ditches is also practicable

where the groundwater table is not too deep. Better soil is filled into these ditches, but not so much that the roots of the cultivated plants cannot reach the capillary fringe of the groundwater table, thus endangering the water supply in the summer. Where this is possible, vegetables and even berries, fruit, and wine grapes can be grown. Trees for windbreaks or wood can also be planted using this method. Suitable are, for example, the salt-tolerant tamarisks, but also *Ulmus*, *Elaeagnus*, *Populus balsamifera* or *Salix* species (Maljugin, 1954). In this manner the scant water reserves in this generally very dry, extensive zone can be used. The question of the profitability of such agriculture, however, has not been raised.

The most important biogeocoenoses of the desert-steppe (semi-desert) of Central Kazakhstan

The biogeocoenoses of the desert-steppe of Kazakhstan are described in the third volume of "Biocomplex Investigations in Kazakhstan" (Anonymous, 1976). The following biogeocoenoses are included:

(1) The complex consisting of the *Stipa lessingiana* + *Festuca sulcata* + *Artemisia gracilescens* biogeocoenosis[1] and the *Artemisia gracilescens* + *Festuca sulcata* + *Stipa sareptana* + *Stipa lessingiana* biogeocoenosis.

(2) The complex consisting of the *Stipa sareptana* + *Artemisia gracilescens* biogeocoenosis and the *Artemisia gracilescens* + *Stipa sareptana* + *Festuca sulcata* biogeocoenosis.

(3) The *Atriplex cana* + *Artemisia pauciflora* biogeocoenosis.

(4) The *Artemisia pauciflora* biogeocoenosis.

The first two complexes represent the zonal vegetation on light chestnut earths with some solonetz characteristics. The biogeocoenoses with *Stipa lessingiana* are found somewhat further north than those with *Stipa sareptana*. A certain micro-mosaic structure is noticeable in these biogeocoenoses: within the generally distributed grass stratum the dominant *Artemisia* occurs in patches. The biogeocoenoses 3 and 4, in contrast, are found with halophilic semi-shrubs in shallow hollows with solonchak-like soils. They are especially characteristic for the northern Turanian Lowland.

The studies were carried out during the period 1957 to 1962 and were supplemented by studies during 1965 and 1968. The experimental areas were located around the experimental station at Zhanaarkin.

Ecology of Complex 1

The experimental area was located at the base of a 2 to 3° southeast slope on the lower Koksengir Hills. It formed an area 350 to 400 m long by 120 to 200 m wide covering five hectares.

The first above-mentioned unit (Complex 1) covered 70% of the area, with 5 to 15 m^2 patches of the second unit (Complex 2) with *Stipa sareptana* scattered throughout. Three species are dominant in the first, more important biogeocoenosis: *Stipa lessingiana*, *Festuca sulcata*, and *Artemisia gracilescens*.

The microclimatic conditions in Complex 1 can be seen in Table 4.3.

The soil profiles were studied in detail. The soil under Complex 1 was somewhat rockier and thus better aerated. The loamy strata, which are slightly saline, are found at a depth of 84 cm in Complex 1 and at 61 cm in Complex 2. The pH is neutral, and the sorbed-sodium content does not exceed 1.5 to 1.7% of the sorption capacity. The slight solonetz characteristics are thus caused by magnesium, of which over 20% has been sorbed.

Detailed data were gathered on the variation of soil water content. The water content of the upper 100 cm under Complex 1 was about 8.5 to 11.5% of field capacity up until May 1959. After that time the soil began to dry out, so that in July only 2.5 to 5.5% of field capacity remained.

The year 1960 was somewhat wetter, for the total supply during the drought period was still 1110 $m^3 ha^{-1}$, as opposed to 743 $m^3 ha^{-1}$ during 1959. The next year, 1961, however, was especially dry. Water available for uptake by plants was completely lacking only at certain depths during the most extreme dry periods. In general, the available water content is always higher in Complex 1 than in Complex 2, especially in the rocky horizons. The somewhat drier conditions in Complex 2 are reflected by the occurrence of *Stipa sareptana*.

The soils are relatively poor in calcium, as is the

[1] Where a community type is designated by the name of several dominant species, the minus sign separates successive strata, while co-dominants within a stratum are linked by the plus sign.

TABLE 4.3

Microclimatic measurements of temperature, humidity, and wind speed on August 2, 1960 in the *Stipa lessingiana* + *Festuca sulcata* + *Artemisia gracilescens* biogeocoenosis near the Zhanaarkin Experimental Station (from Anonymous, 1976, p. 9)

Hour	Temperature (°C)			Relative humidity (%)	Wind speed (m sec^{-1})	Temperature of soil at 20 cm (°C)
	surface	air at 10 cm	air at 2 m			
7	23.5	24.4	24.3	68	1.3	23.9
8	24.0	25.2	25.0	68	–	–
9	26.6	29.4	29.9	57	2.2	23.7
10	29.6	33.0	32.5	49	2.7	23.5
11	34.0	33.7	33.0	42	1.9	23.5
12	34.8	33.2	33.0	42	2.1	23.7
13	41.5	36.9	36.0	32	1.9	24.0
14	43.0	37.4	36.0	30	–	–
15	43.5	37.8	36.4	30	2.0	24.8
16	42.9	38.0	36.7	30	2.1	25.6
17	40.5	37.8	36.9	31	2.1	26.2
18	37.5	35.0	35.3	31	1.0	26.7
19	34.5	33.8	33.5	34	0.9	27.0

substrate material. They effervesce with 10% hydrochloric acid only at the lower limit of moisture in springtime.

Since organic plant remains decompose rapidly, the soils are also poor in humus. This low humus content decreases further during the course of the summer.

The readily soluble salts are sulphates, primarily gypsum.

Although the soils of both biogeocoenoses are light chestnut earths, the soil of Complex 2 clearly exhibits more arid characteristics.

The plant cover is composed of the following numbers of species: 33 flowering plants, 2 mosses, 12 lichens, 26 soil algae (12 Cyanophyta), 4 parasitic fungi, 81 soil fungi, and 105 bacteria and Actinomycetes. The individual species are listed in the original work. The Pannono-Pontic-Kazakh geoelements are predominant, followed by the true Kazakh geoelements. Bunch grasses and semi-shrubs are the main life forms, followed by the ephemeroids.

Spring and early summer represent the main growing periods, with the number of active species decreasing sharply from July on. Phenological spectra on the individual flowering plants are given in the original work (Anonymous, figs. 10–17, pp. 43–50).

Vegetal cover is 40 to 50%. The vegetation reaches 40 cm in height by June and July, and three strata can be distinguished. The depths of rooting for the three dominant species are 20 to 25 cm, 50 to 70 cm, and 80 to 120 cm.

The supply of germinable seeds in the soil varies from year to year depending on the number of seeds produced in the previous year. One can also find seeds from plants which occur in communities higher up the slopes, but these seeds produce plants which die sooner or later after germinating.

The phytomass in Complex 1 is composed of that of the three dominant species, the relative proportions of which are not constant, as seen in Table 4.4.

The largest amount of above-ground phytomass is present in July (180.2 g m^{-2}), while only 133.8 g m^{-2} are present in May. Litter accounts for

TABLE 4.4

Proportions of total phytomass represented by the dominant species (in %) at Zhanaarkin (Anonymous, 1976)

	1961	1965	1968 (after fire)
Stipa lessingiana	33	57.8	57
Festuca sulcata	42	32.8	37.6
Artemisia gracilescens	18.5	9.4	1.8

46 to 57% of the phytomass in July but only 11% in May.

Of the above-ground phytomass, 73 to 95% can be used as feed for grazing cattle.

The synusiae of soil micro-organisms were also studied in detail, and estimation of their phytomass (dry weight) was attempted. In the upper 10 cm of the soil they amount to 99 g m^{-2} of soil surface (volumetric weight of the soil: 1.26). Their mass makes up only 1.3% of the humus mass.

Data on the cycling of ash elements are given in the original work (Anonymous, 1976, pp. 81–84).

The phenological aspects of the vegetation of Complex 1 were described as follows. The snow cover disappears at the end of March or beginning of April. The first species to bloom is *Gagea pusilla*, but it is inconspicuous. The red flowers of *Tulipa patens* appear next. The desert-steppe turns green at the end of May. The lighter patches of *Artemisia* are most conspicuous, but one can also recognize yellow *Erysimum* in places. The aspect at the beginning of June is dominated by the awns of *Stipa*, between which stand the bunches of *Festuca sulcata* and the greenish-yellow inflorescences of *Ferula*. The grasses dry out at the end of June or beginning of July, and the plant cover takes on a straw-yellow color, the general effect of which is rather monotonous. The picture is changed little by the dark yellow color of the blooming *Artemisia* at the end of August and the beginning of September. Winter begins at the beginning of November.

The phenology in Complex 1 differs only slightly from that in Complex 2.

Ecology of Complex 2

The second unit (Complex 2), also consisting of two biogeocoenoses, was studied in the same area (Anonymous, 1976, pp. 93–155). In general the situation is very similar, if one ignores the greater aridity. One finds the effervescing horizon at 30 to 50 cm under Complex 1 and at only 20 to 40 cm under Complex 2, i.e., the soil is not wetted as deeply in the spring as is the case under Complex 1. The soil salinity is also somewhat higher in Complex 2 than in Complex 1.

The number of flowering plants, 26 species, is somewhat similar to that in Complex 1, while the numbers of lichens (15 species) and soil algae (42 species) are somewhat larger in the more open Complex 2. Mosses are absent. The cover is somewhat lower and the mosaic structure more pronounced than in Complex 1. The east European–Kazakh geoelements predominate, but Kazakh–Turanian elements are well represented, for example, *Rheum tataricum*, *Megacarpea megalocarpa*, *Scorzonera tuberosa*, and *Allium pallasii*.

The main growing period is limited to springtime, and the number of active species is already small in June. Phenological spectra are compressed toward the springtime but are in general similar to those of Complex 1.

Of the dominant species, *Stipa lessingiana* is completely replaced by *Stipa sareptana*. The age structure of the population is normal, as in Complex 1, with all age groups represented.

The maximum phytomass is reached in June. It is estimated as 2.03 t ha^{-1} for 1961 on areas which had not been grazed for several years.

The biogeocoenoses with *Artemisia pauciflora*

The third experimental area, with *Atriplex cana* and *Artemisia pauciflora*, was situated on a terrace 1.5 m above a saline river bed, in which *Halocnemum* grew. The relief in the biogeocoenosis can best be described as a nano-relief, since the semi-shrubs stood on hillocks of only 2 to 6 cm. The microclimate is essentially the same as that of Complex 1 and is the normal microclimate in the desert-steppe.

The groundwater in this experimental area was not deep, and the sandy horizon at a depth of 135 to 180 cm was wet. The solonchak-like solonetz soil is salty in all horizons and effervesces on the surface with only 10% hydrochloric acid. The B_1-horizon, at a depth of 10 to 35 cm, has a prismatic structure with sorbed sodium chloride (5.6% of the total sorption).

Only the upper 0 to 5 cm of soil are salt-free in the spring. The quantitative distribution of salts in different soil layers is shown in the tables on pp. 164–166 of the original work (Anonymous, 1976).

Although the soil is very moist, available water is very often lacking in the upper 16 cm of the soil during the period of active plant growth.

The humus content of the upper 10 cm is about 4% and decreases rapidly with increasing depth. The humus content increases in the course of the summer in this biogeocoenosis, since the micro-organisms are inactive during the summer drought.

The plant cover forms a total cover of 50 to 70%, of which 30 to 50% is represented by *Atriplex*.

In this biogeocoenosis one finds 22 flowering plant species, 1 moss species, 19 lichen species, 24 algal species, 71 microscopic fungus species, and 72 species of bacteria and, above all, Actinomycetes.

Most flowering plants belong to the Kazakho-Turanian geoelement and are semi-shrubs, while the number of ephemerals is small.

The roots of *Atriplex* reach the capillary fringe of the groundwater, while those of *Artemisia* do not.

The growing season covers eight months with rather indistinct aspects.

Of the above-ground phytomass, 75 to 91% is contributed by *Atriplex* and only 3 to 11% by *Artemisia*. The total above-ground phytomass in 1961 increased from 311 g m^{-2} at the end of May to 731 g m^{-2} in mid-July. The below-ground phytomass contributes 1920 g m^{-2}. The dry weight of the micro-organisms in the upper 0 to 20 cm of soil amounted to 67.7 g m^{-2} (bulk density of soil: 1.33).

The ash content of annual parts of *Atriplex* reaches over 30% of the dry weight (Na > 7%, Cl > 14%). *Artemisia pauciflora* contains, in contrast, only 10% ash (Na > 2%, Cl > 1.5%).

The fourth experimental area is distinguished from the third by the fact that *Artemisia pauciflora* is absolutely dominant (98% of the above-ground phytomass). *Anabasis salsa* and *Kochia prostrata* occur only individually. This biogeocoenosis covers a depression of roughly 5 ha between the hills of the highly eroded landscape. During the snowmelt period and after heavy rainfall it is well wetted, but is otherwise dry. The soil is transitional from columnar solonetz to light chestnut earth. Available water is found only in the upper 1 m, in which *Artemisia pauciflora* has its roots.

The cover can be very high in favorable years. In the dry year 1961 the above-ground phytomass was 173 g m^{-2}.

The entire area around the experimental station is unsuitable for agriculture and is only grazed, with little rational planning. On the average about 0.2 to 0.4 t ha^{-1} of dry matter can be used for livestock forage of average quality. It would be better here to alternate the grazing between grassier areas and those with halophytes.

Small, meadow-like patches can be mowed. Wet meadows with *Phragmites* yield 6 to 10 t ha^{-1} of hay, but these cannot be mowed every year because the yields then drop to one-half to one-quarter of the maximum amounts. Cyperaceae meadows yield only 0.6 to 3 t ha^{-1} of hay. The total area of mowable meadows is very small.

The unregulated grazing and frequent fires lead to very degraded vegetation covers. The proportion of grasses increases on burned areas, since *Artemisia* is damaged greatly by fire.

INVESTIGATION OF THE ECOSYSTEMS IN THE NORTHERN ARAL LOWLAND

Rodin (1977) has collected the results of the investigations from the period 1965 to 1974, which were performed under the auspices of the I.B.P. at the station Tereskent (investigators: B.A. Bykov and others). The station is located on the railroad line between Aral'sk and Chelkar, 65 km from Aral'sk, in the northern part of the Aral Lowland (Fig. 4.15). The climatic conditions are shown in the climate diagram (Fig. 4.16). The climate is continental, with an annual variation of absolute temperature of 84 to 89°. Total solar radiation during the growing period (mean temperature above 10°) amounts to 572 kJ cm^{-2} (diffuse ra-

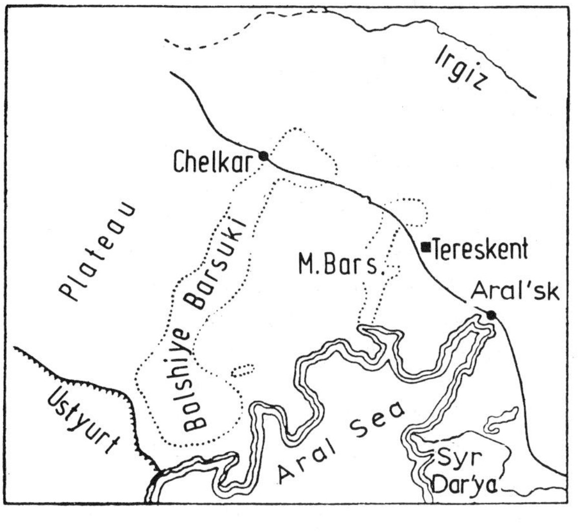

Fig. 4.15. Cartographic sketch of the northern Aral region; with the sand deserts Malyye (M. Bars) and Bolshiye Barsuki. The experimental station of Tereskent is 65 km from Aral'sk (from Rodin, 1977, p. 8). The Aral–Karakumy are east of Tereskent–Aral'sk.

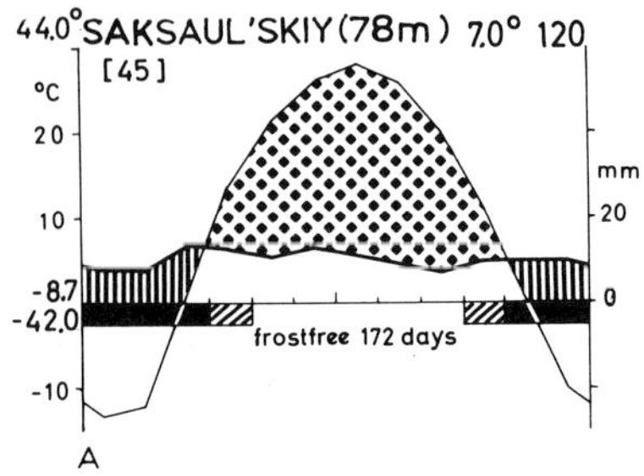

Fig. 4.16. Climate diagram of Saksaul'skiy in the northern Aral region (from Rodin, 1977, p. 8).

diation: 133 kJ cm^{-2}), of which the photosynthetically active radiation is 167 to 188 kJ cm^{-2}. The annual temperature sum (the sum of mean daily temperatures above 10°) is 3647°. A continuous snow cover of 5 to 8 cm is present from the middle of December until roughly the middle of March. The soil is frozen in February to a depth of 1.5 m.

The average annual precipitation is 120 cm, and the average humidity of the air at midday, from April to October, is not over 39 to 42%. The saturation deficit in April is 7 mbar, in July 25 mbar, and in October 5 mbar, with an average of 9.6 mbar for the year.

The average wind speed is 3.5 m s^{-1}; strong winds (> 15 m s^{-1}) occur on 5 to 28 days out of the year.

Surface water is completely lacking, and only the *takyry* and saltpans are inundated in springtime. The groundwater table is deeper than 10 m.

It is the zonal vegetation that was studied. This consists of an *Artemisietum terrae-albae* (with *Anabasis aphylla*, *Rheum tataricum*, *Poa bulbosa*, etc.) on normal burozems, and the slightly halophilic community of *Anabasis salsa*, found in shallow sinks which have no outlet and which have a solonetz horizon at a depth of roughly 2 to 5 m.

The *Artemisietum*

The *Artemisietum* in Kazakhstan covers an area of over 30 million ha. Of the species, 8 to 10% are xeromorphic semi-shrubs with well-developed fragmentation of basal stem, 5 to 10% are perennial herbs, 15% are ephemeroids (perennial geophytes), and 60% are ephemerals (annual therophytes). On the twelve experimental plots 44 species were found, producing a total cover of 37%. These species form particular synusiae. *Artemisia terrae-albae* is dominant, with a cover of 30%, about 80% of the phytomass, and about 150 000 individuals per hectare. The mean distance between actively regenerating individuals is about 25 cm. The age structure of the population was: 19.8% juvenile, 63.7% mature, and 16.6% senescent.

Other semi-shrubs include *Kochia prostrata*, *Ceratoides papposa*, and *Anabasis aphylla*; ephemeroids include *Iris songarica*, *Poa bulbosa*, *Ferula caspica*, *Allium decipiens*, *Tulipa biflora*, *Rheum tataricum*, *Gagea reticulata*, etc. Of the numerous ephemerals, noteworthy are *Leptaleum* (*Tanacetum*) *filifolium*, *Rochelia retorta*, *Lepidium perfoliatum*, *Descurainia sophia* and *Trigonella argenta*. Important also is the soil-lichen synusia containing *Aspicilia terrestris*, *A. aspera*, and *Caloplaca fominii*, whose phytomass makes up 0.67 g m^{-2}.

The vegetation reaches generally 20 to 30 cm above ground, but the rooting zone is about 80 to 100 cm deep. Only the scattered *Anabasis aphylla* reaches 60 cm in height and roots 2.5 and 3.5 m deep.

The phytomass reaches 33.2 t ha^{-1}, but only 18% is found above ground, 82% below ground. Of the total phytomass, *Artemisia* makes up 83%, *Rheum* 14%, *Anabasis* 1.6%, and the other species 1.4%.

Data on the energy content of the phytomass are shown in Table 4.5. The total energy stored in the biogeocoenosis amounts to 600.16×10^6 kJ of which 493.44×10^6 kJ (= 82%) are stored below ground.

The humus content of the soil is 22 t ha^{-1}, which is about three times the above- and below-ground annual litter (including dead plant parts). The total of the dead but not yet decomposed phytomass (*mortmasse*) reaches 9.2 t ha^{-1}.

The phytomass contains 1666 kg ha^{-1} of ash elements, of which *Artemisia* accounts for 1066 kg ha^{-1} (83%). The ash content of *Rheum* is very high (12%), a fact which we must attribute, at least preliminarily, to the many calcium oxalate crystals in the plant tissues. For *Anabasis aphylla*, 50% of the ash component is represented by sodium, chlorine and sulphur, so it is really a halophyte, although not so obviously a halophyte as the xero-halophytic *Anabasis salsa*, which is dominant in the second biogeocoenosis investigated.

TABLE 4.5

Energy content of the phytomass of the three most important species of the *Artemisia terrae-albae* biogeocoenosis (Anonymous, 1976)

Plant part	*Artemisia terrae-albae*		*Anabasis aphylla*		*Rheum tataricum*	
	t ha^{-1}	kJ ha^{-1} ($\times 10^6$)	t ha^{-1}	kJ ha^{-1} ($\times 10^6$)	t ha^{-1}	kJ ha^{-1} ($\times 10^6$)
Total phytomass	28.01	509.71	0.44	7.98	4.73	85.34
Leaves and annual shoots	0.68	12.81	0.04	0.54	0.51	8.56
Perennial shoots	4.35	83.62	0.09	1.68	no data	
Underground phytomass	22.98	413.28	0.31	5.75	4.22	76.77
at 0–25 cm	15.61	290.51	0.11	2.05	0.32	5.50
at 25–50 cm	5.21	99.45	0.07	1.34	2.01	37.12
at 50–75 cm	2.16	23.31	0.05	0.92	1.89	34.14
at 75–200 cm	–	–	0.08	1.42	–	–

1 kJ = 0.24 kcal.

The *Anabasietum salsae*

The *Anabasietum salsae* on solonchak-like solonetz was studied in less detail. This is a very widespread association found in the Caspian Lowland, on the lower course of the Kura in the Trans-Caucasus, and from the plains of Kazakhstan on into China. Thirty species can be identified in this biogeocoenosis: the shrub *Atraphaxis spinosa* (100 individuals ha^{-1}) and four semi-shrubs, among them the dominant *Anabasis salsa* (6700 individuals ha^{-1}) and *Zygophyllum macropterum* (100 individuals ha^{-1}), plus ephemeroids (*Carex pachystylis*, *Colpodium humile*, etc.) and ephemerals (including the halophilic *Petrosimonia brachiata*). Among the lichens, *Aspicilia esculenta* and *A. affinis* dominate.

The cover percentage varies between 12 and 30%, of which amount 95.5% is made up by *Anabasis* (= 6700 individuals ha^{-1}). The floristic composition scarcely varied in the years 1967 to 1970.

The phytomass of *Anabasis* was 7.59 t ha^{-1} (for the other species 0.68 t ha^{-1}), of which only one-eighth is above ground. The annual litter (Russian: *opad*) for *Anabasis* is 0.36 t ha^{-1} above ground and 2.60 t ha^{-1} below ground (about 40% of the phytomass). For the other species (primarily ephemerals), the figures are 0.29 t ha^{-1} above ground and 0.39 t ha^{-1} below ground, giving a total for all species of 3.64 t ha^{-1}. The total dead matter (*mortmasse*) is insignificantly higher at 3.87 t ha^{-1}.

General description

Archaeological findings show that the region north of the Aral Sea was already settled during the Neolithic and during the Bronze Age. Sheep, camel, and horse-raising, as well as hunting, were practised. Conditions were too unfavorable for agriculture. Attempts with special methods (see pp. 67–68) have only been made in this direction in recent times.

Further data were gathered on the number of seedlings, of which almost all died during the first year. The diurnal courses of transpiration, photosynthesis, and respiration were also examined. The productivity of photosynthesis was as follows:

Artemisia terrae-albae	10 g CO_2 (g dry wt.)$^{-1}$ year^{-1}
Kochia prostrata	14 g CO_2 (g dry wt.)$^{-1}$ year^{-1}
Anabasis aphylla	15 g CO_2 (g dry wt.)$^{-1}$ year^{-1}
Anabasis salsa	14 g CO_2 (g dry wt.)$^{-1}$ year^{-1}

The lower value for *Artemisia* is the result of a late summer dormant period.

Quantitative data on the number of seeds produced per unit area during the years 1965 to 1970 were given in a large table (Anonymous, 1976, p. 21).

The following data were presented for the nitrogen cycle of *Artemisia*: 300 kg ha^{-1} of nitrogen are contained in the phytomass; 76 kg ha^{-1} are required annually for growth, and 70 kg ha^{-1} return to the soil with the litter; thus the annual gain is

6 kg ha^{-1} (2 kg ha^{-1} in the above-ground growth). This gain is lost through herbivory.

For phosphorus, 7 kg ha^{-1} are taken up annually, and somewhat over 6 kg ha^{-1} return to the soil.

Of the total amount of rain water reaching the soil in 1966, 19% was used for transpiration and 81% for unproductive evaporation. In the dry year 1967 only 52.2% was used for evaporation. The unproductive evaporation loss was 40 to 60% in the spring but 70 to 80% during the dormant period of *Artemisia*.

Primary production

The primary production is almost completely limited to the first 1 to 1½ months of the growing season; autumn production depends entirely on the abundance of rainfall. Rainfall amounts of less than 5 mm evaporate immediately and have no effect.

The primary production of the *Artemisietum*, as computed from annual growth, was 7.72 t ha^{-1}, that is, about 23% of the phytomass. *Artemisia* represents about 65% of this total. The net increase of phytomass is 0.82 t ha^{-1} year^{-1}. The net increase for *Artemisia*, however, is only 0.25 t ha^{-1}, 10.7% of the total phytomass, a fact which marks the population of this dominant as stable. Production varies more for the ephemeroid *Rheum* and the ephemerals.

Table 4.6 reproduces primary production data for the *Anabasietum*.

Competitive relations in the ecosystem

Competition in the *Artemisietum* is primarily for water. In April and May *Artemisia* transpires, on the average, 0.9 g (g fresh wt.)$^{-1}$ h^{-1}, *Rheum* 0.6 g, *Poa bulbosa* 0.35 g, and the ephemerals 0.5 g. This water consumption speeds up the death of the young seedlings. Of the large herbivores, the most important was the antelope *Saiga tatarica*, but today includes primarily the smaller livestock, sheep and goats. Overgrazing leads to a degradation of the plant cover and the beginning of anthropogenic succession.

The following rodents are also important: *Allocricetulus eversmannii*, *Rhombomys opimus*, *Meriones meridianus*, *Citellus fulvus*, *Allactaga jaculus*, *A. elater*, and *Pygerethmus platyurus*. *Artemisia* suffers considerably from their gnawing. The numbers of the rodents are regulated by the carnivores *Canis lupus*, *Vulpes corsak*, and the rarer eagle *Aquila rapax*.

The plant cover is also greatly influenced by insect outbreaks, which are held in check by reptiles such as *Agama sanguinolenta*, *Phrynocephalus mystaceus*, and also the birds (*Calandrella cinerea* and *Melanocorypha leucoptera*).

The dynamics of the ecosystem

The individual synusiae have different patterns of development: *Artemisia* is active in the spring only, but *Kochia prostrata* and *Ceratoides papposa* have a summer dormancy only in drought years. The shrubs *Athraphaxis spinosa* and *Convolvulus fruticosus* fruit only during the summer and are for the most part dormant during autumn and winter. The xeromorphic *Ephedra distachya* ends its active vegetative phase at the beginning of the summer, whereas the deep-rooted *Anabasis aphylla* shows no dormancy at all and blooms during the driest part of the summer.

TABLE 4.6

Primary production in 1969 of the *Anabasietum* at the Station Tereskent (in t ha^{-1}) (Rodin, 1977, p. 24)

Plant(s)	Primary production			Phytomass increase		
	above ground	below ground	total	above ground	below ground	total
Anabasis salsa	0.43	2.72	3.15	0.07	0.12	0.19
Ephemeroids + Ephemerals	0.28	0.39	0.67	–	–	–
Total	0.71	3.11	3.82	0.07	0.12	0.19

Anabasis makes up 83% of the primary production.

The grasses *Poa bulbosa*, *Colpodium humile*, and *Psathyrostachys lanuginosa* fruit from May to the beginning of June. *Agropyron desertorum*, *Stipa sareptana*, and *Ajania* (*Tanacetum*) *achilleifolia*, fruit early but also often remain active on into autumn.

The ephemeroids *Rheum*, *Ferula*, and others, as well as the ephemerals, develop very quickly in the spring but terminate their development very early. Only the annual *Ceratocarpus arenarius* holds on until autumn.

The xero-halophyte *Anabasis salsa* develops later than *Artemisia* and usually goes through a summer dormancy. Fluctuations in the composition of the plant cover are noticeable from year to year, but the proportions of the dominants *Artemisia* and *Anabasis salsa* change, for the most part, very little.

Fire represents a major disruption. After a fire, 26 to 28 years are required before the normal biogeocoenosis is re-established. Anthropogenic interference also brings about secondary succession.

Biogeocoenosis complexes on sandy soils

The sands of the Malyye Barsuki formed during the Tertiary. They consist of sand ridges and sand hills, often sickle-shaped, wandering barkhans (Russian: *barkhan*). About 97% of the sand is of the grain size 0.25 to 0.05 mm. There is hardly any soil profile, the humus content is under 0.1%, the soil pH is neutral, carbonates are absent to a depth of 2 m, and the Cl^- and SO_4^{2-} content is low.

The biogeocoenosis studied is dominated by *Ammodendron argenteum*, which becomes 80 to 170 cm tall and forms the upper vegetation stratum (root depth over 2 m), along with *Calligonum* spp. and *Eremosparton*. In the second stratum (25 to 40 cm tall) one finds various *Artemisia* species, *Kochia*, grasses, and forbs (root depth up to 2.5 m). The lowest stratum is formed by shallow-rooted ephemerals.

The total vegetal cover reaches 50%. The total phytomass is 34.2 t ha^{-1} (27% above ground and 73% below ground), of which *Ammodendron* contributes 10.7 t ha^{-1}, *Artemisia tomentella* 5.4 t ha^{-1}, *Kochia prostrata* 5.4 t ha^{-1}, and *Agropyron fragile* 4.4 t ha^{-1}. The dead mass was 16.3 t ha^{-1} and the humus mass 5.0 t ha^{-1}. The sandy region is rich in springs. It is grazed by livestock, and *Agropyron* is mowed for hay.

As is always the case on sandy soil, the water availability to the plants is relatively good, so that the deep-rooted species have no summer dormancy.

Ammodendron argenteum is a good sand-binder. The development of the shoot and root systems is described in detail (Anonymous, 1976, p. 32).

The annual primary production (growth) of the biogeocoenosis is 14.0 t ha^{-1} (9 t ha^{-1} below ground). The annual litter production is 12.4 t ha^{-1}, of which 8 t ha^{-1} (65%) consists of root litter, 4 t ha^{-1} (32%) is contributed by annual plant parts, and only 0.4 t ha^{-1} (3%) by perennial parts. The largest part of the total annual litter (3.2 t ha^{-1}) comes from *Artemisia tomentella*, and only 1.3 t ha^{-1} comes from *Ammodendron argenteum*. The net growth of phytomass per year was 1.6 t ha^{-1}. The plants of the biogeocoenosis take up 155.4 kg ha^{-1} of nitrogen and 496 kg ha^{-1} of ash elements annually. Of these amounts, roughly two-thirds are used for building of underground organs. To the soil are returned annually 135 kg ha^{-1} of nitrogen and 417 kg ha^{-1} of ash elements.

The caloric value of the phytomass of individual plant parts varies considerably, from 19 228 J g^{-1} for fruits to 22 521 J g^{-1} for the fine roots of *Ammodendron*.

The total phytomass of the biogeocoenosis has a caloric content of 606.97×10^6 kJ ha^{-1}. There are also 129.78×10^6 kJ ha^{-1} bound in the parts already dead.

Two additional tables show the exact distribution of the ash elements and the nitrogen (Anonymous, 1976, p. 37).

Investigations at the Station Taukum

The northern deserts lie in the endorheic basin between the Kazakh Hills and Betpak-Dala Plateau in the north and the foothills of the northern Tyan Shan (Tien Shan) in the south. Lake Balkhash occupies the lowest part of the basin. To the south of it stretch the vast sand deserts (Sary Ishikotrau) and, further to the south, between the Ili River and the Chu-Ili Mountains, the Taukum (for the location of the station see Fig. 4.17).

Both a summary of results and a bibliography have been published by Rodin (1977). The researchers were Kurochkina and others. Sub-

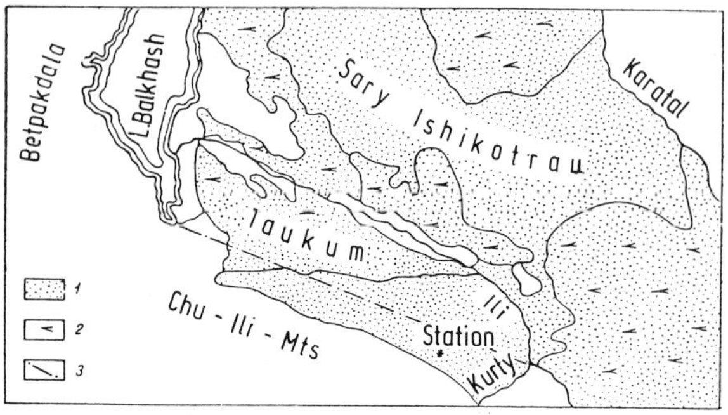

Fig. 4.17. Location of the Taukum experimental station in southern Kazakhstan (from Rodin, 1977, p. 40). Legend: *1* = sand areas; *2* = *Haloxyloneta*; *3* = location of the profile in Fig. 4.18.

divisions based on relief of the study area are shown in Fig. 4.18. The Taukum sands are located 370 to 500 m above sea level and arose out of Pleistocene alluvial deposits. As a result of redistribution by the wind, the present-day surface is composed of fine-grained sands which form ridges with steep north slopes and gentle south slopes. Deep dune valleys are sometimes formed, but there are no typical barkhans.

The mean climatic data are shown in the climate diagram (Fig. 4.19). The climate is somewhat warmer and moister than in the northern Aral region. The average for the years 1955 to 1965 was 218 mm of rainfall annually, but for 1965 to 1972 it was

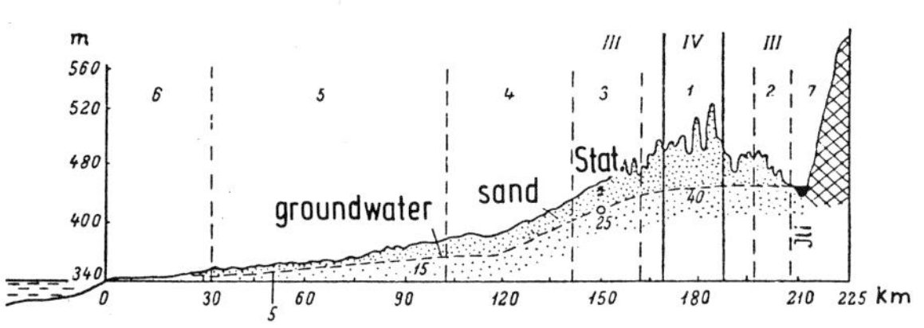

Fig. 4.18. Profile through the Taukum Desert from Lake Balkhash to the Malyye-Sary Mountains. Ordinate: height above sea level (in m). Vegetation: *1* = *Artemisietum* with dwarf-shrubs on high sand ridges (20% mobile, with *Aristida*); *2* = dwarf-shrubs on partly mobile sands, with heavy competition by weeds (*Gypsophila paniculata*, *Secale silvestre*); *3* = *Artemisietum–Agropyronetum* with *Calligonum* and *Kochia prostrata* (around the experimental station); *4* = the same, but with 20% of the area covered with mobile sands: *5* = *Haloxylonetum–Artemisietum* on dust-containing sand; 6 = halophytes on solonchak (*Halostachys caspica*, *Halocnemum strobilaceum*, *Kalidium caspicum*); *7* = meadows in the floodplain of the Ili River. *III* = earlier third terrace of the river; *IV* = earlier fourth terrace. The numbers on the groundwater table (dashed) indicate its depth (in m).

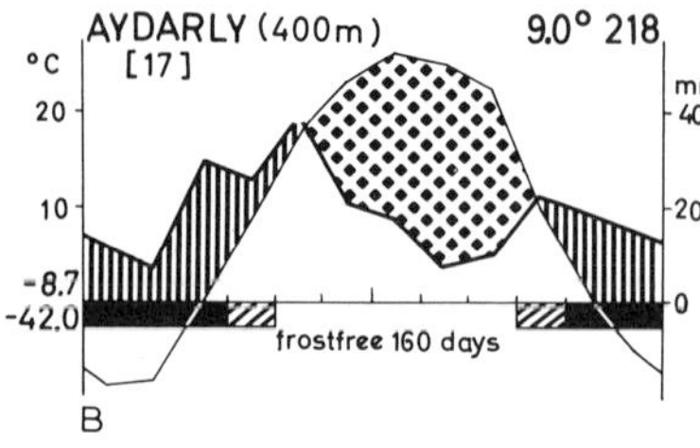

Fig. 4.19. Climate diagram of Aydarly in the Taukum.

232 mm and about 300 mm fell in 1966 and 1972. The maximum rainfall comes some years in the spring, other years in the summer, which is especially good for the rejuvenation of the shrubs and for an increase in the plant cover.

Water for the woody species is most readily available in September–October.

The groundwater level is shown in Fig. 4.18. It lies so deep that it has essentially no effect on the plant cover.

Within the sand areas one can distinguish: (1) recently redeposited sands; (2) sands originating from the nomadic period of the thirteenth to eighteenth centuries, with shallow sandy soils; and (3) old, silty-sandy, firm soils which are, however, still distinguishable from the zonal siero-burozems by their primitive structure.

The vegetation is accordingly very heterogeneous. Four different biogeocoenoses were studied:

(1) The *Agropyronetum fragilis* with *Artemisia terrae-albae* and *Calligonum leucocladum* on gentle south slopes, with *Artemisia* increasing downslope. The number of species is 31. *Agropyron fragile* (12 700 individuals ha^{-1}) is dominant, followed by *Artemisia* (6200 individuals ha^{-1}) and *Kochia prostrata* (6%). All other species are poorly represented. One can distinguish two synusiae, the dominant grasses and the semi-shrubs.

(2) The *Ephedretum lomatolepis* with three *Calligonum* species on flat sand ridges and depressions of the old relief elements. The number of species is 21. *Ephedra* is dominant with 114 100 individuals ha^{-1}. The *Calligonum* species have 800 individuals ha^{-1}, and the other species from (1) are represented.

(3) The *Artemisietum terrae-albae* with *Kochia prostrata* occurs on the final stage of the sands of lower north slopes transitional to the depressions, a location which most closely approximates the euclimatopes (Russian: *plakor*). Total vegetal cover is

50‰, and the number of species is 25. The vegetation has two levels. In the upper layer *Artemisia* (61 000 individuals ha^{-1}) dominates, followed by *Kochia* (26 000 individuals ha^{-1}) and *Agropyron* (9300 individuals ha^{-1}). In the lower level are represented *Carex physodes* (770 500 individuals ha^{-1}), *Colpodium humile* (480 000 individuals ha^{-1}), *Oedibasis apiculata* (24 000 individuals ha^{-1}), *Allium* spp. (8000 individuals ha^{-1}), and also various ephemerals.

(4) The *Ceratoidetum papposae* with *Artemisia terrae-albae* occurs on firm, sandy, carbonate-rich soils of the depressions, with a moss layer composed of *Tortula desertorum* (80% cover). *Ceratoides* (10 000 individuals ha^{-1}) and *Artemisia* dominate. The latter is somewhat more numerous, but the shrubs are very small. There are about 3200 individuals ha^{-1} of *Kochia*. The number of species is 34. Three synusiae can be distinguished: the semi-shrubs, the perennial forbs, and the mosses. Ephemerals are relatively unimportant.

This sand desert still belongs to the Kazakh or north Turanian type, which is floristically clearly distinguished from the Middle Asian and south Turanian deserts. These deserts are also distinguished ecologically by the only slight summer depression in the activity of the plants, the less pronounced winter rain, and the more extreme cold period.

The development of the most important species is treated in detail in the original source (Anonymous, 1976). We limit ourselves more to the quantitative data on production and mineral cycles (Tables 4.7 and 4.8).

Additional tables (Anonymous, 1976, p. 55) show corresponding data for the most important individual species of the individual biogeocoenoses.

If the areas are grazed, the following percentages of total aboveground phytomass of the individual biogeocoenoses can be utilized by livestock: (1) *Agropyretum:* 6.4‰; (2) *Ephedretum:* 4.1‰; (3) *Artemisietum:* 7.3‰; and (4) *Ceratoidetum:* 4.1‰.

The sands are relatively poor in nitrogen, but atmospheric nitrogen is fixed by the root nodules of the many legumes. Table 4.7 gives summary information on the cycling of nitrogen and ash elements.

The nitrogen cycle is especially intensive in the *Artemisietum*. In the phytomass of the *Agropyretum*, 228 kg ha^{-1} of ash elements and 60 kg ha^{-1} of nitrogen are found. For the *Ephedretum* the numbers are 482 and 150, and for the *Artemisietum* 294 and 135 kg ha^{-1}. The chloride content is vanishingly small in all biogeocoenoses, but the calcium content is high. After water, it appears that only nitrogen would limit the productivity of the vegetation.

For interspecific competition, the water conditions are surely the most important factor, but the interrelationships are very complicated. For example, the ephemerals are favored underneath *Calligonum alatiforme* because of the protection from the wind and the nutrients in the litter. On the other hand, the ephemerals inhibit the sprouting of *Calligonum* seedlings and prevent propagation of the fruits.

The relationships between higher and lower plants are also very complicated. Certain species are benefited by mycorrhizal fungi, while others are inhibited by rust and mildew fungi.

TABLE 4.7

Amounts of ash elements and nitrogen accumulated in the annual production and in the annual dying parts (kg ha^{-1}) for communities of the Taukum during 1969–1971 (Rodin, 1977, p. 47)

Fraction	*Agropyronetum*		*Ephedretum*		*Artemisietum*	
	ash elements	nitrogen	ash elements	nitrogen	ash elements	nitrogen
Annual production	160.97	49.0	79.59	25.0	128.77	72.00
above ground	30.90	9.8	40.50	15.0	71.65	17.04
below ground	130.07	39.2	39.09	10.0	87.12	54.96
Annual dying parts	131.25	48.0	57.26	18.0	123.15	60.74
above ground	26.87	48.0	27.84	18.0	37.73	60.74
below ground	104.38		28.43		85.42	

TABLE 4.8

Annual production and dying parts (t ha^{-1} and % of total phytomass) for communities of the Taukum during 1969–1971 (Rodin, 1977, p. 47)

Fraction	*Agropyronetum*		*Ephedretum*		*Artemisietum*		*Ceratoidetum*	
	t ha^{-1}	%	t ha^{-1}	%	t ha^{-1}	%	t ha^{-1}	%
Total phytomass	6.03	100	12.85	100	7.46	100	10.33	100
above ground	1.44	24	3.53	27	2.53	34	1.40	–
below ground	4.59	76	9.32	73	4.93	66	8.93	–
Annual production	4.39	73	2.22	17	4.16	56	6.32	61
above ground	0.67	11	1.16	9	1.41	19	1.01	10
below ground	3.72	62	1.06	8	2.75	37	5.31	51
Annual dying parts	3.60	51	1.54	12	3.57	48	1.28	12
above ground	0.59	1	0.81	6	1.28	17	0.36	3
below ground	3.01	50	0.73	6	2.29	31	0.92	9

Herbivory by caterpillars is often heavy in the spring, both on *Artemisia* and on *Kochia* and *Calligonum*. Certain beetles destroy the fruit of some species. The activity of rodents is also very striking (*Rhombomys opimus*, *Meriones meridianus*, etc.). These not only eat certain plant parts but also, through burrowing, destroy the plant cover and bring about succession. Of the ungulates, *Gazella subgutturosa* occurs mostly in groups of seven to ten animals, but their number is small today and their effect on the vegetation is not of importance.

The effects of man through grazing of domestic animals can be seen in the wind erosion (deflation) and in the increase of weedy species. One can also consider as a partial degradation the disappearance of shrubs and of *Agropyron fragile* and the increase of *Artemisia terrae-albae*.

The number of weeds which have found their way into the region is rather large. Especially striking is *Gypsophila paniculata*, a tumbleweed which spreads out quickly over great distances.

REFERENCES

Anonymous, 1969a. *Biocomplex Investigations in Kazakhstan. Part I: Plant Communities and Animal Populations in the Steppes and Deserts of Central Kazakhstan.* Akad. Nauk S.S.S.R., Nauka, Leningrad, 469 pp. (in Russian).

Anonymous, 1969b. *Biocomplex Investigations in Kazakhstan. Part II: Biocomplex Characteristics of the Main Coenosis-Forming Plants of Central Kazakhstan.* Akad. Nauk S.S.S.R., Nauka, Leningrad, 336 pp. (in Russian).

Anonymous, 1976. *Biocomplex Investigations in Kazakhstan. Part III: Main Plant Communities of the Central Kazakhstan Desert-Steppe.* Akad. Nauk S.S.S.R., Nauka, Leningrad, 292 pp. (in Russian).

Gvozdezkiy, N.A. and Nikolayev, V.A., 1971. *Kazakhstan (Description of Natural Conditions).* Mysl, Moscow, 255 pp. (in Russian).

Lavrenko, E.M. and Sveshnikova, V.M., 1965. Synthetic research on growth forms using the example of *Stipa* tussock grasses. *Zh. Obshch. Biol.*, 26: 261–275 (in Russian).

Maljugin, E.A., 1954. Agricultural methods for cultivation of the sandy semi-deserts of western Kazakhstan. *Pustyn. S.S.R. Ikh Osvoyeniye*, 2: 66–134 (in Russian).

Rodin, L.E. (Editor), 1977. Productivity of vegetation in the arid zone of Asia. In: *Synthesis of the Soviet Studies for the International Biological Program 1965–1974.* Akad. Nauk S.S.S.R., Nauka, Leningrad, 232 pp. (in Russian).

Rustamov, I.G., 1972. Phytomass quantitative characteristics and productivity of subshrublet communities of the Krasnovodsk Plateau. In: *Int. Symp. U.S.S.R., 1972, Leningrad*, pp. 128–132.

Shalyt, M.S., 1952. The underground parts of some meadow steppe and desert plants and phytocoenoses. Part II. *Tr. Bot. Inst. Akad. Nauk S.S.R. Ser. III* (*Geobot.*), 8 pp. (in Russian).

Sveshnikova, V.M., 1962. Eco-physiological peculiarities of dominant *Stipa* species in steppe communities. *Probl. Bot.*, 6: 388–398 (in Russian).

Walter, H., 1968. *Die Vegetation der Erde, II. Die gemässigten und arktischen Zonen.* Fischer-Verlag, Jena–Stuttgart, 1001 pp.

Walter, H., 1974. *Die Vegetation Osteuropas, Nord- und Zentralasiens, VII. Vegetationsmonographien der einzelnen Großräume.* VII. Fischer-Verlag, Stuttgart, 452 pp.

Walter, H., 1979. *Vegetation of the Earth and Ecological Systems of the Geo-Biosphere.* Springer-Verlag, New York, N.Y., 2nd ed., 274 pp.

Walter, H. and Straka, H., 1970. *Arealkunde. Phytologie, Vol. III, No. 2*, Fischer-Verlag, Stuttgart, 2nd ed., 478 pp.

Chapter 5

MIDDLE ASIAN DESERTS[1]

H. WALTER and E.O. BOX

GENERAL GEOLOGICAL HISTORY

The region of the Turanian Lowland, the northern part of the Irano-Turanian deserts (Fig. 2.1), adjoins the previously discussed Kazakhstan deserts to the north and extends to the Kopet Dag (elevations up to 2262 m) along the Irano-Afghanistan border in the south, crossing it in places. It is delimited to the west by the Caspian Sea and in the east by the foothills of the Middle Asian mountains belonging to the Tien Shan and Pamiro-Alay systems. The region lies roughly between latitudes 38° and 44°N and longitudes 55° and 65°E. This corresponds latitudinally to the northern Mediterranean zone.

The large geo-morphological units are the Ustyurt Plateau with the Mangyshlak Peninsula[2] and the southern sand deserts: Kyzylkum (between the Syr-Dar'ya and the Amu-Dar'ya Rivers, primarily in Uzbekistan) and especially the Karakum (between the Amu-Dar'ya and the Caspian Sea, primarily in Turkmenistan). The Karakum (Turkic, meaning black sand) has been ecologically studied in great detail and will be discussed separately in Chapter 6. In addition to these large units, the following can also be distinguished: the loess-covered foothills of the Middle Asian mountains (in the south and east); and many scattered small units such as *takyr* areas, river floodplains with delta formations, and saline soils in depressions or coastal areas.

The broad, rather level Turanian Lowland has a relatively young topography which arose only in the Neogene (Miocene and Pliocene), as the Sarmat Sea receded to the west. The deposits of this sea, mostly limestones and marls, remained only on the Ustyurt Plateau and on the Mangyshlak Peninsula. These deposits were later lifted and never again flooded. They thus form the oldest surface in the entire region, with later small tectonic movements causing only slight changes. Erosion has also had a rather minor effect because the climate of Middle Asia has been distinctly arid since the end of the Miocene. Members of the present flora are descendants of the Tertiary *hammada* (rock desert) desert vegetation.

After the Sarmatian period came a period of massive mountain-building in the south and east, which continues today. The Turanian Lowland was completely isolated from the rest of Asia. At the same time, the weathering products from the erosion of the new mountains came to be deposited by rivers in the Turanian Lowland, which sank, except for the northwestern part, and was flooded anew by the Aktshagyl Sea. Up until the beginning of the Pliocene, the Sarmatian layers in this area were carried off and covered with new marine deposits. These latter remain in the northern part of the Karakum Desert as Trans-Unguz layers and form the second-oldest surface in the region (Fig. 5.1).

The southern part of the Karakum was again flooded by the Apsheron Sea, at the beginning of the Pleistocene, as far as the Kopet Dag. At this time, which coincided with a major glaciation of the mountains, the Amu-Dar'ya River system was formed. The river originally emptied into the Caspian Sea, continuously changing its course through the lower Karakum between the Kopet Dag in the south and the Trans-Unguz area in the north. The Unguz escarpment was the right bank of

[1] Manuscript completed November, 1977.

[2] As a result of the sinking of the Caspian Sea, this is no longer a peninsula and is shown on newer maps as the Mangistau Hills.

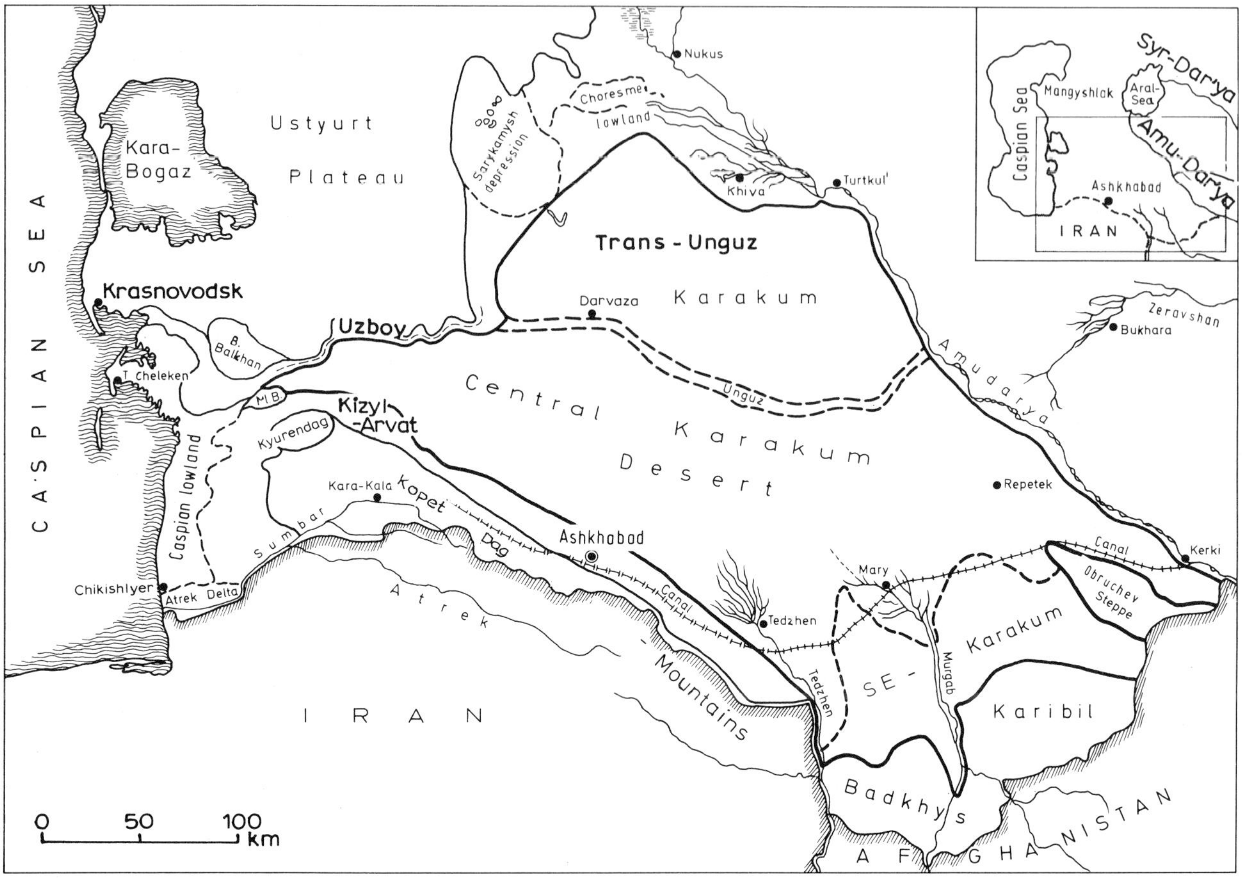

Fig. 5.1. Map of Turkmenistan with the Karakum Desert (outlined) and neighboring regions (from Walter, 1976, fig. 46). *M.B.* = Malyye Balkhany.

the meandering Amu-Dar'ya. At this time the Aral Sea, together with the Sarykamysh Depression, formed a large lake area to the north of the Trans-Unguz.

The entire Karakum lowland was covered with alluvial deposits from the original Amu-Dar'ya. This forms the third-oldest surface today, since the last transgression of the Khvalyn Sea reached to the east only as far as the meridian of Kizyl-Arvat.

At the same time, the lower course of the Amu-Dar'ya was pushed continuously to the east by the delta deposits of the rivers Tedzhen and Murgab coming directly from the south, until finally, in the second half of the Pleistocene, the Amu-Dar'ya broke through to the north to reach the Aral Sea, along roughly its present course. Of course, the new delta area changed places again. At times one arm in the delta emptied into the currently dried-out Sarykamysh Depression, filling it then with water. A freshwater lake formed, which found an outlet to the Khvalyn Sea through the Uzboy River. The water level of the Khvalyn Sea lay considerably above the level of the present Caspian Sea. It sank during the ensuing period, however, so that the Atrek River, coming from the Kopet Dag in the southwest of the region, was able to build a broad delta into the Caspian Sea.

The mouth of the Amu-Dar'ya was later pushed further to the east, to the Aral Sea, by the deposits in the delta area. As a result, the Sarykamysh Depression dried out until only a few salt lakes remained and the western arm of the Amu-Dar'ya dried up. The Uzboy is today a stagnant river. The lakes in its course are not remnants of the river but, rather, depressions without outlets, which are fed laterally and from groundwater and become enriched with salts. An outlet could have existed temporarily during the wet period of the thirteenth to sixteenth centuries (Shnitnikov, 1957).

During the most recent period, the fine soil particles from the broad alluvial areas of the sand deserts have been blown out by the prevailing winds and deposited as loess on the foothills of the mountains to the south and east. It has been estimated that a dust layer of 5 m thickness has been removed from the Karakum with time. The coarser, sandy parts were only moved a short distance, which in the sand desert results in dune formation. These processes are described in more detail in the description of the Karakum Desert (Chapter 6).

CLIMATIC CONDITIONS

The Turanian Lowland lies in the transition zone between the dry, subtropical air masses of the Arabian–South Iranian region and the continental, temperate air masses to the north. This transition zone has, climatically, an arid Mediterranean character. Mediterranean cyclones bring precipitation in the winter and on into the spring (December–April), with a maximum in March. The summer months (May–September) are rainless and very hot. The character of this climate can be seen clearly from the climate diagrams of Nukus in the north, Ashkhabad in the south, Krasnovodsk and Chikishlyer in the west, and Kerki in the southeast (Fig. 5.2).

The aridity of the climate is very pronounced. The annual precipitation is lowest in the delta area of the Amu-Dar'ya, about 80 to 100 mm. Only a little over 100 mm falls along the Caspian coast, and only at the foot of the Kopet Dag does the rainfall climb to something over 200 mm.

The potential evaporation is very high — about 1500 mm annually on the average, over 2500 mm in the interior of the desert, and only on the sea coast somewhat under 1000 mm. The annual evaporation of the Caspian Sea has been estimated as 986 mm, but rises to 1100 mm in the Kara-Bogaz Bay, which extends far inland.

The incident solar radiation is also very high. For Tashkent it is 102.4 kJ cm^{-2} in July and 20.9 kJ cm^{-2} in December, which correspond to 84.4 and 16.3 kJ cm^{-2} for a horizontal surface. The total annual solar radiation is 543 to 585 kJ cm^{-2}. The number of hours of sunshine is 2500 to 3000, 90% of the possible amount in the summer and 50% in the winter. Since the region is completely open to the north, cold air from the Siberian high-pressure system invades roughly 12 to 23 times during the winter, and absolute minimum temperatures reach about −26°C. Thaws also occur intermittently,

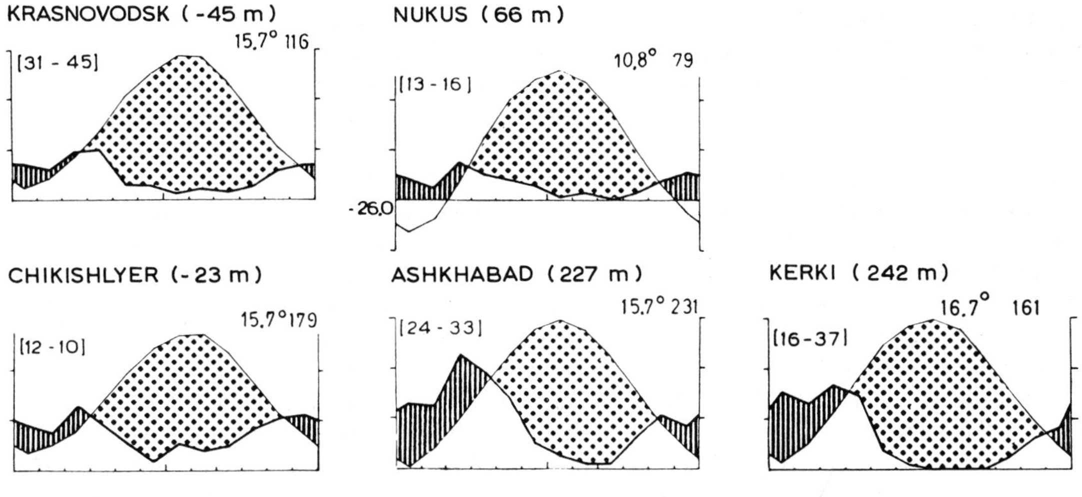

Fig. 5.2. Climate diagrams from the regions bordering the Karakum (from Walter, 1976, fig. 48): Krasnovodsk on the Caspian Sea, Nukus in the Amu-Dar'ya Delta, Chikishlyer in the Atrek Delta, Ashkhabad on the foot of the Kopet Dag, and Kerki on the Amu-Dar'ya (see map in Fig. 5.1).

however, so that, in contrast to Kazakhstan, there is no continuous snow cover. Mean January temperatures are considerably below freezing except in the extreme south, where they are only slightly above freezing.

Sudden changes in weather can also be felt in spring. The temperature can suddenly fall from 25° to freezing, bringing with it heavy snowfall. Late frosts occur until the end of April. These can be very dangerous for cotton crops.

The summers are very unchanging and hot. More than half of the days have mean temperatures above 30°C, with absolute maxima around 50°C. The average diurnal temperature variation is 15 to 20°C, and the difference between the extreme temperatures in summer is 20 to 30°C. Relative humidity can drop to 2 or 3%. Invasions of subtropical air from Iran are especially hot and often connected with duststorms. Drought necessitates a major reduction in leaf area for *Artemisia*, and many *Salsola* and *Astragalus* species in the summer. The *Calligonum* species lose their leaves completely. Only *Haloxylon*, *Aellenia*, and the large *Salsola* species show no essential changes.

Individual cool days begin in autumn at the end of September. These become more frequent during October and November and eventually bring on the winter.

The most favorable season is the spring, when the soil is wetted by the winter precipitation and receives additional spring rainfall. The development of spring ephemerals and ephemeroids (geophytes) is consequently very striking. The length of the spring growing season fluctuates very strongly from year to year, however, depending on the abundance of rain. For *Ceratocephalus orthoceras*, for example, the growing season varies from 75 to 115 days, for *Bromus tectorum* 106 to 185 days, and for *Meniocus linifolia* and *Hypecoum parviflorum* 62 to 162 days.

Depending on temperature and rainfall patterns in the spring, the germination of sometimes one, sometimes another ephemeral is favored, so the composition of the ephemeral cover can vary considerably from year to year. The ephemeroids also come out by no means every year, even if their underground organs are still alive. The annual production of organic matter is subject to similarly large fluctuations, for example, in the range of from 5 to 150 kg ha^{-1} for the grasses, 1 to 54 kg ha^{-1} for the forbs, and 69 to 176 kg ha^{-1} for *Carex physodes*.

Seed germination of perennial species occurs also in the spring, but large numbers of seedlings appear only in wetter years. Their further development requires three consecutive favorable years, which occurs only seldom. It is for this reason that young, perennial plants are found, in general, only infrequently in the desert.

The climate along the coastal strip on the east coast of the Caspian Sea, from Krasnovodsk to the western foothills of the Kopet Dag (Fig. 5.1), is somewhat milder. The compensating effects of the large water surface are noticeable: The winters are not so cold, summer temperatures are 3 to 5° lower, and the humidity in the driest month (August) is 30 to 40% higher than further to the east. The precipitation, however, still remains quite low at 110 to 120 mm. The frost-free period is 200 to 240 days. The growing season is 210 to 275 days, with the same temperature sum of 4000 to 5000° as in the sand desert.

The southwestern district, with the Atrek Delta to the south, has the most favorable climate. It is protected by the Kopet Dag from the northeast winds, so that temperatures do not fall below −10°. The temperature sums reach 5100 to 5500°, and the frost-free period is 270 days. Annual precipitation is around 180 to 280 mm, and the humidity is somewhat higher. The climate at Kara-Kala is thus almost subtropical. Not only olives, figs, almonds, and pomegranates but also date-palms (*Phoenix*) and citrus species can be cultivated (Sharapov, 1954). Of course, in cold winters the leaves of *Phoenix* are killed by frost.

The climate diagram of Chikishlyer (Fig. 5.2) gives an idea of the climatic conditions.

SOIL FORMATION

Because of the aridity, only a small amount of organic matter can be produced by the plants in the desert. Under the prevailing high temperatures, the organic matter which does reach the soil is quickly decomposed, so the humus content of the desert soils is much reduced (1–1.5%) and is only to be found in the uppermost horizons. The soils are relatively inactive, and their carbonate content is high. The readily soluble salts, especially gypsum, are not leached out but rather accumulate quite near the surface.

The high evaporation rate produces a crust on the soil surface. If the groundwater level is high, the surface is kept wet by the capillary fringe. The water in the soil solution evaporates, but the dissolved salts can only accumulate and eventually form a solonchak soil. Only the sandy soils, in which there is almost no capillary movement, do not become saline.

The prevailing physical weathering leads to the formation of primarily coarse-grained soils, mainly rocky, gravelly, and sandy soils.

The following soil types are most often found in Middle Asia:

(1) Dark sierozems (siero-burozems) on the outcropping rock of plateau sites. These have rocky or gravelly upper horizons, compaction and salinization at a depth of 5 to 15 cm, and a gypsum horizon at 50 to 70 cm. The humus content is only 0.7 to 1.5%. Such soils are characteristic of the rock desert (hammada). Gypsophilic plants grow on these soils.

(2) Solonchak soils with a heavy accumulation of salts (NaCl, sulphates) in the upper horizons. These form in depressions with higher groundwater levels. On such sites one finds only salt-tolerant halophytes.

(3) Sandy soils, which cover enormous areas in the desert areas of Middle Asia. These are water-permeable and have low salt content. In arid regions these are among the most favorable sites for plants, since they store water well. Psammophytes must have well-developed root systems, however, because the field capacity of sandy soils is very low. The sand is easily moved when dry, so dune formation can occur as soon as the surface is cleared of its plant cover.

(4) Light sierozems, which arise over loess and carry ephemeral vegetation. These are characterized by calcium deposits in the form of fine, white fibers in the upper horizons. The soil is densely permeated by roots, and the humus content is relatively high at 1 to 1.5%. These soils are thoroughly wetted in the spring but dry out almost completely in the summer. Gypsum crystals accumulate at greater depths.

(5) Takyr soils, which form in shallow depressions or on almost level plains which are flooded temporarily in the spring. These are clayey and, once swollen with water, are water-impermeable. Only the top 5 to 10 cm (maximum 15–20 cm) are wetted. In the dry state these soils crack into polygon-shaped plates. These soils are colonized by algae and lichens.

(6) Alluvial meadow and swampy soils, which are found in the floodplains and delta areas. These carry dense floodplain vegetation and contain more humus (3–4%). They are similar to the soils of the irrigated, cultivated areas.

VEGETATION–SOIL RELATIONSHIPS IN MIDDLE ASIA

The flora of the Turanian Lowland is composed primarily of Turano-Iranian geoelements, which tend also toward the Mediterranean. The number of endemic Turanian genera and species, however, is quite large. In the Kyzylkum, for example, they are 30% of all species. Mongolian elements, on the other hand, are absent, quite in contrast to Kazakhstan, which has a connection to the east through the Dzungarian Gate.

According to Korovin (1961/1962), the Turanian flora contains 246 species of Chenopodiaceae (16%), 160 species of Asteraceae (12%), 148 species of Fabaceae (9%), and 96 species of Brassicaceae (6%). Below 5% are the Liliaceae, Poaceae, Polygonaceae, Apiaceae, Zygophyllaceae, and Plumbaginaceae (this last with only 1.7%).

Endemic genera, mostly monotypic, are: *Spiraeanthus*, *Niedzwedzkia*, *Borszczowia*, *Alexandra*, *Piptoptera*, *Londesia*, *Kirilowia*, *Ormopterum*, *Schumannia*, *Soranthus*, *Iljinia*, *Sympegma*, *Chartoloma*, *Spirorrhynchus*, *Tetracmidion*, *Streptoloma*, *Octoceras*, *Lachnoloma*, *Cithareloma*, and *Sprygima*. Endemism is very well developed in Middle Asia, the result of events in the Late Tertiary (intensive mountain formation, increased aridness, and isolation from neighboring regions).

The following genera are represented by very numerous species: *Calligonum* 67, *Ammodendron* (section of *Astragalus*) 26, *Zygophyllum* 31, *Anabasis* 20, *Salsola* s.l. 54, *Limonium* 22, *Tamarix* 16, and *Suaeda* 22.

The following vegetation types, which correspond to soil types, are taken as the basis for the classification of the vegetation (see also Rodin, 1963):[1]

(1) Gypsophilic dwarf-shrub vegetation
(2) Halophyte vegetation
(3) Psammophyte vegetation of the sand deserts
(4) Deserts with only ephemeral vegetation
(5) Takyr vegetation
(6) Floodplain vegetation

In the desert, the plants diminish in favor of rocks. Closed, above-ground communities are absent, and plants often are found only as individuals on favorable sites. Root competition is certainly not lacking, but the distribution of particular species is influenced, above all, by the locally rapidly changing site characteristics, such as soil catenas on slopes, the presence of soil water only in small pockets or crevices, and rapidly changing salinity. Consequently, it is generally not possible to identify discrete plant communities and it is more practical to undertake a vegetation classification based on substrate material or water availability.

Gypsophilic dwarf-shrub vegetation on dark sierozems of plateau sites

During the Tertiary, the original vegetation of Middle Asia was driven out by increasingly xeric conditions. At the same time, the products of weathering, which came to be deposited in the area, changed also. The earlier, lateritic soils were replaced toward the end of the Oligocene and in the first half of the Miocene by soils rich in gypsum, which was available in abundance as the weathering product of the mountains. This makes the wide expansion of the gypsophilic vegetation understandable. The soils are certainly rich in calcium also, but the presence of gypsum exerts an especially strong influence on the vegetation in arid regions. Thus, according to Korovin (1961/1962), about 400 gypsophilic species can be distinguished in Middle Asia. To this number belong most of the endemic genera already mentioned but also species of the genera *Zygophyllum*, *Limonium*, *Cleome*, *Reaumuria*, *Haplophyllum*, *Ferula*, *Dorema*, *Arthrophytum*, *Anabasis*, *Nanophyton*, and other dwarf shrubs and semi-shrubs, such as *Artemisia terrae-albae* and *A. turanica*, which grow vegetatively in the spring and then bloom and fruit in the autumn after a summer dormancy. (In Middle Asia about 100 *Artemisia* species are distinguished, almost all belonging to the subgenus *Seriphidium*.) A typical species also is *Salsola laricifolia*, whose branch tips become thorny. Thorny or spiny are also *Gaillonia* spp., *Sisymbrium subspinescens*, *Noaea spinosissima*, *Cousinia* spp., *Acanthophyllum pungens*, *Atraphaxis spinosa* and others, as is generally the case with xeromorphic plants. Significant, in addition, are *Caragana grandiflora*, *Convolvulus frutescens*, *Sophora griffithii*, *Ephedra distachya*, and some *Calligonum* species. There are also about 100 annual ephemerals and 30 geophytic ephemeroids.

Especially remarkable are perennial but monocarpic Apiaceae, often 2 m tall (*Ferula assa-foetida*, *F. ferulaeoides*, *Oedibasis apiculata*, *Muretia oeroilanica*, *Hyalolena turcomanica*). Certain badlands composed of soft, saline, often brightly colored Tertiary marls, on which only a few species can occasionally settle, must also be considered to be part of the gypsum desert.

On plateau-like rises in the Tertiary formation one can find, above all, the *Artemisia–Salsola* community. Dominant are *Artemisia terrae-albae* and *A. kemrudica* (on rocky soils, *A. turanica*), which reach 20 to 25 years of age, along with *Salsola rigida* and often also *S. arbuscula* or *S. laricifolia*, two *Aellenia* species, *Arthrophytum* or *Reaumuria*, and four *Anabasis* species. The Turanian variant of this community is clearly different floristically from the Kazakh and has more species (70 were found on 4000 ha in the Kyzylkum area). In southern Kazakhstan one finds this community from the Ustyurt Plateau in the west to the Betpak-Dala Desert in the east. Lichens (*Squamarina*, *Diploschistes*, *Psora*, *Aspicilia*, *Fulgensia*, and *Collema*) are quite widespread, as well as soil algae and the moss *Tortula desertorum*.

Nanophyton erinaceum (see Fig. 4.14) dominates in other open communities with a cover of 25 to 30%. Another important type is the *Artemisia*–ephemeral desert, often with *Poa bulbosa*, *Carex pachystylis*, *C. physodes*, and other annual and summergreen species. This type of desert is widespread in the southern part of Middle Asia, at the foot of the mountains. As the relief becomes flatter, it changes over into halophyte marshes and takyry.

[1]Illustrations of typical species of Middle Asian gypsophytes, halophytes, psammophytes, ephemerals, and geophytes have been published by Walter (1974, figs. 168–172).

As soon as the soils become sandier, *Astragalus squarrosus* and *Convolvulus hamadae* often appear in large numbers. As the soil of the gypsum desert is converted more to solonetz, a community of *Hammada leptoclada*, with *Aellenia*, *Gamanthus*, *Halocharis*, *Salsola lanata*, *S. sclerantha*, *Halimocnemis*, and other Chenopodiaceae, usually appears.

It can often be observed that sand accumulates as a result of the weathering and subsequent blowing out of the finer particles, so that psammophytes appear in new areas. On the other hand, a dominance of *Salsola* species, which contain up to 3% chloride and 10% sodium in their tissues, can with time lead to a salinization of the soil and to the invasion of halophytes.

Halophilic desert vegetation

The halophytes are represented by many genera in Middle Asia (numbers of species in parentheses): *Salsola* (20), *Suaeda* (9), *Atriplex* (4), *Panderia* (2), *Camphorosma* (3), *Echinopsilon* (3), *Kochia* (2), *Kalidium* (3), *Halostachys* (1), *Halocnemum* (1), *Salicornia* (1), *Seidlitzia* (1), *Anabasis* (8), *Haloxylon* (1), *Petrosimonia* (10), *Halocharis* (3), *Halimocnemis* (2), *Halogeton* (2), *Zygophyllum* (14), *Tetradiclis* (1), *Nitraria* (3), *Limonium* (9), and *Reaumuria* (3). Chenopodiaceae predominate heavily.

The halophyte vegetation is restricted to solonchak soils close to groundwater and, thus, accompanies the lower courses of several rivers, occupies depressions into which rainfall runs off, and encircles the salt lakes. The decisive ecological factor here is not water but rather the salt, primarily chlorides. Only a few species are adapted to such high salt concentrations. Especially characteristic are the widespread chenopods *Salicornia herbacea*, *Halocnemum strobilaceum*, and *Halostachys caspica*, while *Seidlitzia rosmarinus* is an endemic. In addition, there are *Haloxylon ammodendron* (*aphyllum*) and other species, especially those from the families Plumbaginaceae, Caryophyllaceae, Boraginaceae, and Tamaricaceae, the presence of which generally is scarcely noticed.

Most species are of the succulent halophyte type, with high osmotic values caused by the chlorides stored in the cell fluids. Some of these are woody plants which reach a considerable age, such as *Halostachys*, *Haloxylon*, *Kalidium*, and *Seidlitzia*. There are also nonsucculent, salt-excreting halophytes, including *Cressa cretica* and the grass *Aeluropus*.

The halophytes with high salt content can be significant as livestock feed when the salts are leached out of the litter in winter. The protein content of the dry matter is relatively high.

An extreme, mostly unvegetated form of the salt desert is that represented by the salt pans (*shory*). These arise in depressions with very high groundwater level. In such situations, where the capillary fringe reaches the surface, continuous evaporation of the salt solution forms swollen crusts which also contain gypsum. The high salt concentration prevents any plant growth. The first pioneer species is *Halocnemum strobilaceum*, which appears on the periphery in the form of low cushions with a cover of about 25% (see central part of Fig. 5.3). If water stands between these cushions in the spring, it gives the impression of a hummock-moor. *Halocnemum* is joined by *Salicornia*, *Halostachys*, *Kalidium*, and *Halopeplis pigmaea*. In the next zone outward are *Suaeda* and *Salsola* species, on some sites also *Petrosimonia*. These are followed, on higher ground, by *Anabasis salsa* and *Artemisia* species (Fig. 5.3).

The stands of *Haloxylon ammodendron* are very characteristic. This species is arborescent and can become 14 m high under optimal conditions, with a trunk diameter of 1 m. It can form very dense, impenetrable stands, which are still very open since the trees have no leaves (illumination at the soil surface is 50% of normal daylight). In spite of this, undergrowth is absent. The ground is bare, but groundwater is always available at a depth of a few meters and is reached by the *Haloxylon*. Seedlings of this *Haloxylon* species can form a taproot of 2.8 m within one year.

One stand of this tree, covering 600 000 ha, exists along the Chu River; other large stands can also be found along the Amu-Dar'ya River, i.e., almost always in alluvial plains. These *Haloxylon* forests are very important as sources of firewood and are heavily exploited, and in some cases have been annihilated (Fig. 5.4).

Other halophyte communities are dominated by dwarf shrubs up to 40 cm tall, such as *Salsola gemmascens*, which also forms pure stands but can be mixed with up to twenty species (all appearing

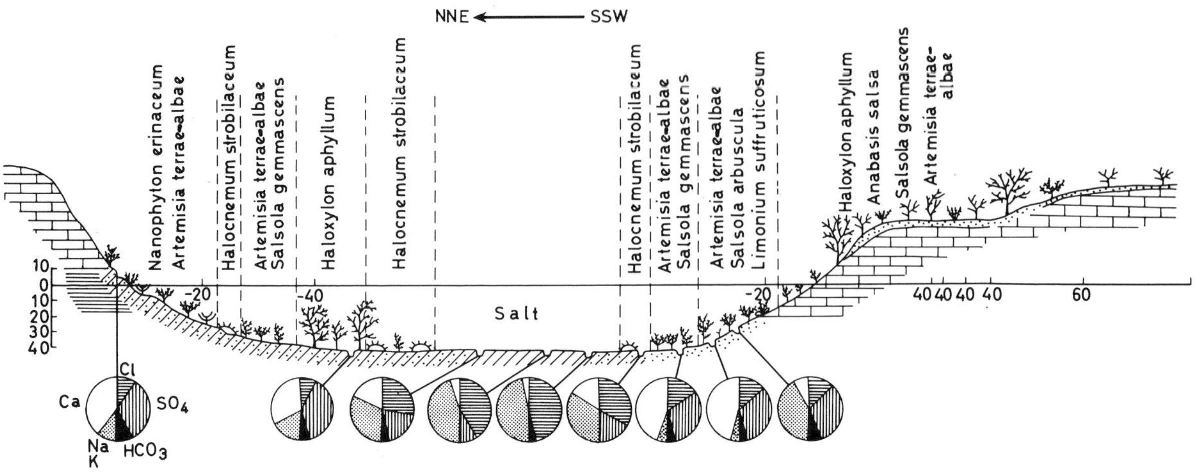

Fig. 5.3. Zonation of the salt pan Kaundy. Height in the profile is magnified twenty-fold. Circles show the percentage of salt ions: Cl^- dominates in the center with SO_4^{2-} increasing toward the periphery (after Viktorov, in Walter, 1968, fig. 543). *Haloxylon aphyllum* = *H. ammodendron*.

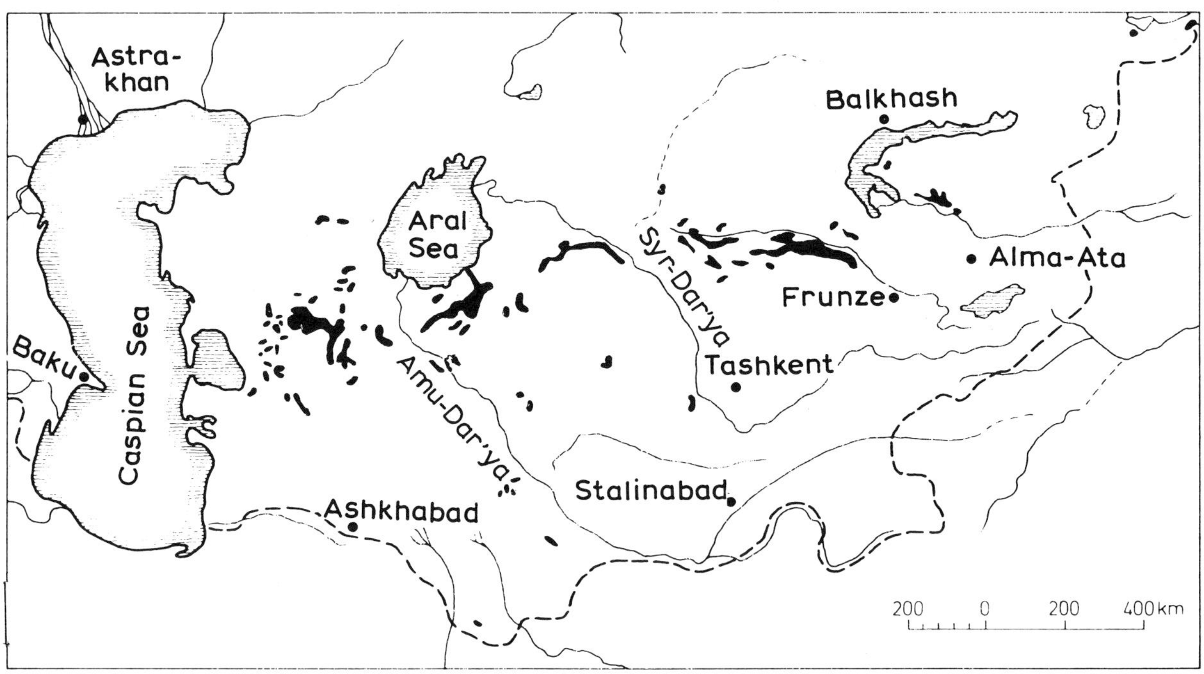

Fig. 5.4. Distribution of extensive *Haloxylon ammodendron* stands in Middle Asia (from *Descriptio Vegetationis U.S.S.R.*, in Walter, 1974, fig. 181).

more or less individually). *Salsola rigida* has a similar sociology.

Psammophilic communities

The flora of the sand deserts includes 350 species, 56% of which are endemic.

The shrubs form an interesting group, which is especially characteristic for Middle Asia. The genus *Calligonum* with thirty species (from the family Polygonaceae) has its center of distribution in Middle Asia and Kazakhstan. These are virgate shrubs, which, according to the form of the fruit, are divided into four sections. Only one species occurs in the Sahara, a few in the Caucasus, Iran, and Mongolia. The two species of the endemic genus *Eremosparton* (Fabaceae) are also virgate shrubs. These are joined by five *Ammodendron* species (Fabaceae), the twenty closely related *Astragalus* species of the section *Ammodendron* (1 m tall, silver-leaved shrubs), *Haloxylon persicum*, three *Salsola* species, the endemic *Smirnovia turkestana* (Fabaceae) with small, cordate leaves, a woody *Artemisia* species, *Ephedra*, and others.

The virgate form seems to be especially characteristic for plants of the sand deserts and for plants of arid regions on sites with relatively good water supplies. Plants with virgate assimilation organs can transpire heavily, but, because of the small surface area, the total water loss is not very large.

Although the annual precipitation in the southern part of the Karakum varies in individual years between 24 and 172 mm, the vegetation, except for the shifting sand dunes, is often so thick that one does not have the impression of a desert at all (see figs. 531 and 535 in Walter, 1968). This is especially true since tree species are present, such as the 7 to 8 m tall *Haloxylon persicum* or the 5 m tall *Ammodendron conollyi* (Fabaceae), as well as the 7 to 8 m tall *Calligonum eriopodum* and *C. arborescens* and the 3 m tall *Salsola richteri* and *S. paletzkiana*. The number of shrubby forms not over 1 m tall is even greater: *Calligonum* spp., *Eremosparton* spp., *Ephedra strobilacea* and *E. lomatolepis*, *Astragalus* spp. from the section *Ammodendron*, and others.

Therophytes are quite numerous: a total of 143 species, of which 98 are spring ephemerals. These are restricted to the areas of stable sand. In the southern Karakum, many of these germinate in November–December at temperatures of 7 to 12°C if the top soil layers are wet. Of the ephemeroids, 31 species are present, including *Carex physodes*, *Poa bulbosa*, 8 *Allium* spp., 4 *Gagea* spp., 2 *Eremurus* spp., 2 *Iris* spp., 2 *Tulipa* spp., 2 *Rheum*, *Leontice* and *Dorema* spp., 4 *Ferula* spp., *Calligonum*, *Orobanche*, and others. Their green growth period covers only 1½ to 2½ spring months.

We return to this psammophyte vegetation in more detail in Chapter 6.

Ephemeral deserts on light sierozems over loess

One must distinguish between the northern and southern loess deserts. The dominant species, *Poa bulbosa* and *Carex pachystylis*, are common to both, but the southern deserts are differentiated by the occurrence of the giant Apiaceae (*Ferula* spp., *Dorema*, and others). Shrubby or arborescent *Pistacia*, *Ficus*, and *Amygdalus* species are also found scattered throughout the southern deserts at particular sites, and these are completely absent from the northern loess deserts.

The *Carex pachystylis* community was first described for the loess plain Mirachul at an elevation of 225 to 360 m. It is quite widespread in the south, however, and extends also over into Afghanistan (Freitag, 1971). The root mass of this community is limited to the upper 20 cm of the soil, with very few species (0.5%) rooting deeper. This comes about as a result of the high field capacity of the loess, which means that the soil is only shallowly wetted. The plants have used up all the water by the end of May, so it is impossible for dwarf-shrubs to survive the summer drought (Fig. 5.5).

Table 5.1 provides some insight into the water conditions in the soil during the spring.

TABLE 5.1

Water content (%) of the soil in the loess desert in the spring (Walter, 1968)

Depth (cm)	19 April	24 April	3 May	19 May
0–2	7.3	23.3	0.7	0.5
8–9	6.3	18.7	8.4	5.3
20–22	9.2	8.0	10.0	8.2
60	8.7	8.7	9.4	8.6
90	4.8	8.7	8.3	7.1
200	2.4	3.7	4.7	–

Fig. 5.5.A. Vegetation profile through the ephemeral desert with very shallowly rooting therophytes and geophytes (from Korovin, in Walter, 1974, fig. 173). B. Ephemeral desert in southwest Turkmenistan, in springtime (photograph by M.P. Petrov; from Walter, 1968, fig. 528).

The depth of thorough wetting varies from year to year, of course, depending on the amount of rainfall. Since it rains roughly every four or five days in March and April, however, the upper soil layers remain wet and dry out quickly only at the end of May when the rain stops and soil temperatures rise to 30°C at a depth of 10 cm.

Conditions for growth are very favorable in the spring. *Carex pachystylis* (*affine C. stenophylla*) and *Poa bulbosa* come out at the beginning of March, when the temperature starts to climb. These are the most important species.

Carex pachystylis (*C. hostii*) is a small plant which reaches a height of 20 cm under good conditions. The leaves are short and narrow. The plant has a rhizome at a depth of about 5 cm which can grow 2 to 8 cm per year and can reach a total length of 6 m. The individual shoots come out about 3 to 5 cm apart. The numerous shoot-producing roots form a thick mass of twisted, tangled roots at a depth of 5 to 15 cm but reach to 60 cm. *Poa bulbosa* (*vivipara*) is also small and has clumps 2 to 3 cm in diameter which are thickened at the base and form several small bulbs. Small bulbs also develop in the inflorescence in place of flowers. Both types of bulbs remain alive for several years, even in the herbarium, and can withstand high temperatures in the dry state.

Carex pachystylis dominates and covers 75 to 80% of the area (with up to 4000 runners m^{-2}). *Poa bulbosa* is well represented and can displace *Carex* under slightly saline conditions.

Ephemeroids (geophytes) include: *Geranium tuberosum*, *Eranthis longistipitata*, *Scorzonera tuberosa*, *Ixiolirion tataricum*, *Bunium capusii*, *Ranunculus severzovii*, *Gentiana olivieri*, *Gagea* spp., and *Tulipa* spp., as well as the perennial, monocarpic *Ferula assa-foetida* and others. These ephemeroids also propagate vegetatively, sometimes primarily vegetatively, which insures the continuation of the species even in dry years with especially short growing seasons. *Ranunculus severzovii*, with its white-pubescent leaves, does bloom regularly but produces few viable seeds. It propagates primarily by means of up to 10 cm long underground rhizomes, from the ends of which new plants arise.

Ephemerals are also found with the ephemeroids. Among the ephemerals are the Brassicaceae *Draba verna*, *Alyssum desertorum*, and *Hymenolobus procumbens*; the Caryophyllaceae *Holosteum umbellatum*, *H. glutinosum*, *Arenaria leptoclados*, and others; the Papilionaceae *Astragalus filicaulis*, *A. campylorhynchus*, and *Trigonella* spp.; and the Poaceae *Agropyron buonapartis*, *Bromus macrostachys*, and *Phleum paniculatum*, as well as species of the genera *Papaver*, *Delphinium*, *Nigella*, *Malcolmia*, *Capsella*, *Veronica*, *Erodium cicutarium*, and others.

The growing season for these species varies from 1 month to 2½ months. The plants reach heights of 10 to 20 cm in good rain years, but either do not germinate at all in bad years or remain dwarfs of 1 to 2 cm. Depending on the amount and distribution of the rainfall and the simultaneously prevailing temperatures, the germination of one species or another will be favored, so that the annual flora varies immensely from year to year and from site to site.

In the spring the vegetation cover is closed and reminds one of a meadow. Those plants which develop under the most favorable conditions are all mesophilic, without xeromorphic characteristics. After the end of May, however, the desert makes a totally dead impression for nine months. One exception is that represented by the two small stenohydric species *Aphanopleura capillifolia* (Apiaceae) and *Diarthron vesiculosum* (Thymelaeaceae). These begin to bloom when the sun has burned everything else and the soil is hard as stone. *Diarthron* lives on until autumn. The ecology of *Diarthron* has been studied by Breckle (1971) in Afghanistan.

The number of species in the ephemeral desert is 45 to 50, but on an area of 10 m² only about 10 to 15 species are usually present, though with a large number of individuals. Near Tashkent the following were found on 1 m²: 1060 individuals of *Festuca myuros*, 545 of *Arenaria serpyllifolia*, 257 of *Galium tenuissimum*, 138 of *Aegilops triuncialis*, and 42 of *Gentiana olivieri*.

The following site sample also serves as an example:[1]

Carex pachystylis	2	*Holosteum umbellatum*	+
Poa bulbosa	4	*Veronica campylopoda*	+
Ranunculus severzovii	3	*Ferula assa-foetida*	r
Malcolmia turkestanica	2	*Haplophyllum pedicellatum*	r
Ixiolirion tataricum	2	*Ziziphora tenuior*	r
Gagea stipitata	1	*Leptaleum filifolium*	r
Gagea reticulata	1	*Bunium capusii*	r

The ephemeral desert on loess-covered slopes (*adyr*) in the foothills of the Middle Asian mountains is very similar. Of course, there are some new species; the geophytes *Valeriana chionophila*, *Anemone petiolulosa*, and *Euphorbia rapulum*, plus some species which root deeper and last all summer, for the annual precipitation here is mostly above 300 mm. There arises a certain stratification of the root systems (Fig. 5.6). In the north the deeper rooting species are the *Artemisia* species, while in the south they are particularly *Cousinia*, *Astragalus*, *Psoralea*, *Haplophyllum*, *Convolvulus*, and *Phlomis* spp. These can take up water from a depth of 2 m. The roots of *Alhagi camelorum* can reach 20 m if the soil is sufficiently moist, but the annual ephemerals still dominate strongly. Especially striking are the southern deserts with the monocarpic *Ferula assa-foetida* and *Dorema aitchisonii*, a 2 m tall umbellifer which blooms only after reaching the age of six to seven years.

On the foothills 6 km from Tashkent, Markova (1973) has studied the development of eighty annual ephemerals. *Aristida adscensionis* and *Tribulus terrestris* (adventive tropical species) germinate only in the spring, while *Eremopyrum hirsutum* and *Hordeum leporinum* germinate only in the autumn and do not bloom if sown in the spring. All others germinate mostly in late autumn, as soon as the soil is wet enough, but bloom also if sown in spring. The development of the ephemerals depends on the water supply. If this is good, then development is prolonged; flowering occurs later but also lasts longer, since secondary flowering shoots may form.

[1] Here and subsequently the Braun-Blanquet indices of cover are used: 5: 76–100%; 4: 51–75%; 3: 26–50%; 2: 6–25%; 1: 1–5%; +: <1%; r: rare.

Fig. 5.6. Vegetation profile through the foothill ephemeral desert in Middle Asia (from Korovin, in Walter, 1974, fig. 186). *1* = *Carex pachystylis*; *2* = *Poa bulbosa* var. *vivipara*; *3* = *Ranunculus pinnatifidus*; *4* = *Scaligeria alloides*; *5* = *Haplophyllum perforatum*; *6* = *Phlomis thapsoides*; *7* = *Ixilirion tataricum*; *8* = *Gentiana olivieri*; and *Eremostachys labiosa* in the middle (without a number).

The ephemeral desert is used for livestock grazing but provides sufficient forage for only three months. It can even be mowed. The best time is the end of April, at which time one can obtain 0.5 to 2.5 t ha^{-1} of dry matter. In dry years, however, the production of plant matter is very low. Hay yields gained by mowing at different times are shown in Table 5.2.

As one can see, *Poa bulbosa* gives the higher yield, while *Carex* hay has double the protein content. *Carex* can competitively exclude *Poa*.

The underground phytomass is also quite large, but 95.5% of this is contained in the top 10 cm of soil, 4% in the next 10 cm, and only 0.5% at depths down to 60 cm. Over 2 kg m^{-2} of rhizomes, bulbs, and roots have been documented.

The level topography and the good soils make the ephemeral desert especially suited to irrigated agriculture, especially cotton. Thus, today only small areas can be found which are still in their original conditions.

As soon as readily soluble salts appear in the soil, the floristic composition of the ephemeral desert changes. At the same time the plant cover takes on the appearance of a mosaic. The slightest unevenness provides small depressions, into which the

TABLE 5.2

Hay yields (100 kg ha^{-1}) in the ephemeral desert (from Korovin, 1961/1962)

Dominant species	19 April	24 April	29 April	4 May	9 May	14 May	19 May
Carex pachystylis	6.4	8.7	4.7	5.0	4.0	3.6	2.4
C. pachystylis and *Poa bulbosa*	12.3	13.9	15.4	12.9	11.6	9.5	7.2
Poa bulbosa	25.2	29.6	40.9	37.8	41.6	24.7	18.8

salts are washed and in which they can accumulate. The vegetation reacts very quickly to the slightest change in salt content.

The first sign of salinization is the appearance of *Salsola sclerantha*, *S. carinata*, or *Halocharis hispida*. These species germinate late and develop well only after the ephemeral vegetation has already dried out. Moreover, *Carex pachystylis* is inhibited by only a very slight salt content, and *Poa bulbosa* begins to take its place. The clumps of *Poa bulbosa*, however, are only weakly anchored, can be easily removed from the soil, and are thus commonly distributed in patches. New annual, slightly salt-tolerant species, some with fleshy leaves, appear simultaneously: *Senecio coronopifolius*, *Lepidium perfoliatum*, *Spergularia marginata*, *Sclerochloa dura*, *Statice* and *Plantago* spp., and others. *Plantago lachnantha* shows the same behavior in the dead, fruiting stage as does the well-known "Rose of Jericho" (*Anastatica hierochuntica*), that is, the plants bend their fruiting stems inward during dryness and extend them again when it is wet.

This slightly halophilic ephemeral desert represents a transition between the takyry, which in the extreme can be completely bare of vegetation, and the typical halophyte desert.

The vegetation of the takyr soils

The takyry were thought for a long time to be totally without vegetation, but, through the studies of Rodin (1954) as well as Kovda (1956), it was shown that they are densely populated by algae and lichens. Only higher plants are either completely absent or remain relatively unimportant, then mostly on takyr-like but not true takyr soils.

The algal vegetation may be a very old vegetation type, perhaps from pre-Devonian times when there were no higher plants. Similar formations may have covered the surface in the perhumid tropics at that time, places where they do not occur today because the soil is shaded by the forests. Lower plants can occur there today only as epiphytes on leaves.

On the clayey, temporarily moistened takyry in the deserts, which are not suitable for the growth of higher plants, however, the algae as *Desertialgeta* or the lichens as *Desertilichineta* gain dominance as the only vegetation.

The takyry are flooded either only in the spring or sometimes several times during the year, but the water is not able to sink very far into the heavy, initially dried-out clay. The water evaporates quickly, and the soil dries out and cracks. These sites are suitable, therefore, only for poikilohydrous plants which can withstand almost complete dehydration.

The river Amu-Dar'ya has two high-water periods, one in the spring and a second in the summer, when the snow in the high mountains melts. The inundated areas, especially in the delta, dry out quickly and break up by cracking into more or less regular polygons with diameters of 15 to 25 cm. The top, crusty layer consists of very dense clay (Fig. 5.7A).

Other large takyr areas in Middle Asia are found at the foot of the mountains, as for example at the foot of the Kopet Dag, in the area of the old delta formations (Fig. 5.7B). These are both deluvial and alluvial young clay deposits originating from the muddiness of the waters flowing down from the mountains. This water also carries a certain amount of soluble salts, such as chlorides and sulphates (up to $2\ g\ l^{-1}$), so that the takyr soils show a light or sometimes distinct salinization. Thus, they are similar to certain, quite dry saline soils. Those higher plants which do appear are ephemerals which germinate from seeds which the water brings in great numbers. The greater the salinity, the smaller is the number of germinating seeds. On low-salt soils the number can be many thousand per square meter.

The takyry were studied by Rodin from 1951 to 1956 in great detail. The blue-green algae of the genus *Phormidium* (32 species) are most important on the clayey takyry, along with *Microcoleus vaginatus*. A total of 147 species were identified: 92 Cyanophyta, 38 Chlorophyta, 5 Xanthophyta, and 12 Bacillariophyta (a more detailed list can be found in Gollerbakh et al., 1956).

Only nineteen species of Cyanophyta and six species of Chlorophyta occurred simultaneously in two different areas. For the most part, one or two species of Oscillatoriales dominate absolutely. (In the semi-deserts of Europe, though, it is primarily Nostocales, which swell up considerably when wetted.) In puddles of standing water it is primarily the *Zygnema* species which develop.

The algae form a film several millimeters in thickness on the surface. The amount of dry organic

Fig. 5.7.A. Dry takyr with algal mat (photo by L.E. Rodin). B (p. 93). Distribution of takyr soils in Turkmenistan (adapted from Rodin, in Kovda, 1956): *1* = algae-takyry (solonchak-like soils); *2* = lichen-takyry (solonchak-like soils); *3* = takyr-like siero-burozems with patches of algae-takyry; *4* = scattered small takyr areas within areas of other soils.

matter formed reaches 0.5 to 1.0, even 1.4, t ha^{-1}. Growth occurs not only as a result of flooding but also after wetting by rainfall.

The microfauna is also ephemeral in character, developing temporarily in the shallow pools. The slightest difference in height means that the water does not distribute itself evenly but overflows from the larger watercourses and remains standing in pools. These are generally very shallow, 5 to 10 cm deep. The water in the pools is often warmed to 20° above the air temperature, providing very good conditions for the development of the Cyanophyta. These form a green, photosynthetic film, which gives off oxygen as small bubbles and, through the consumption of carbon dioxide, lends the water a certain alkalinity during the day. It is known that the algae are able to fix elementary nitrogen. Thus, it is understandable that their dry matter shows a nitrogen content of 4.5%.

Alongside these algae communities (*Desertialgeta*) but on the somewhat higher non-flooded areas, one can also find the lichen communities (*Desertilicheneta*), with primarily *Diploschistes albissimus*, *D. scruposus*, and *Squamarina lentigera*, together with *Collema*, *Aspicilia*, and *Psora* species, and others. The dry matter of the phytomass reaches about 0.5 to 1.0 t ha^{-1}.

The lichens can only survive where there is no sedimentation, since they grow very slowly. These areas are wetted only by rainfall.

Since the surface of the takyry is subject to constant changes resulting from sedimentation, the inundated areas are constantly changing. Thus, the algae and lichen communities are in a constant dynamic state, replacing each other whenever the distribution of the water changes.

Microscopic soil fungi are also widespread on the takyry. As Litvinov (1956) was able to show, these

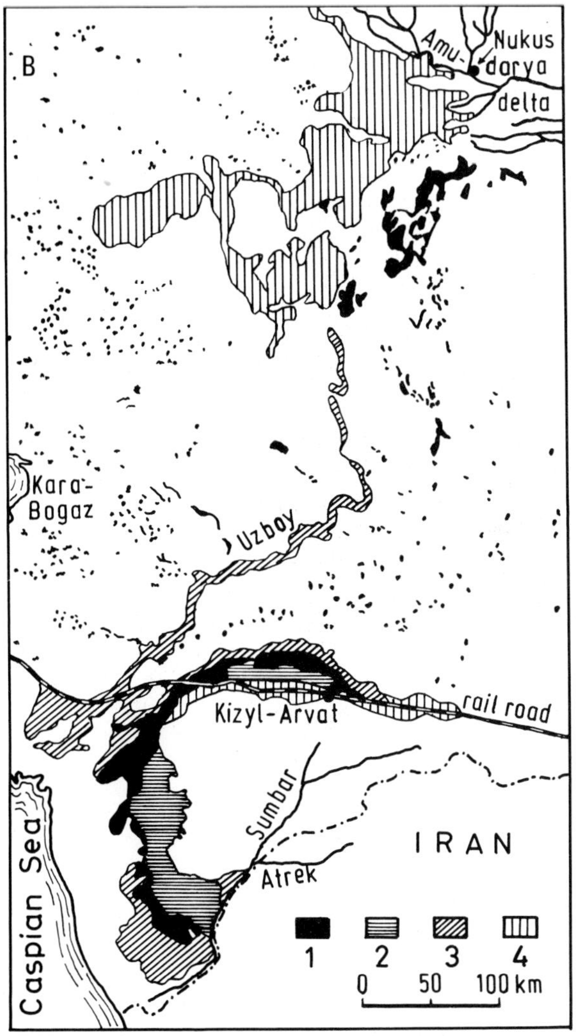

are primarily Dematiaceae, with the genera *Stemphylium*, *Macrosporium*, *Alternaria*, and *Cladosporium*, also *Aspergillus*, but seldom *Penicillium*. The fungus flora, however, shows clear differences on the different takyr sites. The fungi, together with the especially numerous Actinomycetes and the bacteria, take over the role of decomposers in this cryptogam ecosystem.

Of the fungi, one (*Stemphylium algophagum*) was isolated from faded places in the algae film. Laboratory tests showed that the metabolic products of this fungus kill the algae; they also inhibit the growth of *Hordeum* and *Vicia* seedlings (Litvinov, 1956).

It is important that one view the takyry as dynamic, continuously changing patterns, as Rodin (1954) has very graphically depicted them in the Kopet Dag foothills. The components of these dynamic processes are the water, wind, soil salts, lower plants, and higher plants, as well as animals.

The takyry are the lowest, barely inclined flat portions of the rubble fans deposited by intermittent waterfalls at the mouths of the valleys. The episodic water courses meander continuously across these flat areas. One can even observe the small erosion channels formed by the rain falling on the takyry. These *chili* begin on the takyry and disappear again within a short distance. If they run together, then individual hollows remain, in which puddles initially of only a few centimeters depth begin to form. Ephemerals can take hold here, root through the soil, and eventually modify it. Usually they dry out prematurely. Freshly deposited soil is also a suitable place for germination. The seeds are carried along with the water from the mountains and accumulate in the hollows with the soil carried along with them. In such a hollow 31 000 seeds were once found on a single square meter (Rodin and Sukhoverko, 1956).

The most common species are *Lepidium perfoliatum*, *Malcolmia africana*, *Hypecoum trilobum*, *Roemeria hybrida*, *Matricaria lamallata*, the grasses *Hordeum leporinum*, *Eremopyrum orientale*, and *Gagea* and *Allium* spp., just to name a few. As a result of evaporation at the edges of the pools, the salts contained in the water accumulate, especially on the south sides. An embryonic solonchak is formed, on which *Salsola* or *Suaeda* and other Chenopodiaceae settle. The roots of these plants loosen up the crust so that the soil can be wetted somewhat more deeply. This usually means a settling of the soil surface, that is, the hollows become deeper and wider, and the takyr is slowly covered by vegetation. At any time, however, a new deposit of alluvial soil can destroy everything and bring on a new course of succession.

Erosion channels can be merged by the action of the wind or covered up, which also causes individual hollows to form. On the other hand, if the area is inundated, the water can be blown away by the wind. On dry soil, the finer soil particles are piled up around the plants and, in the case of shrubs, actually form significant hillocks (*hummocks*, *nebkas*). Other plants grow on these hillocks, and animals begin to colonize them, so that a definite micro-ecosystem arises, which dies again as

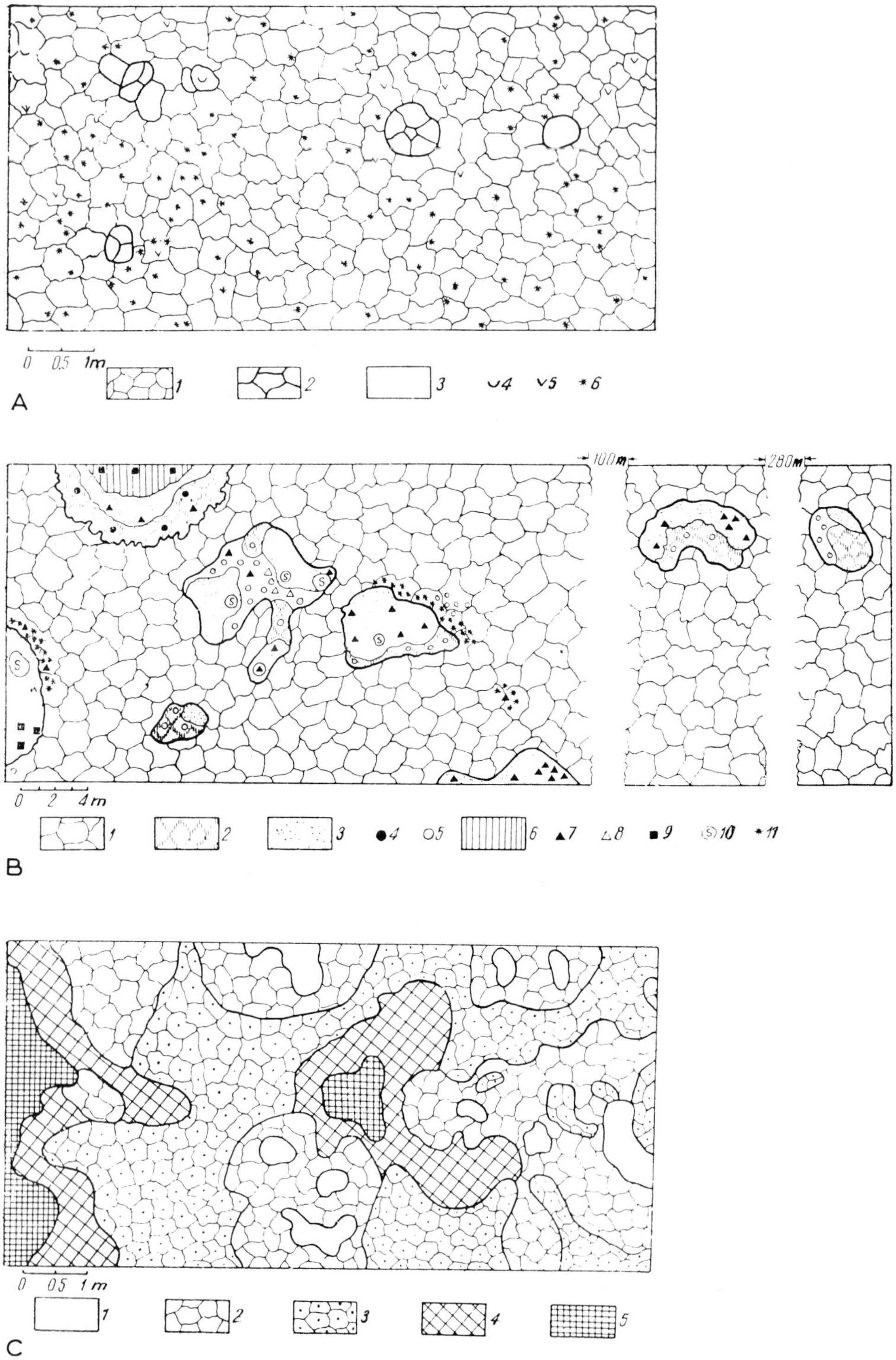

Fig. 5.8. Micro-complexes of takyry (from Rodin, 1954, p. 397). A. Initial stage of an alga-takyr, with polygon structure; *1* = slightly brackish algae areas; *2* = small pools of water with wide cracks and convex polygons; *3* = patches almost totally without algae; *4* = polygon with a scaly algal film; *5* = small individuals of *Lepidium perfoliatum*; *6* = seedling of *Salsola turcomanica* and *Halimocnemis*

soon as the shrub dies or is gathered for wood.

Small hillocks are built also by the rodent *Allactaga elater*, the termites *Hodotermes angerianus* (50–70 cm high, close to hollows with plant growth), and by predatory ants (*Cataglyphis* spp.). Deserted hillocks slowly fall apart, but an uninhabited bare spot may remain for a long time. Tarantulas (*Lycosa* spp.) build cones 3 to 10 cm tall with craters at the tops. These animals can also initiate such micro-successions.

An idea of the structure of the takyry is given in Fig. 5.8.

The vegetation of the floodplains of Middle Asia

Insofar as they have not been cut down, floodplain forests accompany all the larger, permanent streams which rise in the mountains and carry water to the desert. The floodplains are inundated annually in the spring. Some rivers, such as the Amu-Dar'ya and others which rise in the high mountains, are also flooded a second time in the summer, when the snow in the high mountains melts.

The vegetation is quite varied and very dynamic. The level and the duration of flooding vary from year to year. Moreover, the rivers frequently change course. The soils of the floodplains are carried off or covered by new sediments and no longer flooded, which in the arid climate and with high groundwater can bring about salinization.

The following vegetation types should be distinguished (Rodin, 1963); (1) forest; (2) shrub thicket; (3) tall-grass and forb vegetation; (4) meadow; and (5) *ka'ir* (the most recently deposited soil in the floodplains).

(1) Floodplain forests

Floodplain forests are called *tugai* in Middle Asia. The most important tree species is *Populus diversifolia* (*turanga*), with which *Populus pruinosa* (*petta*) is also often found. The trees reach an age of 35 to 40 years and a height of 13 to 15 m. A natural floodplain forest is vertically stratified. Underneath the tree stratum grow the understory trees and shrubs, which are about 3 to 7 m high and include one or two *Salix* species (mostly *S. australior* and *S. wilhelmsiana*) and *Elaeagnus orientalis*. The true shrub stratum is 2 to 4 m high and contains *Tamarix* species (*T. hispida*, *T. leptostachys*, *T. ramosissima*) and *Halimodendron halodendron*. Lianas are richly developed, such as *Clematis orientalis*, *Cynanchum sibiricum*, and *Asparagus persicus*. These can climb to the tops of the trees and then reach over into the neighboring trees.

The herb layer has fewer species and varies with soil wetness, salt content of the soil, and the availability of light on the forest floor. Most herbaceous species propagate vegetatively, since the instability of the soil conditions makes germination of seedlings difficult and since seeds would also be carried away by high water.

Typically floodplain forests grow on slightly raised areas which are under water for only a short time. The soils are usually sandy loams or loamy sands. The wood supply in these forests is 90 to 200 $m^3\ ha^{-1}$. One can distinguish several associations: (a) liana-rich forests; (b) forests with a ground cover of *Calamagrostis pseudophragmites*; and (c) forests with a cover of *Trachomitum scabrum* (Apocynaceae). The liana-rich forests are the most productive, while in the others the wood supply is generally around 90 to 110 $m^3\ ha^{-1}$.

On lower-lying, longer-flooded ground one finds stagnant, swampy forests with a groundwater table normally at a depth of 1 to 2 m. The soil profile shows alternating sandy and clayey strata. The trees are only 7 to 9 m tall and produce a wood supply of only 40 to 70 $m^3\ ha^{-1}$. The canopy is less closed, and the proportion of *Salix* species in the tree stratum is higher. Reed thickets are scattered throughout. These include, in addition to

sp. B. Lichen-alga takyr (solonchak-like soil) above A on a small terrace: *1* = typical formation; *2* = crust-like solonchak; *3* = *Poa bulbosa*-relict stand on meadow-sierozem; *4* = small *Poa* bunches on areas with incipient colonization; *5* = dead *Poa* plants; *6* = *Aeluropus* turf in a hollow with meadow-sierozem; *7* = individual living *Artemisia badhysii* plants; *8* = the same but dead; *9* = *Artemisia scopariformis*; *10* = *Salsola turcomanica*; *11* = individual halophytes on takyr-like solonchak. C. A takyr already covered with some vegetation (above B): *1* = alga-takyr with algae washed off and only young *Microcoleus* colonies remaining; *2* = alga-takyr with *Phormidium* film; *3* = takyr already settled by ephemerals, about 158–502 plants per m^2 (9 species); *4* = primitive, solonchak-like sierozem which developed from a takyr, with a growth of ephemerals, *Artemisia*, and *Poa bulbosa* (cover 40–50%, 25 species); *5* = the same, but already better covered (cover 50–70%, 17 species).

Phragmites communis, also *Typha angustata*, *T. laxmannii*, and *T. pallida* or *Schoenoplectus litoralis*, *S. hippolyti*, *Equisetum ramosissimum* and *Cladium mariscus*.

The stands with *Phragmites* give fuel yields of 40 to 50 m^3 ha^{-1}. The somewhat less wet stands with *Glycyrrhiza glabra*, on the other hand, give higher yields of 50 to 70 m^3 ha^{-1}.

The conditions in the desert-like floodplain forests that are not inundated are less favorable. Groundwater is usually found at a depth of 4 to 7 m. Only *Populus diversifolia* remains in the tree stratum, the canopy is not closed, and the trees are only 10 to 13 m tall. They are often withered at the top or have been attacked by *Fomes* or insects and are sick. As a result of the low shade, primarily drought-resistant shrubs develop, such as *Lycium turcomanicum*, *Halimodendron*, and *Tamarix* species. A slight salinization of the soil surface is also noticeable, on which places annual chenopods develop. In place of the mesophilic grasses, facultative halophytes appear, such as *Zygophyllum fabago*, *Lepidium obtusum*, *Alhagi canescens*, and *Sphaerophysa salsula*.

These stands are transitional to the halophytic floodplain forests, which have slightly brackish groundwater and greater salinization in the upper soil layers. *Populus diversifolia* survives as individuals, *Tamarix* species dominate, and in the herb layer one finds *Aeluropus litoralis*, *Halostachys*, *Limonium*, and *Salsola* species.

If the groundwater level suddenly sinks, for example as the result of a river's change of course, and especially if this is under a heavy soil on which rainwater does not sink in quickly, then the roots of the herbaceous plants lose contact with the capillary fringe and the plants die. The surface is then covered with a thin algal film composed of Cyanophyta (*Microcoleus vaginatus*, *Phormidium* spp.) — that is, *takyr* formation is initiated.

Floodplain forests with psammophytes can also form if, through interference by man (woodcutting, overgrazing), sandy alluvial deposits are set in motion and start to form dunes. Most trees do not tolerate well either the uncovering of their root systems or the covering of their trunks with sand. Only *Tamarix* is insensitive to this and always grows out of the sand again. Typical psammophytes then find their way in, first *Calligonum* spp. and then also *Salsola richteri*, *Ammodendron conollyi*, *Ephedra strobilacea*, and even *Haloxylon persicum*.

Undisturbed floodplain forests are seldom found, since both lumber and firewood are scarce in desert areas. Both are gathered from the floodplain forests. Grazing of floodplain forests is also generally practiced, so one usually finds only degraded forest stands. *Populus* species do easily form rapidly growing thickets, but when old trees fall, clearings arise. The forests take on a park-like character and are permeated by weeds and ruderal species. The tree stratum is destroyed by fire, and *Calamagrostis* and *Glycyrrhiza* occupy the burned areas.

If inundation lasts for a long time, as often happens on level ground between two rivers, and if clay is deposited, then floodplain thickets with *Salix songarica* and *S. wilhelmsiana* form. Such stands ten to twelve years old have a closed shrub layer, with emerging trees 10 to 12 m in height. *Salix songarica* represents two-thirds of the woody plants in the stand. In addition to *Salix*, only single individuals of *Populus* and *Elaeagnus* are found, more seldom *Tamarix* or *Halimodendron*. On one sample area, 1000 *Salix* trunks were counted, 50 of *Populus*, and 10 of *Elaeagnus*.

At high water in these willow thickets, the ground can be under up to 1 m of water. A herbaceous ground-cover is completely absent, and the surface is more often covered with a matted layer of litter composed of branches and leaves, on which *Scytonema* forms an algal film.

(2) Shrub thickets

These are composed of several *Tamarix* species. Most commonly they are the 5 to 6 m tall *T. ramosissima* or *T. hispida* and the shorter *T. leptostachys* or *T. laxa* (and others, in the southwest). These shrub thickets are not inundated, and groundwater is found at a depth of 4 to 5 m, even 6 m. *Tamarix* is able to excrete salt in the form of crystals from its shoots, and these fall to the ground as dust. Thus, the accompanying species must be salt-tolerant or halophytic, such as *Halimodendron* and *Lycium turcomanicum* or, in the grass layer, the salt-excreting grass *Aeluropus litoralis*. Individuals of *Lactuca tatarica*, *Plantago major*, *Zygophyllum fabago*, *Sophora alopecuroides*, *Limonium otolepis*, and some very low, vegetative shoots of *Phragmites* also occur. In dense stands the ground is covered by a layer of *Tamarix* litter 3 to 5 cm thick.

Ephemerals are completely absent. At most, there may occur a few summer annuals (*Launea fallax*, *Chenopodium serotinum*, *Atriplex nitens*, *Salsola lanata*, and others).

On increasingly dry sites, these *Tamarix* thickets form a transition to the desert-like tree vegetation with *Haloxylon ammodendron* (= *aphyllum*), which can be observed in the earlier delta region of the Amu-Dar'ya. Old, dried-out water courses and remnant lakes show that these places were once floodplains. This *Haloxylon* species has a very large root system and can root through a large volume of soil to a depth of 8 m.

If *Tamarix* is cut, it always sprouts again. Good stands yield about $100\,m^3\,ha^{-1}$ of firewood. Degraded stands close to settlements, however, may yield only $7\,m^3\,ha^{-1}$ or even less.

(3) Floodplains with tall-grass and forb vegetation

These are called *jangil* (cf. the Indian "jungle") and are often composed of impenetrable reed thickets (*Phragmites communis* and *P. isiaca*). They can be either pure stands, or other hydrophytes and/or halophytes can be mixed in (Kerbayev, 1964).

These reed thickets cover 300 to 600 thousand hectares in the delta region of the Amu-Dar'ya, along the coast of the Aral Sea, and far out into the shallow water. These areas are the only fuel, fiber and forage resources for the local people. The *Phragmites* serves as year-round pasture, and the young shoots are used as hay. The reeds are also used for lodgings and as firewood (caloric value higher than that of wood), also for reinforcing roads and paths. Even tombstones are made of *Phragmites*. Hunting within these habitats is also very important as a source of meat and furs.

The soils are peaty, muddy, or meadow-like swampy soils, but can occur in both brackish and non-brackish varieties. The reed stands vary from very dense stands, 3 to 4 (up to 6) m tall, with an annual production of $25\,t\,ha^{-1}$ of dry matter, down to low, quite open stands, which are under water for long periods.

The species composition differs accordingly. In addition to the two species of *Phragmites* already mentioned, admixtures of *Typha* spp., *Schoenoplectus tabernaemontani*, and *Bolboschoenus maritimus* can occur. *Carex pseudocyperus*, *Glycyrrhiza glabra*, and *Aeluropus* are important in the meadow-like stands. Other accompanying species include: *Marsilia aegyptiaca*, *Ophioglossum bucharicum*, *Saccharum spontaneum*, the sedge *Dichostylis homulosa*, *Bergia koghanii* (Elatinaceae) and *Oxytropis riparia*, and the saprophytic orchid *Eulophia turkestanica*.

Along the seacoast, *Phragmites* can form mats up to 1 m thick over fluid mud. If carried away by storms, they roam the sea as floating islands until they find a new place to land. On these islands may grow also *Dryopteris thelypteris*, *Calystegia sepium*, *Lycopus europaeus*, *Stachys palustris* and others.

The reed meadows in the Amu-Dar'ya Delta are breeding grounds for grasshoppers and locusts, and control measures are necessary every year.

(4) Floodplain meadows

These occupy large areas in the southwest and in the delta areas of the Amu-Dar'ya and the Atrek but, otherwise, only cover small areas at the foot of the Kopet Dag.

Sward-forming grass species are dominant, such as *Aeluropus litoralis*, *A. repens*, and *Cynodon dactylon*, along with the bunch-grass *Atropis distans*, and also *Calamagrostis dubia*, and the very tall grass *Erianthus purpurascens*. The following stoloniferous forbs occur also: *Glycyrrhiza glabra*, *Karelinia caspia*, *Alhagi pseudoalhagi*, and *A. canescens*. All of these grow on into autumn, root very deeply, and are salt-tolerant.

Especially widespread are *Aeluropus* stands with individual accompanying species, some non-characteristic ruderal species (*Alhagi persarum*, *Peganum harmala*, *Capparis spinosa*), and a number of ephemeroids (*Poa bulbosa*, var. *vivipara*, *Cardaria draba*, *Allium albanum*) and ephemerals, which fluctuate considerably from year to year.

These areas are grazed or mowed and may be used for growing cucurbits.

(5) *Ka'ir* areas

These are areas of recently deposited soil, with very little vegetation, consisting of pioneer plants, which gradually change over into a vegetation of tall grasses and forbs.

The wet surface of fresh sediment deposits is covered for a time by a large number of seedlings ($100–500\,m^{-2}$). On sandy soil one finds primarily *Typha pallida*, on clayey soil *Calamagrostis dubia*. If the vegetation is not destroyed by high water in the next few years, then clearly identifiable groups form around the dominant species of *Typha*, *Calamagrostis*, or *Phragmites*. In the course of natural succession, *Saccharum spontaneum* becomes dominant. This grass holds its place, however, only if the groundwater does not sink below 1.5 to 1.75 m and the surface is inundated annually. Otherwise other species invade, for example, *Equisetum ramosissimum*, *Glycyrrhiza glabra*, *Erianthus purpurascens*, or *Typha elephantina*. *Glycyrrhiza* or *Erianthus* often prevails. Plant communities form in which the tall grasses, such as *Erianthus*, *Phragmites*, and *Saccharum*, constitute an upper stratum 2 to 4 m high, while the understory of *Glycyrrhiza*, *Typha*, or *Alhagi* reaches 1 to 1.5 m. The underground phytomass, however, exceeds that above ground. In dense stands of *Glycyrrhiza*, the below-ground phytomass can be 10 to 15 t ha^{-1}, while the above-ground phytomass is only 3 to 4 t ha^{-1}. For *Typha pallida* the corresponding figures are 25 to 30 t ha^{-1} below ground and 8 to 10 t ha^{-1} above ground. The constant wetness and high temperatures make this abundant growth and high production possible.

One characteristic (probably a requirement) of the species in the flooded zone is that, after heavy sedimentation, they possess the ability to form new roots at arbitrary heights along the main stem. This is true for *Saccharum*, for example, which can be buried in up to 30 to 40 cm of mud (Fig. 5.9) and still survive. *Clematis orientalis*, *Cynanchum sibiricum*, and *Tamarix* behave similarly. The species of later successional stages possess this ability to a lesser extent. For *Erianthus*, *Glycyrrhiza*, and *Equisetum*, the depth of sedimentation cannot exceed 10 to 15 cm. The plants can survive flooding when not more than the lower half of the above-ground parts is covered by water and the water does not remain more than three weeks. The plants are more sensitive, on the other hand, to a sinking of the groundwater table during the dry period. The stands become more open, and another successional stage of the forest (*tugai*) begins.

As soon as the soil becomes saline, various *Tamarix* species and the legume *Halimodendron halodendron* appear. In the understory, the species belonging to the *jangil* stage, including also *Cynodon dactylon* and *Imperata cylindrica*, are replaced by halophytes.

Finally, it is also worthwhile to look briefly at the phenology of this floodplain vegetation. The frost-

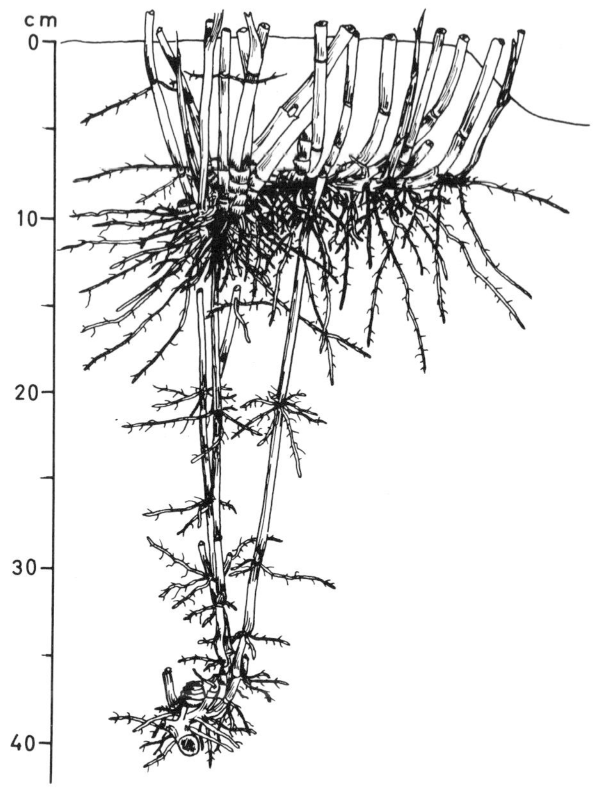

Fig. 5.9. Underground part of *Saccharum spontaneum* (by Kerbayev, from Walter, 1968, fig. 538), showing the new formation of rhizomes and roots at a higher level (depth of 8–10 cm) after deposition of a sediment layer of 30 cm. The remains of older rhizomes can be seen at the bottom.

free period lasts from the end of March to the beginning of November along the middle course of the Amu-Dar'ya River. On the average, it is 216 days, varying from 190 to 246. The floodplain turns green at the beginning of April, and rapid growth begins at the end of April. *Glycyrrhiza* blooms in the first half of May, before it is overgrown by the grasses. A reddish hue prevails at the beginning of June, produced by the blooming of *Tamarix* and the fruiting of *Glycyrrhiza*. The tall grasses begin to bloom in the middle of July, producing a silvery-white color. *Erianthus* and *Phragmites* bloom, finally, in the first half of August. The last aspect of the ripening grasses is characterized by a deep yellow color.

In the floodplain forests, the blooming of the *Populus* species is the first striking aspect. The seasonal development on the wet soils of the floodplain begins a month later than in the sand desert.

Precise phytomass and production assessments were carried out by Gladyzhev (1969) on a *Phragmites* stand along the middle course of the Amu-Dar'ya. The results are contained in Table 5.3. The proportions of rhizomes, coarser roots, and fine roots at different depths are shown in Fig. 5.10. About 45% of the total underground phytomass is made up of rhizomes. The ratio of above-ground and below-ground phytomass is 1:2. The total phytomass is about 120 t ha^{-1}, and the production about 42 t ha^{-1}, of which over 35 t ha^{-1} is above-ground production. This is an extraordinarily high value, even when one considers that it is perhaps 10% lower for absolutely dry matter.

The area of root surface and the length of roots at various depths and in relation to the soil type were also determined (Figs. 5.11 and 5.12).

Production assessments on floodplain forests along the middle course of the Amu-Dar'ya have also been published (Gladyzhev and Rodin, 1977). Studied were:

(1) Stands of *Salix songarica*, which becomes 5.0 to 6.5 m tall with trunk diameters of 6 to 14 cm (a total of 89 trunks ha^{-1});

(2) Stands of *Elaeagnus orientalis* (with *Salix* and *Populus*), which reach 6.5 to 9.5 m in height and trunk diameters of 6 to 20 cm (648 trunks ha^{-1});

(3) *Populus pruinosa* forests, with a height of 6.3 to 8.3 m and trunk diameters of 8 to 18 cm (477 trunks ha^{-1}). *Salix* occupies the lowest ground in the floodplain, *Elaeagnus* prefers drier sites, while *Populus* dominates on sandy soils with groundwater at a depth of 0.5 to 9 m. *Prunus diversifolia* is often admixed.

All forests possess a well-developed herb layer and abundant litter. Rooting in the upper soil layers is very dense.

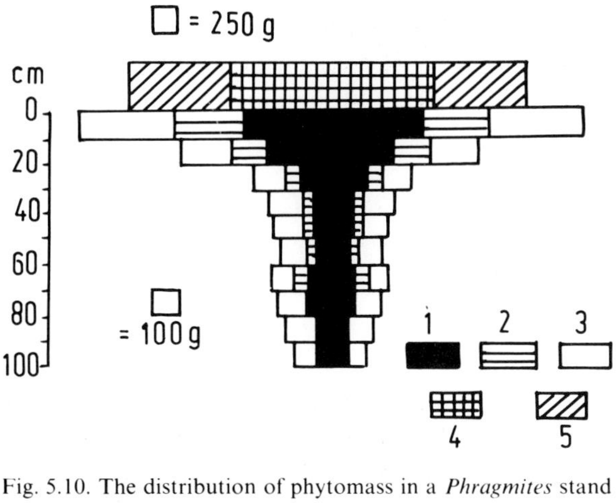

Fig. 5.10. The distribution of phytomass in a *Phragmites* stand (from Gladyzhev, 1969): *1* = thick rhizomes; *2* = thinner rhizomes and coarser roots; *3* = fine roots; *4* = living above-ground phytomass; and *5* = above-ground litter.

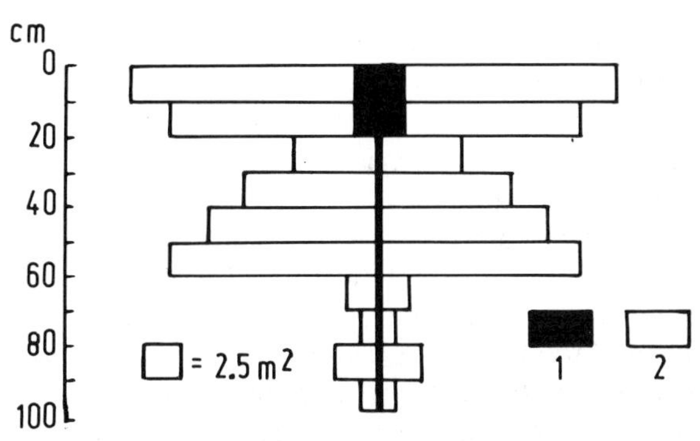

Fig. 5.11. Distribution of root surface in a *Phragmites* stand (from Gladyzhev, 1969): *1* = thick roots and thin rhizomes; *2* = fine roots.

TABLE 5.3

Production of a *Phragmites communis* stand in the floodplain of the middle Amu-Dar'ya River, from Gladyzhev (1969) (I and II are two sample areas; phytomass is measured as air-dry matter)

Number of stems (m^{-2})	Height of plants (cm)	Phytomass in g m^{-2}			Total phytomass (t ha^{-1})	Litter in t ha^{-1} yr^{-1}		
		above-ground	below-ground	fine roots		above-ground	below-ground	total
I 184	220	3800	6330	418	105.5	38.0	7.7	45.7
II 191	230	4100	8440	892	133.9	32.0	6.9	38.9

The quantitative results of the studies are summarized in Table 5.4. Large amounts of nitrogen are returned to the soil with the litter. For the *Populus* stands the amount is around 161 kg ha^{-1}.

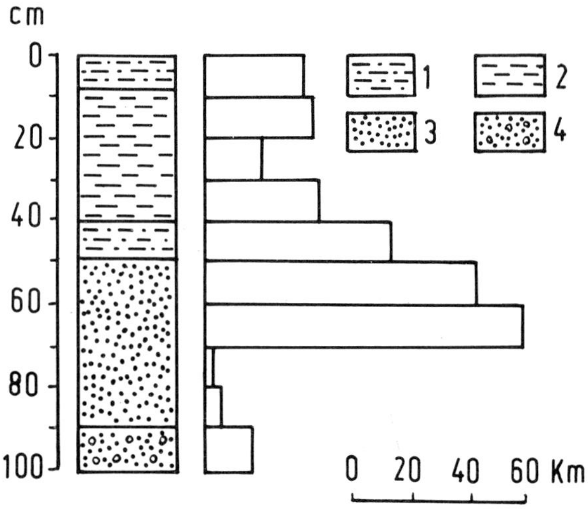

Fig. 5.12. Distribution of root length in a *Phragmites* stand as a function of soil type (from Gladyzhev, 1969): *1* = sandy soil; *2* = sandy loam; *3* = fine sand; and *4* = coarse sand.

A certain amount of sodium and chlorine is contained in the litter also — a sign of some salinization.

The proportion of the total phytomass contributed by the herb layer is 26 to 36%. This is significant since this part of the phytomass turns over very rapidly.

Table 5.5 provides a comparison between the floodplain forests of the Amu-Dar'ya (middle course) and those of the delta area. Striking is the high proportion of underground phytomass in the *Populus* forests of the middle course, a fact which may be due to the greater dryness of these sites.

Table 5.6 shows the composition of the litter at different months of the year.

PRODUCTION ASSESSMENTS IN MIDDLE ASIA

The most detailed studies were carried out in the Karakum Desert. These will be discussed in Chapter 6. We consider here only the studies from the station in the Kyzylkum Desert, located north of the Kul'dzhuktau Mountains and 160 km north-

TABLE 5.4

Distribution of phytomass and of dead organic matter in the floodplain forests along the middle Amu-Dar'ya River, 1977 (from Gladyzhev and Rodin, 1977)

Plant parts	*Salix songarica*		*Elaeagnus orientalis*		*Populus pruinosa*	
	t ha^{-1}	%	t ha^{-1}	%	t ha^{-1}	%
Total phytomass	71.0	–	154.1	–	81.6	–
Tree stratum	45.5	100	114.9	100	57.0	100
leaves	3.2	7	4.9	4	3.1	5
flowers, fruits	2.2	5	18.4	16	2.1	4
branches	8.3	18	28.2	25	7.2	12
trunks	18.2	40	10.7	9	7.4	13
roots	13.6	30	52.7	46	37.2	66
Herb stratum	25.5	100	39.2	100	24.6	100
leaves, stems	9.1	36	6.1	15	1.3	5
rhizomes	11.2	44	14.8	38	3.5	14
roots	5.2	20	18.3	47	19.8	76
Phytomass distribution						
above-ground	41.0	58	68.3	44	21.1	26
below-ground	30.0	42	85.8	56	60.5	74
Mortmass	30.7	100	71.1	100	42.2	100
leaf litter	15.8	52	31.3	44	10.4	25
other above-ground litter	11.7	38	25.7	36	23.1	54
below ground	3.2	10	14.1	20	8.7	21

TABLE 5.5

Distribution of phytomass in different floodplain forests of the Amu-Dar'ya River (in t ha^{-1}) (Rodin, 1961)

Species	Middle course			Delta course		
	phytomass (t ha^{-1})	above-ground (%)	below-ground (%)	phytomass (t ha^{-1})	above-ground (%)	below-ground (%)
Elaeagnus orientalis	154.1	44	56	73.1	51	49
Populus pruinosa	81.6	26	74	77.8	55	45
Salix songarica	71.0	57	43	81.7	45	55

TABLE 5.6

Composition of the litter (average amounts, in g m^{-2}) in ten litter samples of 1 m^2 area in a *Populus pruinosa* floodplain forest along the middle course of the Amu-Dar'ya River (from Gladyzhev and Rodin, 1977)

Month	Litter				
	g m^{-2}	%	fruit parts (%)	leaves (%)	branches (%)
April–May	73.6	4.3	83.9	9.8	6.3
June	106.6	6.3	–	97.3	2.7
July	27.7	1.6	–	98.1	1.9
August–October	292.5	17.2	91.3	4.6	4.1
November–December	759.5	44.8	–	85.1	14.9
January–March	436.5	25.8	–	69.9	30.1
Total	1696.4	100	19.3	64.9	15.8

west of Bukhara (Fig. 5.13) at 40°45′N and 65°45′E (Rodin, 1977).

The Kyzylkum Desert is located between the lower courses of the Syr-Dar'ya and Amu-Dar'ya Rivers. It is largely a sand desert and represents the transition from the northern Turanian to the southern Turanian deserts. The scientific station of the Botanical Institute of the Uzbeki Academy of Sciences (investigator: I.F. Momotov) is located on a plain with slightly sandy siero-burozems of varying salinity.

The climate is extremely continental and arid (Fig. 5.14). The absolute minimum temperature is about −28°C and the absolute maximum 48°C. The mean July temperature is 30.8°C, and the mean for January is −1.5°C.

The temperature sum for the days with means above 10°C is 4560 to 4660°C. Sunshine occurs 2800 to 2900 h in the year. The total annual solar radiation is 635 kJ cm^{-2}.

The mean annual precipitation is only 95 mm, with a maximum in March–April. There is almost never any rain in the summer. The climate is, thus, roughly the same as that of the Turanian Desert.

At the experimental plot, the groundwater table was at a depth of 10 to 15 m. The soils are very rich in carbonates. Sandy-loam intermediate strata at a depth of 65 to 100 cm contain readily soluble salts (chlorides and sulphates), while the upper 0 to 45 cm are salt-free.

Three biogeocoenoses were studied: (1) an *Artemisietum* with ephemeroids; (2) an *Artemisietum* with *Salsola* spp.; and (3) a *Nanophytetum erinacii*.

In the *Artemisietum* with ephemeroids, *Artemisia turanica* (14 200 individuals ha^{-1}) and *A. diffusa* (1600 individuals ha^{-1}) dominate. There are, in addition, 16 600 individuals ha^{-1} of *Poa bulbosa*, 800 individuals ha^{-1} of *Carex physodes*, and 8000 individuals ha^{-1} of other herbs. A total of 23

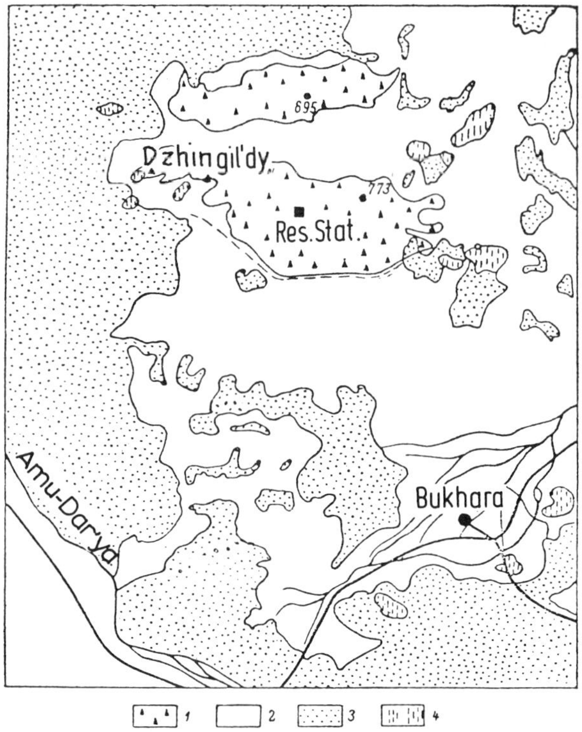

Fig. 5.13. Location of the Kyzylkum Research Station 160 km NNW of Bukhara (from Rodin, 1977; p. 64, with alterations): *1* = plateau, *2* = scree and gravel, *3* = sand areas, and *4* = saline soils.

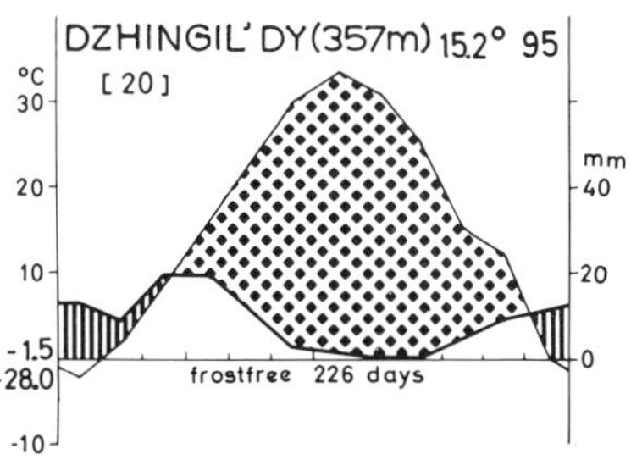

Fig. 5.14. Climate diagram of Dzhingil'dy (from Rodin, 1977, p. 65).

species were determined, of which many were ephemerals. The total vegetal cover was 25 to 30%.

In the *Artemisietum* with *Salsola*, the same two *Artemisia* species dominate along with a third component, *Salsola orientalis*, and a fourth component, *Salsola arbuscula*. The *Salsola* species generally show better vitality than do the *Artemisia* species. The nearly loamy soils contain considerable gypsum at a depth of 25 to 30 cm. A total of 52 species were identified but only 36 in dry years, since many ephemerals do not develop at all in dry years.

Nanophyton erinaceum is absolutely dominant in the *Nanophytetum*, with a cover, though, of only 5 to 15%. This species grows as cushions (see Fig. 4.14). In addition to individual dwarf shrubs and ephemerals, the appearance of annual Chenopodiaceae (*Gamanthus* and *Halimocnemis* spp.) is striking. These live on into the summer.

The soil fungi, algae, and bacteria in the three biogeocoenoses were also studied. Most microorganisms are found in the upper 0 to 12 cm. The bacteria, which require greater wetness, are very active in the thoroughly wetted soil in the spring. In the summer the fungi and Actinomycetes are more important.

All the biogeocoenoses are used as pasture for karakul sheep and for camels. The *Nanophytetum* is grazed only in the autumn and winter. The dominant species does not provide good forage.

In the two *Artemisieta* the rain water penetrates the soil easily. The water stored in the top 2 m exceeds 100 mm but sinks in summer to 60 to 70 mm. The soils of the second *Artemisietum* contain about 1% gypsum in the upper 75 cm and, at greater depths, also chlorides. The soil of the *Nanophytetum* is crusted on top, so a large part of the rain water runs off. Only the top 50 to 60 cm are thoroughly wetted, and the water stored in the top 1 m seldom exceeds 60 mm.

The ecology of seed production and germination was treated in detail.

In the first *Artemisietum* the water consumption during the growing season was 119 mm for *Artemisia diffusa*, 39 mm for *A. turanica*, and 15 mm for the other species.

The transpiration coefficient climbs steadily from spring to autumn, from 1530 to 2082 l water (kg dry wt.)$^{-1}$ for *A. diffusa*, and from 1300 to 1924 l (kg dry wt.)$^{-1}$ for *A. turanica*. In the second *Artemisietum*, *A. turanica* transpires 35 mm, *Salsola orientalis* 28 mm, and *S. arbuscula* 13 mm. The transpiration coefficients are lower for the *Salsola* species (*S. arbuscula* 800 to 2000, *S. orientalis* 100 to 1680). The transpiration coefficient for the *Nanophytetum* was given as 1100 to 1950. The total water consumption for this species is lower since the annual growth is much less.

Various authors have also considered the photosynthesis and respiration. The rates are very different for the different species and vary also depending on the water supply in the course of a day and in the course of the year. The temperature also has a major effect.

Nanophyton possesses a distinct taproot, which can penetrate to the gypsum horizon at a depth of 2 m. At this point it forms smaller absorbing roots.

The primary production is subject to large fluctuations in individual years. This is especially true for the ephemerals, which are completely absent in dry years. The development of the semi-shrubs, however, is also quite different in different years. If one takes the production in a normal year as 100%, then the variation in wet and dry years can be seen from Table 5.7.

The absolute values for individual years obtained for three biogeocoenoses are shown in Table 5.8.

More detailed figures for average production are shown in Table 5.9 (from Anonymous, 1975).

The animal organisms were also treated in detail (Rodin, 1977, p. 82). Of these, 29 rodents have the most important role. They destroy a significant part of the phytomass and eat a large number of the seeds. Their numbers are regulated by the carnivores, of which the fox (*Vulpes vulpes*) is the most common. The rodents make up 70% of the fox's food, while the remaining 30% is made up of insects. The spotted cat (*Felis libyca*) also eats rodents (26% of its diet).

The numerous insects are very important. *Artemisia turanica*, for example, is attacked by *Theone costipennis*. The larvae of this species can eat 50 to 100 g of leaf matter in a single day. Heavy

TABLE 5.7

Production in wet and dry years in the Kyzylkum Desert as percentages of the long-term mean (Rodin, 1977)

Species	Wet year	Dry year
Artemisia turanica and *Artemisia diffusa*	177	40
Nanophyton erinaceum	163	54
Salsola arbuscula	225	13
Salsola orientalis	166	31

TABLE 5.8

Production (t ha^{-1}, above-ground parts) of three biogeocoenoses in individual years in the Kyzylkum Desert (Rodin, 1977)

Community type	1967–68	1968–69	1969–70	1970–72	1971–72
Annual precipitation (mm)	50	111	64	84	83
1. *Artemisietum* with ephemeroids	0.19	0.72	0.27	0.30	0.20
2. *Artemisietum* with *Salsola* spp.	0.02	0.44	0.20	0.21	0.24
3. *Nanophytetum*	0.04	0.19	0.06	0.10	0.11

TABLE 5.9

Phytomass, growth, and litter of the three biogeocoenoses in the Kyzylkum Desert (from Anonymous, 1975)

Biogeocoenosis:	1. *Artemisietum* (ephemeroids)	2. *Artemisietum* (*Salsola*)	3. *Nanophytetum*
Phytomass (t ha^{-1})	3.5	5.7	1.3
leaves (%)	5	6	8
woody tissue (%)	15	13	23
roots (%)	80	81	69
Annual growth (t ha^{-1})	0.37	0.69	0.19
roots (%)	51	48	47
Annual litter			
above-ground (t ha^{-1})	0.28	0.37	0.16
roots (t ha^{-1})	0.48	0.76	0.31
Annual litter as % of total litter	63	68	66

invasions can destroy 60% of the green plant matter. In the first type of *Artemisietum* (with ephemeroids), 200 insect species were identified, of which 60% were phytophagous. In the second *Artemisietum* over 100 species were identified, but in the *Nanophytetum* only about 60.

Great damage is also caused by the destruction of the seeds.

Rodents and arthropods (ants and termites, etc.) are also important to soil-forming processes.

During the course of the growing season the three biogeocoenoses show a number of phenological aspects. These are determined in the spring primarily by the ephemerals but also by *Ferula assa-foetida* and by *Stipa*. The *Artemisia* species bloom at the end of September to the middle of October. In dry years most of these phases do not occur.

Many ephemerals germinate in the autumn and grow during periods of thaw in winter. These form the basis for winter grazing.

The phytomass in these winter pastures fluctuates drastically from site to site. The herbage amounted to 13.2 to 173 g m^{-2} in 1967 to 1968 but only 7.3 to 41.6 g m^{-2} (dry weight) in 1968 to 1969. The production of the above-ground parts was, for the corresponding years, 6.3 to 8.8 g m^{-2} and 0.1 to 0.4 g m^{-2}.

REFERENCES

Anonymous, 1975. *Resources of the Biosphere. Synthesis of the Soviet Studies for the IBP, 1.* Nauka, Leningrad, 288 pp. (in Russian).

Breckle, S.-W., 1971. Ist *Diarthron vesiculosum* (Thymelaceae) ein ökologisches Wunder? *Bot. Jahrb.*, 90: 550–561.

Freitag, H., 1971. Die natürliche Vegetation Afghanistans. *Vegetatio*, 22: 285–344.

Gladyzhev, A.I., 1969. Quantitative characterization of the biological productivity of a reed community in the floodplain of the middle Amu-Dar'ya. *Izv. Akad. Nauk Turkm. S.S.R., Ser. Biol. Nauk (Ashkhabad)*, 4: 68–71 (in Russian).

Gladyzhev, A.I. and Rodin, L.E., 1977. Structure and distribution of phytomass of riverbed forest communities along the middle stretch of the river Amu-Dar'ya (Turkmenian S.S.R.). *Bot. Zh.*, 62: 3–14 (in Russian).

Gollerbakh, M.M., Novichkova, L.N. and Zdobnikova, V.V., 1956. The algae of the takyry. In: V.A. Kovda (Editor), *The Takyry of Western Turkmenistan and the Means of Their Agricultural Utilization.* Nauka, Moscow, pp. 38–54 (in Russian).

Kerbayev, B.B., 1964. The Jangil vegetation in the Amu-Dar'ya floodplains. *Tr. Inst. Biol. Ashkhabad*, 1: 22–74 (in Russian).

Korovin, E.P., 1961/62. *Vegetation of Middle Asia and Southern Kazakhstan.* Nauka, Tashkent, 2nd ed., Vol. I: 452 pp.; vol. II: 577 pp. (in Russian).

Kovda, V.A. (Editor), 1956. *The Takyry of Western Turkmenistan and the Means of Their Agricultural Utilization.* Nauka, Moscow, 736 pp. (in Russian).

Litvinov, M.A., 1956. Biocoenoses of the microscopic soil fungi of takyry. In: V.A. Kovda (Editor), *The Takyry of Western Turkmenistan and the Means of Their Agricultural Utilization.* Nauka, Moscow, pp. 55–74 (in Russian).

Markova, L.E., 1973. Modifications in the development of annual plants of Middle Asia in relation to the water supply. *Probl. Biocoen., Geobot. Geogr. Rast. (Akad. Nauk S.S.S.R.)*, Leningrad, pp. 170–180 (in Russian).

Rodin, L.E., 1954. The role of plants in the formation of takyry and their complexes. *Pustyn. S.S.S.R. Ikh Osvoyeniye*, 2: 392–413 (in Russian).

Rodin, L.E., 1956. The impact of man on the formation of takyry. In: V.A. Kovda (Editor), *The Takyry of Western Turkmenistan and the Means of Their Agricultural Utilization.* Nauka, Moscow, pp. 104–108 (in Russian).

Rodin, L.E., 1961. *Dynamics of Desert Vegetation as Exemplified in Western Turkmenistan.* Akad. Nauk S.S.S.R., Bot. Inst., Leningrad, 227 pp. (in Russian).

Rodin, L.E., 1963. *Desert Vegetation of Western Turkmenistan.* Akad. Nauk S.S.S.R., Bot. Inst., Leningrad, 309 pp. (Russian).

Rodin, L.E. (Editor), 1977. *Productivity of Vegetation in the Arid Zone of Asia. Synthesis of the Soviet Studies for the IBP, 1965–1974.* Akad. Nauk S.S.S.R., Nauka, Leningrad, (Russian).

Rodin, L.E. and Sukhoverko, R.V., 1956. The higher plants and the seed supplies on the takyry. In: V.A. Kovda (Editor), *The Takyry of Western Turkmenistan and the Means of Their Agricultural Utilization.* Nauka, Moscow, pp. 30–37 (in Russian).

Sharapov, N.I., 1954. Basis of plant cultivation in subtropic Turkmenia. *Pustyn. S.S.S.R. Ikh Osvoyeniye*, 2: 135–176 (in Russian).

Shnitnikov, A.V., 1957. Variation of general humidity on the continents of the northern Hemisphere. *Zap. Geogr. Obshch. Soy. S.S.R.*, 16: 337 pp. (in Russian).

Walter, H., 1968. *Die Vegetation der Erde, II. Die gemässigten und arktischen Zonen.* Fischer-Verlag, Stuttgart, 1001 pp.

Walter, H., 1974. *Die Vegetation Osteuropas, Nord- und Zentralasiens. Vegetations-Monographien der einzelnen Großräume.* VII. Fischer-Verlag, Stuttgart, 452 pp.

Walter, H., 1976. *Die ökologischen Systeme der Kontinente (Biogeosphäre).* Fischer-Verlag, Stuttgart, 131 pp.

Chapter 6

THE KARAKUM DESERT, AN EXAMPLE OF A WELL-STUDIED EU-BIOME[1]

H. WALTER and E.O. BOX

THE SAND DESERT KARAKUM AS A LANDSCAPE-ECOLOGICAL UNIT

After the more general description of Middle Asia, it is good to treat a specific example of a biome in more detail. Such large ecological units as the Karakum Desert are called eu-biomes (Walter, 1976; Walter and Box, 1976).

The scientific exploration of the Karakum Desert began at almost the same time as did that of the Sonoran Desert in Arizona. The founding of the Desert Laboratory of the Carnegie Institution in Tucson (Arizona) in 1903 provided the initiative for experimental ecological research in the Karakum. Only nine years later the Sand-Desert Station, Repetek (38°34′N and 63°11′E, elevation 190 m), in the southeastern Karakum region, was opened on the railroad line from Krasnovodsk on the Caspian Sea to Ashkhabad, Bukhara, and Samarkand (Klyushkin, 1955). Ecological work in the Karakum has become more intensive in recent decades as a result of the construction of the large canal from the Amu-Dar'ya River, along the southern edge of the Karakum, to the Caspian Sea. The total length of the canal is supposed to reach 1400 km, of which 1000 km are already finished. The canal is navigable by ships and permits the enlargement of the region's irrigated agricultural lands. The ecological work serves to explore the scientific basis for practical measures. Results have been summarized in the collection edited by Kunin (1955), the geographical volume *Turkmenistan* by Kunin (1969), the book *Middle Asia* by Murzayev (1958), the works of Rodin (1961, 1963), the two-volume work of Korovin (1961/62), and many other publications. All of these works, in Russian, remain almost totally unknown in the west. For this reason, a detailed discussion is attempted here.

The Karakum Desert biome, with an area of about 350 000 km^2, belongs to zonobiome type VII (Walter and Box, 1976), the temperate arid zone. Within this framework it is also ascribed to the subzonobiome constituted by the deserts with cold winters in Eurasia. The desert is a part of the large Aral–Caspian (Turanian) Lowland (Fig. 2.1; see also p. 79).

The largest part of this lowland is occupied by sand deserts, the Karakum between the Caspian Sea and the Amu-Dar'ya River and the Kyzylkum Desert between the Amu-Dar'ya and the Syr-Dar'ya Rivers.

Within these two deserts the Karakum forms a large ecological unit. This is especially true of the central part, a basin filled with up to 500–600 m of river sediments, through which an underground lake flows.

The Karakum Desert is a very large, complex ecological unit, in which interference by man is becoming increasingly noticeable. The entire region is only sparsely inhabited. The sands serve as a year-round pasture primarily for sheep. In the south, however, at the foot of the mountains, there are numerous large oases, around which irrigated agriculture and a large human population have arisen. In the coastal zone of the Caspian Sea, considerable quantities of oil, natural gas, and other natural resources have been found. The interference of man in this biome must also be considered if one wishes to have a well-rounded

[1] Manuscript completed November, 1977.

image of the ecological conditions. We will consider the natural factors in more detail first.

GEOLOGY, SOILS, RELIEF, AND RIVER SYSTEM

The most important events of the geographical history, which began with the recession of the Tethys Sea in the Tertiary (see pp. 10, 45, 79) have already been outlined, for the Karakum is a part of the Turanian Lowland.

The soil types were also identified (see pp. 46–48, 82–83). As far as one can actually speak of soil, of course, the sandy soils are by far the most common in the sand desert. The *takyr* soils and saline soils (solonchak) represent only scattered small areas.

For the most recent, wind-formed relief we must go into more detail. Dry sands are easily blown away, while exposed firmer sand layers are eroded more slowly by sand-blasting. In both cases, individual grains are carried only short distances over the surface. As a result, various types of dunes form, of which the following are generally distinguished:

(1) Low sand hillocks (*nebkas*), which form around shrubs when the wind diminishes momentarily and sand is deposited around them; the plants always grow outwards.

(2) Individual *barkhans* (crescent-shaped dunes), which form on smooth loamy surfaces (e.g., on takyry) when sand is blown across them. These can become 3 to 4 m high. The windward side has a gentle slope of 12 to 15°, whereas the leeward side, down which the sand grains trickle, has a steeper slope of 30 to 35°, depending on the size of the sand grains. Individual barkhans migrate very rapidly, over a hundred meters in a year.

(3) Barkhan fields, such as one finds in large areas of sand, e.g., in the delta region of the Amu-Dar'ya, along its banks, or around oases (Fig. 6.1). The volume of sand is very large and movement is very slow.

(4) Barkhan chains are typical for regions with outcroppings of loose sandstone. The edges of the barkhans interfere with each other laterally, so the crest of the chain which is formed then represents a long, winding line running perpendicularly to the direction of the wind. These dune chains can reach a height of 30 to 40 m but are usually about 5 to 15 m high. If the wind blows continuously from a given direction, then the sand mass will migrate about 5 to 15 m annually. An advancing barkhan chain can cover entire villages. The control of such chains requires that importation of additional sand be prevented by stabilizing the areas from which the sand is blown away. As they continue to migrate,

Fig. 6.1. Barkhan field 20 km from the Amu-Dar'ya, prior to afforestation (photograph of E.A. Bessey, 1902). On the dune slopes are visible the leafless shrubs of *Calligonum* spp., *Salsola arbuscula*, and bunches of *Aristida karelinii*; in the dune valleys is *Tamarix* sp. (from Walter, 1976, fig. 47).

the chains then decrease in height and are finally colonized by plants and thus stabilized. Stabilization can also be accomplished by spraying with bitumen, an oily substance. Usually the wind blows from different directions in different seasons. In this case the base of the dunes remains stable and only the upper portions are rearranged, i.e., the steeper leeward side shifts from side to side.

(5) Sand ridges are especially widespread in the Karakum. Their length is 10 to 100 times greater than their width or height. The absolute length varies from 100 m to several kilometers, the height between 5 and 60 m, and the width from 10 to 1000 m and more. Lower ridges can lie between high ones. The orientation is mostly meridional, so crossing the sand areas in the north–south direction is generally easier than in other directions. These sand ridges often consist of dense sand layers which are not moved around by the wind or possess nuclei of sandstone or loam. They are inhabited by plants and thus immobile, but sand can be blown out of them to form barkhans in surrounding areas. Most sand ridges are asymmetrical, that is, one slope is steeper than the other. The ridge itself is not even, but rather composed of individual hillocks which arose from loose sand deposited around shrubs. These hillocks are also immobile but can be damaged by grazing and then partially eroded by the wind. There then remains a landscape dominated by the irregularly shaped remnants of the former dunes. Almost all immobile sand areas are composed of irregular combinations of sand ridges and hillocks.

Veysov (1976) gives further information on the dynamics of sand. The type of motion of the sand grains depends on their size; sliding or rolling for diameters of 0.5 to 1 mm (dependent on wind speed), bouncing or skipping for diameters of 0.25 to 0.5 mm, and blowing (with no ground contact) for diameters of 0.05 to 0.25 mm. The form of the ripple marks also depends on the grain size, being larger as the size of the grains becomes larger. In the Karakum the ripple marks are 0.5 to 3 cm high and 6 to 20 cm apart, for coarse sands 8 to 10 cm high and 40 cm apart.

The barkhans develop through several stages: (1) regular ripples form; (2) these become irregular, as coarser and finer grains segregate; (3) symmetric, large undulations develop, which are decomposed into egg-shaped parts; (4) the undulations become asymmetrical and take on the shape of the barkhans; (5) true barkhans, with steep leeward slopes, form; (6) the barkhans arrange themselves in pairs or groups; and (7) barkhan chains form, perpendicular to the wind direction.

The barkhans travel faster over a firm surface than over a loose one, for example, faster on wet sand than on dry sand. On clayey takyry the barkhans are blown apart and travel on as individual streams of sand. These settle around obstacles, such as shrubs.

The wind direction in the Karakum is determined by the Pamir–Alay mountain ranges. In summer the wind blows from the northwest toward the mountains, while in the winter it blows from the southeast away from the mountains. Northeast and north winds also occur, however. Barkhan movement resembles that of a pendulum, with net motion toward the northwest (around 4–4.7 m annually at Kerki). Local exceptions also occur.

Since the sand in individual barkhans is constantly being redistributed, it is completely dry. This is not necessarily the case with the barkhan fields or chains, in which moisture from winter rains can be retained in individual intermediate layers. As more water seeps in, from year to year, and only a small part is evaporated at the surface, freshwater lenses eventually are formed over impermeable layers and over saline groundwater. These lenses can be tapped by wells. The completely bare sands are thus very important, since on sands with vegetation the plants use up the rain water and there is no accumulation, as occurs under bare sands, even at annual precipitation amounts of 100 mm and less.

Chemically, the sands are composed of 50% silica, up to 10% feldspar, and up to 4% calcite, with considerable mica as well.

Aerial photographs are especially useful in the analysis of the sand relief, since the different forms can be seen more easily than from the ground.

Of the total area of the Karakum Desert, about 80% is made up of sands with some vegetation, 5 to 10% has little or no vegetation (mobile sands), and the rest is takyry and saltpans, which appear as islands in smaller and larger depressions in the middle of the sands, because the wind can remove the sand only down to wet layers (above groundwater) or to clay layers. Since the groundwater is usually saline, salt crystallizes out of the capillarily wetted surface sand layers as the water evaporates. These crystals tend to loosen up the surface soil grains. In summer the groundwater level sinks, the soil dries out, and the surface disintegrates into dust. The wind blows away the dust with the salt crystals until a constantly wet sand layer comes to the surface. A solid crust of salt then forms on this layer, even when the groundwater is only slightly saline. In this way a saltpan with halophytes develops, and the halophytes root into the only slightly saline groundwater.

Takyry are found on clays which are only temporarily inundated. A capillary rise of the groundwater is generally absent. These form primarily in the delta regions of rivers.

The main river, the Amu-Dar'ya, plays a dominant role in the river system of the region. Its entire length is 2600 km, and its catchment area covers 465 000 km^2. Along its lower course the Amu-Dar'ya has no tributaries but instead loses water to the underground lake under the central Karakum. The last tributary, the Surkhan-Dar'ya, empties into the main river 1276 km above the river's outlet into the Aral Sea. The river Zeravshan, coming from the east, no longer reaches the Amu-Dar'ya because too much water has been taken out of it for irrigation in the Samarkand–Bukhara area.

The source of the Amu-Dar'ya is in the Vrevski Glacier at 5000 m elevation in the southern Pamir. The river first carries the name Vakhan (Afghan: Wakhan), then Pyandzh (Russian: P'andzh, Afghan: Panja), and receives the name Amu-Dar'ya only after confluence with the Vakhsh in southernmost Tadzhikistan. The Pyandzh and the Vakhsh contribute 80% of the total runoff. In the upper catchment area, which is covered by snow in winter, 400 glaciers and perpetual snowfields (firns) are known.

The Aral Sea receives about 42 km^3 of water from the Amu-Dar'ya each year. At Kerki, somewhat north of the Afghanistan border, the mean flow is 2000 m^3 s^{-1}, but at Nukus, just above the delta, only about 1500 m^3 s^{-1}. The loss of 500 m^3 s^{-1} comes from diversions to irrigation and losses to the groundwater.

The water level of the river begins to rise with the beginning of snowmelt in spring, and it reaches its highest level in mid-summer. The high water period, with a series of crests, lasts four to five months. Sometimes a relatively low level intervenes between the spring and summer high-water periods, but usually the first high-water level flows smoothly into the second. The falling of the water level begins at the end of July or beginning of August. The low-water period begins in October and reaches its lowest point in January. At this time the lower course of the river receives water from the groundwater.

The mean width of the river is 1 to 1.5 km, 3 km at high water. During the high-water period, the flow reaches 6000 to 8000 m^3 s^{-1} at Kerki but only 4000 to 6000 m^3 s^{-1} at Nukus. In January the corresponding figures are 400 to 700 m^3 s^{-1} and 400 to 600 m^3 s^{-1}. Riverbank erosion can be considerable. The mean amount of suspended matter in the river is 3.6 kg m^{-3} at Kerki, with a maximum of 150 kg m^{-3} in May. The Indus, Ganges, and Syr-Dar'ya are even muddier.

The amount of solid matter transported by the river is 200 million tons annually at Kerki, but only 10% of this reaches the Aral Sea. Thus, the continuous growth of the delta area is understandable. The present-day watershed boundary between the Aral Sea and the Sarykamysh Lowland was built up by the Amu-Dar'ya, primarily out of loamy deposits. These deposits are useful in the irrigated fields, but they stop up the irrigation channels, which must be cleaned out constantly. These alluvial deposits also make up the floodplain soils, which are still sandy along the middle course of the river. The riverbanks are composed of coarser material. Along the Karakum stretch of the river, the riverbed is being continuously shifted to the east, with sand dunes arising on the left terrace which continues to grow.

The temporary formation of an ice cover in winter is especially dangerous, since it backs up the water and can lead to flooding.

At the beginning of the current century, 7% of the river water was used for irrigation. By 1955 it had increased to 15%, and even more since the construction of the Karakum Canal (see Fig. 5.1).

The infiltration of river water into the groundwater is particularly important for the entire Karakum region (see p. 111).

The second largest river in the Karakum is the Murgab in the southeast. It is fed primarily by melting snow in the spring, that is, at a time when hardly any river water is needed for irrigation. The river carries little water in the summer. The mean flow is 50 $m^3 s^{-1}$. Occasionally it is double this amount, while in summer it is only 10 $m^3 s^{-1}$. The amount of suspended matter is also high, and the river deposits 5 million m^3 of solid particulates every year. The older dams are already partly or fully buried by sedimentation. Through the construction of new reservoirs, the water of the Murgab has been better utilized than that of the other rivers in the Karakum region. The water is used, in particular, for growing cotton. Part of the water in the delta region also makes its way into the groundwater.

The river Tedzhen is longer than the Murgab (1100 km long) but carries less water. The source region is in Afghanistan, in the mountain regions of the Paropamisus and the Siyakh-Kush. A part of the water from the catchment area is used outside the Karakum in the Herat oasis (Afghanistan) and in Iran. The flow depends entirely on the amount of snow which falls in winter along the upper course of the river.

The last large river in the region, which is in the southwest and actually does not touch the Karakum Desert, is the Atrek. Its length is 662 km. Its catchment area lies, for the most part, in Iran. The water of this river is scarcely utilized by the human population. The deposition in the delta area is considerable, and there is constant danger of soil salinization resulting from the high groundwater level.

The tributary Sumbar is more economically important than the Atrek itself. It is used for the irrigation of the small, but subtropical and thus very valuable region of Kara-Kala in the southwest (see p. 82).

Nikitin and Meshcheryakov (1954) have studied the aquatic vegetation in the various reservoirs of the river Murgab.

CLIMATE

Climatic data for the Station Repetek

For the desert experimental station at Repetek, in the southeastern Karakum about 80 km west of the Amu-Dar'ya, observation records covering the years 1913–1917 and 1926–1970 are available (Nechayeva, 1975; Rodin, 1977, p. 118; Fig. 6.2).

The climate diagram of Repetek is almost identical to that of Kerki (Fig. 5.2). The mean annual temperature and average annual precipitation during the period of observation, plus the numbers of years with above- and below-average rainfall, are shown in Fig. 6.2. The following additional information (from Nechayeva, 1975) can also be given.

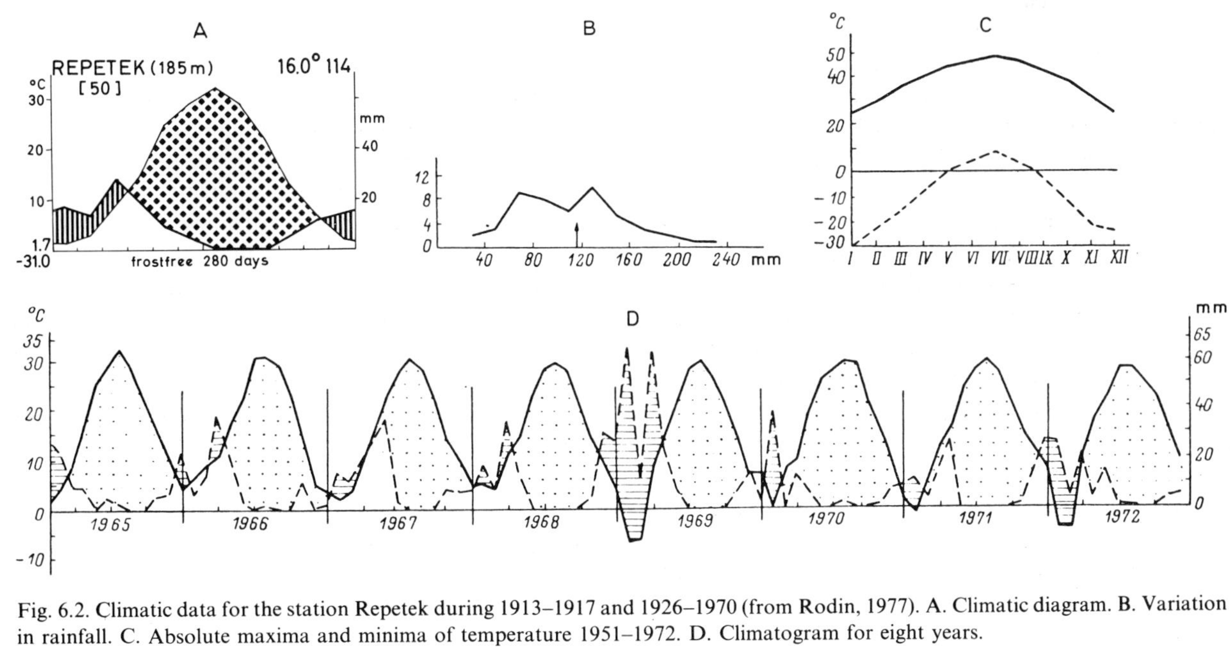

Fig. 6.2. Climatic data for the station Repetek during 1913–1917 and 1926–1970 (from Rodin, 1977). A. Climatic diagram. B. Variation in rainfall. C. Absolute maxima and minima of temperature 1951–1972. D. Climatogram for eight years.

Temperature

The mean annual temperature is 16.2°C, with a maximum year-to-year variation of 2°. There are no months with mean temperature below 0°C, and, from the end of April to the end of October, the monthly mean temperatures lie above 20°C. A number of days with frost occur in the second half of March, and the earliest frosts appear in October. The mean frost-free period is 280 days.

The warm season (mean daily temperature above 5°C) lasts from the second half of February to the first third of December. The annual temperature sum (temperatures above 10°C) reaches 5600°. The difference in mean temperature between the coldest month (January) and the hottest (July) is 30°C. The summer months are characterized by mean temperatures in the range 30 to 32°C. The absolute maximum temperature recorded was 50°C on 3 July, 1915, and the absolute minimum was −31°C on 26 January 1969. The mean monthly temperature is lowest in January at +1.7°C. The mean diurnal temperature variation is around 20 to 25°C.

Precipitation

The average annual precipitation is 114.5 mm, of which 37% falls in winter, 49% in spring, 2% in summer, and 12% in autumn[1].

The rainfall maximum in March–April is caused by heavy downpours. The period from July to September is practically rainless. The variation in rainfall from year to year is very large. In 1917 only 24.3 mm (20% of the average) were recorded, while in 1953 a high of 229.5 mm (200% of the average) was recorded. The average number of rainy days is 32.

Snowfall is rare, and a snow cover of 2 to 5 cm is present for an average of only 8 to 10 days. Only in 1968–69 was there a 25 to 30 cm snow cover, which lasted for 48 days at temperatures around −5°C.

Unmeasurable precipitation in the form of fog, dew, and hoar-frost is seen on 40 to 50 days in winter.

Air humidity

The absolute humidity climbs from winter until June–July and averages 7 to 10 mbar for the year. The highest relative humidity (65–70%) is found in January and the lowest (20–25%) in July–August. For short periods the humidity can fall to only a few percent.

Wind

The wind blows from different directions, from April to October mostly from the north and from December to March mostly from the south. The direction of prevailing winds changes in November and in April. The mean wind speed is 3.6 m s^{-1}.

Groundwater

In dune valleys the groundwater is found at a depth of 4 to 5 m, under the dune ridges at a depth of 25 to 30 m. The groundwater table varies in the course of a year by about 3 to 12 cm. It begins to sink in spring, at the beginning of the growing season, and climbs after the end of the growing season. In addition, over the last forty years, a continuous drop of 1 cm per year has been observed. The salt content of the groundwater is only 1 g l^{-1} under the dunes, whereas it can climb to 15 to 20 g l^{-1} to the west, outside the dune area. The salinization is of the chloride–sulphate or sulphate–chloride type.

MICROCLIMATE

Of the very extensive measurements, only the most important are mentioned.

Incident solar radiation at the surface is 672 kJ cm^{-2} for the year, of which 462 kJ cm^{-2} are contributed by direct radiation and 210 kJ cm^{-2} by diffuse radiation. The albedo for poorly covered sand surfaces is 25 to 26% and about 20 to 22% for surfaces with good vegetation covers.

The outgoing long-wave radiation is 172 kJ cm^{-2} in shifting sands, 159 kJ cm^{-2} in sands with some vegetation, and 134 kJ cm^{-2} (per year) in lightly wooded dune valleys.

The absorbed radiation, on the other hand, is lowest for mobile sand at 500 kJ cm^{-2} annually; for immobile sand it is 508 kJ cm^{-2} and for the hollows 537 kJ cm^{-2}. The mean daily temperature of the air layer near the ground is highest for mobile sand, around 0.5 to 1.0°C lower for immobile sand, and 1.5 to 2.0°C lower for the hollows. Conversely, in summer, the relative humidity is around 5 to 6% higher for the hollows than for the sand biogeo-

[1] In the northern Caspian Lowland the numbers are 17%, 24%, 35%, and 24% (see Miroshnichenko, 1975).

coenoses (corresponding figures are given by Nechayeva (1975), measured at 50 cm and 2 m above the surface). The wind speed at a height of 5 m is also two to three times lower over a shrub cover than over bare sand.

Soil temperatures at various depths show the expected diurnal variation and the corresponding differences between the individual biogeocoenoses (Nechayeva, 1975, p. 46).

The characteristic microclimatic peculiarities are especially distinct during the high insolation at midday.

WATER BALANCE OF THE BIOME, GROUNDWATER AND SALT CONDITIONS

Deserts are generally characterized as lacking sufficient water. In the Karakum Desert, considering the through-flowing groundwater, this is not the case; but this groundwater is often saline.

Requirements concerning the quality of water are not as strict in arid areas as in humid areas. In humid areas drinking water should not contain more than 0.5 $g\,l^{-1}$ of salts. In arid regions water with 2 $g\,l^{-1}$ of salts is still used as drinking water. Even water with 3 to 4 $g\,l^{-1}$ of salt is sometimes drunk.

The drinking-water requirements for livestock accustomed to desert conditions (sheep, goats, camels) are even less strict. The upper limit of salt content is 7 to 8 $g\,l^{-1}$ in summer, 10 to 12 $g\,l^{-1}$ during the cool season, and even 15 to 18 $g\,l^{-1}$ of total salts in the winter.

There are, however, lakes and springs with good freshwater. In order to appreciate this, one must understand the source of the groundwater streams in the Karakum.

The depth of the groundwater table below the surface varies, according to relief, from several hundred meters to 1 to 2 m. If one imagines the overlying rock layers removed, then one can think of the Karakum region as a large sea stretching from the Amu-Dar'ya to the Caspian Sea, into which it merges imperceptibly. Of course the surface of the groundwater table is not perfectly level, showing a slight decline from east to west.

As already mentioned, the water of the Amu-Dar'ya seeps into the loose rock which fills the basin. The amount corresponds to an average of about 150–200 $m^3\,s^{-1}$ of river flow along the entire length of the rivercourse in the Karakum region. The groundwater stream moves very slowly to the west and corresponds, to a certain extent, to the original Amu-Dar'ya, which emptied into the Caspian Sea. The groundwater lake is also fed by water from the Murgab and the Tedzhen, which were earlier tributaries of the Amu-Dar'ya. In addition, there is the underground flow from the rubble strata at the foot of the southern mountains bordering the Karakum, above the Kopet Dag. There is also overland flow from the slopes after heavy downpours. This seeps into the rubble or into the dry riverbeds.

One must also consider the rainwater which falls on the 350 000 km^2 of the Karakum Desert. Even at a mean annual precipitation of 100 mm, this is a large amount, namely 35 km^3. Of course the largest part of this water evaporates from the surface or is lost by the transpiration of the plants, but a certain, not precisely calculable part does sink into the ground and reach the groundwater. This is especially true of rainfall on bare sands.

On the basis of the existing observations, one can make the following very rough computation of the annual runoff to the groundwater of the Karakum Desert (Kunin, 1955):

infiltration from the Amu-Dar'ya riverbed	150 $m^3\,s^{-1}$
rainwater seepage in barkhan areas	30 $m^3\,s^{-1}$
underground runoff from the Kopet Dag	20 $m^3\,s^{-1}$
infiltration from the Murgab River	17 $m^3\,s^{-1}$
infiltration from the Tedzhen River	4 $m^3\,s^{-1}$
seepage from hillocks and takyry	1 $m^3\,s^{-1}$
total average runoff to groundwater	222 $m^3\,s^{-1}$

The following computation of losses can also be made. It is even less reliable, but does provide a certain orientation point:

evaporation from wet saltpans	165 $m^3\,s^{-1}$
groundwater losses through transpiration	57 $m^3\,s^{-1}$
total losses from the groundwater to the atmosphere	222 $m^3\,s^{-1}$

According to this very crude balance, no appreciable amount of groundwater would flow into the Caspian Sea, which is very probable from the observations and experience which exist.

The alternating alluvial materials are permeable to different degrees. In general, however, the groundwater moves very slowly from east to west, and losses through capillary rise and evaporation occur where the groundwater reaches to within a meter of the surface. Roots of plants can, under some conditions, reach groundwater at a depth of 5 to 10 m and cause it to be lost to the air through transpiration.

Since river sediments were deposited under arid conditions, smaller or larger temporary salt lakes were continuously being formed. These could dry out partially and have been covered by new sediments. As a result, varying amounts of different salts are contained in the loose rocks. The most common salts are sodium chloride and sodium sulphate. The groundwater dissolves these salts as it moves along and thus has a certain salt content, which varies greatly, both quantitatively and qualitatively, from place to place. A mixing of the different groundwater streams does not occur or occurs only to a limited degree. Consequently, the salt content of the water from individual springs can be very different and unpredictable. The same is true of well water.

Springs around saltpans are marked by high salt content, since the salt accumulating at the surface from evaporation is dissolved in the spring, when the saltpans are full of water, and seeps into the ground to reach the groundwater. On the other hand, as we already emphasized, the groundwater under barkhan areas is always very low in salts and serves as good drinking water.

The two to three spring months bring 50 to 70% of the rainfall in the Karakum region. In the barkhan areas the rain water sinks into the loose sand completely. Only a small part evaporates from the wet surface; another fraction is held within the sand as capillary soil water. The largest fraction sinks gravitationally to the groundwater level and collects as a freshwater lens on top of the groundwater, since it has a lighter specific gravity. This freshwater lens is asymmetrical, rising only slightly above the groundwater level but sinking more deeply into the groundwater, just as an iceberg retains 90% of its volume under water. Mixing with the saline groundwater occurs only to a very limited extent.

These freshwater lenses can be reached by wells. If water is removed too quickly, salty groundwater can flow into the wells.

The formation of freshwater lenses benefits from overgrazing, since if plant cover is completely destroyed, open sand areas with small dunes arise.

Artificial freshwater lenses are also deliberately produced by the inhabitants on the larger takyry. The takyry correspond to the deltas of small watercourses; they are almost level and are covered with a thin layer of clay. After the spring rains, the water runs off from the slopes and distributes itself over the existing plain. The clay particles which were carried along are also deposited. The water remains as a thin layer over the impermeable clay and evaporates. There is no storage of water in the soil. Algae develop in the water temporarily, then dry out and survive the drought period in a latent state. The clay layer dries out and cracks, but these cracks close again after the first 3 to 4 mm of rainfall. Underneath the clay strata of the takyry there are usually sand strata. If one breaks through the clay layer artificially, then the rainwater can penetrate into the sand and be stored.

In order to do this on a large scale, the broad layer of flowing water is led by shallow channels to the lowest point in the takyry. At this point an artificially larger depression is created, with several rings of small well shafts which are dug out down to the sand layers. The water gathering in the depression during the rainy period is swallowed by the wells and penetrates from the well shaft laterally into the sand above the saline groundwater. This forms a freshwater lens which is protected from evaporation. The freshwater can then be used during the drought period for drinking and watering purposes, possibly for small vegetable gardens. The water is regained from the same wells which were used to collect it.

A water layer of 10 mm forming over a takyr during the rainy season represents 10 000 m^3 of relatively salt-free water per square kilometer, which can be stored in the wells. Since a karakul sheep needs only 1.5 m^3 of water in a year, such a water supply would be enough for a herd of 6000 sheep. In the sand desert, each sheep needs a grazing area of about 6 ha. This means that, for a watering place which can hold 10 000 m^3 of stored water, 36 000 ha of grazing land can be made accessible. The storage area of a takyr should, of course, be kept as free of plants as possible, for around every larger plant a sand dune will form, which will reduce the flow of water to the wells and cause losses by transpiration.

The freshwater lens formed under the wells will be shifted laterally somewhat by the movement of the groundwater. The distance is usually a few meters per year. If the volume of sand above the groundwater is small, as when the groundwater is at a depth of only 3 to 4 m, then the collecting wells are laid out in a row parallel to the direction of groundwater flow. The water is brought first to the first well, which is filled, then to the second, and so on. When the wells are used during the dry period, they are also used in sequence, as long as usable water can be obtained. Adaptation to the particular local conditions is necessary in every detail.

No accumulation of fresh groundwater occurs under sand with a vegetation cover. The surface of these sands is somewhat compacted; the upper sand layers (with a slight humus content) have a greater field capacity and hold more water back. There are indeed years with especially high rainfall, but in these years the ephemeral vegetation develops more abundantly and the perennial species form larger shoots, so the water consumption rises accordingly. When a certain water surplus still remains, it merely replenishes the water supply of the deeper layers, which is then available to deeper-rooting, woody species in dry years.

A striking phenomenon on the northwestern border of the Karakum is the presence of freshwater lakes, on the surface, in the western part of the Uzboy Valley, in which otherwise only bitterly salty lakes would be found. The shores of the salt lakes are desert-like, for only a few hygro-halophytes are able to survive. These lakes are produced by standing groundwater, which reaches the surface and evaporates. The level of the lake is maintained by the throughflow of groundwater, but the salt content continually increases. Along the shore, the ground surface is wetted by the capillary fringe, and white salty crusts form where the last patches of water evaporate.

The situation in the freshwater lakes is more complicated. Reeds, tamarisks, and poplars grow along their shores (Fig. 6.3A and B). These lakes are found in the valleys of earlier tributaries of the Uzboy. They often extend for 10 km or more and are also fed by groundwater, but not by the salty groundwater from the Karakum side. Since these lakes are not far from the Ustyurt Escarpment and the Great Balkhan, one can assume that the higher rainfall of the mountains finds its way to the freshwater lakes through fissures. Some water from these fissures also runs off, above or below ground, into the Uzboy Valley. Thus, these lakes are formed not by standing but by slowly running water. Although one must expect an evaporation of about 2 m per year, there is still a certain amount of runoff, which prevents an increase in the salt concentration.

Relatively well-supplied with water is the lowland just below the Kopet Dag in the south. Numerous springs are to be found at the foot of the mountains. Water can also be obtained out of the rubble fans by means of gallery wells (*kanats*).[1] The outcropping rocks of the mountains are highly fissured limestones of the Lower Cretaceous, into which the rain water penetrates easily and through which it feeds the springs at the foot of the mountains. Many small and larger oases have existed here for a long time.

The situation apparently becomes less favorable to the west of Kizyl-Arvat. There are here the sort of loamy outcroppings into which rain does not easily penetrate. The landscape is in essence "badlands," and there is no vegetal cover. Numerous small erosion channels furrow the slopes of the cone-shaped mountains. Contrary to the widespread opinion, such slopes are hardly eroded at all. The dried-loam surface shows itself to be very resistant against erosion. The rainwater runs off superficially, is only slightly muddied by the loam, and sinks in only when it reaches the plain. It is for this reason that the neighboring Karakum region has such large freshwater lenses above the groundwater. The lack of vegetation on the slopes increases the runoff tremendously. This landscape, thus, does not suffer from lack of water.

More efficient utilization of the water has been attained in more recent times through the construction of numerous dams. Groundwater should be used as close to the foot of the mountains as possible, as long as the water has not yet mixed with the saline groundwater of the Karakum.

There are in the Karakum region also very deep groundwater streams carrying water under signi-

[1] These are nearly-horizontal wells used to intercept at a higher level the downward flowing groundwater on slopes, especially in valleys. The water can be used to irrigate terraces built just below the wells.

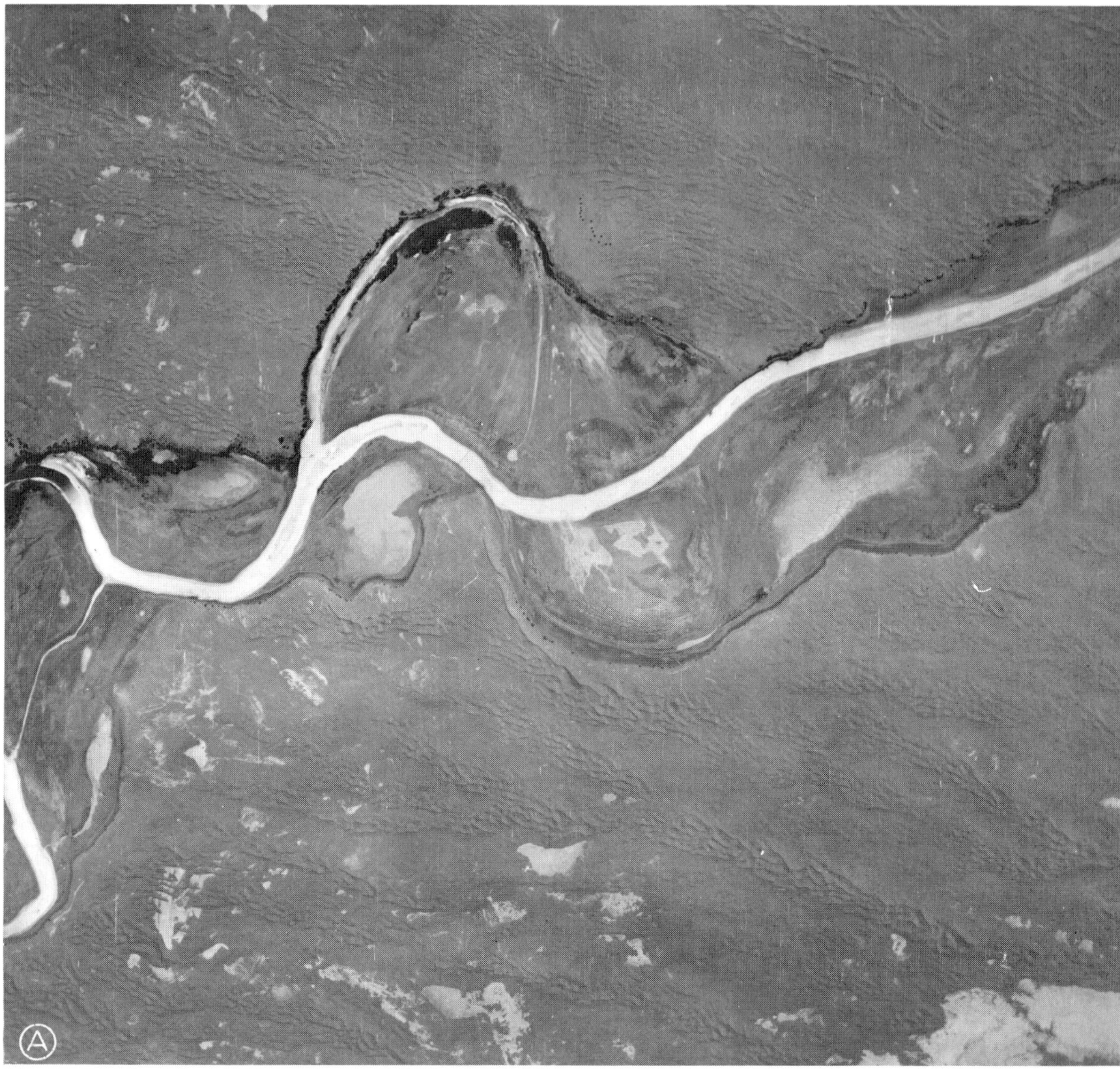

ficant pressure. Artesian wells, however, have not yet played a significant role. Much more important for the opening up of the land to exploitation is the huge Karakum Canal, which runs roughly in the direction of the original Amu-Dar'ya at the foot of the southern mountains. This is being continually expanded.

The canal uses the widened Bosaga–Kerki Canal for its first 37 km, then unites after another 9 km with the Kelif-Uzboy, an old bed of the Afghan river Balakh. It continues for 80 km in this channel, forming a reservoir with a volume of 350 million m^3. It then cuts across a section of the Karakum and reaches the Murgab Delta at the railroad station Zakhmet. The canal continues by way of Mary to the city Tedzhen and is there channeled into the river of the same name. The length of the canal extends 600 km up to this point.

The drawing of water from the Amu-Dar'ya into the canal takes place without any dams. The maximum water outtake amounts to 300 $m^3 s^{-1}$.

The new oasis on the sandy-loam plain of the Obruchev Steppe permits the growing of fodder for intensive cattle-raising. The irrigation in the delta

Fig. 6.3.A. The dry Uzboy bed covered with a white salt crust. Single trees and marshes (dark spots) on the northern river side with a freshwater seepage from the dune area (air photo received from W.K. Leimbach). B. Freshwater lake in the Uzboy Valley, with reed stands and individual poplars (*Populus diversicolor*) as the remains of an earlier floodplain forest (photograph Petrov, from Walter, 1968, p. 770).

area of the Murgab and the Tedzhen makes possible better utilization of the arable areas, since the Amu-Dar'ya has its high-water level in the summer, at a time when the Murgab and Tedzhen are at their lowest levels. The canal guarantees a supply of drinking water and provides a water route for transporting goods. The same is true of the other sections along the Kopet Dag.

THE NATURAL VEGETATION

Floristic relations

After looking at the environmental factors in the Karakum Desert we should now look at the biotic component.

Fossil findings show that the flora of the Karakum is very old (Korovin, 1961/62). This is strongly suggested also by the many endemic, often monotypic, genera. The flora of the sand desert includes 350 species, of which 56% are endemic. The evolution of the plants in the Turanian region took place, as already mentioned, under arid conditions and in nearly complete isolation since the Tertiary. This led to the development of an independent flora and to the selection of characteristic life forms, which we will discuss in some detail. The Karakum Desert is perhaps the most interesting of all temperate cold-winter deserts.

A list of plant species in this region is given by Nechayeva et al. (1973). It includes data on biotope, growing season, type of phytocoenosis and life form (tree, shrub, semi-shrub, herb, each with several classes, groups, and subgroups). The development, morphology, ecological significance, and other details, are described for the most important representatives, based on lengthy observations in the field and under cultivation. One peculiarity of the Karakum Desert is the occurrence of trees as dominants. These reach a height of 9 m, are not limited to groundwater supplies, and grow to be more than fifty years old. The differentiation of the Raunkiaer life forms is not sufficient for the special situation here. In addition to abundant microphylly and aphylly, small, short-lived shoots are often formed. These are photosynthetic and fall off during drought or in the winter. In other cases, the green, assimilating shoot tips may die back, so the production of new shoots is mostly sympodial. The wide spectrum of life forms is unusual for a low-rainfall desert. In the sand desert three biogeocoenoses are distinguished:

(1) Mobile sands with *Aristida karelinii*, *Ammodendron conollyi*, *Calligonum arborescens*, *Eremosparton flaccidum*, etc. This biogeocoenosis makes up 25% of the protected area.

(2) Immobile sands: sandy areas on which *Haloxylon persicum* and other woody plants grow. These also carry an undergrowth of various herbs, especially *Carex physodes*, and numerous ephemerals and ephemeroids in the spring. This biogeocoenosis, with 34%, covers the largest fraction of the protected area around Repetek.

(3) Hollows: valley-like depressions between dunes, which have slightly saline groundwater at a depth of 5 to 10 m and light, forest-like stands of *Haloxylon ammodendron*. As accompanying species there appear other woody plants and herbs, with an undergrowth of *Carex physodes*. The areal fraction is, along with the takyry and halophyte areas, hard to estimate individually.

With decreasing mobility of the sands, the dust content of the soil increases. The dust which is blown away with the sand is carried farther and can be deposited locally in wind-protected parts of the relief. The local dust deposition is shown very graphically in Fig. 6.4.

Table 6.1 gives numbers of species and the percentage fractions of the individual life forms in the entire flora. We must limit ourselves to a few typical examples of characteristic forms.

Life forms

Trees with definite trunks

Ammodendron conollyi (Fabaceae), the sand acacia, is an important tree on slightly mobile sands. It grows 4 to 9 m tall, with a trunk diameter of 6 to 30 cm and trunk height of 1 to 3 m. The upper parts

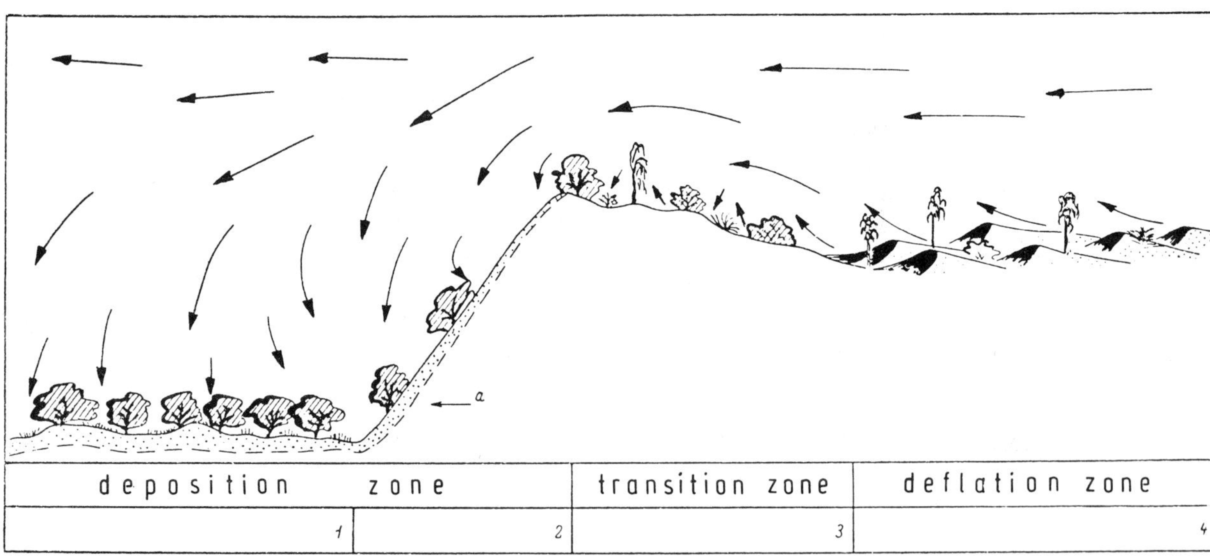

Fig. 6.4. Deflation and deposition of dust in the dunes near Repetek and its significance for the distribution of the vegetation (from Rodin, 1977, p. 138). Legend: *1* = *Haloxylon ammodendron–Carex physodes* biogeocoenosis in a hollow formed by deposition; *2* = a *Haloxylon persicum–Carex physodes* + *Bromus tectorum* biogeocoenosis at higher elevations; *3* = a transition area occupied by *Haloxylon persicum* + *Ammodendron conollyi–Calligonum* spp. biogeocoenosis; *4* = blown-out dunes, *Ammodendron conollyi–Calligonum* spp. biogeocoenosis; *a* (dotted) = with dust particles of accumulated fine sand.

TABLE 6.1

Life-form spectrum (numbers of species and corresponding percentages) for the three main biogeocoenoses (see p. 116) in the Karakum Desert (from Nechayeva et al., 1973)

Life form	I: mobile sand		II: immobile sand		III: hollows		Total	
	species	%	species	%	species	%	species	%
Trees and shrubs	5	26.3	9	10.2	12	11.9	17	12.7
Semi-shrubs and dwarf-shrubs	2	10.5	2	2.3	1	1.0	4	2.9
Perennial herbs	4	21.1	13	14.8	9	8.9	24	17.9
Annual herbs	6	31.6	59	67.1	71	70.3	80	59.7
Fungi (large)	2	10.5	4	4.5	7	6.9	8	6.0
Mosses (*Tortula desertorum*)	–	–	1	1.1	1	1.0	1	0.8
Total	19	100.00	88	100.0	101	100.0	134	100.0

of the lateral, 2nd to 6th order branches die back annually and are replaced by new branches from renewal buds.

The silvery pinnate leaves are reduced to a pair of leaflets 30 mm long and 3 mm wide. The tip of the rachis is spiny. Long, 1st to 3rd order shoots are produced only in favorable years. These reach a length of 150 cm and form the crown. Long trunk shoots can be produced from dormant buds. The showy flowering racemes are reddish-violet.

The formation of the root system can be seen from Fig. 6.5. While the shoot of a first-year seedling grows to only 10 to 20 cm in height, the root penetrates 70 to 100 cm deep into the ground. In the second year, the root increases its length to only 120 cm but forms long, horizontal, lateral roots, which supply the plant with water after each rainfall. Even in old plants, the main root does not reach deeper than 3 m. The trees live to be fifty years old.

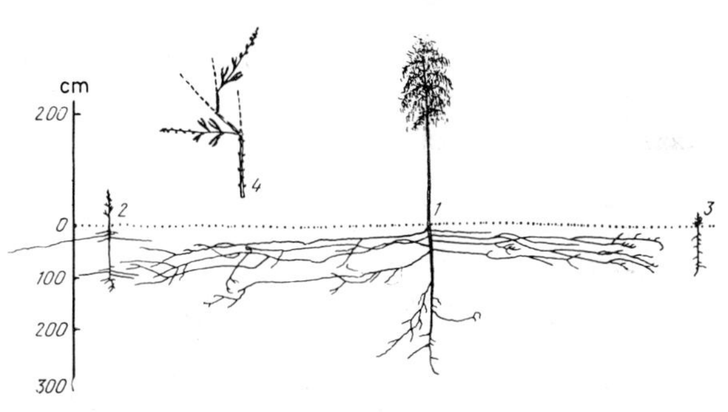

Fig. 6.5. *Ammodendron conollyi*: *1* = 10 year old plant; *2* = 2 year old plant; *3* = 1 year old plant; *4* = shoot branching (from Petrov, in Nechayeva, 1973; and Walter, 1976, fig. 49).

In the open stands of sand acacia one finds about 75 trees per hectare. The above-ground phytomass is about 0.57 t ha^{-1}. The age-structure of such a stand is as follows:

Age (years)	Number of trunks (%)
1	6
2	2
3–5	11
6–10	18
11–15	9
16–20	6
21–25	3
26–30	2
> 30	1
dead	42

This species tolerates both covering and uncovering by sand. In such cases, it forms either long trunk shoots or adventitious roots.

Eremosparton flaccidum (Fabaceae) is another endemic of similar significance, but this tree grows to only 5 m with a trunk height of 2 m. In a normal stand one finds about 40 small trees per hectare, with a phytomass of 60 kg ha^{-1}.

Calligonum eripodum (Polygonaceae) grows to 6 m (trunk 1.5 m) and is the only tree belonging to an otherwise well-represented genus, to which we will return.

Haloxylon ammodendron (*aphyllum*) (Cheno-

Fig. 6.6. *Haloxylon ammodendron* (*aphyllum*) and, on the left, *Salsola arbuscula*, 5 km from the Amu-Dar'ya River (photograph by E.A. Bessey, from Walter, 1976).

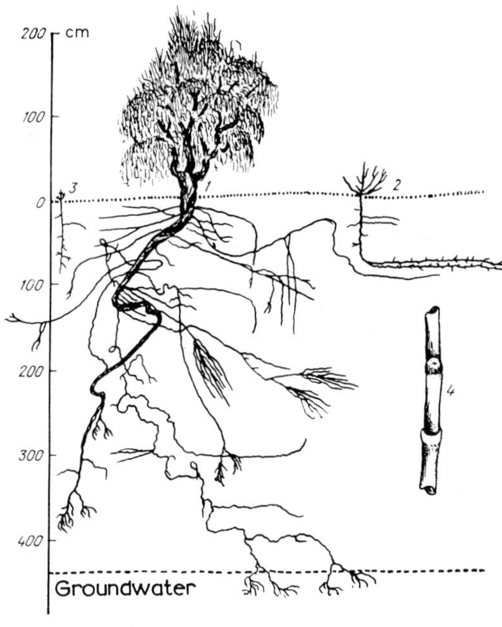

Fig. 6.7. *Haloxylon ammodendron*, with root system: *1* = a plant about 10 to 15 years old; *2* = 2 year old plant; *3* = 1 year old plant; *4* = part of a vegetative shoot (from Petrov, in Nechayeva, 1973; and Walter, 1976, fig. 51).

podiaceae), the black *saksa'ul* ("with black wood"), is the most important tree, growing 5 to 9 m high. Its trunk diameter reaches 24 to 38 cm at an age of 50 to 70, sometimes 100, years (Fig. 6.6). The tree is leafless, and the assimilating shoots die at the end of the growing season. Branching is sympodial.

The root system is shown in Fig. 6.7. The growth form may be shrubby or arborescent, depending on site conditions. Large individuals develop when the roots reach groundwater at a depth of 4 to 5 m, sometimes even more than 8 m. In such cases there arise dense thickets, which are used for firewood. This species tolerates slightly saline water. The young shoots provide a good fodder for camels and sheep throughout the year.

The distribution of the phytomass is shown in a profile (Fig. 6.8) taken from Miroshnichenko (1975a).

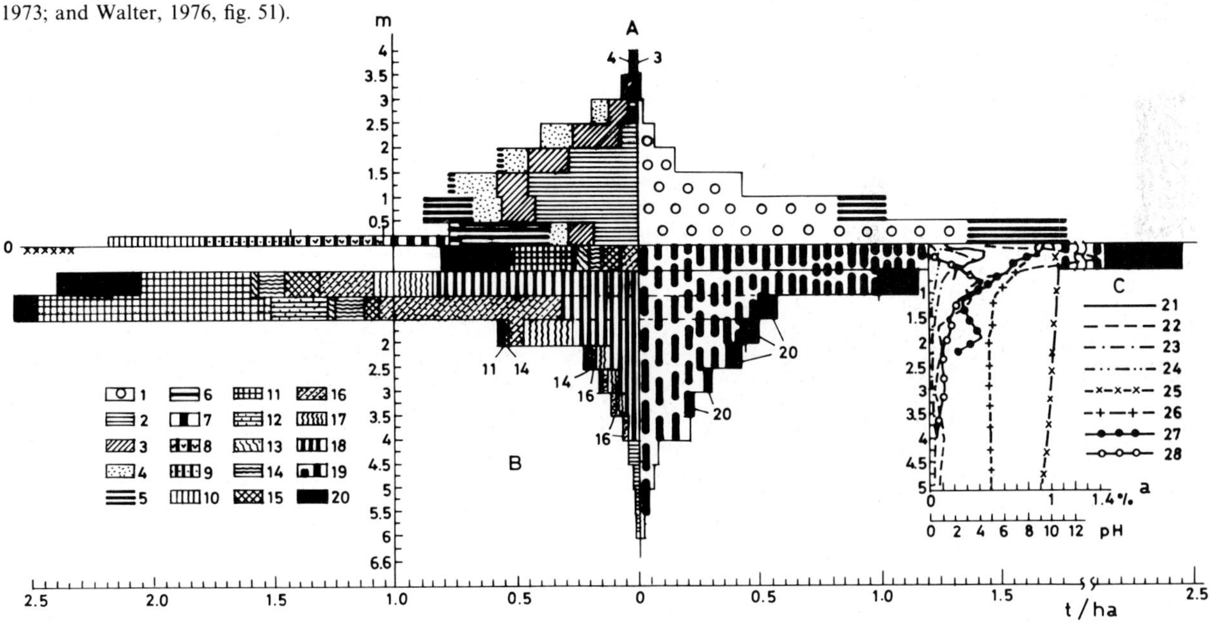

Fig. 6.8. Quantitative distribution of the phytomass of *Haloxylon ammodendron* in the biogeocoenosis with *Carex physodes*, above groundwater at a depth of 7 m (from Miroshnichenko, 1975b). A. Above-ground phytomass: *1* = trunk; *2* = branches over 1 cm thick; *3* = thinner branches; *4* = shoot from the last year; *5* = dead branches in the crown; *6* = dead wood on the ground, including: *7* = over 7 cm in diameter, *8* = 3–5 cm in diameter, *9* = 1–3 cm in diameter, and *10* = under 1 cm in diameter. B. Below-ground phytomass: lateral roots, including: *11* = less than 0.5 mm in diameter; *12* = 0.5–1 mm in diameter; *13* = 1–2 mm in diameter; *14* = 2–4 mm in diameter; *15* = 4–10 mm in diameter; *16* = 10–20 mm in diameter; *17* = 20–40 mm in diameter; *18* = more than 40 mm in diameter; *19* = taproot; and *20* = dead roots. C. Composition and moisture content of the soil: *21* = humus percentage; *22* = dry-weight percentage; *23* = Na percentage; *24* = Cl percentage; *25* = pH value; *26* = CO_2 percentage; *27* = water content on May 5th; and *28* = water content on July 2nd. The taproot of *Haloxylon* reaches the capillary fringe of the groundwater.

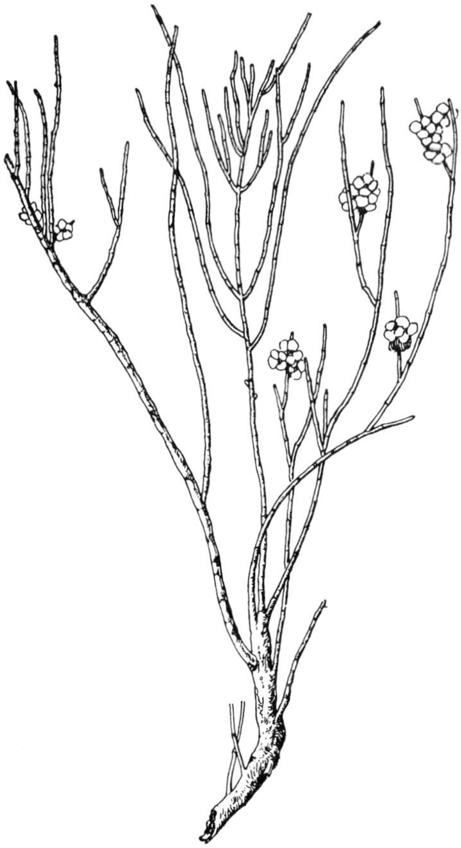

Fig. 6.9. *Haloxylon persicum*: part of a vegetative branch (middle) and fruiting branches (on both sides) (from Petrov, in Walter, 1976, fig. 52).

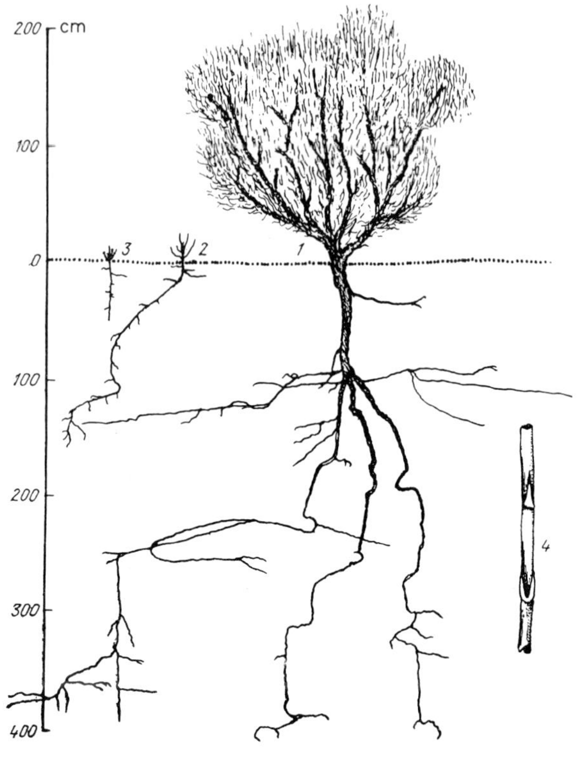

Fig. 6.10. *Haloxylon persicum*: *1* = plant 10–15 years old; *2* = 2 year old plant; *3* = 1 year old plant; and *4* = part of a vegetative shoot, with scaly leaflets (from Petrov; in Walter, 1976, fig. 53).

Shrubs (up to 6 m tall, without trunk)

Haloxylon persicum (Chenopodiaceae), the white saksa'ul ("with white wood"), is the character plant for the Karakum Desert. This species, along with *Carex physodes*, forms the most common plant community on immobile sand. The shrub grows to 3 to 5 m tall (trunk height only 10 to 20 cm). The long shoots, with small scale-like leaves, have a length of 50 cm, of which the outer 10 to 20 cm die back annually. The flowers are found on short shoots only 1 to 3 cm long (Fig. 6.9). The root system is shown in Fig. 6.10. The shrubs reach an age of over 30 years. They are eaten readily by sheep but rejuvenate easily from stump sprouts. This saksa'ul forms open stands with 120 to 150 individuals per hectare and produces 0.4 to 0.5 t ha^{-1} of firewood and up to 0.2 t ha^{-1} of green fodder.

Calligonum (Polygonaceae) is a genus which, in Turkmenistan, is represented by 55 species, only one of them a tree. Their cylindrical leaves fall off early, leaving the shrubs bare (Fig. 6.11). Photosynthesis is carried on by annual, densely clustered shoots (Fig. 6.12). These develop from axillary buds, which branch even before elongation and are thrown off during the dry season. The sympodial shoot system is shown in Fig. 6.13. The taproot can reach groundwater if it is no deeper than 3.5 m. Water is obtained otherwise by means of horizontal roots which can reach a length of up to 20 m.

The number of shrubs on firm sand is 100 to 300 per hectare, on mobile sand only 50 to 100 per hectare. The annual growth of the shoots amounts to 25 to 100 kg ha^{-1}. The most important species are *C. caput-medusae* (height: 2.5 m) and *C. setosum* (height: 1.2 m). The age-structure of a stand of *C.*

Fig. 6.11. *Calligonum arborescens* along the railroad embankment 5 km from the Amu-Dar'ya River (photograph by E.A. Bessey, in Walter, 1976, fig. 54).

setosum on drifted sand in the southern part of the central Karakum was as follows:

Age (years)	Number (%)
1	8
2	3
3–5	1
6–10	14
11–15	20
16–20	41
>20	11
dead	2

One can see that reproduction by seeds occurs only in favorable years. A shrub produces about 14 200 nutlets.

Salsola richteri (Chenopodiaceae) forms shrubs 1.5 to 2 m tall, of which the upper 20 to 30 cm long, flower-bearing parts die back annually. Photosynthesis occurs by means of brachyblasts which develop at the base of perennial shoots and have already been dropped before the fruit are ripe. In good years the shrubs grow primarily vegetative shoots, in dry years primarily generative shoots. The roots reach a depth of 3 to 4 m.

This species is one of the subdominants. One

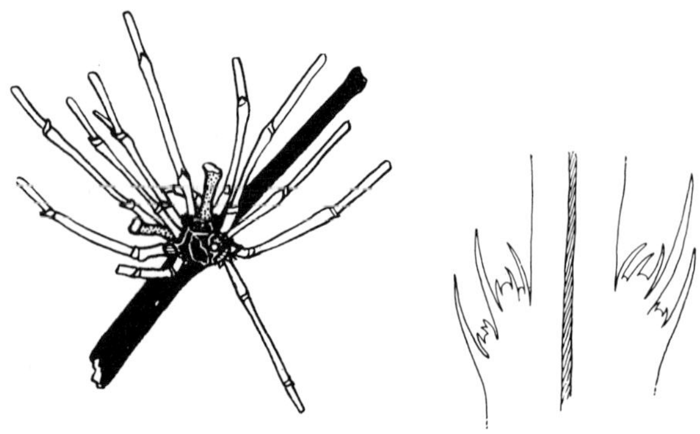

Fig. 6.12. *Calligonum rubens*: 1 year old, green, dense shoots on a 4 year old branch; the leaves are only small scales. Right: longitudinal section through a shoot bud, with primordium for lateral shoots (from Nechayeva et al., 1973, in Walter, 1976, fig. 55).

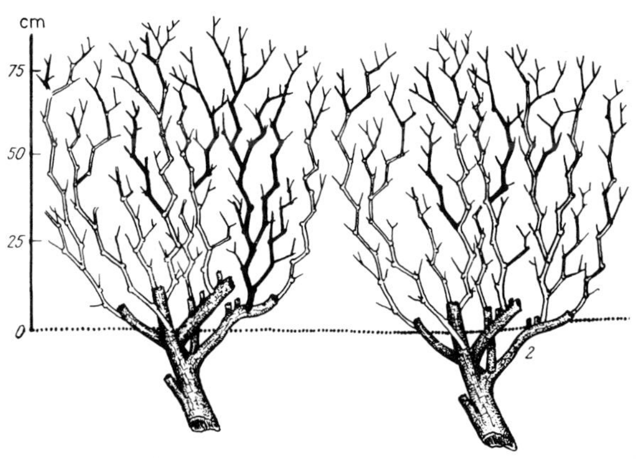

Fig. 6.13. *Calligonum setosum*: formation of new shoots (black) after a moist spring (old branches removed); notice the sympodial shooting structure (from Nechayeva et al., 1973, in Walter, 1976, fig. 50).

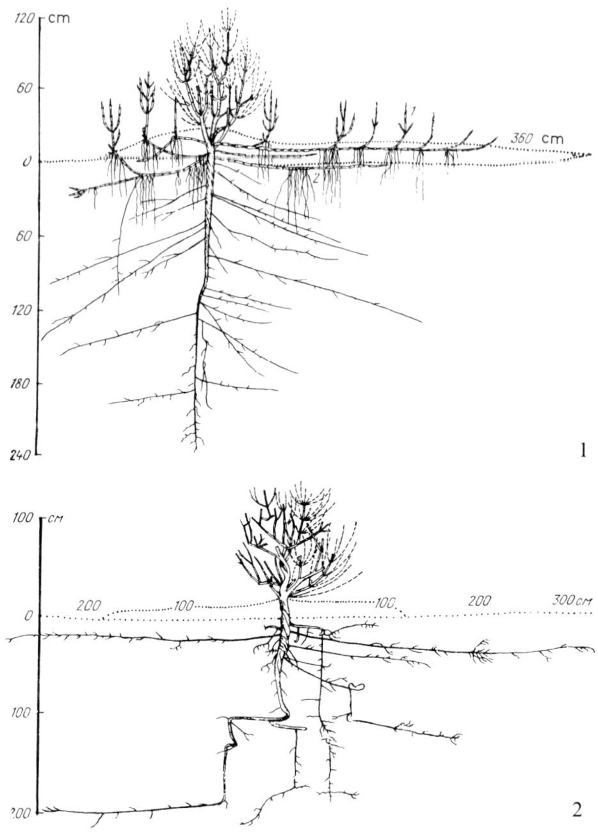

Fig. 6.14. *Ephedra strobilacea*. Above: root system and formation of runners on a 10 year old plant, the sand substrate of which has been blown out (*1* = 1 year old shoot on a young runner; *2* = 1–3 year old runners). Below: trunk-forming, roughly 20 year old plant, the sand substrate of which was blown away during the juvenile stage (from Nechayeva et al., 1973, in Walter, 1976, fig. 57).

finds 25 to 100 individuals per hectare. The annual above-ground growth can be 30 kg ha^{-1}.

Other important shrubs are the following: *Salsola arbuscula* (height: 1 m), with a hint of trunk (10 cm high), and other *Salsola* species; *Aellenia subaphylla* (Chenopodiaceae); the lightly-foliated *Astragalus paucijugus*; and *Ephedra strobilacea* (height: 1–2 m). The last species can become 100 years old and occurs either as individual shrubs or in stoloniferous form (Fig. 6.14). On slightly mobile sands one can find dense stands of 400 to 500 individuals per hectare with an annual growth of 0.4 to 0.5 t ha^{-1}. This is a valuable fodder plant.

Semi-shrubs

These include the endemic *Smirnovia turkestana* (Fabaceae) with small, summer-deciduous leaflets; various *Artemisia* spp.; many *Astragalus* spp.; and small *Salsola* spp.

Herbaceous species (forming several synusiae)

Carex physodes, with its long underground rhizomes, is the most important turf-forming species on immobile, hillocky sand areas of the Karakum. The number of shoots reaches 600 to 700 per square meter. The total phytomass, with rhizomes, is 360 to 780 g m^{-2} or 3.6 to 7.8 t ha^{-1}. The underground

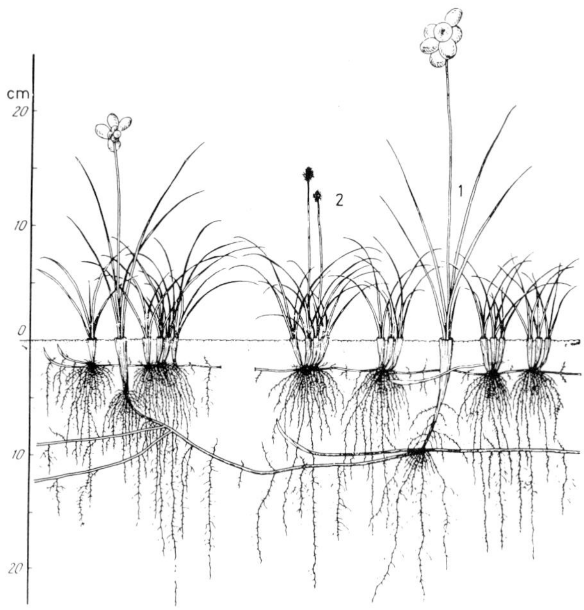

Fig. 6.15. *Carex physodes* with vesicular utriculi (*1*) and, for comparison, *Carex pachystylis* (*2*) in the *Poa bulbosa–Carices* biogeocoenosis of the Badkhys region (from Nechayeva, 1973, in Walter, 1976, fig. 58).

mass, is, of course, 10 to 18 times greater than the above-ground mass. This species, called *rang* and also *ilak*, is the most important forage for sheep and is eaten readily in both the green and dried state. Its mode of growth is shown in Fig. 6.15. *C. physodes* on sandy soil is closely related to *C. subphysodes* on sandy gray-earths and *C. pachystylis* (*C. hostii*) on loess soils of the foothills.

C. physodes was considered to be a poikilohydrous species (see Vasilevskaya, 1954, 1965) but Vasilevskaya has shown this assertion to be false. The leaves die in May but remain on the plant until September or even the next spring. Only the rhizomes, with the active meristems at a depth of 2 to 3 cm, survive the drought period and the winter. The plants grow to a height of 15 to 40 cm. The scarcely branched rhizomes lie at a depth of 12 to 20 cm. The main root mass is found 7 to 22 cm below the soil surface, but some roots penetrate to 120 cm. In April, after it rains, numerous 0.3 to 1 cm long, white ephemeral rootlets form around the base of the runners. When the soil dries out again after 5 to 15 days, they die. There are, however, also longer, thin, but also short-lived roots.

C. physodes comes out immediately after the first rain in autumn (earliest date: October 20; latest date: November 20) and holds on through the winter. The growing season varies from 90 to 215 days. The yield of dry mass is strongly correlated with the amount of rainfall in November to May: 70 kg ha^{-1} for 50 mm rain, 160 kg ha^{-1} for 140 mm. Even under the most favorable conditions, however, the highest yield does not exceed 200 kg ha^{-1}. After one dry year, the yield in the next year is relatively good; after two consecutive dry years it is low. As a result of its very extensive rhizomes, this species is a very successful competitor with the spring ephemerals and the seedlings of woody species. In dry years it definitely suppresses the others. It is a very useful species for sand stabilization. On slightly mobile sand it can advance 0.5 to 1 m in a year. On highly mobile sand the primary colonizers are the perennial species *Heliotropium arguzoides* and *Tournefortia sogdiana* (both Boraginaceae). Their production reaches 45 to 94 g m^{-2}, which is probably related to the lack of competition.

The most important stabilizer of shifting sand is *Aristida karelinii*. The spring geophytes are also represented by many species in sand areas but are not as important as in loess soils. The most noteworthy are *Allium* spp., *Gagea* spp., *Tulipa sogdiana*, *Rhinopetalum arianum*, *Eremurus inderiensis* (all Liliaceae), *Iris longiscapa*, *I. songarica*, the araceous *Eminium lehmannii*, and some dicotyledons with bulbs (*Rheum turkestanicum*; the umbellifer *Schummania karelinii*; and the meter-tall *Dorema sabulosa*). *Cistanche flava* (Orobanchaceae), which parasitizes *Calligonum* species on the shifting sands, is also interesting. A plant which weighs 7 to 10 kg may consist of up to 94% water.

The phenology of *Carex physodes* is shown in Fig. 6.22.

Annual species

These are represented by 143 species, of which 98 are spring ephemerals and 45 are summer annuals. Only 24 occur in large numbers. The winter annuals in this number (species of the genera *Ceratocephalus*, *Alyssum*, *Meniocus*, *Veronica*, etc.) are actually spring ephemerals. Others hold on to May, such as various Poaceae, Caryophyllaceae, Papaveraceae (*Hypecoum*, etc.), Brassicaceae, Fabaceae, Geraniaceae, Boraginaceae and Asteraceae. We have already mentioned the fluctuation of the species composition from year to year. The longer the species need for their fruit to mature in the spring, the deeper their roots penetrate, though a depth of 20 cm is seldom exceeded. Only the summer annuals, which germinate in the spring and grow on into the summer, are able to penetrate with their roots to a depth of 70 cm. In unfavorable years the summer annuals scarcely develop. The chenopods *Agriophyllum latifolium* and *Horaninovia ulicina* can expand on highly mobile sand — that is, in the absence of competition. They germinate between March 1st and April 6th, produce three long roots, and, as a result of the xeromorphic structure, hold on for six months.

In summary, one can say that the most important component of the vegetation is that formed by the woody plants. These are characterized by long annual shoots (1–2.5 m) and by aphylly or microphylly, which reduces the transpiring surface to a very small area, despite the denseness of the vegetation. The leaf-area index lies considerably below one. This is understandable because the woody species must survive the drought season. They can further reduce water losses by dropping the small leaves or by the dying-back of transpiring shoots (suffrutescence). The fact that transpiration values, both per square decimeter of leaf area and per gram of fresh weight, are sometimes very high for desert plants (only during the favorable season, of course) does not imply anything at all about the degree of adaptation of these plants to their environments. For the water balance of plants it is more important to compute the transpiration per unit of ground surface covered by the individual plant or by plant communities. Only then can the water loss be related to precipitation, considering, of course, that, depending on soil type, a varying fraction of the precipitation is lost to the plants through evaporation from the soil surface.

For fully-grown plants of the individual species average values of above-ground phytomass (dry weight) and percentage fraction of annual shoots are given in Table 6.2. These figures should at least provide a general idea of the size of the plants and the approximate above-ground production.

The weights for individuals of winter annuals lie under 1 g, for summer annuals between 0.3 and 32.6 g, and only for the long-lived annuals, *Salsola paulsenii* and *Horaninovia ulicina* (both Chenopodiaceae), does the phytomass (dry weight) reach 128 to 500 g.

TABLE 6.2

Above-ground phytomass (kg) and percentage as annual shoots (* = major species collected in the vicinity of Repetek Station by Togyzayev, unpublished) (NW = NW Turkmenistan)

	Above-ground phytomass	Percentage annual
Trees		
**Ammodendron conollyi*	8 kg	6
**Eremosparton flaccidum*	2 kg	6
**Haloxylon aphyllum*	152 kg	8
Shrubs		
**Haloxylon persicum*	18 kg	8
Haloxylon persicum (NW)	5.9 kg	8
Salsola arbuscula	0.8 kg	12
Calligonum rubens	3.8 kg	15
C. setosum	2.3 kg	15
**Ephedra strobilacea*	8.6 kg	19
Ephedra strobilacea (NW)	3.8 kg	20
Dwarf-shrubs		
Limonium suffruticosum	550 g	20
Semi-shrubs		
Convolvulus erinaceus	650 g	31
Astragalus longipetiolatus	330 g	32
Salsola orientalis	170 g	32
S. gemmascens	160 g	33
Mausolea eriocarpa	280 g	34
Smirnovia turkestana	545 g	35
Small semi-shrubs		
Artemisia kemrudica	120 g	39
A. kemrudica (NW)	90 g	37
Reaumuria turkestanica	120 g	62
Aellenia glauca	90 g	65
Acanthophyllum stenostegium	40 g	65
Convolvulus korolkovii	30 g	70

Ecological studies on individual plants

Transpiration measurements by the quick-weighing method, with cut branches, were performed in the Karakum Desert by Vasilyev (1931) and by Arcichovski and Opsipov (1931). The experiments of Vasilyev, which were published in German, aroused considerable attention because of

TABLE 6.3

Transpiration values [g (g fresh wt.)$^{-1}$ h^{-1}] in June, in the morning, according to Kokina (1935) and Vasilyev (1931)

Species	Kokina	Vasilyev
Smirnovia turkestana	1.075	2.93
Ammodendron conollyi	0.876	1.99
Haloxylon aphyllum	0.305	0.87
Haloxylon persicum	0.281	0.43
Salsola subaphylla	0.073	0.20

the high values, especially the value for the endemic *Smirnovia turkestana*. Kokina (1935) checked these values a few years later and was able to show that they were two to three times too high. Kokina assumed the reason for the over-estimation to be the use of twigs of only 0.5 g and an insufficiently long exposure period. The most important table from Kokina's results, published in Russian (Kokina, 1935), is reproduced here as Table 6.3 (see also Walter, 1968, pp. 771–774). The highest transpiration values are attained in June, but the daily maximum is reached by 10 to 11 h. With increasing drought in the summer, a reduction in transpiration becomes noticeable. The leaf area is also reduced. Setting the June transpiration values of the different species equal to 100%, one sees that the transpiration at the end of July has dropped to 76.6 to 95.8% and in September to 42.2 to 66.6%.

In spite of this, one must characterize the water supply of the Karakum plants as relatively good, even during the drought period. This is substantiated also by the measurements of water deficit by Bobrovskaya (1969a, b). The deficits obtained are lower than those typical for steppe species. The deficits for desert species, measured at the sites, do not exceed 25%, while sublethal values are in the range 43 to 53%. Thus, the threat of water shortage does not normally exist.

The potential osmotic pressure increases from spring through the drought period by about 1.5 to 2 times (i.e., the osmotic potential decreases), which leads to the formation of smaller, more xeromorphic summer leaves or to the dropping of transpiring organs (see Walter, 1972; Walter and Stadelmann, 1974).

It is the view of the Russian researchers that the scant precipitation in the Karakum Desert is not sufficient, at these high transpiration rates, to guarantee the plants the necessary water supply. It is, therefore, assumed that additional water finds its way into the soil through condensation of water vapor out of the atmosphere and/or through the groundwater. The condensation is presented to take place in the temporarily cooler soil layers.

Blagoveshchenskiy (1954a) carried out the necessary experiments in the Karakum Desert and, further north, on Lake Balkhash. He was successful after five years in showing small amounts of water uptake (a few tenths to 1%) which could be the result of condensation (see Walter, 1968, pp. 154–156).

These experiments are not convincing. There can be no doubt that a temperature-conditioned vapor pressure gradient in the soil will cause diffusion and condensation of water. With sufficiently sensitive methods this can surely be shown. An ecologist, however, is more interested in how large this condensation would be in relation to the water needs of the plants. All condensation amounts which have been measured are too small to be of any ecological significance.

Indeed, Blagoveshchenskiy (1954b) does show chrono-isopleths of soil water content in the upper 6 m of a sandy soil which seem to show that the change in water content at a depth of 3 to 5 m is not related to the water content of the higher layers. But even for a sandy soil, gradations of water content of 0.5% must lie within the limit of error, which is surely greater than 1%. The samples were taken only once per month. After heavy rains water can also sink in very quickly along old root channels and be stored at greater depths. No information, however, was given on rainfall amounts.

If his assertion were correct, Blagoveshchenskiy would be right that plants could grow in the rainless southern Libyan Desert from water condensed in the soil out of the air. But the assertion is wrong. The authors know the stretch from Asyut on the Nile to the oases Al-Khargha and Baris. Along this entire stretch of over 200 km, not a single plant was sighted except around oases with artesian water.

Our position on water condensation in the soil holds also for the question of water uptake from humid air at night by cut plant branches, which Sveshnikova (1972) observed in the Karakum Desert. She found the greatest water uptake, 12.5%

of the original weight, to be by *Tamarix*, with lower values of 2 to 4% or even water losses for the other species. This is a purely physical phenomenon which is almost always without ecological significance. *Tamarix* excretes NaCl and other salts by means of special salt glands, so the shoots appear a dusty gray. These salts must absorb water from wet air and then run off — it is not at all a question of dew. Concentrated solutions of sodium chloride have an osmotic potential of about −360 bar. Plant cells can obtain no water from such solutions. The solution usually drips off. It has no ecophysiological significance.

The outer layers of the shoots of non-salt excreting species lose considerable water (from epidermis, bark, and cork), due to the dry desert air. At night these layers take up water again from the wetter air and swell. The amount of water uptake depends entirely on the structure of the outer walls of these shoots and is also of little ecological significance. Only when the water potential of the cells inside the plant organs is very low (high suction) can water uptake into the interior of these organs be shown. This is a very slow process, whereas, in the experiments mentioned above, the water uptake took place during the first three hours. It has not yet been shown in any case that this process has practical ecological significance. Plants which transpire slightly at night have a vapor-pressure gradient toward the atmosphere — water uptake cannot take place.

Miroshnichenko (1975b) did not consider, during the assessment of water content in sandy soils, that the distribution of water under a plant cover is very heterogeneous.

The question arises whether the assumption of a special water source for the water supply of the plants is even necessary. One can by careful investigation show that there is water in the root zone of desert plants everywhere. This can come from rainfall which fell in wet years and was protected in the soil from evaporation, remained stored for many years, and is sufficient for the water consumption of the few plants that are present. The high transpiration values of the leaves, measured per square decimeter of leaf area, do not by themselves imply anything about the total water needs of the plant cover. One must at least consider the very low leaf-area index in the desert.

In fact, more recent studies (see Nechayeva, 1975) have shown that the amount of water consumed by the vegetation, in millimeters per unit area, is significantly lower than the amount actually present in the soil.

The plants of biogeocoenosis I, growing on drifted sand, transpired during their growing season a total of only 37.4 mm. The small trees of *Ammodendron conollyi* showed the highest water consumption (21.6 mm, 57% of the total), followed by the shrubs of *Calligonum arborescens* (13.7 mm, 36.6%). All the plants of *Aristida karelinii* together used only 2.1 mm (5.7%), although their daily transpiration per gram of fresh weight is, at 20–21 g, very high.

In the upper 2 m of sand in biogeocoenosis I, there were 83 mm of water stored at the beginning of the growing season. During the growing season, 49 mm of the stored water was lost to evapotranspiration, leaving 34 mm in the autumn. The rain water stored in the soil is, thus, fully sufficient to cover the transpiration of the plants.

The situation in biogeocoenosis II, with *Haloxylon persicum* and *Carex physodes*, is somewhat less favorable. The amount of transpiration here was 47 mm. Of this, the shrubs used 30 mm (64%), *Haloxylon* alone used 16 mm. The transpiration of *Carex physodes* required a similar amount (17 mm, 35%).

The water supply in the top 2 m at the beginning of the growing season was 62 mm, the decrease was 54 mm, so the remainder in the autumn was only 8 mm. The water supply is, thus, even in this case, just enough.

The situation in biogeocoenosis III is different. This biogeocoenosis includes the *Haloxylon ammodendron* stands in the dune valleys, which also have a light undergrowth of *Carex physodes* and some halophytic summer annuals. The daily transpiration rate per gram of fresh weight is very high here also: *Haloxylon ammodendron* 7 to 9 g, *Ephedra strobilacea* 4 to 4.5 g, *Carex physodes* 15 to 16 g, summer annuals such as *Suaeda* 2 to 3 g or *Kochia* 5 to 6 g. The computation of transpiration per unit of surface area shows that the greatest losses are caused by the shrubs. The total transpiration was 149 mm, that of the shrubs 138 mm (92.5%). *Haloxylon* alone accounted for 108 mm, with the light *Carex* stand needing only 10.6 mm (7%).

At the beginning of the growing season there

were only 76 mm of water stored in the upper 2 m of the sandy soil. This is far short of the transpiration need. In the dune valleys, though, *Haloxylon* can reach the capillary fringe of the groundwater with its deep roots and, thus, largely cover its transpiration needs through uptake from the groundwater. If 100% of its water needs came from groundwater, then the transpiration of the other plants (41 mm) could easily be covered by the water stored in the upper 2 m of soil. There is still enough left that *Haloxylon* can use some of this.

These studies show incontestably that the water supply of all sand-desert plants is guaranteed and that the assumption of special condensation processes or other uptake of water, as by shoots directly from the air, is totally unnecessary.

That the plants of the Karakum Desert do not suffer from water shortage is corroborated by the recent explanations of Sveshnikova (1975).

We refer the reader to this English-language publication and also take some important values (Table 6.4) from the summary of the studies at Repetek during 1965–69 (Anonymous, 1975, table on p. 143).

The water deficits of the species in the Karakum are always low and lie far below the experimentally determined lethal values. Moreover, the potential osmotic pressure (−osmotic potential) is continuously high only for the halophytes (*Haloxylon* spp., *Salsola*). For the other species, only the maximum value during the drought is higher. The transpiration rates, per gram of fresh weight, are quite different. As is normal, the values are lower for the larger woody plants than for semi-shrubs and herbs. The values are reduced during drought. The values of relative transpiration (related to the evaporation from an open water surface) are usually given as 1/4 to 1/18. The ephemerals, which only root in the top soil layers where there is sufficient water in spring, show no special adaptations (Sveshnikova, 1975).

Numerous transpiration experiments were also performed by Nechayeva and Savinkin at the experimental station Karrykul near Ashkhabad (Anonymous, 1975, pp. 145–56). These gave similar results.

For ecosystem research, however, the water losses of the entire soil-water supply, in mm, are more important. These are low, as we have seen, because the density of the plant cover is low and the plants have adapted to the meager water supply by reducing their transpiring leaf areas. Many species

TABLE 6.4

Data on the water budgets of dominant species in the Karakum Desert, Repetek (Sveshnikova, 1975)

Growth forms and species	Transpiration level [mg (g fresh wt.)$^{-1}$ h^{-1}]		Water content (% of fresh weight)	Osmotic potential (bar)		Water deficit (% of water saturation)	
	maximum	typical		minimum	maximum[1]	natural	sub-lethal
Trees and shrubs							
Haloxylon ammodendron	670	100–400	62–80	−58	−46	14	47–50
Haloxylon persicum	850	100–700	50–66	−58	−38	21	48–53
Calligonum caput-medusae	780	100–500	60–76	−41	−16	19	45–48
Salsola richteri	760	100–400	55–76	−56	−31	24	46–48
Ammodendron conollyi	810	100–600	51–76	−50	−24	19	43–46
Semi-shrubs							
Astragalus paucijugus	1540	680–770	63–83			23	45–49
Smirnovia turkestana	2600	600	57–81			33	50–66
Herbs							
Aristida karelinii	930	240					
Heliotropium arguzioides	1750	780	52–89			33	50–67
Carex physodes	2760	600					

[1] Values at optimal water availability (juvenile values still lower).

are leafless, such as *Calligonum* species, *Eremosparton*, etc. *Salsola* and *Astragalus* species, among others, have very small leaves. *Acanthophyllum* has thorns in place of leaves, while *Smirnovia* drops its leaves early. Vasilevskaya (1954) has concerned herself, in more detail, with the development of xeromorphic leaves.

It was determined for *Haloxylon ammodendron* that, on the average, 19% of the rain water is intercepted by the crown and 14% runs on down the trunk. (The corresponding values for *H. persicum* are 16% and 9%.) The water which runs down the trunk can penetrate along the tap root more deeply into the ground than otherwise. This makes it easier for roots to grow deeper, usually in wetter years, and to reach the groundwater. The continued existence of the tree is then assured even in drought years.

Kostychev and Kardo-Syssoleva (1930) studied the photosynthesis of the Karakum plants, but their results cannot be evaluated because the error due to overheating of leaves in the measuring chamber was at that time not yet appreciated. Reviews in English of newer photosynthesis studies can be found in Rodin (1972, pp. 8–27). Zakharyants studied 225 species in the Kyzylkum Desert. Nasyrov, Lebedeva, and Kichitov refer to the fact that, in the ecosystems of the Karakum, organic matter is produced all through the year because of the constant changing of the synusiae. Fazylova gives values of 200 to 700 lux for the light-compensation point and 20 000 to 70 000 lux for the light-saturation point of the species there. Naaber considered the chlorophyll content of the various desert plants. Aleksandrov, Durdyev, Shukhtina, and Yaskulyev studied the problem of heat-hardening associated with a certain degree of inactivity. Reinus showed that the difference between air temperature and the temperature of assimilating virgate shrubs is relatively small, even under high insolation, and not more than 2.7°C for *Aristida karelinii*.

Summaries of the photosynthesis studies of 1965–1972 in the Karakum and southern Kyzylkum were given by Zalenski and Voznesenskiy (Anonymous, 1975, pp. 148–152). The main result was that the desert plants show positive photosynthesis all through the summer, even during the drought. This was to be expected from the balanced water budget. Only on very hot days and only for ephemerals and only slightly xerophytic species of the Kyzylkum, could a severe depression or cessation in the assimilation of carbon dioxide be ascertained.

The highest photosynthesis levels were observed in the spring, but for some species the maxima occurred in the autumn, always at the time of flowering. The most active were the ephemerals and some semi-shrubs. The enduring but slowly developing chenopods showed low photosynthetic rates. Light saturation occurred for ephemerals at 50% of direct sunlight but at 66% for perennial species. The temperature optimum for net photosynthesis was around 25 to 30°C and the maximum around 40 to 45°C, both of which were somewhat lower in the spring. *Haloxylon persicum* showed a rather even photosynthetic rate over the entire range from 10° to 45°C. The lowest optimum temperature was found to be for *Ephedra strobilacea*, a species which can probably assimilate also in winter.

The highest rate and productivity of photosynthesis are shown by those species which only develop in the spring. This is compensated by the short duration of their growing seasons.

The highest potential respiration is found among the ephemerals and the legumes, always at the time of flowering. The values in Table 6.5 and 6.6 should give an idea of the quantitative relations. The reference unit was different for species of the Karakum and for those of the Kyzylkum.

It is assumed that most of the woody species have C_4-photosynthetic systems, but there is no information concerning this question.

Michelson (1955a, p. 156) portrayed, in a phenological diagram, the spectra of 48 species of different life forms. Perskaya (1955) considered the morphology and ecology of the seeds and fruits of sand-binding tree and shrub species. Germination capacity is retained by legumes for many years under laboratory conditions. The germination capacity of *Calligonum* species was retained for only two years, and that of *Haloxylon* species and *Salsola richteri* decreased considerably within the first year. Preliminary treatment was necessary in order to induce the germination of the hard-shelled legumes.

TABLE 6.5

Data for the gas-exchange of Karakum Desert plants (Voznesenskiy, 1975)

Species	Intensity of apparent photosynthesis [mg CO_2 (g dry wt.)$^{-1}$ h^{-1}]		Productivity of photosynthesis [mg CO_2 (g dry wt.)$^{-1}$ day^{-1}]		Optimum temperature for apparent photosynthesis (°C)	Respiration intensity	
	average	maximum	average	maximum		(°C)	[mg CO_2 (g fresh wt.)$^{-1}$ h^{-1}]
Ammodendron conollyi	7.6	20	110	160	22–35	30	1.2
Haloxylon ammodendron	7.1	10	80	100	10–30	33	1.2
Haloxylon persicum	6.1	10	55	85	10–45	30	1.1
Salsola richteri	6.0	18	120	160	20–30	30	0.6
Calligonum caput-medusae	5.8	12	80	125	25	30	0.7
Ephedra strobilacea	2.3	4	30	35	5–20	–	–
Astragalus paucijugus	14	22	170	240	15–23	35	0.7
Aristida karelinii	21	52	250	460	30–40	32	1.6
Heliotropium arguzioides	8.6	24	170	215	15–30	35	1.1
Carex physodes	14	29	205	220	17–30	–	–

TABLE 6.6

Data for the gas-exchange of Kyzylkum Desert plants (Voznesenskiy, 1975)

Species	Intensity of apparent photosynthesis (mg CO_2 dm^{-2} h^{-1})		Productivity of photosynthesis (mg CO_2 dm^{-2} day^{-1}) (maximal)	Optimum temperature for apparent photosynthesis (°C)	Respiration intensity	
	average	maximum			(°C)	[mg CO_2 (g fresh wt.)$^{-1}$ h^{-1}]
Salsola orientalis	10	23	200	15–35	22–25	1.1
Calligonum aphyllum	25	32	300	–	22–25	1.3
Atraphaxis seravschanica	–	24	–	20–35	22–25	1.6
Aellenia subaphylla	20	22	270	15–35	22–25	0.5
Eurotia eversmanniana	–	40	–	15–35	22–25	0.9
Artemisia turanica	30	50	400	–	22–25	0.4
Alyssum desertorum	–	47	200	20–30	15	0.9
Malcolmia africana	–	27	230	25–30	15	0.6
Astragalus filicaulis	20	51	310	15–30	–	–
Ferula assa-foetida	28	30	340	–	22–25	0.8
Veronica campylopoda	–	25	–	30	15	0.3
Lamium amplexicaule	–	24	150	25	22–25	0.9

The plant cover as the basis for ecological subdivision

In the Karakum Desert one can distinguish between the vegetation of the sand regions, the clay patches, and the saline soils, each of which consists of a series of plant associations which correspond to biogeocoenoses. These can be made to comprise three large biogeocoenosis complexes: the psammophyte, takyr, and halophyte biogeocoenosis complexes. It is, thus, the edaphic characteristics, which are always especially important in the desert, that are characteristic for the individual biogeocoenosis complexes. One could also call these pedobiomes, but, as already mentioned, a clear differentiation between zonal and azonal vegetation, that is, between zonobiomes and pedobiomes, is very difficult in deserts.

Psammophyte biogeocoenosis complexes

The sand sea of the Karakum, with its characteristic vegetation, stretches in endless monotony over hundreds of kilometers from the Unguz to the Kelif

Uzboy (Fig. 6.16). The comparison with the sea is very accurate, since the surface is not smooth but instead seems to be undulating. The sand waves are either elongated low sand ridges or irregularly distributed dunes with sparse plant cover. The changing relief, with raised areas and with depressions or lower-lying plains between them, implies a certain classification of the vegetation ranging from the more poorly covered hills to the densely covered low areas. We can thus speak of a biogeocoenosis complex determined by soil catenas (Fig. 6.17):

(1) The highly mobile tops of the dunes carry only a few pioneer species.

(2) On the upper part of the slopes the sand is already less mobile, so a number of typical woody plants can form sparse stands and stabilize the sand.

(3) The sand of the lower part of the slopes is immobile, and the white saksa'ul, with many accompanying species, forms denser vegetation.

(4) In the lower parts of the relief the sand is covered, in addition, by the sand sedge (*Carex physodes*).

(5) In addition, there are the valley-like hollows, which we will discuss separately.

The boundaries between 1 and 2, as well as between 3 and 4, are indistinct. Thus, it is more practical to distinguish only two biogeocoenoses: an *Ammodendretum aristidosum* on mobile sand (I) and a *Haloxyletum persici* on immobile sand (II). The open and scattered vegetation of the desert is, in general, very difficult to subdivide into sharply delimited communities[1]. The varying relief causes a certain mosaic structure.

The areas with shifting dunes (individual barkhans and barkhan fields), which one finds around the station at Repetek and along the Amu-Dar'ya, form a special vegetationless biogeocoenosis with a psammofauna which has not yet been studied in any detail[2].

[1] A number of more narrowly conceived associations of the sand vegetation are described by Michelson (1955a, b) and by Blagoveshchenskiy (1955).

[2] A similarly rich fauna based on organic material blown in from the neighboring areas was identified in the extensive, vegetationless dune region of the southern Namib Desert in southwestern Africa (Koch, 1961; Kühnelt, 1965). The situation there is more favorable, however, as a result of the very frequent fog which wets the top sand layers, giving the microfauna the possibility to take up the necessary water. In the Karakum Desert this is possible only in the spring. In such ecosystems without producers, the organic material blown in by the wind accumulates in the dune valleys.

Fig. 6.16. Typical picture of the vegetation of the central Karakum Desert, with *Haloxylon persicum* shrubs. *Carex physodes*, in the herb layer, is not evident during the dry period (photograph by Petrov; from Walter, 1968, fig. 535).

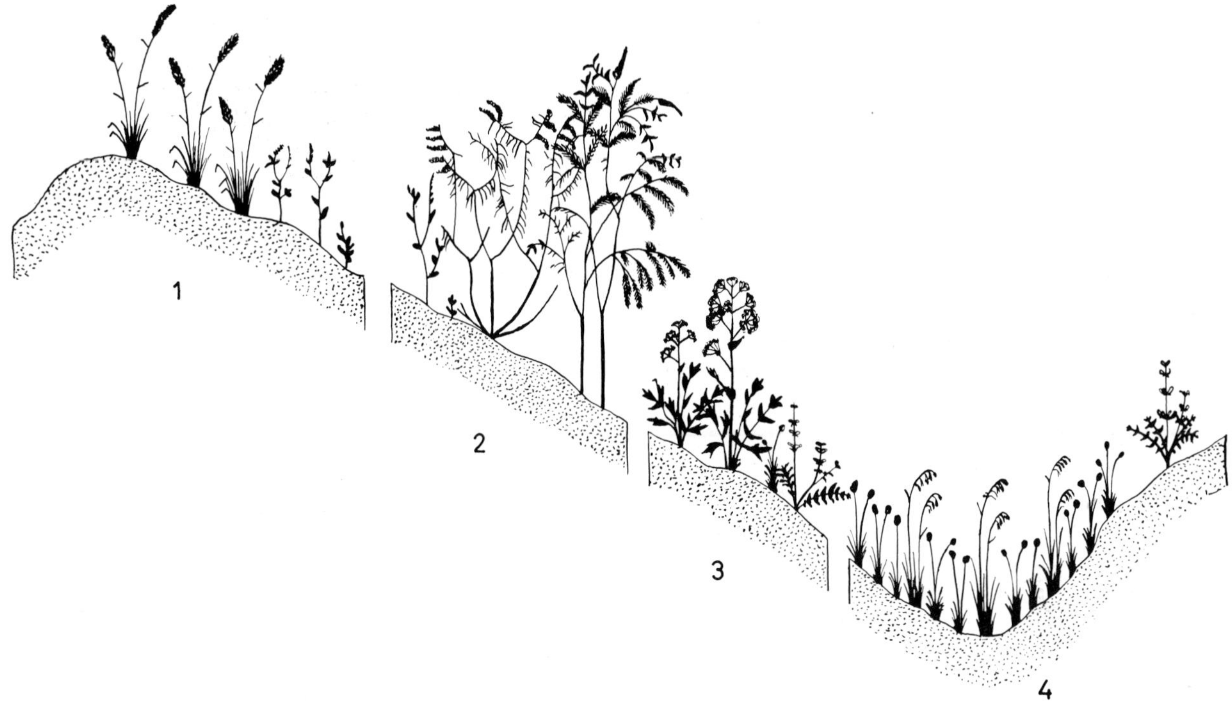

Fig. 6.17. A catena (ecological gradient, biogeocoenosis complex) in the hilly sand-desert region of the central Karakum (from Korovin; in Walter, 1968): *1* = pioneer stand on the mobile sand of dune ridges, with *Aristida karelinii*, etc.; *2* = upper dune slope, with still mobile sand and shrubs (*Ammodendron conollyi*, etc.); *3* = lower slope, with little mobile sand and with *Haloxylon persicum* (not shown) accompanied by various herbs (*Dorema*, *Astragalus*, etc.); and *4* = sand completely covered by a closed turf of *Carex physodes* (see also Fig. 6.19).

The series *1* to *4* in Fig. 6.17 represents an ecological gradient or successional sequence which one can observe with decreasing mobility of the sand.

The first pioneer species to appear on highly mobile sand in the Karakum region is *Aristida karelinii*. (The larger grass *Elymus giganteus* has its southern limit around the Aral Sea.)

Aristida karelinii forms clumps 1 to 1.35 m tall with diameters from 0.8 to 2.6 m. The orthotropic shoots grow from the end of February to the beginning of December; blooming continues from the end of April until October, but with a minimum during the summer drought. The fruiting stems and leaves die in autumn; the main stalks survive the winter and develop again in the spring. The stalks grow for two or three years, the entire clump for fifteen to twenty years. This species tolerates being covered by blown sand and always grows outwards (Fig. 6.18). The root system is very extensive in the sand. The 40 to 350 adventitious roots grow out radially and reach a length of 10 to 15 m. They penetrate up to 2.2 m deep into the ground, but most are found in the first 15 to 60 cm. The root-

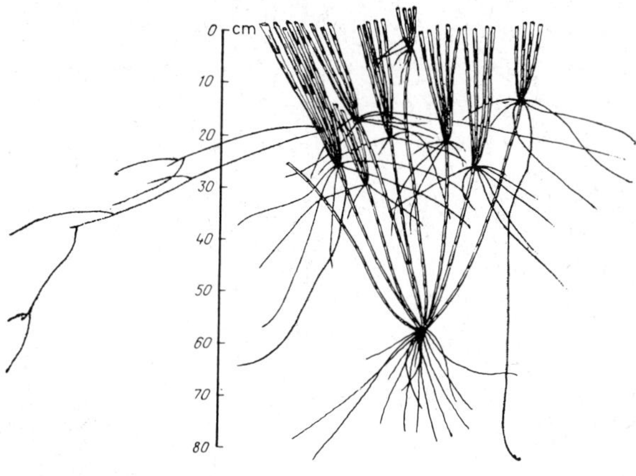

Fig. 6.18. Formation of adventitious roots at a higher level after being covered by sand: *Aristida karelinii*. Earlier level of the sand surface was at a depth of 55 cm, a later level at 20–30 cm (from Nechayeva et al., 1973, in Walter, 1976, fig. 61).

tips are thickly covered with root-hairs, to which the sand grains adhere. The root-cortex dies early, and only the guiding central cylinder remains. It is surrounded by a loosely fitting sheath, which is made up of the dead rhizodermis, with the root-hairs, and the attached sand grains. One can find the same phenomenon among the *Aristida* species in South-West Africa (Namibia) (Walter, 1939, pp. 756–757). This structure forms a sort of protective sheath for the roots once the sand dries out.

The mobility of the sand is reduced by the extensive root system and by the sand-covered shoots, so that other species can also colonize.

The above-ground phytomass of the *Aristida* stands reaches 6.3 to 8.9 t ha^{-1}, and the total phytomass reaches 12.2 to 17.3 t ha^{-1}. The number of clumps can reach 3800 to 6400 per hectare. Table 6.7 shows the age-structure of populations in protected and grazed areas. This grass is not preferred, but can be used as an emergency forage for sheep in the winter.

A second smaller species is *Aristida pennata* (height 40–65 cm, clump diameter 40–80 cm). It is found more on small, mobile sand patches in the hilly parts of the Karakum. The occurrence of this species on grazed areas is a sure sign of overgrazing.

The still smaller *Aristida plumosa* grows only on immobile sand.

The next pioneer species, after *Aristida karelinii*, are *Heliotropium arguzioides* and *Tournefortia sogdiana* (both Boraginaceae). These spread very quickly, the first by means of horizontal roots with adventitious shoots at a depth of 20 cm, and the second by means of underground runners at a depth of 10 cm. These species also tolerate covering by sand. They are especially typical in blown-out depressions. Their roots reach 1 to 2 m deep. Both species hold the sand together in summer, after the annual species of the open sands, *Agriophyllum latifolium* and *Horaninovia ulicina* (both Chenopodiaceae), have died. *Heliotropium* and *Tournefortia* provide summer forage. The phytomass of such stands is 1.2 to 1.4 t ha^{-1}, of which 18 to 33% of the total is represented by accompanying species.

If the sand movement is diminished, then seedlings of woody species can often gain a foothold in the shelter provided by the grass clumps. With this begins the sere represented by open shrub and tree stands on only slightly mobile sands. The following species are important:

Woody plants: *Ammodendron conollyi, Eremosparton flaccidum, Salsola paletzkiana, S. richteri, Smirnovia turkestana, Ephedra strobilacea, Calligonum arborescens, C. caput-medusae, C. eriopodum, C. densum, C. erinaceum, C. microcarpum, Astragalus ammodendron*, and singly *Haloxylon persicum*.

Perennial herbs: *Aristida karelinii, Convolvulus erinaceus, C. divaricatus, Acanthophyllum elatius, A. korolkovii, Artemisia diomoana, Heliotropium arguzioides, Tournefortia sogdiana, Dorema sabulosum*.

Annual herbs: *Agriophyllum latifolium, A. minus, Horaninovia anomala, H. ulicina, H. minor, Corispermum lehmannianum, C. papillosum*, and *Cutandia memphitica*.

Usually only 10 to 12 species appear on a given site.

One can also speak of two synusiae in this biogeocoenosis, one on highly mobile sand with *Aristida*, etc., and a second on slightly mobile sand with *Ammodendron*, etc. Possibly one can distinguish a third consisting of ephemerals.

The sand is so well bound by the shrubs that the white saksa'ul (*Haloxylon persicum*) becomes dominant and *Carex physodes* also invades.

TABLE 6.7

Age-structure of *Aristida karelinii* stands on mobile barkhan stands in the central Karakum Desert (from Nechayeva et al., 1973)

Age (in years):	1	2–3	4–5	6–10	11–15	dead
Protected area						
Number of plants	1843	493	345	588	193	290
Percentage	49	13	9	16	5	8
Grazed areas						
Number of plants	2583	1620	655	768	440	283
Percentage	41	26	10	12	7	4

Haloxyletum persici is the most important plant community (Rodin, 1963). It is limited to deep, slightly hilly sands with little or no sierozem formation. This is the most structured community in the Karakum, and the following strata and synusiae can be distinguished:

(1) **Upper stratum:** composed of low trees and tall shrubs, with a height of 2.5 to 4.5 m and a cover of 20 to 30%. The dominant species is *Haloxylon persicum* (300 to 600 ha^{-1}). Subdominants are *Salsola richteri, Calligonum arborescens, C. eriopodum, Ephedra strobilacea*, and *Ammodendron conollyi* (only on blow-out scars).

(2) **Middle shrub stratum:** usually 1 to 1.5 m tall, composed of individual *Astragalus paucijugus, A. unifoliatus, Smirnovia turkestana*, and various *Calligonum* species (in individual stands only two or three of these species).

(3) **Semi-shrub stratum:** 30 to 50 cm high, with 4 to 5 of the following: *Convolvulus erinaceus, C. divaricatus, Haplophyllum* spp., *Mausolea eriocarpa, Artemisia badhysi* var. *arenicola, A. lobulifolia*, and *Acanthophyllum* spp.

(4) **Herb stratum:** Korovin (1961/62) identified a total of 130 to 140 herbaceous species for the entire Karakum region, of which 46.5% are ephemeral and 19% are perennial. The most important perennial is *Carex physodes*, generally considered to be an ephemeroid. It has a cover of 30 to 40% (up to 60 to 70%). Ephemerals are very numerous, often appearing in groups which represent the seedling foci of the previous year. Especially common are *Arnebia decumbens, Koelpinia linearis, Papaver pavonium, Senecio subdentatus, Cutandia linearis, Bromus tectorum, Eremopyrum buonapartis, E. orientale, Schismus arabicus, Trisetum cavanillesii, Amberboa turanica, Erodium oxyrrhynchum, Isatis emarginata, I. trachycarpa, Acantholepis orientalis, Euphorbia densa, Scabiosa oliveri, Astragalus* spp., *Spirorrhynchus sabulosus*, and *Malcolmia* spp.

In individual stands there occur usually a total of 30–35 species (see site samples in Nechayeva, 1954a).

With a greater plant cover on the sands, the water situation for the plants deteriorates. The vegetation-less barkhan fields store about 50% of the rainwater, so, as mentioned, a freshwater water-table forms under them. Where groundwater reaches close to the surface, at the base of barkhans, one can even find reed stands (*Phragmites*) in the middle of the dunes.

The pioneer species on the tops of the dunes also do not suffer from lack of water. The pure sand contains at most 1.5 to 2% of grain sizes under 0.01 mm. It has a wilting point at less than 1% water content, while the water content at a depth of 75 cm is always 1 to 3%, at a depth of 2 m even 3 to 5%, sometimes 5 to 7%. The clumps of *Aristida karelinii*, whose roots permeate a large volume of soil, can easily survive the summer drought.

As a result of decomposing litter, the percentage of finer particle sizes reaches 3% during the sere represented by woody plants on easily mobile sand. In addition, there is 0.5% humus. Under the shrubs a certain activity within the soil becomes noticeable.

In the next successional stage on the stable sands, the proportion of finer particles is already 4 to 4.5% and the humus content is also higher. An actual soil (a loose sierozem) is beginning to form. This is true to an even greater extent in the *Haloxylon persicum* community (Fig. 6.19). The soil receives calcium, in particular, from the ash component of the litter. The ash content of *Haloxylon* is 17.3%, of which only 6.7% is water-soluble. Similar ash content is found in the other woody species. The litter of the herbs also contributes considerable organic material. The nitrogen content of the soil climbs to 5 to

Fig. 6.19. *Haloxylon persicum* with a very dense undergrowth of *Carex physodes* (with highly swollen utriculi recognizable) near Repetek (photograph by Petrov; in Walter, 1968, fig. 533).

6 times that of the dune sand, while the phosphorus content doubles. The soil surface becomes somewhat compacted, and the water-holding capacity of the upper soil layers increases. A larger fraction of the rain water is stored at lesser depth, and less water, especially in rainy years, is stored at greater depths. The annual precipitation varies between 24 and 172 mm, while the potential evaporation is 1500 to 2000 mm. Since the water in the deeper layers is taken up by the roots, without adequate replacement, the soil underneath dries. Its water content becomes only 1.5 to 2 times the hygroscopic amount and is thus not usable. Even so, the precipitation should be sufficient to cover the needs of the vegetation (see p. 126).

In dry years the woody plants and perennial herbs reduce their transpiring surfaces considerably, and the ephemerals barely develop at all. They exercise a certain buffering effect on the water balance, which, cybernetically considered, represents a regulatory circuit with negative feedback.

Where especially dense *Carex physodes* stands develop, very little water penetrates to deeper layers. This is especially true where a thick cover of moss (*Tortula desertorum*) occurs. This happens mostly outside the central Karakum. In such cases, the summer water supply of the higher plants is endangered, and they usually die. The opinion is often held in the Russian literature that succession is proceeding toward this as a final lichen-dominated stage, but these purely theoretical ideas are seldom accurate under natural conditions. The activities of the rodents and their burrowing are not considered, nor those of the earlier large herds of wild grazers, now replaced by herds of sheep. A closing of the plant cover is prevented by these animals (see p. 138, 142).

Biogeocoenosis III, with *Haloxylon ammodendron*, is scattered throughout the sandy regions and limited to the valley-like, deeper depressions and hollows (Fig. 6.20). It is encountered more often outside the central Karakum, when layers of loam intervene between the sand strata. When these layers are not at too great depth and water accumulates above them, the black saksa'ul, *Haloxylon ammodendron* (= *aphyllum*), often occurs alongside *Haloxylon persicum* (Fig. 6.21). *H. ammodendron* has a wide ecological range, occurs mostly on loamy soils, and is much less sensitive to salt, though it is absent from solonchak soils (Kalenov, 1972, pp. 7–19). *Haloxylon ammodendron* forms dense thickets on soils with shallow groundwater tables.

Haloxylon ammodendron is an alkali halophyte, that is, the plants have a high ash content (22.5% of the dry weight) with considerable sodium and relatively little chloride and sulphate. The soil receives from the litter (4.5 t ha^{-1}) about 252 kg ha^{-1} of ash elements per year, which leads to a certain salinization. As a result, the accompanying plants are often halophytes, for instance *Tamarix* spp., which are also bound to groundwater situations. In the central Karakum one can find only small stands of the black saksa'ul in the depressions of the dune regions. In the Trans-Unguz Karakum, *Haloxylon ammodendron* stands with dense covers of *Tortula desertorum* occur.

The biogeocoenoses with *Haloxylon ammodendron* are less common in the sand desert. They develop much better when their roots reach groundwater. In the river valleys one can find dense forest-like stands with heights up to 14 m and with trunk diameters of 1 m. The number of trees and larger shrubs can reach 2500 to 5000 ha^{-1}. The phytomass is estimated as 55 t ha^{-1} and the primary production as 10.5 t ha^{-1} (Rodin, 1961). Most trees are destroyed for use as firewood but are partially replaced.

The *Haloxyletum ammodendrii* is, thus, clearly related to the floodplain biogeocoenosis complex, on the one hand, and to the takyr biogeocoenosis complex or halophyte biogeocoenosis complex on solonchak with high groundwater, on the other hand. The *Haloxylon ammodendron* stands are characterized by an especially distinct mosaic structure. Underneath the tree crowns one finds specific synusiae which have been studied, above all, by Rotshild (1960) in the Kyzylkum Desert. The microclimate under the trees is much milder and less extreme because of the shading, the wind-protection, and the accumulation of snow in winter. The soil is also changed. It contains more humus and nitrogen, as a result of continuous litter availability, but also contains more salts which form during the decomposition of the litter. This leads to formation of solonetz and to compaction. Species such as *Carex physodes* and *Artemisia* are absent under tree crowns. In their places appear certain character species (chenopods, etc.) which are more hygrophilic, nitrophilic, and halophilic — that is,

Fig. 6.20. Dense stand of black saksa'ul (*Haloxylon ammodendron*) in depressions of the sand desert with brackish soil. (The groundwater is reached by the roots of *H. ammodendron*.) The large bunch grass in front of the tree is *Lasiagrostis splendens*. (Photograph by Petrov; from Walter, 1976, fig. 63.)

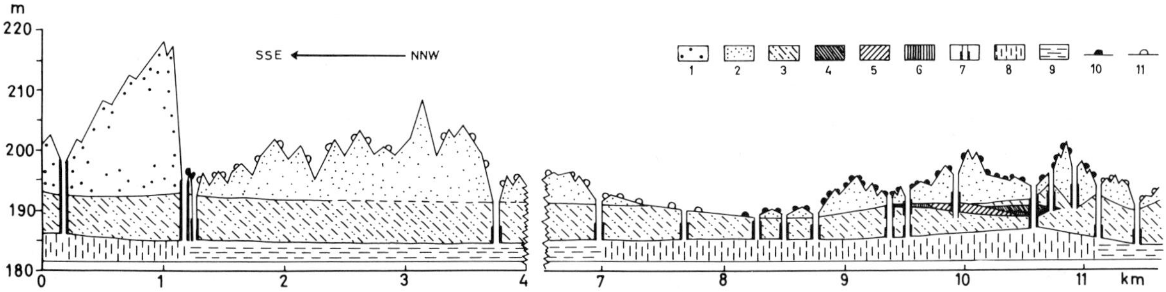

Fig. 6.21. Typical profile through the Karakum during the dry period (from Blagoveshchenski; in Walter, 1968, p. 763): *1* = mobile dune sand; *2* = old, stabilized sand hills, with a slight soil profile; *3* = gray, biotite-containing sands; *4* = gray loamy sand; *5* = yellow loamy sands; *6* = loam; *7* = sand stratum with a water content above 1%; *8* = fresh groundwater; *9* = saline groundwater; *10* = *Haloxylon ammodendron*; *11* = *H. persicum*. The missing section of the profile (4–7 km) is similar to neighboring parts.

which show a more ruderal character. Burrows of *Rhombomys opimus* are common under the trees and make the mosaic structure more complicated. Ephemerals are often especially dense where water drips down from the tree canopy.

The special characteristics of the microbiotopes under *Haloxylon persicum* and other shrubs are not so distinct but are still clearly visible. These also represent especially suitable ecological niches for animals.

Takyr biogeocoenosis complexes

Only small takyr areas occur in the true sand desert, where the wind has removed the sand and uncovered firm, clayey, geologically older strata. Water collects on these areas after heavy rains, and they are covered shallowly before the water evaporates. The temporarily wetted soil is a suitable biotope for the takyr algal communities.

These communities can represent the starting point for an ecological gradient which can also correspond to an ecological succession:

raw soil with algae → takyr-like sierozems with lichens → typical sierozems with seed-plants

This gradient or succession is a typical example of a biogeocoenosis complex.

Table 6.8 provides information on the change in phytomass, annual production, and chemical composition.

Precise observations were made concerning the runoff from Karakum takyry on an area of 1200 m^2. Table 6.9 shows the results for this takyr for five rainy days.

The runoff is greater when the snow melts. The melting of a 33 mm thick snow cover in March 1948 yielded a runoff of 16.6 mm.

Runoff from a dry takyr begins always after 3 mm rainfall, from a wet takyr somewhat sooner. Before runoff begins, however, the cracks in the 2 to 3 cm thick surface crust must close completely. On the takyry at the foot of the Kopet Dag, which are less clayey, the runoff is somewhat less even though the slope is slightly greater.

The computations yielded the result that, from a takyr area of 27 600 k m^2, a total of 514.5 million m^3 of water run off.

Halophyte biogeocoenosis complexes

In the lowest-lying parts of the sand-desert relief, the groundwater can be so near the surface that the surface is always kept wet by the capillary fringe. As a result of the high evaporation, the salts contained in the groundwater then accumulate and form a solonchak soil. Saltpans (*shory*) form, around

TABLE 6.8

Phytomass of takyr vegetation and its chemical composition (after Rodin, 1954)

Zone	Dry organic matter present (phytomass) (kg ha^{-1})	Annually decomposed organic matter (kg ha^{-1})	Nitrogen contained therein (kg ha^{-1})	Amount of ash components (kg ha^{-1})
I Lowest-lying areas with clay deposits, saline soils, very few ephemerals	100	100	1.5	6–7
II Middle levels with soil algae or soil lichens, few ephemerals	300	300	9–10	20–22
III (a) patches of ephemeral vegetation (transition zone)	1200–1600	1200–1600	20–30	90–140
(b) *Artemisia* semi-desert with ephemerals on sierozem soils	12 000	10 000	90	475

TABLE 6.9

Observations of runoff and soil-wetting on takyry (area 1200 m^2) in the Karakum (Bogdanov, 1954)

Days	Rainfall (mm)	Surface condition (prior)	Runoff after rain (mm)	Depth of water layer (mm)	Runoff into basin (m^3)	Depth of soil wetting (cm)	Amount of water going into soil (mm)
13 March	1.9	dry		no runoff		1.5	–
20 March	3.6	dry	3.0	0.13	0.16	4	0.47
21 March	8.8	wet	2.8	4.0	4.8	6	2.0
17 April	15.7	dry	2.9	9.6	11.6	9	3.2
18 April	2.7	wet		no runoff		–	–

which there is a clear zonation of plant communities. We can combine these into a biogeocoenosis complex corresponding to an ecological series along a gradient of decreasing salt concentration (with increasing depth to groundwater), from the center of the saltpan to the periphery.

The innermost part of the saltpan is mostly covered with a salt crust and carries no vegetation. Proceeding outward, concentric rings with very few species follow. In the southeastern Karakum these are the following: (1) *Halocnemum strobilaceum*, the first pioneer species; (2) low trees of *Seidlitzia rosmarinus* (Chenopodiaceae); (3) finally, *Halostachys caspica* with *Limonium suffruticosum*; these can be joined by *Kalidium caspicum* and other Chenopodiaceae (site sample by Michelson, 1955b). Sometimes, salt-excreting grasses also occur, usually *Aeluropus litoralis*.

Blown sand from the surrounding areas is dumped around all shrubby species of the saline soils, so that small sand hummocks (nebkas) form. Only the tips of the shoots stick out of these small dunes.

The dunes around *Halocnemum* remain small (0.5 m high), whereas, for the quickly-growing *Tamarix* shrubs, considerable amounts of sand are dumped. The same is true for *Nitraria*, which is native to the region (Blagoveshchenskiy, 1955). Thus, a nebka landscape can form around the saltpans (see Walter, 1968, pp. 705–707). The dunes rise at regular intervals above the solonchak soil, each with only the ends of the branches of shrubs sticking out. The shrubs themselves root into the groundwater. If the wind blows from a single direction, as it does between the Balkhan Mountains, then the dunes gradually take on the shape of long ridges roughly parallel to the direction of the wind.

The small dunes are a favorable site for non-halophilic herbs. They also harbor a certain fauna and, thus, form microecosystems.

Haloxylon ammodendron stands often grow around the edge of such saltpan depressions (Fig. 6.20).

Characteristic for the true halophytes is the constantly high content of sodium and chlorine in their transpiring parts (Bazilevich et al., 1972) and in their litter, as shown in Table 6.10. These are, thus, typical halophytes.

The phytomass in this biogeocoenosis is not high: *Halocnemum* 1.76 t ha^{-1}, of which 59% is roots (litter: 0.515 t ha^{-1}); *Kalidium* 1.38 t ha^{-1}, of which 75% is roots (litter: 0.7 t ha^{-1}).

Sodium and chlorine, in the form of sodium chloride, return to the soil with the litter and can lead to salinization of the soil. The main salinization on these wet sites, of course, occurs through the evaporation of the capillary fringe at the surface, leaving the salts behind. These salts contain a much larger fraction of sodium chloride than does the litter, in which more potassium, calcium, and magnesium are found. Thus, the salts coming from the litter are more favorable for plant growth than are those coming from the groundwater. One can even speak of a certain, relative desalinization by the litter. The halophytes also actually prepare the soil for the ephemerals by their ability to collect windblown, salt-free sand.

TABLE 6.10

Content of sodium and chlorine (in percent) in the dry matter of various portions and litter of *Halocnemum strobilaceum*, *Kalidium caspicum*, and *Salsola turcomanica* (from Bazilevich et al., 1972)

Parts studied	*Halocnemum strobilaceum*		*Kalidium caspicum*		*Salsola turcomanica*	
	Na	Cl	Na	Cl	Na	Cl
Annual shoots	5.37	10.05	10.05	8.63	5.90	11.06
Older shoots	2.08	1.68	1.46	1.28	–	–
Roots	1.65	2.09	2.84	1.93	1.04	1.19
Litter	3.10	5.18	–	–	–	–

THE ANIMALS

Consumers: general description

Turkmenistan is a region which had human settlement by at least the Neolithic. Great empires were repeatedly founded during later history. Under these conditions the original animal populations could not remain undisturbed. Animals which were important in hunting, as well as the large carnivores which threatened their numbers, were depleted from the settled fringes on into the desert. The thousands of saiga (wild goat-antelopes), tarpans (wild horses), and kulans (wild asses), which were reported in the old chronicles, have disappeared. In order to protect the herds, the carnivores were exterminated, destroying the natural equilibrium in the process. The rodents profited from the extermination of carnivores, including carnivorous birds, and are today the most numerous mammals. Arthropods, as a result of their small size, were not greatly disturbed, but we have no comprehensive studies on this especially varied group.

On hot summer days the sand desert gives a totally dead impression as far as animals are concerned. Animals are neither seen nor heard. Except for the noise of the wind, the desert is silent. But the appearance is deceiving because life awakens only at night. Since the ground surface heats up to 60°C, in extreme cases even 80°C, during the day, and since the air is so dry, almost all animals seek shelter underground, in the many underground burrows, where it is cool and the air is almost saturated with water. As a result the animals' bodies lose almost no water, which is important for survival in an area with so few watering places. Many animals have no means of taking up water but instead cover their water needs by eating water-containing plant parts. Metabolic water, which is formed during respiration, is also of significance. Since it is cool at night and the air is relatively moist, the total water consumption above ground is relatively low. Fluid excretions of these animals are quite small. Thus, the animals show special adaptations to the rigorous conditions to which they are subject in the desert, just as do the plants.

The ecological roles of the animals in the Karakum Desert are highly varied (Kunin, 1955; Rodin, 1961):

(a) The herbivores, as consumers, feed themselves at the cost of the plant matter produced, while the carnivores eat the herbivores and thus regulate their numbers.

(b) Animals, e.g., hoofed animals, break up the sand surface and thereby prevent crusting and compaction.

(c) Through burrowing, the animals redistribute the upper sand layers (e.g., rodents, ants).

(d) Animals affect the generative propagation of the plants by distributing seeds and fruits (epizoochorously and endozoochorously) and by pressing them into the ground under foot, all of which favor germination.

The significance of animals in an ecosystem, as consumers, is less quantitative than qualitative. The total amount of consumed plant matter is small, but, by damaging certain species through the destruction of their seeds and fruit, the animals affect the competitive relations of the plants and prevent any single species from attaining absolute dominance. On the other hand, the destructive animals are held in check by the carnivores. There are, thus, many interdependent regulatory loops by which the equilibrium of the system is maintained. Rodin (1961) has summarized the animal-caused changes in the vegetation of western Turkmenistan, as far as they are known.

The vertebrates in the protected area at Repetek are classified according to their feeding habits in Table 6.11.

The feeding habits of the invertebrates are equally varied. The ants of the genus *Messor* concentrate totally on the fruits of *Carex physodes*, *Bromus tectorum*, *Kochia schrenkiana*, and *Microcephala lamellata*. The coprophages are represented by dipterous larvae and *Scarabaeus sacer*.

The animals are limited to particular biogeocoenoses; for instance, one reptile, one mammal, and three bird species occur only with *Haloxylon ammodendron*. These relations are depicted very graphically by Zaletayev (1976), especially for the colonization by animals under the canopy of *Haloxylon ammodendron*. He draws attention also to the fact that 65% of the seedlings of this species are found on the mounds of soil left by burrowing rodents. The loosened soil is probably a good germination site, and there is no competition from other plants. On the other hand, the larvae of *Turcmenigena varentzovi* are quite destructive, since they live in the lowest part of the trunks of old trees and make it easier for the fungus *Poria desertorum* to infect the wood. This fungus destroys the wood so that trunks 30 to 45 years old usually fall over. The hummock which arose around the tree is blown out and forms a hollow instead. *Carex physodes* settles on such sites, and then a leveling of the ground follows.

Animals often effect a series of micro-successions in this way, causing continuously changing microbiotopes.

Mammals in the protected area at Repetek (southeastern Karakum)

The area at Repetek, which has been protected since 1928, covers an area of 34 000 ha. Of these, 18 000 ha have *Haloxylon persicum–Carex physodes* sand vegetation on small, only slightly mobile sandhills, on which *Aristida karelinii* and *Ammodendron conollyi* also grow. The region of shifting dunes with dune valleys makes up 14 000 ha, and the remaining 2000 ha represent lower-lying sand areas occupied by *Haloxylon ammodendron* stands.

In this relatively undisturbed area about 200 species of vertebrates were identified (29 mammals, 23 reptiles, and over 140 bird species). The invertebrates have not yet been sufficiently studied, but over 1000 species of insects have been counted (Kuznetsov et al., in Anonymous, 1975, p. 142). A more detailed description of the animals is given by Stalmakova (1955). The species were cited in the order of the individual animal groups:

Insectivora

The most important representative is the eared hedgehog (*Hemiechinus auritus*). This fairly common species inhabits old rodent burrows under blown-out shrubs. In search of food it can cover a distance of 5 km in a single night. Its prey consists of reptiles and various arthropods.

Shrews (*Diplomesodon pulchellum* and *Crocidura suavolens*) also belong to this group and have a similar way of living.

Chiroptera

The bats (*Vespertilio* spp.) need a shelter during the day, but this is lacking in the desert. They live,

TABLE 6.11

Vertebrate species by feeding category, for the protected area at Repetek (Rodin, 1977)

Animal group	Carpophages	Anthophages	Phyllophages	Xylophages	Rhizophages	Saprophages	Zoophages	Necrophages
Reptiles	–	–	5	–	–	–	19	–
Birds	17	–	–	–	–	–	46	1
Mammals	9	4	11	2	6	3	15	4
Total	26	4	16	2	6	3	80	5

therefore, in the vicinity of the station and can be found under the roofs of the houses and in other parts of the buildings. Five hundred individuals were counted in 1948.

Rodentia

The rodents are the largest group in the sand desert and are of great ecological importance.

The local ground squirrel, *Spermophilopsis leptodactylus*, is everywhere, though not in great numbers. Its diet consists of graminoids and forbs, but it disdains the halophytes. It generally does not forage more than 100 to 150 m from its burrow, 200 to 300 m if the plant cover is very sparse. This rodent tolerates relatively high temperatures and can leave its burrow during the day. It seeks shelter underground only when the heat becomes very great, or during rainfall. Its diet consists mainly of the fruits of *Calligonum*, the herbaceous ephemerals, and the bulbs of the ephemeroids. The latter are found without error by smell, even at a depth of 7 to 9 cm. The same is true for other rodents. Rodin mentions that one often finds places which are completely free of *Allium*, *Tulipa*, *Iris*, etc., although one would expect to find them there. One notices from the many holes, in which remains of bulb shells can be found, that the bulbs have been dug up and eaten by the rodents.

The porcupine *Hystrix hirsutirostris* lives in dense *Haloxylon ammodendron* stands and often covers up to 9 km per night in search of food. The diet consists of *Nitraria* fruit on saline areas, otherwise of bulbs of *Tulipa* and the roots of *Ferula*.

The jerboas are represented by three species. *Dipus sagitta* leaves its burrow only at night and closes the opening with sand during the day. This rodent ranges up to 1 km from its burrow. When in flight it can jump up to 2 m and then quickly bury itself in the sand. Their main enemies are owls. The jerboa diet consists of fruits and assimilating branches of shrubs, fruits and shoots of herbs, and insects. This is also the diet of the desert jerboa (*Eremodipus lichtensteinii*), which one finds sporadically everywhere. On the other hand, *Paradipus ctenodactylis* is a widespread species appearing in larger numbers but is always restricted to a shrub vegetation, provided that slightly mobile sands are present for building its burrow. The burrow reaches down into the wet sand layers, while the entrance is always covered with dry sand. Food is provided by the shrubs which grow on such open sand areas, such as *Ammodendron* and *Calligonum* species, *Salsola richteri*, *Astragalus paucijugus*, and *Haloxylon persicum*. This species gnaws off branches from the bottom of the shrub on up to a considerable height. The damage to the shrubs then leads to the blowing away of the sand. The winter dormancy of this species lasts 37 to 78 days, at a time when there are no *Haloxylon* fruit available. The animals awake when *Haloxylon* comes out in the spring, at a time when the temperature in the burrow 2 to 2.5 m deep begins to climb.

The vole (*Ellobius talpinus*) occurs only in the southeastern Karakum close to oases. It was, however, also observed in 1940–1943 in the protected area, where it spread over an area of 150 m^2 and fed itself on the roots of the shrubs *Haloxylon* and *Ammodendron*.

Rhombomys opimus is probably the most important and widespread rodent. In September 1938, 1600 individuals per km^2 were counted in the *Haloxyletum ammodendri*, but only 39 in the *Haloxyletum persici* and only 2 in the blown-out stands. On a sample area of 5 ha in 1936–1950, however, the number fluctuated from 0 to 135. The areas with *Carex physodes* are most densely settled. On such areas, 7000 entrances to the underground passageways were counted per hectare. These passages run at a depth of 50 cm directly under the rhizomes of *Carex physodes*, where the water content of the sand does not fall below 0.5% and the temperature varies only between 18 and 35°C. The nesting chamber is at a depth of 80 to 120 cm. In these rooms the humidity is over 80% and the temperature never exceeds 30°. Food-storage chambers are also built (70 cm long by 50 cm wide), in which the imported food is eaten. *Rhombomys* settles where there is food, in families which consist of the two parents and the 13 to 16 young from the last two litters. When foraging, the animals go only a few meters from the burrows. The burrows of individual families are connected with those of other families over an area of 1.5 to 2 ha. If food is plentiful, the young remain with their families for up to a year, otherwise they wander off after about four months. This rodent is active all year and stores supplies for the winter. In summer the animals leave the burrows early in the morning and in the evening; in winter, on nice days, they leave around midday. The diet consists of various parts of the saksa'ul, in summer (when the branches contain more salt) only the fruit. In addition, both fresh and dried parts of the herbs are eaten, above all the rhizomes of *Carex physodes*. In the dune valleys one can find one burrow per square meter and can recover 1 kg of hay from each burrow. The hay consists of up to 90% *Carex physodes* rhizomes. *Carex physodes* disappears on such areas and is replaced by annual grasses, such as *Bromus tectorum*, *B. macrostachys*, and *Eremopyrum orientale* (Rodin, 1961). Up to 1.2 t of soil are thrown up by the burrowing. This loose soil is easily blown away by the wind, which prevents a covering of the surface by *Carex* and mosses (see p. 134).

In addition to *Rhombomys*, *Meriones meridianus* is also common. These rodents inhabit plant-covered sands in the winter but move in the spring into the dune regions, where they build burrows in dune valleys under plants, less frequently on open sands. Their diet is concentrated and consists of fruits, seeds, and to a lesser extent herbaceous plant parts. They do considerable damage to the afforestation attempts with saksa'ul seeds since they know how to find the fruit with great certainty.

Meriones erythrourus is only found as individuals in the protected area. It seems to be extending its range, however, since it has been noticed since 1942 in the station's storage sheds, from which it has carried away seeds and fruit which were destined for sowing. In the process it has adapted itself to new species for food, such as *Elaeagnus* fruit and cotton seeds. The animals reproduced so rapidly that it was necessary to undertake control measures.

Even though the animals are mobile, they are often strictly bound to particular phytocoenoses. Although *Meriones erythrourus* and *Rhombomys opimus* occur in the same part of the sand desert, Rodin (1961) emphasized that they never settle in the same community. This is also true for other groups of rodents.

The soil is enriched with various organic materials as a result of the rodents' burrowing (urine, feces, corpses, plant remains). The soil structure is also destroyed and loosened up. The piles of earth thrown up around burrows increase the microrelief and soil erosion. The redeposited soil often has a higher salt content, having come from deeper soil layers.

On takyr areas the clay crust is broken up by the rodents so shrubs can colonize, where it would otherwise not be possible.

On sandy areas around abandoned rodent colonies the first plants to reinvade after abundant rains are ruderal species, followed by *Bromus tectorum*.

The summer, because of lack of water-rich food, is an especially difficult period for the rodents. The juicy organs of the succulent halophytes contain so much salt at this time of the year that the rodents will not eat them. The animals must subsist on rather dry food, such as seeds, dry leaves and stems, and a few somewhat juicier bulbs and roots or rhizomes of *Carex physodes*. The rodents' thirst in summer is so great that they come to the experimental station at night in order to drink from a mud-hole at the pump station.

Precise experiments to determine the water balance of the rodents were carried out by Stalmakova (1954) on the following species: *Spermophilopsis leptodactylus*, *Dipus sagitta*, *Meriones meridianus*, and *Rhombomys opimus*. A precise, complete water balance was established. The water budget was balanced in the spring, when juicy vegetation is available, and the budget of *Meriones* was balanced in the autumn, since this species feeds on juicy *Salsola* fruit at this time. The balance in summer was constantly negative, that is, the animals must use metabolic water gained from the respiration of the carbohydrates in the food. If this water is not sufficient, then body fat or protein is respired and the animals lose weight.

As a result of the strained water balance, the rodents in the Karakum reproduce only during the short favorable period in March–April. If these species occur in wetter steppe areas, they generally have two or three reproductive periods per year. *Citellus fulvus*, which is found only in the northwestern Karakum, becomes dormant as soon as the herbaceous plants dry up in summer.

Lagomorpha

As an important species in the protected area one must mention the desert hare (*Lepus tolai*). It is very mobile, even though it has its storage chambers mostly in plant-covered sands. In the spring it feeds on the fresh vegetative parts of the herbs, otherwise on the bark of the large shrubs *Ammodendron conollyi*, *Eremosparton flaccidum*, *Calligonum ar-*

borescens, and *Astragalus paucijugus*. It does not eat the *Haloxylon* or *Salsola* species. Even at the low level of two to three animals per square kilometer it can cause great damage. In the vicinity of *Haloxylon ammodendron* stands, all *Ammodendron* and *Eremosparton* shrubs are gnawed on. Young shrubs are bitten off completely. The stabilization of the dune sands is impeded because it is always precisely the animals' favorite species which are used for the stabilization attempts.

Carnivora

Of the Mustelidae, the mouse-weasel (*Mustela nivalis*) was found only once, next to the burrow of a ground squirrel (*Spermophilopsis leptodactylis*). The polecat (*Vormela peregusna*) on the other hand, is generally distributed. This species seems to feed primarily on the large *Rhombomys opimus*.

Among the canines, the wolf (*Canis lupus desertorum*) was never seen in the protected area, although it does occur in winter in the Karakum. This is probably because watering places are too infrequent. The jackal is also found only close to the oases along the Murgab and Amu-Dar'ya. Within the protected area the most common carnivore is the fox (*Vulpes vulpes flavescens*). It eats small mammals, birds, reptiles, and also sand-cockroaches (*Anisogamia tamerlana*), scorpions (*Liobuthus kessleri*), *Galeodes fumigatus* and tarantulas (*Tarantula bergsonii*). The stomach of one fox killed in the dune region contained twelve tarantulas and two scorpions. In autumn and winter the foxes try to catch chickens from the settlement around the experimental station.

Of the cats, the spotted wildcat (*Felis ocreata*) is very common. These are found in blown-out sand areas near *Haloxylon* stands, where many birds make their nests. In 1940 it was established that the cats destroyed one-half of the nests of the warbler (*Sylvia curruca*) and the nightingale (*Erythropygia*), as well as *Rhodespiza obsoleta*, the dark broad-tail (*Scotocerca platyura*) and the turtle-dove (*Streptopelia turtur*). During the day the cats lie in shallow ditches next to shrubs. The dune cat (*Felis margarita thinobius*) was observed only once, as were the tracks left by a desert lynx (*Lynx caracal*).

Artiodactyla and Perissodactyla

When the ephemerals develop in the spring, wild pigs (*Sus scrofa*) come from the floodplain forests of the Amu-Dar'ya far into the sand areas. Their tracks were observed in the protected area, where they had dug up the bulbs of *Eminium lehmannii*. They do not remain continuously in the region.

Another ungulate is the Zheiran gazelle (*Gazella subgutturosa*), of which one continuously finds a few within the protected area. Most come in October, when the gazelles migrate. They stay in the *Haloxyletum persici*, where they eat the dry leaves of *Carex physodes*, or in the dunes, where they eat *Heliotropium arguzioides*. At the beginning of March they usually leave. The remainder seek shelter in the dense *Haloxylon* stands or sleep during the day under shrubs in the dune valleys. They do not need water. They live from dry sedge leaves and do not eat the juicy shoots of the saksa'ul.

Just 100 to 200 years ago, before firearms came into general use, there were large herds of antelope (including also the saiga antelope, *Saiga tatarica*) and also wild sheep (*Ovis orientalis arkal*). Herds of 500 to 1000 animals were not uncommon. The same was true for the kulan (*Equus hemionus*). Even at the beginning of the century, herds of up to 1000 of these wild asses were present in southern Kazakhstan in the Lake Balkhash region. Today they are probably extinct.

These large wild herds certainly had a major ecological significance. The running of such herds causes the sand surface to break up and prevents growth of *Carex physodes* and mosses, decreasing the production of the sand vegetation. As a result, the theoretically assumed final successional stage (see p. 134) should not be reached, since it is precisely the *Carex*-covered areas which were the main grazing grounds for the wild herds. Today these wild grazers are replaced by man's domesticated animals. These are primarily flocks of sheep. Turkmenistan is the home of the karakul sheep, which yields the well-known black skins. Karakul skins have become today one of the most important export commodities in this dry region.

Up to 1973 a total of 29 mammals had been identified in the protected area, out of 49 in the whole of Turkmenistan.

Other vertebrate groups in the protected area

The other vertebrate groups, as well as the invertebrates, have now also been studied in more detail (Nechayeva, 1975). Of the birds, 25 nesting species were named, of which 13 were nesting migratory birds. Sixty species were identified for all of Turkmenistan. Eighteen species of reptiles were identified (40 for all of Turkmenistan).

Twenty-one bird species nest in the forest-like *Haloxylon ammodendron* stands. One can also find 11 reptile and 19 mammal species here. In the areas of mobile sands with partial plant cover, the numbers are 12 bird, 11 reptile, and 15 mammal species.

Four reptile species inhabit all three biogeocoenoses: two geckos (*Gymnodactylus russowi* and *G. caspius*), the "arrow-snake" (*Psammophis lineolatus*), and *Varanus griseus*. Seven reptile species are found only in the mobile sand: *Phrynocephalus interscapularis*, *P. mystaceus*, *Crossobamon eversmanni*, *Teratoscincus scincus*, *Eryx miliaris*, *Eremias scripta*, and *E. grammica*. The following are found in the stands of the two *Haloxylon* species: *Agama sanguinolenta*, *Eremias intermedia* and *E. lineolata*, *Echis carinata*, *Coluber tyria*, *C. karelini*, and the tortoise *Testudo horsfieldii*. The tortoises are among the most important herbivores because there can be as many as one hundred per hectare. A tortoise eats several hundred grams per day of young shoots in inflorescences of the ephemerals. Even though these animals are only active for 2.5 to 4 months, the damage which they cause is clearly visible. During their active season the tortoises lay 10 to 15 eggs in the sand. Embryonic development lasts three months, but the young tortoises do not appear on the surface until the next spring and are then the prey of many carnivores, including carnivorous birds. As soon as the herbaceous vegetation dries out, the tortoises disappear into the sand for eight to ten months.

Most of the other reptiles are carnivorous, especially the snakes (*Thaphlometopon lineolatum*, *Eryx miliaris*, *E. tataricum*, etc.), which greatly reduce the rodent populations. The snakes use the underground tunnels of the rodents for access to their prey, but also for winter dormancy. The same is true for *Varanus griseus*, which grows to a length of 1 m and a weight of 2.5 kg. It hunts young tortoises and lizards. These latter are also carnivorous, for example the extremely quick species of the genus *Eremias*.

Agama lizards often sit high in the branches of the shrubs in summer but quickly descend and disappear into rodent burrows when danger comes. Their most striking characteristic is the ability to change their skin color (blue when caught).

Of the bird species which are constantly found in the region and in all three biogeocoenoses, the following are cited: *Corvus corax*, *Dendrocopos leucopterus*, *Aquila chrysaetus*, *Buteo rufinus*, and *Athene noctua*. The most common migratory birds are *Lanius excubitor*, *Sylvia curruca*, *Oenanthe isabellina*, and *Circaetus gallicus*. Only one species (*Parus major*) can always be found in the *Haloxylon ammodendron* stands, but five species nest there (*Hippolais languida*, *H. caligata*, *Rhodospiza obsoleta*, *Sylvia cantillans*, and *Upupa epops*). *Caprimulgus aegiptius* nests in the *Haloxyletum persici*, *Passer simplex* in the mobile sand. Nesting in both *Haloxylon* stands are *Scotocerca inquieta*, *Galerida cristata*, *Passer ammodendron*, *Streptopelia turtur*, *Cercotriches galactotes*, and *C. europaeus*. *Podoces panderi* and *Sylvia nana* nest in both mobile and plant-covered sands.

Of the polyphagous and herpetophagous birds, one can find all species in all three biogeocoenoses. After the breeding period, and especially in winter, many entomophagous birds eat the fruits and seeds of the desert plants. This is especially true for the species in the mobile sands. Carnivorous birds are significant in limiting the numbers of the rodents.

Invertebrates of the protected area

Only those species which are ecologically closely related to the woody plants were studied in any detail. Even for these, though, we must limit the amount of information given.

In the crowns of *Haloxylon ammodendron* one can find many groups of invertebrates. Of these, the majority are phytophagous: Trombiformes, Collembola, Orthoptera, Psocoptera, Auchenorrhyncha, Psylloidea, Aleuroidea, Aphidoidea, Coccidea, Hemiptera, Thysanoptera, Coleoptera, Lepidoptera, and Cecidomyiidae. One group, the spiders (Araneae), is carnivorous. In July the spiders represent 35.8% of the zoomass of all invertebrates in the canopy. Insect galls are very numerous. *Mesopsocus hyemalis* (Psocoptera) feeds on the fungi and algae on the tree trunks and in February constitutes 9% of the total invertebrate zoomass on *Haloxylon*. The number of cicadas climbs in October to 130 000 to 150 000 per hectare. Very common also are butterfly caterpillars, which often make up 50 to 80% of the zoomass.

Dangerous to man are the mosquitos, which carry infectious parasites (e.g. *Leishmania*) from the rodents. Camping close to areas inhabited by rodents is not to be encouraged. The danger of malaria, on the other hand, exists only around the oases. Scorpions, tarantulas, etc., are poisonous; ticks often carry viruses.

The invertebrate soil fauna is also very rich. Carnivores include the scorpions and pseudoscorpions, as well as spiders. Phytophagous and detritivorous groups include the Trombiformes, the Thysanoptera, the larvae and imagines (adults) of beetles, and the larvae of the Diptera.

Plant litter is a main source of food for the soil fauna. As a result, the soil fauna underneath tree crowns is quite different from that between trees, where *Carex physodes* grows and produces almost no litter. Many invertebrates are restricted to the litter of *Haloxylon*, especially the larvae of many beetle species which feed on detritus. Their numbers can reach 40 000 per hectare. The larvae alone make up 30 to 40% (3–6 kg ha^{-1}) of the invertebrate zoomass of the edaphon. They are most commonly found at a depth of 50 to 100 cm. Many larvae feed on the finer roots of the trees.

Among animals which spend most of their time underground, the most noteworthy are the bright-colored ants of the genus *Camponotus* and the soil cicadas of the genus *Pentasteridius*. Termites, on the other hand, are rare in *Haloxylon* stands. The large termites (*Hodotermes angerianus*), whose mounds reach 50 to 70 (100) cm in height, are restricted more to the foothills of the Kopet Dag. Termites do appear on takyry, mostly close to depressions with dense marsh vegetation. Ants also inhabit takyry. They are partially seed-eaters but also carnivores. In general, the number of animal species on the takyry is relatively small because of the low plant productivity.

Information on the zoomass of the various animal groups can be found on page 156.

THE DECOMPOSERS

Little is known about the decomposition and mineralization of the organic remains of plants and animals in the sand desert.

Decomposition is initiated by the microfauna, by saprophages which break up plant parts and partially eat them, also by coprophages and necrophages, which do the same to the excrement and corpses of the animals. The mineralizing activities of the fungi and bacteria begin simultaneously. Because of the high summer temperatures, there is hardly any humus formed. Since the litter of the halophytes (above all the Chenopodiaceae) always contains large amounts of soluble salts (ash content: 22–32%), these salts accumulate in the soil and can result in salinization. For the true halophytes, the salts are usually sodium chloride and sulphate; for the alkali-halophytes, of which *Haloxylon ammodendron* (ash content: 22.5%) is one, the sodium ions outweigh those of chloride and sulphate together by a wide margin. These are compensated in the cell-solution by ions of the organic acids, often oxalic acid. Since the organic acids are oxidized to carbon dioxide, the sodium in the soil accumulates in the form of carbonate and the soil becomes strongly alkaline. This can result in the formation of solonetz.

Rodin (1961) performed a number of ash analyses (see Walter, 1968, pp. 784–787).

Within the protected area of Repetek, thirteen species of fungi from the rhizosphere and fourteen species from the surrounding soil were isolated. Half were *Penicillium* species.

Soil bacteria are very numerous: 3 to 8 million (g dry soil)$^{-1}$. *Pseudomonas* and *Bacillus masentericus* were especially common, but *Azotobacter* and *Clostridium* occurred also, as well as Actinomycetes.

INTERFERENCE BY MAN

Of the 366 000 km^2 of the Karakum Desert, 250 000 km^2 are natural rangeland which can be grazed throughout the year, since there is no snow cover in winter. Of course, the pastures are sufficient for only the undemanding goats, and, most importantly, camels and the karakul sheep. The latter species' number was over three million in Turkmenistan in 1968.

The most important interferences by man in the Karakum biome are those connected with grazing, since grazing is seldom carried on rationally. Overgrazing leading to the destruction of the plant cover and to the formation of barkhan fields is common. In other cases the pastures are not sufficiently used, also presenting problems. Therefore, in order to find out the most rational methods of using the Karakum rangeland, experiments were carried out in 1940–1943 on an experimental area in the northern part of the southeastern Karakum with a herd of 700 to 1000 sheep (Nechayeva, 1954).

Of course, in a desert with a good growth of woody species, the direct intervention of man in cutting firewood is also important. Since man lives generally close to watering places, the effect is especially noticeable there. Wood-cutting within five kilometers of water places should be forbidden.

One might hold the opinion that the effects of grazing by sheep are the same for the vegetation as were the effects of grazing by the previous wild herds. The domesticated herds move slowly, however, and do not stray a great distance from the watering places. They stress the pastures around the watering places, while the areas further away from the water remain almost totally undisturbed. The result is that concentric annular vegetation zones form around the watering places, with increasingly degraded vegetation as one approaches the center. Around the waterhole itself, where the animals remain the longest, there is no vegetation at all, and the soil is overfertilized by the animals' excrement. The negative effect of this overfertilization remains noticeable for many years after the watering places have been abandoned. There is too little precipitation for leaching to take place. The exposed sand around the wells is often set in motion, so former watering places are recognizable from a great distance as barkhan fields. That this may lead to the formation of freshwater lenses above the salty groundwater was mentioned on page 107.

Around the barkhan field in the center, one finds a ring of *Aristida karelinii*, *A. pennata*, *Dorema sabulosum*, *Astragalus chivensis*, etc., all plants which the livestock do not like to eat. Continuing outward from the center, there follows a ring with tree-shrubs such as *Eremosparton flaccidum*, *Salsola richteri*, *Ammodendron conollyi*, *Astragalus paucijugus*, etc. Still farther away come *Haloxylon persicum* and *Calligonum* species, as well as herbs, though *Carex physodes* is still only poorly represented. The normal composition of the vegetation is reached only at a greater distance from the well.

Typical pasture weeds also appear, mostly nitrophilic species. On loamy soil the weed *Peganum harmala* (Zygophyllaceae) is found, which can also be found around watering places as far west as the Ebro Basin in Spain.

The experimental area consisted, primarily, of a *Haloxyletum caricosum* with variations. The experiments were carried out with varying grazing densities. The average was one sheep per 6.2 ha; the minimum was one per 9.3 ha, the maximum was twice the average density. The grazing seasons used were: autumn, 94 days; winter, 62 days; spring, 91 days; and summer, 118 days, on the average.

Grazing can exert both positive and negative effects on the vegetation. Positive effects are: (1) a slight loosening of the soil surface; (2) promotion of seed-setting by the plants and the pressing of seeds into the soil under foot; and (3) fertilization of the soil through excrement. These positive effects are noticeable only with an intermediate grazing intensity. At the maximum grazing intensity the soil is loosened too much and can be blown away to start forming barkhan fields. At minimum grazing intensity, the loosening is too little and unwanted mosses take over the area. Similarly, the pressing of seeds under foot to the optimal depth of 0.5 to 2.5 cm occurred only with intermediate numbers of animals. At maximum numbers the seeds were buried too deep, and at minimum numbers they were not buried at all but were eaten by the birds and rodents instead.

The effect of trampling is always a negative one for the higher plants. At maximum occupancy the loss of forage was 75%; at medium occupancy it was still 67%, and at minimum occupancy 54%.

The change in the composition of the plant cover resulting from varying occupancy levels is shown in Table 6.12. One sees from the numbers that the intermediate density has a positive effect after just four years, not only quantitatively but also qualitatively, since the best forage plant, *Carex physodes*, is promoted.

Complete protection of the grazing areas does not have a positive effect because *Carex physodes* is gradually replaced by annual grasses and forbs, and the surface becomes compacted.

After removal of the carnivores, the many rodents presented major problems. In places there can be 20 to 75 rodents per hectare, which can eat up to 78% of the edible matter present. Thus, foxes, wildcats, and other carnivores should be protected.

The experiments showed that very early grazing,

TABLE 6.12

Numbers of plants and their dry weights per square meter after four years of grazing at different levels of occupancy (from Nechayeva, 1954a)

Occupancy level	*Carex physodes*		Grasses		Forbs		Total	
	number	g m^{-2}	number	g m^{-2}	number	g m^{-2}	number	g m^{-2}
Minimum	342	15.8	242	6.4	88	2.6	672	24.8
Intermediate	347	17.5	256	6.4	82	2.9	685	26.8
Maximum	243	12.7	343	5.3	93	2.3	679	20.3

in March–April when the soil is still wet, is well tolerated. The plants grow back quickly, especially *Carex physodes* and the grasses, less well the forbs. The annuals are not able to bloom, however, and are reduced. The phenology of *Carex physodes* feed is shown in Fig. 6.22.

Grazing the same area twice in the spring proved to be very damaging, especially when it is done every year, as is often the case. The pastures become degraded after only two or three years. *Carex physodes* is also reduced and rarely reaches the blooming stage. Certain annual species which were not eaten were able to increase, such as *Ceratocephalus falcatus*, but the desirable species disappeared because they were not able to bloom. This can be seen clearly from Fig. 6.23 and, even more clearly, from Table 6.13.

Grazing during both spring and autumn on the same area is equally damaging. When a degraded pasture is completely protected, it recovers. The recovery is very slow, however, whereas degradation proceeds very rapidly.

Continuous grazing in summer is also damaging because the dry sand is so loose and the hooves of the sheep sink in so deeply that even the rhizomes of *Carex physodes* are damaged. In such cases the annual grasses increase considerably, especially *Bromus tectorum*. One may find up to 600 plants per square meter on such areas, but the individuals are so feeble that their grazing value is much reduced (Fig. 6.24). Grazing during the summer every four years had no negative effects.

Shrubs can also serve as forage, but only part of their photosynthetic tissue is eaten off and the

TABLE 6.13

Change in species composition after three years of grazing twice during the spring (after Nechayeva, 1954a)

Species	Number of shoots m^{-2}		Weight in g m^{-2}	
	1940	1943	1940	1943
Carex physodes	290	150	15.2	6.7
Bromus tectorum	82	345	3.1	2.8
B. macrostachys	7	5	0.2	0.1
Trisetum cavanillesii	20	50	0.2	0.3
Eremopyrum orientale	1	–	0.1	–
Ceratocephalus falcatus	12	770	0.1	0.2
Astragalus argilobus	5	2	0.1	0.1
Hypecoum parviflorum	18	5	0.3	0.1
Streptoloma desertorum	10	11	0.3	0.2
Malcolmia grandiflora	3	–	0.2	–
Senecio subdentatus	5	–	0.1	–
Scorzonera hemilasia	4	–	0.2	–
Meniocus linifolius	10	7	0.1	0.1
Total	467	1345	20.2	10.6

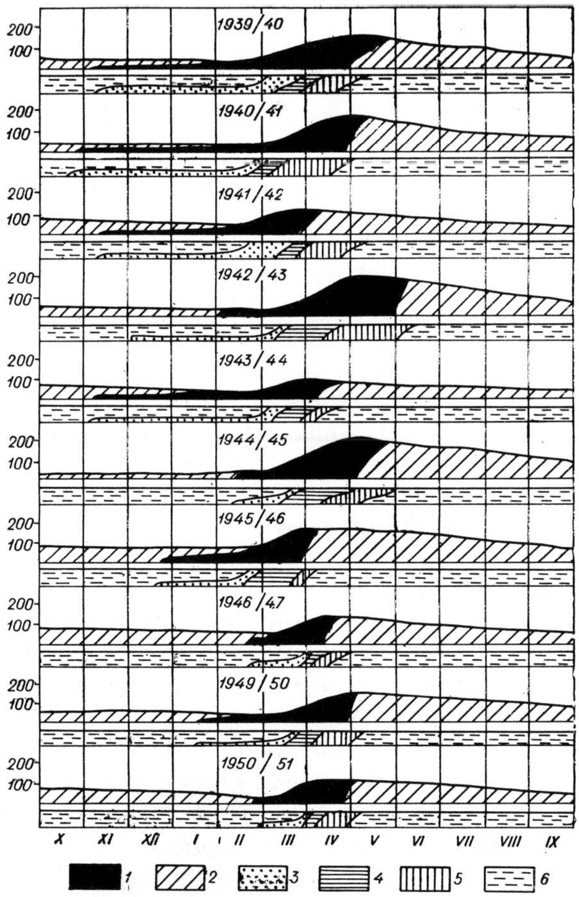

Fig. 6.22. Phenology of amounts ($kg\ ha^{-1}$) of livestock feed provided by *Carex physodes* during the period 1939–40 to 1950–51 in the *Haloxyletum caricosum* of the southeastern Karakum (from Nechayeva, in Beydeman, 1974): *1* = green feed; *2* = dry feed; *3* = vegetative stage; *4* = flowering period; *5* = fruits; *6* = plants dried out.

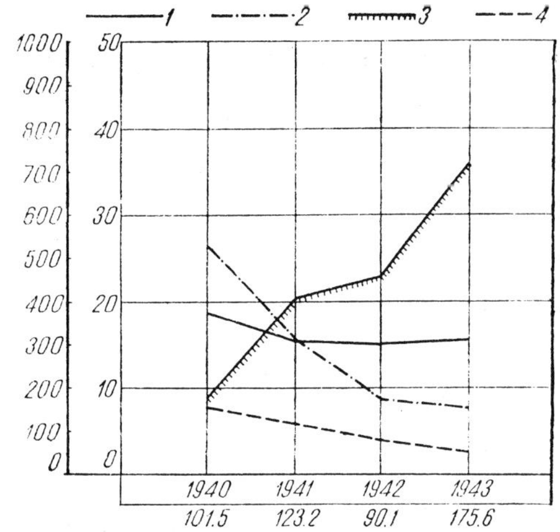

Fig. 6.23. Modification of shoot density and biomass of *Carex physodes* and the annuals by grazing twice a year (in spring) and by maximum grazing stress (from Nechayeva, 1954b). *Carex physodes*: *1* = number of shoots per m^2, *2* = their dry weight in $g\ m^{-2}$; ephemerals: *3* = number of plants per m^2, *4* = their dry weight in $g\ m^{-2}$. Ordinates: number of plants per m^2 (left row) and dry weight in $g\ m^{-2}$ (right row). The lower row of numbers shows the precipitation during 1940–43.

plants are not damaged. Of the total plant matter eaten, that from shrubs made up the following amounts: spring, 25%; summer, 13%; autumn, 33%; winter, 48%. The corresponding fractions of photosynthetic tissue lost are as follows: early spring, 5%; late spring, 12%; summer, 9%; autumn 18%; winter, 22%. In autumn and winter it is more the branch litter which is eaten by the sheep.

It can be seen from these figures that a certain degree of rotation is necessary in order to achieve the best results. A wide variety of grazing systems was tried.

It was found that the pasture areas suffer the least when they are divided into many lots, each of which is grazed alternately for a short period during each season (rotation grazing). That presupposes, however, that all the paddocks have similar plant covers, which is seldom the case. With heterogeneous pastures the rotation must be adapted to the particular local conditions. One must also consider the locations of the watering places. One variation, that of alternate grazing of a paddock in the spring and summer one year and in autumn and winter the next year, proved to be quite useful.

Under rational use with moderate occupancy, optimal utilization could be sustained without damaging the plant cover of the pastures or altering the composition in a given direction. Short grazing during each season most closely resembles the original use by the wild herds, which constantly changed location and never remained for very long at any one place. This is, of course, more difficult to carry out.

Within the framework of these experiments it was also possible to study the seed supply in the sands and the distribution of seeds of individual species throughout the different biotopes (Nechayeva, 1954b). A good production of seeds and a sufficient supply in the soil are important

Fig. 6.24. *Bromus tectorum*: a, on an ungrazed area; and b, stunted individuals on an area which is grazed continuously in summer (from Nechayeva, 1954b).

factors for the rejuvenation of the stands when they are grazed by the rotation method.

Other than grazing, the interference of man in the desert has been minimal. The population of Turkmenistan (Turkmen S. S. R.) has increased from about 1 million in 1913 to 1.5 million in 1959 and 2.2 million in 1970, but most of this population is in the southern fringe along the Amu-Dar'ya, in the delta area and along the Caspian Sea. Even here it is concentrated in the few cities. Irrigated agriculture on about one million hectares is also very important. Irrigation serves cotton plantings primarily, and all of the agricultural areas lie outside the Karakum.

Industrialization and the quick growth of the cities has been brought on by the rich oil and gas fields in the area. The oil production for 1967 was given as 12 million tons, primarily in the region between the Balkhan Mountains and the Caspian Sea.

The salt deposits around the Kara-Bogaz Bay are also important. The bay is connected to the Caspian Sea only by a channel which is only 10.5 km long and 100 to 300 m wide. The bay covers 13 000 km^2 and is, because of the high rate of evaporation, about 4 m lower than the Caspian Sea. As a result, a continuous stream of saltwater (200 to 300 m^3 s^{-1}) flows from the sea into the bay. With time, an accumulation of 48×10^9 tons of various salts of sodium, magnesium, potassium, chlorine, bromine, and iodine has taken place. Two-thirds of this amount is available today in solid form and is being processed chemically.

As a result of the industrialization of the border areas, the influence of man in the Karakum Desert will certainly become more noticeable, especially if oil is also found in the delta area of the original Amu-Dar'ya. Oil has already been discovered at one place near Darvaza, on the border between the central Karakum and the Trans-Unguz section of the Karakum.

PRODUCTION STUDIES IN THE KARAKUM[1]

Studies at the Station Karrykul

The station at Karrykul (38°14′N and 58°16′E) was founded in 1960 in connection with the grazing experiments. Here, 90 km from Ashkhabad, an experimental area 80 km long and 30 km wide was marked off. Of this area, 50 ha are protected from grazing (Fig. 6.25 and 6.26).

This is the original delta region of the river

[1] See also data in Nechayeva et al. (1979).

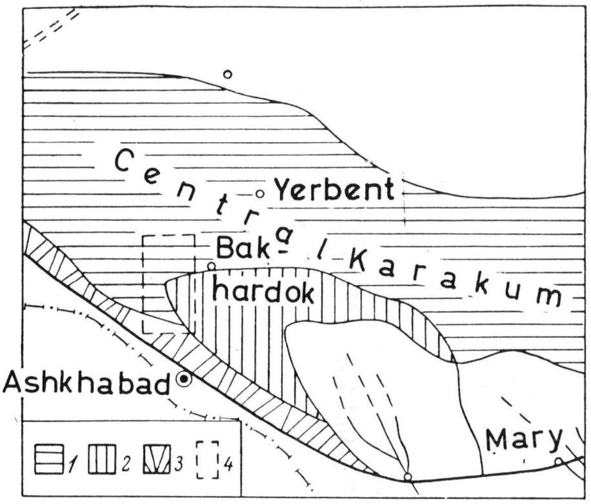

Fig. 6.25. Location of the experimental area at Karrykul, north of Ashkhabad: *1* = old alluvial deposits in the Karakum sand region; *2* = old delta deposits of the Tedzhen; *3* = alluvial deposits at the foot of the Kopet Dag; *4* = experimental area (from Rodin, 1977, with alterations).

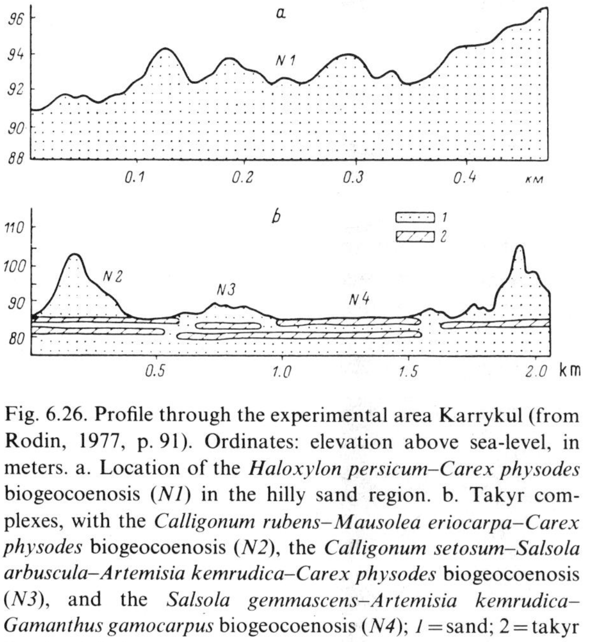

Fig. 6.26. Profile through the experimental area Karrykul (from Rodin, 1977, p. 91). Ordinates: elevation above sea-level, in meters. a. Location of the *Haloxylon persicum–Carex physodes* biogeocoenosis (*N1*) in the hilly sand region. b. Takyr complexes, with the *Calligonum rubens–Mausolea eriocarpa–Carex physodes* biogeocoenosis (*N2*), the *Calligonum setosum–Salsola arbuscula–Artemisia kemrudica–Carex physodes* biogeocoenosis (*N3*), and the *Salsola gemmascens–Artemisia kemrudica–Gamanthus gamocarpus* biogeocoenosis (*N4*); *1* = sand; 2 = takyr (horizontal clay).

Tedzhen. It is a slightly inclined, clayey plain, on top of which were deposited aeolian sands. The original vegetation was replaced over the last 80 to 100 years by a highly anthropogenic vegetation. Of the large species *Haloxylon ammodendron* and *H. persicum*, only scattered patches remain.

Four biogeocoenoses were studied:

(1) The *Haloxylon persicum–Carex physodes* biogeocoenosis. This covers the sand ridges, which make up 70% of the total area of plant-covered sands. Only at the tops of the ridges do fragments with *Aristida karelinii* occur. A total of 50 species are present; the cover is 30%.

(2) The *Calligonum rubens–Mausolea eriocarpa–Carex physodes* biogeocoenosis, often with *Salsola richteri* and *Astragalus longipetiolatus*, on the slopes of sand ridges with sand layers of at least 3 m in depth. The sand is immobile. The number of species is 60 and the cover 35%.

(3) The *Calligonum setosum* + *Salsola arbuscula–Artemisia kemrudica–Carex physodes* biogeocoenosis on thinner sand areas and on lower slopes underlain by clay. The water supply is greater. The number of species is 78 and the cover 45%. Blowouts occur in some places and are colonized by *Tournefortia* and *Heliotropium.*

(4) The *Salsola gemmascens* + *Artemisia kemrudica–Gamanthus gamocarpus* biogeocoenosis on clayey takyr-like areas. *S. gemmascens* often forms pure stands. The number of species is quite low at 10 to 15, and the cover is only 25%. In addition to *Gamanthus* there occur also other annual, late-blooming Chenopodiaceae. Spring ephemerals are unimportant. On less dense, loamy areas or on areas with shallow sand deposits one finds *Artemisia kemrudica* dominating.

The precipitation at the Karrykul station is 148 mm, lower than in Ashkhabad (220 mm), but the seasonal distribution is the same — winter rainfall with a maximum in the spring.

In the description of the life-cycle of the plants, some interesting facts about the root systems came to light. The *Aristida* species have, here also, the sand sheaths which were first described for *Aristida* in southern Africa (see p. 132). This protects them from drying out in the dry sand and protects them from being blown away. The roots are numerous (40–450) and up to 15 m long; they penetrate downward to a depth of about 2.2 m. Water uptake occurs through whorls of thin absorbing roots which live only 2 to 13 weeks.

Heliotropium, *Tournefortia*, and *Jurinea* have rhizomes up to 5 to 7 m long, from which vertical roots go down 1 to 2.5 m. *Heliotropium* and *Tournefortia* also form two types of short-lived absorbing roots: (1) 3 to 4 cm long and over 1 mm thick; and (2) 7 cm long but thin. *Jurinea* has only the short roots. *Tournefortia* has thickenings on its vertical roots, which serve to store water.

The rhizomes of *Carex physodes* have fine rootlets with thick tomentum which holds the sand together. This species also has about 30 (up to 70) cm long roots, which live 7 to 8 years, as well as the

TABLE 6.14

Phytomass and annual dying parts for communities around the Station Karrykul (t ha^{-1}, air-dried weight): means for 1960–1967 (from Rodin, 1977, pp. 106–107)

	Trees and shrubs	Semi-shrubs	Perennial herbs		Annual herbs		Total (rounded)
			spring/ summer–fall	winter–spring	spring/ summer–fall	winter–spring	
1. *Haloxylon persicum–Carex physodes*							
Total phytomass	6.035	0.015	0.014	1.890	0.009	0.110	8.1
green parts	0.270	0.004	0.006	0.090	0.003	0.100	0.5
wood	3.280	0.001	–	–	–	–	3.3
roots and rhizomes	2.485	0.010	0.008	1.800	0.006	0.010	4.3
Annual litterfall	0.664	0.005	0.004	0.442	0.008	0.090	1.2
green parts	0.190	0.003	0.003	0.082	0.003	0.100	0.4
wood	0.328	–	–	–	–	–	0.3
roots and rhizomes	0.146	0.002	0.001	0.360	0.006	0.010	0.5
Litterfall as % of phytomass	11	30	30	23	100	100	–
2. *Calligonum rubens–Mausolea eriocarpa–Carex physodes*							
Total phytomass	3.355	1.698	0.015	2.226	0.007	0.110	7.4
green parts	0.186	0.108	0.007	0.106	0.005	0.100	0.5
wood	1.480	0.609	–	–	–	–	2.1
roots and rhizomes	1.689	0.981	0.008	2.120	0.002	0.10	4.8
Annual litterfall	0.249	0.214	0.005	0.522	0.006	0.092	1.1
green parts	0.138	0.084	0.004	0.098	0.005	0.100	0.4
wood	0.026	0.032	–	–	–	–	0.1
roots and rhizomes	0.085	0.098	0.001	0.424	0.002	0.010	0.6
Litterfall as % of phytomass	7	13	26	23	100	100	–
3. *Calligonum setosum + Salsola arbuscula–Artemisia kemrudica–Carex physodes*							
Total phytomass	3.437	1.250	0.033	1.869	0.018	0.130	6.7
green parts	0.128	0.106	0.015	0.089	0.006	0.118	0.5
wood	1.578	0.766	–	–	–	–	2.3
roots and rhizomes	1.731	0.378	0.018	1.780	0.012	0.012	3.9
Annual litterfall	0.204	0.139	0.009	0.440	0.016	0.107	0.9
green parts	0.087	0.066	0.008	0.080	0.006	0.118	0.3
wood	0.030	0.035	–	–	–	–	0.1
roots and rhizomes	0.087	0.038	0.001	0.360	0.012	0.012	0.5
Litterfall as % of phytomass	6	11	27	24	100	100	–
4. *Salsola gemmascens + Artemisia kemrudica–Gamanthus gamocarpus*							
Total phytomass	0.486	4.409	–	0.042	0.180	0.008	5.1
green parts	0.012	0.369	–	0.002	0.060	0.007	0.4
wood	0.313	2.489	–	–	–	–	2.8
roots and rhizomes	0.161	1.551	–	0.040	0.120	0.001	1.9
Annual litterfall	0.028	0.501	–	0.010	0.160	0.007	0.7
green parts	0.012	0.369	–	0.002	0.060	0.007	0.4
wood	0.006	0.127	–	–	–	–	0.1
roots and rhizomes	0.010	0.120	–	0.008	0.120	0.002	0.3
Litterfall as % of phytomass	5	11	–	23	100	100	–

TABLE 6.15

Changes in green phytomass over several years for communities near the Station Karrykul (t ha^{-1} air-dried weight) (from Rodin, 1977, p. 107)

	1960	1961	1962	1963	1964	1965	1966	1967	Mean for eight years
1. *Haloxylon persicum–Carex physodes*									
Green phytomass	0.258	0.378	0.566	0.402	0.924	0.300	0.657	0.268	0.473
% of mean	55	80	120	85	195	63	139	57	100
2. *Calligonum rubens–Mausolea eriocarpa–Carex physodes*									
Green phytomass	0.327	0.265	0.508	0.522	0.909	0.343	0.918	0.313	0.512
% of mean	64	52	99	102	178	67	180	61	100
3. *Calligonum setosum + Salsola arbuscula–Artemisia kemrudica–Carex physodes*									
Green phytomass	0.265	0.201	0.498	0.491	0.859	0.289	0.842	0.253	0.462
% of mean	57	43	108	106	186	63	182	55	100
4. *Salsola gemmascens + Artemisia kemrudica–Gamanthus gamocarpus*									
Green phytomass	0.217	0.078	0.439	0.353	0.846	0.216	0.153	0.389	0.450
% of mean	48	17	96	79	188	48	234	87	100

short-lived roots (6–15 days), only 1 cm long. These latter are formed on the rhizomes when the upper soil layers are thoroughly wetted and warm.

In each biogeocoenosis several synusiae can be distinguished, based on growth forms and ecological relations. These permit full utilization of the environmental possibilities for production.

Information on the phytomass and annual litter is given in Table 6.14. The formation of green assimilating parts fluctuates considerably from year to year and contributes to variation in production. This can be seen in Table 6.15.

As a result of heavy earlier use, the biogeocoenoses are not in equilibrium. Succession leading to a nearly complete plant cover over sand can be observed quite clearly on the protected area. Stump-sprouts could be seen on the *Haloxylon ammodendron* stumps but not on *H. persicum*.

Studies at the Station Repetek

The most complicated of the biogeocoenoses, that of the groundwater-bound *Haloxylon ammodendron–Carex physodes* system, was studied within the 34 000 ha of the protected area at the station Repetek (Fig. 6.27). The summary of the results covers pp. 118–201 in Rodin (1977). We limit ourselves to the most important data.

The climatic conditions during the study-years 1965–72 are shown in the climatogram (Fig. 6.2).

The *Haloxylon ammodendron–Carex physodes* biogeocoenosis has a canopy cover of 30% and a total cover of 90%. *Tortula desertorum* appears in patches, covering 2 to 3% of the total area. The number of species is 60.

Carex physodes can survive only between the crowns of *Haloxylon*, since it is killed underneath the crowns by the saline litter. The result is that the water content of the soil under the crowns of the trees is 4 to 5 times higher in summer than under the *Carex physodes*. Young *Haloxylon* plants can thus establish themselves best just under the *Haloxylon* crowns, where there is no competition from *Carex*. The crown of *H. persicum* is not as large, and its litter is only barely salty, so *Carex physodes* can grow underneath this species.

The phytomass of the *Haloxylon ammodendron–Carex physodes* biogeocoenosis is described by Miroshnichenko (1974) by data reproduced here as Table 6.16.

The average annual litter production is described in Table 6.17.

The total, above-ground dead organic mass in this biogeocoenosis was 3.42 t ha^{-1} (trunks 0.64, branches within the canopy 0.82, litter on the ground 1.96 t ha^{-1}). Of this total, 76% was from

TABLE 6.16

Amounts of phytomass (t ha^{-1} dry weight) in *Haloxylon ammodendron–Carex physodes* communities (from Miroshnichenko, 1974, p. 333)

Fraction of phytomass	*Haloxylon ammodendron*	*H. persicum*	*Ephedra strobilacea*	*Calligonum setosum*	*C. caput-mudusae*	Other shrubs	*Carex physodes*	Ephemerals and forbs	Total phyto mass	% of total
Green shoots	0.72	0.05	0.24	0.007	0.012	0.01	0.2	0.04	1.279	4.12
vegetative	0.63	0.025			0.01					
growing	0.06	0.018	0.24	0.007		0.01	0.02	0.04		
flowering	0.03	0.07			0.002					
Perennial parts	5.14	0.12	0.51	0.10	0.133	0.03	0	0	6.033	19.44
biennial branches	0.34	0.02	0.1	0.06	0.01		0	0		
branches <1 cm diam.	0.51	0.03			0.025		0	0		
branches >1 cm diam.	1.38	0.03	0.2	0.03	0.021	0.03	0	0		
trunks	2.91	0.04	0.21	0.01	0.077		0	0		
Live above-ground phytomass	5.86	0.17	0.75	0.107	0.145	0.04	0.2	0.04	7.312	23.56
Roots and rhizomes	9.53	0.27	0.84	0.24	0.22	0.05	8.23	0.044	19.424	62.57
very fine roots (<0.5 mm)	1.17	–	0.05	0.004	0.01		1.54			
fine roots (0.5–4 mm)	0.5	–	0.05	0.006	0.02		5.29	0.044		
medium roots (4–20 mm)	1.28	–	0.14	0.04	0.02	0.05	0			
large roots (>20 mm)	6.58	–	0.60	0.19	0.17					
rhizomes	0	0	–	0	0	0	1.4	0		
Total living phytomass	15.39	0.44	1.59	0.347	0.365	0.09	8.43	0.084	26.736	86.13
Dead branches in crown	0.57	0.01	0.04	0.04	0.16	0.01	0	0	0.83	2.67
Total above-ground phytomas (alive+dead)	6.43	0.18	0.79	0.147	0.305	0.05	0.2	0.04	8.142	26.23
Annual above-ground litter	0.72	0.037	0.02	0.005	0.014	0.01	0.2	0.04	1.046	3.37
Perennial branches	0.05	0.002	0.01	0.001	0.002	0.01	0	0		
Annual shoots	0.67	0.035	0.01	0.004	0.012		0.02	0.04		
Litter (total lying)	1.42	0.04	0.28	0.03	0.06	–	0.01	0	1.830	5.89
small twigs (1 mm)	0.38	0.01	0	0	0.02	–	0.01	0		
medium twigs (1–3 mm)	0.04	0.01	0	0.01	0.005	–	0	0		
large twigs (>3 mm)	0.36	0.01	0.25	0.01	0.025	0	0	0		
branches	0.27	0.01	0.03	0.01	0.009	0	0	0		
Dead wood (lying)	0.59	–	0.01	0	0	0	0	0	0.600	1.94

TABLE 6.17

Annual litter (t ha^{-1}) for the *Haloxylon ammodendron–Carex physodes* biogeocoenosis (Repetek, 1969–1972) (from Rodin, 1977)

Plant parts	*Haloxylon ammodendron*	Shrubs	*Carex physodes*	Ephemerals	*Tortula desertorum*	Algae	Total	
							t ha^{-1}	%
Annual shoots and leaves	0.67	0.176	0.20	0.040	0.09	0.03	1.206	16
Branches	0.21	0.090	–	–	–	–	0.300	5
Trunks	0.24	0.100	–	–	–	–	0.340	5
Underground parts	2.10	0.420	2.90	0.044	0.10	–	5.564	74
Total (t ha^{-1})	3.22	0.786	3.10	0.084	0.19	0.03	7.410	
Total (%)	43	11	42	1	3	1	100	

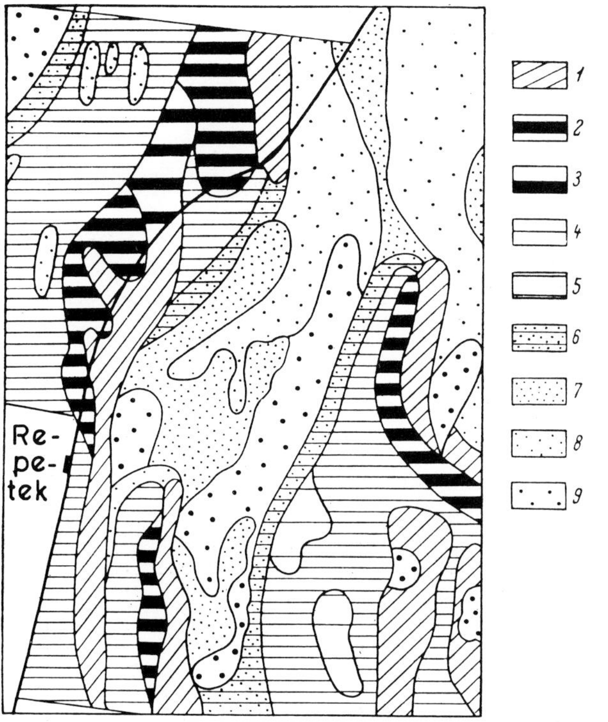

Fig. 6.27. Subdivisions of the vegetation around the experimental station at Repetek on the railway Mary–Chardzhou (from Rodin, 1977, p. 119); *1* = sand ridges with *Haloxylon persicum* and *Ammodendron*; *2* = depressions with *Haloxylon*, *Ammodendron* and *Carex physodes*; *3* = the same, but also with *H. persicum* on the uneven sand hillocks, *4* = typical *H. persicum–Carex physodes* biogeocoenosis on stable sand; *5* = the same, but with *Tortula*; *6* = more mobile sands; *7* = *Aristida karelinii–Ammodendron–Aristida* biogeocoenosis on highly mobile low dunes; *8* = the same on dunes of intermediate height; *9* = the highest dunes, with *Aristida* and *Eremosparton flaccidum*.

Haloxylon, 20% from other shrubs, 1% from *Carex physodes*, and 1% from the moss *Tortula*. About 77% of the litter lay under the crowns of the shrubs, 92% of this fraction under *Haloxylon* and half within 1 m of the trunks. This non-uniform distribution of litter produces a mosaic structure in the plant cover.

The annual production (growth) is estimated as 7.47 t ha^{-1}, hardly more than the annual litter (7.41 t ha^{-1}). Of this production, the assimilating organs make up 1.27 t ha^{-1} (17%), the woody parts (branches and trunks) 0.64 t ha^{-1} (8%), and the roots 5.56 t ha^{-1} (75%).

The main producer is *Haloxylon ammodendron* (64% of the total); in second place is *Carex physodes* with 3.1 t ha^{-1}. These two species together represent 76% of the production. The fraction represented by assimilating parts of *Haloxylon* decreases with age from 83% to 56%, while the woody fraction increases with age from 17% to 44% of the aboveground production.

The lower plants are also of some importance here. The moss synusia, which also always contains algae, was studied in special detail by Novichkova-Ivanova (1976). The gametophytes grow for at least three years. The development cycle is shown in Fig. 6.28. In addition to reproduction by spores, there is also vegetative reproduction by means of gemmae on the rhizomes, which represents an adaptation to the adverse conditions in the desert.

Mosses grow poorly in the hollows, since they are easily covered by drifting or blown sand and are destroyed by rodents. The production of the mosses

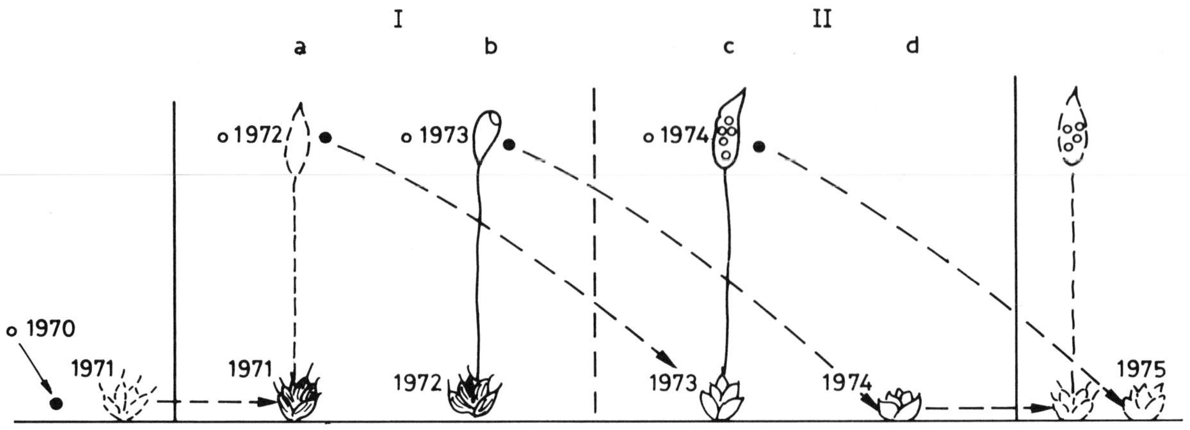

Fig. 6.28. *Tortula desertorum* synusia with plants of different ages (*a–d*) in the spring 1974; ○ before the number of the year means in the fall of that year; --- are parts already dead or not yet formed (1975); *I* = slowly growing plants, *II* = faster growing (from Novichkova-Ivanova, 1976).

is estimated as 0.2 t ha^{-1}, while the phytomass of the algae is estimated as an average of 0.03 t ha^{-1}. These algae are primarily Cyanophyta, which form a film on the soil surface. These synusiae on open sites without *Carex physodes* can produce a dry weight of 20 g m^{-2}. Of course the weather has a very strong effect, as shown in Table 6.18.

The phytomass of the algae is turned over many times during the course of a growing season, so the production can be enormous. These algae can also fix nitrogen from the air.

The production of the algae and mosses together is estimated as 3% of the total for the biogeocoenosis.

The ash components and their cycles were also studied. With the decomposition of the litter, these elements go back into the soil and can lead to salinization of the soil, as already mentioned

TABLE 6.18

Phytomass of algae in g m^{-2} at Repetek, after Novichkova-Ivanova (1976)

Year	Weather	Under crown of *Haloxylon ammodendron*	Between *Carex physodes*
1969			
April 12	cool and wet	0.42	0.24
April 18	cool and wet	0.24	0.12
1970			
April 19	warm and	20.33	11.18
April 26	moderately	0.34	0.42
May 3	wet	0.96	8.16
1972			
April 8	cool and	0.06	5.28
April 11	dry	0.07	0.06
April 14		0.07	2.64
1973			
April 26	cool and wet	0.12	3.60
May 3	for some time	0.03	4.00

(p. 137). *Haloxylon ammodendron* has a very high sodium content (6.19% of dry weight) and somewhat lower chlorine content (0.69% of the dry weight). It is a typical alkali-halophyte and causes an increase in the alkalinity of the soil. This is true to a much lesser extent for *H. persicum*: sodium 1.32% and chlorine 0.27% of the dry weight. In this respect the litter of *Carex physodes* is insignificant: sodium 0.06% and chlorine 0.18% of dry weight.

The ash analyses of Rodin (from Walter, 1968, pp. 785 and 787) are shown in graphic form in Fig. 6.29 and 6.30.

The large tables 65 to 69 of Rodin (1977) contain detailed data on the chemical analyses of soil samples from various depths, the nitrogen content

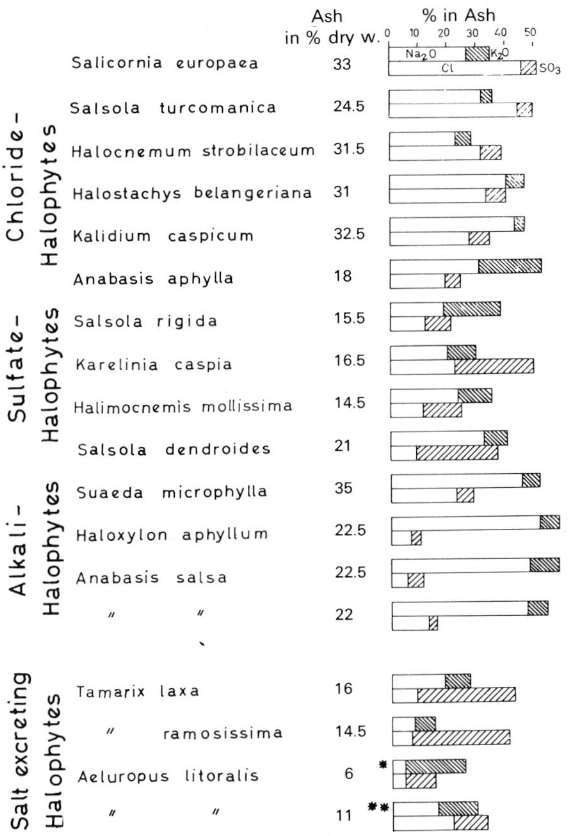

Fig. 6.29. Ash analyses of halophytes, computed and graphed from data of Rodin (1961) (from Walter, 1968, p. 785). The meanings of the different columns is explained under *Salicornia*. * = computed on the basis of SiO_2-free ash (65% SiO_2 in the ash); ** = same (36% SiO_2 in the ash).

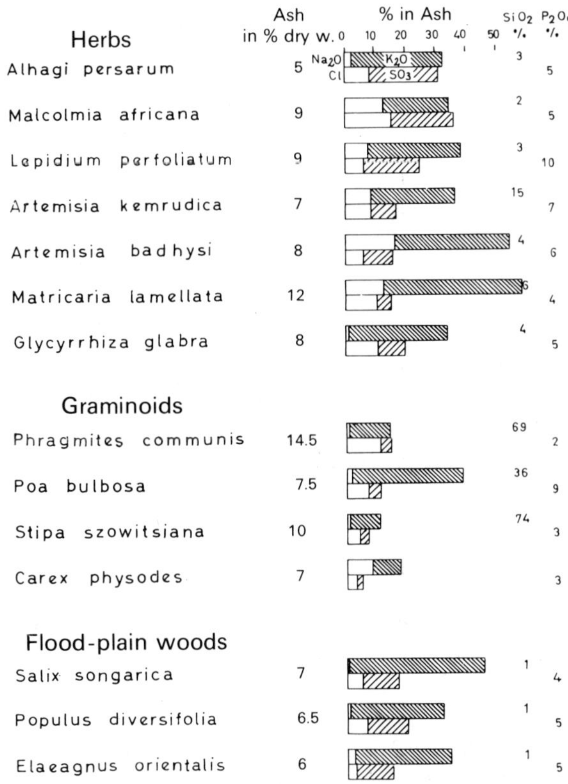

Fig. 6.30. Ash analyses of Middle Asian non-halophytes, computed from data of Rodin (1961) (from Walter, 1968, p. 787). The legend is given under *Alhagi*.

and ash-element content in the litter, in the dead plant parts, in the annual litter, and in the plants themselves for the *Haloxylon ammodendron–Carex physodes* biogeocoenosis. From these figures certain retrospective conclusions can be drawn concerning the mineral cycling and the salinization of the soil (see also the English summary by Cherpurko et al., 1972).

About 0.5 kg of litter with 0.07 kg of ash elements falls annually on each square meter under *Haloxylon ammodendron*. From this amount the following chemicals are formed during decomposition:

Na_2CO_3	42 g	$MgCO_3$	26 g
K_2CO_3	32 g	NaCl + KCl	4 g
$CaCO_3$	27 g	$CaSO_4$	4.4 g

This leads to the salinization of the soil.

This is much less true under *H. persicum*, where

only 0.04 kg m^{-2} of ash elements reach the soil. The amounts of substances formed are:

Na_2CO_3	8 g	$MgCO_3$	18 g
K_2CO_3	18 g	$NaCl + KCl$	1.7 g
$CaCO_3$	40 g	$BaSO_4$	3.9 g

In this case it is primarily calcium which accumulates. The mineralization of the soil by *Carex physodes* litter can be ignored.

The phenological spectra of the most important perennial species in the Karakum Desert are shown, for two years with very different weather patterns, in Fig. 6.31.

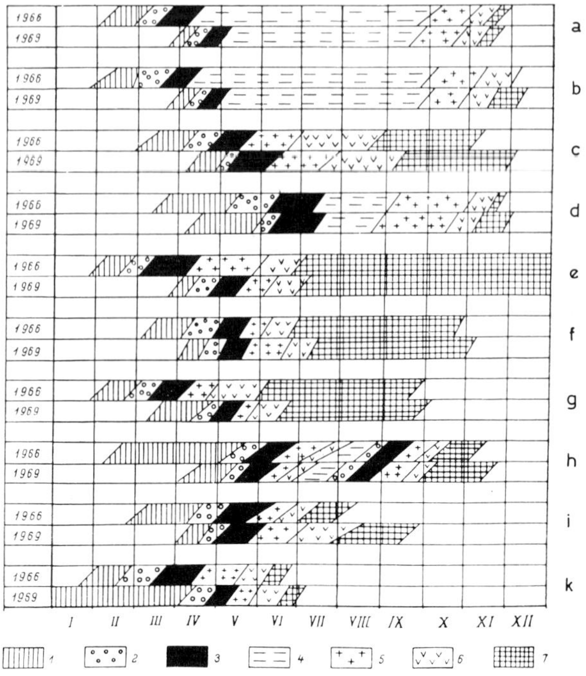

Fig. 6.31. Phenological spectra of the most important perennial species of the Karakum Desert, covering two years with different weather patterns (from Rodin, 1977, p. 156): *a* = *Haloxylon ammodendron*; *b* = *H. persicum*; *c* = *Ammodendron conollyi*; *d* = *Salsola richteri*; *e* = *Ephedra strobilacea*; *f* = *Calligonum caput-medusae*; *g* = *Astragalus paucijugus*; *h* = *Aristida karelinii*; *i* = *Heliotropium arguzioides*; *k* = *Carex physodes*; *1* = vegetative growth; *2* = formation of flowering buds; *3* = flowering; *4* = fruit-formation not yet discernible; *5* = green fruits; *6* = seed propagation; *7* = vegetative growth after seed propagation.

Zoomass

Compared with phytomass the zoomass is almost negligible, but it was determined when possible. The data given represent numbers of individuals per hectare in the sand area and zoomass in kg ha^{-1}. Table 6.19 summarizes the results.

Data were also given for invertebrates. The invertebrate zoomass in biogeocoenosis III was 10.1 kg ha^{-1} in the tree canopy and 15.0 kg ha^{-1} in the soil. Spiders (Araneae) increased from 7800 ha^{-1} in February (zoomass <0.1 kg ha^{-1}) to 66 000 ha^{-1} in July (zoomass 0.53 kg ha^{-1}), then decreased to 19 600 ha^{-1} in October. The number of insects is very small.

Various groups of invertebrates alternate in the canopy of *Haloxylon ammodendron*, and corresponding quantitative fluctuations in the zoomass can be seen. The same is true for the organisms in the soil, both under tree crowns and between them (see Table 6.20).

Production figures for regions neighboring the Karakum

One can obtain the values shown in Table 6.21 for typical takyr vegetation.

TABLE 6.19

Number of individuals per hectare and zoomass in kg ha^{-1} for vertebrates in the protected area (III represents the *Haloxylon ammodendron–Carex physodes* biogeocoenosis of Kuznetzov, in Rodin, 1977)

Biogeocoenosis	Mammals		Birds		Tortoises		Other reptiles	
	number	mass	number	mass	number	mass	number	mass
I (mobile sand)	4.3	0.3	0.8	0.02	–	–	41.0	0.21
II (immobile sand)	7.0	0.7	1.4	0.05	0.4	0.3	19.0	0.10
III (hollow)	16.4	1.4	2.6	0.06	0.6	0.5	24.8	0.8

TABLE 6.20

Seasonal variation in zoomass (kg ha^{-1}) in the canopy of *Haloxylon ammodendron* and in the soil (from Rodin, 1977)

	February	April–May	July	October
Invertebrates in tree crowns	2.8	10.1	1.2	1.9
Invertebrates in the soil under crowns	7.0	15.0	?	8.5
Invertebrates in the soil between crowns	?	1.5	?	?

TABLE 6.21

Production of takyr vegetation (after Rodin, 1954)

Zone	Phytomass (t ha^{-1})	Primary production (t ha^{-1} yr^{-1})	Nitrogen (kg ha^{-1})	Ash (kg ha^{-1})
I	0.1	0.1	1.5	6–7
II	0.3	0.3	9–10	20–22
IIIa	1.2–1.6	1.2–1.6	20–30	90–140
IIIb	12	10	90	475

I = Lowest-lying areas, with clay deposition; soils saline; only soil algae.
II = Somewhat higher areas, with soil algae or lichens; few ephemeral seed-plants.
IIIa = Transition zone with patchy ephemeral vegetation.
IIIb = *Artemisia* semi-desert with rich ephemeral vegetation on sierozem.

TABLE 6.22

Productivity of succulent halophytes on solonchak with a cover of 10% (after Bazilevich et al., 1972)

Species	Phytomass (t ha^{-1})	Roots (t ha^{-1})	Primary production (t ha^{-1} yr^{-1})
Halocnemum strobilaceum	1.76	1.04	0.5–0.7
Kalidium caspicum	1.38	1.04	0.5–0.7

For the halophilic plants of the wet soils in southwestern Turkmenistan a few data are also available (Table 6.22). These are chloride-halophytes. The biohalogens make up 66 to 75% of the total ash (67 to 72% for annual shoots), but the salt content of the soil, in spite of the salt-containing litter, is lower under the plants than in the vegetationless areas between plants.

Very detailed tables are published by Rodin and Basilevich (1965) for other desert communities on the edge of the Karakum. When one consults the original, one must remember that the floodplain forest and reed communities receive considerable additional water. These resemble forest communities more closely in productivity.

REFERENCES

Anonymous, 1969a. *Biocomplex Investigations in Kazakhstan. Part I: Plant Communities and Animal Populations in the Steppes and Deserts of Central Kazakhstan.* Akad. Nauk S.S.S.R., Nauka, Leningrad, 496 pp. (in Russian).

Anonymous, 1969b. *Biocomplex Investigations in Kazakhstan. Part II: Biocomplex Characteristics of the Main Coenosis-Forming Plants in the Vegetation of Central Kazakhstan.* Akad. Nauk S.S.S.R., Nauka, Leningrad, 336 pp. (in Russian).

Anonymous, 1975. *Resources of the Biosphere: Synthesis of the Soviet Studies for the IBP, 1.* Nauka, Leningrad, 288 pp. (in Russian).

Anonymous, 1976. *Biocomplex Investigations in Kazakhstan. Part III: Main Plant Communities of the Central*

Kazakhstan Desert-Steppe. Akad. Nauk S.S.S.R., Nauka, Leningrad, 292 pp. (in Russian).

Arcichovski, V.M. and Opsipov, A., 1931. Die Saugkraft der zentralasiatischen Wüsten nebst Transpirationsmessungen am Saksaul. *Planta* (Berlin), 14: 552–565.

Bazilevich, N.I., Rodin, L.E. and Gorina, A.I., 1972. Productivity and biochemistry of succulent communities on solonchaks. In: L.E. Rodin (Editor), *Int. Symp. Ecophysiological Foundations of Ecosystems Productivity in Arid Zones*. Nauka, Leningrad, pp. 203–207.

Beydeman, I.N., 1974. *Methoden zur Untersuchung der Phänologie von Pflanzen und Pflanzengesellschaften*. Nauka, Novosibirsk, 155 pp.

Blagoveshchenskiy, E.N., 1954a. The function of the Kara-Kum Station in relation to afforestation in deserts. *Pustyn. S.S.S.R. Ikh Osvoyeniye*, 2: 181–190 (in Russian).

Blagoveshchenskiy, E.N., 1954b. On the water condensation in desert soils. *Pustyn. S.S.S.R. Ikh Osvoyeniye*, 2: 508–519 (in Russian).

Blagoveshchenskiy, E.N., 1955. The desert Cheleken and the possibility of its development. In: M.P. Petrov (Editor), *Papers of the Sand-Desert Station Repetek*. Ashkhabad, pp. 180–203 (in Russian).

Bobrovskaya, N.I., 1969a. New data on the water deficits of desert plants. *Bot. Zh.*, 54: 1054–1058 (in Russian).

Bobrovskaya, N.I., 1969b. On the water balance of tree and shrub species in the sand desert Kara-Kum. *Bot. Zh.*, 56: 361–368 (in Russian).

Bogdanov, V.V., 1954. Periodical runoff on the takyry. *Pustyn. S.S.S.R. Ikh Osvoyeniye*, 2: 488–507 (in Russian).

Bykov, B.A., 1973. *Introduction to Phytocoenology*. Nauka, Alma-Ata, 319 pp. (in Russian).

Cherpurko, N.L., and others, 1972. Biogeochemistry and productivity of *Haloxyloneta ammodendroni* in the south-eastern Kara-Kum deserts. In: L.E. Rodin (Editor), *Int. Symp. Ecophysiological Foundations of Ecosystem Productivity in Arid Zones*. Nauka, Leningrad, pp. 198–203.

Kalenov, G.S., 1972. The ecology of *Haloxylon ammodendron* Bge. in the Kara-Kum desert. In: L.E. Rodin (Editor), *Int. Symp. Ecophysiological Foundations of Ecosystem Productivity in Arid Zones*. Nauka, Leningrad, pp. 7–19.

Klyushkin, E.A., 1955. The Repetek Natural Reserve. *Tr. Inst. Biol. Askhabad*, (1): 20–30 (in Russian).

Koch, C., 1961. Some aspects of abundant life in the vegetationless sand in the Namib Desert dunes. *J. S.W. Afr. Sci. Soc.* (*Windhoek*), 15: 8–34; 79–82.

Kokina, S.L., 1935. Water-economy and inner factors of resistance in plants of the sand desert Kara-Kum. *Probl. Rasteniyevod. Osvoyeniye Pustyn*. 4: 99–196 (in Russian).

Korovin, E.P., 1961/62. *Vegetation of Middle Asia and Southern Kazakhstan*. Tashkent, 2nd ed., Vol. I: 452 pp.; vol. II: 577 pp. (in Russian).

Kostychev, S. and Kardo-Syssoleva, H., 1930. Untersuchungen über den Tagesablauf der Photosynthese in Zentralasien. *Planta* (*Berlin*), 11: 117–143.

Kühnelt, W., 1965. Nahrungsbeziehungen innerhalb der Tierwelt der Namibwüste (Südwestafrika). *Sitzungsber. Österr. Akad. Wiss.* (*Math.-Naturw. Kl.*), Abt. I, 1974: 185–190.

Kunin, V.N. (Editor), 1955. *Description of the Nature of the Kara-Kum Region*. Akad. Nauk S.S.S.R., Geogr. Inst., Moscow, 407 pp. (in Russian).

Kunin, V.N., 1969. *Soviet Union: Turkmenistan*. Mysl, Moscow, 277 pp. (in Russian).

Lavrenko, E.M. and Sveshnikova, V.M., 1965. Synthetic research on growth-forms using the example of *Stipa* tussock grasses. *Zh. Obshch. Biol.*, 26: 261–275 (in Russian).

Maljugin, E.A., 1954. Agricultural methods for cultivation of the sandy semi-deserts of western Kazakhstan. *Pustyn. S.S.S.R. Ikh Osvoyeniye*, 2: 66–134 (in Russian).

Michelson, E.G., 1955a. Dynamics of the vegetation of the Kelif-Uzboy. *Tr. Repetekskoy Pestchano-Pustyn. Stantsii III*. Akad. Nauk Turkmen. S.S.S.R., Ashkhabad, pp. 75–140 (in Russian).

Michelson, E.G., 1955b. Dynamics of the vegetation of the eastern Kara-Kum (Repetek). *Tr. Repetekskoy Pestchano-Pustyn. Stantsii III*. Akad. Nauk Turkmen. S.S.R., Ashkhabad, pp. 141–175 (in Russian).

Miroshnichenko, Yu.M., 1974. Biological productivity of the association *Haloxylon ammodendron–Carex physodes* in the eastern Kara-Kum. *Rastit. Resursy*, 10: 329–337 (in Russian).

Miroshnichenko, Yu.M., 1975a. Peculiarities of seasonal dynamics of productivity in phytocoenoses of the Afro-Asian arid area. *Bot. Zh.*, 60: 1164–1178 (in Russian).

Miroshnichenko, Yu.M., 1975b. Root systems of trees and bushes and their ecology in the eastern Kara-Kum. *Bot. Zh.*, 60: 1776–1196 (in Russian).

Miroshnichenko, Yu.M., 1975c. Water supply of plants and their influence on the water content of the soil under Saksaul communities. In: *Vodnyi obmen v osnovnykh tipakh rastitel'nosti S.S.S.R.* Nauka, Novosibirsk, pp. 228–232 (in Russian).

Murzayev, E.M. (Editor), 1958. *Middle Asia*. Nauka, Moscow, 648 pp. with a vegetation map (in Russian).

Nechayeva, N.T., 1954a. The influence of grazing on Kara-Kum pasture-land as a basis for grazing rotation. *Pustyn. S.S.S.R. Ikh Osvoyeniye*, 2: 303–369 (in Russian).

Nechayeva, N.T., 1954b. Seeds stored in the soil of Kara-Kum pastures and the influence of grazing on the trampling of seeds into the soil. *Pustyn. S.S.S.R. Ikh Osvoyeniye*, 2: 370–391 (in Russian).

Nechayeva, N.T. (Editor, with 14 collaborators), 1975. *Biocoenoses of the Eastern Kara-Kum*. Akad. Nauk Turkmen. S.S.R. (Exp. Station Repetek), Ashkhabad, 75 pp. (in Russian).

Nechayeva, N.T. and Kaplina, V.G. (Editors), 1978. *Biocoenosis Research in the East Karakum*. Ylym, Ashkhabad, 138 pp. (in Russian).

Nechayeva, N.T., Vasilevskaya, V.K. and Antonova, K.G., 1973. *Life-forms of the Plants in the Kara-Kum Desert*. Akad. Nauk S.S.S.R., Nauka, Moscow, 243 pp. (in Russian).

Nechayeva, N.T., Antonova, K.G., Karshenas, S.D., Mukhamedov, G. and Nurberdeyev, M., 1979. *Production of Central Karakum Vegetation*. Akad. Nauk, Turkm. S.S.R., Ashkhabad, 255 pp. (in Russian).

Nikitin, S.A. and Meshcheryakov, A.A., 1954. The water and shore vegetation in the Murgab Reservoir. *Pustyn. S.S.S.R. Ikh Osvoyeniye*, 2: 414–429 (in Russian).

Novichkova-Ivanova, L.N., 1976. Moss synusiae of the Kara-Kum Desert (population structure and productivity). *Bot. Zh.*, 61: 1168–1179 (in Russian).

Perskaya, A.D., 1955. Fruits and seeds of sand-stabilizing tree and shrub species in Middle Asia. In: M.P. Petrov (Editor), *Papers of the Sand-Desert Station Repetek.* Ashkhabad, pp. 235–293 (in Russian).

Petrov, M.P. (Editor), 1955. *Papers of the Sand-Desert Station Repetek, III.* Ashkhabad, 358 pp. (in Russian).

Rodin, L.E., 1954. The role of plants in the formation of takyry and their complexes. *Pustyn. S.S.S.R. Ikh Osvoyeniye*, 2: 392–413 (in Russian).

Rodin, L.E., 1961. *Dynamics of Desert Vegetation As Exemplified in Western Turkmenistan.* Akad. Nauk S.S.S.R., Bot. Inst. Moscow–Leningrad, 227 pp. (in Russian).

Rodin, L.E., 1963. *Desert Vegetation of Western Turkmenistan.* Akad. Nauk. S.S.S.R., Bot. Inst. Moscow–Leningrad, 309 pp. (in Russian).

Rodin, L.E., 1972. *Eco-physiological Foundations of Ecosystem Productivity in Arid Zones. Int. Symp. U.S.S.R., 7–19 June 1972.* Nauka, Leningrad, 232 pp.

Rodin, L.E. (Editor), 1977. *Productivity of Vegetation in the Arid Zone of Asia: Synthesis of the Soviet Studies for the International Biological Program 1964–1974.* Akad. Nauk S.S.S.R., Nauka, Leningrad, 232 pp. (in Russian).

Rodin, L.E. and Bazilevich, N.I., 1965. *Dynamics of Organic Matter and Biological Circulation of Ash-Elements and Nitrogen in the Main Vegetation Types of The Earth.* Akad. Nauk S.S.S.R., Bot. Inst., Moscow–Leningrad, 253 pp. (in Russian).

Rotshild, E.B., 1960. The role of trees and shrubs in the formation of the mosaic pattern of vegetation in the deserts. *Bot. Zh.*, 45: 1750–1770 (in Russian).

Stalmakova, V.A., 1954. Rodents of the Kara-Kum and their economic importance. *Pustyn. S.S.S.R. Ikh Osvoyeniye*, 2: 756–782 (in Russian).

Stalmakova, V.A., 1955. Mammals of the sand-desert natural reserve Repetek and the neighboring Kara-Kum Desert. In: M.P. Petrov (Editors), *Papers of the Sand-Desert Station Repetek.* Ashkhabad, pp. 307–349 (in Russian).

Sveshnikova, V.M., 1972. On the absorption of water vapor by the above-ground parts of the Kara-Kum Desert plants. *Bot. Zh.*, 57: 880–888 (in Russian).

Sveshnikova, V.M., 1975. The water regime of desert plants in Middle Asia. *Ekologiya*, 4: 35–42 (in Russian; English translation in *Sov. J. Ecol.*).

Vasilevskaya, V.K., 1954. *Leaf Development in Drought-Resistant Plants.* Akad. Nauk Turkmen. S.S.R., Ashkhabad, 183 pp. (in Russian).

Vasilevskaya, V.K., 1965. Anatomical adaptations of plants in hot and cold deserts of Middle Asia and Kazakhstan. *Probl. Sovrem. Botan.*, 2: 5–18 (in Russian).

Vasilyev, I.M., 1931. Über den Wasserhaushalt von Pflanzen in südostlichen Karakum. *Planta (Berlin)*, 14: 225–309.

Veysov, S., 1976. *Dynamics of the Dune Sand Relief, as Exemplified by the Southeast Kara-Kum.* Ylym, Ashkhabad, 195 pp. (in Russian).

Voznesenskiy, V.L., 1975. Photosynthesis and respiration. In: Anonymous, *Resources of the Biosphere: Synthesis of the Soviet Studies for the IBP, 1.* Nauka, Leningrad, pp. 148–152.

Walter, H., 1939. Grasland, Savanne und Busch der ariden Teile Afrikas in ihrer ökologischen Bedingtheit. *Jahrb. Wiss. Bot.*, 87: 750–860.

Walter, H., 1968. *Die Vegetation der Erde, II. Die gemässigten und arktischen Zonen.* Fischer-Verlag, Jena, Stuttgart, 1001 pp.

Walter, H., 1972. Der Wasserhaushalt der Pflanzen in kausaler und kybernetischer Betrachtung. *Ber. Dtsch. Bot. Ges.*, 85: 301–313.

Walter, H., 1976. *Die ökologischen Systeme der Kontinente (Biogeosphäre).* Fischer-Verlag, Stuttgart, 131 pp.

Walter, H., 1979. *Vegetation of the Earth and Ecological Systems of the Geo-Biosphere.* Springer-Verlag, Berlin, 274 pp.

Walter, H. and Box, E., 1976. Global classification of natural terrestrial ecosystems. *Vegetatio*, 32: 75–81.

Walter, H. and Stadelmann, E., 1974. A new approach to the water relations of desert plants. In: *Desert Biology*, II. Academic Press, New York, pp. 213–310.

Zaletayev, V.S., 1976. *Life in Deserts (Geographical-Biocoenological and Ecological Problems).* Mysl, Moscow, 271 pp. (in Russian).

Chapter 7

THE OROBIOMES OF MIDDLE ASIA[1]

H. WALTER and E.O. BOX

GENERAL DESCRIPTION

Up until now we have only looked at the horizontal demarcation of Middle Asian biomes. In the southeast of this region, however, there rises a series of mountain systems reaching heights of over 7000 m. These are orobiomes with very complicated vertical zonations which must be considered three-dimensionally. We will follow, to a large extent, the presentation by Stanyukovich (1973) on the vegetation of all orobiomes (mountain ranges) of the Soviet Union, although we are interested at present only in the orobiomes of the desert zone. Other data on the individual mountain ranges of Middle Asia and their altitudinal belts can be found in Korovin (1961/62) and in various individual publications (see Walter, 1974, pp. 256–274).

The Middle Asian mountain ranges form a very complicated unit which one can ascribe to the interaction of three systems (Fig. 7.1): (1) the Tien Shan (Tyan Shan); (2) the Pamir–Alay system; and (3) the Turkmen–Khurasan system. This last reaches into Middle Asia only at its northernmost edge, the Kopet Dag and the Balkhans. These ranges are completely isolated from the other systems and are related orographically to the Elburz Mountains of northern Iran.

The Kopet Dag is built up of layers of the Cretaceous group and consists of three longitudinal ridges: the northern (up to 1000 m), the middle (up to 2000 m), and the southern, which forms the boundary with Iran (up to 3000 m). The climate of these orobiomes is extremely arid. Precipitation at the base, as well as at an altitude of 1500 m, is only slightly over 200 mm, but the temperature and thus the potential evaporation decreases with altitude. The sequence of altitudinal belts is very simple but is somewhat different in the eastern and western Kopet Dag.

In the east one finds, up to 800 to 1200 m, a slightly sloping ephemeral desert with *Agropyron* and *Artemisia* spp. In the ravines there grow also woodlands or thickets of *Celtis caucasica* up to 4 to 6 m high at the lower locations, *Acer turcomanica* somewhat higher, and *Berberis turcomanica* and *Lonicera bracteolaris* at still higher locations.

Above this belt lies the montane steppe, extending up to 1600 to 1800 m. It consists primarily of *Festuca sulcata*, *Stipa lessingiana*, *S. turcomanica*, and *Agropyron trichophorum*, with *Poa bulbosa* and *Carex pachystylis* as ephemeroids.

Above 1800 m these are joined, especially on rocky soil, by more and more dwarfed cushion-plants, as well as by creeping *Juniperus* krummholz. *Juniperus* can reach a height of 6 to 12 m on protected sites in ravines. The cushion-plants are primarily *Acanthophyllum glandulosum*, *Astragalus* (*Tragacantha*) spp., *Onobrychis cornuta*, and *Acantholimon khorasavicum*. If overgrazed, the cushion-plants drive out the junipers.

In the western Kopet Dag, the areas up to 400 m are covered by an ephemeral desert with *Carex pachystylis* and *Poa bulbosa*.

Artemisia ciniformis dominates between 400 and 600 m along with dwarf-shrubs (*Kochia prostrata*, *Ceratoides*, *Noaea*, etc.).

From 600 m up to the lower peaks the ephemerals are joined by various composites (*Codonocephalum*, *Cousinia*, *Centaurea*), *Convolvulus subhirsutus*, or *Crambe*. Along small brooks one finds thickets of *Rubus turcomanicus* with *Ulmus campestris*, *Malus pumila*, *Prunus cerasifera*, *Crataegus*, *Rosa*, *Ficus*

[1] Manuscript completed November, 1977.

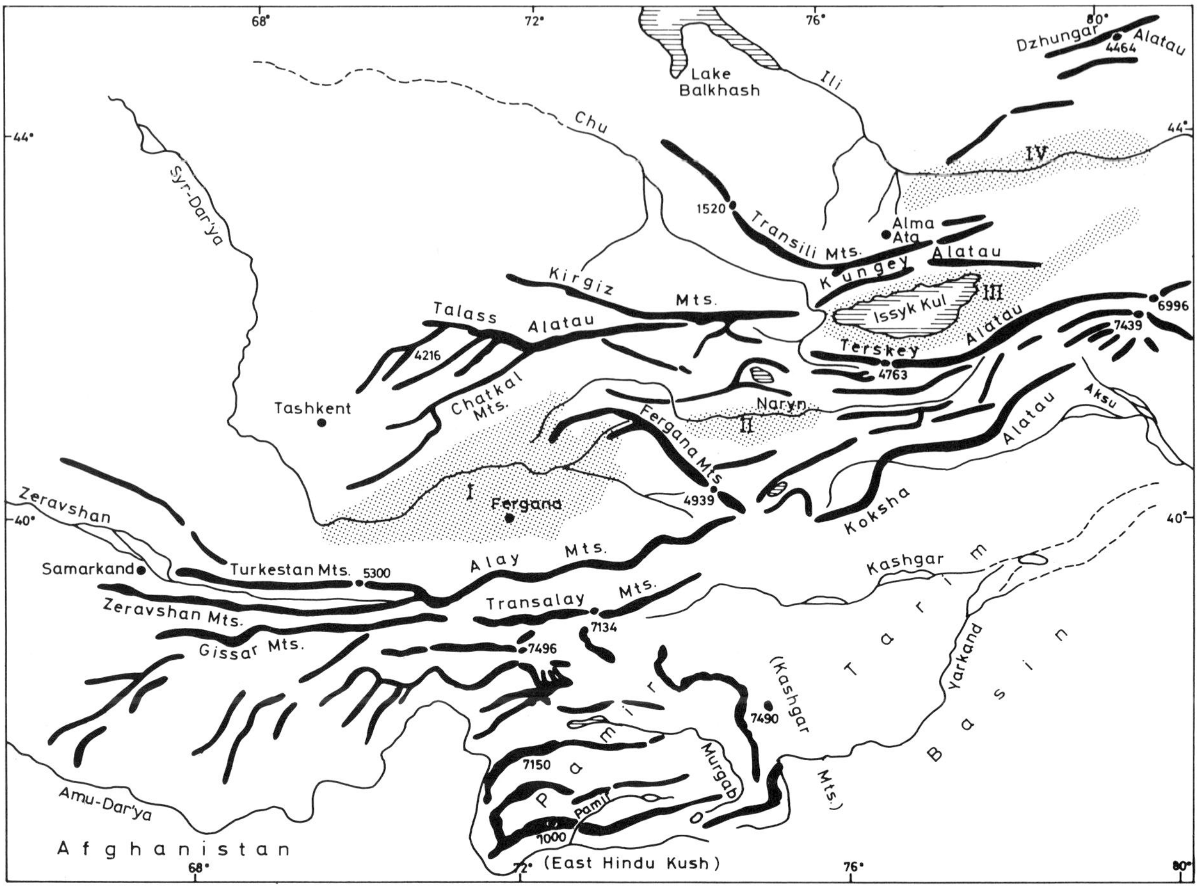

Fig. 7.1. Cartographic sketch of the Middle Asian mountain ranges (explained in the text), with summit lines. Basins are indicated by gray shading: *I* = Fergana Basin; *II* = Naryn Basin; *III* = Issyk-Kul' Basin; and *IV* = Ili Basin. (Adapted from Korovin, from Walter, 1974, Fig. 187.) North of the basins is the Tien Shan system (see Fig. 7.2), south, the Pamiro-Alay system (see Fig. 7.19).

carica, and *Vitis sylvestris*. *Juglans regia* and, less often, *Pistacia vera*, *Rhus coriaria*, or *Mespilus germanica* grow at the bottoms of the larger ravines.

All of these mountain ranges are heavily overgrazed and eroded.

Somewhat further to the north are the Small (Malyy) Balkhan (up to 800 m) and the Great (Bol'shoy) Balkhan (up to 1700 m). They are also dry, and the altitudinal belts are similar.

The extensive Pamir–Alay system occupies the entire southeast of Middle Asia. To this system belongs the Pamir, with over 7000 m high peaks, to which we will return, however, in connection with Tibet in Central Asia. This region is separated from the Tien Shan by the Fergana Basin (Fig. 7.1). Only the most northern mountain ranges belong to Middle Asia. The order from north to south includes the Turkestan (up to Samarkand), the Zeravshan (ending south of Samarkand), and the Gissar mountain ranges, each running due west from the main system, and a few additional ranges with more southwestern orientation. Running off to the east are the Alay (south of Fergana) and the Trans-Alay ranges, the last of which directly borders the Pamir in the south. The Alay Range represents the direct continuation of the Turkestan Range to the east. The steepness of these ranges depends entirely on the rock types which compose them. Shale forms a more gentle relief, while limestones form very rugged relief. The Turkestan and Zerevshan Ranges fall off steeply to the south, the Gissar Range to the north.

All mountain ranges to the north of the Fergana Basin, to and including the Tarbagatay, which is transitional to the Altai, are considered to be part of the Tien Shan. The central Tien Shan reaches heights of over 7000 m. To the Tien Shan, with eastern orientation, belong the Kungey Alatau[1] north of Lake Issyk-Kul' and the Terskey Alatau to the south. The Koksha Alatau runs off to the southwest. The western border of the central Tien Shan is formed by the Fergana Range, which separates the Fergana Basin from the Naryn Basin. Characteristic for this mountain area are broad valleys and high plateaus (3000–4000 m) called *syrts*, above which the mountains rise only slightly higher. As a result, there arises along the upper course of the Syr-Dar'ya River a landscape, resembling the eastern Pamir, which has slowly flowing rivers in very shallow riverbeds. The mountain range arose by block-lifting, whereas the other ranges with steeper relief arose from folding.

The western Tien Shan, also called the Outer Tien Shan, consists of the Talass Alatau and the many parallel ranges which go off from it toward Tashkent to the southwest. These chains decrease steadily in altitude, from 4000 m down to low, loess-covered, rather flat ridges, which are called *adyry* (singular: *adyr*). The northern Tien Shan consists of the Kirgiz ranges (continuation of the Kungey Alatau west of Issyk-Kul'), the Chu-Ili, and the Dzungarian Alatau with the Tarbagatay. These ranges also reach 4000 m or more. A graduated structure is often dominant. Between the Kungey Alatau and the Dzungarian Alatau lies the Ili Basin, through which the river Ili flows from Dzungaria into Lake Balkhash. In the north, the Dzungarian Alatau, separated from the Tarbagatay Range and the Altai Mountains of Siberia, is the basin of Lake Zaysan, through which the upper course of the Irtysh River flows. This does not belong to Middle Asia but more correctly to Siberia.

The eastern Tien Shan stretches far into Dzungaria, separating it from the Tarim Basin, and with its last low foothills even reaches Mongolia and the Gobi Desert. It is thus Central Asian.

This very schematic description should show how complicated the entire mountain region actually is. This complexity is reflected in the great variety of climatic conditions and the corresponding vegetation and altitudinal subdivisions. The variety and the contrasts are even greater since this mountainous area lies around latitude 40°, a zone which, in arid regions, is characterized by especially great contrasts on northern and southern exposures (Troll, 1972).

The subdivision of the Tien Shan mountain system is seen more clearly in Fig. 7.2. Although all the various chains rise out of the desert zone, the sequences of altitudinal belts are very different in the different chains. Depending on geographical location and exposure to prevailing winds, one can distinguish a more moist sequence with forest belts, a drier sequence with only one wooded belt, or a very dry sequence with no wooded belt. The first type of sequence is found on mountain ranges whose slopes face the west wind, which by blocking the path of the prevailing west winds force the air masses to climb and produce clouds and orographic rainfall at the higher levels. The last type of sequence of altitudinal belts, on the other hand, lies in the rain shadows. On these slopes there is no significant increase in precipitation with altitude, though with increasing altitude the temperature and thus potential evaporation do decrease and the water situation of the plants is improved somewhat. Since winds in the mountains can be easily deflected and cause local vertical as well as horizontal wind patterns, even neighboring mountains can have greatly differing altitudinal patterns, to say nothing of the differences between north and south slopes (Vykhodtsev, 1956).

The northern Tien Shan, whose boundaries can be seen in Fig. 7.2, reaches heights of from 5000 to 5400 m. The snow line is between 3000 and 3900 m. During the Ice Ages the snow line was 600 to 800 m lower on the outer ranges and 400 to 500 m lower on the inner, drier ranges. Today the glaciers are in the process of receding.

The mountains with western exposure, which receive more rainfall, possess the moist pattern of altitudinal belts, including a broad-leaved belt (*Malus*, *Prunus*, *Sorbus*) and above it a needle-leaved belt with meadows. The ridges oriented toward the east, on the other hand, are very dry, so the montane steppes flow directly into the alpine belt. Even the alpine belt has alpine mats; there is no true alpine tundra in Middle Asia. Pure ephemeral communities, mediterranean elements, and the shrubby cushion-plants are also absent from the northern Tien Shan.

---

[1] Alatau (Turkic) = mountain range.

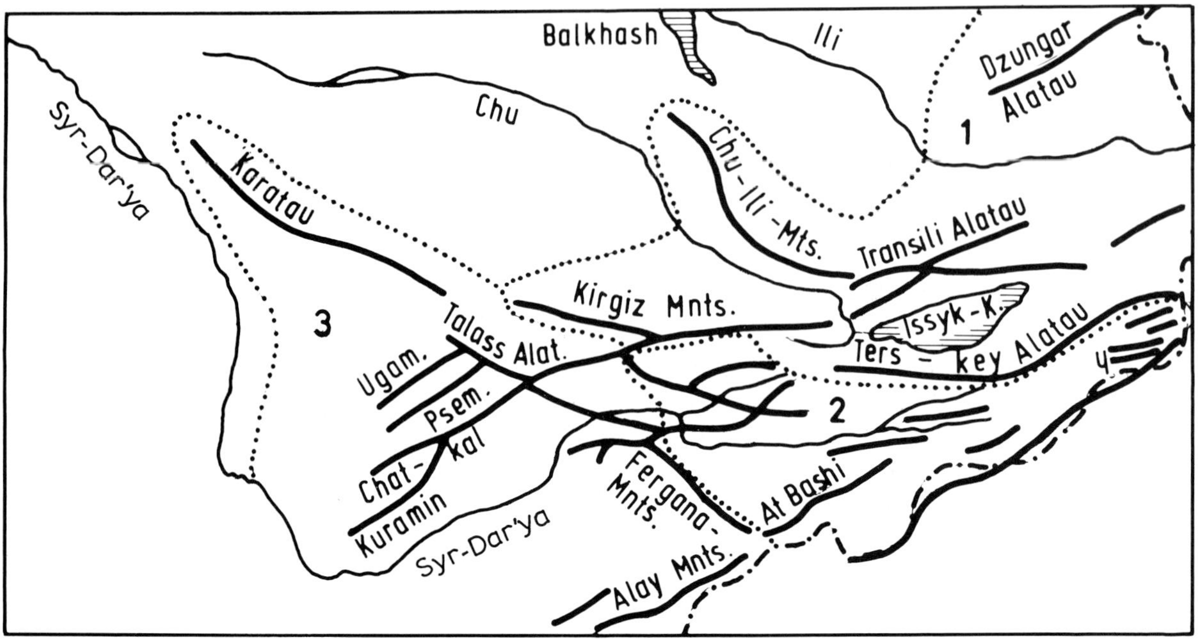

Fig. 7.2. Subdivisions of the Tien Shan system (by Rubtzov, from Stanyukovich, 1973): *1* = northern Tien Shan; *2* = central Tien Shan; *3* = western Tien Shan (border with *2* formed by the east slope of the Fergana Range). The border between *1* and *2* is the Terskey Alatau. The eastern Tien Shan belongs to Central Asia.

THE ALTITUDINAL SEQUENCE OF THE NORTHERN TIEN SHAN

Stanyukovich (1973) gives the following general pattern of altitudinal belts for northern slopes of the moist type:

Up to 800–850 m	desert belt with considerable *Artemisia sublessingiana* and *Poa bulbosa*
800–1500 m	montane steppe with *Stipa* spp., *Festuca sulcata*, *Agropyron sibiricum*, and *Artemisia* spp., as well as forbs; upward more meadow-like, with *Malus sieversii*, *Padus racemosa*, *Ribes meyeri*, *Crataegus*, and *Rosa* also appearing
1500–2300 m	forest-meadow belt with quite European-looking meadows (p. 167) and forests of *Picea schrenkiana* and *Abies semenovi* (*A. sibirica* only in the Dzungarian Alatau, Fig. 7.3), with an understory composed of boreal elements; *Populus tremula* occurs and forms, under the *Picea* belt, a belt 300 to 600 m wide. *Picea* becomes lower-statured at timberline, with *Juniperus sibirica* and *J. turkestanica* krummholz on the ground (Fig. 7.4)
2300–3000 m	alpine mats (*Alchemilla*, *Geranium*, *Phleum alpinum*) with reptant *Juniperus* and steppe elements (*Phleum phleoides*, *Koeleria gracilis*, *Poa angustifolia*, *Festuca sulcata*) in the lower part
Above 2800 m	*Kobresia* swards between snow patches, with small sedge moors and alpine elements (*Arctous alpina*, *Gentiana algida*, *G. falcata*, *Leontopodium*, etc.).

The southern slopes are much drier:

1200–2800 m	montane steppes, in the upper parts with open *Picea* woodland patches and, above that, creeping *Juniperus* with steppe elements
Above 2800 m	*Kobresia* swards which, however, cover only 15 to 20% of the area; rubble areas are widely distributed.

We can also look at some concrete examples. Especially interesting is the broad-leaved forest belt with wild fruit species.

Forests of wild fruit trees

Three Middle Asian species of *Malus* are cited, of which *Malus sieversii* is closely related to our cultivated apples, but is very polymorphous.

The wild apple blooms at the beginning of May

Fig. 7.3. Dzungarian Alatau with *Picea schrenkiana* with *Abies sibirica* in the Koksu Valley; *Betula pendula* and *Salix* species along the bank. (Photograph by V.N. Pavlov, University of Moscow.)

but ripens at different times in June–August or in September–October. Reproduction can occur vegetatively as well, through root shoots or through rooting of branches which are in contact with the ground.

The wild apple occurs as an understory tree in open *Juglans* and *Acer* forests or in stands of *Crataegus* species, which are widespread in the mountains as far as the Dzungarian Alatau and Tarbagatay and in the southern Pamir–Alay. They replace the walnut forests where these no longer survive.

The wild apple stands on the north slopes of the Trans-Ili Alatau above Alma-Ata (Fig. 7.5) are well known. Even today, after five-sixths of the forests have been destroyed, they still cover 5000 ha. Local people wander into the forest when the apples are ripe and collect hundreds of tons of fruit. Since the fruit are only distantly related to the cultivated apples, they are usually dried as slices in the sun or cooked for jam, as are also the wild apricots.

The wild apricots are not as important, but are no less interesting. The species (*Armeniaca vulgaris*) occurs at various places in the Tien Shan. Worth mentioning is the 17 m tall stand with *Juglans*, *Malus*, etc., in the Itokar Valley. The opinion has been expressed that these very xerophilic apricots have escaped from cultivation with no assistance by man and have become fully acclimatized.

Other xerophilic woody species include the five wild pear species (*Pyrus*), of which two have been identified only in the Kopet Dag.

Crataegus azarolus, with large fruit (up to 3 cm in diameter), is also an important species of wild fruit. It is a drought-resistant shrub which forms dense stands on loess and on rocky slopes of the semi-desert belt (Fig. 7.6). The same is true of the wild almond (*Prunus dulcis* = *Amygdalus communis*). Its seeds are usually bitter, but varieties with sweeter seeds do occur. Species of the *Prunus* subgenus *Cerasus* also have edible fruit.

The Alych plum (*Prunus divaricata*) is found more as an understory or undergrowth in the walnut and wild fruit forests. It is one of the two species from which our cultivated plums are derived. The fruit, 2 cm in diameter, are variously colored but have a good taste and are suitable for jam. This species even occurs in steppes in the northern Caucasus.

All of these types of wild fruit have certainly represented, for millennia, the starting point for unconscious or conscious breeding of fruit trees by man in Middle Asia. This is also true of the wild grape, which, according to Popov (1940), is *Vitis vinifera* and not the European–Caucasian *Vitis sylvestris*.

Meadow–needle-leaved forest belt

The needle-leaved forests of the Middle Asian mountains are limited to the Tien Shan system and are composed primarily of *Picea schrenkiana* (Fig. 7.7). This species has cones up to 15 cm long with rounded scales. At the upper limit of its distribution the crown is often broader and the top flattened as a result of the freezing of the main shoot. In contrast, the spruces in the Alps often have more pointed crowns at their upper limits. At 2900 to 3100 m, *Picea schrenkiana* forms low, wind-sheared krummholz. Short-needled forms with more pointed crowns have also been described. The trees can

Fig. 7.4. Upper limit of *Picea*, with cushions of *Juniperus turkestanica*, in the Trans-Ili Alatau (2100 m altitude). (Photograph by V.N. Pavlov, University of Moscow.)

reach a height of 50 m and a trunk-diameter of 120 cm. Most trees 100 to 150 years old, however, are only 20 to 30 m tall (Fig. 7.8). The roots reach lengths of up to 30 m, but only in the upper 10 to 15 cm of the soil. The spruce occurs, therefore, only where the uppermost soil layers remain wet from frequent rainfall even in summer. Below the spruce belt is, as already mentioned, the aspen belt.

In addition to the spruce, the fir (*Abies semenovii*) is also important in Middle Asia. *Abies sibirica* has its southernmost limit in the Dzungarian Alatau, but there only in the western part. *Abies semenovii* is actually an ecotype of *Abies sibirica*. It extends to the northern edge of the Talass and Chatkal Alataus and is very similar to *Picea schrenkiana* in its ecological requirements.

Together with these two needle-leaved trees there occur also small-leaved broadleaf species: *Populus tremula*, *P. laurifolia*, *Betula tianschanica*, and *Sorbus tianschanica*, plus 38 species of shrubs. Of the herbaceous species the ferns are most noteworthy, also some Pyrolaceae, *Goodyera repens*, *Corallorhiza*, *Adoxa*, *Arctous alpina*, and Middle Asian species of *Carum*, *Aegopodium*, *Aquilegia*, *Galium* and *Astragalus*, and also *Impatiens parviflora*, *Galium boreale*, *Chelidonium majus*, *Geum urbanum*, *Poa nemoralis* and *Coeloglossum viride*. Tall forbs also appear in openings: *Aconitum*, *Doronicum*, *Trollius*, etc. The usual mosses also belong to this forest vegetation.

Firs occur only together with spruces and do not extend as far upward. All needle-leaved forests prefer north slopes and wet ravines. Stands without undergrowth ("nudum") have also been described.

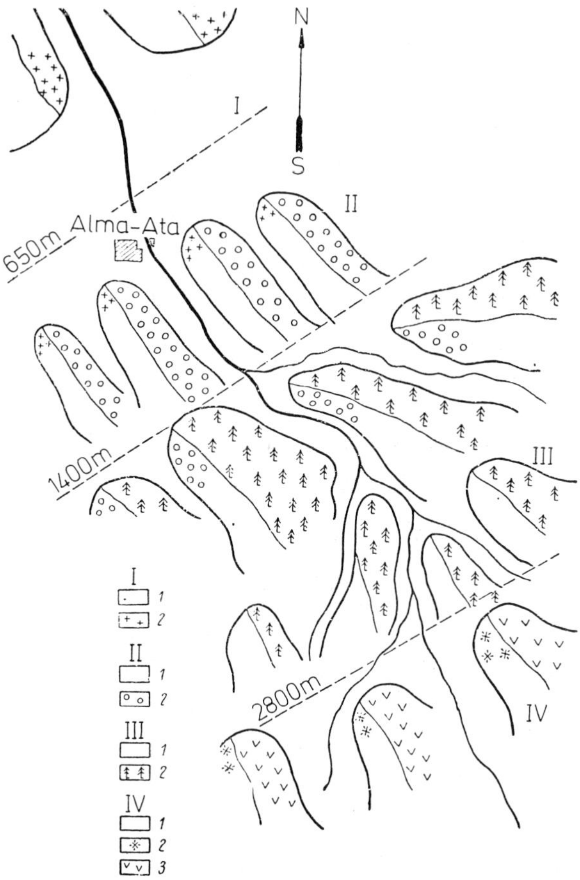

Fig. 7.5. *Malus* forests occurring above Alma-Ata (by Popov, from Walter, 1974): *I* = semi-desert belt (*1* = ephemerals on SW slope, *2* = with *Artemisia* on NE slope); *II* = desert-steppe belt (*1* = with *Agropyron*, *2* = with wild apple woodlands on NE slopes); *III* = Steppe and forest belt (*1* = *Festuceta-Stipeta*, *2* = *Picea* forest on NE slopes); *IV* = High-mountain belt (*1* = alpine steppes and meadows, *2* = *Juniperus* krummholz, *3* = *Salix* species on NE slopes).

Fig. 7.6. Open *Crataegus* woodland in the western Tien Shan (photograph by V.N. Pavlov, University of Moscow). The herb stratum is composed of *Hordeum bulbosum* and *Eremurus regelii* (both ephemeroids, i.e. geophytes).

Fig. 7.7. *Picea schrenkiana* at the upper forest limit (2000–2500 m) on rock and rubble slopes of the Trans-Ili Alatau (photograph by V.N. Pavlov, University of Moscow).

This phenomenon is probably the result of the complete utilization of the water supply by the tree stratum, which is typical of all forests near their drier boundaries.

Meadows occur also within the various more humid belts but only locally, on continuously wet sites, in basins or on gentle north slopes, on black earth-like humus-rich mountain soils. These are regularly mowed once a year. Arystangaliyev (1956) studied them in the Kungey Alatau of the central Tien Shan north of Lake Issyk-Kul'.

At intermediate-to-lower altitudes he distinguished the *Brachypodium pinnatum* and the *Dactylis glomerata* meadows, which are very rich in species and often possess an upper and a lower herb stratum. The upper is 50 to 150 cm tall (cover 90–100%), and the number of species on 200 m^2 was 20 to 55 (200 in all 48 samples). The productivity was 3.5 t ha^{-1} of dry matter (Fig. 7.9).

Dactylis meadows are often found on *Läger* (the German word for sites rich in animal excrement, especially in the Alps). Their height is 50 to 60 cm (cover 90–100%) and the productivity is 2.2 to 5.2 t ha^{-1} of dry matter or 18.0 to 20.0 t ha^{-1} of

Fig. 7.8. *Picea schrenkiana* forest along a brook in the Trans-Ili Alatau (photograph by V.N. Pavlov, University of Moscow).

fresh weight. In order to obtain the best hay, the meadows should be mowed two weeks before grass flowering (Fig. 7.10).

If the meadows are grazed, then the forbs gain the upper hand. The forb-rich *Phlomis* meadow-steppe occurs on black-earth soils of north slopes. Its cover is also 100%, with 172 species found in 12 samples. The productivity is 1.05 to 1.55 t ha^{-1} of dry matter (Figs. 7.11 and 7.12). In the higher regions, around 2800 to 3500 m, *Kobresia capilliformis* meadows occur in patches on sunny slopes with good water supplies and fine-grained soils. The dominant species forms a cover of 40 to 80% (total cover: 60–100%). Sward height is 15 to 45 cm. The number of species on 17 sample plots was 154, about 25 to 60 species per 200 m^2. Productivity of dry matter was 0.6 to 1.6 t ha^{-1} (Fig. 7.13).

The *Kobresia humilis* meadows on rocky soils or southern slopes are much less tall (10–15 cm). *Kobresia* dominates even more strongly, there being only 15 to 25 species on 200 m^2.

THE SEQUENCE OF ALTITUDINAL BELTS WITHOUT FORESTS

As an example, we have taken an arid mountain chain from the Tien Shan (42°50′–41°50′N and 76°15′–75°10′E) in which the steppe elements, otherwise considered to be xerothermic, continue

Fig. 7.9. *Brachypodium* meadow-steppe in the Kungey Alatau (by Arystangaliyev, from Walter, 1974, fig. 199): *1 = Brachypodium pinnatum*; *2 = Phleum phleoides*; *3 = Ranunculus polyanthemus*; *4 = Pedicularis dolichorrhiza*; *5 = Campanula glomerata*; *6 = Lathyrus pratensis*; *7 = Dactylis glomerata*; *8 = Helictotrichon (Avena) pubescens*; *9 = Thalictrum simplex*.

Fig. 7.10. *Dactylis* meadow-steppe in the Kungey Alatau (by Arystangaliyev, from Walter, 1974, fig. 200): *1* = *Dactylis glomerata*; *2* = *Brachypodium pinnatum*; *3* = *Silene commutata*; *4* = *Pedicularis dolichorrhiza*; *5* = *Thalictrum simplex*; *6* = *Stachyopsis lamiiflora*; *7* = *Ranunculus polyanthemus*.

Fig. 7.11. *Phlomis* meadow-steppe in the Kungey Alatau (by Arystangaliyev, from Walter, 1974, fig. 201): *1* = *Phlomis oreophila*; *2* = *Trollius dschungaricus*; *3* = *Helictotrichon pubescens*; *4* = *Euphorbia alatavica*; *5* = *Polygonum nitens*; *6* = *Aegopodium alpestre*; *7* = *Papaver croceum*; *8* = *Brachypodium pinnatum*; *9* = *Alchemilla sibirica*.

Fig. 7.12. Subalpine *Phlomis oreophila* meadow (2400 m) in the Kirgiz Mountains (photograph by V.N. Pavlov, University of Moscow).

into the alpine belt and mix there with the hygro-cryophilic alpine elements. A sharp boundary between the alpine belt and the lower subalpine and high-montane belts is not present in this case. This seems paradoxical when one knows only the humid sequence of belts, but can be explained by the special ecological conditions. Before we do that (p. 172), the actual sequence of belts must be described in more detail (Abramov, 1949).

Between 2000 and 4000 m the following belts were distinguished:

semi-desert	below 2000 m
montane steppe	2000–2600 m
subalpine belt	2600–2900 m
alpine belt	2900–3500 m (lower)
	3500–4000 m (upper)
nival belt	above 4000 m

Semi-desert belt

Tertiary loams with "badlands" characteristics are present below 2000 m. The vegetation is desert-like and is composed of *Artemisia* and *Salsola* species. Shrubs are present only in the river flood-plains (*Hippophaë rhamnoides*, *Nitraria schoberi*) and on rocky slopes [*Dasiphora* (*Potentilla*) *fruticosa*, *Lonicera microphylla*]. Wet soils in the valleys are covered with a meadow-like vegetation which, along the rivers, is composed of the following (with cover indices — see footnote, p. 89): *Phragmites* (4; up to 2.5 m tall), *Alopecurus ventricosus* (1), *Poa pratensis* (1), *Sium latifolium* (+), etc.

Alopecurus ventricosus dominates somewhat further from the river. In the second stratum are *Rumex* spp. and *Sium latifolium*, and in the lowest stratum *Myosotis palustris*, *Blysmus compressus*, *Triglochin palustre*, etc. The total cover reaches 70%.

Fig. 7.13. *Kobresia* high-mountain meadow in the Kungey Alatau (by Arystangaliyev, from Walter, 1974, fig. 202): *1* = *Kobresia capilliformis*; *2* = *Allium atrosanguineum*; *3* = *Myosotis suaveolens*; *4* = *Euphorbia alatavica*; *5* = *Erigeron aurantiacus*; *6* = *Primula algida*; *7* = *Potentilla nervosa*; *8* = *Geranium saxatile*; *9* = *Cerastium tianschanicum*; *10* = *Papaver croceum*; *11* = *Leontopodium campestre*.

Montane steppe

The dominating species are *Stipa caucasica*, *Festuca sulcata*, *Artemisia compacta*, and *A. tianschanica*. The ephemerals develop in early spring. The cover is 45 to 50% on steep southwest and west slopes. The number of species on 100 m^2 is around 15 to 20. Three strata can be distinguished:

I. 40–50 cm tall *Elymus* spp. and *Agropyron dshungaricum*
II. 10–35 cm tall *Stipa caucasica*, *Agropyron cristatum*, *Festuca sulcata*
III. 3–5 cm tall *Androsace villosa*

On northwest and northeast slopes with 18 to 30° inclination, the cover reaches 60 to 75% and the number of species 25 to 30 in 100 m^2. The stratification is as follows:

I. 20–50 cm tall *Festuca sulcata*, *Helictotrichon avenastrum*, *Avena desertorum*, and *Artemisia* spp.
II. 10–20 cm tall *Astragalus* spp., *Leontopodium alpinum*, *Potentilla conferta*, etc.
III. 3–5 cm tall *Androsace villosa*

Shrubs grow on rocky slopes: *Caragana jubata*, *Dasiphora* (*Potentilla*) *fruticosa*, *Rosa webbiana*, *Lonicera microphylla*, *Clematis songarica*, *Convolvulus spinifera*, *Kochia prostrata*, *Ceratoides papposa* (*Eurotia ceratoides*).

A site sample at 2600 m gives a more precise impression of the composition of the vegetation:

Artemisietum-stiposo-festucosum, ESE slope with 18° inclination

I. *Artemisia compacta* (3), *A. tianschanica* (2), *Festuca sulcata* (2), *Stipa caucasica* (1), *Helictotrichon desertorum* (1), *Koeleria gracilis* (1), *Carex* spp. (2, 3), *Elymus* spp. (+), *Saussurea salicifolia* (+), *Astragalus* sp. (+), *Bupleurum densiflorum* (+), *Jurinea lanipes* (+), *Phlomis oreophila* (+)
II. *Potentilla conferta* (2), *Lagochilus platyacantha* (Labiatae) (1), *Allium korolkovi* (1), *Tulipa dasystemon* (+), *Gentiana kaufmannia* (+), *Aster altaicus* (+)

Flowering period: July (beginning of August); for *Artemisia*, *Saussurea*, *Bupleurum*, and *Phlomis*: middle of August. *Gentiana kaufmannia* can be considered to be an alpine species.

Subalpine belt

This belt at 2600 to 2900 m is characterized by the increasing admixture of alpine species to the still dominating species of the meadow-steppe, with considerable *Phlomis oreophila* on southwest exposures.

As an example, a site sample for an area at 2800 m on a north slope with 22° inclination is shown. This location means that the true steppe elements must be present in smaller numbers. The cover reaches 100%. Three strata can be distinguished (I, 35–70 cm; II, 20–35 cm; III, less than 20 cm, with a ground cover of moss almost completely absent). The number of species is 35 to 40 in 100 m^2.

Herbetum phlomidosum

I. *Helictotrichon pubescens* (2), *Poa pratensis* (2), *Phleum phleoides* (1), *Koeleria gracilis* (1), *Roegneria* (*Agropyron*) *schrenkiana* (1), *Festuca kirilovii* (+), *F. rubra* (+), *Carex* sp. (2), *Neogaya mucronata* (Umbelliferae) (2), *Polygonum bistorta* (2), *Campanula glomerata* (+), *Solenanthus stylosus* (Boraginaceae) (+).
II. *Phlomis oreophilia* (3), *Trollius altaicus* (3), *Galium verum* (2), *Kobresia humilis* (2), *K. capillifolia* (2), *Cerastium arvense* (2), *C.* sp. (+), *Geranium collinum* (2), *Potentilla nivea* (2), *P.* sp. (1), *Artemisia sericea* (1), *Euphorbia alatavica* (1), *Dianthus hoeltzeri* (1), *Thalictrum foetidum* (1), *Leontopodium alpinum* (1), *Aconitum napellus* var. *tianschanicum* (1), *Polygonus viviparum* (+), *Melandrium apetalum* (+), *Allium* sp. (+)
III. *Astragalus* (2), *Parnassia palustris* (2), *Gentiana kaufmannia* (2), *Thalictrum alpinum* (1), *Ligularia altaica* (+), *Tulipa dasystemon* (+)

Flowering period: almost completely limited to July.

The following alpine species occur: *Kobresia* spp., *Potentilla nivea*, *Thalictrum alpinum*, *Aconitum napellus*, *Polygonum viviparum*, and *Leontopodium alpinum*, which also appears in the montane steppes. Another sample plot at 2700 m was characterized by the appearance of nitrophilic species (*Rheum wittrockii* up to 1 m tall, *Taraxacum officinale*, *Tragopogon dubius*, *Erysimum marschallianum*). This may have been an earlier läger area. The steppe species *Festuca sulcata*, *Agropyron dshungaricum*, and *Poa relaxa* were present also.

On a third sample plot, also at 2700 m, there appeared, in addition to the meadow-steppe grasses, the steppe species *Stipa caucasica*, *Festuca sulcata*, *Potentilla bifurca*, *Pulsatilla albana*, and *Saussurea salicifolia* alongside the alpine elements *Gentiana* spp., *Potentilla nivea*, *Leontopodium alpinum*, etc.

The alpine belt

The belt above 2900 m, in which steppe elements are present only as scattered individuals and *Kobresia* meadows dominate, is designated as the alpine belt. In addition, *Carex stenocarpa* often occurs as a dominant. This is a species which one also finds in the dark Siberian taiga composed of *Pinus cembra* ssp. *sibirica*. In this belt the soil is composed of a dense network of *Kobresia* and *Carex* roots. As a result of solifluction, the ground surface has become humped, around low-growing *Caragana jubata* and *Dracocephalum imberbe*, with *Polygonum viviparum*, *Minuartia verna*, etc., between humps. *Saxifraga hirculus* and *S. sibirica* grow on wet sites; between large blocks of conglomerate one finds *Chrysosplenium nudicaule*, *Cortusa matthioli*, *Festuca coelestis*, etc. The genus *Dracocephalum* is also represented in the meadow-steppe; *Festuca coelestis* and *Poa relaxa* are species of the alpine steppes of Pamir, above 4000 m. Steppe elements still occur individually: *Stipa*, *Ptilagrostis*, *Roegneria*. One sees how difficult it is to delimit the alpine belt. The transitions from the montane steppe to the subalpine and alpine are very gradual.

The following stand on a sample plot at 3200 m on a 24° east slope can serve as an example of the alpine belt. The stratification of the stand is not distinct. Only a few species reach above the closed plant cover, and even in the ground layer there are only a few, individual *Polytrichum alpinum* cushions.

Kobresietum of the lower alpine belt

Kobresia capillifolia (3), *K. humilis* (2), *Carex stenocarpa* (2), *Festuca tianschanica* (1), *F. coelestis* (1), *F. kryloviana* (+), *Stipa regeliana* (1), *Ptilagrostis mongolica* (1), *Roegneria schrenkiana* (+), *Trisetum spicatum* (1), *Astragalus* sp. (2), *Polygonum viviparum* (2), *Potentilla nivea* (2), *Thalictrum alpinum* (2), *Oxytropis* sp. (1), *Primula algida* (1), *Aster alpinum* (1), *Gentiana kaufmanniana* (1), *G.* spp. (+), *Allium monadelphum* (+), *Trollius altaicus* (1, 2), *Melandrium apetalus* (1), *Dracocephalum imberbe* (1), *Polygonum bistorta* (+), *Pulsatilla albana* (+), *Papaver nudicaule* (+), *Parrya* sp. (+), *Erigeron surantiacus* (+), *Bupleurum densiflorum* (+), *Phlomis oreophilia* (+), *Ligularia altaica* (+), *Tulipa dasystemon* (+), *Corydalis gortschakovii* (+), *Aconitum rotundifolium* (+).

The vegetation on läger is very different because of the nitrogen fertilization. Nitrophilic species gain dominance: *Alchemilla retropilosa* (3, 4), *Poa pratensis* (3), *Carex melanantha* (3), *Taraxacum officinale* (2, 3), *Geranium collinum* (2, 3), *Myosotis sylvatica* (2, 3), *Polygonum aviculare* (1, 2), alongside many others.

At the same time the production increases to three or four times that on non-fertilized alpine mats, reaching almost 10 t ha^{-1} of fresh weight and 3.5 t ha^{-1} of dry matter. So much for the facts, now we will try to explain them.

ECOLOGICAL INTERPRETATION OF THE ALTITUDINAL SEQUENCE WITHOUT FORESTS

In order to understand the ascent of the steppe plants into the alpine belt in continental regions, one must know the ecological requirements for the spread of steppes in the European–North Asian region (Walter, 1975). These are as follow:

(1) The rainfall, or more properly the water reserves in the soil, must lie below the level necessary for the survival of forest.

(2) The growing season for the grasses (good conditions for development of grasses: wet soil and relatively high summer temperatures) must be about four months.

(3) The remaining eight months can be either cold or dry.

(4) The temperature minima are unimportant, for the grasses retract and their reproductive buds are protected from frost and/or dehydration by even a thin snow cover.

Important for our considerations is that the soil be thoroughly wetted in the spring and that the fine-grained soil of the steppe sites store so much water that drought for the plants only begins when water reserves are used up in late summer. Using the phenological phases which were observed over the summer at the sample areas in the Tien Shan, we can depict the situations in the various altitudinal belts as shown in Fig. 7.14.

In the semi-desert belt below 2000 m the cold season lasts four months, from November to the beginning of March. In March the soil is thoroughly wetted, so the development of ephemerals (therophytes and geophytes) begins and lasts into April. The semi-desert chamaephytes require higher temperatures for their growth and do not come out until later. Their assimilating leaf-surface increases

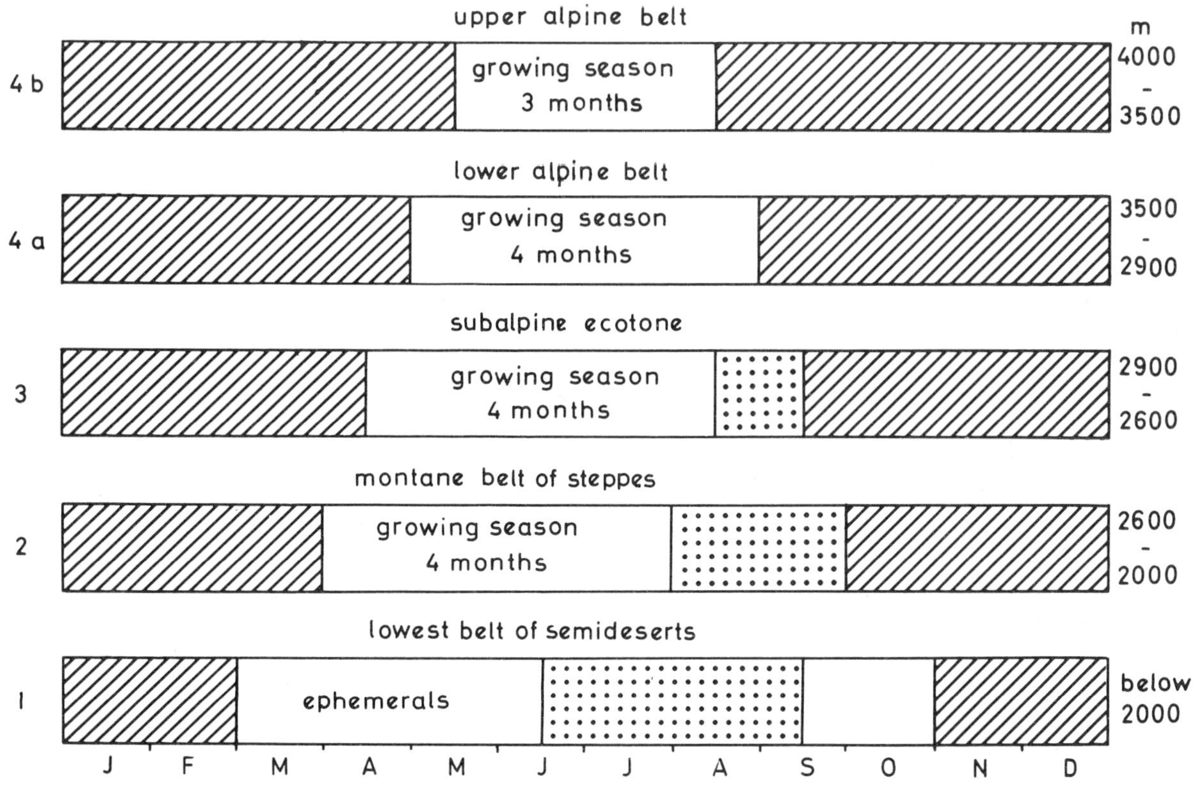

Fig. 7.14. Schematic representation of the arid sequence of altitudinal belts in the Tien Shan and of the decrease in length of growing season with increased altitude (from Walter, 1975): white = growing season; dotted = drought period for plants; lined = winter-cold dormancy (further details in text).

slowly, so the accumulation of the assimilates necessary for flowering and fruiting occurs over the course of several months. Because the water consumption of the chamaephytes is also low, however, the water reserves are used up only gradually and the drought period is not a completely dormant period. Flowering and fruiting occur in late summer and autumn. Of the total growing season of eight months, the first two months are exploited by the ephemerals and the remaining six months by the chamaephytes. At the prevailing low levels of precipitation, the lower soil layers are too dry for steppe plants.

In the montane steppe belt the cold season lasts six months. Precipitation, especially in the form of snow, is higher. In spring, the soil is thoroughly wetted, and the temperature permits a rapid development of leaf area among the grasses and the beginning of intensive assimilation and storage of assimilates. In contrast to the slow pace of development among the semi-desert plants with later flowering periods, the development of the montane grasses proceeds very rapidly, and flowering begins early in the summer. With the high rate of water consumption, however, the water reserves in the soil are exhausted by the end of July. The steppe dries out in the autumn, after the fruit are ripe. The growing season is four months and the dormant period eight months (two months of drought and six months of cold). The good cover by perennial plants represses the ephemerals.

In the subalpine belt the cold season lasts seven months or a bit more, but the drought period is decreased accordingly, so the steppe plants still have a growing season of four months. The summer is wetter and not as hot as at lower elevations. This gives the alpine plants with their similarly short growing seasons the possibility of competing successfully with the steppe plants. Both ecological types occur side by side. Of the steppe plants, the more hygrophilic meadow-steppe plants are better represented than the thermophilic steppe plants.

In the lower alpine belt the growing season of four months, contrasted with a cold season of eight months, is still just long enough for the development of the meadow-steppe plants. The alpine species, however, which grow more quickly under the given moisture and temperature conditions, gain the upper hand in competition. In the upper alpine belt the growing season is finally only three months or less and, even on sunny southern exposures, is too short for the development of the steppe plants.

Above 4000 m one finds alpine steppes only in the Pamir with extremely low precipitation and almost no cloudiness. Alpine meadows are limited there to wet soils on spring sites, in stream channels, and under melting snow patches (see Ch. 9).

In the mountains of more maritime climates, including the Alps, the ascent of steppe plants to great heights is prevented primarily by the cloudiness. On clear days the temperatures just above the surface reach equally high levels on southern slopes, but the number of such days is minimal and the temperature sum for the whole growing season is too low. Only on sites with especially favorable microclimates, as for example on those mentioned by Meusel (1952), can steppe elements ascend to the higher belts.

The temperature sums in the arid mountains of the Old and New Worlds, areas with cloudless summers, are very high in the air-layer near the surface of south slopes. This is especially true of mountains around the 40th parallel, since southern slopes here are struck almost perpendicularly by the sun for a large number of hours each day (see Ch. 9). At the same locations the north slopes receive only the weak, diffuse radiation from the dry air and cloudless sky. At Mount Evans (Colorado, U.S.A.), for example, this diffuse radiation made up only 0.05% of the total radiation at 4335 m (Troll, 1972). The differences between northern and southern exposures in these mountains are so extreme that the sequence of vegetation belts on the northern and southern slopes cannot be compared. Troll (1972) already pointed this out in his studies in the Nanga Parbat (Kashmir), and we can confirm the same situation for the mountains in Utah and Nevada (U.S.A.). One often has a well-developed forest belt on the north slope, while this is only hinted at on the south slope by an open woodland (often *Juniperus* species) or is lacking completely and is replaced by desert or steppe vegetation.

THE ALTITUDINAL SEQUENCE OF THE CENTRAL TIEN SHAN

It is appropriate to look at this region more closely than was done in Fig. 7.2, but without the Issyk-Kul' Basin in the north and the Fergana Range in the west. This is an extremely high-mountain landscape with high plains (*syrts*) at around 3000 m in mountain chains which can reach heights of 7000 m. Only in the border areas, which are all still above 1500 m, are there any intermediate elevations. The climate in the center is extremely continental, and precipitation reaches only 200 to 300 mm even at great altitudes. The snow line lies at about 4200 m. The general conditions remind one very much of the Central Asian Pamir (see Ch. 9) with high-mountain deserts and desert-steppes. Needle-leaved forests appear only on the northern and western margins.

The following series of altitudinal vegetation types are given:

I. Subhumid sequence on northern slopes above the Naryn Basin (southern slopes completely devoid of forest)

1. 1900–2000 m	semi-desert belt with *Salsola* and *Artemisia* spp.
2. 2000–2500 m	dry *Stipa–Artemisia* montane steppe
3. 2500–3000 m	semi-open needle-leaved woodland (*Picea schrenkiana*)
4. 3000–3200 m	*Juniperus* krummholz with *Caragana* and meadow-steppes
5. 3200–3500 m	*Kobresia* swards and fragments of meadow-steppe

II. Dry sequence in the western part

1. Montane desert, at very high altitudes
2. Dry montane steppes
3. Montane meadow-steppes
4. Hemicryophilic steppes and meadows with reptant *Juniperus*
5. Eucryophilic steppes, meadow-steppes, and alpine cushion-plants.

III. Very dry sequence beginning above 2800 m

1. Hemicryophilic *Festuca* steppes and montane semi-deserts
2. Eucryophilic alpine cushion-plants and low swards, also with fragments of steppe between rubble patches

Central Asian elements are floristically very important in the central Tien Shan, especially cushion-plants such as *Dryadanthe tetrandra*, *Saussurea alpina*, and *Oxytropis immersa*.

ALTITUDINAL SEQUENCE OF THE WESTERN TIEN SHAN

The western Tien Shan includes the Fergana Range between the Naryn and Fergana Basins, which slopes to the south toward the Fergana Basin, and the Talass Alatau with the associated chains connected to it toward Tashkent in the southwest.

The climate in the entire area is almost subtropical at the lowest altitudes. This mountain range receives west winds which result in orographic rainfall at the higher elevations. Precipitation corresponds to the mediterranean pattern and falls in the winter half-year. The maximum is in the spring, and the summers are hot and dry (Fig. 7.15).

As a result, ephemerals are very important at lower elevations. The snow line is at 3600 m in general, at 4000 m on south slopes. The vegetation differences also distinguish the western Tien Shan from the central and northern sections: Boreal elements are completely missing in the west, and Mediterranean elements play the leading role. Broad-leaved forests are common at middle elevations, especially those with *Juglans* and wild fruit trees. Needle-leaved forests are found only in the Fergana Range. Mountain steppes and cryophilic cushion-plants dominate at the higher elevations.

The individual sections of the Tien Shan differ significantly from each other. The Fergana Range has an especially complicated vertical zonation. The Talass Alatau is essentially dry, but the low Kuramin Chain, which extends furthest to the south, is still drier.

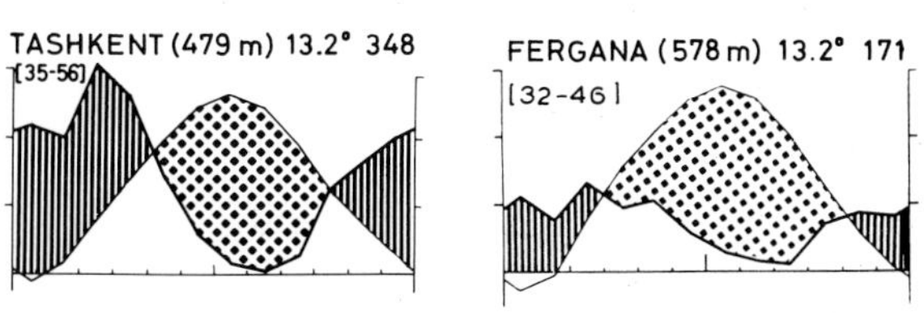

Fig. 7.15. Climatic diagram of Tashkent and Fergana, with dry mediterranean climate (rainfall maximum in spring).

Fergana Range (south slope)

The mountains rise out of the Fergana Basin, which lies at 700 m altitude. This was earlier an ephemeral–*Artemisia* desert but is today irrigated agricultural land. The use of too much irrigation water at times caused the groundwater table to rise, causing secondary marshiness in many places (Levina, 1944). The marsh vegetation is composed mostly of widespread species.

Growing in the water are *Myriophyllum spicatum* and *M. verticillatus*, as well as the usual *Potamogeton* spp., *Utricularia vulgaris* and *Salvinia natans*. The algae *Chara*, *Spirogyra*, *Cladophora* and *Enteromorpha* are represented. The reed thickets consist of *Phragmites*, *Typha*, and *Scirpus* spp., plus *Cladium mariscus*. These are joined by the marsh plants *Sparganium*, *Butomus*, *Alisma*, *Sagittaria*, *Lythrum*, *Lysimachia*, *Sium*, *Mentha*, etc. More Middle Asian elements occur also: *Arthraxon langsdorfii*, *Apluda inermis* (both tropical grasses), *Erianthus purpurescens*, *Imperata cylindrica*.

In marshy meadows one finds many *Carex* spp. and also *Dryopteris thelypteris*, *Equisetum ramosissimum*, *Marsilea*, etc.

Phragmites peat has also been found in part with woody *Alnus glutinosa* remains. This is a species of *Alnus* which hardly occurs in Middle Asia today. The age of the peat deposits, however, was not determined.

The sequence of altitudinal belts on the south slopes was described as follows (Stanyukovich, 1973):

700–900 m — Ephemeral desert (*Poa bulbosa*, *Carex pachystylis*, and many annual ephemerals, plus the higher-growing *Leontice*, *Ferula*, and *Cousinia*), with *Artemisia* and *Salsola* spp. but also *Kochia prostrata*, *Ceratoides papposa*, etc. *Pistacia vera* (see Fig. 7.20) and *Cerasus erythrocarpa* are scattered throughout.

900–1000 m, on *adyry* — Dominant *Hordeum bulbosum*, *Agropyron* (*Elytrigia*) *trichophorum*, *Andropogon ischaemum*, and *Inula grandis*, in addition to the ephemerals. Especially striking are the tall *Eremurus*, *Ferula*, and *Prangos* species. The number of trees and shrubs increases on eroded soils [in addition to those named, *Pyrus regelii*, *Crataegus* spp., *Prunus divaricata* (*cerasifera*), *Amygdalus communis*, and *A. ulmifolia*], with even *Acer* and *Malus* on northern exposures.

1000–1300 to 2000 m — Broad-leaved forest belt, with *Juglans regia* only in ravines and on northern sites in the lower part of the belt but, above 1400 m, on all sites except extreme southern exposures (which are covered by ephemerals and shrubs only). The *Juglans* forests have three strata, with a shrub stratum and a herbaceous stratum (*Brachypodium sylvaticum*, etc.). There are often *Acer turkestanicum* (*affine A. laetum*) forests with an admixture of *Juglans* and *Malus*, plus *Rosa kokanica* and *Lonicera*, and *Rhamnus*. On steep slopes one more often finds stands of *Exochorda* (Rosaceae). *Abelia corymbosa* also occurs.

Above 2000 (2200) m — In the next higher belt, which is difficult to delimit, there grow light stands of *Acer*, or one finds a *Piceto-Abietetum nudum* with a canopy cover of about 0.5% and an undergrowth of shrubs and herbs or creeping *Juniperus* (Fig. 7.16). On flatter slopes, where *Betula turkestanica* occurs alongside *Picea schrenkiana*, *Juniperus* can become 8 m tall.

The next higher belt consists of subalpine meadows and bushes, with the *Juniperus* krummholz decreasing. In these meadows one finds especially large amounts of *Polygonum coriaceum*, with *Iris ruthenica*, *Geranium collinum*, *Allium monadelphum*, and the grasses *Dactylis glomerata*, *Alopecurus songoricus*, *Phleum phleoides*, *Hordeum turkestanicum*, and *Poa* species. *Festuca sulcata*, *Koeleria gracilis*, *Eremurus altaica*, etc., grow on southern exposures. Within the *Geranium* meadows there is often an admixture of *Anemone protracta*, *Aquilegia karelinii*, *Myosotis alpestris*, *Polemonium coeruleum*, etc.

Above 3000 m — Eucryophilic, low mats and *Kobresia* swards, as well as cushion-plants (*Dryadanthemum*, *Oxytropis*). On dry, rocky slopes there appear, on southern exposures, *Festuca sulcata*, *Potentilla hololeuca*, *Geranium collinum*, *Cousinia caespitosa*, etc.

Talass Alatau

This area has been described by Kultiasov (1955). The mountains reach heights of over 4000 m and have a few permanent snow fields and small glaciers, so the altitudinal belts extend into the nival belt. The south slope is considerably steeper than the north slope. Springs are present at middle elevations, fed by the water from melting snow and ice. Desert lies at the base of the mountains, the Muyunkum in the north and the Kyzylkum in the south. The annual precipitation is low and distributed as follows: winter 21%, spring 42%, summer 13%, and autumn 24%. The absolute temperature variation reaches 66°C. Summer is hot (up to 36°) and dry (humidity 13%).

The Talass Alatau is especially exposed to the dry winds from the desert. The orographic snow line lies above 3000 m, but even at 3500 m there is no closed snow cover. The snow remains only on north slopes and in the shadows of rocks in August.

The upward sequence of altitudinal belts is as follows: (1) ephemeral desert; (2) forb-rich steppes

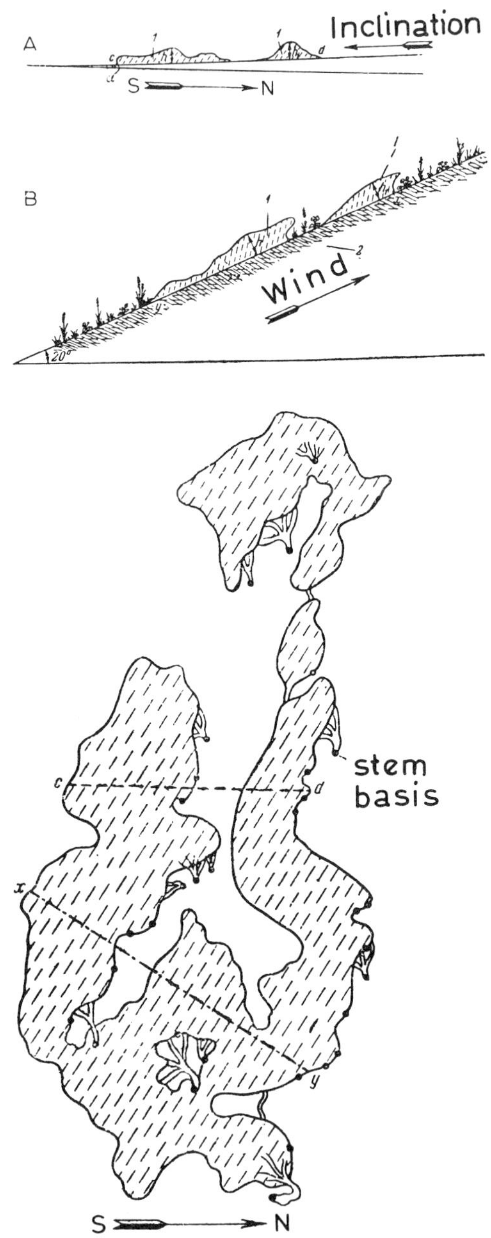

Fig. 7.16. *Juniperus turkestanica* in the high mountains of Middle Asia, growing as a creeping dwarf-shrub (by Kultiasov, from Walter, 1974): basal outline (below) and vertical profile (above). A. Profile along c–d (height: 50–65 cm). B. Profile along x–y (height: 75–90 cm). *1* = *Juniperus*; *2* = substrate.

with individual shrubs; (3) *Juniperus* woodland (open); (4) subalpine belt with *Prangos* and *Ferula* species; (5) alpine steppes, and *Kobresia* swards. There is no forest belt. The arborescent *Juniperus semiglobosa* grows at 1700 to 2000 m, but the trees are scattered or occur at best in very light stands.

The situation at 3000 m is described as follows: High-mountain steppe on 30° northwest slope.

Soil profile (lower limit of horizon is given)

A_0 2–3 cm	steppe turf
A_1 10–12 cm	dark gray-brownish, loosely crumbling, loamy, heavily rooted
B_1 35–40 cm	gray-brownish, dense
B_2 80–100 cm	brownish-gray, whitish calcium efflorescences, highly effervescent, very dense and compacted at depth of 35 to 45 cm.

The water content of the soil falls in the summer from 20% to 10% in August. The ground is wetted primarily by melting snow in the spring.

Vegetation: bunch grasses 20 to 25 cm tall are most important, such as *Festuca sulcata*, *Helictotrichon desertorum*, *Koeleria gracilis*, and *Bromus turkestanicus*; these are joined by 10 to 20 cm tall forbs, such as *Potentilla hololeuca*, *Artemisia aschurbajevii*, *Geranium saxatile*, *Veronica campylopoda*, etc. The number of species on 100 m^2 is 20. *Kobresia* forms thick swards with 85% cover on fine-grained soils with ample water. In such cases the number of species drops to 12 to 18.

The following plot sample shows the composition of an alpine steppe-meadow at 2950 m on a 20° southwest slope.

Upper herb stratum (20–40 cm): *Festuca sulcata* (3), *Helictotrichon desertorum* (3), *Hordeum brevisubulatum* (1), *Poa attenuata* (1).

Middle herb stratum (10–20 cm): *Potentilla hololeuca* (+), *Artemisia aschurbajevii* (1), *Rhodiola heterodonta* (1), *Myosotis suaveolens* (r), *Potentilla tephroleuca* (r).

Lower herb stratum (up to 10 cm): *Kobresia schoenoides* (3), *Veronica campylopoda* (+), *Tulipa dasystemon* (1), *Oxytropis immersa* (r), *Aconitum rotundifolium* (r), *Pachypleurum alpinum* (r).

The mixed composition of the high-mountain steppes, derived from typical steppe elements and true alpine species, as well as montane xerophytes, is very characteristic.

The subalpine meadows with alpine elements are found on well-wetted places, mostly underneath snow patches and in snow valleys.

Clear horizons are apparent in the soil profile. There is no effervescence. The water content varies only a little and lies between 5 and 15%. Herbs growing to 65 cm are dominant: *Polygonum hissaricum*, *Geranium saxatile*, *Alopecurus songoricus*, *Bromus turkestanicus*. The lower stratum is composed of species up to 20 cm tall: *Nepeta pulchella*, less often *Rhodiola heterodonta*, the 15 cm tall and more common *Oxytropis immersa*, etc. The cover is 75% and the number of species is 22. Variants with different dominants occur also: with *Polygonum hissaricum* and *Bromus*, or with *Geranium* or *Alopecurus*, often with considerable *Trollius asiaticus* or *Allium monadelphum* (cover 100%). *Polygonum nitens*, with numerous alpine species (*Doronicum oblongifolium*, *Aster alpinus*, *Aconitum rotundifolium*, *Polemonium coeruleum*, etc.), dominates immediately next to snow valleys. The number of species on 25 m^2 can exceed 50.

The snow valleys are characterized by leached, acidic (pH = 3.5), continuously wet soils, which contain 10 to 12% humus and 25 to 30% organic matter. The roots go down 25 to 30 cm (some 1 m), but the upper 20 cm dry out considerably, so the situation is different from that of the snow valleys in the Alps, where the plants root only a few centimeters deep.

Dominant are: *Allium monadelphum* with *Potentilla tephroleuca* or with *Carex melanantha*, or both of these; or *Oxytropis immersa* with *Poa alberti;* also *Oxyria elatior* with *Ranunculus rubrocalyx*. Other characteristic species are: *Cerastium cerastioides*, *Minuartia biflora*, *Primula algida*, *Pedicularis amoena*, *Lloydia serotina*, *Myosotis suaveolens*, *Gagea* spp., *Draba talassica*, *Pachypleurum alpinum*, *Saxifraga sibirica*, and less often *Gnaphalium supinum*.

The mountain xerophytes, with the shrubby cushions of *Onobrychis echidna*, occur around 2900 m on rocky soils without snow cover in winter. The leaf surface is reduced in these xerophytes, and the tips of the rachises are often transformed into thorns. *Ziziphora clinopodiodes*, *Richteria* (*Chrysanthemum*) *pyrethroides*, *Cousinia caespitosa*, etc., also possess low leaf-surface areas. There are also scattered steppe grasses.

At the upper limit of the open *Juniperus* woodland (parkland), *Juniperus turkestanica* often takes on the reptant (creeping) growth form. The stem with branches, up to 20 m long, lie along the ground covering an area of 4 to 5 m width. They rise to a height of only 0.5 to 1.5 m and possess a clearly wind-sheared shape (Fig. 7.16). *Berberis oblonga*, *Lonicera humilis*, steppe grasses, *Galium simile*, *Geranium*, *Ziziphora*, *Gypsophila cephalotes*, *Rosa*, etc., are established on the protected sides of the *Juniperus* bushes. This plant community is found mostly on southern exposures on rocky soil where it has not been burned off in order to enlarge the livestock grazing areas.

At 3000 m the soil temperature on north slopes at the end of July is almost constant at a depth of 15 cm. On a corresponding south slope the soil temperature, at comparable depth, is 3.7°C higher in the morning, 6.3°C higher at noon, and 8.8°C higher in the evening. From July 19 to August 31 surface frost was recorded 24 times on the north slope and only four times on the south slope. The maximum temperature on the north slope was 34.5°C and on the south slope 42.5°C. The soils of the high-mountain steppes are, at 3000 m, 5.8°C warmer than those of the snow valleys, at 2225 m around 7.0°C warmer. The development of the snow-valley plants begins under the snow.

The situation around 1500 m is somewhat wetter on the mountain chains stretching on from the Talass Alatau to the southwest (Ugam, Pakem, and Chatkal). Well-developed *Juglans regia* forests can be found here on the eastern slope. These have been studied in some detail (Korovin, 1961/62). Table 7.1 gives climatic data for such forests (at 1433 m).

The soils are brown forest soils, rich in humus but sometimes with a bleached horizon. They remain moist until the beginning of July, then dry out quickly so that the underbrush wilts. There is a snowdrop (*Galanthus*) phenophase in the spring. The walnut species is identified as *Juglans fallax* or *Juglans regia* ssp. *fallax*. It is polymorphous and variable but is cold-resistant to a certain degree. Reproduction by seeds is good (840–6000 seedlings ha^{-1}), but reproduction by stump-sprouting is also possible. The roots penetrate to 3.5 m; trunks can become 1 m thick. The forests are shady, have only one tree stratum, and undergrowth is strikingly poorly developed.

What undergrowth there is consists of: *Brachypodium sylvaticum*, *B. pinnatum*, *Melica altissima*, *Festuca gigantea*, *Agropyron caninum*, *Poa nemoralis*, *Bromus benekeni*, *B. inermis*, *Carex polyphylla*,

TABLE 7.1

Climatic data for *Juglans regia* forests (from Korovin, 1962)

Month	Jan.	Feb.	Mar.	April	May	June	July	Aug.	Sept.	Oct.	Nov.	Dec.	Year
Mean temperature (°C)	−3.4	−2.0	1.9	8.7	12.1	17.5	17.5	20.6	14.0	8.4	3.3	1.9	9.5
Minimum temperature (°C)	−20.7	−15.1	−13.5	−1.6	−2.9	−2.2	−2.2	5.2	0.7	−10.0	−13.4	−16.2	−20.7
Precipitation (mm)	56	113	203	193	92	32	32	0.3	13	143	69	75	995

Fig. 7.17. Walnut stand in the Fergana Mountains (photograph by V.N. Pavlov, University of Moscow): old *Juglans regia* ssp. *fallax* (left), *Lonicera* and *Rosa* bushes (middle), with small *Prunus divaricata* (*cerasifera*) trees (to the right). In the background is the Baubash Alatau.

Archangelica decurrens, *Conioselinum latifolium*, *Ligusticum discolor*, *Aegopodium podagrarica*, *Crepis sibirica*, *Alliaria officinalis*, *Impatiens parviflora*, *Lamium album*, and Middle Asian *Allium*, *Arum*, *Cerastium*, *Melandrium*, *Polygonatum*, and *Senecio* species.

Usually only a few of these species are present. The poor development can be due to antibiotic effects (allelopathy) of the *Juglans* litter.

In addition to these shady forests on protected north slopes there are also more open stands on drier biotopes and in floodplains. These floodplain stands often have grapevines as lianas and also considerable *Impatiens parviflora* undergrowth (Zaprayagayeva, 1964).

Most stands are 60 to 90 years old and have trunk diameters of 35 to 65 cm (Figs. 7.17, 7.18). The number of trunks can be 120 to 150 ha^{-1}. The yield of nuts varies from 3 to 32 kg per tree (Popov, 1940).

The Pamir–Alay System (Fig. 7.19)

One must also distinguish different subsections of this mountain system:

(1) The northern ranges with the Alay Range, which delimits the Fergana Basin in the south and which forks in the west into the Turkestan Range to the north of the Zeravshan River, ending near Samarkand, and the Zeravshan Range south of the river, ending south of Samarkand.

(2) The southwestern ranges with the Gissar and Darvaz Ranges, together with the ranges branching off from them, which become lower and lower toward the south but reach to the Pyandzh and the Amu-Dar'ya Rivers.

(3) The central mountain system south of the Trans-Alay Mountains, with the 77 km long Fedchenko Glacier (the longest outside the polar regions), the highest peak (Kommunizm, 7495 m), and additional mountains around 7000 m.

Fig. 7.18. Belt of broad-leaved forest in the western Tien Shan (photograph by V.N. Pavlov, University of Moscow): individual groups of *Juglans*, *Malus sieversii*, *Crataegus* spp., *Prunus divaricata* in the middle of steppe communities (with *Agropyron trichophorum* and *Hordeum bulbosum*), and large umbellifers (*Ferula tenuiseta* in the foreground).

(4) The part of Badakhshan or west Pamir with the ranges Vanch, Yazgulem, Rushan, and Shugnan, which connects with the main system further to the south.

(5) East Pamir, the 4000 m high plateau surrounded by mountain ranges still 1000 m higher.

The subsections 3 to 5 belong to Central Asia and will be discussed later in connection with Tibet. The entire region outlined above belongs to the Tadzhik S.S.R.

The northern ranges

The individual mountain ranges run parallel to the latitudes which is not a favorable orientation for high precipitation. The climate is mediterranean and dry. The ephemeral vegetation is only weakly developed at lower elevations, forest belts are missing at middle elevations, and these are replaced by mountain steppes and open *Juniperus* stands, which become creeping juniper krummholz at higher elevations. Cushion-plants appear, particularly in the western part.

Precipitation at low elevations totals around 220 mm per year and increases only slightly with increasing elevation.

We shall describe the vegetation of the Turkestan Range, which is similar to the others.

The open *Juniperus* stands cannot be considered part of a forest belt, for they are all extremely open. The desert, semi-desert, steppe, and alpine communities all flow gradually into each other. Because of the steepness of the slopes here, it is very difficult to delimit the various belts. Edaphic conditions are often more important than elevation. The vegetation is also considerably degraded as a result of grazing and the cutting of wood.

The mountains decrease in height to the west, from 5400 m to 2600 m. The plain at the foot of the mountains is at 500 m. The glaciers stop today at 3020 to 3480 m on the north slope and 3200 to 3500 m on the south slope. During the Pleistocene they extended down to 2100 to 2200 m. The climate as a whole is arid, and the continentality increases toward the east. Precipitation may reach a maximum at an elevation of around 2500 m. At 3000 m, 714 mm were measured from September to June 13, that is, in $9\frac{1}{2}$ months.

The soil changes from sierozem in the desert belt to black-earth-like soil at middle elevations, lessivized[1] brown soil at higher elevations, and finally to alpine turf soil at the highest elevations. As a result of the steepness, however, the soils are either rocky or fully eroded.

Afanasyev (1956) described the vegetation as follows. At the foot of the mountains is the desert called *chul*, with *Artemisia* and *Salsola* species and low *Stipa caucasica* f. *desertorum*. On the loess-covered foothills, called *adyry*, one finds *rang* vegetation, consisting primarily of *rang* (*Carex pachystylis*) and *Poa bulbosa* (see p. 89). The elevation is about 600 to 1400 m, and the soil is a typical sierozem. Woody plants occur on rocky soil: *Amygdalus spinosissima*, *Cerasus verrucosa*, and also *Pistacia vera*. The belt above this, up to 2700 m, is covered with ephemerals in the spring.

[1] Leached, especially referring to the leaching of calcium, as is common in Europe (from French: *sols lessivés*).

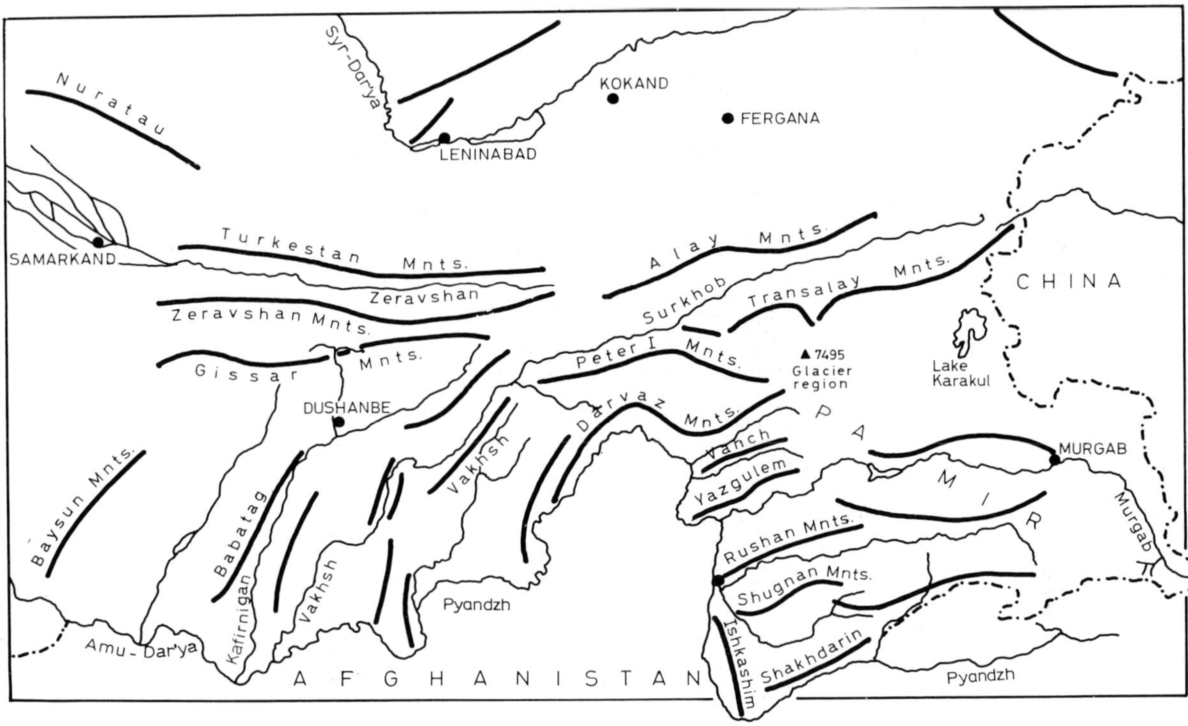

Fig. 7.19. Summit line of the Pamiro-Alay system in Tadzhikistan (adapted from a map in Anonymous, 1968).

One sample yielded the following: *Agropyron trichophorum* (4), *Festuca sulcata* (2, 3), with *Bromus oxyodon*, *Koelpinia linearis*, *Rochelia peduncularis*, *Linum corymbulosum*, two or three *Astragalus* species, *Senecio subdentatus*, and many other ephemerals, including *Eremurus*.

The largest part of this area has been converted to cultivated land. On sites with better water supplies, namely in the upper part of this belt, woody plants can become established. These include *Berberis*, *Rosa*, *Ephedra*, as well as wild apples (*Malus sieversii*) and wild walnuts (*Juglans regia* ssp. *fallax*) along the streams (Fig. 7.18).

These scattered or single-standing woody plants simulate a landscape that remotely suggests a savanna. This led to the term "semi-savanna" used by some geobotanists, but this is not really appropriate because the vegetation has nothing to do ecologically with the tropical or subtropical savannas. The woody plants are limited to local, edaphically different rocky or wetter sites.

The high-mountain steppe above 2700 m is called *yaila*, which is a mountain-side pasture land (= the German *Alm*) used only in the summer. The soils are burozems or chernozems. The most important grass species is *Festuca sulcata* found with *Poa bulbosa* or *Agropyron trichophorum*, *Colpodium parviflorum*, and also *Carex* species, with many forbs. On north slopes *Agrostis canina* is important. Afanasyev (1956) has described numerous plot samples.

Within this belt, mostly on north slopes, junipers (*Juniperus turkestanica* and *J. seravschanica*) up to 4 to 5 m high occur as single-standing trees. The crown closure is normally 10 to 20% but can reach 30%. Under the crowns are often found *Lonicera bracteolaris* with *Poa nemoralis*, *Campanula glomerata*, *Geranium collinum*, etc. The question is still in dispute whether this is a special juniper-parkland belt. The soils are mostly rocky. One often finds treestumps, which suggests that the stands were once denser and that the steppe vegetation has been brought about in part by the cutting of trees for wood. It can often be shown, however, that the density of the stands is clearly dependent on the water content of the soil.

We made some observations in the analogous juniper–pinyon belt in the Intermountain Region of the western United States. As soon as the water supply becomes more favorable, as in an erosion channel or a small valley, the stands become denser. These can form woodlands, which, because of the branching all along the trunks, can be almost impassable. On dry, rocky slopes the trees are widely spaced, but one has the impression that their root systems are in contact with each other — that is, that the stand is closed below ground. The accompanying and sometimes dominant pines of North America (*P. monophylla* and *P. edulis*) are absent in Asia.

In the Zeravshan Valley there is, according to Zakirov (1955, 1962), no *Juniperus* belt, but instead all woody stands are limited to locally wetter sites. In contrast, Afanasyev (1956) mentions dense stands in places on the north slope of the Turkestan Range from 1900 to 2300 m up to 2800 to 3100 m. These stands have a crown closure of 70% and good rejuvenation. The height of the stand is mostly 5 to 6 m to 10 to 12 m. Scattered *Sorbus tianschanica* occur in the tree stratum and, in one case, *Betula*. Mentioned in the shrub layer were: *Lonicera microphylla*, *L. bracteolaris*, *L. persica*, five *Rosa* species, *Cotoneaster multiflora*, *Spiraea hypericifolia*, and *Ribes meyeri*. Of these, usually only two or three occur together (cover 20–30%). *Juniperus seravschanica* is most common in the lower part of the belt, with *J. semiglobosa* in the middle section and *J. turkestanica* most common toward the top. These are joined at 2300 to 2500 m in the western part by *Acer turkestanicum*.

On skeletal soils one finds thorny mountain xerophytes and cushion-plants (*Onobrychis echidna*, *Cousinia* species, *Arenaria griffithii*, *Scorzonera scanthoclada*, and less often *Acantholimon alatavicum*).

This vegetation is joined above 3000 m by more and more alpine elements. At this height, however, the mountains are so steep and rocky that a true altitudinal belt is no longer formed. The plant growth depends entirely on the site characteristics, especially on local water availability.

At 3100 to 3600 m, in the headwater areas of the rivers, a cryophilic vegetation with a similar composition to that in the Talass Alatau is encountered. It includes, for example: *Oxytropis immersa*, *Potentilla pamiroalaica*, *P. hololeuca*, *P. flabellata*, *Leontopodium ochroleucum*, *Pachypleurum alpinum*, *Veronica biloba*, *Gnaphalium supinum*, *Geranium saxatile*, *Cerastium cerastioides*, *Myosotis suaveolens*, *Allium oreophiloides*, and *Polygonum biaristatum*. These are joined, in definite communities, by *Poa litwinowiana*, *Festuca alaica*, *Atropis subspicata*, or *Carex melanantha* and *Sibbaldia tetrandra*.

The *Kobresia* mats are less common in this mountain system. The alpine vegetation here has still not been studied in detail because of the inaccessibility of the region.

Southwestern mountains (Gissar–Darvaz region)

The sequences of altitudinal belts on the Gissar and Darvaz Ranges are of the moderately humid type. The ranges, together with the mountain Peter I, form a high arc open to the west which catches the air-masses coming from the west and causes them to rise, producing orographic rainfall on the southern side of the Gissar Range and the western side of the Darvaz Range (Fig. 7.19). The result is that the middle elevations receive 1000 mm (up to 1800 mm) of rainfall and broad-leaved forests are able to form a distinct belt. The snow line is at about 3700 to 4000 m, and the glaciated area covers several hundred square kilometers.

The climate is mediterranean and almost subtropical at lower elevations, with mean January temperatures above freezing. Heavy frosts can, however, occur occasionally.

The corresponding altitudinal belts, on the favorable slopes, are the following:

700–1500 m The original ephemeral-desert belt with considerable *Agropyron* (*Elytrigia*) *trichophorum* and scattered wooded patches of *Amygdalus bucharica*, *Crataegus pontica*, *Cotoneaster racemiflora*, *Pistacia vera* (Fig. 7.20), and *Cercis griffithii* plus *Ziziphus jujuba* with *Punica granatum*. Through the interference of man, however, the woodlands have been reduced to a few scattered remains. The plant cover today consists mainly of *Agropyron* with *Hordeum bulbosum* and *Poa bulbosa*, *Cynodon dactylon* and *Andropogon ischaemus* on unplowed ground. These are joined by the especially widespread ephemerals *Vulpia myuros*, *Aegilops triuncialis*, *Bromus macrostachys*, and *B. oxyodon*. *Juglans regia*, *Platanus*, *Populus*, with *Erianthus* and *Phragmites* grow in floodplains.

1500–2800 m Broad-leaved forest belt with many tree and shrub species, including more warmth-loving species in the lower part (*Juglans regia*, *Acer turkestanicum*, and *Vitis hissarica* in the upper stratum; in the second stratum *Prunus divaricata*, *Crataegus tur-*

Fig. 7.20. Open tree woodland with *Pistacia vera* surrounded by an ephemeroid carpet in the Chaltan of southern Tadzhikistan (from the photograph collection of E.M. Lavrenko).

kestanica, *Cerasus mahaleb*, *Malus sieversii*, *Celtis caucasica*, on rocky sites *Juniperus seravschanica*; underneath many shrubs; and in the herb layer *Dactylis glomerata*, *Aegopodium tadzhicorum*, *Impatiens parviflora*, etc., and on open sites *Inula grandis* and *Prangos pabularia*). The forests have also suffered greatly from man's presence, but closed 15 to 20 m tall *Juglans regia* forests with many accompanying species do remain. *Acer turkestanicum* (13–15 m tall) dominates above 2300 m. On rocky sites it is no longer *Juniperus seravschanica* but *J. semiglobosa* and *J. turkestanica*. *Betula* and *Populus* are found along watercourses, and *Rosa divina* is abundant on open sites. Meadows with *Polygonum coriarium*, *Ligusticum discolor*, etc. also cover large areas.

2800–3400 m Subalpine meadow belt, with mountain steppes, *Juniperus* krummholz and cushion-plants on rocky ground. Precipitation drops rapidly at the highest elevation. The moist meadows consist of *Alopecurus seravschanicus* with *Poa pratensis*, *Festuca rubra* and many forbs (*Geranium saxatile*, *Ligularia thomsonii*, *Astragalus andaulgensis*, often *Rumex ulsenianus*). *Polygonum coriarium* dominates among the tall forbs. Cushion-plants are not widespread (*Astragalus lasiosemius*, *Onobrychis echidna*) but rather accompany the steppes.

3400–3800 m Eucryophilic mat belt, often with *Polygonum hissaricum*, plus *Kobresia* swards with *Carex* species. *Festuca coelestis*, *F. sulcata*, and *Poa litwinowiana* occur in the steppe patches.

Koroleva (1958) has published numerous plot samples and a vegetation map (1.5 cm = 1 km).

We do not need to go into great detail in discussing the low mountains south of Dushanbe. These are drier and their vegetation is similar to that of the Kopet Dag:

300–900 m desert with dwarf-shrubs and ephemerals

900–1700 m considerably degraded vegetation with *Poa bulbosa* and *Carex pachystylis*, and individual *Amygdalus* and *Pistacia*.

1700–2300 m belt with open *Juniperus* stands and *Rosa kokanica* (after removal of the *Juniperus*), plus small fragments of broad-leaved woodland

We add to this overview a complete list of all wild fruit species which occur in Tadzhikistan (Zaprayagayeva, 1964, with many maps of distributions). The species are the following: *Juglans regia*, *Pistacia vera*, *Amygdalus spinosissima*, *A. bucharia*, *A. vavilovii*, *Celtis caucasica*, *Ficus carica*, *Berberis heterobotrys*, *B. oblonga*, *B. multispinosa*, *B. integerrima*, *B. nummularia*, *B. kaschgarica*, *Ribes meyeri*, *R. janczewskii*, *R. heterotrichum*, *R. villosum*, *R. malvifolium*, *Cotoneaster hissarica*, *Pyrus bucharica*, *P. regelii*, *P. tadshikistanica*, *P. cajon*, *Malus sieversii*, *Sorbus tianschanica*, *S. turkestanica*, *S. persica*, *Crataegus pontica*, *C. turkestanica*, *C. altaica*, *C. hissarica*, *C. songorica*, *Rubus caesius*, *Fragaria bucharica* (without any economic importance), *Prunus sogdiana* (related to *P. cerasifera*, the *Alych* plum), *P. divaricata*, *P. domestica* (probably escaped from cultivation), *Aflatunia* (related to *Cerasus*) *ulmifolia* (no economic importance), *Cerasus verrucosa*, *C. tadshikistanica*, *C. erythrocarpa*, *Padus mahaleb*, *Rhus coriaria*, *Ziziphus jujuba*, *Vitis vinifera*, *Hippophae rhamnoides*, *Elaeagnus angustifolia* (unimportant), *E. orientalis*, *E. songarica*, *Punica granatum* (wild, related to the garden form), *Diospyros lotus*, *Vitex agnus-castus*.

Middle Asia is a very impoverished refuge for the Tertiary forests. Only in the wettest parts of individual mountain ranges were some species able to hold on until today. The number of these tree and shrub species is rather large. We identify the genera and include the number of species in parentheses: *Acer* (5), *Betula* (10), *Crataegus* (11), *Lonicera* (15), *Malus* (3), *Pyrus* (5), *Juglans* (2), *Prunus* (2), *Cerasus* (4), *Rhamnus* (1), *Armeniaca* (8), *Mespilus* (1), *Jasminum* (1), *Berberis* (6), *Cotoneaster* (5), *Padus* (1), *Alnus* (1), *Aflatunia* (Rosaceae) (1), *Platanus* (1), *Viburnum* (1), *Euonymus* (3), *Fraxinus* (3), *Sorbus* (2), *Abelia* (Caprifoliaceae) (1), *Populus* (9), *Thelycrania* (Cornaceae) (2), *Atraphaxis* (Polygonaceae) (1), *Ribes* (1), *Grossularia* (1), *Sorbaria* (1), *Rubus* (4), *Vitis* (2), *Exochorda* (Rosaceae) (2), *Spiraea* (1). In addition there are many *Rosa* and *Salix* species.

There are many endemic species, often with very limited distributions.

PHYTOMASS AND PRODUCTION OF THE HIGH-MOUNTAIN VEGETATION

Zlotin (1971) has presented interesting assessments of the phytomass and primary production (= annual litterfall) of the syrts (high plateaus) at 3000 to 4000 m in the arid central Tien Shan. The following biogeocoenoses of the individual altitudinal belts up to 4000 m were studied:

Plot No.	Elevation (m)	Cover (%)	Vegetation, dominants, and their fraction of total phytomass in %
(1)	4000–3000	10	cryophilic cushion-plants: *Dryadanthe tetrandra* (90%)
(2a)	3800–3500	60	moist mats: *Kobresia capilliformis* (85–90%)
(2b)	3800–3500	60	moist mats: *Kobresia* + *Carex stenocarpa*
(3a)	3600–3300	30–40	steppe-meadows: *Kobresia humilis* (*c*. 50%), *Festuca kryloviana* + *Ptilagrostis purpurea* (*c*. 50%)
(3b)	3600–3300	15	steppe-meadows: *Festuca tianschanica* (60–80%)
(4a)	3700–3300	5–20	high-mountain steppes (cold): *Saussurea leucophylla* (50–80%) with *Oxytropis* spp.
(4b)	3700–3300	5–20	like 4a, but with *Ptilagrostis*
(5a)	3700–3300	30	high-mountain steppe: *Ptilagrostis subsessiliflora* (90%)
(5b)	3700–3300	30	like 5a, but with *Ptilagrostis purpurea* + *Festuca sulcata* (90%)
(6a)	3400–3200	5–15	dry steppes: *Ptilagrostis subsessiliflora* (90%)
(6b)	3400–3200	20	*Leucopoa* steppes: *Leucopoa olgae* (87%), *Hordeum turkestanicum* (11%), *Roegneria czimganica* + *Poa litwinowiana* (2%)
(7a)	3300–3100	30	semi-desert: *Artemisia rhodantha* (50%), *Oxytropis* (36%), *Ptilagrostis subsessiliflora* (13%)
(7b)	3300–3100	10–20	*Artemisia* semi-desert: *A. rhodantha* (50%), *Oxytropis* (36%), *Ptilagrostis subsessiliflora* (13%)
(8a)	3200–3000	5	*Artemisia* desert: *A. rhodantha* (85%), *Ptilagrostis subsessiliflora* (13–14%)
(8b)	3200–3000	<5	halophyte desert: *Limonium hoeltzeri* (40%), *Reaumuria kaschgarica* (38%), *Kalidium schrenkianum* (5%)
(9)	3700–3300	*c*. 100	moors: *Carex melanantha* + *C. oxyleuca* (98%)

Table 7.2 contains the estimated values for phytomass and production.

TABLE 7.2

Phytomass of biogeocoenoses of the syrts of the central Tien Shan in t ha^{-1} (air-dry weight) (after Zlotin, 1971)

Plot No. (see p. 184)	Number of species	Phytomass					Annual green parts (litter)	Annual litter total	Lying litter	Zoomass	Phytomass/ zoomass
		green parts	above-ground perennial parts	roots		total					
				living	dead						
1	14	0.8	9.4	5.9	0.7	16.8	0.8	2.9	0.21	0.03	550
2a	18	2.2	–	19.6	21.6	43.4	2.1	8.7	0.4	3.00	14
2b	30	1.4	0.8	14.4	17.3	33.9	1.2	5.4	2.92	3.00	11
3a	14	1.7	–	7.8	4.9	14.4	1.1	3.8	0.2	0.70	20
3b	15	1.0	–	7.6	0.2	8.8	0.9	3.5	1.50	3.00	2.9
4a	12	0.8	0.4	4.8	0.3	6.3	0.7	2.0	0.009	0.01	630
4b	7	0.5	–	6.3	1.0	7.8	0.4	2.1	0.006	0.07	110
5a	6	1.4	–	4.7	1.6	7.7	1.4	2.4	0.20	0.07	110
5b	6	1.7	–	3.8	1.5	7.0	1.7	2.6	0.08	0.07	100
6a	5	0.2	–	1.5	2.0	3.7	0.2	0.7	0.02	0.07	53
6b	6	0.5	–	2.2	3.3	6.0	0.5	1.2	1.2	0.13	46
7a	10	1.0	1.1	3.9	0.9	6.9	1.0	1.7	0.22	0.02	350
7b	12	0.6	0.3	2.4	0.9	4.2	0.6	1.3	0.12	0.03	140
8a	4	0.13	0.2	0.4	1.2	1.9	0.13	0.3	0.38	0.007	270
8b	6	0.4	0.3	1.9	0.4	3.0	0.4	0.9	0.12	0.003	1000
9	11	1.8	–	21.4	9.9	13.3	1.8	8.8	1.70	0.07	190

In addition, data are available on phytomass and production of a wild-fruit forest with the edible *Malus sieversii* and for other mountain communities of the Tien Shan (Malinovsky, 1975).

The *Malus* forests in Kazakhstan cover an area of 11 370 ha from the Tarbagatay Mountains to the Pamir–Alay Ranges. In the study area of the Trans-Ili Alatau they cover 6560 ha. The crown closure is 40 to 60%.

Alongside the apple *Malus sieversii* there is a rich admixture of *Crataegus songorica*, *Armeniaca vulgaris* (the apricot), *Populus tremula*, etc. In the herbaceous stratum are *Brachypodium pinnatum*, *Digraphis* (*Phalaris*) *arundinacea*, *Urtica dioica*, *Polygonum coriarium*, *Gagea pseudorubescens*, *Leontice altaica*, and others.

The age of the stand, studied at 1500 m on a 25° southeast slope, was 80 to 100 years. The energy content of the *Malus* phytomass was 1235 million kJ ha^{-1}, of which 1023 million were in the above-ground mass and 212 million were below ground. Details are shown in Table 7.3.

The low net growth of phytomass (annual increase) of 1 t ha^{-1} above ground shows that the stand is already approaching its equilibrium condition.

The primary production has been estimated from the annual wood growth and annual leaf and fruit litter to be 4 t ha^{-1} above ground for *Malus* and 9 t ha^{-1} for the entire stand. It is, of course, very difficult to determine wood growth and litter of short-lived absorbing roots.

Nitrogen, phosphorus, and potassium balances were also determined for this stand (Table 7.4).

Especially interesting also are the data for eighteen biogeocoenoses (BGC) on the syrts at 3000 to 4000 m in the mountains south of Lake Issyk-Kul' (41°30′N, 77°30′E). The extreme climatic conditions at these elevations are as follows. The mean temperature in July does not exceed 10°C, and that of January is around −20°, so the mean annual temperature is below freezing. The net annual energy income is 190 kJ cm^{-2} at 3600 m and is negative at 4000 m. The mean annual ground-surface temperature corresponds to the air temperature; at a depth of 5 to 10 cm the temperature is 3°C lower than air temperature; in summer the surface temperature is 2 to 3°C (up to 5°C) higher than air temperature.

Precipitation falls primarily in summer and amounts to 160 to 450 mm annually. Moisture is slight at 3000 to 3200 m, still insufficient for plant-cover formation at 3200 to 3500 m, moderate at

TABLE 7.3

Phytomass, annual growth, annual litter, and net annual growth (in t ha^{-1}) for a wild apple (*Malus sieversii*) forest in the Trans-Ili Alatau (after B. S. Rodinov, R. Zh. Argimbayeva, G. Sh. Ayazbayeva, and K. Kazantseva, in Malinovsky, 1975)

Fraction	Phytomass	Annual growth	Annual litter	Net growth
TREE STRATUM				
Malus				
Wood	30.67	0.43	0.33	0.10
Bark	1.51	0.01	0.01	0.01
Thick branches	15.11	1.14	0.88	0.26
Thin branches	2.99	0.78	0.60	0.18
Annual shoots	0.07	0.07	–	0.07
Leaves	0.86	0.86	0.86	–
Underground parts	36.66	1.77	1.36	0.41
Fruit	0.76	0.76	0.76	–
Total	88.63	5.82	4.79	1.03
Crataegus				
Wood	6.09	0.14	0.07	0.07
Bark	0.61	0.01	0.01	0.01
Thick branches	2.40	0.18	0.09	0.09
Thin branches	0.84	0.07	0.03	0.04
Annual shoots	0.05	0.05	–	0.05
Leaves	0.22	0.22	0.22	–
Underground parts	7.36	0.33	0.16	0.17
Fruit	0.22	0.22	0.22	–
Total	17.79	1.22	0.79	0.43
Total for tree stratum	106.42	7.04	5.58	1.46
SHRUB STRATUM				
Lonicera, Rubus idaeus, Rosa, Ribes, Euonymus				
Above-ground parts	0.42	0.17	0.17	–
Underground parts	0.35	0.15	0.15	–
Total	0.77	0.32	0.32	–
HERB STRATUM				
Forbs (above ground)	2.71	2.71	2.71	–
Grasses				
Above-ground parts	0.84	0.84	0.84	–
Miscellaneous	0.21	0.21	0.21	–
Underground parts	12.54	3.74	3.74	–
Total	16.30	7.50	7.50	–
Total for entire stand	123.49	14.86	13.40	1.46

3500 to 3750 m, sufficient at 3750 to 3950, and 1.5 times the necessary amount above 3950 m. In general, precipitation at 4000 m is about three times higher than at 3000 m, but local circulation patterns cause considerable variations.

Three altitudinal belts were distinguished: (1) alpine (3000–4000 m); (2) subnival (4000–4400 m); and (3) nival–glacial (over 4400 m). In the alpine belt the eighteen biogeocoenoses (BGC) were studied, beginning at 3000 m and going upward:

Desert BGC's with mountain burozems (humus 1.5–2.5%, water content 2–5%):

1. BGC with *Artemisia rhodantha*
2. BGC with halophilic *Kalidium schrenkianum*, *Limonium hoeltzeri*, and *Reaumuria kaschgarica*

TABLE 7.4

Balance of nitrogen, phosphorus, and potassium cycles in a *Malus sieversii* forest (kg ha^{-1}) in the Trans-Ili Alatau (from Malinovsky, 1975)

Fraction	Phytomass			Annual growth			Annual litter			Net growth		
	N	P	K	N	P	K	N	P	K	N	P	K
TREE STRATUM												
Malus												
Structural organs	71.87	10.04	218.90	6.49	0.60	5.76	4.46	0.44	4.30	2.03	0.16	1.46
Leaves	12.27	0.93	20.77	12.27	0.98	20.77	5.88	0.90	3.73	6.39	0.08	17.04
Roots	76.56	42.81	67.01	3.68	2.06	3.22	2.36	1.33	2.08	1.32	0.73	1.14
Total	160.70	53.78	306.68	22.44	3.64	29.75	12.70	2.67	10.11	9.74	0.97	19.64
Crataegus												
Structural organs	21.62	9.10	25.07	1.16	0.08	1.06	0.27	0.03	0.35	0.89	0.05	0.71
Leaves	3.26	0.40	2.37	3.26	0.40	2.37	0.16	0.15	0.88	3.10	0.25	1.49
Roots	27.32	3.24	50.14	1.23	0.14	2.26	0.60	0.07	1.10	0.63	0.07	1.16
Total	52.20	12.74	77.58	5.65	0.62	5.69	1.03	0.25	2.33	4.62	0.37	3.36
Total for tree stratum	212.90	65.52	384.26	28.09	4.26	35.44	13.73	2.92	12.44	14.36	1.34	23.00
SHRUB STRATUM												
Lonicera, Rubus idaeus, Rosa, Ribes, Euonymus												
Above-ground organs	1.88	0.43	2.01	0.67	0.18	0.83	0.41	0.12	0.43	0.26	0.06	0.40
Underground organs	1.23	0.37	0.70	0.52	0.15	0.29	0.27	0.04	0.28	0.25	0.11	0.01
Total	3.11	0.80	2.71	1.49	0.33	1.12	0.68	0.16	0.71	0.51	0.47	0.41
HERB STRATUM												
Above-ground organs	59.46	8.62	67.48	59.16	8.62	67.68	59.16	7.66	67.68	–	–	–
Underground organs	164.48	15.29	58.80	54.83	5.10	19.57	54.83	5.10	19.57	–	–	–
Total	223.64	28.91	126.28	113.99	13.72	87.25	113.99	12.76	87.25	–	–	–
Total for biogeocoenosis	439.65	95.23	513.26	143.27	18.31	123.81	128.40	15.84	100.40	14.87	1.51	23.41

Semi-desert BGC's with high-mountain desert-steppe soils (humus 2–3%, water content 10–15%):

3. BGC with *Artemisia rhodantha* and graminoids
4. BGC with *Artemisia rhodantha* and *Oxytropis* spp.

Dry-steppe BGC's with chestnut soils (humus 1.5–6%, water content up to 20%):

5. *Festuca sulcata–F. kryloviana* BGC
6. *Ptilagrostis subsessiliflora* BGC
7. *Leucopoa olgae* BGC

BGC's of the cold steppe with takyr-like soils (humus 3–6%):

8. *Saussurea leucophylla–Oxytropis chionobia–O. globifolia* BGC
9. BGC with *Saussurea* and graminoids (*Festuca kryloviana, Ptilagrostis, Calamagrostis tianschanica*).

Mountain meadow-steppe BGC's with corresponding soils (humus 4–6% to 8–11%, water content in spring up to 30%, then decreasing):

10. *Kobresia capilliformis–K. humilis* BGC
11. *Festuca tianschanica–Kobresia* spp. BGC
12. forb-rich *Festuca tianschanica* BGC

Wet meadows with humus-rich (up to 10%) soils (water content 30–50%):

13. forb-rich *Kobresia* BGC
14. *Caragana–Kobresia* BGC

Alpine cushion-plants with mountain raw-humus soils under the cushion-plants (humus content 24%, water content 25%) and bare polygon soils between plants (humus content 0.5–2.5%, water content very low):

15. *Dryadanthe tetrandra* with *Oxytropis platysoma* BGC
16. *Dryadanthe tetrandra* with *Carex stenocarpa* BGC

Moors with 5 to 15 cm of peat over gley soils; permafrost at depths of 40 to 70 cm (humus 35–40%, ignition loss 50–60%, water-saturated):

17. *Carex melanantha* BGC with four other *Carex* spp. and mosses (*Hypnum, Tortularia, Stegonia, Grimmia*), plus *Nostoc*, etc.
18. The same, but in floodplains, periodically wetted.

The results of these studies are summarized in Table 7.5.

One recognizes from the Table that the largest amounts of phytomass are found in the wet biogeocoenoses (13, 14, 17, 18). The meadow-steppes have intermediate values (10, 11, 12), the cold and dry steppes lower values, and the very dry biogeocoenoses have very low values. The temperature factor plays almost no role at all.

Zoomass was also determined. It represents less than 1% of the total dry weight. Zoomass increases rapidly from low values in the deserts to highest values in the moist meadows. It is as low in the cushion-plant belt, however, as in the cold steppes, and is also relatively low in the moors. The largest part of the zoomass is made up of invertebrates (76–99.7%). The vertebrates play a noticeable role only in the dry biogeocoenoses, in which the fraction of above-ground species is also relatively high. The number of endemic animal species is very large, especially in the dry biogeocoenoses (see Table 7.6).

In the Kirgiz Mountains of the northern Tien Shan (42°48′N, 75°43′E) the phytomass of the following stands of meadow communities at various altitudes were determined:

1. A *Bothriochloa ischaemum* stand in the semi-desert (1200 m) on light chestnut earths
2. A *Stipa capillata–Agropyron repens–Festuca sulcata* stand of the mountain steppe (1500 m) on dark chestnut earths
3. Tall herbs with dominant *Aegopodium alpestre, Ligularia heterophylla,* and *Dactylis glomerata* (total of 40 species) at 2000 m in the montane meadow-steppe on deep chernozem-like soils
4. A mountain meadow with *Phlomis oreophila, Alchemilla retropilosa,* and *Poa angustifolia* (total of 70 species) on a deep mountain meadow soil
5. A subalpine meadow with *Alchemilla retropilosa, Aegopodium alpestre,* and *Phleum phleoides* (total of 66 species) on a less deep, subalpine meadow soil

The results are shown in Table 7.7.

In stand 1, on the average, 68% of the above-ground phytomass is due to *Bothriochloa* (*Andropogon*) *ischaemum* (range of variation: 61–85%), although 91 species occur in these stands. The maximum phytomass is measured in July. Of the total phytomass, 91% is under ground. Unregulated grazing causes a reduction in the phytomass present.

In stand 2 the above-ground phytomass varied over the course of six years between 1.2 and 3.4 t ha^{-1}. The maximum phytomass here was also measured in July, during flowering. The meadow vegetation possesses some vertical zonation.

Stand 3 has a very distinct maximum of phytomass at the end of July. The amount varies from year to year between 3.8 and 4.4 t ha^{-1}. The ratio of underground biomass to above-ground biomass is 8:1.

Stand 4 has a cover percentage of 70 to 80% and is relatively rich in species (a total of 71 species). Three strata could be distinguished. The total height is 65 to 80 cm, but most of the biomass is found within 30 cm of the surface. Phytomass

TABLE 7.5

Phytomass, annual growth, and annual litter of the high-mountain biogeocoenoses (BGC) 1–18 (see text) in t ha^{-1} (after R.I. Zlotin, in Malinovsky, 1975)

BGC	Above-ground parts				Roots			Total mass			Annual growth			Annual litter
	annual	perennial			living	dead	total	living	dead	total	above-ground	roots	total	
		living	dead	total										
1.	0.13	0.44	0.16	0.60	0.38	1.22	1.60	–	–	2.29	0.14	0.14	0.28	0.26
2.	0.43	0.16	0.38	0.54	1.66	0.64	2.30	2.57	0.76	3.33	0.45	0.41	0.86	0.83
3.	0.63	0.31	0.12	0.43	2.35	0.92	3.27	3.31	1.03	4.34	0.64	0.80	1.44	1.43
4.	1.02	0.44	0.16	0.60	4.38	0.90	5.28	5.84	1.08	6.92	1.03	0.80	1.83	1.82
5.	1.54	–	–	–	4.25	1.65	5.90	5.79	1.83	7.62	1.52	1.42	2.94	2.94
6.	1.10	–	–	–	2.81	3.04	5.85	3.95	3.11	7.06	0.87	0.90	1.77	1.77
7.	0.49	–	–	–	2.20	3.30	5.50	2.69	4.51	7.20	0.49	0.71	1.20	1.20
8.	0.83	0.40	–	0.40	4.19	0.98	5.17	5.40	1.01	6.41	0.83	1.13	1.96	1.93
9.	0.51	–	–	–	5.18	2.22	7.40	5.70	2.21	7.92	0.47	1.70	2.17	2.17
10.	1.61	–	–	–	13.11	16.69	29.80	14.71	16.90	31.61	1.25	4.40	5.65	5.65
11.	1.71	–	–	–	7.75	4.85	12.70	9.51	5.06	14.57	1.02	2.78	3.80	3.80
12.	1.49	–	–	–	10.10	6.73	16.83	11.59	8.41	20.00	1.40	3.30	5.70	5.50
13.	2.43	–	–	–	21.42	19.78	41.20	24.03	19.99	44.02	2.14	7.20	9.34	9.34
14.	1.40	0.76	–	0.76	14.60	17.14	31.74	16.60	20.22	36.82	1.31	4.81	6.12	5.98
15.	0.70	0.82	0.97	1.79	3.92	0.44	4.36	5.46	7.69	13.15	0.61	1.33	1.94	1.92
16.	0.80	2.49	3.57	6.06	5.13	1.05	6.18	8.36	13.75	22.11	0.87	1.60	2.47	2.40
17.	5.04	–	–	–	31.56	34.24	65.80	37.40	39.00	76.40	2.90	10.60	13.50	13.20
18.	3.04	–	–	–	2.24	0.96	3.20	25.44	11.60	37.04	2.00	7.50	9.50	9.40

Note: The figures show some inconsistencies which were already present in the original, without any explanation.

TABLE 7.6

Zoomass and its composition in the high-mountain ecosystems 1–18 (see text) (from Malinovsky, 1975)

Subdivisions of alpine belt from bottom to top	Living zoomass (kg ha^{-1})	Composition by groups in % of total zoomass						
		systematics		habitat		trophic type		
		invert.	verteb.	under-ground	above-ground	sapro-phages	phyto-phages	carni-vores
Desert BGC's	2.2	96.9	3.1	76.6	23.4	53.8	34.4	11.8
Semi-desert BGC's	7.7	75.9	24.1	68.9	31.1	31.2	65.8	3.0
Dry steppes	12.1	79.7	20.3	75.7	24.3	43.4	51.7	4.9
Cold steppes	26.8	95.5	4.5	93.6	6.4	87.7	10.4	1.9
Meadow-steppes	215.2	99.5	0.5	99.1	0.9	96.7	3.0	0.3
Wet meadows	542.0	99.7	0.3	99.5	0.5	97.6	1.9	0.5
Alpine cushions	25.8	99.7	0.3	98.6	1.4	82.2	13.9	3.9
Moors	109.5	99.5	0.5	98.8	1.2	98.2	1.1	0.7

TABLE 7.7

Phytomass and litter layer (t ha^{-1}) of meadow biogeocoenoses (see text for meaning of stand numbers) in the Kirgiz Mountains of the northern Tien Shan (after L.I. Popov, L.P. Lebedeva, R.I. Ionov, and I.G. Korneyeva, in Malinovsky, 1975)

Stand No.	Elevation (m)	Phytomass			Total organic matter
		above-ground	below-ground	litter layer	
1.	1200	4.34	64.84	1.40	70.58
2.	1500	2.78	46.20	1.58	50.56
3.	2000	2.89	42.36	2.95	48.20
4.	2300	2.15	30.59	2.30	35.04
5.	2400	2.68	29.54	3.43	36.65

increases rapidly in May–June and reaches a maximum in the middle of July. The range of variation is 2.2 to 3.5 t ha^{-1}. The ratio of below-ground to above-ground biomass is high here also at 7:1.

Stand 5 has a cover of 80 to 90% and consists of 66 species. The plant development begins as early as the end of April. The variation of phytomass in individual years is very low. The relation of below-ground to above-ground phytomass is 5:1.

The phytomass of these five stands studied decreased with increasing altitude, while the amount of litter and the annual litterfall increased.

The primary production in this region was also assessed, namely at 1900 to 1950 m (montane meadow-steppe belt) in a *Trifolium*-rich meadow. The main species were: *Festuca sulcata*, *Trifolium repens*, *Taraxacum officinale*, *Poa pratensis*, *Phleum phleoides*, *Agropyron repens*, *Agrostis alba*, and *Koeleria gracilis*. The soil was rich in skeletal materials and had a humus content as high as 5.77%. The water content in May–June was 29.5% at a depth of 70 to 80 cm. The nutrient content per 100 g of soil, in extractable amounts, was 11.5 to 14 mg nitrogen, 0.25 to 1 mg phosphorus, and 16.7 to 34.4 mg potassium. The primary production was determined only in the years 1969–1972, in the last year with a comparison to an *Artemisia–Ziziphora* pasture in a semi-desert area (see Table 7.8).

TABLE 7.8

Primary production of air-dried biomass in t ha^{-1} for stands from the Kirgiz Mountains at 1900–1950 m (from Malinovsky, 1975)

	1969	1970	1971	1972
Precipitation (mm)	680	450	307	–
Primary production (t ha^{-1})				
Trifolium meadow	0.89	0.99	0.87	1.46
Artemisia pasture	–	–	–	1.87

REFERENCES

Abramov, T. G., 1949. From results of the stationary observations in the basin of the Ukok River in Kirghiz SSR. *Tr. Leningr. Ova. Estestovoispyt.*, 6: 32–62 (in Russian).

Afanasyev, K.S., 1956. *Vegetation of the Turkestan Mountain Range*. Akad. Nauk S.S.S.R., Moscow, Leningrad, 278 pp. (in Russian).

Arystangaliyev, S., 1956. Meadow vegetation of the Kungey-Alatau in the Kazakhstan region. *Roboty Bot. Inst. Akad. Nauk Kazakh. S.S.R.*, 4: 130–175 (in Russian).

Chumichev, D.A., 1968. *Soviet Union: Tadzhikistan*. Nauka, Moscow, 238 (with map) (in Russian).

Korovin, E.P., 1961/62. *Vegetation of Middle Asia and Southern Kazakhstan*. Tashkent, 2nd ed., Vol. I: 452 pp.; Vol. II: 577 pp. (in Russian).

Koroleva, A.S., 1958. Sketch of the vegetation on the central south slope of the Gissar Mountains. *Tr. Tadzhik. Otdel. Akad. Nauk S.S.S.R.*, 8: 1–140 (in Russian).

Kultiasov, I.M., 1955. *Peculiarities of the Ecology of High-Mountain Plants in the Western Tien Shan*. Akad. Nauk S.S.S.R., Moscow, 127 pp. (in Russian).

Levina F.Ya., 1944. Contemporary inundation processes and the peat quantities of the Fergana Valley. *Bot. Zh.*, 29: 192–203. (in Russian).

Malinovsky, K.A., 1975. Productivity of mountain communities of the U.S.S.R. *Resursy Biosfery* (*Leningrad*), 1: 167–198 (in Russian).

Meusel, H., 1952. Über die Elyneten der Allgäuer Alpen. *Ber. Bayer. Bot. Ges.*, 29: 47–55.

Pavlov, V.N., 1980. *Vegetation of West Tien-Shan*. Moscow University Press, Moscow, 245 pp.

Popov, M.G., 1940. The vegetation of Kazakhstan. *Tr. Kaz. Fil., Akad. Nauk*, 18: 215 pp. (in Russian).

Stanyukovich, K.W., 1949. *Plant Cover of Eastern Pamir*. Nauka, Moscow, 159 pp. (in Russian).

Stanyukovich, K.W., 1973. *Mountain Vegetation of the U.S.S.R.* Akad. Nauk Tadzhik S.S.R., Dushanbe, 310 pp. (in Russian).

Troll, C., 1972. The three-dimensional zonation of the Himalayan system. *Erdwiss. Forsch.* (*Wiesbaden*), 4: 264–275.

Vykhodtsev, I.V., 1956. *Altitudinal Vegetation Belts in the Kirghiz* (*Tien-Shan and Alai*). Akad. Nauk S.S.S.R., Moscow, 84 pp. (in Russian).

Walter, H., 1974. *Die Vegetation Osteuropas, Nord- und Zentralasiens. Vegetationsmonographien der einzelnen. Großräume, Band VII.* Gustav Fischer-Verlag, Stuttgart. 452 pp.

Walter, H., 1975. Ecological relations between plants of the steppe and those of the alpine belt. *Flora*, 164: 339–346.

Zakirov, K.S., 1955. *Flora and Vegetation in the Basins of the River Zeravshan, I. Vegetation*. Akad. Nauk Uzbek. S.S.R., Tashkent, 207 pp. (in Russian).

Zakirov, K.S., 1962. *Flora and Vegetation in the Basins of the River Zeravshan*, II. Flora. Akad. Nauk Uzbek. S.S.R., Tashkent, 446 pp. (in Russian).

Zaprayagayeva, V.I., 1964. *The Wild Fruit-Tree Species of Tadzhikistan*. Tadshik. Akad. Nauk, Bot. Inst., Nauka, Moscow, Leningrad, 695 pp. (with 319 figures and 81 tables) (in Russian).

Zlotin, R.I., 1971. Production and phytomass distribution in the syrts of the Tien Shan. In: *Biological Productivity and Mineral Cycling*. Leningrad, pp. 145–150 (in Russian).

Chapter 8

THE DESERTS OF CENTRAL ASIA[1]

H. WALTER and E.O. BOX with coverage of the Gobi by W. HILBIG

GENERAL DESCRIPTION

Central Asia consists of a series of large basins separated by high mountains. It is a huge area with, in general, no external drainage. Only in the south and east do the upper courses of the great river systems of India, Southeast Asia, and China reach into this area (see Fig. 2.1).

During the Early Tertiary there were two centers of aridity in Eurasia, the Sahara and Central Asia. The old Tethys Sea extended across Iran, Turania, and into the West Siberian Lowlands. After the Tethys Sea disappeared in the Late Tertiary as a result of massive orogenesis, the region was colonized by plants from both the west and the east. Thus, in eastern Kazakhstan and in Dzungaria, mixed floras arose (Grubov, 1959, 1972).

The Mongolian element in Central Asia is represented by a series of old taxa which stand quite isolated in Asia, for example *Potaninia mongolica*, a member of the family Rosaceae related to the South African genus *Cliffortia*; legumes which show close relationships with the South African genus *Podalyria*; and the taxonomically isolated species *Zygophyllum xanthoxylon* (Fig. 8.1), *Z. kaschgaricum*, *Tetraena mongolica* and *Brachanthemum gobicum* (Fig. 8.2). Very characteristic of Central Asia are the shrub-shaped *Caragana* species and *Hedysarum* species (both Fabaceae).

Typical species which occur only in Central Asia are the following:

Sand-deserts: *Caragana microphylla*, *C. bungei*, *C. korshinskii*, *Hedysarum scoparium*, *H. mongolicum*, *Atraphaxis frutescens*, *Oxytropis aciphylla*, *Salix flavida*, and the herbs *Artemisia ordosica*, *A. sphaerocephala*, *Pugionium cornutum*, *Agriophyllum gobicum*, and *Psammochloa villosa*.

Sandy-gravelly deserts on high plateaus of Cretaceous and Tertiary formations: *Zygophyllum xanthoxylon* (Fig. 8.1), *Z. kaschgaricum*, *Potaninia mongolica*, *Ephedra przewalskii*, *Nitraria sphaerocarpa*, *Reaumuria soongorica*, *Calligonum mongolicum*, *Anabasis brevifolia*, *Salsola passerina* (Figs. 8.3 and 8.4), *Caragana pygmaea*, *C. frutescens*, *Asterothamnus* spp., *Artemisia xerophytica*, *Ammopiptanthus mongolicus*, *Stipa glareosa*, and *S. gobica*. Species which occur in Kashgaria–Dzungaria are: *Sympegma regelii*, *Iljinia regelii*, *Convolvulus tragacanthoides*, and *C. gortschakovii*.

Rock and rubble deserts on higher ground: *Gymnocarpus przewalskii*, and the already-mentioned *Nitraria*, *Zygophyllum*, *Tetraena*, and *Ammopiptanthus* species, plus *Iljinia*, *Sympegma*, and *Salsola laricifolia*.

Fig. 8.1. *Zygophyllum xanthoxylon* in the northern Gobi Desert (photograph by Z.G. Bespalova).

[1] Manuscript completed November, 1977.

Fig. 8.2. *Brachanthemum gobicum* in the dwarf-shrub desert of the Gobi Desert (photograph by Z.G. Bespalova).

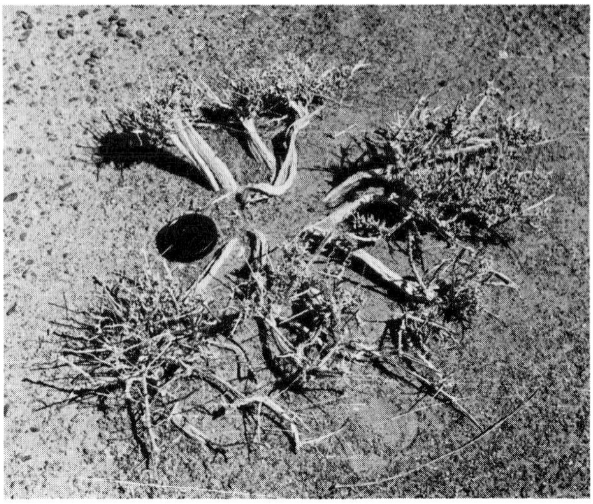

Fig. 8.4. Very old *Salsola passerina*, the main stem of which has split at the base into many parts (particularization, or vegetative segmentation). (Photograph by Z.G. Bespalova.)

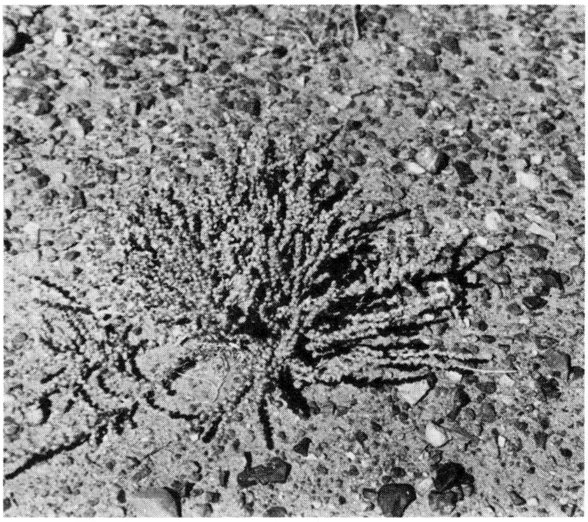

Fig. 8.3. A young *Salsola passerina* plant in the sandy gravel desert of the Gobi Desert (photograph by Z.G. Bespalova).

Fig. 8.5. *Kalidium foliatum* on brackish sand in the Gobi Desert (photograph by Z.G. Bespalova).

Salt-deserts: *Kalidium gracile* and *K. cuspidatum*, while *K. foliatum* (Fig. 8.5) is more Turano-Gobic; *Tamarix chinensis*, plus others which occur also in Middle Asia.

Other species characteristic of Central Asia are *Tanacetum* (*Hippolytia*) *achilleoides*, *T. trifidum*, *T. gobicum*, as well as *Allium mongolicum* and *A. polyrrhizum*.

The gypsum, clay, and loess deserts which cover such large areas in Middle Asia are hardly represented in Central Asia.

The northern part of Central Asia belongs to the Mongolian floristic province, the southern part to the Tibetan, while Dzungaria represents a transition region and, by its flora, can be considered to be part of the Kazakh-Dzungarian province (see Fig. 2.1).

Since we have treated Kazakhstan in some detail, we only need to look briefly at Dzungaria.

DZUNGARIA (see Fig. 8.7)

Just as was the case in eastern Kazakhstan around Lake Balkhash, the precipitation in Dzungaria is also rather evenly distributed throughout the year (Fig. 8.6). The annual amount at Ching-ho (Dshiho, Jinghe, at elevation 318 m) on the lake Ebi Nor (Ai-pi Hu) is 96 mm, at Wu-lu-mu-ch'i (Urumchi, at 903 m) at the north foot of the Tien Shan, 247 mm, and at Ch'i-t'ai (Jitai, Tsitai, at elevation 795 m) further to the east, 193 mm. The central part of Dzungaria is certainly even drier, but data are lacking. The flora of Dzungaria reminds one more of eastern Kazakhstan, while the biotopes, on the other hand, are more similar to those of other parts of Central Asia.

Typical species are: *Anabasis salsa*, *Nanophyton erinaceum* (see Figs. 4.14, 8.18), *Haloxylon persicum*, and also *H. ammodendron* and *Salsola arbuscula*. Ephemerals are much less important. Characteristic communities are those with *Salsola dshungarica*, *Stipa glareosa*, *Arthrophytum*, etc.

The largest part of the area is taken up by the central basins. The northern part is a rock and rubble desert at the foot of the southern Altai Mountains. The central part is a sand desert with saksa'ul. The eastern part is a high plateau with an *Artemisia–Salsola* desert, and the southern part is again a rock desert on the north slope of the eastern Tien Shan. The mountains themselves have forest belts at higher elevations, the Altai in the north with *Larix sibirica* and the Tien Shan in the south with *Picea schrenkiana* up to the alpine belt.

The saksa'ul desert is, according to Murzayev (1971), the most widespread vegetation type in Dzungaria, not only on sand but also on the loamy *hammadas* in the east and west, from Ebi Nor to the rivers Chyorny Irtysh and Urungu. Both saksa'ul species occur, but *Haloxylon persicum* reaches its eastern limit on sand in Dzungaria. *Haloxylon ammodendron*, the Gobi variety (earlier separated from the Turanian *H. aphyllum*), is more widespread on rubble and on gray-brown, skeletal soils, together with the xerophytes *Ephedra przewalskii* and *Iljinia regelii* on gypsum-rich soils.

Tulips and *Rheum nanum* occur as ephemeroids, as well as the parasites *Orobanche salsa* and *Cynomorium songoricum*.

At the border between the sand areas and alluvial areas, where water collects, salinization of the soil often occurs. *Haloxylon ammodendron* dominates here, along with wide expanses of *Reaumuria soongorica* or *Tamarix* and *Ceratoides papposa* (*Eurotia ceratoides*). In the north one finds communities reminiscent of Kazakhstan: in addition to the saksa'ul considerable *Nanophyton erinaceum*, *Artemisia schischkinii*, *A. gracilescens*, *A. terrae-albae* var. *massagetovii*, *Salsola rigida*, and on sand *Artemisia arenaria*, *Ferula* sp., *Ephedra*, and ephemerals. But on the Tertiary plateau of eastern Dzungaria there occur also the true Mongolian elements, such as *Brachanthemum mongolicum*, *Zygophyllum xanthoxylon*, *Caragana leucophloea*, on sand *Atraphaxis compacta*, and *Calligonum mongolicum* as well as *Hedysarum mongolicum*. The lowest part of Dzungaria is the Ebi Nor Basin with extensive sandy soils and solonchak. Found here are *Haloxylon persicum*, *Aristida pennata*, *Ephedra przewalskii*, *Anabasis aphylla*, and *Artemesia borotalensis*, with *Anabasis brevifolia* on rocky ridges. Thus, one sees how Middle Asian and Central Asian (Mongolian) elements interdigitate in this transition region.

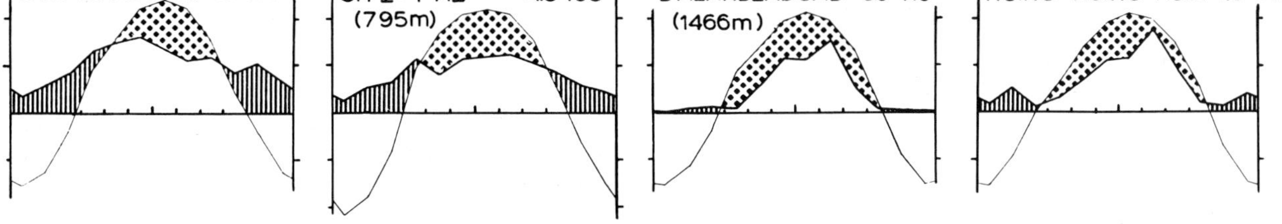

Fig. 8.6. Climate diagrams for stations in Dzungaria (Urumchi, Ch'i-t'ai), north of the Gobi–Altai (Dalandzadgad), and south of the Gobi in the Pei Shan Desert (Hsing-hsing-hsia). The latter two have summer rainfall.

GEOBOTANICAL SUBDIVISIONS OF CENTRAL ASIA

Central Asia is normally divided into two floristic-ecological provinces, provided that one excludes Dzungaria as a transition zone toward eastern Kazakhstan. The two actual provinces (Grubov, 1959; see also Lavrenko, 1965) are the Mongolian and the Tibetan (see Fig. 2.1).

The Mongolian province includes the entire Mongolian People's Republic (Outer Mongolia), as well as eastern Mongolia (the Chinese Inner Mongolia) (Fig. 8.7, region *6*) and the Cis-Hsing-An region (Fig. 8.7, region *4*). This last region, however, is steppe with some forest and will not be treated here (see Walter, 1974). The same is true of the northern part of Outer Mongolia (Fig. 8.7, regions *2*, *3*, *5*), with the exception of the Large Lake District (*7*). Only the southern part (Fig. 8.7, regions *9*, *10*, and the northern part of *11*) belongs to the Gobi, roughly as far as the Gashuun Nuur (Ka-shun Hu).

We separate off the southern part of region *11*, the Pei Shan Desert, since this region is an orographic region in itself.

Bordering the Pei Shan region to the west is the huge Tarim Basin between the eastern Tien Shan in

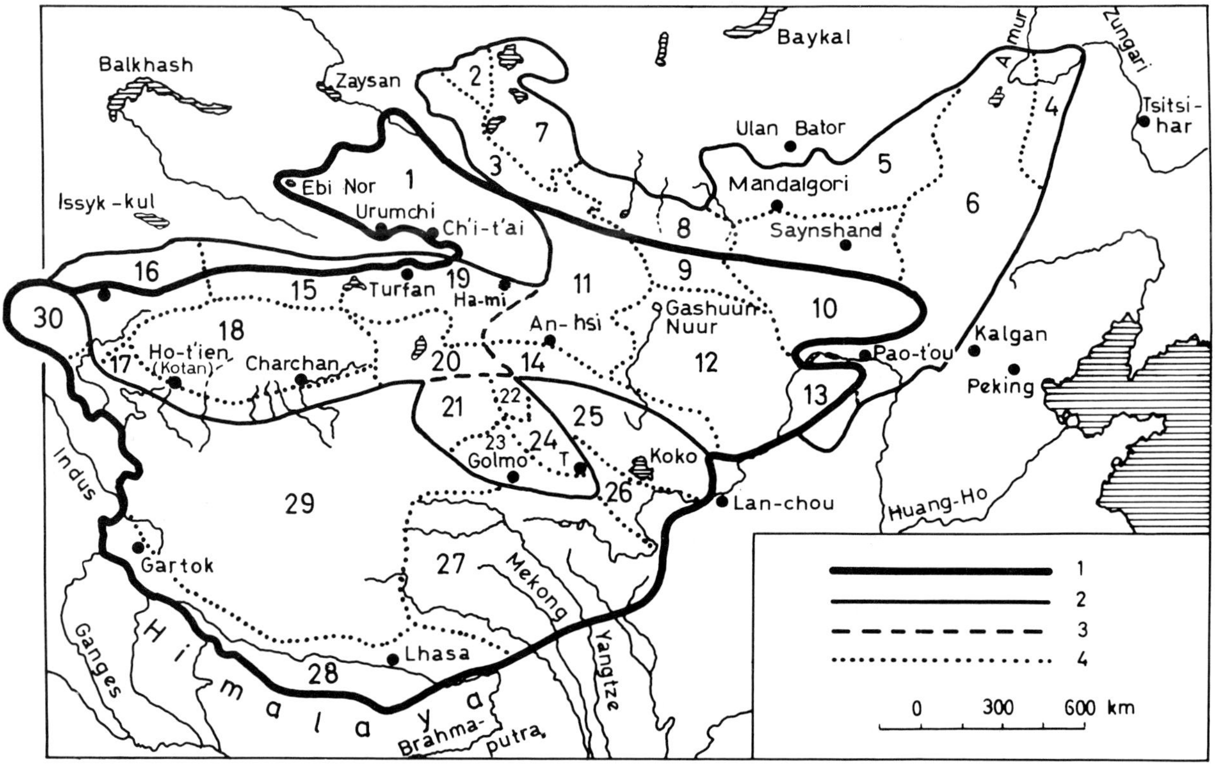

Fig. 8.7. Geobotanical subdivision of Central Asia (after Grubov, 1959); the Dzungaro-East Kazakh Province belongs more to Middle Asia. Outer borderline shows the region of Central Asia. *1* = boundary of the deserts; *2* = boundaries of provinces; *3* = boundaries of subprovinces; *4* = boundaries of regions.

Dzungaria. *1* = Dzungaria.

Mongolian Province (*2–14*). Steppes and semi-deserts: *2* = Kobdo Basin; *3* = Mongolian Altai; *4* = Cis-Hsing-An region (in the east); *5* = Middle Chalcha; *6* = eastern Mongolia; *7* = Basin of the large lakes; *8* = valley of lakes. Deserts: *9* = Gobi–Altai; *10* = eastern Gobi (steppe in the north); *11* = western Gobi (southern half = Pei Shan Desert); *12* = Ala Shan Desert; *13* = Ordos (steppes in the southeast); *14* = Kansu (Hessi) Corridor.

Kashgaria and Tarim Basin (*15–20*). *15* = northern Kashgaria; *16* = western Kashgaria (desert only at low elevations); *17* = southern Kashgaria; *18* = Takla Makan; *19* = eastern Kashgaria; *20* = Lop Nor Depression.

Tsaidam (*21–24*). *21* = northwestern Tsaidam; *22* = northern Tsaidam; *23* = southeastern Tsaidam; *24* = eastern Tsaidam; *T* = Tsagan.

Tibetan Province (*25–30*). *25* = Nan Shan Mountains; *26* = Chamdo; *27* = Hsikiang region of deep river valleys with forest; *28* = southern Tibet (longitudinal valley); *29* = Chang Tang (High Tibet); *30* = Pamir.

the north and the K'un Lun Shan (Kun Lun Mountains) in the south. The extreme Takla Makan Desert is in the center of the Tarim Basin (Fig. 8.7, regions *15–20*). To the east stretches the huge Ala Shan Desert, with the Ordos Desert on the other side of the river Huang-Ho (Fig. 8.7, regions *12–13*). The region is delimited in the south by the Kansu Corridor (Fig. 8.7, region *14*).

The last part of the Mongolian province, the Tsaidam region (Fig. 8.7, regions *21–24*), forms the transition to the Tibetan province. It lies 1500 m higher than the other Mongolian deserts but still 1500 m lower than the Tibetan province with its very cold deserts. At just over 2000 m the winters are quite cold but the summers distinctly hot. It is, thus, a region of extremely continental conditions.

In addition to Tibet (Fig. 8.7, regions *25–29*), the Tibetan province includes only the Pamir (Fig. 8.7, region *30*), the node of the massive Asian mountain ranges.

The desert areas named will be discussed in sequence. These are:

I Deserts with hot summers

A. Gobi and Large Lake District	E. Ala Shan
B. Pei Shan	F. Ordos
C. Tarim Basin with Takla Makan	G. Tsaidam (transition)
D. Kansu Corridor (Hessi Corridor)	

II. Deserts with cool summers

A. Tibet	B. Pamir

Table 8.1 contains comparative climatic data for Middle Asian and Central Asian deserts with mean annual temperatures above freezing.

Deserts with hot summers

The Gobi and the Large Lake District (W. Hilbig)

The Gobi. This desert and desert-steppe region is divided, after Grubov (1963), into the Gobi–Altai (Fig. 8.7, region *9*), the eastern Gobi (Fig. 8.7, region *10*), and the western Gobi (Fig. 8.7, the northern half of region *11*). In Grubov's (1955) phytogeographical subdivision of Outer Mongolia (Mongolian People's Republic), the western Gobi corresponds to the district of Transaltai–Gobi. The easternmost part of Dzungaria, belonging to the Mongolian People's Republic, is separated as the Dzungarian Gobi (Fig. 8.18). The northern border areas of the Ala Shan Desert are separated as the Ala Shan–Gobi.

Fig. 8.8 shows a very schematic subdivision of the vegetation regions of Outer Mongolia. One sees that the real desert zone only occupies a relatively narrow belt (*c.* 15% of the whole territory) in comparison with the semi-desert or desert-steppe zone located to the north. A large part of the desert-steppe zone, constituting 27% of the territory of the Mongolian People's Republic, also belongs to the Gobi. The term Gobi in the Mongolian language includes desert and desert-steppe areas. Therefore, geographical names with the term Gobi are often found also in the desert-steppe areas north of the Mongolian and Gobi–Altai. The continuous desert zone is situated south of the Mongolian and Gobi–Altai and begins in the western part of Mongolia at about 46°N, in the east at about 43°N.

The Gobi has an average altitude from 1000 to

TABLE 8.1

Climate data for the deserts of interior Asia (from Petrov, 1966/67)

Station	Elevation (m)	Mean temperatures (°C)			Precipitation (mm)
		January	July	annual	
Turanian Lowlands	22 to 242	−6.9 to 4.8	27.1–32.3	11.0 to 17.1	82–244
Iranian Plateau	516 to 1773	1.0 to 6.0	27.9–33.0	15.4 to 20.0	76–250
Kazakhstan	−22 to 396	−13.6 to 6.0	25.1–28.6	7.5 to 12.1	103–184
Dzungaria	318 to 903	−22.0 to −15.2	23.9–25.4	4.3 to 6.8	96–247
Gobi Desert	991	−18.1	23.3	3.8	107
Ala Shan	1240 to 1562	−11.2 to −6.8	22.6–24.5	7.0 to 9.6	39–219
Pei Shan	1122 to 1776	−15.6 to −9.8	21.1–26.3	4.3 to 9.6	42–74

1300 m. Extensive depressions from 700 to 1000 m can also be found.

The rainfall is nearly completely restricted to the summer, and amounts to 100 to 150 mm, but can be as low as 40 mm. In the driest parts, in the west, there can be entire years with no rainfall at all. The typical distribution of temperature and precipitation in the course of the year is shown in the diagram of Dalandzadgad (Fig. 8.6). At Saynshand (45°N, 110°E, 991 m in elevation) in the desert-steppe zone (northern part of region *10* of Fig. 8.7), 107 mm of precipitation have been measured, with a mean annual temperature of 3.8°C (January −18.1°C, July 23.3°C). Winters are cold, cloudless, and very dry. As a result of the high insolation, daily maximum temperatures can reach above freezing even in winter. The very thin and discontinuous snow cover usually evaporates without melting. The dry soil remains unfrozen.

Extensive areas with loess-like soils or with sand, as were found in Middle Asia, are missing in the Mongolian deserts. Most soils are rocky, similar to hamadas, and covered with a coating of broken stone and gravel, which accumulates on the surface (desert pavement) as the finer materials are blown away. Through desert-varnishing these stones have taken on a dark appearance. Gypsum crystals are also found. Their surface is often cork-like and porous, stratified, and brick-red (iron oxides) in color, as in subtropical soils.

The compact soil surface and precipitation falling in short heavy rains causes high runoff and remarkable water erosion. In the smaller areas of sand deserts the sand shifts off because of the continuous and strong wind (also sand storms). One can see shifting dunes. The more extensive sand areas exist in the desert-steppe zone (Fig. 8.8).

Important components of the Mongolian desert vegetation are low shrubs and semi-shrubs of the Chenopodiaceae (*Anabasis*, *Haloxylon*, *Salsola*, *Kochia*, *Sympegma*, *Iljinia*, *Nanophyton*), Tamaricaceae (*Reaumuria*), Zygophyllaceae (*Zygophyllum*, *Nitraria*), Fabaceae (*Oxytropis*, *Caragana*), Polygonaceae (*Atraphaxis*, *Calligonum*), Convolvulaceae (*Convolvulus*), and Asteraceae (*Artemisia*, *Brachanthemum*, *Tanacetum*).

An especially characteristic plant of the Central Asian desert regions is *Anabasis brevifolia*, the so-called "baglur". The *Anabasis brevifolia* desert is widespread on gravel sites in all parts of the Gobi.

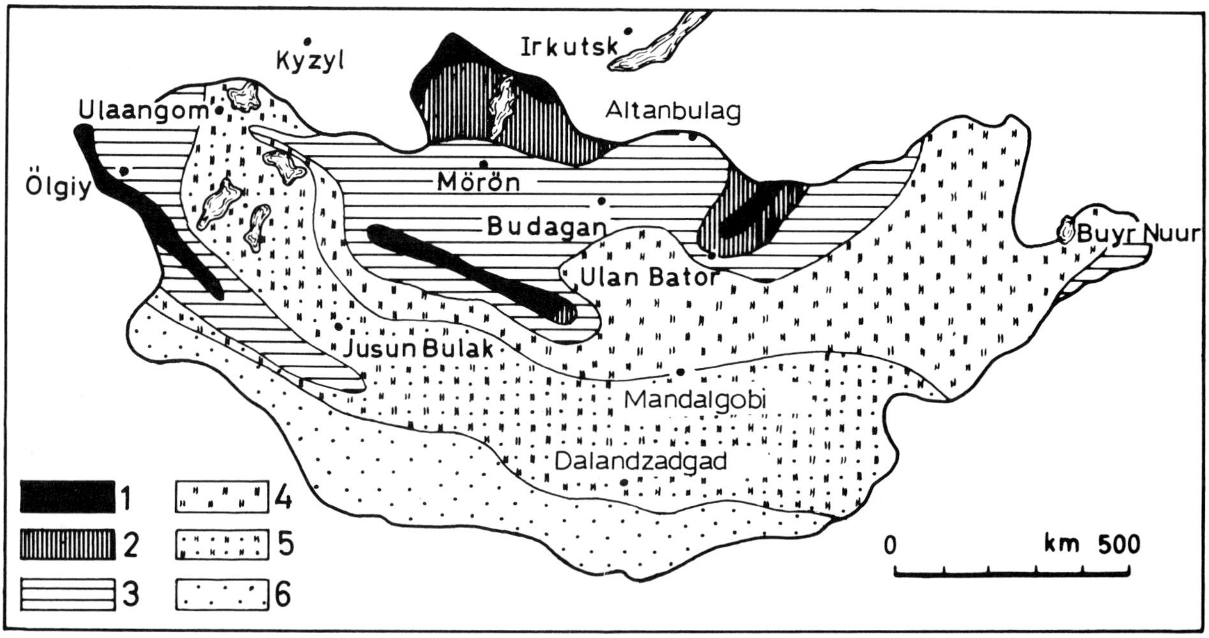

Fig. 8.8. Schematic subdivision of the vegetation of Outer Mongolia (after Yunatov, 1950): *1* = alpine belt; *2* = montane coniferous forest belt; *3* = montane forest-steppes and steppes; *4* = grass steppes; *5* = semi-desert; *6* = desert (see note, p. 236).

Extrazonal desert vegetation within the desert-steppe zone is dominated in most cases by the same species, as in the large lake district. One can distinguish a more extreme form of this community with *Reaumuria soongorica*, *Salsola passerina*, and *Convolvulus gortschakovii* and a form with species from desert-steppe vegetation.

The saksa'ul (*Haloxylon ammodendron*) is the characteristic plant of sandy areas in the desert region, but can also be found on clay and gravel (both on salty and non-saline soils). On sandy sites the yield of green assimilating branches of *Haloxylon* amounts to 0.3 to 0.5 t ha^{-1}, on gravel only to 0.08 to 0.15 t ha^{-1} (Yunatov, 1954). Often the plants show a low shrub shape and hedging as a result of overgrazing. Similar potential osmotic pressures observed on undamaged luxuriant growth as well as in heavily damaged stands of *Haloxylon* verify that damage is often caused by overgrazing (Helmecke and Schamsram, 1979a).

The zone of desert-steppes or semi-deserts north of the desert zone shows a gradual transition to the steppe zone in the eastern parts of Mongolia. One main type of the desert-steppe, showing the most steppe-like character, is the *Stipa* community. It is dominated by short *Stipa* species (*S. glareosa*, *S. gobica*, *S. orientalis*). In the southern part of Mongolia the desert-steppes are characterized by a *Stipa–Allium* community (Figs. 8.9, 8.10). *Allium polyrrhizum* and *Allium mongolicum* are the most common plants in this community. Notable forbs of both desert-steppe communities are *Lagochilus ilicifolius*, *Zygophyllum rosovii*, *Gypsophila desertorum*, *Ptilotrichum canescens*, *Scorzonera divaricata*, *Tanacetum achilleoides*, *Convolvulus ammanii*, *Rheum nanum*, and on overgrazed areas *Peganum nigellastrum*.

Another important type of desert-steppe vegetation on sandy gravel and rubble sites in the Gobi and in the Large Lake District is the shrub desert steppe (Fig. 8.11) with *Caragana* (*C. leucophloea*, *C. pygmaea*, *C. bungei*), *Atraphaxis* (*A. frutescens*, *A. pungens*), *Amygdalus* (*A. pedunculata*), *Oxytropis* (*O. aciphylla*), and semi-shrubs of *Artemisia* (*A.

Fig. 8.9. Desert-steppe near Bulgan (Gobi Desert, northwest of Dalandzadgad) at the end of June: fruiting *Stipa glareosa*, with *Allium polyrrhizum*, *A. mongolicum*, etc. (photograph by P. Hanelt).

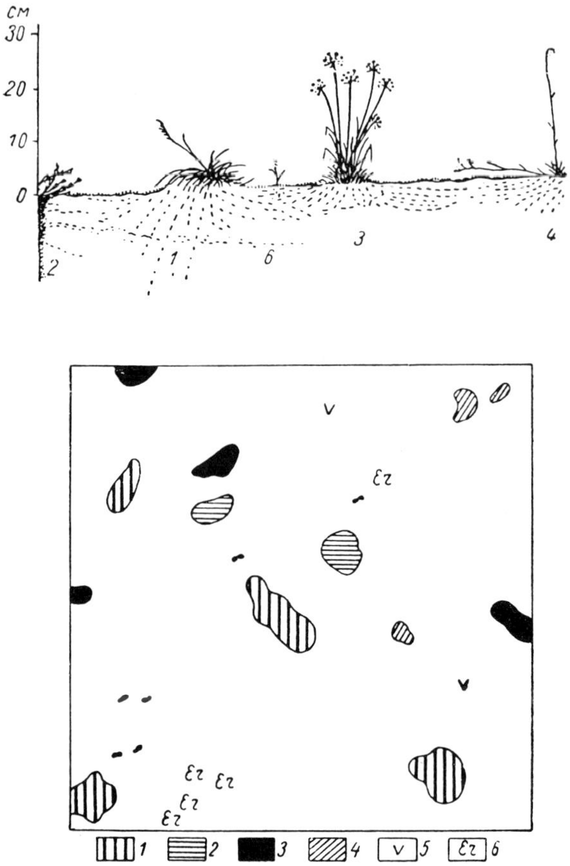

Fig. 8.10. *Stipa-Allium* biogeocoenosis: profile (above) and plan of 1 m^2 (below), from Kalinina (1954): *1* = *Stipa glareosa* + *Stipa gobica*; *2* = *Anabasis brevifolia*; *3* = *Allium polyrrhizum*; *4* = *Cleistogenes mutica*; *5* = *Carex duriuscula*; *6* = *Eragrostis minor*.

xanthochroa, *A. xerophytica*, *A. rutifolia*). Considering the structure and combination of species in this case, they should be considered semi-deserts. Summer annuals appear in the desert and desert-steppe communities at the beginning of summer precipitation (*Aristida adscensionis*, *Setaria viridis*, *Pappophorum boreale*, *Eragrostis minor*, *Tribulus terrestris*, *Bassia dasyphylla*, *Plantago minuta*, *Agriophyllum*, *Halogeton*, *Salsola*, *Corispermum*). Seeds of annuals predominate in the seed-bank of the soil. Generally the plants in deserts and desert-steppes show highly variable seed production from year to year.

Mosses are almost completely absent from these communities. Lichens can be found to a small extent. Klement (1966) stated that nearly all species of the Irano-Turanian lichen association *Parmelietum vagantis* occur in the Mongolian deserts and desert-steppes. Lichens on rocks, such as *Xanthoria*, *Acarospora*, and *Omphalodina*, are spread from deserts to mountain steppes. Nevertheless, the deserts in the lee of the huge mountain ranges are not empty of lichens. Crustose lichens of the genera *Sphaerothallia* (*S. lacunosa*, *S. desertorum*, *S. terrestris*) and *Amphoridinum* (*A. caesiopsilon*) appear as "window lichens" at the edges of stones near the ground surface (Fig. 8.20). These also occur on the bottom of transparent pebbles (Schubert and Klement, 1971). Dew and condensation of soil humidity at night after rains effect the accumulation of some water at these points, enough to wet the lichen thalli thoroughly and to enable assimilation at daybreak.

Skalon (1977) presented first results of investigations on soil micro-organisms of different desert-steppe and desert communities of Mongolia. His investigations led to different species numbers and quantities in dependence on soil and vegetation. In the *Stipa gobica–Allium polyrrhizum* community he found 79 species, in the *Reaumuria soongorica–Brachanthemum gobicum* community 32 species, of which 10 were *Actinomyces*, 3 were *Azotobacter* and 1 was *Clostridium*. The micro-organisms are most dense in the rhizospheres of the plants.

The Mongolian deserts and desert-steppes are refuge areas of very rare, and highly protected wild animals, the Mongolian kulan or Dshiggetai (*Asinus hemionus*), the Saiga antelope (*Saiga tatarica*), the Zheiran gazelle (*Gazella subgutturosa*), and the wild camel (*Camelus ferus*). The wild Przewalski horse (*Equus przewalskii*) lived in natural habitat in the Gobi Desert. The horse, however, has not been seen there in the last few decades and may be extinct in its natural habitat.

The most common mammals are rodents. Typical rodents in the Mongolian desert-steppe and desert are representatives of the jerboas (*Allactaga*, *Alactagulus*, *Dipus*, *Salpingotus*, *Cardiocranius*), gerbils (*Meriones*, *Rhombomys*), and dwarf-hamsters (*Allocricetulus*, *Cricetulus*, *Phodopus*). Among the genera mentioned *Rhombomys* is closely linked to the stands of *Haloxylon* and *Nitraria*. The souslik (*Citellus erythrogenys*) and the pika (*Ochotona pallasii*) are characteristic animals of the desert-steppe region. In desert vegetation there are far fewer rodents than in the desert-steppe. The

A

B

Fig. 8.11.A. Sandy semi-desert zone of the Gobi Desert east of Bulgan, with dense shrub vegetation of *Caragana bungei*. B. *Caragana leucophloea* in shallow dry valleys (wadis) in the desert along the Mongolian–Chinese border, with a view of the Dzagan Bogd (photographs by P. Hanelt).

Fig. 8.12.A. Dzun Mod oasis with *Populus diversifolia* in the Transaltai–Gobi north of Dzagan Bogd (photograph by P. Hanelt). B. The watering place in the depression of the Dzun Mod oasis (photograph by P. Hanelt).

average quantity of soil dug out by rodents generally amounts to 22 to 52 $dm^3\ ha^{-1}$ annually.

Reptiles of the genus *Eremias* and the species *Phrynocephalus versicolor* are common, both in the desert-steppe and desert vegetation. *Agama stoliczkana* is clearly linked to granite massifs in the desert region (Stubbe and Chotolchu, 1968; Zevegmid et al., 1974).

The Gobi is one of the least settled regions in the world because oases are very rare. Lavrenko and Yunatov (1960) gave some impressions of these small natural oases around springs at the bases of hills. They can also be created as the result of the damming up of groundwater, when rock layers cross larger runoff channels or at sites where dry rivers surface. Sparse stands of *Populus diversifolia* locate these oases from a distance (Fig. 8.12). At such sites one finds solonchak and solonetz soils with stands of *Phragmites australis*, and *Lasiagrostis splendens* with *Glycyrrhiza uralensis*, *Sphaerophysa salsula*, and *Alhagi sparsiflora*. Saline pastures with *Potentilla anserina*, *Glaux maritima*, *Hordeum brevisubulatum*, *Halerpestes*, and *Triglochin* also occur, as do stands of annual halophyte species like *Suaeda* and *Salicornia* with *Chenopodium glaucum*. A salt crust can form as the result of evaporation from the wet soil surface but, in spite of this, *Populus diversifolia* and *Salix* root into the fresh groundwater. On such sites in the Ala Shan–Gobi one can find *Elaeagnus moorcroftii*. The vegetation of these oases is often heavily disturbed by cattle. *Lasiagrostis splendens* occurs in the transition from the true desert to the groundwater-influenced vegetation. It is limited to the higher unflooded border areas. Around the edges of the oases are stands of *Tamarix ramosissima* (Fig. 8.13), *Lycium ruthenicum*, *Haloxylon ammodendron*, and *Nitraria sibirica* (often with the parasite *Cynomorium songaricum*, Fig. 8.14) on drifted sand-hillocks (*nebkas*). On sites with groundwater flowing closely under the soil surface or along erosion channels, one occasionally finds trees of *Ulmus pumila*.

The altitudinal belts of the Gobi–Altai are shown in Fig. 8.15. The mountain massifs south of the Gobi–Altai show the same altitudinal transition of the vegetation belts with different proportions of vegetation types on the northern and southern slopes. The alpine *Kobresia* belt and the sub-nival belt situated above it are only rarely developed. The whole sequence can only be found in the Arts Bogd (3802 m).

The eastern Gobi (Fig. 8.7, region *10*) is steppe-like in the north where *Stipa gobica*, *Stipa*–

Fig. 8.13. *Tamarix ramosissima* on a takyr area in the Transaltai–Gobi, south of the Tost Uul Mountains (photograph by P. Hanelt).

Fig. 8.14. *Cynomorium songaricum* (Cynomoriaceae), a parasite on *Nitraria* species (photograph by P. Hanelt).

Cleistogenes or *Stipa–Tanacetum* desert-steppes occur and *Caragana bungei* and *Artemisia* species form extensive shrubby semi-deserts (Fig. 8.11A). More to the south, on loamy or rocky soils, one finds desert communities with *Anabasis brevifolia* or *Salsola passerina*, plus *Zygophyllum xanthoxylon* and *Convolvulus gortschakovii*. More common on sandy, rubble soils are *Potaninia mongolica*, *Brachanthemum gobicum*, *Salsola arbuscula*, and *Ceratoides papposa*. If small sand dunes occur, they are covered by sparse stands of *Haloxylon ammodendron* (Figs. 8.16 and 8.17B).

The Ala Shan–Gobi, the northern border part of the Ala Shan Desert, is more or less a plain sloping

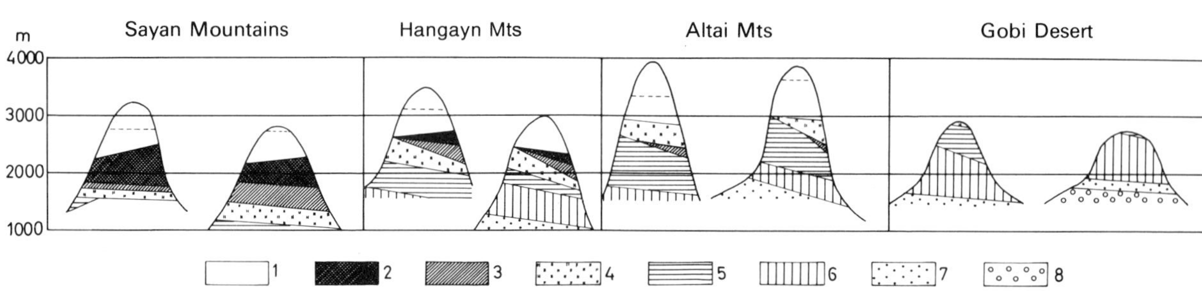

Fig. 8.15. Schematic diagram of the different altitudinal belts in the mountains of Mongolia and the neighboring regions (from Yunatov, 1950): *1* = high-mountain belt (with upper limit of closed turf vegetation indicated); *2* = montane taiga; *3* = larches and some mixed forest; *4* = montane steppes; *5* = rocky, dry *Stipa–Artemisia* steppes; *6* = montane desert-steppes with *Stipa* and *Salsola*; *7* = *Salsola* deserts; *8* = *Haloxylon ammodendron* deserts.

Fig. 8.16. *Haloxylon ammodendron*. A. In a sandy depression northeast of Bulgan. B. A young plant in the southern Gobi (photographs by P. Hanelt).

to the salty lake Gashuun Nuur (Ka-Shun Hu). The zonal vegetation is a desert dominated by *Anabasis brevifolia* with participation of *Reaumuria soongorica*, *Salsola passerina*, and *Nitraria sphaerocarpa* (Fig. 8.17A). Often *Salsola passerina* is the most dominant species. In the most extreme areas, on gravel and sand material, the *Ephedra przewalskii* community occurs. Along larger erosion channels one can find the endemic *Gymnocarpus przewalskii* and the also endemic evergreen shrub *Ammopiptanthus mongolicus*, as well as *Astlerothamnus* and *Caragana leucophloea* (Fig. 8.11).

The Transaltai–Gobi, especially in its western parts, is among the driest regions of Central Asia. The precipitation per year south of the mountain massif of the Arts Bogd amounts to 18 mm. It is common there to experience years without any rainfall. Temperatures show very high fluctuations, up to 30°C per day, and to 70°C per year. The Transaltai–Gobi is geographically more varied than the other districts of the Gobi. Hilly landscapes predominate. Numerous mountain ranges, spurs of the Gobi–Altai and the eastern Mongolian Altai, appear. One finds, however, also plains of gravel, rock and stone deserts. Sandy deserts occur relatively infrequently. There are three vegetation belts from north to south: desert-steppes, deserts, and extremely arid deserts. The desert-steppes are characterized by a *Stipa glareosa–Allium mongolicum* community with participation of *Anabasis brevifolia*. This zone occupies more extensive territories north to the Gobi–Altai. In the desert zone, perennial grasses and *Allium* species are largely lacking. Semi-shrubs and shrubs predominate. The very scant cover commonly reaches values of less than 10%, and commonly only 1–2%; *Haloxylon ammodendron* is a widely distributed shrub. On extreme sites, *Haloxylon* has low stature (1–1.5 m) and is accompanied by annual or biennial plants, e.g. *Halogeton arachnoides*, *Artemisia scoparia*, *Bassia*, and *Agriophyllum* (Fig. 8.17B). In dry river valleys (*sairs*) one can find stands of *Haloxylon* with shrubs to 6 m in height. The holoparasite *Cistanche salsa* parasitizes on roots of *Haloxylon*. The *Anabasis brevifolia* desert with *Reaumuria soongorica* and *Sympegma regelii* has also a wide distribution. Sandy deserts are characterized by *Zygophyllum xanthoxylon* and *Ephedra przewalskii*. *Nitraria sibirica* forms little sand dunes at the edges of saline depressions. Gypsum deserts are character-

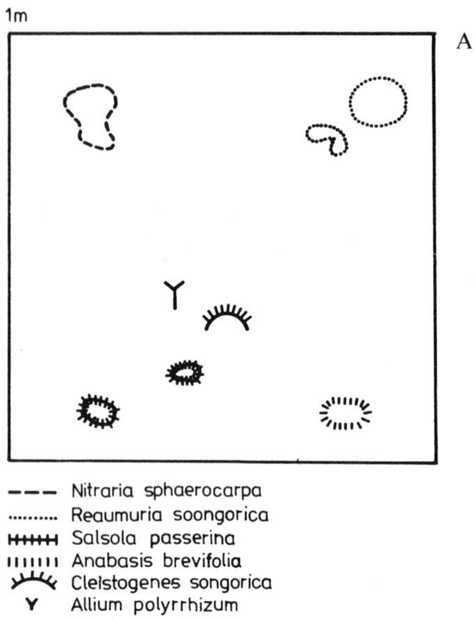

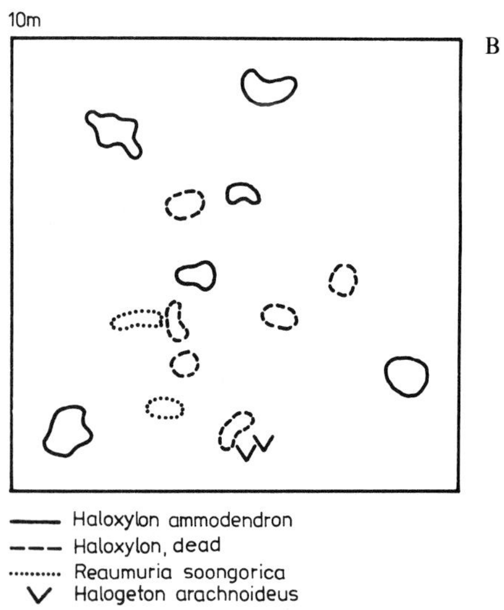

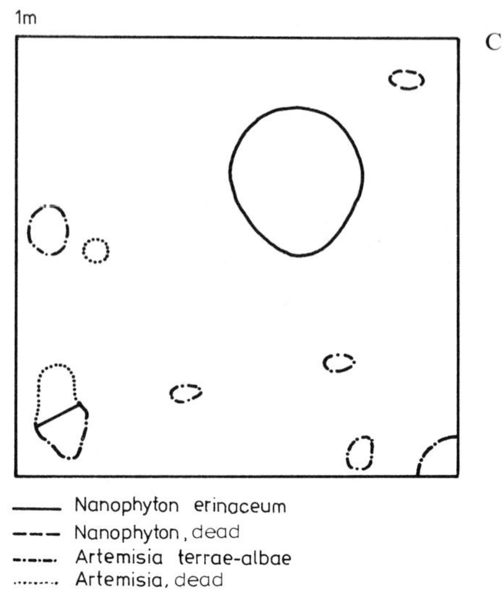

Fig. 8.17.A. Projection drawing of a *Reaumuria soongorica–Salsola passerina* community, Ala Shan–Gobi (from Hanelt, 1970). B. Projection drawing of a *Haloxylon ammodendron* community, Transaltai–Gobi (from Hanelt, 1970). C. Projection drawing of a *Nanophyton erinaceum* community, Dzungarian Gobi (from Hanelt, 1970).

ized by *Nitraria sphaerocarpa*. All the species mentioned accumulate sand in hummocks around themselves.

The zone of extremely arid deserts occupies the largest part of the Trans-Altai–Gobi. The border to the real deserts runs NW to SE from 44°50′N to 43°20′N. The floristic composition is the same as mentioned above, but the vegetation is contracted, i.e. restricted to the erosion channels. The area between these channels is nearly empty of vegetation.

Lavrenko (1957) has shown, by means of profile samples, how relief influences the distribution of various species. Yevstifeyev and Rachkovskaya (1976) have shown the coincidence between the zonal soil and vegetation types. The vegetation map of the Transaltai–Gobi by Rachkovskaya and Volkova (1977) makes clear how dependent the highly variable desert vegetation is on soil conditions.

The westernmost part of the Gobi, the Dzungarian Gobi, is a plain with many dried-up beds from salty lakes. The salt lakes get their water from seasonal rivers from the Mongolian Altai. The Dzungarian Gobi occupies a special geobotanical position because of the appearance of Kazakho-Dzungarian elements in the flora, penetrating from Middle Asia (*Nanophyton erinaceum*, *Artemisia terrae-albae*, *Anabasis aphylla*, *A. elatior*, *Dodartia orientalis*, *Halimodendron halodendron*). They, together with typical Central Asian species, determine the character of the vegetation in this district. The Dzungarian Gobi is, according to Grubov (1963), a part of the phytogeographical subprovince Dzungaria.

A typical element of the desert vegetation is *Nanophyton erinaceum*, the so-called *tar* (Fig. 8.18). This species usually forms compact hemispherical

Fig. 8.18. *Nanophyton erinaceum* desert in the Dzungarian Gobi (photograph by W. Hilbig). (See Fig. 8.17C.)

cushions 15 to 35 cm wide and 5 to 10 cm high. In consequence of the extraordinarily slow growth, larger plants have an age of 50 to 100 years. Associated species are *Reaumuria soongorica*, *Anabasis aphylla*, *Artemisia terrae-albae*, and *Stipa glareosa* and others. Very salty takyr-shaped areas in the central part of the Dzungarian Gobi are occupied by *Salix ledebourana*, *Tamarix ramosissima*, *Haloxylon ammodendron* and *Halimodendron halodendron*. Along the rivers shrub-like *Salix* species develop (*S. viminalis*, *S. triandra*, *S. ledebourana*). At upper elevations, near the Mongolian Altai, gallery forests of *Populus pilosa* occur. These communities are often highly disturbed by overgrazing.

The Large Lake District. Between the Mongolian Altai and the mountain massif of the Hangayn Nuruu (Changai), a large closed depression with a series of freshwater lakes (Har Us Nuur, Har Nuur) and salt lakes (Uvs Nuur, Hyargas Nuur, Döröö Nuur) is situated. The Large Lake District (Fig. 8.7, region *7*), is surrounded by mountains. The area is linked with the desert-steppe areas in the southeast of Mongolia only by a narrow rift valley between the Hangayn Mts. and the Gobi–Altai (Fig. 8.7, region *8*). The climate occurring in the lee of the mountain ranges causes this extent of the desert-steppe belt and the extensions of desert vegetation to the Uvs Nuur depression (latitude of 50°30′N) (Karamysheva and Banzragh, 1976). Some important climatic values collected at two meteorological stations from the Large Lake District may be compared with values of Dalandzadgad (Table 8.2).

In the Uvs Nuur depression one can find extensive areas of the desert and desert-steppe vegetation with the characteristic Kazakho-Dzungarian elements. The principal species is *Nanophyton erinaceum*, already mentioned for Dzungaria and the Dzungarian Gobi. It can have a phytomass of 0.3 to 0.8 t ha^{-1}. The *Nanophyton* communities have *Stipa glareosa*, *Cleistogenes squarrosa*, *Artemisia*

TABLE 8.2

Climatic values of Ulaangom and Hovd (Large Lake District) in comparison with values of Dalandzadgad (from Hanelt, 1970)

	Mean annual temperature (1941–60) (°C)	Mean annual precipitation (1955–60) (mm)	Absol. max. temperature (1941–60) (°C)	Absol. min. temperature (1941–60) (°C)
Ulaangom (936 mm)	−4.0	148.3	35.1	−48.2
Hovd (1395 m)	−0.4	135.3	35.4	−46.0
Dalandzadgad (1470 m)	+3.9	160.5	37.5	−36.5

pectinata, and the Turanic grass *Psathyrostachys juncea* associated with them. In the depressions of Hyargas Nuur and Har Us Nuur, to the south, there are deserts dominated by *Anabasis brevifolia* and *Reaumuria soongorica*, with *Convolvulus gortschakovii* and *Stipa glareosa*.

Desert-steppes often show more frequent occurrence of short bunch grasses and low forbs (*Convolvulus ammanii*, *Potentilla acaulis*, *Ptilotrichum canescens*). *Stipa* species, of the series Barbatae (*S. glareosa*, *S. gobica*, *S. orientalis*) dominate. *Cleistogenes* can also sometimes dominate here.

As salinities become more extreme, the *Stipa* species disappear and *Tanacetum achilleoides* or *T. trifidum* becomes dominant. The structure of these desert-steppes is clearly seen in the projection drawings (1 m^2) of Kalinina (1954).

The *Stipa–Allium polyrrhizum* association reaches only to the southeastern part of the Large Lake District, the Chargin Gobi. Farther to the west *Allium polyrrhizum* disappears. Shrub desert-steppe is very widespread in the Large Lake District. Shrubs (*Caragana*, *Atraphaxis*, *Amygdalus pedunculatus*) and semi-shrubs (*Artemisia*) dominate the different communities. Some summer annuals also exist in the desert-steppe communities but have highly variable frequency depending on precipitation.

Ecological studies in the Gobi. The climatic values of different meteorological stations in the Gobi show the variable precipitation. Considerable differences in climate and vegetation occur over remarkably short distances. The climatic values in a sequence of years at Dalandzadgad (Fig. 8.19) illustrate the fluctuation of summer precipitation according to the amount and the mean time of precipitation. The station Bulgan is only about 70 km NW of Dalandzadgad but shows a more arid situation. The incomplete climatic values of 1971 and 1972 at Nomgon, about 100 km SSE of Dalandzadgad, in the true desert zone, show the distinctly smaller amount of precipitation and the increase of temperature extremes in comparison with the aforementioned stations within the desert-steppe zone.

Microclimatological investigations of scientists from the universities of Halle/Saale, German Democratic Republic, and Ulaan Bataar (Ulan Bator), Mongolian People's Republic, in 1973 show distinctly high amplitudes of air and soil temperature during cloudless days in different types of desert-steppe and desert vegetation (Helmecke and Schamsran, 1979a). At 1 m above soil surface a temperature plateau of 32 to 35°C forms. The air temperature near the soil surface and the temperature of the uppermost soil layer drop greatly below the values of the deeper soil layers at night. Daily temperature amplitudes of 45 to 50°C were measured on the soil surface in Bulgan (Beresneva, in Rodin, 1977).

Condensation of dew is a possible means of accentuating cooling during periods of high relative atmospheric humidity after precipitation. Condensation of moist soil air may occur by means of this deep cooling during periods of high relative atmospheric humidity after precipitation and even with low relative atmospheric humidity if the soil is moist. However, in Bulgan only one to two days

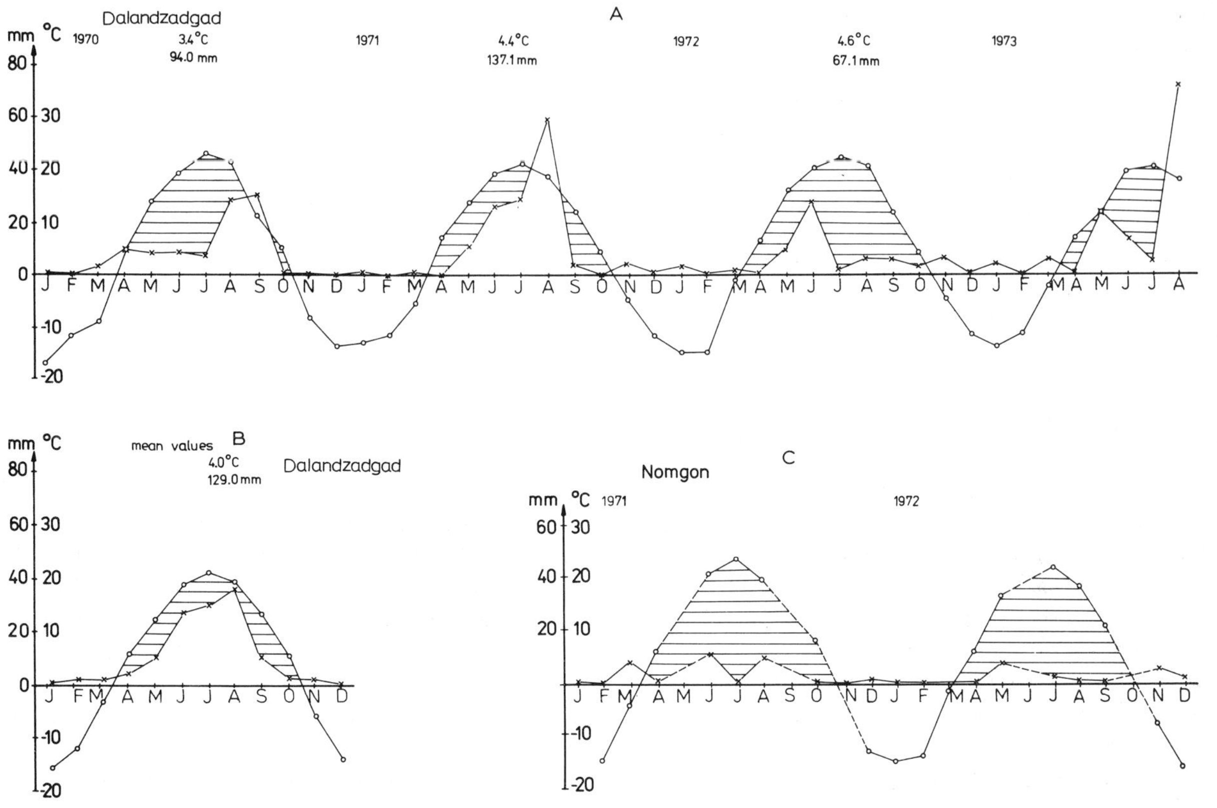

Fig. 8.19.A. Climatograms of Dalandzadgad for 1970–1973. B. Climate diagram, mean values. C. Two-year diagrams for Nomgon, south Gobi (incomplete) (from Helmecke and Schamsran, 1979a).

with dew were observed from July to September on the average over some years. Water condensation on the soil surface has a great importance for desert and desert-steppe vegetation in the average season. The "window-lichens" (Fig. 8.20) in particular may improve their water budget in this way.

The relative atmospheric humidity drops to less than 10% even shortly after precipitation events. The relative humidity does not increase higher than 40 to 45% at night. Evaporation generally reaches high values (>3.5–4 ml h^{-1}; Piche evaporimeter). Evaporation can also occur at night. The great increase in evaporation during clear days becomes milder with hazy weather and at somewhat lower temperatures. This effects an improvement of the water budget for some plants. The relative atmospheric humidity can average 30%. The values of Table 8.3 make this clear. The potential osmotic pressure of different plants is lower than on clear

Fig. 8.20. *Sphaerothallia desertorum* on the bottom of stones on the soil surface in the desert steppe (from Helmecke and Schamsran, 1979b) (photograph by S. Rodemann; ×0.7).

TABLE 8.3

Evaporation, relative atmospheric humidity, air and soil temperature during a clear day (June 7, 1973) and a hazy day (July 7, 1973) in the *Stipa–Allium polyrrhizum* desert-steppe near Bayan Ondor, SW Dalandzadgad (from Helmecke and Schamsran, 1979a)

	June 7, 1973		July 7, 1973	
Height above surface:	10 cm	100 cm	10 cm	100 cm
Max. evaporation (ml h^{-1})	2.3	2.55	1.3	1.5
Evaporation sum from 7 to 17 h (ml h^{-1})	18.75	20.4	12.55	14.45
Min. relative atmospheric humidity (%)	12	12	28	28
Min. temperature (°C)	16.2	16.8	23.0	23.1
Max. temperature (°C)	33.5	31.9	30.2	28.5
	min.	max.	min.	max.
Soil temperature at 2 cm (°C)	14.0	40.0	21.5	34.5

Evaporation was measured by means of standard Piche evaporimeter.

days. The microclimatic conditions in scattered stands of *Haloxylon ammodendron* with a vegetal cover of 5 to 15% were more moderate than in the low-growing *Anabasis brevifolia* desert and the *Stipa–Allium polyrrhizum* desert-steppe.

Scientists from the Botanical Institutes of the Academy of Sciences in Leningrad and of the Mongolian Academy of Sciences in Ulaan Bataar have carried out ecological studies in the northern Gobi near Bulgan, south Gobi Aimak, at an experimental station located at an elevation of 1000 m (44°12′N, 103°10′E). These studies were performed during the period 1970 to 1974 (Rodin, 1975, 1976, 1977) under the leadership of T.K. Gordeyeva. The precipitation in this desert-steppe region is 130 to 150 mm annually. The mean annual temperature is 3.8°C (July 21.4°C, January −15.9°C). The annual sum of temperatures above 10° is 2200°. Sunshine was recorded during 3200 h per year, and the annual radiation was 504 kJ cm^{-2}. During a clear and sunny day 2843 J cm^{-2} were measured. The frost-free period lasts 150 days, with the growing season beginning in April–May. The mean wind speed is 3.5 m s^{-1}. High wind speed in spring and summer is characteristic for the Mongolian desert and desert-steppe regions. It can exceed 10 m s^{-1} and reach 16 to 18 m s^{-1}.

The soils are sandy burozems, 20 to 70 cm deep, with high runoff. The small accumulation of phytomass and the slow decomposition processes are reasons for the very low humus content (0.26%).

The vegetal cover of 7 to 10% in the desert community of *Reaumuria soongorica* and *Brachanthemum gobicum* investigated had the following densities: *Reaumuria soongorica* 3400 plants ha^{-1}, *Brachanthemum gobicum* 1980 ha^{-1}, *Salsola passerina* 480 ha^{-1}, *Zygophyllum xanthoxylon* 320 ha^{-1}, plus scattered *Nitraria sibirica*, *Potaninia mongolica*, *Stipa glareosa*, *Convolvulus ammanii*, and others. In 1972 there were a total of 45 species; in the drier two years before only 25 and 35 species were found respectively.

The total phytomass for the steppe-like *Agropyron fragile* community was 6 t ha^{-1}, for a *Reaumuria* community 4.5 t ha^{-1}, and for a *Brachanthemum* community 2.4 t ha^{-1}. The annual production of the *Agropyron fragile* community amounts to 4.4 t ha^{-1}, of which 85% is underground. If we only compare the above-ground phytomass, we obtain the following differences. The total above-ground phytomass of the *Reaumuria-Brachanthemum* community (794 kg ha^{-1}), with the main portion perennial shoots, had very low fluctuations from year to year (between 92 and 100%). The phytomass of annual shoots and leaves changed in dependence on meteorological conditions (fluctuations between 48 and 130%):

1970 (59.4 mm annual precipitation): 64 kg ha^{-1} annual shoots and leaves
1973 (292.8 mm annual precipitation): 137 kg ha^{-1} annual shoots and leaves

As in all deserts, by far the largest fraction of phytomass is underground. Miroshnichenko (1967) investigated the productivity and stratification of phytomass in the *Stipa–Allium polyrrhizum* desert-steppe near Dalandzadgad (10% vegetal cover). The phytomass of green shoots and leaves was

120 kg ha^{-1} (air-dried). The fraction of *Stipa* (*S. glareosa*, *S. gobica*) was 65% and of *Allium polyrrhizum* 25%. The roots concentrate in the upper soil layers because the periodical wetting of the compact soil only reaches to 30 cm. The main fraction of underground phytomass (83% of the total phytomass) is in the upper 20 cm. Only 3% goes deeper than 40 cm.

The water budget of different plants near Bulgan was studied during a dry year, 1970 (very high fluctuation of values) and during a moist year, 1971 (more balanced transpiration values) by Sveshnikova and Bobrovskaya and by Tsend (both in Rodin, 1975) (Table 8.4).

The difference between maximum and minimum transpiration rates was twice as high in 1970 as in the wetter 1971. The water supply was relatively good in both years, as one can see from the relatively low water deficits of the plants. This seems to contradict the high suction values (i.e., negative of the water potential), but these values may be exceeded, as is almost always possible under arid conditions. This can be seen even more clearly among the succulent halophytes, which grow on sandy soils and show very slight water deficits. They have high suction values in spite of that (Table 8.5) since they depend on the high salt concentration in the cell sap.

During the same years, Slemnev and Bold (see Anonymous, 1975) studied the photosynthesis of these plants. Both the potential photosynthesis under optimal external conditions (increased CO_2 concentration), according to the C^{14} and manometer methods, and the apparent photosynthesis at the site (titration or conductivity methods) were studied. The maximum apparent photosynthesis was only a small proportion of the potential (Table 8.6).

Two photosynthetic maxima during the growing season are characteristic for *Reaumuria*. These occur during vegetative growth and during flower formation. For *Salsola* there is only one maximum during fruiting, and for *Brachanthemum* only one maximum at the beginning of the growth of generative shoots. Although water conditions are very unfavorable in the Gobi, a positive photosynthesis could always be found during the growing season.

The potential osmotic pressure of most desert plants (*Reaumuria*, *Kalidium*, *Nitraria*, *Zygophyllum*, *Anabasis*, *Haloxylon*) is very high. The species, often

TABLE 8.4

Transpiration, water content and water deficit (in %), and suction force (bar) for species of the northern Gobi Desert (after Rodin, 1977)

Species	Year	Transpiration rate [mg (g fresh weight)$^{-1}$ h^{-1}]			Water content (% fresh weight)			Water deficit (% of water saturation)	Suction (bar)	
		max.	min.	diff.	max.	min.	diff.		min.	max.
Reaumuria soongorica	1970	1450	165	1285	72.0	41.0	31.0	8.0–24.0	–	100
	1971	450	80	370	71.0	48.0	23.0	11.1–26.1	24	30
Zygophyllum xanthoxylon	1970	640	100	540	86.0	65.0	21.0	5.0–18.0	–	90
	1971	320	40	280	89.0	78.0	11.0	2.8–5.8	25	37
Salsola passerina	1970	760	60	700	79.0	52.0	27.0	19.0–32.0	60	85
	1971	370	60	310	73.0	58.0	15.0	18.0–33.7	23	45
Brachanthemum gobicum	1970	680	120	560	68.0	48.0	20.0	15.0–27.0	70	100
	1971	380	40	340	73.0	62.0	11.0	12.3–17.7	24	42

TABLE 8.5

Transpiration, water content and deficits (% of water content of water-saturated plants) and suction forces for plants from the northern Gobi Desert (from Rodin, 1977)

Species	Year	Transpiration rate [mg (g fresh wt.)$^{-1}$ h^{-1}]				Water content (% fresh weight)				Water deficit (%)		Suction force (bar)	
		mean	max.	min.	diff.	mean	max.	min.	diff.	mean	max.	min.	max.
Haloxylon ammodendron	1970	240–340	430	145	285	69–77	84	60	24	3.5	4.0	–	–
	1971	110–270	430	30	400	75–80	84	72	12	5.5	7.5	30	43
Nitraria sibirica	1970	260–620	960	160	800	75–82	89	60	29	9.5	14.0	20	80
	1971	140–460	740	60	680	76–82	89	73	16	8.2	11.0	25	30

shrubs and semi-shrubs, show high fluctuations between the maximum and minimum values. The high pressures of some desert plants are caused by the very high salt content in the cell sap (*Reaumuria soongorica*, *Kalidium gracile*). Chlorides were detected in all species, and sulphates were detected in some (Fig. 8.21).

The salinity of desert soils in the Gobi is mentioned by several authors. Petrov (1966/67) especially refers to the large extent of solonchak deserts in the depressions of the Gobi.

Species primarily found in desert steppe (*Stipa gobica*, *Tanacetum achilleoides*, *Allium polyrrhizum*, *Lagochilus ilicifolius*) show less fluctuation between the maximum and minimum values of potential osmotic pressure. Their maximum values are also below the values of true desert plants, when they grow together.

The Pei Shan Desert (see Fig. 8.7, region *11*, and Fig. 8.22)

Petrov (1966/67) reported on this southernmost part of western Gobi in great detail (see his pp. 231–255). Since this is a region which had not been studied botanically to any great extent, we include some more detailed site samples.

Pei Shan is a landscape of low mountains which stretches from the Jo Shui (Edsin Gol) River in the east over about 500 km to the Lop Nor (Lo-Pu Po) Depression in the west; it is about 200 km wide. The northern boundary runs through the tectonic depression of Ha-mi to the Gashuun Nuur (Ka-shun Hu). It is the oldest part of the Central Asian continental area. It has not been covered by the sea since the Carboniferous and has been under the influences of heavy erosion and denudation ever since. The mountains are composed of Precambrian metamorphic rocks which cover more than half of the region. Jurassic and Cretaceous deposits cover only small areas, and Tertiary deposits (sandstones, clays, and conglomerates) even less. Pleistocene deposits, on the other hand, cover somewhat more than 40% of the area.

In the dry river valleys are considerable amounts of angular rocks which were only transported a short distance. All steeper slopes are heavily weathered and crumble when walked upon. The dust particles have been removed from this desert by wind or water; the lower areas of the relief are covered over with rubble. Where sand drifts occur,

TABLE 8.6

Maximal photosynthesis values [mg CO_2 (g dry weight)$^{-1}$ h^{-1}] for plants of the northern Gobi Desert (see text) (from Rodin, 1977)

Species	1971		1970	
	potential	apparent	potential	apparent
Nitraria sibirica	63	14	64	20
Zygophyllum xanthoxylon	50	9	33	10
Brachanthemum gobicum	33	9	40	8
Haloxylon ammodendron	31	7	23	16
Reaumuria soongorica	29	7	25	11
Salsola passerina	16	8	13	6

they are formed of primary sands with angular grains, a sign that they arose directly from weathering and were not transported any significant distance.

The central mountain chain reaches a height of 2800 m, a height of 1000 m relative to the surrounding plateaus. It is furrowed by numerous erosion channels. Fig. 8.22 gives an idea of the relief with the profile from An-hsi (1100 m) to Ha-mi. The profile contains many mountain chains and hills made up of metamorphic rocks and rocks of the Carboniferous and Cretaceous, with numerous rubble-filled depressions between the chains.

The climate diagram of Hsing-hsing-hsia (1776 m) between An-hsi and Ha-mi gives some information on the climate (Fig. 8.6). In winter the region is entirely within the Siberian–Mongolian high-pressure area; in the summer the East Asian

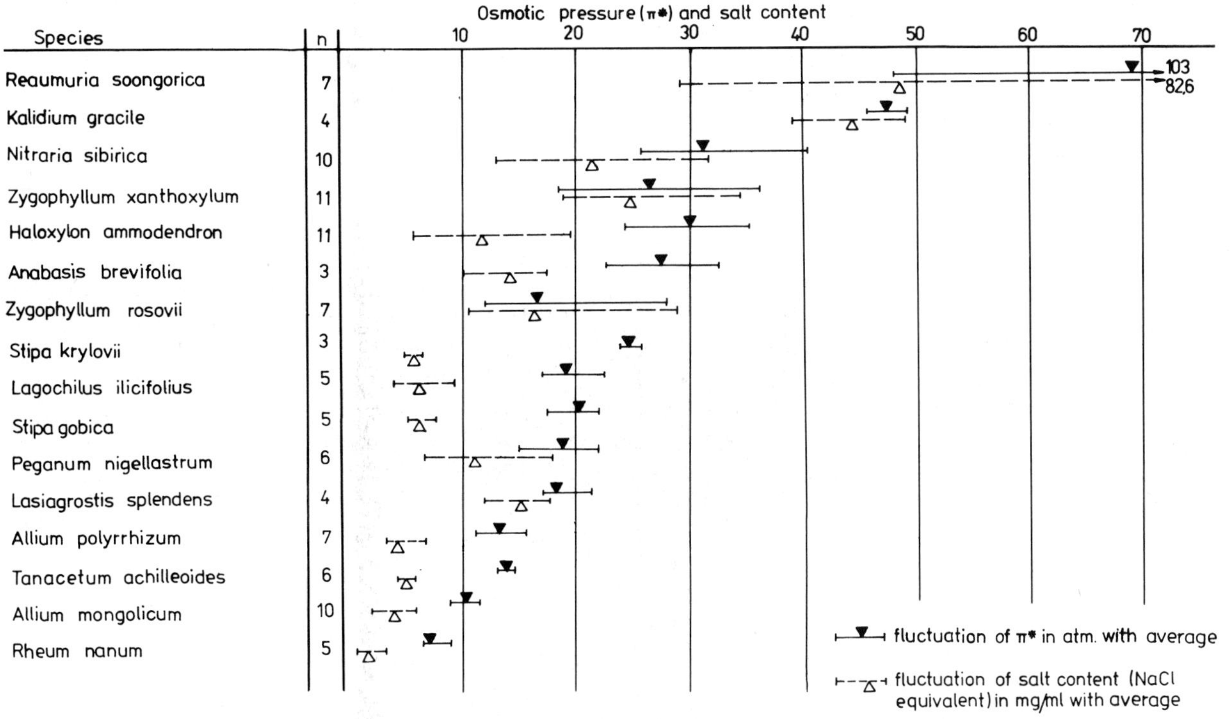

Fig. 8.21. Relative osmotic pressure (π^*) and salt content (NaCl equivalents) of selected desert and desert-steppe plants of the Gobi (after Helmecke and Schamsran, 1979a).

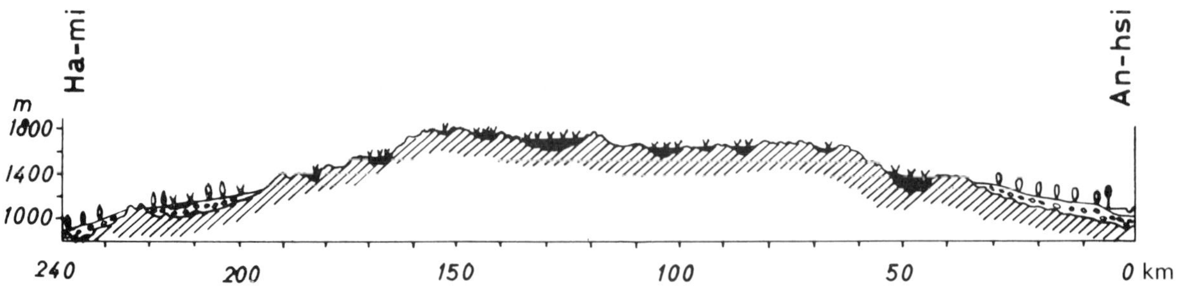

Fig. 8.22. Profile through the Pei Shan Desert from the northwest (Ha-mi) to the southeast (An-hsi) (from Petrov, 1966/67); black = rubble-filled basins with dwarf-shrub vegetation and with groundwater near the surface in places (wells, settlements); hatching = outcroppings of old, hamada-like rock layers with very sparse vegetation (water only in fissures); at the base of the mountains are found almost level areas with dry riverbeds, along which there is a somewhat richer shrub vegetation (groundwater very deep).

monsoon has no effect so far inland. Rainfall is brought by dust storms from the west and occurs in the form of individual downpours.

The mean wind speed is 3.5 m s^{-1}. The temperature extremes are −35.1°C and +42.5°C. The annual precipitation is 39 to 85 mm, but on one day 20 mm was recorded. Precipitation may reach 500 mm only in high-mountain areas. The potential evaporation at An-hsi is over sixty times greater than the precipitation.

Judging from the erosion channels and valleys, more rain fell during the Pleistocene than at present. Even today, though, there is some runoff from the wetter central area toward the periphery. The water sinks into the rubble masses, so the depressions remain dry. In the mountains groundwater collects in shallow, rubble-filled hollows and can be reached by wells. There are also springs with slightly brackish water (2 g l^{-1} of salts) at the foot of the mountains; the springs and wells in the mountains offer less saline water. It is from this water that the inhabitants live. One finds watering places every 100 to 200 km along the caravan routes.

Soils with stone pavement are characteristic in the flat areas of the Pei Shan Desert. Underneath the pavement one can find up to one meter of fine earth. Calcium and gypsum crusts are common. The depressions and hollows contain larger or smaller areas with saline soils.

The sierozems of the lower areas change over into chestnut earths in the mountains and probably into dark-gray forest soils at the highest elevations.

Petrov (1966/7) gives tables with site samples from various desert communities. We limit ourselves to showing the lists only and denote the more common species by means of a plus sign (+).

(a) Vegetation of sandy gravel deserts in lower-lying areas (1100–1400 m) of the broad plains (cover: seldom 20% and often only 1%; 9 samples, with 5–15 species per sample).

Dwarf-shrubs: *Ephedra przewalskii* (+), *Nitraria sphaerocarpa* (+), *N. schoberi*, *Zygophyllum xanthoxylon*, *Calligonum mongolicum* (+), *Atraphaxis mongolica*, *Caragana leucophloea*, *Haloxylon ammodendron*, *Salsola laricifolia*, *Sympegma regelii*, *Anabasis brevifolia*, *Tamarix ramosissima*, *Reaumuria soongorica* (+), *Artemisia salsoides* (+), *Asterothamnus centralasiaticus* (+), *Lycium ruthenicum*, *Caryopteris mongolica*, *Clematis songarica*, *Gymnocarpos przewalskii*.

Perennial herbs: *Limonium aureum*, *Haplophyllum dauricum*, *Tournefortia sibirica*, *Zygophyllum mucronatum*, *Scorzonera divaricata*, *Tanacetum fruticulosum*, *Artemisia anethifolia* (+), *Jurinea* sp.

(b) Rubble desert of lower mountains (up to 2000 m) with cover of 1 to 5% (five samples, 8–13 species per sample).

Dwarf-shrubs: The first fourteen under (a) above, but with *Nitraria sibirica* and *Atraphaxis buxifolia* instead of *N. schoberi*; also *Amygdalus mongolica*, *Kalidium foliatum*, and *Artemisia sphaerocephala*.

Perennial herbs: The first six in (a) above, plus *Stipa glareosa*, *S. bungeana*, *Allium mongolicum*, *Rheum nanum*, and *Aster altaicus* in crevices.

(c) Halophytic meadow communities of moister depressions (4 samples, 4–12 species per sample, cover 25–30%).

Woody plants: *Tamarix ramosissima*, *Nitraria schoberi*, *N. sibirica* (+), *Lycium ruthenicum*, *L. turcomanicum* (+), *Clematis songarica*.

Perennial herbs: *Phragmites communis* (abundant), *Lasiagrostis splendens*, *Aneurolepidium dasystachys*, *Aeluropus litoralis* (abundant in patches), *Carex* sp., *Iris ensata*, *Peganum harmala*, *Glycyrrhiza uralensis*, *Alhagi sparsifolia*, *Limonium aureum*.

(d) The same species are found along most rivers which contain water; in addition there are *Populus diversifolia*, *Elaeagnus moorcroftii*, and *Halimodendron halodendron*.

(e) In dry lowlands between mountain chains one can find *Haloxylon ammodendron* stands.

(f) The species composition on sands in the northeast is different. On mobile sands one finds species of *Hedysarum*, *Caragana*, and *Psammochloa*, on sand drifts *Nitraria schoberi* and individuals of *Kalidium*.

(g) The following were determined for the highest peaks (up to 2800 m) by means of binoculars: a steppe-shrub belt (*Stipa bungeana* with *Amygdalus mongolica*, *Cotoneaster*, *Rhamnus*, *Rosa*, *Spiraea*); and a spruce-birch belt probably with *Picea asperata* and an undergrowth of *Sorbus*, *Lonicera*, *Rhamnus*, *Cotoneaster*, etc. The closest similar montane forests are in Nan Shan (350–400 km away) and in the Tien Shan (300 km away, with *Picea schrenkiana*).

Pei Shan is distinguished from the other Central Asian deserts by the heavy predominance of rock deserts (hammadas).

Kashgaria, Tarim Basin and Takla Makan

The tectonically derived Tarim Basin covers an area of almost 400 000 km^2 and is 1300 km long from east to west and 500 km wide from north to south. It is delimited on three sides by a horseshoe-shaped arc of high mountains made up of the eastern Tien Shan in the north, the Pamir and Karakoram Ranges in the west, and the Kun Lun Mountains and Astin Tagh (A-erh-chih Shan) in the south (Figs. 8.7 and 8.23). The narrowed basin is open to the Lop Nor depression to the east. The elevation varies from 780 m in the east to 1400 m in the west. The basin is filled with pre-Pleistocene weathering products, which come to the surface around the edges but are otherwise covered by Pleistocene layers, primarily aeolian sands. These form the pathless, uninhabited, and in Asia the most extreme sand desert, the Takla Makan (see Fig. 8.28) with dunes as high as 300 m.

The Tarim Basin is the most arid part of Central Asia, but at the same time is rich in groundwater. The glacier-fed rivers from the south carry water to the southern edge of the desert, where they form large rubble fans and disappear into the sand. Only the Khotan River manages to cross the most extreme part of the desert, thereby dividing the western sand region from the actual Takla Makan. The western and northern edges of the desert are watered by the Tarim River system. The Tarim River arises at the confluence of the Yarkand, Kyzylsu, and Aksu rivers. Its length is about 2000 km, and its width varies from 80 to 1000 m, with an average depth of 10 m. It carries an average of about 1200 to 1300 m^3 sec^{-1} of water from west to east, about twice this amount at high water, forms wide floodplains, and constantly shifts its bed toward the sand desert, into which part of its water disappears (Fig. 8.24). It ends in the endorheic Lop Nor Depression forming a salt lake of about 100 km diameter. This lake has lain dry at times, as the water was diverted in another direction. The

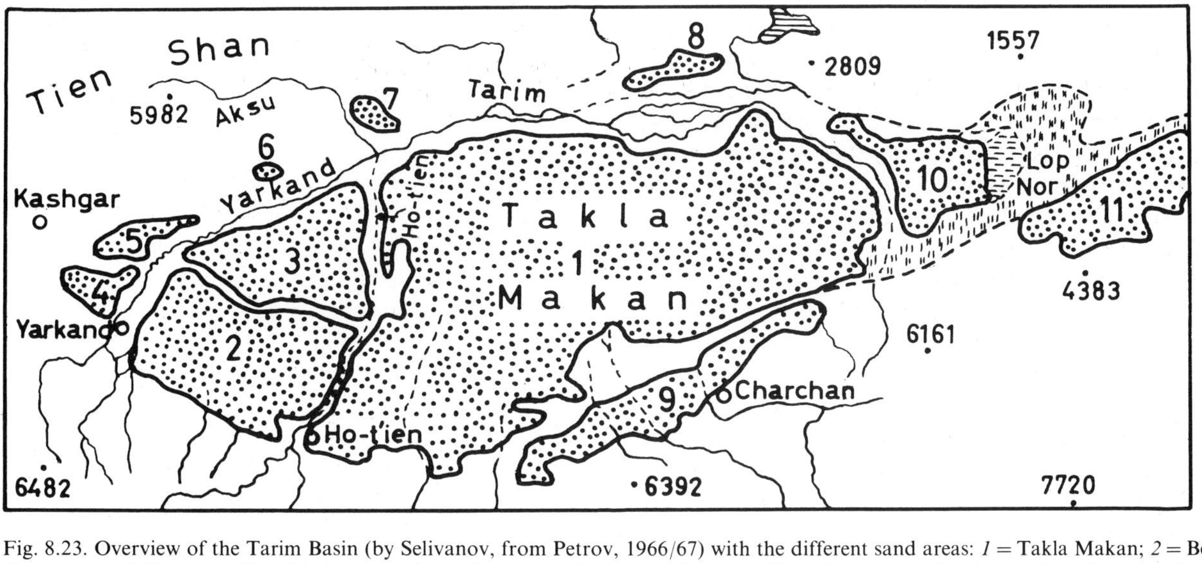

Fig. 8.23. Overview of the Tarim Basin (by Selivanov, from Petrov, 1966/67) with the different sand areas: *1* = Takla Makan; *2* = Bel Kum; *3* = Tagh Kum; *4* = Yangishahr; *5* = Faizabad; *6* = Tum Shuk; *7* = Ak Su; *8* = Chiglyk; *9* = Kum Katti; *10* = Lop Nor; *11* = Kum Tagh; --- = earlier boundary of the Lop Nor.

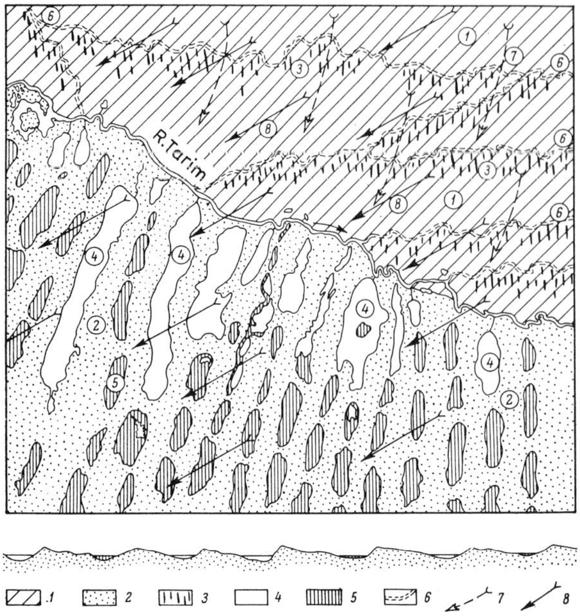

Fig. 8.24. Northeastern edge of the Takla Makan Desert, which also forms the right bank of the Tarim River (by Sven Hedin, from Petrov, 1966/67): *1* = floodplain forest in the Tarim Valley; *2* = dune sands of the Takla Makan; *3* = sand ridges on the right bank of earlier rivercourses; *4* = lakes lying between the dunes; *5* = dry dune valleys; *6* = old rivercourses of the Tarim; *7* = direction of shifting of the Tarim riverbed; *8* = direction of prevailing winds; --- = earlier boundary of the Lop Nor lake.

term "nomadic river" has been applied to such rivers which constantly shift their riverbeds and deltas.

As a result of the scant precipitation the sand desert Takla Makan is generally completely devoid of vegetation. The sand, however, "floats" on water-saturated alluvial deposits, and a through-flowing groundwater table is assumed to exist at no great depth underneath the desert. It was reported that Hedin, during his crossing of the eastern part in 1904, was able to find fresh or saline groundwater almost daily in wells of 1.2 to 1.5 m in the deepest dune valleys. One can also find *Tamarix* shrubs and reeds only on such sites. The dunes are otherwise devoid of vegetation, since the precipitation probably does not exceed 25 mm annually. The salinity of the groundwater increases as one approaches the Lop Nor Depression.

The overall depth of the Tarim Basin reaches 10 km, and both oil and natural gas deposits have been found there. The depth of the Pleistocene deposits is about 400 to 500 m. Loess-like loamy silts are widespread around the periphery and form a cover of 1 to 15 m, 25 to 30 m around the oases Kashgar and Aksu. Marine sediments with insular remains of earlier river and lake terraces are common in the Lop Nor Depression. These have been heavily attacked by the frequent sand-blasting.

Most weathering products deposited in the region came from the Kun Lun Mountains to the south, since their rivers carry more water than do those from the drier Tien Shan south slopes. These proluvial deposits, which are periodically carried out of the mountains to the foothills by intermittent storm runoffs, form alluvial fans with a characteristic structure. In the upper parts a section through the fans is convex, but in the lower parts, where the river empties and cuts into the lowlands, the section is concave (Fig. 8.25). The longitudinal section is shown in Fig. 8.26. If the river begins to form a delta along its upper course and to extend it over a wide plain, then the cross-section is flat, as is the case with the Kyzylsu around Kashgar.

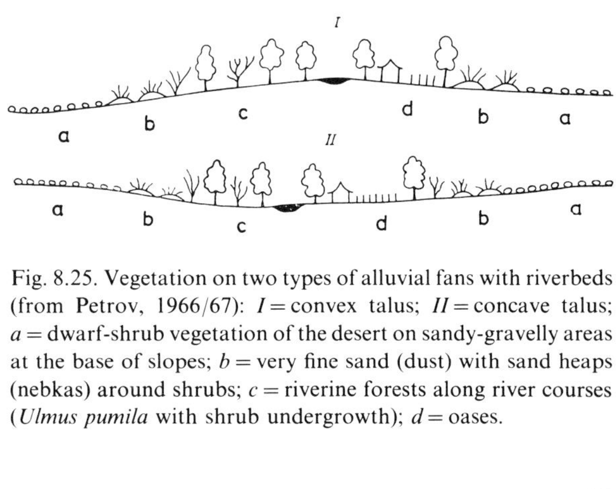

Fig. 8.25. Vegetation on two types of alluvial fans with riverbeds (from Petrov, 1966/67): *I* = convex talus; *II* = concave talus; *a* = dwarf-shrub vegetation of the desert on sandy-gravelly areas at the base of slopes; *b* = very fine sand (dust) with sand heaps (nebkas) around shrubs; *c* = riverine forests along river courses (*Ulmus pumila* with shrub undergrowth); *d* = oases.

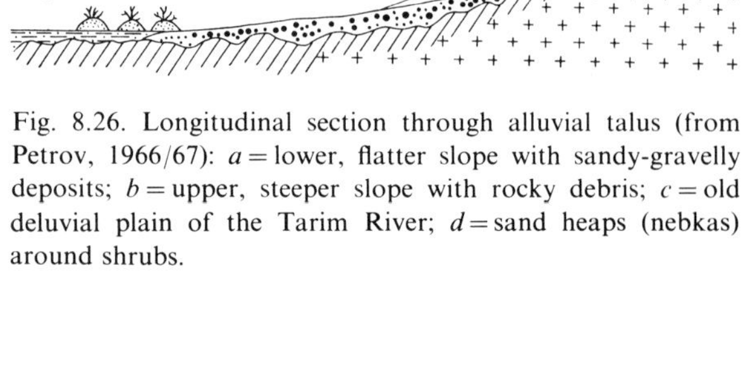

Fig. 8.26. Longitudinal section through alluvial talus (from Petrov, 1966/67): *a* = lower, flatter slope with sandy-gravelly deposits; *b* = upper, steeper slope with rocky debris; *c* = old deluvial plain of the Tarim River; *d* = sand heaps (nebkas) around shrubs.

The valley of the Tarim has three clearly distinct terraces at the confluence of its three constituent tributaries (Fig. 8.27). This number decreases along the lower course until the terraces finally disappear into the plain of the delta. At the confluence with the Khotan the Tarim forms a floodplain half a kilometer wide. From this point on it continuously shifts its bed to the right (Coriolis effect) toward the sand desert, thus constantly increasing the width of the floodplain on the left bank. The interesting transition from the Tarim Valley into the sand desert is shown in Fig. 8.24.

It is the mobile sands, however, which cover the greatest areas in the Tarim Basin. These are mostly barkhans, but in some places also drifts arise from the influence of woody plants (nebka formation): small dunes around *Ephedra* or *Nitraria*, larger dunes around *Tamarix* (Fig. 8.28). Parabolic dunes are absent, because these require that plants grow over the sands; under extremely arid conditions this is only possible when the plants are able to root into the groundwater. In this case the shrubs form the nebkas, as already mentioned. The most important sand areas in the Tarim Basin are shown in Fig. 8.23.

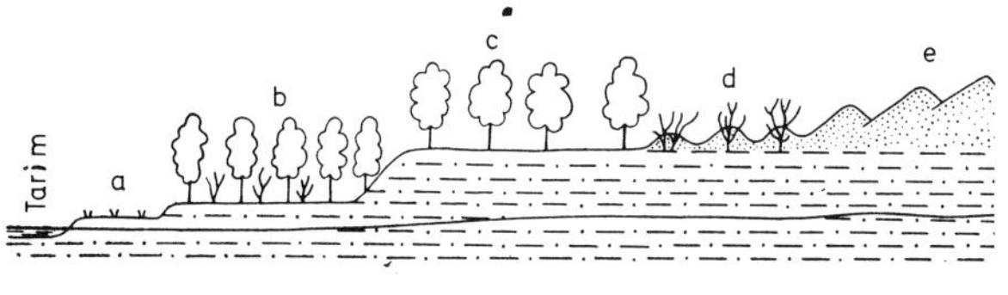

Fig. 8.27. Profile through the Tarim Valley with three terraces (from Petrov, 1966/67): *a* = lower terrace along the floodplain with marshy meadow vegetation; *b* = higher terrace with young *Populus diversifolia–P. pruinosa* forest; *c* = upper terrace with an old forest; *d* = sand heaps (nebkas) with *Tamarix ramosissima*; *e* = shifting dunes (barkhans) of the Takla Makan Desert.

The climate is extreme (Fig. 8.29). The annual precipitation at Kashgar (1306 m) in the west is only 63 mm, and this decreases toward the east to 9 to 11 mm. The potential evaporation is 2500 to 3000 mm. Mean temperatures are 9.8 to 11.6°C for the year, −5.6 to −10°C for January, and 24.4 to 26.8°C for July. Since the basin is open to the east, there is no protection against invasions of cold Siberian–Mongolian air masses in winter, and absolute minima lie below −27°C. Maxima in summer are over 40°C, and the relative humidity in July is often only 2 to 3%. Because of the dust-filled air, however, insolation is not very high. Dust clouds can hang over the area for weeks or even months and reach up to 4500 m in height. Visibility is usually less than 1 to 1.5 km, often only 50 to 100 m. The dust is deposited as loess on the slopes in the west.

All of these peculiarities of the Tarim Basin are reflected in the vegetation. Next to vast areas with no vegetation one finds the rich floodplain forests of the Tarim Valley. The flora is low in species (120 species), and endemics are lacking.

Fig. 8.28 shows a schematic classification of the vegetation. The upper parts of the slopes of the Tien Shan are not part of the desert and can be forested. The lower parts (Fig. 8.7, regions *15–17*), however, are rock, rubble or gravel desert.

It is striking that, in the Takla Makan sand

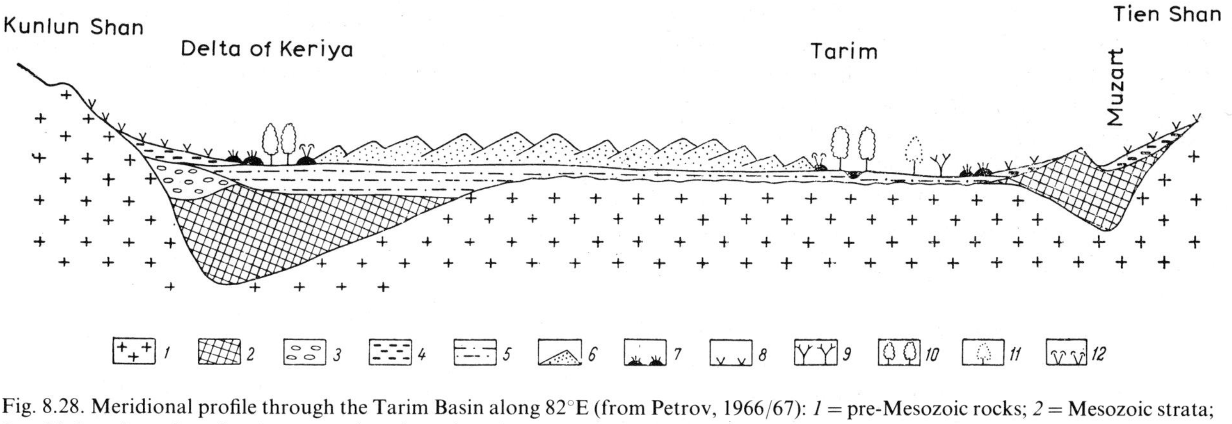

Fig. 8.28. Meridional profile through the Tarim Basin along 82°E (from Petrov, 1966/67): *1* = pre-Mesozoic rocks; *2* = Mesozoic strata; *3* = old slope-base deposits; *4* = young slope-base deposits; *5* = alluvial lake deposits; *6* = barkhans of the Takla Makan; *7* = small dunes with plant covers; *8* = dwarf-shrub biogeocoenosis of sandy-gravelly Gobi areas; *9* = *Haloxylon ammodendron* stands; *10* = floodplain forests with *Populus* and *Elaeagnus*; *11* = dying floodplain forests; *12* = stands of *Tamarix* and shrubby halophytes.

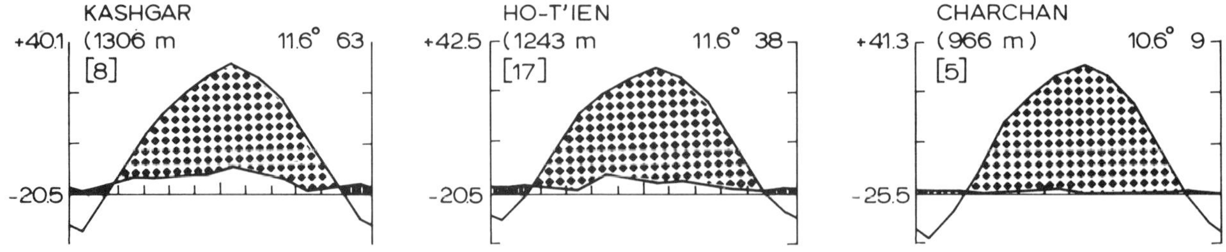

Fig. 8.29. Climate diagrams for stations in the Tarim Basin (Walter, 1974, p. 307).

desert, there are none of the psammophytes of the Mongolian sand deserts (*Caragana*, *Psammochloa*, *Pugionium*) but rather there occur the Middle Asian species (*Aristida pennata*, with *Calligonum* spp. and *Tamarix ramosissima*). After rain, the annuals *Agriophyllum arenarium* and *Corispermum* spp. can sometimes be found, however. The eastern part of the Tarim Basin (Fig. 8.7, region *19*) is separated from Dzungaria by desert-like mountains (Barkol Tagh, 4867 m above sea level) and by the eastern spurs of the Tien Shan system. The large oasis Hami is located on the south slope above a depression and, further to the west, the oasis Turfan lies in a graben 154 m below sea level. Both have extensive irrigated agricultural lands.

The Gashun-Gobi (Fig. 8.7, region *20*), stretching in front of the oases, leads over into the Lop Nor Depression (780 m). The Gashun-Gobi consists of sand and salt deserts with scattered dried-loam deserts.

Petrov (1966/67) distinguished the following biogeocoenosis complexes:

(1) Those with few dwarf-shrubs (*Ephedra*, *Zygophyllum*, *Gymnocarpus*, *Sympegma*, *Anabasis*, *Iljinia*, *Reaumuria*, etc.) on the slopes of the encircling mountains.

(2) The very stunted psammophilic vegetation of the Takla Makan Desert and smaller sandy areas between floodplains. Widespread are drifts bearing *Tamarix ramosissima*, *Nitraria schoberi*, *Haloxylon*

Fig. 8.30. Nebka landscape in the Tarim Basin with *Nitraria schoberi* on an area with slightly brackish groundwater near the surface (photograph by M.P. Petrov).

Fig. 8.31. *Haloxylon ammodendron* tree in the Tarim Basin (photograph by M.P. Petrov).

ammodendron, *Ephedra strobilacea*, *Calligonum roborowkii*, *Halostachys caspica*, *Lycium aphyllum*, *Anabasis aphylla*, and the herbs *Aristida pennata*, *Karelinia caspica*, *Scorzonera divaricata*, *Alhagi sparsiflora*, *Glycyrrhiza inflata*, *Phragmites communis*, and some annuals (10 samples with 2 to 9 species per sample, depending on the salinity of the groundwater in the root zone). The species mentioned extend also into the sand desert.

(3) The halophilic vegetation of old river channels and delta formations, which are composed of annuals (*Salsola*, *Suaeda*, *Halogeton*), dwarf-shrubs (*Kalidium cuspidatum*, *Halocnemon*), shrubs such as *Tamarix ramosissima*, *T. hispida*, *Halostachys caspica*, *Nitraria schoberi* (Fig. 8.30), and a few stands of *Haloxylon ammodendron* (Fig. 8.31).

(4) The floodplain vegetation of rivers and oases around the edges of the Tarim Basin (Fig. 8.32), which is very characteristic of the basin. These river floodplains are very reminiscent of the *tugai* vegetation of Middle Asia, but in this almost uninhabited region they have not been altered so much by man. As a result the stands of *Populus* or *Ulmus pumila* along smaller rivers cover almost the entire floodplain, and open grass or reed areas are less important. *Hippophaë rhamnoides*, as well as *Tamarix ramosissima* and *Elaeagnus*, are also very common. The actual floodplain forests, with a crown closure of 40 to 60%, are composed of

Fig. 8.32. *Populus diversifolia* floodplain forest in the Tarim Basin (photograph by M.P. Petrov).

Populus diversifolia and, to a lesser extent, *P. pruinosa*, *Halimodendron halodendron*, *Hippophaë rhamnoides*, *Tamarix ramosissima*, *Lycium turcomanicum*, *Elaeagnus moorcroftii*, *Myricaria alopecuroides*, and *Clematis orientalis*, with an herb layer of *Glycyrrhiza inflata*, *G. uralensis*, *Poacymum hendersonii*, *Oxytropis glabra*, *Goebelia alopecuroides*, *Calamagrostis pseudophragmites*, etc.

In standing water one sees *Phragmites communis*, *Typha angustata*, *T. minima*, *Scirpus mucronatus*, *Potamogeton perfoliatus*, etc. In general, however, the floodplains have fewer species than in Middle Asia.

Kansu Corridor (Hessi Corridor, Jümönn Passage) (Fig. 8.7, region *14*)

This corridor opens to the west into the Lop Nor Depression. The term "hessi" refers to the asymmetric graben which arises as the result of a chain of narrow depressions. These depressions extend for a distance of 1000 km along the north base of the Astin Tagh (A-erh-chih Shan) and the Nan Shan. They are separated from the Ala Shan Desert to the north by chains of low mountains (over 2000 m). The corridor is 100 km wide south of An-hsi but narrows to only 20 km further to the east. The floor of the graben climbs from 800 m in the west to 1500 m in the east, where it leads over into the Ala Shan Desert (Fig. 8.33).

A series of rivers rising in the Nan Shan forms the basis for the oases within the corridor. The largest river is the Su-lo Ho, on which An-hsi is situated. So much water is taken from the river at An-hsi for irrigation that its lower course now consists of only a series of salt lakes. It no longer reaches the Lop Nor. Because of the great east–west length of the corridor, the climate changes considerably from one end to the other. One can observe here the transition from the desert climate of the Lop Nor Depression to the steppe climate at Lan-chou on the Huang Ho (Fig. 8.34). Precipitation increases from 45 mm to over 100 mm, while the temperature

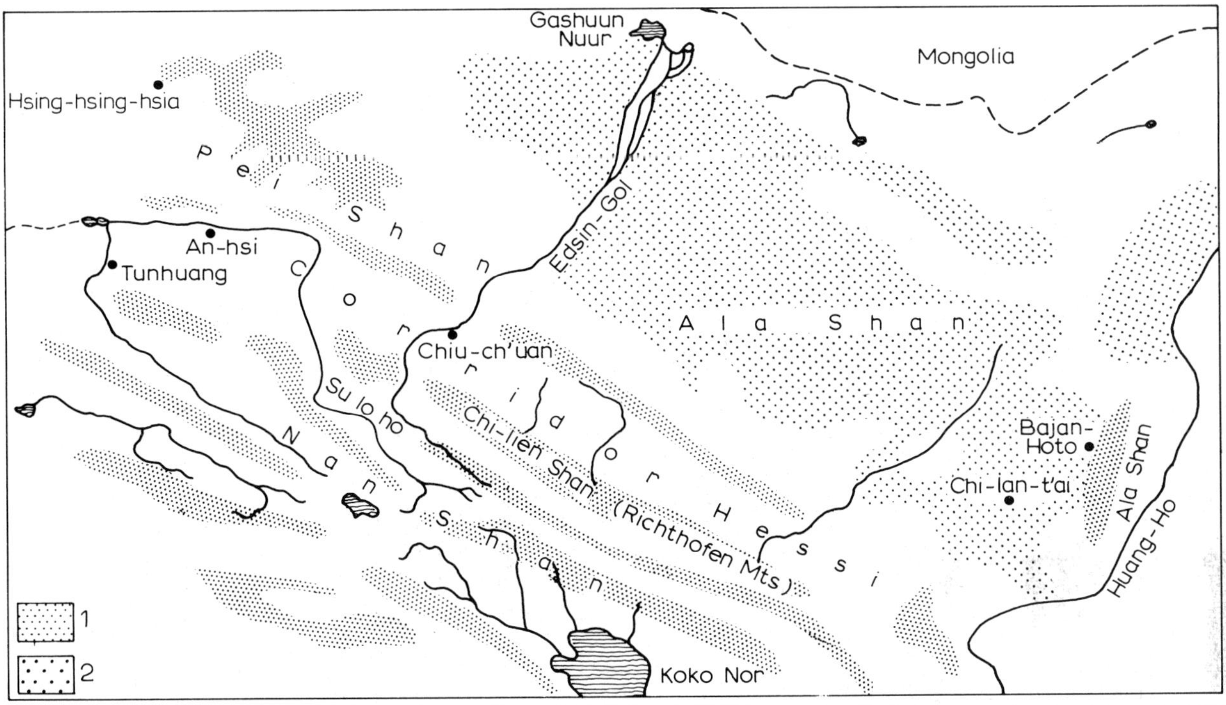

Fig. 8.33. Cartographic sketch of the location of the Kansu (Hessi) Corridor and the Pei Shan and Ala Shan Deserts (from Petrov, 1966/67): *1* = mountain chains; *2* = mainly sand areas. The Richthofen Range is the northernmost ridge of the Nan Shan system. The smaller Ala Shan Range is to the east along the Huang Ho. The Hessi Corridor opens to the west into the Lop Nor Depression.

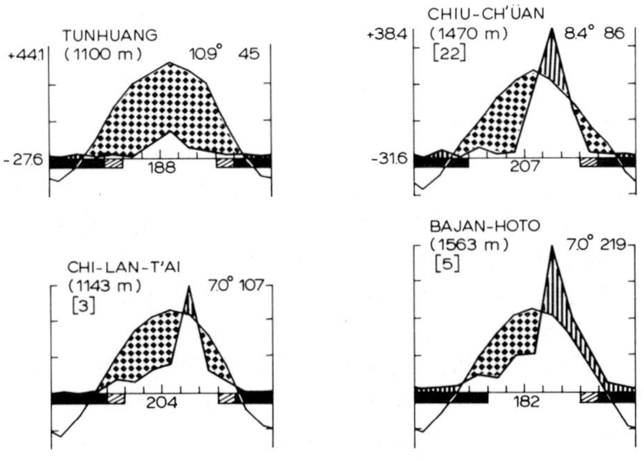

Fig. 8.34. Climate diagrams for stations in the Kansu Corridor (above: Tunhuang and Chiu-ch'üan) and in the southeastern Ala Shan Desert [below: Chi-lan-t'ai and Bajan-Hoto (Pa-yen-hao-t'e)].

decreases with the increasing altitude toward the east: Annual means decrease from 10.9° to 7.1°C, January means from −6.3° to −8.9°C, and July means from 26.1° to 22.4°C. The potential evaporation at An-hsi is 3200 mm.

Looking at the vegetation, one can distinguish a series of altitudinal belts on the north slope of the Nan Shan: (1) oases at 1500 m; (2) the sandy gravel deserts at the foot of the slopes (1500–2000 m); (3) the dry montane steppe of the foothills (2000–2300 m); (4) the shrub steppe (2300–2500 m); (5) spruce forests with forest-meadows (2500–3200 m); (6) high-montane meadow steppe (3200–3500 m); (7) alpine meadows (3500–3600 m); and (8) the nival belt above about 3600 m (Fig. 8.35).

In the lowest-lying areas of the graben one finds halophilic meadows around the oases. These are composed of *Schoenoplectus tabernaemontani*, *Bolboschoenus affinis*, *Triglochin palustre*, and considerable *Carex* spp. On solonchak one finds *Halocnemum*, *Halostachys belangeriana*, *Kalidium foliatum*, *Tamarix laxa* or *Haloxylon ammodendron* and *Nitraria schoberi*.

On the slightly inclined sandy gravel areas there grow, alongside the usual Mongolian shrubs, *Stipa glareosa* and *Artemisia capillaris*. Sand areas are dominated by *Caragana korshinskii*, *Hedysarum scoparium*, *Artemisia sphaerocephala*, *Psam-*

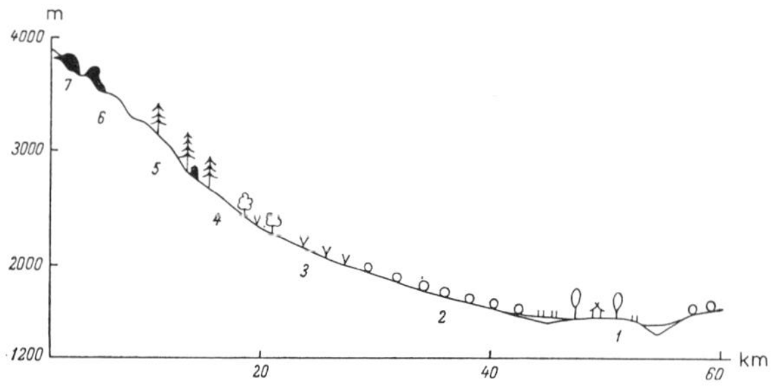

Fig. 8.35. Altitudinal belts on the north slope of the eastern Nan Shan Mountains (from Petrov, 1966/67): *1* = central depression in the Kansu Corridor with the river Hei Ho (marsh-meadows and oases); *2* = sandy-gravelly dwarf-shrub desert on sierozems and burozems at the base of the slopes; *3* = forb-rich grass steppes of the higher foothills on skeletal chernozem; *4* = complex composed of open bushlands and montane steppes on brown soils; *5* = forest-meadow belt with islands of *Picea* on north slopes and forest-meadows on south slopes on dark-brown, calcium-rich soils; *6* = subalpine meadows; and *7* = alpine meadows (from Walter, 1974, fig. 260).

mochloa villosa, and *Pugionium cornutum*; mobile sands by *Calligonum mongolicum*, *C. zeidamense* and *Aristida pennata* (see Lavrenko, 1966).

Characteristic dominants of dry montane steppe are *Stipa bungeana*, *S. brevifolia*, *Agropyron* spp., *Aneurolepidium dasystachys*, *Amygdalus mongolica*, *Dasiphora fruticosa* s.l., *Ephedra equisetina*, *Zygophyllum xanthoxylon*, etc. The total vegetal cover can reach 60 to 70%.

The shrub steppe is even better developed, reaching 100% cover on moist sites. Among the shrubs [*Cotoneaster*, *Lonicera*, *Rhamnus*, *Rosa*, *Berberis*, *Dasiphora* (*Potentilla*), *Spiraea*, *Clematis*] are steppe communities composed of *Stipa* species and other grasses and scattered with colorful forbs such as *Adonis*, *Pedicularis*, *Oxytropis*, *Stellaria*, *Torularia*, etc.

The spruce belt is very rich in species. *Picea asperata*, very closely related to *P. schrenkiana*, is a tree which can reach a height of 40 m and a trunk diameter of 1 m (Fig. 8.36). In the eastern Nan Shan, where the precipitation reaches 1000 mm, the forests can become closed. For the most part only more xerophilic shrubs grow on the south slopes: *Dasiphora fruticosa*, *Rosa omeiensis*, and *Caragana* species. Many shrubs are found as undergrowth in sparse spruce forests (woodlands). In addition to

Fig. 8.36. Needle-leaved forest belt with *Picea asperata* in the eastern Nan Shan (photograph by M.P. Petrov).

Fig. 8.37. Tengri sand region in the southeastern Ala Shan Desert (photograph by M.P. Petrov).

those of the shrub-steppe there are *Sorbus*, *Juniperus*, and *Salix* shrubs. On open sites one notices *Lilium tenuifolium*, *Iris ensata*, *Thalictrum* and *Galium* species. *Populus davidiana* (aff. *P. tremula*) establishes itself on areas of felled forest. Patches of forest alternate with meadows. In the western Nan Shan Mountains the dryness is too great and the forest-meadow belt disappears.

The Ala Shan Desert (see Fig. 8.7, region *12*)

This desert area lies between the rivers Jo Shui (Ochina-Ho, Edsin-Gol, Ruoshui), which empties into the Gashuun Nuur, and the Huang Ho (Fig. 8.33). Immediately to the west of the Huang Ho is the Ala Shan mountain range with peaks up to 3600 m. The region borders the Kansu (Hessi) Corridor in the south, through which the railroad from Lan-chou to An-hsi runs. From an airplane the region appears as a complex of large areas of mobile dunes with dark patches of saksa'ul and small salt flats or broad gravel plains with inselbergs. Aeolian sand covers 65% of the total area, and half of this is shifting barkhans (Fig. 8.37).

In contrast to Pei Shan, therefore, the Ala Shan Desert is mainly a sand desert. The climate is continental but does receive the full effect of the East Asian summer monsoon. Winter is always dominated by the Mongolian–Siberian high-pressure cell. The precipitation in the western part is 39 to 68 mm, but this climbs to 220 mm at the foot of the Ala Shan Mountains in the east. The potential evaporation is around 2360 to 3790 mm annually. Two climate diagrams from the eastern part are shown in Fig. 8.34. Precipitation is sufficient in this area for the sands to carry some vegetation, as one can see from the water content of the soil (Fig. 8.38). Vegetation can occur in the west only on sites with shallow groundwater.

The vast expanse of shifting dunes is, as almost always, the result of irrational grazing and destruction of the woody plants by man. *Tsaidams* (moist depressions) are scattered throughout the sand region. These are the remains of an old river system which was covered by the sands. The depressions are mostly surrounded by shifting dunes, since the effects of overgrazing were especially

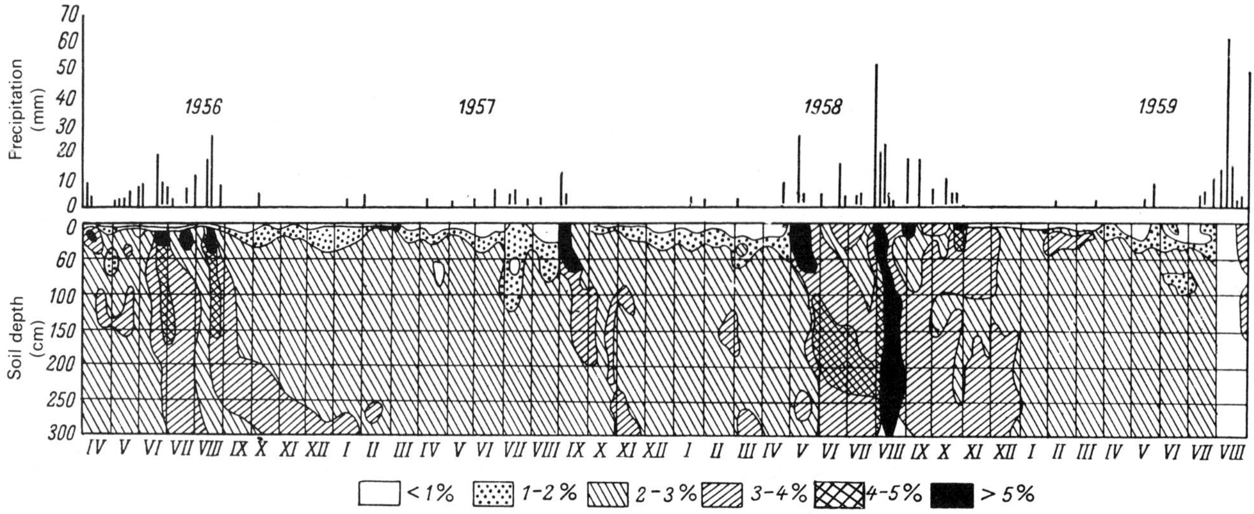

Fig. 8.38. Annual course of the water content of the sand soil of dune valleys (Tengri Sands, southeastern Ala Shan). At the bottom is given the water content in the upper 3 m of the sand soil, of which only the upper 40 cm dried out completely, even in the dry year 1957 (by Chdhi-bi-Zhenyu et al., from Petrov, 1966/67).

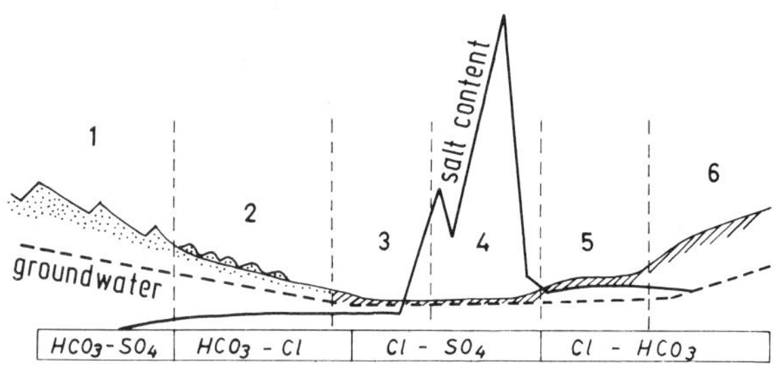

Fig. 8.39. Saltpan (Tsaidam Tudao-Chu) in the Ala Shan region (by Chdzi-Mao-Zhenyu, from Petrov, 1966/67), with data on the groundwater level and the relative salt content of the groundwater: *1* = barkhans; *2* = dunes half-covered by plants; *3* = sulphate–chloride solonchak; *4* = chloride solonchak; *5* = carbonate–chloride solonchak; *6* = slope of the Holan Shan high plateau. The dominant anions in the groundwater are given at the bottom: carbonate salinization around the periphery and chloride salinization in the center.

severe here. Figure 8.39 shows the course of the groundwater level as well as the type and degree of salinization. *Nitraria schoberi* is quite widespread on the smaller drifted dunes with high groundwater. It is accompanied by individuals of *Kalidium* and *Lycium* species, on moister places by *Lasiagrostis splendens*, and on swampy sites by *Phragmites communis* and *Typha minima*. Along the dry river beds one can find the pitiful remains of a floodplain forest composed of *Populus diversifolia* and *Ulmus pumila* (Fig. 8.40). On sandy alluvia one also finds stands of *Haloxylon ammodendron*.

Some water-containing basins in the central part of the Ala Shan dry out in the summer. They are fed by groundwater and have their highwater levels in the winter, since the groundwater coming from the southern margins in the summer requires half a year to reach the central areas of the desert. The groundwater often flows in old river courses.

As always, the bare sand dunes store the rainfall and groundwater tables form, which can reach the surface in the dune valleys to be used for drinking water.

The vegetation of Ala Shan is rather varied and somewhat richer in species. It consists of typically Mongolian–Central Asian geoelements, as can be seen from the following plot samples:

Vegetation of rocky inselberge (9 samples, 12–23 species per plot, cover is low; see Fig. 8.41):

Shrubs: *Zygophyllum xanthoxylon*, *Nitraria sphaerocarpa*, *N. sibirica*, *Caragana stenophylla*, *C. tragacanthoides*, *Ephedra przewalskii*, *E. equisetina*, *Haloxylon ammodendron*, *Atraphaxis* sp., *Convolvulus fruticosus*, *C. tragacanthoides*, *Calligonum mongolicum*, *Gymnocarpos przewalskii*, *Potaninia mongolica*, *Ammopiptanthus mongolicus*, *Lespedeza dahurica*.

Semi-shrubs: *Reaumuria soongorica*, *R. trigyna*, *Artemisia frigida*,

Fig. 8.40. *Ulmus pumila* in earlier riverbed of the Ala Shan Desert (photograph by M.P. Petrov).

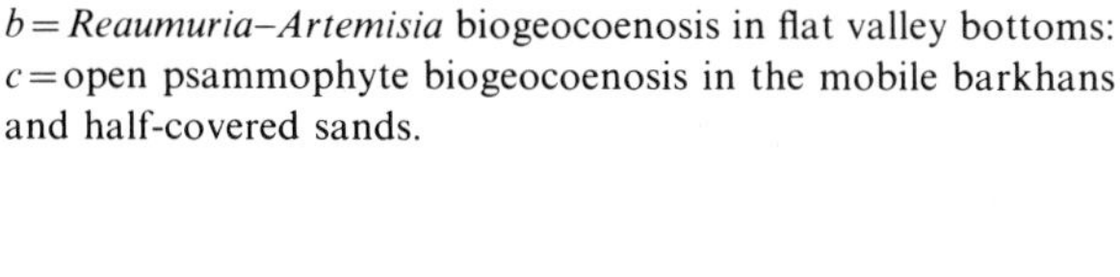

Fig. 8.41. Vegetation subdivision in the eroded, hilly landscape of the Ala Shan Desert (after Petrov, 1966/67; from Walter, 1974, fig. 261): *a*=dwarf-shrub vegetation on inselberge; *b*=*Reaumuria–Artemisia* biogeocoenosis in flat valley bottoms: *c*=open psammophyte biogeocoenosis in the mobile barkhans and half-covered sands.

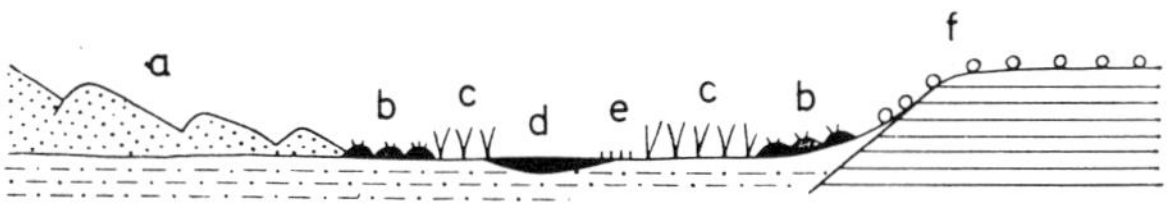

Fig. 8.42. Brackish depression in the Ala Shan Desert (by Petrov, 1966/67; from Walter, 1974, fig. 262): *a* = psammophytes in the barkhan region; *b*=*Nitraria schoberi* sand heaps (nebkas); *c*=*Lasiagrostis splendens* stands; *d*=lake with reed thicket; *e*=*Kalidium*–halophyte stands; *f*=sandy-gravelly plains with psammophilic dwarf-shrubs.

Salsola laricifolia, *S. passerina*, *Anabasis brevifolia*, *Sympegma regelii*, *Ceratoides papposa*, *Caryopteris mongolica*, *Asterothamnus centraliasiaticus*, *Clematis songarica*.

Perennial herbs: *Stipa gobica*, *S. breviflora*, *Cleistogenes mutica*, *Pennisetum flaccidum*, *Carex duriuscula*, *Iris ensata*, *Asparagus gobicus*, *Allium mongolicum*, *Phleum nanum*, *Oxytropis aciphylla*, *Astragalus melilotoides*, *A.* sp., *Zygophyllum mucronatum*, *Arnebia saxatilis*, *Limonium aureum*, *L. bicolor*, *Artemisia* sp., *Aster altaicus*, *Jurinea* sp., *Scorzonera divaricata*, *S. pseudodivaricata*, *Tanacetum fruticulosum*, *Echinops gmelini*, *Ixeris chinensis*.

The plant cover is similar on the surrounding sandy-gravelly flats with *Reaumuria* dominating and *Stipa gobica* abundant. The saksa'ul is typical of extensive sandy depressions.

Haloxylon ammodendron stands in depressions, often between dunes (5 samples, 3–9 species per plot, cover up to 50%):

Shrubs: *Haloxylon ammodendron* (2), *Nitraria schoberi* (1–2), *Oxytropis aciphylla* (1), *Caragana stenophylla* (+), *Zygophyllum xanthoxylon* (+), *Kalidium foliatum* (+).

Semi-shrubs: *Artemisia frigida* (+), *Potaninia mongolica* (+), *Reaumuria soongorica* (+).
Perennial herbs: *Peganum nigellastrum* (+), *Cleistogenes mutica*, *Stipa gobica*, *Saussurea salsa*, *Corispermum mongolicum*.

If the groundwater table is higher, then halophytes such as *Kalidium* occur on solonchak and *Nitraria* grows on nebkas.

Vegetation of solonchak soils (2 samples, 4–5 species per plot):

Shrubs: *Nitraria schoberi*, *N. sibirica*, *Lycium turcomanicum*, *Kalidium gracile*, *K. foliatum* (Fig. 8.5).
Perennial herbs: *Lasiagrostis splendens*, *Glaux maritima*.
Annual herbs: *Salicornia herbacea*, *Suaeda przewalskii*, *S. glauca*.
These are joined on halophilic meadows by: *Phragmites communis*, *Iris ensata*, *Tournefortia sibirica*, *Glycyrrhiza uralensis*, *Periploca graeca*, *Cynomorium*, *Triglochin palustre*, *Cynanchum sibiricum*, *Taraxacum chinense*, *Potentilla anserina*, *Thermopsis lanceolata*, and *Puccinellia distans*.

It is, however, the psammophyte vegetation which is most widespread, both on mobile and immobile sands. The barkhans are completely devoid of vegetation.

Vegetation of dunes and sand flats (13 samples, 3–8 species per plot):

Shrubs: *Caragana korshinskii*, *C. microphylla*, *Hedysarum mongolicum*, *H. scoparium*, *Nitraria schoberi*, *Atraphaxis frutescens*, *Calligonum mongolicum*.
Semi-shrubs: *Artemisia ordosica*, *A. sphaerocephala*.
Perennial herbs: *Psammochloa villosa*, *Phragmites communis*, *Arnebia saxatilis*, *Stilpnolepis centifolia*, *Sophora alopecuroides*, *Orobanche* sp.
Annuals: *Pugionium cornutum*, *P. dolabratum*, *Agriophyllum arenarium*, *Corispermum mongolicum*, *Artemisia anethifolia*, *Astragalus* sp.

The distribution of the various communities in relation to the relief is shown in Fig. 8.42.

The psammophytes of Central Asia are distinguished from those of Middle Asia by a later development in the spring (necessitated by the dry spring) and by a shallower root system in the top 1 m of soil, which in Central Asia is wetted by summer rain. The cover on immobile sands can reach 40 to 60%. The most common tree in the Ala Shan Desert is *Ulmus pumila*, which can be found singly or in groups on sites with shallow groundwater tables. It is thus found around wells and springs. Old trees attain a trunk diameter of 60 cm and often indicate the locations of earlier floodplains. *Populus diversifolia* can often be found in the southern part of the region, sometimes also in dune valleys.

The Ala Shan Mountains in the eastern part of the region, over 3000 m high, run north and south, that is, across the path of the eastern monsoon winds. They thus receive higher precipitation, and the altitudinal belts are more distinct than in the Nan Shan Mountains.

Petrov (1966/67) gave the following classification of the vegetation in northern Ala Shan:

1500–1600 m: Surrounding plains, with *Reaumuria soongorica*–*Artemisia* communities with *Stipa glareosa*, *Allium mongolicum*, etc.
1600–1900 m: Montane steppe of the foothills, with *Stipa brevifolia*, *Caragana* spp., *Ephedra*, *Atraphaxis*, *Allium mongolicum*, etc.
1900–2500 m: Shrub belt with forb-steppe: *Lonicera* sp., *Rosa* sp., *Rhamnus erythroxylon*, *Dasiphora fruticosa*, *Ephedra equisetina*, *Cotoneaster* sp., *Spiraea trilobata*, *Ziziphus jujuba*, with individuals of *Ulmus glaucescens* in ravines.
2500–3000 m: Spruce–pine forest (canopy closure 60–70%): *Picea asperata*, *Pinus tabulaeformis*, *Juniperus rigida*, *Lonicera* sp., *Rhamnus parvifolia*, *Euonymus nana*, *Cotoneaster* sp., *Ribes pulchellum*, etc. In dry ravines *Ostryopsis davidiana* and *Syringa oblata*. Tall forbs on open sites: *Valeriana officinalis*, *Thalictrum* sp., *Aquilegia* sp., *Astragalus* sp., *Sanguisorba* sp., etc. On sunny rocky slopes individual *Lilium tenuifolium*, *Rhaponticum uniflorum*, *Hedysarum* sp. On felled areas closed stands of aspen and birch (*Betula platyphylla*–*Populus davidiana*).
Above 3000 m (Subalpine–alpine belt): individual *Caragana jubata*, *Spiraea trilobata*, *Dasiphora* (*Potentilla*) *fruticosa*, *Salix* shrubs in meadows with *Polygonum viviparum*, *Saussurea pygmaea*, *Dianthus* sp., *Ranunculus* sp., *Delphinium* sp., etc.

The Ordos region (Mu Us Shamo) (Fig. 8.7, region *13*)

The Ordos Desert area occupies the northwest half of the region within the great bend of the Huang Ho, while the southeastern half, which is separated by the Great Wall, is a loess plain to the steppe region (Fig. 8.43). The outcropping rocks in the desert are primarily Cretaceous and Jurassic rocks. Their weathering products produced the raw materials for vast expanses of aeolian sands, with alluvial deposits in the depressions. The size of the entire area is 95 000 km^2. The elevation varies between 1150 and 1500 m; only in the west do mountains rise significantly higher (the Arbisso Mountains, Zhuozi Shan, with steep slopes and peaks as high as 3015 m). The heavily furrowed loess plains in the southeast lie about 150 to 200 m

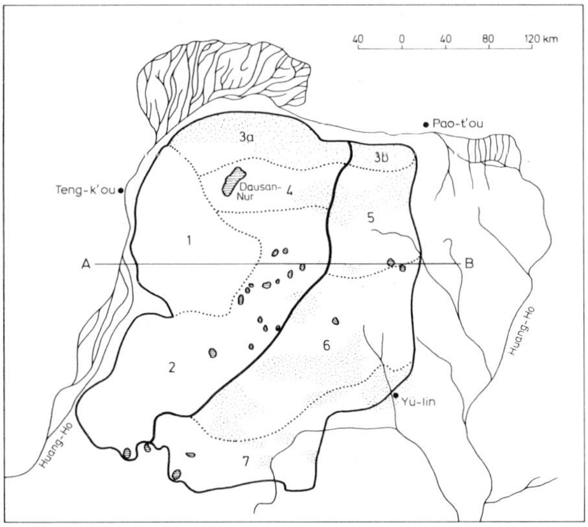

Fig. 8.43. Subdivisions of the landscape of the Ordos region (from Petrov, 1966/67, adapted by Walter, 1974, fig. 263): *1*, *2*, *3a*, 4 = desert; *3b*, *5*, *6*, *7* = desert-steppe; *1* = Arbisso Mountains; *2* = raised central Ordos; *3* = sand-dune region; *4* = sandy, brackish depression with Lake Dausan; *5* = sandy-gravelly heights; *6* = depression with lakes and meadows; *7* = eastern barkhan region. *A–B* is the vegetation profile shown in Figure 8.45.

higher than the bordering Ordos region. This loess represents dust blown out of the sand areas of the Ordos.

A total of seven landscape units can be distinguished: (1) the heavily eroded Arbisso Mountains, built up of old strata; (2) the high plain with outcropping, only slightly eroded Cretaceous and Jurassic rocks and scattered, endorheic depressions with soda salinization; (3) the northern region of shifting dunes, which reaches all the way to the wide, green Huang Ho Valley; (4) the lower-lying basin around the lake Dausan Nur, with chloride salinization; (5) the northeastern, raised plain with immobile sands on top of Cretaceous strata; (6) the endorheic lowland with slightly brackish lakes or freshwater basins and sandy soils with about 20% plant cover; and (7) the southeastern area of shifting dunes, with continuously flowing rivers cut deep into the solid rock; the mobile sands here are primarily the result of overgrazing (Li Hsiao fang, 1979) and store the extra summer rainfall as groundwater, which then flows off in the rivers.

The continentality of the climate decreases markedly within the Ordos as one goes from west to east. Dry, icy weather dominates the winter here also, but the monsoon brings so much rain in the eastern part (405 mm at Yü-lin) that it can no longer be called a desert (Fig. 8.44).

The varying salinization of the lakes is less the result of the climate than it is the result of the salt content of the particular substrata. Both climatic and substrate conditions determine the quite varied vegetation of the Ordos region. The east–west profile in Fig. 8.45 shows the subdivisions schematically.

Petrov (1966/67) gives the following additional data on the species composition of the plant cover:

(1) In the mountains one finds only individual plants of *Amygdalus mongolica*, *Zygophyllum xanthoxylon*, *Reaumuria trigyna*, etc.

(2) On sandy gravelly flats there grow, alongside *Reaumuria* and *Caragana*, *Tetraena mongolica*, *Asparagus gobicus*, and on sandy sites *Ammopiptanthus mongolicus*.

(3) The cover reaches 40 to 50% on drifted sands. A sample yielded: *Artemisia ordosica* (2), *A. frigida* (1–2), *Caragana microphylla* (2), *C. stenophylla* (2), *Stipa glareosa* (1), *Pycnostelma lateriflorum* (1), *Carex duriuscula* (2), *Allium mongolicum* (1), *Peganum nigellastrum* (1), *Lasiagrostis splendens* (+), *Panzeria* (*Lycium*) *lanata* (+).

(4) The halophilic meadows (slight soda-salinization) consist of *Carex* species with *Potentilla anserina*, *Silene repens*, *Anemone* sp., *Oxytropis glabra*, plus an *Orchis* species and a *Spiranthes* species; noteworthy from other places were *Taraxacum chinense*, *Glaux maritima*, *Triglochin palustre*, *Saussurea glomerata*, *Suaeda prostrata*, *Heleocharis*, *Cyperus* and *Phragmites*.

(5) The dry *Artemisia* steppes have a denser plant cover. Cited in addition to the dominant species mentioned in (3) are: *Artemisia capillaris* (1), *Salsola collina* (1), *Cleistogenes mutica* (+), *Scutellaria grandiflora* (+), *Aster altaicus* (+), *Thymus serpyllum* s.l. (+), *Echinops gmelini* and others (Fig. 8.46).

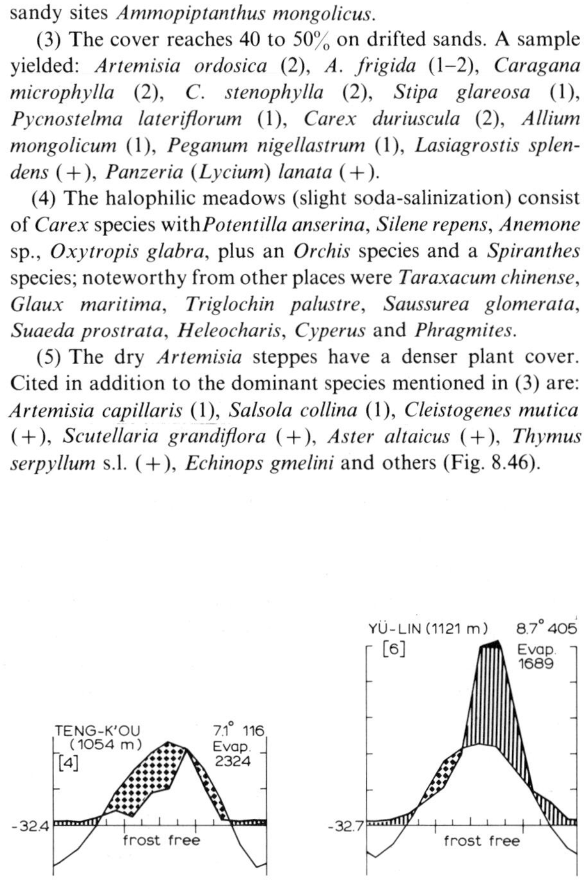

Fig. 8.44. Climate diagrams from Teng-k'ou in the western Ordos (desert) and Yü-lin in the southeast (steppe).

Fig. 8.45. East–west profile through the Ordos region (compare Fig. 8.43, *A–B*) (from Petrov, 1966/67; in Walter, 1974, fig. 265): *1* = dwarf-shrub desert of the Arbisso Mountains; *2* = *Reaumuria soongorica–Caragana tragacanthoides–Stipa glareosa* semi-desert; *3* = *Artemisia ordosica* semi-desert (Fig. 8.46); *4* = hydrohalophyte biogeocoenosis; *5* = steppe stands with *Artemisia ordosica*; *6* = grass steppes; *7* = calciphilic vegetation of outcropping Cretaceous sandstone with carbonate-rich soils (*Euphorbia*, *Gypsophila*, *Haplophyllum*); *8* = steppes of the eroded loess plateau.

(6) The following occur in the steppe-meadows: *Carex duriuscula* (4), *Aneurolepidium dasystachys* (4), with *Astragalus melilotoides*.

(7) Calciphilic vegetation with *Gypsophila*, *Euphorbia*, *Haplophyllum*, etc.

(8) The following are given for the steppes on the loess plains around the margins of the Ordos: *Stipa grandis* (2), *S. bungeana* (2), *Agropyron mongolicum* (1), *Pappophorum boreale* (1), *Cleistogenes mutica* (+), *Artemisia frigida* (1), *A. pectinata* (1), *Thymus serpyllum* s.l. (1), *Oxytropis psammocharis* (1), *Astragalus adsurgens* (+), *Stellera chamaejasme* (+), *Pycnostelma lateriflorum* (+), *Euphorbia kansui* (+).

The vegetation on blown sands is very sparse. Noteworthy are the shrubs *Hedysarum mongolicum*, *H. scoparium*, *Caragana microphylla*, *C. korshinskii* (Fig. 8.47), as well as other psammophytes (*Artemisia ordosica*, *A. sphaerocephala*, *Atraphaxis frutescens*, *Psammochloa villosa*, *Agriophyllum arenarium*, *Corispermum gmelini*, *Pugionium cornutum*). In dune valleys one finds *Salix cheilophila*, *Hippophaë rhamnoides*, with *Typha*, *Phragmites*, *Calamagrostis epigeios*, or in small freshwater lakes, *Scirpus maritimus*, *Trapa natans*, *Alisma plantago*, *Nymphoides peltata*, *Potamogeton pectinatus*.

Halophytes grow on solonchak around salt lakes: *Kalidium gracile*, *K. foliatum*, *Nitraria tangutorum*, *Lasiagrostis splendens*, etc. (see figs. 549–550 in Walter, 1968, pp. 790–791).

The Tsaidam Basin (see Fig. 8.7, regions *21–24*)

Moist, mostly brackish depressions (inland salt marshes) are generally called *tsaidam* in Central Asia. The Tsaidam Basin, however, is a raised, endorheic basin, 100–200 km wide and 600 km

Fig. 8.46. Sands half-covered by *Artemisia ordosica* in the semi-desert of the Ordos region (photograph by M.P. Petrov).

Fig. 8.47. *Caragana korshinskii* on mobile sand soil in the Ordos region (photograph by M.P. Petrov).

long, once filled by an earlier sea. It is orographically a part of Tibet but belongs floristically and by its vegetation to the Mongolian province. It is separated from Mongolia by the Astin Tagh (the north branch of the Kun Lun Mountains) and by the Nan Shan Mountains. The mean elevation of the region is around 3000 m, about 1500 m higher than the Mongolian Plateau but about the same amount lower than Tibet itself. It is thus a transitional level, both geographically and climatically. The mean annual temperatures are between 2.3 and 3°C. The mean temperature in July does not reach 18°C, and the very scant precipitation amounts to about 13 to 26 mm. The desert conditions are easily recognized in the climate diagram from Ge'ermu (Golmo, Fig. 8.48) at the southern edge of the basin. The higher-lying station at Tsagan Usa in the east receives somewhat more rain.

The uplift of Tsaidam, together with the uplift of Tibet, occurred no earlier than Early–Middle Pleistocene. Before that the flora was certainly that of Mongolia, with few exceptions. As a result only

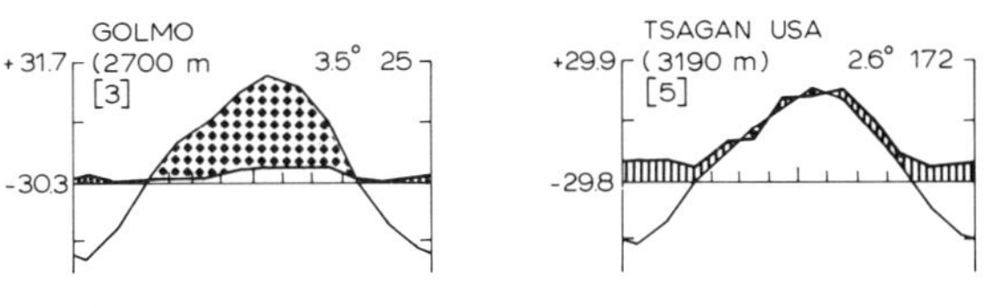

Fig. 8.48. Climate diagrams from the Tsaidam region. The locations of the stations Golmo and Tsagan Usa are shown in Fig. 8.7, regions *23* and *24*.

a very few endemic species have developed (*Calligonum zaidamense*, *Salsola zaidamensis*).

The western part of Tsaidam (region *21* in Fig. 8.7) is an elevated plain with a very light plant cover composed of complexes of gypsophilic, halophilic, and psammophilic shrub vegetation. The northern part (region *22*) includes the foothills of the Nan Shan, consisting in part of sandy hills, and carries an *Artemisia–Salsola* vegetation. The lowest lying parts in the center and southeast (region *23*) are vegetationless salt flats, to which the entire region owes its name (Fig. 8.49), plus *Tamarix–*

Fig. 8.49. Extensive unvegetated salt flats in the center of the Tsaidam Basin (photograph by M.P. Petrov).

Nitraria saline meadows in river valleys, their delta formations, and salt lakes. Sandy areas with barkhans are present only in the south at the base of the Kun Lun Mountains (Fig. 8.50). The eastern part is the highest and most mountainous (region *24*). The wet (though vegetationless) central part and the many salt lakes receive their water from the numerous rivers coming down from the snowy Kun Lun in the east.

In the Tsaidam Basin itself there are no representatives of the Tibetan flora, though certainly there are on the surrounding mountains. These include *Microula sikkimensis*, *Rumex nepalensis*, *Rheum spiciforme*, *Saussurea sorocephala*, *Thylacospermum rupifragum*, *Lagotis brachystachya*, *Geranium pylzovianum*, *Salvia roborovskii*, *Adenophora himalayana*, *Aconitum gymnandrum*, and *Dracocephalum tanguticum*.

The proximity of Dzungaria is expressed through the presence of such species as *Poacynum hendersonii*, *Trachomitum lancifolium* and many *Calligonum* species. The Mongolian elements, however, are the most widespread. Around 200 species have been collected.

Six plot samples on sandy-gravelly and rubble-covered areas in the foothills showed the following species: *Tamarix ramosissima*, *Myricaria alopecuroides*, *Ephedra przewalskii*, *Nitraria schoberi*, *Calligonum zaidamense*, *Calligonum* spp., *Salsola abrotanoides*, *Reaumuria soongorica*, *Sympegma regelii*, *Ceratoides papposa*, *Artemisia* spp., *Halogeton glomeratus*, *Cynomorium songarium*, and *Lepidium perfoliatum*. *Haloxylon ammodendron* also forms scattered stands in places.

On partly blown-out sand hillocks one finds mainly *Calligonum zaidamense*, also *Tamarix ramosissima* and *Nitraria* if the groundwater table is not deep, and *Populus diversifolia* and *Phragmites* if the groundwater is almost at the surface. One often finds dying trees, which indicates a sinking groundwater table due to deeper cutting by rivers.

On immobile sands with a 40% plant cover the dominants are: *Artemisia arenaria*, *A. xerophytica*, *Oxytropis aciphylla*, *Aneurolepidium dasystachys*, *Thermopsis lanceolata*, with *Stipa glareosa*, *Salsola paulsenii*, *Corispermum mongolicum*, *Echinops gmelini*, *Potentilla bifurca*, etc. (Fig. 8.50).

In the eastern part with more rainfall the cover in the *Artemisia–Salsola* semi-desert reaches 70 to 80%.

Fig. 8.50. *Artemisia* semi-desert at the foot of the Kun Lun Mountains in the Tsaidam region. Just in front of the mountains there is also a region of sand dunes (photograph by M.P. Petrov).

On solonchak there is either no vegetation or one finds individuals of *Kalidium gracile* with *K. foliatum*, *Salicornia europea*, *Lycium turcomanicum*, *Salsola praecox*, and rarely *Phragmites communis*.

The plant cover of the marshy, halophilic meadows is the richest. These receive water continuously from the rivers. The following plants can be named: *Phragmites*, *Atropis*, *Carex* spp., *Calamagrostis macileuta*, *C. pseudophragmites*, *Heleocharis atropurpurea*, *Juncus gerardi*, *Trichophorum pumilum*, *Triglochin striatum*, *T. maritimum*, *T. palustre*, *Glaux maritima*, *Youngia paleacea*, *Saussurea salsa*, *Mulgedium tataricum*, *Thermopsis lanceolata*, *Sphaerophysa salsula* and, in places, *Tamarix* with *Myricaria* and *Nitraria* on river terraces.

Around salt lakes, where the groundwater is probably relatively fresh, there form stands (total vegetal cover 80–90%) of *Psathyrostachys (Hordeum) juncea*, *Aneurolepidium dasystachys*, *Phragmites communis*, *Bolboschoenus affinis*, *Typha angustata*, *Saussurea glomerata*, etc.

In the shifting sands of Tsaidam one can encounter a rare form of dune chains which run parallel to the wind direction and whose length can reach 4 to 5 km and height 15 to 16 m. These are fairly widespread in the areas around the central depression, and their direction is determined by the locally prevailing north or southwest winds. These dunes form on level, brackish surfaces.

The barkhans in Tsaidam are, as always, devoid of vegetation. Only between the sickle-shaped dunes do individuals of the following species occur: *Artemisia arenaria*, *Psammochloa villosa*, *Corispermum mongolicum*, *Agriophyllum arenarium*, and *Thermopsis lanceolata*.

The sands have formed as a result of the blowing out of weathered Tertiary sandstones and conglomerates, as well as river and marine deposits.

The geological structure of the basin is shown in Fig. 8.51.

Deserts with cool summers

To the cold-winter deserts of Central Asia belongs also the "roof of the world", the world's highest mountain plateau, Tibet with Pamir in the west. The plateau lies between the Kun Lun and Nan Shan Mountains in the north and the Himalaya in the south. Pamir (see Fig. 8.7, region *30*) forms the subject of a separate chapter (Ch. 9).

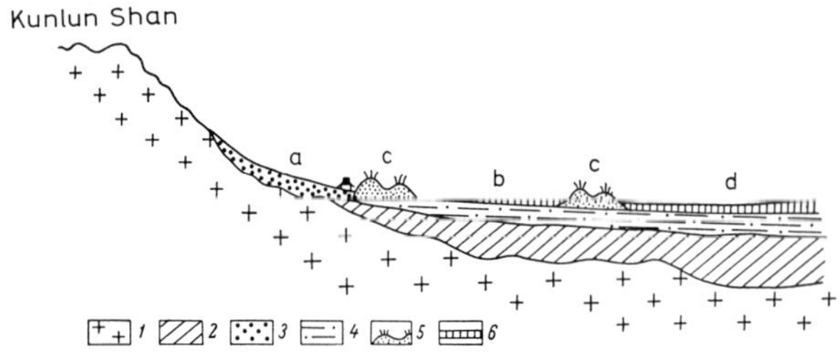

Fig. 8.51. Geomorphological profile through the southern part of the Tsaidam Basin (from Petrov, 1966/67, in Walter, 1974, fig. 267): *1* = pre-Tertiary strata; *2* = Tertiary rocks; *3* = gravelly-sandy rubble fans; *4* = old marine deposits; *5* = large sand dunes; *6* = recent, saline deposits of Lake Dausan Nur; *a* = base of slopes; *b* = delta of Kun Lun River; *c* = sands with *Tamarix*; *d* = solonchak vegetation.

Tibet (see Fig. 8.7, regions *25–29*).

Tibet stretches from west to east over 2000 km and from north to south over 1200 km. Its mean elevation is 4200 to 4800 m, and many mountain chains which cross the plateau rise another 1000 m above it. These chains dissect the region into numerous individual endorheic basins, into which the snowmelt from the mountains flows and forms saturated rubble-strewn marshes and salt lakes, in spite of the low precipitation and as the result of the low evaporation. Only the easternmost and southernmost parts have an outlet to the ocean. These are the sources of the great rivers of India and Southeast Asia.

The Tibetan floristic province is divided into four spheres (Fig. 8.7): (1) the Nan Shan and Chamdo (Amdo) in the northwest (region *25*); (2) the region of the deep river valleys in the southeast (region *26*); (3) the deep longitudinal Valley of Tibet in the south (region *27*); and (4) the vast High Tibetan or Chang Tang Desert (regions *28* and *29*), the most extreme highland area in the world, with the exception of the polar ice caps. The climate of the Tibetan High Plateau is hostile to life in general. The atmospheric pressure is only half that at sea level, and the mean annual temperature is −5°C. Only July has a mean temperature above freezing (8°C); the mean diurnal temperature variation is 37°C. In addition, there are the stormy winds of 15 m s^{-1} on the average plus the sparse precipitation of less than 100 mm annually, with a maximum in the summer, mostly in the form of hail, the very high insolation, and the extreme dryness of the air.

Accordingly, the plant cover is poor, somewhat like that of the polar tundra. It is composed overwhelmingly of forbs with large underground storage organs and small above-ground shoots, usually in the form of cushions or rosettes. These are joined by small grasses and especially Cyperaceae, plus mosses and lichens.

Fig. 8.52 gives an overview of the relief and the altitudinal subdivisions.

The flora of the high plateau consists of about fifty species. The flora of the entirety of Tibet, including the relatively species-rich margins, includes about 1000 species (Ward, 1935; Grubov, 1959; Wissmann, 1960/61).

These plants migrated into Tibet's highlands after the Ice Ages, since the entire region was glaciated by ice caps. Nothing is known of the earlier Tertiary flora, since no fossils have been found. At that time the elevations were also quite different. The flora immigrated from the neighboring regions which were not covered by ice. The plants were thus Central Asian desert species, East

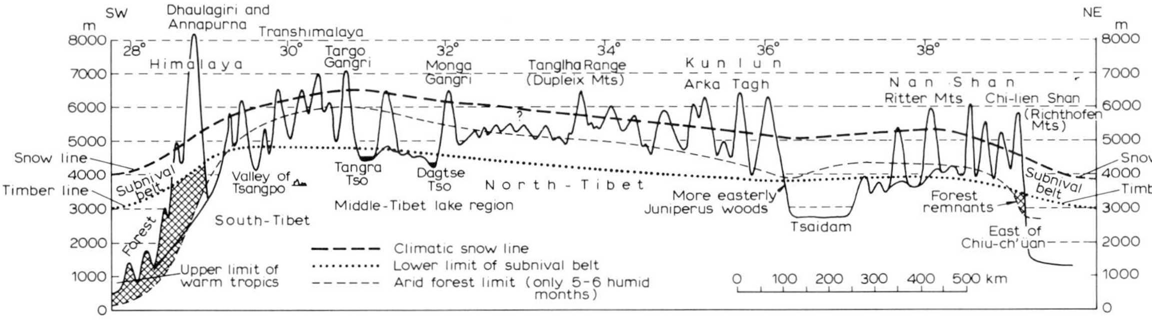

Fig. 8.52. Climate profile through Tibet (with Tsaidam Basin) from SW to NE, exaggerated 7.5 times (from H. von Wissmann, 1961, in Walter, 1974, fig. 268). The tree line in Tibet and Tsaidam is only theoretical, since tree growth is precluded by the extreme dryness.

Asian temperate species, and especially Himalayan mountain species.

The most widespread are the Central Asian geoelements, with a few neo-endemics: *Ceratoides papposa*, *Kochia prostrata*, *Microula tibetica*, *Myricaria prostrata*, *Reaumuria trigyna*, *Rheum spiciforme*, *Thylacospermum rupifragum*, *Chamaerhodos sabulosa*, *Oxytropis aciphylla*, *Thermopsis alpina*, *Astragalus* spp., *Acantholimon diapensioides*, *Tanacetum tibeticum*, *Artemisia wellbyi*, *Ptilagrostis concinna*, *P. purpurea*, *Agropyron arnoldianum*, *Avenastrum* spp., *Festuca* spp., *Elymus junceus*, *E. gerardianus*.

The East Asian geoelements are more important only in the eastern and southeastern mountains. To this group belong species of the Ranunculaceae, as well as *Gentiana*, *Pleurogyne*, *Swertia*, *Pedicularis*, *Meconopsis*, *Pegaeophyton scapiflorum*, *Primula*, *Androsace*, *Rhododendron*, *Corydalis*, *Saxifraga*, *Rhodiola*, *Cremanthodium*, *Ligularia*, *Saussurea*, and others.

The number of holarctic elements is small, but these show dispersal on a large scale: species of the genera *Carex*, *Eleocharis*, *Eriophorum*, *Juncus*, and *Kobresia* form extensive swards and marshes around lakes and in lowlands between mountain chains.

The eastern part, which is separated from the rest of Tibet by the Tsaidam Basin, is very different from the main area. It is known as Tsinghai (Koko Nor Lake) and includes the regions *25* (forested Nan Shan) and *26* (the montane steppe-covered plain of Chamdo) in Fig. 8.7. This region has a less extreme climate: the mean temperature of the warmest month is 10 to 15°C (17°C), the absolute minimum −26°C, and the absolute maximum 34°C. Precipitation amounts are 150 to 380 mm, and the stormy winds are absent.

The alpine meadows contain species of the genera mentioned above plus species of *Anemone*, *Trollius*, *Paraquilegia*, *Crepis* and *Leontopodium*, among others.

The subalpine tall forbs are species of the genera *Polygonum*, *Aconitum*, *Delphinium*, *Rheum*, *Ligularia*, etc. Endemics of Himalayan affinity include *Przewalskia tangutica*, *Pomatosace filicula*. *Nannoglottis carpesioides*, *Xanthopappus*, and *Littledalea racemosa*. Of course species of *Astragalus*, *Oxytropis*, *Allium*, and *Potentilla* (some endemic) suggest Central Asian affinities. The number of shrub species is relatively large: *Rhododendron* (4), *Acer* (2), *Philadelphus* (1), *Hydrangea* (1), *Sorbus* (2), *Prunus* (3–4), *Sorbaria* (1), *Lonicera* (6), etc. These also continue as undergrowth into the spruce and mixed forests.

The richness in species can be attributed to the only slight glaciation of the region during the Pleistocene. Compared with the flora of the mountains further to the east, however, the flora is still very impoverished (over 200 *Rhododendron* species in Eastern China and the Himalayas).

Region *25* of Fig. 8.7 (the Nan Shan) represents a system of parallel mountain chains running west-southwest to east-northeast with a mean elevation of 4000 m and valleys which go down to 2000 to 2500 m. The entire region has no external runoff. The eastern part is lower and moister, and the north slopes are covered with *Picea asperata* forests; *Pinus tabulaeformis* occurs less often, and in clearings also *Betula japonica*, *B. alba* var. *sinensis*, and *Populus davidiana* (*P. tremula* var.). At lower elevations the following appear in the shrub or understory–tree stratum: *Acer davidii*, *A. tetramerum*, *Corylus mandshurica*, *Philadelphus kansuensis*, *Daphne girardii*, *Rhododendron agglutinatum*, *R. anthopogonoides*, four *Lonicera* species, four *Prunus* species, three *Ribes* species, a number of *Salix* species, *Spiraea*, *Rosa*, *Euonymus*, *Berberis*, and others. This is a foreign flora, pointing more toward East Asia. Still further to the east there occur purely East Asian, species-rich forests with *Picea purpurea*, *Abies faxoniana*, *Arundinaria*, and many other species. The spruce forests remind one of the European forests but are much richer in ferns.

Around the high lake Koko Nor (3200 m) in the southeast there extend *Artemisia–Festuca–Stipa* steppes with *Artemisia desertorum*, *A. macrocephala*, *Festuca rubra*, *Stipa capillata* s.l., *S. sibirica*, *Poa bulbosa*, *Astragalus* spp., *Oxytropis* spp., etc.

The southern half of Tsinghai, a fairly uniform, high-mountain level steppe, drops off gradually from 4000 m to 3000 m toward the east. It is drained by the Huang Ho and is drier. The mountain swards are composed of *Kobresia*, *Carex*, *Festuca*, *Deschampsia*, *Poa*, *Avenastrum*, *Stipa*, etc. The most important East Asian–Himalayan forbs were already mentioned. There are no vegetation studies published for Tsinghai.

The next region (region *27*, Fig. 8.7), the region of the deep river valleys, is a high plateau which falls off slightly to the east (mean elevation 3800 m). The Huang Ho, Ya-lung, Yangtze Kiang, Mekong, and Salween have all cut into the plateau with their upper courses. The mean annual temperature here is 0 to −4°C; precipitation climbs from 100 mm in the west to 400 mm in the east. This is a region of high-mountain meadow steppes dominated by *Kobresia royleana* and *Carex moorcroftii*. Grasses (*Festuca*, *Ptilagrostis*, *Poa*, *Agropyron*) are also well represented. *Polygonum sphaerostachyum* and *Poa vivipara* form stands on moist sites. Common also are cushion-plants (*Thylacospermum rupifragum*, *Stellaria decumbens*, *Androsace tapete*, *Arenaria musciformis*), hemicryptophytes, and geophytes (*Astragalus*, *Oxytropis*, *Pedicularis*, *Gentiana*, *Primula*, *Rheum*, *Stellera*, *Meconopsis*, *Allium*, *Fritillaria*, etc.). *Eleocharis*, *Eriophorum*, and *Carex* grow in marshes and bogs.

In the western part the flora is almost as impoverished as in the high plateau of Tibet. Low *Saussurea*, *Cirsium*, *Stipa*, etc. grow on rubble and in crevices; in ravines one can find *Phlomis rotata*, *Pedicularis rupicata*, and *Meconopsis integrifolia* (Walter, 1968, fig. 576, p. 827).

In the east, on the other hand, there appear meter-tall but mostly cushion-shaped or rounded shrubs with small leaves: *Juniperus squamata*, *J. tibetica*, *J. semiglobosa*, *J. przewalskii*, *J. convallium*, with *Rhododendron fastigiatum*, *Salix biondiana*, *S. variegata*, *Dasiphora fruticosa* s.l., *Caragana jubata*, *C. tibetica*, *Spiraea*, and *Cotoneaster*. *Hippophaë* spreads across rubble slopes.

Southern Tibet (region *28*, Fig. 8.7) is the large, hilly longitudinal valley between the highest mountains of the Himalaya and those of the Trans-Himalaya. It arose tectonically, falls off gradually to the east from 4500 m to 3500 m, and is drained by the upper courses of the Brahmaputra, the Indus, and the tributaries of the Ganges. This part of Tibet is the warmest but is still dry, even though precipitation can reach 500 mm (Lhasa: 437 mm and mean annual temperature 9.8°C). The climate has a very distinct summer monsoon pattern, and winters are dry and mild, so tree plantations are possible around Lhasa (*Populus alba*, *P. cathayana*, *Ulmus pumila*, *Corylus heterophylla*, *Prunus ansu*). The main vegetation types, however, are rubble communities with *Oxytropis aciphylla*, *Astragalus tibeticus*, *Thermopsis alpina*, *Artemisia salsoloides*, *Stellera chamaejasme*, and *Taraxacum* spp. Xerophilic shrubs appear in the east, such as *Sophora viciifolia*, *Lonicera spinosa*, *Berberis* spp., *Ceratostigma griffithii*, and *Buddleia tibetica*. In river valleys are found *Buddleia lindleyana* plus *Elaeagnus pungens* and *Hippophaë rhamnoides*, also *Caragana* spp., *Rosa sericea*, *Cotoneaster acutifolia*, and *Dasiphora fruticosa* s.l. Typical of the upper mountain slopes are stands of *Juniperus pseudosabina*, for alluvial soils of river valleys communities of *Imperata cylindrica* and *Stipa inebrians*. East of Lhasa *Sophora moorcroftiana* grows on mobile dunes. Halophytes (*Salsola* spp., etc.) occupy the shores of the lakes at the north base of the Himalaya, since these lakes partially dry out. Agriculture is possible in this part of Tibet, and for this reason it is the most densely populated region.

The situation further to the west is quite different and includes badlands with canyons up to 600 m deep. Still further west, along the upper courses of the Indus and Sutlej Rivers, one again finds meadow-steppes and even *Salix* and *Populus*. This region has not been studied botanically.

The largest part of High Tibet (region *29*, Fig. 8.7) is a high plateau (5000 m) divided by ridges, 300 to 400 m high, which fan out toward the east. The entire region has no external runoff except for the small section which is drained by the Indus. Potential evaporation is low, but many salt lakes and salty marsh areas have formed, in contrast to the better-drained regions *27* and *28* which contain only a few freshwater lakes. The relief is formed by glaciation and frost-weathering. Rocky slopes and rubble-strewn valleys dominate the landscape. Soils are shallow and mostly brackish.

Herbaceous cushion-plants are very important, including *Artemisia wellbyi*, *Astragalus malcolmii*, *A. arnoldii*, *Acantholimon diapensioides*, *Thylacospermum rupifragum*, *Saussurea tridactyla*, *S. wellbyi*, and *Thermopsis alpina*. Also important are dense bunch-grasses (*Ptilagrostis concinna* and *purpurea*) and dwarf-shrubs (*Ceratoides papposa*, *Tanacetum tibeticum*, *Ephedra distachya*, *E. gerardiana*, *Myricaria prostrata*, and *Reaumuria trigyna*).

The plant cover is closed only in lower areas along rivers and lake shores. This consists of swards of *Kobresia bellardii* and *Carex moorcroftii*, with *Juncus thomsonii*, *Triglochin palustre*, *Eleocharis* spp., *Ranunculus tricuspis*, etc. The only larger woody plant is *Juniperus squamata* (closely related to *J. pseudosabina*), except for *Myricaria elegans* in the river valleys (Walter, 1968, fig. 581. p. 830).

Especially typical are the hill or *naka* moors on north slopes (Fig. 8.53). These are composed of *Kobresia tibetica* and are so hard that the camels injure not only their lips but also their feet. The vegetation is, in general, so sparse that it is not sufficient for the nomads' herds, only for the wild herds of antelopes, kyang, and yaks.

The climate diagrams in Fig. 8.54 show how different the climates of the different regions are in their details.

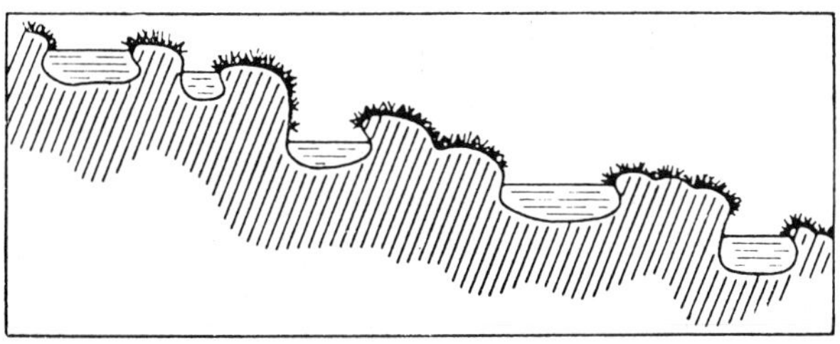

Fig. 8.53. *Naka* moors on a north slope, composed of low dams and pools resulting from solifluction. The pools dry out and/or freeze periodically (from A. Tafel, in Walter, 1968, fig. 582).

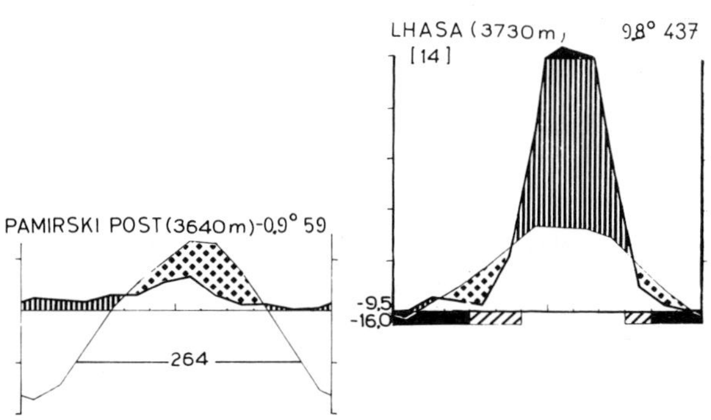

Fig. 8.54. Climate diagrams of Pamirski Post or Murgab (with a climate similar to that in the desert Chang Tang of western Tibet, where there are no stations) and Lhasa in southern Tibet, which is under the influence of the Indian monsoon and receives summer rainfall.

REFERENCES

Anonymous, 1975. *Resources of the Biosphere; Synthesis of the Soviet Studies for the IBP, I*. Nauka, Leningrad, 288 pp. (in Russian).

Grubov, V.I., 1955. *Conspectus of the Flora of the Mongolian People's Republic*. Nauka, Moscow, 302 pp. (in Russian).

Grubov, V.I., 1959. *Attempted Botanical-Geographical Classification of Central Asia*. Akad. Nauk S.S.S.R., Leningrad, 77 pp. (in Russian).

Grubov, V.I., 1963. *Plants of Central Asia, I*. Nauka, Moscow, 168 pp. (in Russian).

Grubov, V.I., 1972. The relation between the floras of Central Asia and Eastern Asia. *Akad. Nauk, S.S.S.R., Komarov, Lektsii. (Leningrad)*, 24: 86–93 (in Russian).

Hanelt, P., 1970. Vorkommen und Vergesellschaftung von *Nanophyton erinaceum* (Pall.) Bge. in der Mongolischen Volksrepublik. *Arch. Natursch. Landschaftsforsch.*, 10: 19–40.

Helmecke, K. and Schamsran, Z., 1979a. Ergebnisse ökologischer Untersuchungen in der Gobi der Mongolischen Volksrepublik. 1. Untersuchungsgebiet, Vegetationseinheiten und Ergebnisse der mikroklimatischen Untersuchungen. *Arch. Natursch. Landschaftsforsch.*, 19: 1–22.

Helmecke, K. and Schamsran, Z., 1979b. Ergebnisse ökologischer Untersuchungen in der Gobi der Mongolischen Volksrepublik. 2. Untersuchungen über die osmotischen Verhältnisse ausgewählter Pflanzenarten. *Arch. Natursch. Landschaftsforsch.*, 19: 81–95.

Hilbig, W. and Schamsran, Z., 1977. Beitrag zur Kenntnis der Vegetation in Chovd Aimak (Mongolische Volksrepublik). *Arch. Natursch. Landschaftsforsch.*, 17: 35–82.

Ismailova, N.N., 1963. Water and temperature conditions in the soil and transpiration of plants in the alpine belt of Pamir. *Tr. Pamirsk Biol. Stantsii*, 1: 172–180 (in Russian).

Ismailova, N.N., 1965. Transpiration intensity of high-mountain plants in relation to the altitude of their biotopes. *Probl. Bot.*, 7: 205–212 (in Russian).

Ismailova, N.N., 1974. Transpiration and water loss of the plant communities in Eastern Pamir. *Ekologiya*, 1: 95–97 (in Russian).

Ismailova, N.N. and Litvinova, N.P., 1969. On the productivity of green matter and the water expenditure for transpiration in some wormwood communities (Artemisieta) of Eastern Pamir. *Bot. Zh.*, 54: 1586–1590 (in Russian).

Ismailova, N.N. and Litvinova, N.P., 1972. Water regime and productivity of main dominant plants of the high-mountain deserts (Eastern Pamir). In: *Int. Symp. U.S.S.R.: Eco-Physiological Foundations of Ecosystem Productivity in Arid Zones*. Nauka, Leningrad, pp. 76–79.

Kalinina, A.V., 1954. Stationary investigation of the pastures in the Mongolian People's Republic. *Tr. Mongolsk. Komm.*, 60: 128 (in Russian).

Karamysheva, S.V. and Banzragh, D., 1976. Vegetation of the mountain range Chan-Chuchijn Ula and of the southern part of the Uvs Nuur depression. In: *Biological Resources and Natural Conditions of the Mongolian People's Republic*, 8. Nauka, Leningrad, pp. 99–124 (in Russian).

Klement, O., 1966. Flechten aus der Mongolischen Volksrepublik. *Feddes Repert.*, 72: 98–123.

Lavrenko, E.M., 1957. Vegetation of the Gobi Desert in the Mongolian People's Republic and its relation to contemporary geological processes. *Bot. Zh.*, 42: 1361–1382 (in Russian).

Lavrenko, E.M., 1965. Subdivision into provinces of the Central Asian and Irano-Turanian subregions of the Afro-Asian desert region. *Bot. Zh.*, 50: 3–15 (in Russian).

Lavrenko, E.M., 1966. Botanical geological observations in the deserts of the Kansu Corridor and on the north slope of the Nan Shan Mountains. *Bot. Zh.*, 51: 1816–1823 (in Russian).

Lavrenko, E.M. and Nikolskaya, N.I., 1965. The distribution of some western *Stipa* species in the Mongolian Altai, in Dzungaria, and in the eastern Tien Shan. *Bot. Zh.*, 50: 1419–1429 (in Russian).

Lavrenko, E.M. and Soczava, V.B., 1956. *Descriptio Vegetationis U.R.S.S.* Akad. Nauk S.S.S.R., Moscow, 971 pp. (with map in 8 sheets).

Lavrenko, E.M. and Sveshnikova, V.M., 1965. Synthetic research on growth-forms using the example of *Stipa* tussock grasses. *Zh. Obshch. Biol.*, 26: 261–275 (in Russian).

Lavrenko, E.M. and Yunatov, A.A., 1960. Natural oases in the Transaltai-Gobi (Mongolian People's Republic). In: *Questions of Evolution, Biogeography, Genetics, and Selection.* Akad. Nauk S.S.S.R., Moscow, pp. 125–136 (in Russian).

Li Hsiao fang, 1979. A discussion of the problem of desertification in China's semi-arid region – Ordos. In: J.R. Goodin and D. K. Northington (Editors), *Arid Land Plant Resources.* Texas Technological University Press, Lubbock, Texas, pp. 678–683.

Miroshnichenko, J.M., 1967. Biological productivity and vertical structure of the plant biomass of some phytocoenoses in the Mongolian People's Republic. *Bot. Zh.*, 52: 1162–1169 (in Russian).

Murzayev, E.M., 1971. The deserts of Dzungaria and the Tarim Basin. In: S.R. Eyre (Editor), *World Vegetation Types.* London, pp. 241–246.

Petrov, M.P., 1966/67. *Deserts of Central Asia.* Nauka, Leningrad, Vol. I, 274 pp.; Vol. II, 288 pp. (in Russian).

Rachkovskaya, E.I. and Volkova, E.A., 1977. The vegetation of the Transaltai-Gobi. In: *Biological Resources and Natural Conditions of the Mongolian People's Republic*, 7. Nauka, Leningrad, pp. 46–74 (in Russian).

Rodin, L.E., 1975. Productivity of desert communities. *Resursy Biosfery*, 1: 128–165 (in Russian).

Rodin, L.E., 1976. Primary productivity of desert communities in North Africa and Asia. In: *Problems of Desert Development*, pp. 55–64 (in Russian).

Rodin, L.E. (Editor), 1977. *Productivity of Vegetation in the Arid Zone of Asia: Synthesis of the Soviet Studies for the International Biological Program 1965–1974.* Akad. Nauk S.S.S.R., Nauka, Leningrad, 232 pp. (in Russian).

Schubert, R. and Klement, C., 1971. Beitrag zur Flechtenflora der Mongolischen Volksrepublik. *Fedd. Repert.*, 82: 187–262.

Skalon, I.S., 1977. Species composition and peculiarities of the development of the microflora of plant communities of zonal stations in the MPR. In: *Biological Resources and Natural Conditions of the Mongolian People's Republic*, 7. Nauka, Leningrad, pp. 157–192 (in Russian).

Stubbe, M. and Chotolchu, N., 1968. Zur Säugetierfauna der Mongolei. *Mitt. Zool. Mus. Berlin*, 44: 5–121.

Walter, H., 1968. *Die Vegetation der Erde, II. Die gemässigten und arktischen Zonen.* Gustav Fischer-Verlag, Jena, Stuttgart, 1001 pp.

Walter, H., 1974. *Die Vegetation Osteuropas, Nord- und Zentralasiens. Vegetationsmonographien der einzelnen Großräume, Band VII.* Gustav Fischer-Verlag, Jena, Stuttgart, 452 pp.

Ward, F.K., 1935. A sketch of the geography and botany of Tibet, *J. Linn. Soc. Lond.*, 50: 239–265.

Wissmann, H., 1960/61. Stufen und Gürtel der Vegetation und des Klimas in Hochasien und seinen Randgebieten. *Erdkunde*, 14: 249–272; 15: 19–44.

Yevstifeyev, J.G. and Rachkovskaya, E.I., 1976. To the question of correlation between soil and vegetation cover in the southern part of the MPR. In: *Biological Resources and Natural Conditions of the Mongolian People's Republic.* 8. Nauka, Leningrad, pp. 125–144 (in Russian).

Yunatov, A.A., 1950. Main features of the plant cover of the Mongolian Republic. *Tr. Mongolsk. Komm.*, 39: 223 pp. (in Russian).

Yunatov, A.A., 1954. Food plants of pastures and meadows of the Mongolian People's Republic. *Tr. Mongolsk. Komm.*, 356 pp. (in Russian).

Zevegmid, D., Stubbe, M. and Dawaa, N., 1974. Das neue mongolische Jagdgesetz vom 6. Januar 1972, die Naturschutzgebiete und Wirbeltierarten der MVR. *Arch. Natursch. Landschaftsforsch.*, 14: 3–36.

NOTE

After completion of the manuscript, a very detailed vegetation map of the Mongolian People's Republic (Outer Mongolia) was published in colour at a scale of 1:1 500 000 (in four sheets measuring 70 × 90 cm). It was prepared by A.A. Yunatov, with B. Dashnyan and A.A. Helbich: *Karta Rastitel'nosti Mongol'skoy Narodnoy Respubliki.* (Editor E.M. Lavrenko) Bot. Inst. Acad. Sci. U.S.S.R., Leningrad, 1979.

Subsidiary maps (at a scale of 1:10 000 000) cover climate, soils, zonal vegetation, distribution of certain plant species, and relief. On the main map, 104 cover types are distinguished: 8 of alpine and sub-alpine vegetation, 5 of boreal coniferous forest, 4 of steppe woodlands, 3 of shrublands, 14 of grass steppes, 12 of dry steppes, 16 of desert-steppes, 22 of deserts, 14 of wetlands, and 6 of glaciers, lakes, rocks, etc.

Chapter 9

THE PAMIR — AN ECOLOGICALLY WELL-STUDIED HIGH-MOUNTAIN DESERT BIOME

H. WALTER and E.O. BOX

GENERAL SUBDIVISION

The region of Pamir connects with Tibet in the east. It has been studied in great detail for several decades from the Pamir Biological Station at 3864 m, about 220 m above Murgab. We can thus now return to this biome and discuss it in more detail. We refer to the presentations by Walter (1968, pp. 809–826; 1974, pp. 320–327) and supplement them with the results of more recent investigations. A color map of Pamir (1:3 000 000) is included in the geographical series *Soviet Union*, volume Tadzhikistan (Chumichev, 1968).

High-mountain ranges radiate from the high plateau of Pamir in all directions. Pamir forms the junction of the Hindu Kush from the southwest, the Alay system and adjoining Tien Shan from the north, the Kun Lun ridges from the east, and the Karakoram from the southeast, which connects with the Himalaya. The highland of Pamir extends outside the Soviet Union and consists of two parts (Agakhanyantz, 1964; Agakhanyantz and Yusufbekov, 1975; Stanyukovich, 1973; see also Fig. 9.1):

(1) Eastern Pamir, the more plateau-like part, with elevations between 3500 and 4000 m and with wide, shallow valleys, which are filled with alluvial deposits and separated from each other by mountain chains. Some especially high ranges (over 7000 m) occupy a more marginal position. These hold the moist air-masses away from the interior, which despite its elevation is quite dry. The meandering rivers flow very slowly and become stagnant in places or are accompanied by meadows. The underground rivers are very interesting. These flow through rubble masses and layers, come to the surface when blocked by obstacles, and then sink quickly into the rubble again. There are numerous lakes, some formed by landslides.

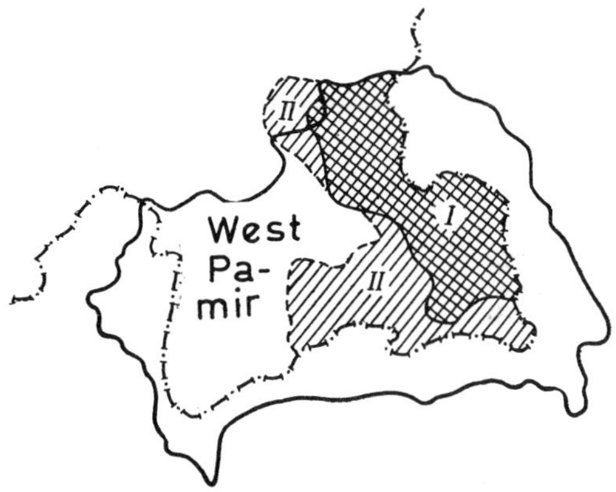

Fig. 9.1. Pamir (dark outline) and the portion within the U.S.S.R. (–·–·–), with its subdivisions: *I* = eastern Pamir (65% Central Asian flora elements), white = western Pamir of Badakhshan (primarily Mediterranean–Turanian flora elements); *II* = transition region (30% Central Asian flora elements) (from Ikonnikov, in Agakhanyantz, 1964). The regions outside the U.S.S.R. (white) are in the east and south. Kashgaria and northernmost Afghanistan are high, inaccessible mountains.

(2) Western Pamir (Badakhshan), a wild mountainous land with high ridges over 5000 to 6000 m and deep valleys reaching down to below 2500 m. This is the less well-known part, reaching over into Afghanistan. Precipitation falls primarily in the winter and spring. The snow line in the region is very high because of the aridity, indeed somewhat higher as the precipitation decreases. Where 500 mm of precipitation falls, the snow line can be found at about 4000 m; where 100 mm falls, it is around 5500 m. At only 40 mm of precipitation there is no snow line because the snow evaporates

[1] Manuscript completed November, 1977.

directly under the combined influence of dry air and high insolation.

CLIMATE

The meteorological measurements of the station at Murgab yielded, for 1941–1951, an average annual precipitation of 103 mm (maximum 145 mm, minimum 40 mm). Its distribution over the course of the year is shown in Table 9.1. The summer months May–August receive the most rain. The climate diagram (see Fig. 8.54) from Murgab shows the desert character of the climate clearly, while the diagram from the Fedchenko Glacier (Fig. 9.2) shows the climate of the surrounding mountains.

Precipitation falls in small events (half in amounts between 0.2 and 0.5 mm), and most of it falls as sleet or snow, seldom as rain. Only on August 8, 1952, did as much as 22 mm fall on one day. Such heavy rainfall occurs probably once every 15 to 20 years, but it is very significant for the vegetation. The light precipitation has no effect. In the winter the monthly precipitation seldom exceeds 2 to 3 mm. The rocky skeletal soils are thus never wetted below a depth of 1 m. The individual precipitation amounts for one year are shown in Table 9.2.

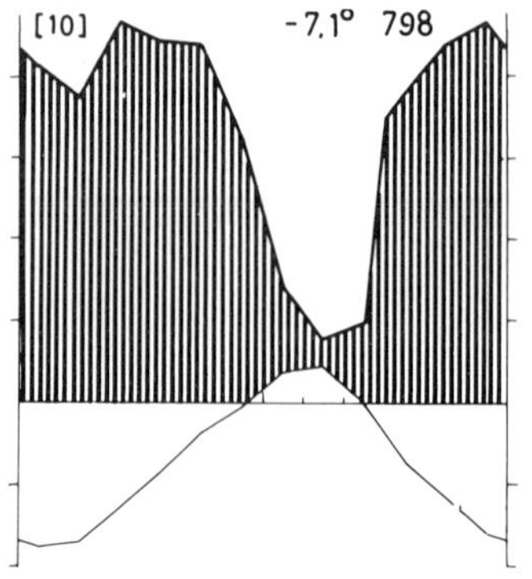

Fig. 9.2. Climate diagram of Fedchenko Glacier situated north of Pamir (from Breckle, 1973).

The relative humidity in Pamir is extremely low, in winter 50 to 70% and in summer around 20% as monthly averages. Because of the extreme elevations and the dry air, the insolation during the day and the outgoing radiation at night are extremely high. Solar radiation is 90% of the solar constant. The mean cloudiness is 39%, as opposed to 65% in the Alps at 3100 m. There are an average of only 45 gray days in the year but an average of 130 cloudless days. The total hours of sunshine for the year are very high (around 3000 h), but because of the elevation the active temperature sum for the

TABLE 9.1

Distribution of precipitation in spring and summer (mm) at the Pamir Biological Station (3864 m elevation) (from Sveshnikova, 1962)

Year	April	May	June	July	August	September	Total May–Sept.
1939	–	0	0	18.0	36.6	0	65.4
1940	–	0	23.6	32.7	0	0	56.3
1941	7.5	0	42.0	28.1	0	5.9	83.5
1942	6.0	30.0	73.8	4.0	2.5	0	116.3
1943	15.3	36.4	10.6	12.0	22.8	8.7	105.3
1944	9.4	20.2	8.6	0	24.1	4.3	66.6
1945	18.1	15.0	19.9	3.4	23.5	6.0	85.9
1946	0.0	16.2	0.6	0	0	15.6	32.4
1947	7.0	3.6	10.5	35.2	4.5	0	60.8
1948	0.6	20.3	29.7	12.6	19.4	0.1	82.7
1949	5.0	6.6	30.6	2.0	26.7	0	70.9
1950	18.2	19.5	8.1	0	20.7	18.5	85.0
1951	–	17.5	4.8	0.8	6.0	0.2	29.1
1952	0.1	13.7	2.8	24.5	59.5	1.7	102.3
1953	–	40.7	36.8	19.4	11.0	24.0	131.7
1954	–	0.3	0.3	32.7	29.8	8.2	71.5
1955	–	37.8	6.9	6.2	0	0	51.0

TABLE 9.2

Precipitation amounts during the period September 1, 1951 to August 31, 1952 at the Pamir Biological Station (from Sveshnikova, 1962)

Date	Amount (mm)	Date	Amount (mm)	Date	Amount (mm)	Date	Amount (mm)
Sept. 10	2.0	Feb. 18	0.9	May 28	5.7	July 20	4.6
Sept. 11	3.1	March 4	2.0	May 29	3.2	July 26	0.5
Sept. 24	0.9	March 6	0.2	May 31	4.3	July 27	0.5
Oct. 10	0.1	March 8	9.5	June 1	2.8	July 30	4.8
Oct. 26	0.2	March 13	0.3	July 4	0.5	July 31	0.9
Dec. 26	0.1	March 27	0.2	July 15	1.0	Aug. 8	22.2
Jan. 13	0.3	April 8	0.1	July 16	2.8	Aug. 23	2.5
Feb. 4	0.4	April 27	0.1	July 18	1.6	Aug. 24	5.7
Feb. 10	0.8	May 18	0.2	July 19	2.8	Aug. 25	3.0
						Aug. 26	5.0

year is not over 1500°C, and there are areas where lasting temperatures over 10°C are completely lacking. The mean January temperature is −20°C or lower and only as high as −16°C in specially favored sites. The absolute minimum is around −40° to −45°C. The mean July temperature is 8 to 12°C and the absolute maximum around 25 to 30°C.

The evaporation from the surface can exceed 5 mm per day on wet areas in the summer. From a free water surface evaporation should be about 8 mm per day. The annual potential evaporation could be estimated roughly as over 1000 mm, but there are no measurements. The mean annual temperature varies between −1.0° and −2.8°C, but air temperatures at 2 m above the ground in summer reach 20 to 22°C, about 28°C near the surface, and 52°C on the surface. The temperature falls below freezing at night, and only ten to thirty nights during the year are frost-free. The thermal conditions at the soil surface are reminiscent of those in the high Alps. The temperature extremes are ameliorated perhaps by the very strong wind, but diurnal extremes of 68°C have been measured. The average afternoon wind speed is around 15 m s^{-1}.

The temperature variation in the upper soil layers can be 25 to 30° daily and over 100° in the course of a year (Fig. 9.3). Even at a depth of 1 m the temperature varies between 13.0° and −8.5°C. The dry soil is seldom frozen at a depth of 4 m. Permafrost occurs thus only locally, in valley sites or cold hollows. The mean monthly maxima and minima in the upper soil layers are shown in Table 9.3.

As a result of the arid conditions there arise on the broad alluvial rubble flats peculiar gray-brown, very humus-poor skeletal soils, which are closely related to the sierozems of the desert. They tend toward soda-salinization.

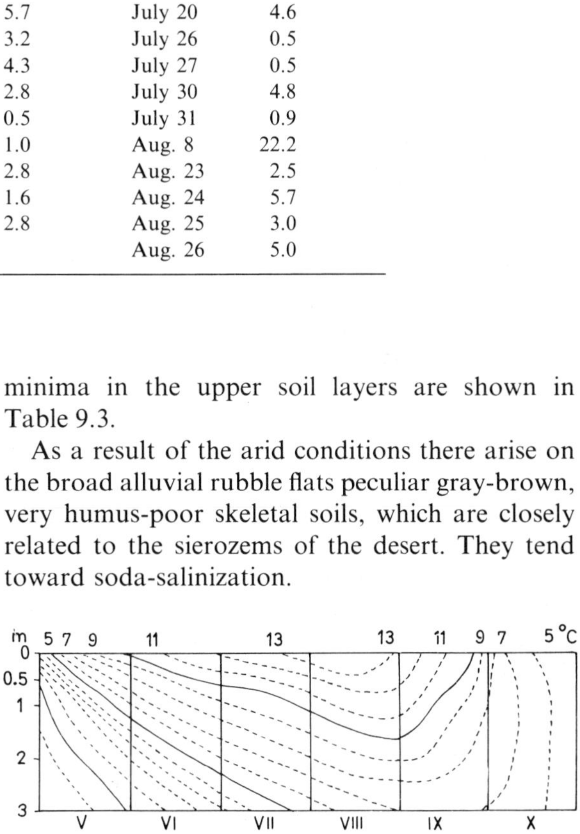

Fig. 9.3. Soil temperatures (chrono-isopleths) near the Pamir Biological Station in the summer of 1940 (from Sveshnikova, 1962).

THE VEGETATION

Eastern Pamir[1]

The vegetation of the Pamir region varies in certain respects. In the southern part of Pamir the vegetation has the character of alpine dry mea-

[1] Plant names are those given by Ikonnikov (1963). Only for *Eurotia ceratoides* is the newer name used: *Ceratoides papposa*. The book covers the flora of eastern Pamir, with its 251 genera and 636 species.

TABLE 9.3

Monthly soil temperature (°C) extremes to a depth of 80 cm at the Pamir Biological Station (from Sveshnikova, 1962)

Depth (cm)	December		January		February		March		April	
	max.	min.	max.	min.	max.	min.	max.	min.	max.	min.
0	+11.9	−30.5	−2.3	−42.7	+7.8	−45.7	+30.5	−39.4	+37.1	−18.2
20	−2.7	−17.3	−11.3	−29.4	−8.4	−28.3	+2.1	−14.9	+11.4	−4.3
40	−3.0	−10.1	−7.9	−22.1	−9.9	−20.7	−1.1	−11.3	+9.1	–
80	−0.9	−4.5	−3.7	−12.8	−8.1	−13.3	−3.8	−9.0	+3.4	−3.6

Depth (cm)	May		June		July		August	
	max.	min.	max.	min.	max.	min.	max.	min.
0	+35.4	−11.0	+50.0	−4.3	+53.3	−0.6	+56.8	−4.2
20	+16.9	−2.3	+22.4	+7.6	+24.8	+10.1	+24.3	+10.1
40	+13.0	+2.4	+19.0	+9.4	+21.3	+10.4	+20.2	+14.1
80	+6.9	+2.7	+12.3	+7.0	+14.9	+12.1	+16.3	+13.1

dows, with *Festuca sulcata* and *Poa relaxa* as well as alpine forbs. Forb-poor *Stipa* steppes prevail in the central part (Fig. 9.4). The eastern part, around the biological station, is still drier and is dominated by *Ceratoides papposa*, *Artemisia rhodantha* (*skorniakovii*) or *Xylanthemum* (*Tanacetum*) *pamiricum*. Meadows occur along rivers, with shrubs also at lower elevations. Of the 150 to 200 plant species which occur, no fewer than 100 are restricted to moist sites. One seldom finds more than five to seven species on the dry sites. These are 10 to 15 cm tall dwarf-shrubs, which reach a cover of 10% at most. Photographs of the vegetation can be found in Walter (1968, figs. 567–573) and could not be reproduced here (see also the photographs in Ikonnikov, 1963).

Fig. 9.4. High-mountain steppe with *Stipa kirghisorum* in the Alay Valley between the Alay and Trans-Alay Ranges (northern boundary of Pamir) at 3500 m (photo by O.V. Zalenski and E.M. Lavrenko).

In contrast to his earlier work (1949), Stanyukovich distinguished only two altitudinal belts in eastern Pamir in his latest contribution (1973). These were called subalpine and alpine: (1) 3500–4200 m: subalpine hemicryophytes; (2) 4200–4700 m: alpine eucryophytes. Above this level lies the nival belt, at least as far as one can distinguish or speak of it when there is no distinct snow line. This belt is almost completely without vegetation.

The vegetation of each belt is xerophilic and can sometimes be called desert-steppe. Mesophilic belts disappear completely in the Pamir and adjoining Tibet.

(1) **Lower belt**: This is dominated by *Ceratoides* desert (Fig. 9.5), often with *Artemisia rhodantha*, less frequently with grasses or cushion-plants, with meadows only directly next to water; *Xylanthemum* (*Tanacetum*) *pamiricum* prevails on sandy flats at the base of rubble slides; *Artemisia rutifolia* is scattered on steep rock walls.

(2) **Upper belt**: The precipitation is roughly

Fig. 9.5. Desert-steppe in Pamir with *Stipa glareosa* and *Ceratoides papposa*, roughly 4000 m above sea level (photo by I.A. Raykova).

double that of the lower belt, especially in the south. The eucryophilic *Ajania tibetica* (= *Tanacetum xylorrhizum*) dominates, accompanied by *Dracocephalum paulsenii*; cushion-plants appear as precipitation increases and, in the more moist southern part, also *Poa litwinowiana* and *Festuca sulcata*. The *Kobresia* swards cover large areas.

The only woody plants in all of eastern Pamir are found in the lowest-lying river valleys. These are *Betula turkestanica* and *Salix* spp., plus *Berberis kaschgarica* and *Hippophaë rhamnoides*.

At the higher elevations the temperature conditions for plants are best near the ground surface. Therefore, the plants do not rise very far above the surface and their roots spread out at relatively shallow depths — though not as shallow as in the Alps, because of the dryness.

The low, padded cushion-plants (German: *Kissenpolsterpflanzen*) are important in Pamir, as in all windy regions (for example, Patagonia, where there are almost no plants except cushion-plants — see Chapter 17). The hard cushions easily absorb the rain water which falls on them and store up to five to six times their own weight in water (e.g. *Sibbaldia tetrandra*). The water is protected from evaporation and is absorbed by adventitious roots inside the cushion. The temperature conditions within the cushion are also ameliorated. Growth of the cushion is very slow, for *Acantholimon diapensioides* only 2 to 5 mm annually, for *Sibbaldia tetrandra* only 3 to 4 mm in height but 16.5 mm radially, on the average.

Herbaceous cushions can be composed of many individual plants which grow densely together and form the cushion. The seeds of older plants find good conditions for germination between dead branches of the cushion and around the periphery. A cushion with a diameter of 1 m can consist of up to 600 individual plants of different ages. The young shoots of *Sibbaldia tetrandra* grow plagiotropically on the ground surface and force the older branches upward, continuously producing a greater height and curvature (Steshenko, 1973).

One can distinguish several types:

1. Hard cushions with a single tap-root:
Gypsophila capituliflora, flat cushions 10 to 15 cm in diameter
Acantholimon hedini, convex cushions 15 to 25 cm high and up to 70 cm in diameter
Potentilla pamiroalaica, very spread-out, flat cushions with the same form as *P. pamirica* and *Plantago arachnoidea*.

2. Solid cushions with rooted lateral branches:
Acantholimon diapensioides, very drought-resistant, flat cushions 20 to 50 cm in diameter; grows centrifugally and forms a "witches' ring", in which the middle is dead
Sibbaldia tetrandra, up to 1 m high and over 1 m in diameter, often wind-sheared (dead on the windward side); prefers moister climates with snow deposits at elevations of 4400 to 4600 m.

3. Loose cushions without adventitious roots (very common in Pamir):
Oxytropis poncinsii, 25 to 30 cm high, 50 to 60 cm in diameter; on sandy soils also three additional *Oxytropis* species, *Hedysarum cephalotes*, *Astragalus myriophyllus*, *Saussurea gnaphalodes*, *Braya pamirica*, *Erysimum humillimum*, *Draba korschinskyi*, *Smelovskia calycina*, and *Macrotomia euchroma*.

4. Loose cushions with adventitious roots:
Artemisia viridis and *Androsace akbaitalensis*

5. Air-pillow plants without adventitious roots:
Acantholimon pamiricum, hemispherical thorn-cushion, with radial growth of 1.5 to 2 cm annually

According to Stanyukovich and Krivonogova-Stanyukovich (1973), six *Acantholimon* species occur in Pamir, of which only three vicarious species are able to dominate:

Acantholimon pamiricum forms 10 to 15 cm high cushions (10–120 cm diameter, usually 30–40 cm), covering 20 to 30% (7–60%) of the ground, at middle elevations of 3300 to 3600 m (limits: 2300–4300 m).
Acantholimon parviflorum prefers the lower elevations (2300–3500 m), is 10 cm tall (diameter 25–60 cm), and covers 15 to 20% of the ground.
Acantholimon korolkovii is a species of higher elevations (3700–4300 m), is more limited to the west, more mesophilic and less prickly, and prefers rocky soils or rubble.

In addition to these cushion-plants, semi-shrubs and the various forms of rubble plants are common. Annuals, however, are extremely rare.

Deserts with semi-shrubs dominating

The high-mountain deserts of Pamir have a special type of vegetation which occurs completely isolated from the deserts of the lower elevations. They cover by far the greatest area of all the vegetation types in Pamir. The cover is 8 to 15%, and the plant communities are quite poor in species (3 to 7 species in 100 m^2). Semi-shrubs and cushion-plants are the most common life forms. The following communities are found:

Ceratoides papposa communities: subalpine deserts on sand with *Christolea pamirica* or *Stipa orientalis*; with *Artemisia rhodantha* or *Hordeum turkestanicum* on rubble with fine-earth; with *Zygophyllum rosovii* or *Acantholimon diapensioides* on brackish soil; with *Ajania tibetica*, *Acantholimon hedini* or *Oxytropis immersa* in the alpine belt.

If permafrost is shallow, these species combinations are all joined by a higher stratum of *Elymus dasystachys*; if spring is wetter, *Carex duriusculiformis* joins. Additional species are *Stipa glareosa*, *Macrotomia euchroma*, *Crepis flexuosa*, *Poa litwinowiana*, *Potentilla moorcroftii*, *Smelovskia calycina*, *Dracocephalum heterophyllum*, and *Ephedra regeliana*.

Artemisia rhodantha communities: with *Stipa orientalis* and *S. glareosa* or *Ptilagrostis subsessiliflora* on sandy soils. These are joined by the species mentioned above plus *Braya pamirica*, *Roegneria czimganica*, etc.

Less common are communities with two other *Artemisia* species or with *Xylanthemum pamiricum*.

One more species-rich community with 50 to 60% cover is that with numerous cushion-plant species: *Artemisia leucotricha* (2), *Sophiopsis annua* (1), *Stipa orientalis* (1), *S. turkestanica* (+), *Poa relaxa* (+), *Carex duriusculiformis* (+), *Cousinia* sp. (+), *Astragalus* spp. (1), *Acantholimon pamiricum* (1), *Lindefolia pterocarpa* (+), *Arenaria griffithii* (+), *Lactuca tatarica* (+), *Hedysarum cephalotes* (+).

The alpine deserts are found between 4300 and 4700 m in elevation. *Ajania* (*Tanacetum*) *tibetica* communities prevail on all types of soil with cover percentages of 3 to 10%. These also include the constantly accompanying *Parrya excapa*, plus *Oxytropis poncinsii*, *Braya pamirica*, *Pyrethrum pyrethroides*, *Hordeum turkestanicum*, or *Melandrium apetalum* and two *Rhodiola* species in the upper part of the belt.

Desert-steppes, poorer in semi-shrubs

Desert-steppes with very low cover form the transition from the deserts. Grass steppes and forb-rich steppes can be distinguished.

The forb-rich (desert) steppes include the following: (1) the *Christolea pamirica* community on moraines (3–10% cover) with *Oxytropis chiliophylla* and *Stipa glareosa* or with *Braya pamirica*, *Ajania tibetica*, and *Oxytropis poncinsii* (total of 13 species) on sand; and (2) the *Allium polyphyllum* community at 3900–4500 m (often extending over several kilometers) with *Ptilagrostis subsessiliflora* associated and also *Iris loczyi*, *Roegneria schugnanica*, *Oxytropis* spp., and *Potentilla moorcroftii*; on rocky soils of southern exposures also with *Artemisia*, *Dracocephalum*, and *Ajania tibetica*.

The grassy (desert) steppes are best developed on sands in valleys (Fig. 9.5). The dominants, in order of increasing hygrophily, are *Stipa orientalis*, *S. glareosa*, and *Ptilagrostis subsessiliflora*; these are joined by *Ceratoides*, *Artemisia*, *Ajania*, and other species already named. *Halogeton* spp., *Echinopsilon divaricatus* and *Trigonella pamirica* could also be named.

True sub-alpine steppes with 50% cover (remaining under 4200 m) are formed by *Festuca sulcata* with the following composition: *Festuca sulcata* (3), *Hordeum turkestanicum* (1), *Poa litwinowiana* (+), *Puccinellia subspicata* (+), *Potentilla pamiroalaica* (1), *Saussurea leucophylla* (2), *Oxytropis immersa* (1), *Smelovskia* (1), *Potentilla moorcroftii* (1), *Erysimum humillimum* (+).

The steppes prefer north slopes with dark chestnut-soils. The following grasses prevail in the alpine steppes with increasing wetness: *Poa litwinowiana*, *Festuca alaica*, and *Poa relaxa*.

The accompanying species are once again the species already mentioned for the alpine belt, plus *Colpodium leucolepis*, *Androsace*, *Carex pseudofoetida*, *Pyrethrum*, and, with better snow cover, also *Nepeta kokanica*.

With high levels of precipitation the *Artemisia viridis* community also joins the steppes. It has a cover of 40% on loamy soils. The accompanying species are similar.

Cushion-plant communities

These alpine communities have a cover percentage of 30 to 40%. With increasing moisture the following species dominate: *Acantholimon diapensioides* (rubble soils), *Saussurea sorocephala* (sand), *Oxytropis poncinsii* (sand), *O. tianschanica* (sand), *Potentilla pamiroalaica*, *P. pamirica*, *Oxytropis immersa* (rubble), *Sibbaldia tetrandra*.

Only *Saussurea*, *Oxytropis immersa*, and *Sibbaldia* go on up into the nival belt.

Similar cushion-plant communities are known also from the central Tien Shan and from the Caucasus.

Between cushions, and often on the cushions themselves, there grow *Ceratoides*, *Stipa glareosa*, *Carex* spp., *Pedicularis cheilanthifolia*, *Polygonum molliiformis* or *P. mezianum*, *Ajania*, *Elymus*, *Allium*, *Iris*, *Braya*, *Roegneria*, *Smelovskia*, *Macrotomia*, etc.

With greater moistness one can find, at 4200 m (cover 35–40%), the following community: *Sibbaldia tetrandra* (3), *Kobresia stenocarpa* (2), *Saxifraga hirculus* (1), *Leontopodium ochroleucum* (+), *Potentilla pamiroalaica* (+), *Primula nivalis* (2), *Erigeron heterochaeta* (+), *Puccinellia subspicata* (+), *Alopecurus mucronatus* (+).

Alpine meadows

Alpine meadows are almost completely absent in the lower belt and are present in the upper belt only as small patches along brooks or in depressions. Snow valleys are not known. High groundwater is always to be associated with salinization in arid Pamir, and this certainly affects the vegetation.

The following dominant biotopes can be distinguished:

Low soil moisture:
moderately saline — *Carex duriusculiformis*
very saline — *Elymus dasystachys* or *Hordeum turkestanicum*

Moderately moist meadows:
non-saline — *Kobresia stenocarpa*, *Swertia marginata*, *Kobresia capilliformis* or *Calamagrostis anthoxanthoides*
on running water with low salinity — *Carex pseudofoetida*, *Kobresia pamiroalaica*, *Polygonum viviparum*
slightly brackish groundwater — *Puccinellia haeckeliana*, *P. pamirica*

Wet meadows
fed by melting snow — *Carex melanantha* and *Primula nivalis*
near running, slightly brackish water — *Carex orbicularis*
slightly brackish groundwater — *Carex parva*

Halophytes such as *Glaux maritima* and *Triglochin maritimum* occur on brackish soil. Dominant *Carex pseudofoetida* and *Hordeum turkestanicum* are joined by *Poa tibetica*, *Taraxacum dissectum*, and *Geranium collinum*.

On sites saturated by melting snow, on the other hand, one finds a cover of 100% composed of *Carex melanantha* and *Primula nivalis*, plus *Cerastium cerastioides*, *Saxifraga hirculus*, *Leontopodium ochroleucum*, *Trisetum spicatum*, *Thalictrum alpinum*, *Swertia*, *Gentiana*, etc.

Western Pamir (Badakhshan)

Up to now we have considered only the eastern part of Pamir, that part east of about 72°45′E which belongs to Tadzhikistan. Despite the heavy glaciation during the Ice Ages and despite the elevation of 4000 m, the region has a relatively level relief. The reason for this may be that the ice evaporated in the dry climate. Agakhanyantz (1965) mentions that P.V. Gubski measured snow sublimation of 43.7 mm near Murgab at 3860 m from December 1963 to March 1964, while the precipitation during this period was only 22.3 mm. During eighteen days in March, on some areas, an amount of snow corresponding to 80 mm of precipitation sublimated.

In contrast to eastern Pamir, the region west of 72°45′E and south of 38.30°N is a wild mountainous area with ranges reaching up over 5000 to 6000 m and with deep valleys lying below 2500 m. Annual precipitation amounting to 90 to 217 mm has been measured in this area. Precipitation comes from the west and falls primarily in winter. The mountain ranges south of the Trans-Alay run east–west and include the following: Vanch, Yazgulem Mountains, Rushan (Roshan), Shugnan (Shighnan), Shakhdarin Range and Vakhan in the south, reaching into Afghanistan. Only the Ishkashim Range in the southwest runs north–south. This western part of Pamir also includes the northeastern part of Afghanistan (Badakhshan). The mountains mentioned are shown in Fig. 7.19.

West Pamir is also characterized by extreme temperatures. Raw, primitive soils dominate the steep slopes and rugged relief. Agakhanyantz (1958) depicts the quite varied sequences of altitudinal belts for the Tadzhik part. Aridity is greatest in the south.

In the Vakhan Mountains the desert reaches up to 3500 m. Very open *Hammada wakhanica* stands prevail, with some *Echinops nanus*, *Lepidium latifolium*, etc. (on north slopes more *Artemisia* spp., *Ephedra intermedia*, etc., as well as ephemerals). Above the desert on south slopes lies a montane-xerophyte belt, which reaches up to over 4000 m and goes over into the subnival–nival belt. It consists almost exclusively of thorny cushion-plants (*Acantholimon* spp. and *Onobrychis echidna*). These are joined in the Shugnan and Rushan Mountains by *Artemisia* species and even scattered steppe grasses.

In these less arid northern ranges, *Artemisia* species prevail in the desert belt along with *Cousinia rubiginosa*. These are joined in the Yazgulem Mountains by xerophytic *Stipa* species, which together with *Festuca sulcata* dominate in the montane-steppe belt just above 3000 to 4000 m. On north slopes in the lower part of the montane steppe one can find fragments of open parklands with *Juniperus serawschanica* and *J. semiglobosa*.

In the subnival–nival belt the vegetation is only able to form closed mats on small areas in wet hollows, otherwise it consists of open stands (7–15% cover) on rubble.

Woody plants are restricted throughout the entire region to valleys. These are floodplains with *Salix*, *Populus*, *Betula*, *Hippophaë*, *Myricaria*, *Lonicera*, *Rosa*, etc., with an undergrowth of grasses and forbs. *Juglans* only goes up to 2300 m. Small floodplain meadows of *Kobresia* mats can still be found at 3000 m.

In Pamir the woody plants go up higher in narrow ravines than in broad valleys (Agakhanyantz, 1970). There are several causes for this. As shown by microclimatic measurements, one cause is the formation of cold-air lakes (cold-air drainage) in the broad valleys, while the steeper ravines receive more wind to break up accumulations of cold air. In addition, more snow accumulates in the ravines so the young plants are protected by a better snow cover. There is also no soil salinization in the ravines, as is very noticeable in the valleys. Finally, woody growth in the ravines is less vulnerable to livestock grazing and to interference by man. Individual trees could be found in ravines as high as 4000 m and some shrubs even as high as 5000 m.

It is noticeable in western Pamir that the boundaries between altitudinal belts occur higher as one goes from west to east. This entire region of western Pamir, the eastern part of Afghan Badakhshan, and eastern Pamir forms a phytogeographical unit which is sharply distinguished from central Tadzhikistan and from the western part of Afghan

Badakhshan. The region definitely belongs to Central Asia (Agakhanyantz and Yusufbekov, 1975).

BIOLOGICAL–ECOLOGICAL STUDIES IN PAMIR

First of all, some lists of species should give an idea of the vegetation in the vicinity of the biological station at Murgab (3864 m). The deserts occupy most of the region, and the following species, among others, occur:

Lower desert belt (3860 m)	Upper desert belt (4700 m)
Ceratoides papposa	*Ajania tibetica* (*Tanacetum xylorrhizum*)
Artemisia rhodantha	*Oxytropis poncinsii*
Artemisia pamirica	*Oxytropis immersa*
Artemisia macrocephala	*Pyrethrum pyrethroides*
Stipa glareosa	*Dracocephalum paulsenii*
Zygophyllum rosovii	*Hedysarum cephalotes*
Christolea crassifolia	*Smelowskia calycina*
Oxytropis chiliophylla	*Braya pamirica*
Potentilla pamiroalaica	*Parrya excapa*
Potentilla moorcroftii	*Acantholimon hedinii*
Astragalus borodinii	*Saussurea glacialis*

On somewhat less dry sites one can find steppes of *Poa litwinowiana* and *P. relaxa*, in which also *Allium polyphyllum*, *Christolea pamirica*, *Dracocephalum heterophyllum*, *Iris loczyi*, and *Lindefolia pterocarpa* occur.

In valleys along brooks, whose water supplies decrease drastically in August, there are meadows with *Elymus dasystachys*, *Elymus nutans*, *Puccinellia haeckeliana*, *Hordeum turkestanicum*, *Poa pamirica*, *P. tibetica*, *Carex duriusculiformis*, *C. melanantha*, *Kobresia capillifolia*, *Polygonum viviparum*, *Gentiana karelinii*, *Swertia marginata*, *Ranunculus* spp., *Clematis tangutica*, *Pedicularis ludwigii* and many others.

Finally we mention also the vegetation of snow patches, that is, continuously wet sites. This is composed primarily of *Primula nivalis*, *Androsace akbaitalense*, *Ranunculus rufosepalus*, *Thalictrum alpinum*, *Waldheimia glabra*, *Rheum reticulatum*, *Rhodiola pamiroalaica*, *Oxygraphis humifusa*, *Braya pamirica*, *Parrya excapa*, *Smelowskia calycina*, and *Hymenolaena nana*.

These lists show that such sites do not really have a desert flora. Steppe elements associate themselves with alpine elements, and representatives of the semi-desert are found only on the driest sites (e.g. the xeromorphic *Artemisia rhodantha* and the chenopod *Ceratoides papposa*, with *Nitraria schoberi* and *Sympegma regelii*; these last two are halophytes, for their normal osmotic values are around 33 to 48 bar and the maxima only 42 to 58 bar). Although precipitation is extremely scant, the dryness is reduced by the low temperatures and consequent low evaporation, and by the wetting of the soil by snowmelt during the first half of the summer.

Biology of the Pamir species

The biology of the vegetation's most important structural species (Russian: *edifikatory*), *Ceratoides papposa* and *Artemisia rhodantha* (=*skorniakovii*), was studied by Steshenko et al. (1969) and by Sveshnikova (1962).

The first species is found in the dry regions of Asia and Europe, and even in the northernmost parts of Africa. Its range extends from the Iberian Peninsula to the Sayan Mountains in Siberia and from northern Kazakhstan to Tibet. *Ceratoides* is found in Pamir between 1800 and 4500 m, while in Kazakhstan it extends down to sea-level.

The community type (*Eurotietum ceratoides*) dominated by this species contains 35 species, of which 15 can be co-dominants (e.g., *Artemisia rhodantha*, *Stipa glareosa*, *Acantholimon diapensioides*, *Xylanthemum pamiricum*). Usually only 2 to 12 species occur simultaneously in a stand; the cover is 10 to 12% (5–40%). The community has two strata, the upper 15 to 20 cm tall and the lower (cushion-plants) only 1 to 3 cm. The annual production is 0.02 to 0.03 t ha^{-1}.

Artemisia rhodantha is a Central Asian element, occurring in the Pamir at 3600 to 4200 m, in the Alay Mountains at 2800 to 3400 m, and in the Tien Shan on the high plateaus (*syrts*) at 2500 m. In Pamir the species prefers loamy-rocky soils on the lower parts of gentle slopes or the dry valley or lake biotopes.

The community type dominated by *Artemisia rhodantha* (*Artemisietum*) contains 50 species, of which 23 can be co-dominant (e.g., *Ceratoides*, *Stipa orientalis*, *S. glareosa*, *Hordeum turkestanicum*, *Acantholimon diapensioides*). The cover is 7 to 50% and the number of species per stand 5 to 15 (20). Vertical stratification is noticeable only if cushion-plants are present. The annual production is 0.04 to 0.48 t ha^{-1}.

The seeds of both *Ceratoides papposa* and *Artemisia rhodantha* germinate 90 to 100% within 3 to 15 days after maximal soaking. The seedlings reach a height of 2 to 3 cm by autumn, while the main root is growing 15 to 20 times as fast. Most of the young plants die in the winter. The phenological spectra of the two species are shown in Fig. 9.6.

Very interesting is the lateral branching at the base of the vertical shoot, the dying of branches in the center, and the splitting of the older shoots (particularization), as shown in Fig. 9.7 for *Ceratoides*. The lignifying parts of *Artemisia* grow about 2 to 3 cm annually, the generative shoots die back to the base, and the splitting is not so striking. The root system is well developed. From the tap-root 1 to 1.5 m long, lateral roots extend to a length of 1.5 to 2 m at a depth of 10 to 15 cm. The ratio of above-ground to below-ground parts is 1:10 for *Ceratoides* and 1:15 (20) for *Artemisia*. The habit of the plants changes radically with changing environmental conditions. At low elevations *Ceratoides* is 85 to 150 cm tall, but at its upper altitudinal limit it forms small cushion-like plants of only 8 to 10 cm in height. These produce few fruit but can become very old (maxima 200–300 years), while the plants in more favorable conditions usually die after 15 to 30 years. The same is true of *Artemisia*. After seeding at Tashkent the plants remained alive only 3 to 4 years, but they live 100 years in the high mountains.

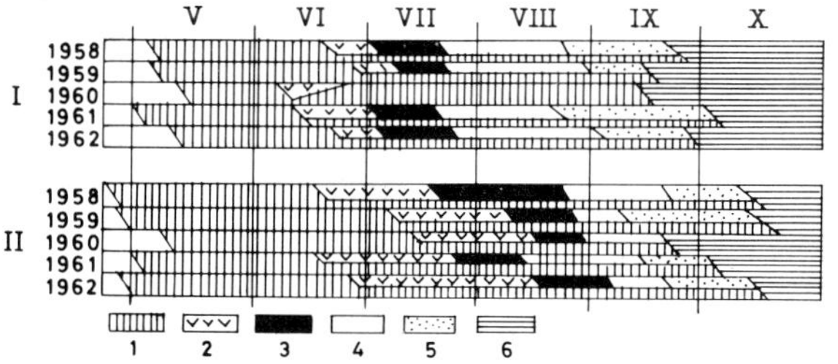

Fig. 9.6. Phenological spectra of *Ceratoides papposa* (*I*) and *Artemisia rhodantha* (*II*) in different years: *1* = vegetative growth; *2* = flower buds; *3* = flowering; *4* = fruiting; *5* = propagation of seeds or fruits; *6* = beginning of winter dormancy (from Steshenko et al., 1969).

Frost resistance of young leaves of *Ceratoides* is around −13°C and about −16°C for older leaves; the corresponding values for *Artemisia* are −20° and −25°C.

Both species are very important forage plants with high protein content. The raw-protein content of young leaves of *Ceratoides* can be 30.2% during their formative stages and decreases to 18.8% at the end of the growing season. The dead leaves in winter, however, still contain 9.7% raw protein.

The growing season in Pamir begins as early as May, but, in spite of this, the annual growth on dry sites is extremely low, at most a few millimeters for cushion-plants. It is a consequence of this that the prevailing life forms are dwarf-shrubs, cushion-plants, and small rosettes. It is a long time before the plants reach the flowering stage: 25 years for *Ceratoides*, 15 years for *Stipa glareosa* and *Acantholimon diapensioides*, up to 8 years for *Kobresia pamiroalaica* and *Carex orbicularis*. On the alpine meadows, on the other hand, *Trisetum spicatum* and *Puccinellia pamirica* bloom within a year, and *Blysmus compressus*, *Astragalus hetero-*

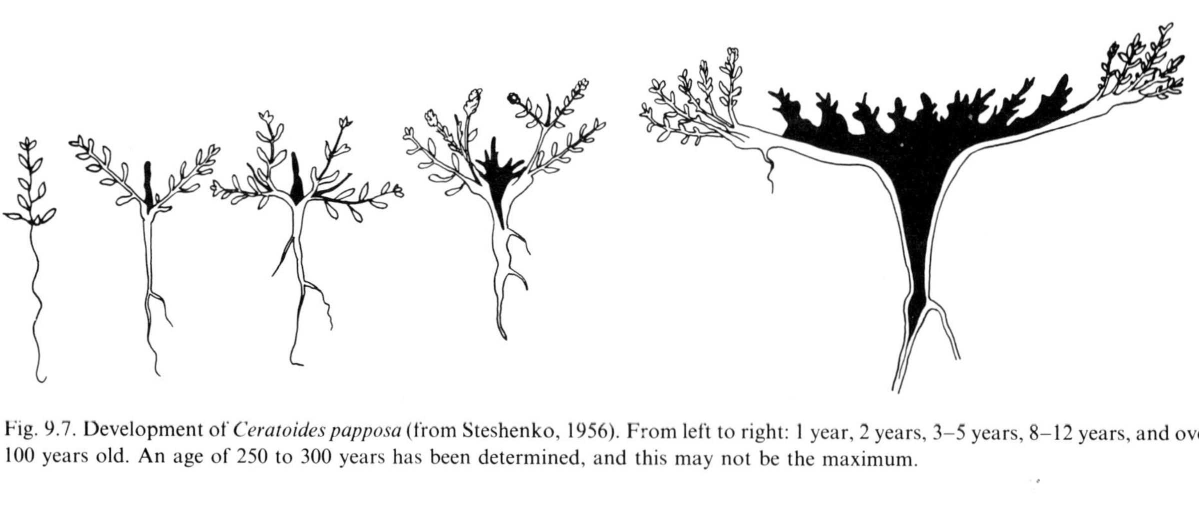

Fig. 9.7. Development of *Ceratoides papposa* (from Steshenko, 1956). From left to right: 1 year, 2 years, 3–5 years, 8–12 years, and over 100 years old. An age of 250 to 300 years has been determined, and this may not be the maximum.

dontus, etc. within three or four years. Because of their slow growth, however, the desert, steppe, and cushion-plants reach ages of 100 to 300 years, apparently 1000 years for *Acantholimon*. The species of the alpine meadows, such as *Trisetum*, *Puccinellia*, *Primula algida*, and *Swertia marginata*, only become 10 to 15 years old.

We refer also, briefly, to a comparative ecological study of the interesting composite genus *Cousinia*, which is widespread around the Pamir–Alay region (Kuprevich et al., 1949; Grigoryev, 1955). *Cousinia* is closely related to the burdock (*Arctium*), of which the mesomorphic *C. umbrosa* and *C. pulchella* are reminiscent. The genus is otherwise noteworthy in that it exhibits all transitional stages between the mesomorphic species and the extreme cushion-plants. In Central Asia we can distinguish: (1) species with large, soft leaves, which grow more under trees at intermediate elevations; (2) other species with prickly leaves, which form the transition to the next group; and (3), the thorny cushion-plants of high mountains, such as *C. splendida*, *C. fedtschenkoana* and *C. macilenta*.

The species of the first group are characterized by high transpiration levels, a high water content, low sap concentrations, and low water deficits. The leaves wither at the beginning of the dry season. The third group, in contrast, shows low transpiration levels, low water content, and high potential osmotic pressure. Water deficits can reach very high values. At the beginning of the drought season the water turnover is sharply limited, so the plants survive the drought period in an inactive state. The species of this group have either converted almost their entire leaf area into thorns or have thickly pubescent, prickly leaves. These thorny cushion-plants possess very deeply penetrating tap-roots.

The second group occupies an ecologically intermediate position. Their transpiration levels are very high in spring but are then restricted, which causes the sap concentration and water deficits to rise. These species grow on sunny sites and survive the drought for some time before finally withdrawing from activity. The drought adaptation of this group, therefore, is mainly morphological–anatomical in nature. It consists of an increasing reduction of leaf area and better development of the root system, which permits the plants to maintain their water balances, even under extreme conditions, by reduction of water consumption.

The water budget of the Pamir vegetation

Winter precipitation has an almost negligible effect on the water supply in the soil. The snow falls on ground which is already frozen and is easily blown away or evaporated. There is no closed snow cover; a cover of a few decimeters can form only where snow is blown together, as in depressions. Much of this is lost to evaporation even before spring thawing begins. During snowmelt, which begins in March, the water comes in contact with the cold ground and freezes in the upper layers of soil, making the soil impermeable. The snowmelt thus runs off, for the most part, and penetrates only on those sites where the soil thaws early. The soil, which has only been very shallowly wetted, dries out very quickly during the spring. Consequently, the plants develop very slowly.

The water supply for the plants is augmented only after the summer rains begin to fall. One peculiarity of the soils is a certain flow from below of water which condensed in the extremely cold soil layers during the winter. One can see in Fig. 9.8 how the water content at a depth of 1 m in the rubble soil increases in autumn, especially in November–December. This increase may proceed upward to a greater or lesser extent, but the water content falls off again toward springtime. If one looks at the soil temperatures, one sees that the condensation begins only after soil temperatures have fallen below freezing down to a depth of 1 m but still remain above freezing at greater depths. The process must, therefore, be a sort of hoarfrost-forming process within the loose rubble. If the deeper layers, down to about 4 m, also freeze (January or February), then the water content at 1 m decreases again. In our opinion this condensation can be of little significance to the plants, since the roots are not active in winter. One must also consider that the differences in water content shown in Fig. 9.8 are only about 1%, which easily lie within the range of variation of water content in the highly heterogeneous rubble soil. In summer it is the uppermost layers that are wetted most, but in this case by rainfall. They dry out quickly again as a result of surface evaporation and plant transpiration. The water content of the soil is usually 2 to 3%, barely above the wilting point and often less.

Sveshnikova (1962) refers to the experience of Blagoveshchenskiy in the sand desert (see p. 125)

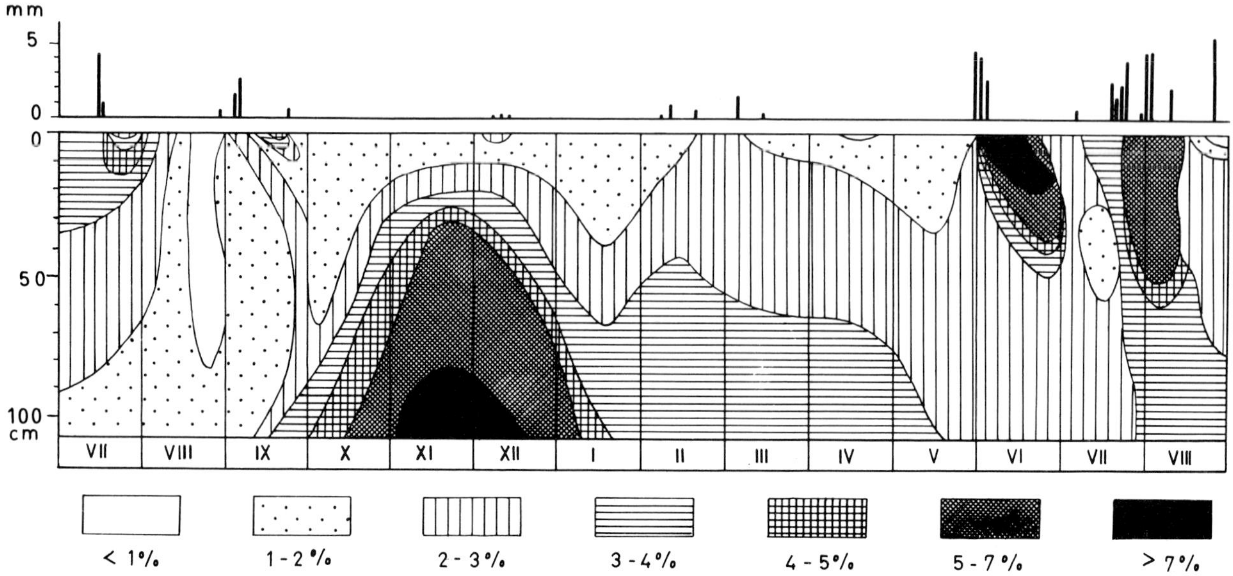

Fig. 9.8. Change in soil water content in the cold desert of Pamir (from Sveshnikova, in Walter, 1968, fig. 575). Water content (%) is given underneath, precipitation (in mm) at the top. Further details are given in the text.

and accords the water condensation in the soil a major ecological significance. Under the prevailing large diurnal temperature variations in the upper soil layers, the hygroscopicity of the soil can indeed cause water to be taken up at night. Quantitatively, however, the amounts concerned are equivalent to fractions of a millimeter of rainfall. Sveshnikova presumed that the plants take up this water immediately. She multiplied the daily values by the number of days in the growing season and came up with about 35 mm of water, about 30% of the annual precipitation. It is much more probable, however, that the small amounts of water taken up at night in the soil are lost by evaporation the next day, for the air humidity near the surface is only 5% at midday. It can be 100% at sunrise after rainy days but is usually 50% during the dry season, even at sunrise, so there is no dew formation. These variations in water content of the uppermost layers can have little significance for the water supply of the plants. The same is true of condensation in deeper soil layers. Under extreme dryness this condensation can be shown to occur (on the average about 0.16 mm per day), but one cannot say whether this water is taken up by plant roots during the winter months.

In general, the water supply in the skeletal soils is very irregular because of the low water-holding capacity and the irregularity of the rainfall. Even so, the water supply in the upper 1 m of soil under desert vegetation varies from 24 to 104 mm, depending on rainfall in individual years. During 1940–1945 and 1950–1955 the water supply during the growing season varied from 53–55 mm to 28–30 mm. This corresponds to a maximum of about 26 mm of available water and a minimum of about 5 mm. This is very little when one considers that the water supply in the upper 1 m of soil in the forest zone is 500 mm on the average and amounts to 70 to 120 mm even in the southern steppe.

Quantitative data on the root systems of Pamir plants are shown in Figs. 9.9, 9.10, 9.11 and 9.12, and in Table 9.4. A total of 310 root profiles (excavation method) for 33 species were made. One finds most roots at a depth of 0 to 40 cm. This layer has temperatures over 10°C for about three months in summer. Of course it can dry out completely for varying periods, since the variations in soil water in the upper layers are always very large. The deeper layers (to 180 cm) show a fairly uniform moisture content, but they are colder. In general, and as is always the case with perennial plants in dry areas, the underground parts are much more massively developed than the above-ground parts. The depth of the root system can be 25 to 45 times greater than the height of the shoots above ground. The dry

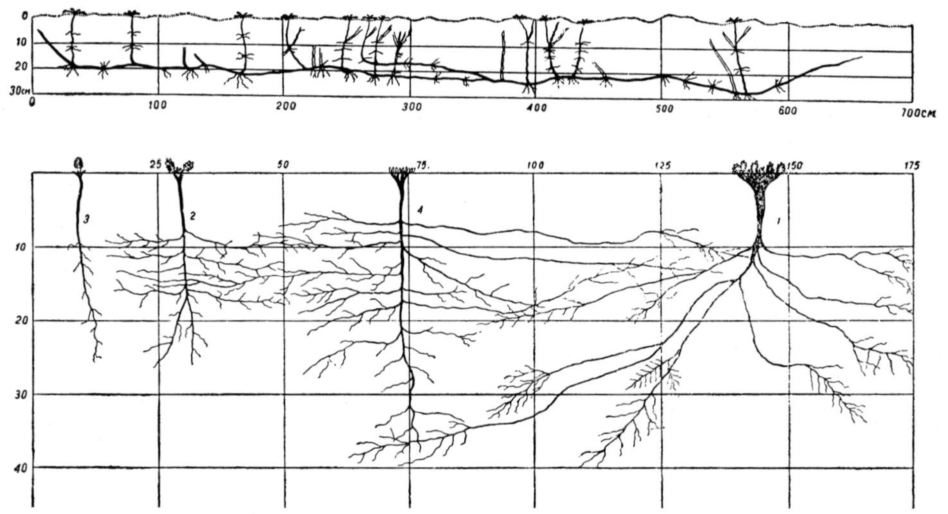

Fig. 9.9. Root systems of sand plants in Pamir (from Stanyukovich, 1949). Roots of *Polygonum pamiricum* (above) and of *Oxytropis poncinsii* (below) are shown: *1* = several years old; *2* = 2-year old plant; *3* = 1-year old plant; *4* = roots of *Braya pamirica* (several years old). Numbers represent dimensions in centimeters.

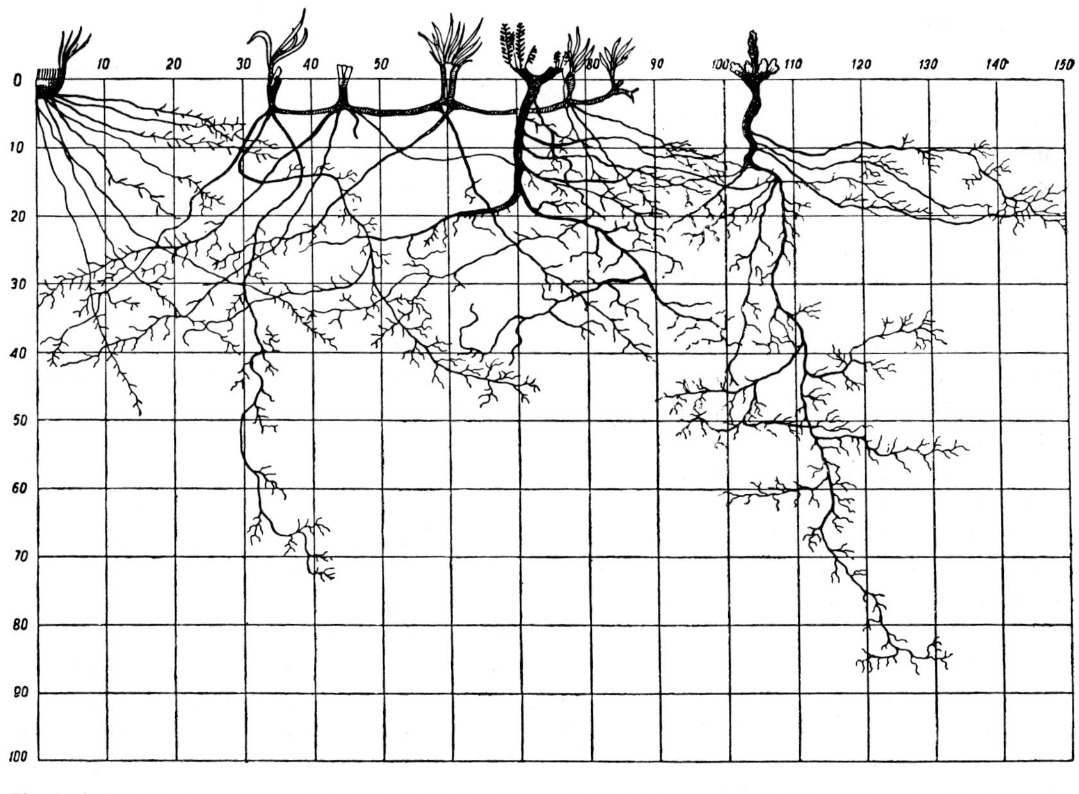

Fig. 9.10. Root systems of the *Artemisia rhodantha–Stipa orientalis* biogeocoenosis (from Stanyukovich, 1949). From left to right: *Stipa orientalis, Carex stenophylloides, Oxytropis chiliophylla, Artemisia rhodantha.*

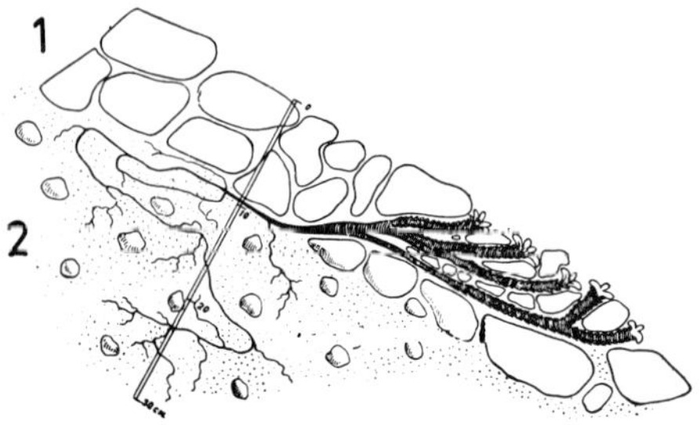

Fig. 9.11. Root system of *Saussurea sorocephala*, a species found on rubble slopes (from Stanyukovich, 1949): *1* = soil profile with coarse debris; *2* = rubble mixed with fine earth.

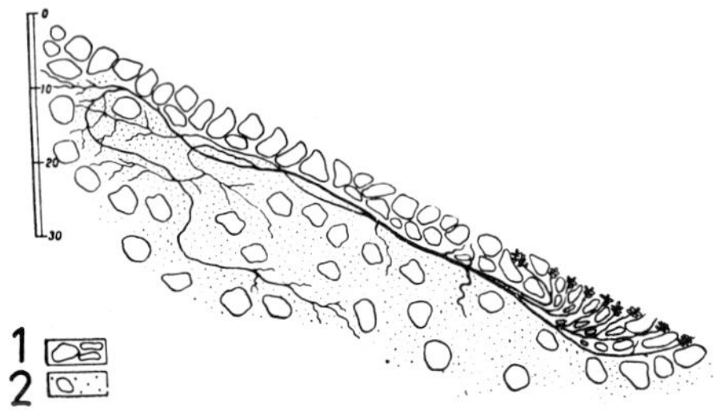

Fig. 9.12. The same as in Fig. 9.11 but with the root system of *Didymophysa fedtschenkoana*.

TABLE 9.4

Extent of root systems (maxima in parentheses) for individual species in the Pamir Desert (from Sveshnikova, 1962)

Species	Horizontal extent (cm)	Vertical extent (cm)
Ceratoides papposa	160–240	120–140 (180)
Artemisia rhodantha	120–220	50–60 (120)
Zygophyllum rosovii	180–220	90–120
Christolea crassifolia	180–200	100–120

mass can be 10 to 12 times greater below ground, and the volume covered by roots can exceed the space occupied by above-ground shoots by factors up to 500 to 800.

Stipa glareosa roots only 50 to 80 cm deep, the meadow plants on moist soils root only 20 to 60 cm deep.

The water consumption of plants in the *Ceratoides–Artemisia* desert was determined by measuring the water content in the top 1 m on a sample plot in each of the following situations: (1) with natural vegetation (*Ceratoides*, *Artemisia*, *Stipa*); (2) with all vegetation removed from the area, the soil remaining open; and (3) on sample areas with the surface covered to prevent evapotranspiration. The plants grew poorly on one part of the sample area. On this area the evapotranspiration in 1950 for 45 to 50 summer days was about 4 mm. The plants were better developed in the other parts where the soil was moister; evaporation was 35 mm and transpiration 16 mm. Depending on the development of the plants, 4 to 35 mm were transpired during the same period in 1951, but in the cold summer of 1952 only 3 to 6 mm during the same period. The water uptake by the roots took place mainly at depths between 20 and 50 cm, where most roots are found. At a depth of 1 m the water content remained almost unchanged.

These results show that, in spite of the extreme dryness, the water supply of the very sparse vegetation is assured. It is, thus, not so astounding that the potential osmotic pressures determined by Sveshnikova for 113 species were higher only on extreme sites. The samples were taken: (1) from desert species at 3860 m; (2) from desert species at 4300 m; (3) from steppe species at 4700 m; and (4) from plants of the nival belt at 4700 m. The highest sap concentrations are shown by the species on the driest sites (see Table 9.5). Otherwise, only two values above 30 bar were obtained, even under the most extreme conditions.

Looking at all the measured values, one can say that:

(1) The snow plants have the lowest average values: 10.5 to 18.8 bar.

TABLE 9.5

Potential osmotic pressures of Pamir Desert plants (from Sveshnikova, 1962)

Species	Osmotic pressure		
	minimum	average	maximum
Stipa glareosa	23.6	34.7	37.7
Artemisia rhodantha	19.5	32.5	54.3
Pyrethrum pyrethroides	19.0	24.6	30.8
Potentilla pamiroalaica	20.9	26.9	41.6
Christolea crassifolia	15.0	22.5	43.7
Oxytropis chiliophylla	16.6	24.2	34.2
Astragalus chadjanensis	19.2	22.6	32.4
Zygophyllum rosovii	19.4	25.5	38.4

(2) Meadow plants have the next lowest average values: 12.3 to 22.9 bar.

(3) Steppe representatives are next in line: 16.2 to 22.2 bar.

(4) The species of the upper desert belt are next: 13.8 to 24.6 bar.

(5) The species of the lower desert belt have the highest values: 17.1 to 34.7 bar.

If there is a good water supply, however, even the last group shows significantly lower values (9.4–23.6 bar). In general, the values are somewhat higher than those of plants of the Alps, among which 20 bar is never attained.

Especially striking are the large diurnal variations in potential osmotic pressure caused by the high insolation during the day and the effects of high altitude. The variations for species on the drier sites can be 12 to 14 bar. (If the plants are watered at the site, then the diurnal variations drop to 0.3 to 3.0 bar). One can see that these species are actually hydrolabile, malacophyllous species, and that real sclerophyllous species do not occur. There are also no evergreen species. The Ericaceae of the Alps and the Arctic, and *Dryas*, are completely absent. *Leontopodium ochroleucum* (aff. *alpinum*) occurs instead. It has a mean potential osmotic pressure of 18.4 bar (12.3–23.2 bar at extremes). The numerous *Artemisia* species are also very typical. *A. macrocephala* shows the lowest average osmotic value at 15.8 bar (8.7–21.6 bar). The diurnal variations are not, however, as Sveshnikova assumed, caused by the accumulation of assimilates. The amounts of sugar formed are much too small for this to happen. The parallel courses of the diurnal curves of osmotic pressure and sugar content, which she mentions, are obviously purely as a result of the concentration of the cell fluid caused by water deficits. The increase in sugar concentration through accumulation of assimilates, expressed in bars, is very low.

Species with succulent leaves or water-storing roots also have low diurnal variations in the Pamir. These species are hydrostable, and the osmotic values remain rather constant and relatively low. They vary for *Zygophyllum rosovii* between 19 and 24 bar and for *Christolea crassifolia* and *Gypsophila capituliflora* between 16 and 20 bar. This is true to an even greater extent for the species of the alpine meadows, which grow on continuously moist sites. Their osmotic values hardly change at all during the entire growing season.

The potential osmotic pressure of all species increases in the course of the summer, especially the values for the species on drier sites. In general, the lowest values are found in May and the highest in August and September. This seems to contradict the measured values for plant water budgets, since it is characteristic of the Pamir plants that the highest water content of the leaves is observed in summer and the lowest water content in the dry spring.

Table 9.6 shows the extreme values of water content under natural conditions. Depending on the site, the diurnal variations in water content were 1.2 to 9.7%. The lowest water-content values were found in the over-wintering parts of plants during the winter months (roots, regenerative buds).

One must consider, however, that the water content expressed as a fraction of fresh weight depends on the amount of dry matter and, thus, on the age of the leaves. Since the Pamir plants do not come out until summer, the summer values of water content represent very young leaves with thin cell walls and not yet fully developed mechanical tissue. The osmotic values, on the other hand, are not affected by the amount of cell-wall tissue.

If one compares the potential osmotic pressure for plants of the same species but on different sites, one finds, as always, a clear relationship between osmotic values and soil moisture. The drier the site, the higher is the sap concentration. Comparison of species of the same genus also show those on drier sites having higher values. The osmotic values for given species increase with increasing altitude. It is not easy to decide whether the determining factor is

TABLE 9.6

Extreme values of water content (% of fresh weight) under natural conditions (Sveshnikova, 1962)

Species	Water content
Desert plants (dry sites)	
Ceratoides papposa	54.5–75.2
Artemisia rhodantha	49.9–72.7
Stipa glareosa	46.5–67.0
Alpine meadow plants (moist sites)	
Polygonum viviparum	73.4–82.5
Potentilla multifida	71.0–78.0
Hedysarum cephalotes	68.3–73.7

the decrease in temperature with altitude or the greater dryness of the sites.

The water relations of the high-mountain plants of the Pamir are more similar to steppe plants than to species of the hot deserts. The dryness in the Pamir is indeed very great, but the drought periods are also relatively short, for the long, cold winter is a period of absolute dormancy. Even in summer the low temperatures at night cause transpiration to cease.

Suction-pressure measurements according to the method of Shardakov (1953, 1956) were also carried out. Tyurina (1957) obtained very high negative turgor values for *Artemisia*, but this is hardly possible for a malacophyllous species. The Shardakov method is inaccurate.

The transpiration studies showed no peculiarities. The stomata of all plants are open during the day, and the daily transpiration curves show a midday peak for most species. The absolute values are relatively high, but there are, as always, species with high rates and others with low rates. The maximum transpiration rates correspond to those determined in the temperate semi-desert regions of Middle Asia.

The daily water consumption by transpiration, in grams per gram of fresh weight, is substantial. The highest values were found for *Polygonum viviparum* (22–34 g), and similarly high values were found for other plants of alpine meadows on continuously moist soils (23–31 g). The succulents (*Christolea* and *Zygophyllum*) have the lowest values (3.4–4.2 g). The most important desert species, such as *Ceratoides papposa* and *Artemisia rhodantha*, had values up to 6.8 g, but these rose to 16 to 20 g under artificial watering. The water contained in the leaves is thus turned over several times in the course of a day, sometimes even in the course of an hour. The transpiring mass of *Ceratoides* per plant increases (from growth) during the growing season from 3 to 10 g (to 60 g if watered additionally). For *Artemisia rhodantha* the corresponding values are similar (up to 33 g, if watered), but for *Stipa glareosa* the values are much lower, from 0.8 to 2.5 g (up to almost 4 g, if watered).

The water loss of one plant per day increases as a result of the increase in transpiring mass and the increasing potential evaporation from May to July or August. The average amounts are given in Table 9.7.

TABLE 9.7

Increase in daily water loss per plant between May and August for plants of the Pamir (Sveshnikova, 1962)

Species	Water loss ($g\ day^{-1}$)
Desert plants	
Ceratoides papposa	2.5–50
Artemisia rhodantha	12–80
Stipa glareosa	0.6–17
Succulent species	
Christolea crassifolia	16–75
Zygophyllum rosovii	2.5–45
Meadow plants	
Artemisia pamirica	23–320
Potentilla multifida	12–140

One sees that the succulents, as a result of their greater fresh weight and despite their low transpiration rates (per gram of fresh weight), consume just as much water as the desert species. If the desert plants are watered so that they can develop fully, the water consumption per plant climbs to over 300 g daily — that is, they exceed the values of the well-watered plants of the alpine meadows.

If one considers the number of plants of a given species on a given area, one can compute the water consumption of the desert communities on the rubble fans in millimeters per year. This was done for the years 1949–1952. The values varied in the upper part of the fan between 13 and 26 mm, in the dry middle part between 7 and 14 mm, in the moist lower part between 20 and 30 mm, and reached 220 mm (maximum 325 mm), as an average, when additional water was provided.

Of this water loss, *Artemisia* accounted for 45 to 70%, *Ceratoides* 20 to 30%, *Stipa* 5 to 10%, *Christolea* 5 to 8%, and *Zygophyllum* 2 to 3%.

If one considers the large experimental error in the computation of the total water consumption, one can say that the figures given here agree well with those obtained from measuring the decrease in soil water content (for 45–50 days).

Sveshnikova (1962) provided the following annual water balance:

water gains:		water losses:	
average precipitation	102 mm	surface evaporation	120 mm
		transpiration	21 mm
totals	102 mm		141 mm

The deficit was held by Sveshnikova to be proof that some other source of water was available, in particular the condensation in the soil (39 mm year^{-1}) and additional condensation on the soil surface. It appears to us, though, that the values are far too uncertain to perform such a precise computation of the water balance. The surface evaporation could just as easily be only 80 mm or the precipitation on the experimental area 140 mm. These numbers would also balance the budget. Besides, the water condensed in the soil must ultimately come from precipitation, since groundwater on a local perched groundwater horizon must also be fed by the precipitation. The deficits could also be covered by the precipitation in wetter years. This extra water probably penetrates deeper into the soil and is stored there for many years, as has been found in many deserts of the world.

The studies of the water budgets of Pamir plants which were made in 1962 have more recently been summarized in English (Sveshnikova, 1973). It is emphasized here that the individual ecological types of plant regulate their water budgets in different ways in order to adapt to the general water shortage in this high-mountain desert.

These studies of Sveshnikova have been continued since by Ismailova (1963, 1965, 1974; Ismailova and Litvinova, 1969, 1972; see also Litvinova and Ismailova, 1968). She did not limit herself to the determination of transpiration rates in grams per gram of fresh weight, but also computed the water losses of the different communities.

The comparison of transpiration rates at various elevations in Pamir (3860–4760 m) showed that the same species, under otherwise equivalent conditions, transpired less with increasing altitude. This is related to the decreasing temperatures and lower vapor pressure near the ground and in the root zone as altitude increases. This decrease in transpiration rate is more obvious among plants of moist sites than among those of drier biotopes.

The water budgets of the dominant species of the various communities were studied in special detail. These species included *Ceratoides papposa*, *Artemisia rhodantha*, *Stipa glareosa*, and *Acantholimon diapensioides*. The results are summarized in Table 9.8.

Underground phytomass makes up 84 to 91% of the total weight of *Ceratoides*, and the transpiring and assimilating organs make up a very small

TABLE 9.8

Production (in g; between brackets % of total) and water expenditure for evaporation by a middle-aged plant (from Ismailova and Litvinova, 1972)

Species	Production					Water expenditure	
	above-ground parts			roots	total weight	g day^{-1}	liters over a growing season
	annual	perennial	dead				
Ceratoides papposa	3–10 (1–2)	10–57 (5–10)	3–18 (1–3)	174–425 (91–84)	190–510 (100)	17–90	2.5–13.5
Artemisia rhodantha	1–5 (1–3)	2–14 (3–8)	1–3 (1–2)	79–146 (95–87)	83–168 (100)	14–37	2–6
Stipa glareosa	0.6–3 (1–3)	2–6 (4–5)	6–8 (12–16)	42–83 (84–74)	50–111 (100)	6–21	0.8–3
Acantholimon diapensioides	23–41 (7–6)	100–243 (35–41)	50–122 (17–20)	111–200 (41–33)	284–606 (100)	130–394	13–40

Note: Estimation of production values for plants was performed over the period of maximal development (July–August) in ten representative individuals. The ranges of values obtained under different ecological conditions are shown.

fraction. The plant thus has the possibility of surviving with very little water, but grows very slowly and becomes very old. *Ceratoides* is a relatively stenohydric xerophyte. *Artemisia*, on the other hand, must be considered a malacophyllous xerophyte (potential osmotic pressures up to 54 bar). This species prefers loamy soils, on which the individual plants can attain quite large dimensions. During the drought period both the transpiring area and the transpiration rate are reduced. *Stipa glareosa* is a xeromorphic grass which tends to be more similar to the steppe species. On favorable sandy soils it can transpire at higher rates than can the other plants. The thorn-cushion species *Acantholimon* shows the greatest water loss because of its greater green area. The tap-root system is well developed; the growth of the cushions is very slow (1 mm per year). These cushion-plants are thought to live as long as 1000 years. They accumulate 35 to 51% of their total phytomass in their extremely dense branching structure.

Once the water losses of individuals of various sizes have been determined for the entire growing season, the water consumption of the entire community can be estimated. Newer results are summarized in Table 9.9 and substantiate what we have just said.

One sees that, despite the relatively high transpiration rates (Table 9.9), the total water consumption (expressed in mm) is so small as a result of the low cover that it can be covered by the scant precipitation that does fall. The cushion-plants grow near streams and receive additional water.

The water content of the soil in the upper 1 m does not exceed 3 to 7% and is even as low as 1 to 3% under *Ceratoides* communities. Under herbaceous cushion-plant communities, however, it can be 15 to 20%. The water loss by the dry *Ceratoides* communities is correspondingly low, 7.7 mm with a green phytomass of 0.14 t ha^{-1} on sandy rock soils but 21 mm on sandy alluvial fans.

The *Artemisia* community consumes more water: 27 to 40 mm, and up to 78 mm on moister sites.

The cushion-plant communities behave very differently. The water consumption of *Ajania tibetica* communities is very low at 30 to 60 mm, since they grow on the tops of dry moraines with southern exposures. The consumption of *Sibbaldia tetragona* communities on rocky, somewhat clayey soils of northern exposures is even lower at 25 to 52 mm. The consumption of the herbaceous *Oxytropis immersa* and *Potentilla pamirica* communities on the other hand, is very high, 117 to 446 mm. These communities are limited to sites with water subsidies.

The annual curve of water consumption reaches a maximum during the flowering season in the summer, when vegetative growth is also most intense. After the middle of July the consumption drops, and is 20 to 30% in August. Water consumption is not the same in different years, being about half as much in dry years as in wet. In general, the available water within the root zone, which corresponds to about one-third to one-half of the annual precipitation, is completely consumed by transpiration.

Photosynthesis and production analysis

Blagoveshchenskiy (1935) studied the photosynthesis of the Pamir plants. *Ceratoides papposa*, *Elymus dasystachys*, *Gentiana algida*, and *Lepidium* assimilate very actively, though only in the morning. The daily gain is given as 25 mg of dry weight per dm^2 leaf surface. The low night-time temperatures prevent large respiratory losses, but the plants must be able to survive the nightly frosts without damage to the photosynthetic apparatus.

Further studies by Glagolyeva (1962), using high

TABLE 9.9

Water consumption of the biogeocoenoses of eastern Pamir (Ismailova, 1974)

	Desert	Steppe	Cushion-plants
Cover (%)	5–18	15–20	15–30
Phytomass (t ha^{-1})	0.14–0.54	0.09–0.48	0.4–0.89
Transpiration g (g fresh wt.)$^{-1}$ h^{-1}	0.3–0.9	0.1–0.7	0.1–0.9
Transpiration of the biocoenosis during growing season (mm)	8–40	6–87	25–446

carbon dioxide concentrations, cannot be used for computation of normal production rates[1]. They also show, however, that all Pamir species are capable of intensive photosynthesis. Other studies have shown that the carbon dioxide assimilation of plants of alpine meadows is especially high. The leaves of these plants possess a multi-layered mesophyll. The photosynthesis of Chenopodiaceae, however, which have only two chlorophyll layers in the leaves, is only half as high (Filipov, 1959; Zalenski, 1963, 1973).

The phytomass of the various biogeocoenoses at 3850 m to 4800 m was studied in great detail. The studies included the following communities:

1. Pure *Ceratoides papposa* (3850 m)
2. *Ceratoides papposa–Stipa glareosa–Oxytropis chiliophylla* (3800 m)
3. *Artemisia rhodantha–Stipa glareosa–Ceratoides papposa* (3850 m)
4. *Stipa glareosa–Ceratoides papposa–Astragalus chomutovii* (3820 m)
5. *Festuca valesiaca* (*sulcata*)–*Carex stenophylloides* (3900 m)
6. *Hordeum turkestanicum–Hedysarum cephalotes + Potentilla pamirica* (4300 m)
7. *Oxytropis immersa + Draba altaica + Leontopodium ochroleucum* (4760 m)
8. *Potentilla pamirica + Oxytropis immersa + Calamagrostis anthoxanthoides* (4650 m)
9. *Sibbaldia tetrandra–Colpodium leucolepis* (4800 m)

The actual results are shown in Table 9.10, in which the annual growth and litterfall are also given. The annual growth of the roots could not be determined directly but was estimated (Ismailova et al., 1975).

If one compares the annual litter production, one sees that the growth is not significantly higher, i.e., there is hardly any net annual accumulation of phytomass. The biogeocoenoses are, thus, stable and in equilibrium with their environments.

In general, the organic mass of the Pamir biogeocoenoses is low. It is lowest for the desert communities (6–20 t ha^{-1}), somewhat higher for the steppes (10–30 t ha^{-1}), and highest for the cushion-plants (19–42 t ha^{-1}). In each case, the proportion of dead organic matter is very high (36–70%).

The roots represent by far the largest fraction of the organic matter, especially in the desert communities. The photosynthetically active parts represent only 1 to 9% of the living mass.

The organic mass increases with increasing elevation and increasing moisture, especially among the cryophilic cushion-plants.

The primary production (=annual growth) is not high: 0.6 to 2 t ha^{-1} in the deserts, 2 to 4 t ha^{-1} in the steppes, and 5 to 12 t ha^{-1} among the cushion-plants. The parts which die are generally not discarded but remain attached to the plants.

In summary, the studies discussed here have provided us for the first time with insight into the ecophysiological and biological relationships in this region of high-mountain deserts above 3500 m, which is unique within the temperate zone.

At the same time it was possible to distinguish two vegetation types: (1) the desert vegetation of dry sites on rubble deposits, the main species of which are *Ceratoides papposa* and *Artemisia rhodantha*; and (2) the eucryophilic vegetation of moist sites at the highest elevations, above 4500 m, of which the characteristic plants are the cushion-plants.

The desert vegetation migrated or wandered into the mountains from the semi-deserts of lower elevations. The vegetation was able to adapt itself to a similar length of growing season, the main difference being that the dormant period in the high mountains is forced by the long, cold winter instead of the long summer drought and short cold winter of the lower elevations. Since the plants were already cold-resistant, the difference for them was not great. Their well-developed root systems gave them the ability to occupy the dry sites. The vegetation takes advantage of the short growing season as best it can. It transpires and assimilates at high rates, as far as water availability permits, and makes allowances for the high water deficits and increased osmotic pressure. In autumn, when the lasting cold begins, the vegetation discards its leaves just as it would in the desert at the beginning of the drought. By this time the main branches are sufficiently lignified to survive the winter. *Stipa glareosa* behaves similarly but also regulates its transpiration during the day by closing its stomata early. The transpiration values are thus lower, but the production of dry matter is lower also. *Stipa* is a more sclerophyllous species, in contrast to the

[1] The partial pressure of CO_2 is greatly reduced at high elevations. The CO_2 content per liter is only about half the normal amount.

TABLE 9.10

Phytomass of nine biogeocoenoses of the Pamir, plus annual growth and litterfall (t ha^{-1}) (from Ismailova et al., 1975)

Biogeo-coenesis (see text)	Above-ground parts: annual parts				Above-ground parts: perennial parts			Roots			Total phytomass			Annual growth			Annual litterfall	
	inflores-cences	leaves	stems	total	living	dead	total	living	dead	total	living	dead	total	above-ground	roots	total	above-ground	roots
1.	0.02	0.05	0.03	0.10	0.23	0.31	0.54	4.00	3.75	7.75	4.33	4.06	8.39	0.11	1.26	1.37	0.10	1.24
2.	0.01	0.05	0.02	0.08	0.49	1.12	1.61	4.19	8.81	13.00	4.76	9.93	14.69	0.10	0.73	0.83	0.08	0.72
3.	0.01	0.07	0.02	0.10	0.33	0.30	0.63	6.47	4.77	11.24	6.90	5.07	11.97	0.12	1.63	1.75	0.10	1.42
4.	0.004	0.08	0.006	0.09	0.53	0.93	1.46	3.82	9.53	13.35	4.44	10.56	15.00	0.11	0.76	0.87	0.09	0.63
5.	0.02	0.29	0.05	0.36	1.21	1.73	2.94	5.12	5.94	11.06	6.69	7.67	14.36	0.39	1.70	2.09	0.37	1.22
6.	0.04	0.70	0.06	0.80	1.96	1.76	3.72	9.74	8.09	17.83	11.95	10.39	22.34	0.86	2.91	3.77	0.84	2.90
7.	0.02	1.92	0.14	2.08	1.52	1.00	2.52	18.23	9.40	27.63	21.83	11.47	33.30	2.11	7.93	10.04	2.08	7.80
8.	0.06	0.81	0.08	0.95	0.85	1.68	2.53	20.36	18.74	39.10	22.16	20.42	42.58	0.98	10.61	11.59	0.95	9.53
9.	0.01	0.91	–	0.92	2.04	3.50	5.54	15.12	20.96	36.07	18.08	24.43	42.51	0.97	4.23	5.20	0.95	4.03

Note: The figures for "Total phytomass" show certain discrepancies, which were in the original table.

malacophyllous species mentioned above. *Stipa* is approaching its physiological limits in the Pamir. Some of the leaves dry up during the summer, and fruiting is very poor. *Zygophyllum* and *Christolea* resemble more the succulent halophytes with compensated water balance. These also come from the lower-lying semi-deserts but, without noticeable salinization in the mountains, are not as important. The semi-desert shrubs *Artemisia pamirica*, *Astragalus borodini*, and *Dracocephalum heterophyllum* and the cushion-plants *Gypsophila capituliflora* and *Acantholimon diapensioides* form a certain transition to the second vegetation type (alpine species with higher water requirements). The shrubs are probably more like steppe elements; the last two are characteristic for the near-Asian highlands which are not as dry as Pamir and receive more winter precipitation. The real alpine meadow plants find suitable temperature conditions in the Pamir region and a sufficiently long period without snow cover, since there is generally no snow cover at all. The species normally occur more in mountains with more moist climates, but in Pamir they can occur on soil which remains constantly moist. This second vegetation type is, thus, a more extrazonal occurrence also, following the law of relative site-constancy (Walter, 1974, p. 15). Each of these two vegetation types corresponds to a zonal vegetation type and occurs extrazonally in Pamir on corresponding sites.

AGRICULTURAL USE

The meadows of the river valleys of the Pamir have a special significance as pastures and actually gave the area its name (*pamirs* = meadows). Personnel of the Pamir Biological Station have studied these meadows in detail (Ladygina and Litvinova, 1974).

The high-mountain meadows cover an area of 200 000 ha. They are restricted to moist sites, that is, riverine floodplains and lake shores, less frequently to outcroppings of aquifers on slopes or to seepage from melting snowfields. The wetness of the floodplains has two maxima during the year, one in spring when the snow melts and the other in summer, when the glaciers melt back and the rivers have their high-water levels.

The upper soil horizons are very rich in humus,

since the low temperatures delay decomposition. These layers are also often slightly saline because of the arid climate.

The dominant plants are *Carex* and *Kobresia* species (*Carex pseudo-foetida*, *C. orbicularis*, *C. melanantha*, *Kobresia capilliformis*, *K. pamiroalaica*, etc.). Grasses (*Hordeum brevisubulatum*, *Poa pamirica*, etc.) and forbs (*Primula algida*, *Oxytropis hirsutiuscula*, etc.) are less important. The cover is 70 to 100%. The height of the plants is 20 to 30 cm, and the number of species per stand varies from 5 to 30.

The investigations of Ladygina and Litvinova (1974) were carried out in 1961–1966 at elevations between 3800 and 4800 m. Temperatures and precipitation levels during the summer months are given in Table 9.11. The phenospectra for elevations 3800 m (Fig. 9.13) and 4350 m (Fig. 9.14) show the course of development of the vegetation. The years 1961 and 1963 show the most favorable weather conditions, while the other years were colder and wetter.

Carex pseudo-foetida occurs at altitudes from 3500 to 4800 m, while the other dominant species do not survive above 4400 m. Only a few species are able to prevail at great heights, at 4200 to 4800 m, mainly *Carex melanantha*, *C. stenocarpa*, and *Primula moorcroftiana*.

The moistness of the biotope is also important for the composition of the particular meadow. *Carex melanantha* and *C. orbicularis* tolerate the greatest wetness and grow in the floodplains, where they are often inundated. *Carex pseudo-foetida* and *Kobresia pamiroalaica* occupy the intermediate biotopes, which are still often marshy and saline. The higher-lying biotopes in the floodplains and on the first terraces are covered by *Kobresia capilliformis*.

If one compares Figs. 9.13 and 9.14, one sees that the life cycles of the plants at higher elevations are completed in shorter periods of time. Most forbs

TABLE 9.11

Air temperature (in meteorological shelters) and precipitation at various elevations (up to 4760 m) in the Chechekty River Valley and, at 3640 m, near Murgab (from Ladygina and Litvinova, 1974)

Elevation (m)	Temperatures (°C)					Precipitation (mm)					
	June	July	August	Sept.	mean	June	July	August	Sept.	Sum	annual total
1963											
4760	−1.0	2.8	2.6	−1.5	0.7	21.9	13.6	0	0	35.5	204.0
4350	2.0	4.6	4.3	1.0	3.2	17.8	12.7	0	0	30.5	175.0
4100	5.0	7.4	6.5	3.0	5.5	4.0	12.1	0	0	16.1	94.2
3860	9.4	11.7	11.2	7.9	10.0	2.1	11.8	0	0	13.9	66.7
3640	9.9	12.5	12.0	6.1	10.1	4.0	8.8	0	0	12.8	35.8
1964											
4760	−3.8	2.1	2.5	−2.6	−0.4	26.3	72.0	43.0	15.0	156.3	340.0
4350	−1.0	4.0	4.5	−0.2	1.8	21.3	62.6	23.5	14.0	121.4	287.4
4100	3.5	6.4	7.7	2.4	5.0	15.0	60.0	6.0	9.0	90.0	209.5
3860	6.2	9.6	11.5	5.4	8.2	10.2	42.8	0	0.3	53.6	135.9
3640	7.9	11.5	12.5	6.6	9.6	12.8	21.9	1.3	0	36.0	81.1
1965											
4760	−3.0	2.0	0.5	−2.9	−0.7	13.8	31.5	93.2	8.0	146.5	324.8
4350	1.0	4.1	1.7	−1.5	1.3	11.1	30.9	87.2	8.0	137.2	292.2
4100	4.0	7.0	5.0	1.8	4.4	7.5	20.6	74.2	3.5	105.8	254.8
3860	8.1	10.9	8.5	4.4	7.9	4.2	5.2	47.2	2.6	60.3	196.8
3640	8.4	11.6	8.8	5.3	8.5	2.8	6.0	56.5	9.2	75.3	135.8

The wetting of the measuring vessels for precipitation was taken into account by means of a correction factor (Nechayev, 1966). The rainfall-measuring method of Tretyakov was employed during the summer and that of Kosarev at 4350 and 4760 m during the winter.

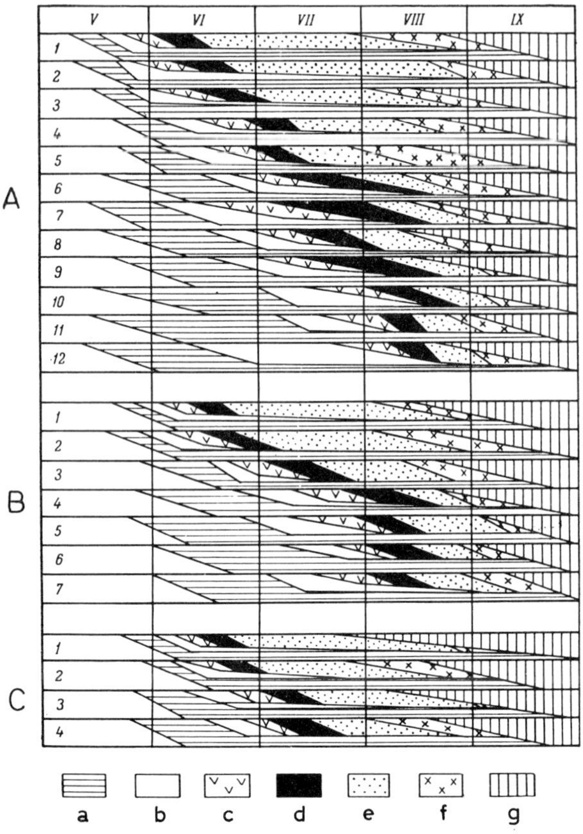

Fig. 9.13. Phenology of the high-mountain pastures at 3800 to 4060 m (from Ladygina and Litvinova, 1974): *a* = vegetative growth; *b* = formation of inflorescences; *c* = formation of flower buds; *d* = flowering; *e* = fruiting; *f* = seed propagation; *g* = drying out. Above: the months, May to September.
The species of three biogeocoenoses are as follow (*A–C*):
A: *Kobresia capilliformis* + *Hordeum brevisubulatum–Oxytropis hirsutiuscula* (at 3900 m): *1* = *Kobresia capilliformis*; *2* = *K. stenocarpa*; *3* = *Carex alexeenkoana*; *4* = *Brayia rosea*; *5* = *Taraxacum officinale*; *6* = *Oxytropis pagobia*; *7* = *Potentilla multifida*; *8* = *Hymenolaena nana*; *9* = *Oxytropis hirsutiuscula*; *10* = *Parnassia laxmannii*; *11* = *Gentianella pygmaea*; *12* = *Hordum brevisubulatum*.
B: *Carex pseudo-foetida–Puccinellia tenuiflora–Herbosae* (at 4060 m): *1* = *Kobresia capilliformis*; *2* = *Carex pseudo-foetida*; *3* = *Taraxacum pamiricum*; *4* = *Oxytropis hirsutiuscula*; *5* = *Poa pamirica*; *6* = *Gentiana leucomelaena*; *7* = *Puccinellia tenuiflora*.
C: *Carex orbicularis* + *Polygonum viviparum* (at 4060 m): *1* = *Carex orbicularis*; *2* = *C. microglochin*; *3* = *Kobresia stenocarpa*; *4* = *Polygonum viviparum*.

bloom in July, whereas the anthesis of the grasses begins relatively late. In cold summers the buds and flowers may be frozen or the fruit may not have time to ripen.

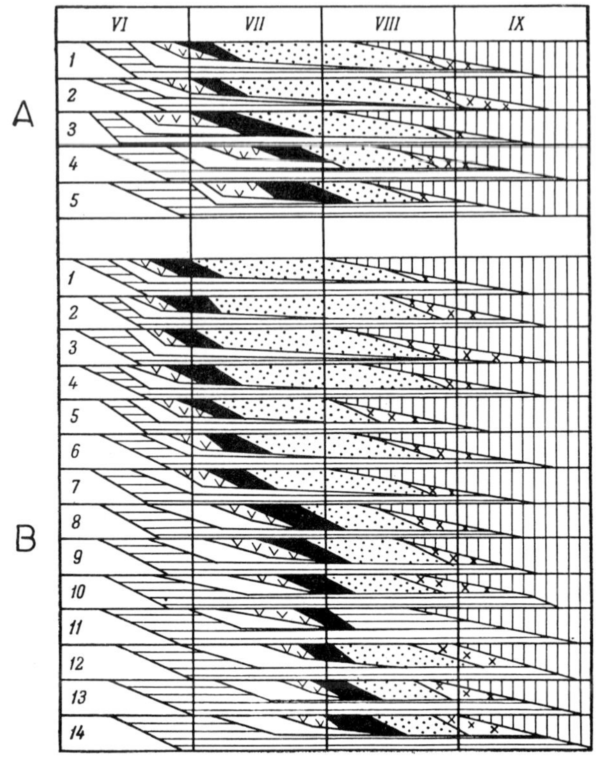

Fig. 9.14. Phenology of the high-mountain pastures at 4350 m (legend as in Fig. 9.13). The species of the two biogeocoenoses (*A–B*) are as follow:
A: *Kobresia capilliformis–Herbosae* (at 4350 m): *1* = *Kobresia capilliformis*; *2* = *Carex melanantha*; *3* = *Primula algida*; *4* = *Erigeron heterochaeta*; *5* = *Peducularis rhinanthoides*.
B: *Carex pseudo-foetida* + *Carex orbicularis–Herbosae* (at 4350 m): *1* = *Kobresia capilliformis*; *2* = *Carex pseudo-foetida*; *3* = *Carex orbicularis*; *4* = *C. melanantha*; *5* = *C. microglochin*; *6* = *C. oxyleuca*; *7* = *Kobresia pamiroalaica*; *8* = *Oxytropis platysema*; *9* = *Potentilla multifida*; *10* = *Leontopodium ochroleucum*; *11* = *Calamagrostis anthoxanthoides*; *12* = *Trisetum spicatum*; *13* = *Festuca coelestis*; *14* = *Oxytropis hirsutiuscula*.

The meadows have their greatest amounts of standing fresh weight from July to the beginning of August (Table 9.12). Since decomposition is greatly delayed at these elevations, the harvested plant matter always contains many dead plant parts: averages of 70% in the spring, 35 to 26% at the end of July to beginning of August, and again 80% in September (Tables 9.12 and 9.13). When harvested for scientific purposes the plant shoots are cut off at ground level, whereas they are cut at a height of 5 cm when harvested for agricultural purposes. This

TABLE 9.12

Relationship of air-dried phytomass (100 kg ha^{-1}) of various botanical groups of Pamir meadow vegetation at the time of maximum development (from Ladygina and Litvinova, 1974)

Biogeocoenosis	Year		Botanical groups				Total weight
			Carex spp.	grasses	legumes	other forbs	
I	1963	living	19.7	9.2	5.7	1.3	35.9
		dead	13.0	3.0	1.5	0.5	18.0
	1964	living	13.8	3.0	2.3	1.0	20.1
		dead	13.3	1.8	0.7	0.6	16.4
II	1963	living	10.8	–	–	4.5	15.3
		dead	6.3	–	–	2.0	8.3
	1964	living	5.6	–	–	2.2	7.8
		dead	5.4	–	–	0.8	6.2
III	1963	living	18.8	5.8	1.0	2.6	28.2
		dead	6.6	2.0	0.2	1.4	10.2
	1964	living	11.3	1.0	0.3	1.0	13.6
		dead	7.3	0.6	0.1	0.4	8.4
IV	1963	living	15.2	0.7	1.0	1.4	18.3
		dead	10.0	0.3	0.3	0.7	11.3
	1964	living	15.6	0.4	0.7	1.3	17.0
		dead	6.6	0.2	0.1	0.6	7.5
V	1963	living	17.1	2.0	0.9	2.2	22.2
		dead	7.5	1.0	0.4	1.1	10.0
	1964	living	16.1	1.2	0.3	2.0	20.0
		dead	6.1	0.9	0.3	1.0	8.3

I: *Kobresia capilliformis + Hordeum brevisubulatum–Oxytropis hirsutiuscula* BGC at 3900 m. II: *Kobresia capilliformis + Herbosae* BGC at 4350 m. III: *Carex pseudo-foetida + Puccinellia tenuiflora–Herbosae* BGC at 4060 m. IV: *Carex pseudo-foetida + C. orbicularis–Herbosae* BGC at 4350 m. V: *Carex orbicularis + Polygonum viviparum* BGC at 4060 m.

sort of agricultural harvesting includes only 45 to 63% of the total above-ground phytomass. Experimental plots showed that, compared with protected areas, the irregularly grazed plots produced a yield of only 28 to 32% under the biological (ground-level) cut and only 17 to 20% under the agricultural cut at 5 cm (Table 9.14). On the protected plots, on the other hand, so much dead organic matter accumulated over a few years that the effect was clearly negative.

The meadow soils are extraordinarily densely rooted, mainly in the upper horizons (Table 9.15), and imposing quantities of peat accumulate.

The underground organic matter makes up 94.6 to 98.5% of the total, and this is especially noticeable in cold summers, when the development of aerial shoots is inhibited (Table 9.16). It is interesting to compare these figures with corresponding values from other regions, where the above-ground phytomass is only slightly less than the underground phytomass (Table 9.17). The decomposition of dead plant parts and roots in the soil is greatly inhibited in Pamir by the low temperatures, and this leads to the accumulation of organic matter in the soil.

The meadows of Pamir have the highest productivity of all the biogeocoenosis types in the region, a fact which can be traced to the better water supplies which the meadows have available. The economic importance of the meadows, therefore, is also high.

An experimental garden of 93 ha was established at 2320 m (1500 m lower than the biological station) in order to test the possibilities of cultivating field

TABLE 9.13

Seasonal change of biological yields in 1963–1964 (100 kg ha^{-1} of air-dried matter) of meadow biogeocoenoses protected from grazing since 1960 (I–V = biogeocoenoses as in Table 9.12) (from Ladygina and Litvinova, 1974)

Biogeocoenosis	Sampling date	Total weight	Green parts	Dead parts	Dead fraction (%)
I	1963 24.VI	43.6	14.8	28.8	66
	14.VII	50.0	28.8	21.2	42
	5.VIII	53.9	35.9	18.0	33
	8.IX	40.2	10.2	30.0	74
	1964 25.VI	30.2	6.1	24.1	80
	20.VII	35.3	17.3	18.0	50
	10.VIII	36.5	20.1	16.4	45
	14.IX	33.0	4.5	28.5	86
II	1963 23.VI	14.6	3.8	10.8	74
	24.VII	18.8	9.3	9.5	50
	26.VIII	23.6	15.3	8.3	35
	3.IX	18.8	4.3	14.5	77
	1964 28.VI	10.5	1.6	8.9	84
	22.VII	11.0	4.0	7.0	63
	30.VIII	14.0	7.8	6.2	44
	4.IX	10.0	2.0	8.0	80
III	1963 25.VI	24.2	10.0	14.2	58
	28.VII	36.2	24.0	12.2	33
	20.VIII	38.4	28.2	10.2	26
	4.IX	26.1	10.1	16.0	61
	1964 20.VI	23.0	8.2	14.8	64
	15.VII	25.4	13.9	11.5	45
	16.VIII	28.0	19.6	8.4	30
	10.IX	22.5	8.5	14.0	62
IV	1963 29.VI	18.0	6.0	12.0	66
	29.VII	29.6	18.3	11.3	38
	10.VIII	23.6	15.5	8.1	34
	3.IX	19.3	6.0	13.3	68
	1964 15.VI	15.2	5.0	10.2	67
	21.VII	21.4	11.8	9.6	45
	13.VIII	24.5	17.0	7.5	30
	5.IX	12.6	2.6	10.0	80
V	1963 24.VI	22.5	6.8	15.7	69
	15.VII	28.2	17.0	11.2	40
	5.VIII	32.2	22.2	10.0	31
	4.IX	27.1	7.3	19.8	73
	1964 28.VI	20.0	5.1	14.9	74
	15.VII	21.8	9.8	12.0	55
	20.VIII	28.3	20.0	8.3	29
	5.IX	20.1	6.0	14.1	70

crops, fruit and other trees under the extreme conditions of Pamir. It was found that potatoes, various types of cabbage and other green vegetables and root crops, as well as forage crops previously unknown to the region, produce very good yields under artificial irrigation. Tomatoes must be grown under glass or in greenhouses before being planted outside. Alfalfa (lucerne) and espar-

TABLE 9.14

Influence of height of cutting and of grazing on the maximum hay yields (air-dried matter in t ha^{-1}) of meadow biogeocoenoses in Pamir (from Ladygina and Litvinova, 1974)

		Protected since 1960		Grazed area	
	Height of cut:	0 cm	5 cm	0 cm	5 cm
Kobresia capilliformis + Hordeum brevisubulatum–Herbosae BGC at 3900 m					
August 5, 1963	living parts	3.59	1.21	1.43	0.93
	dead parts	1.80	0.92	0.32	0.20
	total	5.39	2.13	1.75	1.13
	% of potential[1]	100	39(61)	32(68)	20(80)
August 10, 1964	living parts	2.01	0.80	0.89	0.61
	dead parts	1.64	0.57	0.28	0.10
	total	3.65	1.37	1.17	0.71
	% of potential	100	37(63)	30(70)	19(81)
Carex pseudo-foetida–Puccinellia tenuiflora–Herbosae BGC at 4060 m					
August 20, 1964	living parts	2.82	1.30	0.87	0.43
	dead parts	1.02	0.73	0.25	0.10
	total	3.84	2.03	1.12	0.53
	% of potential	100	50(50)	29(71)	13(87)
August 16, 1964	living parts	1.96	0.84	0.60	0.40
	dead parts	0.84	0.70	0.20	0.08
	total	2.80	1.54	0.80	0.48
	% of potential	100	55(45)	28(72)	17(83)

[1] Biological cut on protected areas = 100%. Numbers in parentheses: percentage lost in comparison with the biological cut.

TABLE 9.15

Root weight (g m^{-3} of soil) in meadow soil during the period of maximum development of the meadow vegetation (from Ladygina and Litvinova, 1974) (I–V are as in Table 9.12)

Soil horizon (cm)	Biogeocoenoses									
	I		II		III		IV		V	
	1963	1964	1963	1964	1963	1964	1963	1964	1963	1964
0–10	4250	3080	4832	4144	4040	3620	4070	3802	3266	2950
10–20	2330	2015	2680	1915	1360	1002	1770	1540	1285	1300
20–30	984	1044	1934	1608	644	570	675	703	985	840
30–40	910	955	560	761	240	225	194	175	833	620
40–50	402	552	230	291	136	131	137	120	720	515
50–60	292	120	100	94	98	100	117	105	276	250
60–70	120	108	90	84	91	80	49	60	82	76
70–80	48	56	40	52	70	65	35	28	50	51
80–90	39	40	18	17	35	38	20	16	17	20
90–100	25	32	15	12	30	26	12	14	13	15
Total	9400	8002	10 499	8978	6744	5857	7079	6563	7527	6637

sette (*Onobrychis sativa*) thrive. Grapevines bear fruit if they are covered in winter. Apricots, peaches, cherry plums, plums, sour cherries, red and black currants, and raspberries produce large quantities of fruit. Sweet cherries, pears, and apples also produce yields, but walnut and mulberries do not

TABLE 9.16

Phytomass (air-dried) of various parts of meadow biogeocoenoses in Pamir during the period of maximum development of above-ground matter (1964 = cool year); I–V are as in Table 9.12 (from Ladygina and Litvinova, 1974)

Biogeocoenosis	Year	Above-ground parts				Underground parts		Total organic matter (t ha^{-1})	Ratio of above-ground/below-ground parts
		living		dead					
		t ha^{-1}	%	t ha^{-1}	%	t ha^{-1}	%		
I	1963	3.59	3.6	1.80	1.8	94.00	94.6	99.39	1:17
	1964	2.01	2.4	1.64	2.0	80.02	95.6	83.67	1:22
II	1963	1.43	1.3	0.83	0.8	104.99	97.9	107.25	1:46
	1964	0.78	0.9	0.62	0.6	89.78	98.5	91.18	1:64
III	1963	2.88	4.0	1.02	1.4	67.44	94.6	71.28	1:17
	1964	1.96	3.2	0.84	1.3	57.85	95.5	60.61	1:20
IV	1963	1.83	2.3	1.13	1.7	70.79	96.0	73.75	1:22
	1964	1.70	2.5	0.75	1.1	65.63	96.4	68.08	1:26
V	1963	2.22	2.8	1.00	1.4	75.27	95.8	78.51	1:23
	1964	2.00	2.9	0.83	1.4	66.37	95.9	69.20	1:27

TABLE 9.17

Above- and below-ground phytomass (in t ha^{-1} air-dried matter) for various meadow biogeocoenenoses in Central Asia (Ladygina and Livinova, 1974)

Region	Phytomass			Soil sample (cm)	Above-ground/below-ground	Source of data
	total	above-ground	below-ground			
1. Tien Shan						
Forb-rich alpine meadow at 2850 m	38.75	0.70	38.05	0–80	1:54	Nikitina (1962)
Kobresia mats	75.80	1.20	74.60	0–100	1:62	Mamytov (1963)
Kobresia capilliformis	51.52	0.46	51.06	0–100	1:111	Popova (1963)
Phlomis–Iris brevituba meadow, 2600 m	40.71	3.15	36.56	0–70	1:12	Blagoobrazov (1966a, 1966b)
Geranium–Alchemilla meadow, 2550 m	32.12	3.74	28.38		1:8	
Floodplain meadows with *Carex melanantha*, 2550 m	149.20	3.17	146.03		1:46	
Alpine mats with *Trisetum spicatum*, 3300 m	37.84	3.32	34.52		1:10	
2. Gissar Range						
Allium–Puccinellia subspicata mats, 3360 m	26.13	0.49	25.65	0–100	1.52	Kaletkina (1966)
3. Eastern Pamir						
Carex pseudo-foetida meadows, 3600 m	43.27	1.40	41.87	0–40	1:29	Sveshnikova (1962)
Poa tibetica meadows, 3600 m	63.43	1.85	61.58		1:33	
Brookside meadows of *Blysmus compressus*, 3600 m	42.11	1.35	40.78		1:33	

fruit. Most of the broad-leaved foliage trees of the temperate zone can be cultivated, as well as *Robinia pseudoacacia*, but some produce shoots several times per year and have abnormal crown forms. Barley and turnips can be grown at much higher altitudes.

The ecological conditions and relationships which we have depicted for Pamir are probably also accurate for the Tibetan high-mountain deserts.

In connection with this, we look also at the latest summary of grazing conditions in all of the Pamir (Agakhanyantz, 1975). The economically usable phytomass in air-dried state was determined for the pastures over the course of twenty years (1949–1968) on transects through the experimental areas in the individual provinces of Pamir and, for comparison, in the pre-Pamir region. Each area covered 30 to 150 thousand hectares. In total, about 600 000 ha in western Pamir (about 25% of the total area) and about 550 000 ha in eastern Pamir (about 20% of the total area) were studied. The individual provinces (shown in Fig. 9.15) were the following:

Pre-Pamir:	0 Darvaz (control area)
Western Pamir:	1 Yazgulem
	2 Shugnan
	3 Vakhan
Transition:	4 Sarez (less studied)
Eastern Pamir:	5 Karakul'
	6 Murgab
	7 Southeast

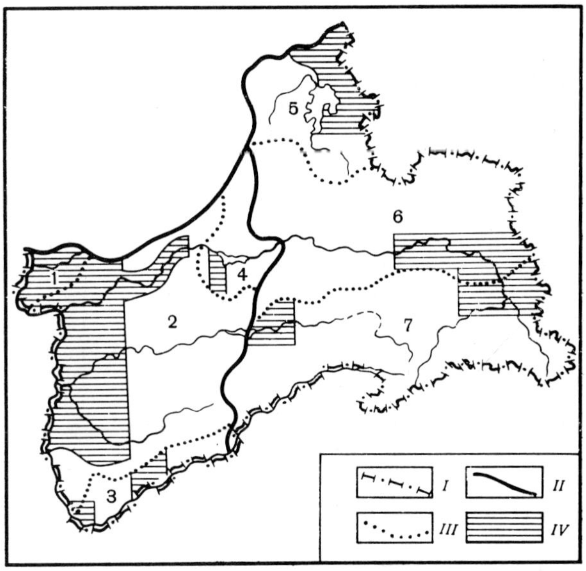

Fig. 9.15. Geobotanical pasture-land provinces according to Agakhanyantz (1975): *I* = Boundary of the U.S.S.R.; *II* = boundary between eastern and western Pamir; *III* = boundary of the pasture-land provinces (see text); and *IV* = the region studied.

Basic environmental conditions are given in Table 9.18.

Within the individual provinces the following altitudinal belts can be distinguished (from bottom to top).

1. **Yazgulem**: total grazing area of 162 600 ha

1740–3100 m: montane desert with *Artemisia*
3100–3500 m: mountain xerophytes with *Acantholimon korolkovii* and *A. pamiricum*
3500–4200 m: montane steppes with *Stipa*, *Poa*, and *Roegneria*
4200–4500 m: cryophilic vegetation with mats

2. **Shugnan**: total grazing area of 1 487 400 ha

2000–3300 m: montane desert with *Artemisia*
3300–3800 m: montane xerophytes with *Acantholimon* and *Onobrychis*
3800–4000 m: fragments of montane steppes with *Stipa*, *Cousinia*, etc.
4000–4600 m: cryophilic vegetation

3. **Vakhan**: total grazing area of 249 200 ha

2600–3400 m: saksa'ul desert with *Haloxylon* and *Artemisia*
3400–4100 m: montane xerophytes with *Acantholimon pamiricum* and *A. diapensioides*
4100–4700 m: cryophilic vegetation
The steppe belt is almost completely lacking

4. **Transition** (Sarez): total grazing area of 569 100 ha
Above the montane desert with *Artemisia* there follow immediately the montane steppes (inversion) with *Stipa* and white-pubescent *Artemisia*, then the montane xerophytes (as in province 3), and then the cryophilic vegetation; this region has not been completely studied

5. **Karakul'** (northern part of eastern Pamir): total grazing area 789 700 ha

3900–4200 m: *Ceratoides* desert
4200–4500 m: montane xerophytes (as in province 3)
4500–5200 m: cryophilic vegetation with *Oxytropis immersa*
Montane steppes are almost completely lacking

6. **Murgab**: total grazing area 1 034 700 ha

3600–4100 m: *Ceratoides* desert with *Stipa glareosa* and *Artemisia rhodantha*
4100–4600 m: montane xerophytes with *Acantholimon diapensioides*
4600–4800 m: cryophilic vegetation

7. **Southeast**: total grazing area 1 015 300 ha

3000–4200 m: *Ceratoides* desert with *Christolea pamirica* plus *Artemisia rhodantha* and *Tanacetum pamiricum*
4200–4600 m: montane xerophytes with *Acantholimon diapensioides*
4600–4800 m: fragments of montane steppe with *Festuca sulcata*, *Stipa orientalis*, plus *Artemisia*
4800–5000 m: cryophilic vegetation

TABLE 9.18

Environmental data for the individual provinces of Pamir (1–3 western Pamir, 5–7 eastern Pamir) (from Agakhanyantz, 1975)

Province No.	Latitude (center)	Range of elevation (m)	Desert (%)	Temperature (mean, °C)	Precipitation (annual, mm)
0	38°45′N	3100	51.7	8.7°	900
1	38°13′N	2860	73.2	9.9°	300
2	37°45′N	2800	56.9	9.1°	260
3	36°53′N	2400	41.4	6.8°	100
5	39°01′N	1300	88.1	−3.0°	70
6	38°10′N	1200	42.0	−1.7°	75
7	37°30′N	1100	59.4	−3.5°	120

TABLE 9.19

Mean yields (air-dried matter) of natural pastures on eu-climatopes in Pamir (kg ha^{-1}), plus pre-Pamir (high precipitation) for comparison (from Agakhanyantz, 1975)

Province		Organic matter eaten		Matter not eaten		Total matter (kg ha^{-1})
		kg ha^{-1}	%	kg ha^{-1}	%	
Pre-Pamir	0	423	56.6	324	43.4	747
Western Pamir	1	72	29.2	174	70.8	246
	2	47	23.5	153	76.5	200
	3	18	17.8	83	82.2	101
Eastern Pamir	5	31	40.8	45	59.2	76
	6	46	44.2	58	55.8	104
	7	76	46.6	87	53.4	163

The provinces 5 to 7 in eastern Pamir replace each other in sequence from north to south. The precipitation increases in this direction, while the vertical extent of the altitudinal belts decreases from 1300 m to 1100 m (see Table 9.18).

The mean yields per hectare of the pastures in the various provinces are given in Table 9.19.

One sees from the table that the yields decrease with precipitation both in western Pamir and in eastern Pamir. In western Pamir the yields decrease also from north to south, while in eastern Pamir the decrease is from south to north. This is especially noticeable in the values for organic matter eaten, the percentages of which also change in the same direction. In the wetter pre-Pamir area the yields are many times higher, and the percentage of usable matter is above 50%. The greater the aridity, the poorer are, in general, both the yields and the quality of the forage.

The effect of mean annual temperature, on the other hand, is hardly noticeable, although the low temperatures of eastern Pamir (see Table 9.18) do result in much lower yields in comparison with western Pamir.

Table 9.20 permits one to compare the mean yields on eu-climatopes for the individual altitudinal belts. The highest yields are found in the montane steppes, when steppes occur in the altitudinal sequence. The production of the cryophilic mats is reduced by the short growing season, that of the extreme deserts of eastern Pamir by aridity.

From these values for mean yield and from knowledge of the pasture areas, the entire feed supply of each province can be computed. In round numbers, Table 9.21 gives the following figures: in western Pamir the usable phytomass is 40 000 t (22%), the unusable phytomass is 140 000 t (78%), and the total phytomass is 180 000 t annually. In

TABLE 9.20

Mean pasture yields of the various altitudinal belts of Pamir in kg ha^{-1} (measurements on eu-climatopes, after Agakhanyantz, 1975)

Altitudinal belt	Western Pamir			Eastern Pamir		
	1	2	3	5	6	7
Cryophilic vegetation	220	90	50	90	110	140
Montane steppes	410	220	–	–	–	300
Montane xerophytes	310	200	110	110	170	250
Montane deserts	370	260	160	40	60	120

TABLE 9.21

Pasture areas and mean yields for the individual provinces of Pamir (from Agakhanyantz, 1975)

Province	Total area	Pasture area		Feed supply (in tons)		
		ha	%	usable	unusable	total
Western Pamir						
1	162 600	43 570	26.8	3 140	7 580	10 720
2	1 487 800	641 240	43.1	30 140	98 110	128 250
3	249 200	146 030	58.6	2 630	12 120	14 750
4	569 100	282 550	50.0	5 080	23 450	28 530
Total	2 468 700	1 113 390		40 990	141 260	181 250
in %				22.4%	77.6%	100.0%
Eastern Pamir						
5	789 700	93 100	11.9	2 880	4 190	7 070
6	1 034 700	599 600	58.0	27 580	34 780	62 360
7	1 015 300	407 200	40.6	30 940	35 430	66 370
Total	2 839 700	1 099 900		61 400	74 400	135 800
in %				45.3%	54.7%	100.0%
Total Pamir	5 308 400	2 213 290		102 390	215 660	317 050
in %				31.9%	68.1%	100.0%

eastern Pamir the usable phytomass is 60 000 t (45%), the unusable phytomass is 70 000 t (55%), and the total phytomass is 130 000 t annually.

In spite of the greater area in eastern Pamir, the total phytomass of the pastures is lower but the usable amount somewhat higher than in western Pamir. The seasonal distribution of precipitation is more favorable in eastern Pamir for the development of the phytomass, since the maximum precipitation falls during the growing season.

The exclusion of pastures which are not located on eu-climatopes and which, thus, are not included in these calculations can be justified because their total area is very small.

Of the total pasture area in Pamir, 92 to 98% has an annual yield (on eu-climatopes) in the range 100 to 400 kg ha^{-1}.

PRODUCTION ESTIMATES FROM THE HIGH-MOUNTAIN DESERT

The following production figures are taken from the latest available papers (Ladygina and Litvinova, 1971). The phytomass of the *Ceratoides–Artemisia rhodantha* biogeocoenosis on eu-climatopes is not large, about 2.2 to 15.2 t ha^{-1}, of which 90 to 96% is in the root system and only 1 to 3% in the annual assimilating organs. The longevity of the plants requires a low annual litterfall (branches and leaves) of 20 to 40%. Since decomposition is greatly inhibited by the low temperatures, however, considerable amounts of dead organic matter do accumulate.

Thc following two biogeocoenoses are discussed in more detail:

(1) *Ceratoidetum papposae purum*, which covers very large, rocky areas at 3800 to 3900 m and represents the most xerophilic formation, with less than 5% cover. The mean height is 12 cm. In other

TABLE 9.22

Productivity in the desert biogeocoenoses (t ha^{-1}) of eastern Pamir (from Ladygina and Litvinova, 1971)

	Ceratoidetum papposae		*Artemisietum rhodanthae*	
	var. *purum*	+ *Artemisia*	+ *Stipa* and *Ceratoides*	+ *Stipa* and forbs
Phytomass, t ha^{-1}	5.85	13.26	8.17	15.21
annual parts, t ha^{-1}	0.10	0.28	0.10	0.20
% of total phytomass	1.8	2.1	1.0	1.3
perennial parts, t ha^{-1}	0.23	0.89	0.33	0.36
% of total phytomass	3.9	6.7	4.0	2.3
roots, t ha^{-1}	5.52	12.09	7.74	14.65
% of total phytomass	94.3	91.2	94.7	96.4
Dead matter, t ha^{-1}	2.53	6.81	3.80	8.24
% of total organic matter	30.0	34.0	32.0	34.0
Total organic matter, t ha^{-1}	8.38	20.07	11.97	23.45
annual litter, t ha^{--1}	2.39	4.21	2.38	5.48
% of phytomass	41.0	31.0	29.0	36.0
Annual litter as fraction of dead matter	0.9	0.6	0.7	0.6

TABLE 9.23

Productivity of cryophilic cushion-plant biogeocoenoses from 4100 m (4300–4800 m) in Pamir (t ha^{-1}); for definitions of the communities I–IV, see text (from Ladygina and Litvinova, 1971)

	Potentilletum		*Ajanietum*	*Sibbaldietum*
	I	II	III	IV
Phytomass, t ha^{-1}	18.00	23.50	7.89	16.12
annual parts, t ha^{-1}	0.92	0.95	0.37	0.92
% of total phytomass	5.1	4.0	4.8	5.6
perennial parts, t ha^{-1}	0.85	0.85	0.53	2.04
% of total phytomass	4.7	3.6	6.7	12.0
roots, t ha^{-1}	16.23	21.70	6.99	13.16
% of total phytomass	90.2	92.2	89.5	82.4
Dead matter, t ha^{-1}	15.04	18.90	11.71	26.40
% of total organic matter	46.0	44.0	64.0	62.0
Total organic matter, t ha^{-1}	33.04	42.40	19.60	42.52
annual litter, t ha^{--1}	9.34	10.22	3.22	4.86
% of phytomass	51.0	43.0	41.0	30.0
Annual litter as fraction of dead matter	0.6	0.5	0.3	0.1

TABLE 9.24

Annual plant growth of the main biogeocoenoses of Tadzhikistan (t ha^{-1} air-dry weight) (after Stanyukovich, 1971)

Vegetation communities	Community types[1]	Herbs			Semi-shrubs			Trees and shrubs				Total				
		Annual shoots and leaves	Roots	Total	Annual shoots and leaves	Woody above- and below-ground parts	Total	Annual shoots and leaves	Woody above-ground parts	Roots	Total	Annual shoots and leaves	Woody above-ground parts	Woody above- and below-ground parts, semi-shrubs	Roots	Total
I. *Vegetation receiving only natural precipitation*																
A. Desert vegetation																
a. lowland	1	–	–	–	0.43	0.16	0.59	–	–	–	–	0.43	–	0.16	–	0.59
	2	0.12	0.52	0.64	0.44	1.29	1.73	–	–		–	0.56	–	1.29	0.52	2.37
	3	0.16	1.05	1.21	0.18	0.62	0.80	–	–	–	–	0.34	–	0.62	1.05	2.01
b. mountain	4	0.008	0.35	0.35	0.09	0.37	0.46	–	–	–	–	0.09	–	0.37	0.35	–
	5	0.008	0.24	0.24	0.17	0.12	0.29	–	–	–	–	0.17	–	0.12	0.24	0.53
c. high mountain	6	0.88	0.61	1.49	0.55	0.13	0.68	–	–	–	–	1.43	–	0.13	0.61	2.17
B. Ephemeral vegetation	7	4.50	4.39	8.89	–	–	–	–	–	–	–	4.50	–	–	4.39	8.89
C. Low shrub vegetation	8	3.35	3.15	6.50	8.00	4.50	12.50	–	–	–	–	11.35	–	4.50	3.15	19.00
D. Shrublands (*shiblyak*)																
a. shrublands	9	1.45	8.75	10.20	–	–	–	2.26	2.73	4.50	9.49	3.71	2.73	–	13.25	19.69
(mountain)	10	3.62	6.29	9.91	–	–	–	6.00	3.00	4.49	13.49	9.62	3.00	–	10.69	13.31
b. shrublands	11	2.87	11.14	14.01	–	–	–	1.29	1.08	0.49	2.86	4.16	1.08	–	12.53	16.87
(high mountain)	12	5.06	2.30	7.36	–	–	–	0.90	0.63	6.81	7.34	5.96	0.63	–	9.11	15.60
	13	0.77	0.53	1.30	–	–	–	28.87	6.67	3.40	38.94	29.64	6.67	–	3.93	40.24
II. *Vegetation receiving only ground water*																
A. Forests	14	16.00	1.04	17.04	–	–	–	1.74	0.60	1.75	3.89	17.74	0.60	–	2.79	21.03
	14	25.00	3.54	28.84	–	–	–	1.74	1.20	2.00	4.94	26.74	1.20	–	5.54	33.48
B. Shrublands	15	4.93	40.33	45.26	–	–	–	7.54	37.53	13.78	58.85	12.47	37.53	–	54.11	104.11
C. Herbaceous	16	6.25	11.50	17.75	–	–	–	–	–	–	–	6.25	–	–	11.50	17.75
	16	7.43	11.50	18.93	–	–	–	–	–	–	–	7.43	–	–	11.50	18.93
	17	7.00	33.40	40.40	–	–	–	–	–	–	–	7.00	–	–	33.40	40.40
	17	9.75	33.40	42.97	–	–	–	–	–	–	–	9.75	–	–	33.40	42.97

[1]Community types: 1 = *Artemisia scotina* + ephemerals; 2 = *Artemisia namanganica* + ephemerals; 3 = *Salsola richteri* + *Poa bulbosa* + ephemerals; 4 = *Artemisia glaucina* + ephemerals; 5 = *Artemisia porrecta* + ephemerals; 6 = *Artemisia lehmanniana* + *Poa relaxa* + *Acantholimon diapensioides*; 7 = *Poa bulbosa* + *Agropyron trichophorum*; 8 = *Perovskia scrophularifolia* + *Agropyron trichophorum*; 9 = *Amygdalus ulmifolia*; 10 = *Crataegus turkestanica*; 11 = *Rosa ovczinnikovii*; 12 = *Malus sieversii*; 13 = *Juglans regia*; 14 = *Populus diversifolia*; 15 = *Salix iranica*; 16 = *Calamagrostis pseudophragmites* + *Equisetum ramosissimum*; 17 = *Carex* sp. + *Phragmites communis* + *Equisetum arvense*.
Some of the figures in the original table were clearly misprints, and have been corrected.

variants there occur individuals of *Artemisia rhodantha*, *Torularia korolkovii*, and *Veronica biloba*.

(2) *Artemisietum rhodanthae* with *Ceratoides*, which grows on fine-grained rubble areas at 3900 to 3950 m on north exposures. The cover is 15% and the height of the plants 20 cm. Individual companions are *Stipa glareosa* and *Carex stenophylloides*, etc. The results of the production measurements are given in Table 9.22.

Of the cryophilic cushion-plant communities from greater altitudes the following were studied:

(1) The herbaceous *Potentilletum pamiricae* in two forms, a variation with *Poa glauciglumis* and *Calamagrostis anthoxanthoides* (I) and a variation with *Oxytropis immersa* and *Smelovskia calycina* (II); cover is 25%.

(2) The *Ajanietum tibeticae* (III) with *Colpodium leucolepis*, etc.; cover is 40%.

(3) The *Sibbaldietum tetrandrae* (IV) with *Colpodium* and *Rhodiola gelida*; cover is 25 to 30%.

The production values are shown in Table 9.23.

Sveshnikova (1971) continued the studies of Ladygina and Litvinova by determining the densities of the plant stands on various biotopes and computing the fraction of the total above-ground biomass represented by the leaves. The differences are very large, depending on the water supply. The less water is available, the smaller the number of leaves produced.

Finally we look at a preliminary overview of the growth of the individual biogeocoenoses in all of Tadzhikistan, which reaches well beyond the borders of Pamir (Stanyukovich, 1971). The studies were carried out on small experimental plots, on which the ages of the woody plants were determined. The annual growth of wood and of roots was roughly estimated by dividing the entire woody mass or root mass of the upper two meters of soil by the age in years of the individual plant. The growth figures obtained seem to be too high in some cases. The translated but otherwise unaltered table of Stanyukovich (1971) is reproduced as Table 9.24.

REFERENCES

Agakhanyantz, O.E., 1958. Short survey of the Badakhshan vegetation. *Sb. Statei Tadzhik. Filial. Geogr. Obshch. S.S.R.*, 99: 47–67 (in Russian).

Agakhanyantz, O.E., 1964. Pamir and Central Asia. *Izv. Vses. Geogr. Obshch.*, 2: 103–108 (in Russian).

Agakhanyantz, O.E., 1965. On the new paleography of Pamir. *Izv. Akad. Nauk S.S.R., Ser. Geogr.*, 3: 84–88 (in Russian).

Agakhanyantz, O.E., 1970. *Life on Mountain Slopes*. Priroda, Nauka, Moscow, pp. 89–95 (in Russian).

Agakhanyantz, O.E., 1975. Above-ground phytomass of the Pamir pastures. *Izv. Akad. Nauk S.S.R., Ser. Geogr.*, 5: 43–50 (in Russian).

Agakhanyantz, O.E. and Yusufbekov, K.Y., 1975. *Vegetation of West Pamir and Attempts of Reconstruction*. Donesh, Dushanbe, 310 pp. (in Russian).

Blagoveshchenskiy, V.A., 1935. Über den Verlauf der Photosynthese im Hochgebirge des Pamir. *Planta*, 24: 276–287.

Breckle, S.-W., 1973. Mikroklimatische Messungen und ökologische Beobachtungen in der alpinen Stufe der Afghanischen Hindukusch. *Bot. Jahrb.*, 93: 25–55.

Chumichev, D.A., 1968. *Sovietskiy Soyuz. Tadzhikistan*. Nauka, Moscow, 238 pp.

Filipov, A.A., 1959. Daily and annual variations of photosynthetic intensity of plants in East Pamir. *Tr. Bot. Inst. Akad. Nauk S.S.S.R.*, Ser. IV, 13 (in Russian).

Glagolyeva, T.A., 1962. Photosynthesis of plants in the upper alpine belt of East Pamir. *Bot. Zh.*, 47: 11 (in Russian).

Grigoryev, Yu.S., 1955. *Comparative Ecological Studies of Xerophilization in Higher Plants*. Akad. Nauk S.S.S.R., Leningrad (in Russian).

Ikonnikov, S.S., 1963. *Key to the Plants of the Pamir*. Akad. Nauk Tadzhik. S.S.R., Dushanbe, 282 pp. (in Russian).

Ismailova, N.N., 1963. Water and temperature conditions in the soil and transpiration of plants in the alpine belt of Pamir. *Tr. Pamirsk. Biol. Stantsii*, 1: 172–180 (in Russian).

Ismailova, N.N., 1965. Transpiration intensity of high-mountain plants in relation to the altitude of the biotopes. *Probl. Bot.*, 7: 205–212 (in Russian).

Ismailova, N.N., 1974. Transpiration and water loss of the plant communities in Eastern Pamir. *Ekologiya*, 1: 95–97 (in Russian).

Ismailova, N.N. and Litvinova, N.P., 1969. On the productivity of green matter and the water expenditure for transpiration in some wormwood communities (*Artemisieta*) of Eastern Pamir. *Bot. Zh.*, 54: 1586–1590 (in Russian).

Ismailova, N.N. and Litvinova, N.P., 1972. Water regime and productivity of main dominant plants of the high-mountain deserts (Eastern Pamir). In: L.E. Rodin (Editor), *Int. Symp. U.S.S.R.: Eco-Physiological Foundations of Ecosystem Productivity in Arid Zones*. Nauka, Leningrad, pp. 76–79.

Ismailova, N.N., Ladygina, G.M., Litvinova, N.P. and Steshenko, A.P., 1975. East Pamir. In: *Resources of the Biosphere, 1*. Nauka, Leningrad, pp. 181–186 (in Russian).

Kuprevich, V.F., Grigoryev, Yu.S. and Viskovakaya, E.K., 1949. Water economy of some species of the genus *Cousinia* in the Pamiro–Alai. *Bot. Zh.*, 34: 16–33 (in Russian).

Ladygina, G.M. and Litvinova, N.P., 1971. Productivity of above-ground and below-ground phytomass of some high-mountain plant communities in Pamir. In: L.E. Rodin and N.N. Smirnov (Editors), *Biological Productivity and Mineral Cycling*. Nauka, Leningrad, pp. 150–156 (in Russian).

Ladygina, G.M. and Litvinova, N.P., 1974. Productivity of

some meadow communities in the Eastern Pamir. *Probl. Bot.*, 12: 275–285 (in Russian).

Litvinova, M.A. and Ismailova, N.N., 1968. Annual above-ground shoots and the transpiration of *Ceratoides* communities in Pamir. *Izv. Otd. Biol., Akad. Nauk Tadzhik., S.S.R.*, 4: 29–34 (in Russian).

Rodin, L.E. (Editor), 1972. *Eco-Physiological Foundations of Ecosystem Productivity in Arid Zones.* Int. Symp. U.S.S.R., 7–19 June 1972. Nauka, Leningrad, 232 pp.

Rodin, L.E. and Smirnov, N.N. (Editors), 1975. *Resources of the Biosphere: Synthesis of the Soviet Studies for the International Biological Program, I.* Nauka, Leningrad, 288 pp. (in Russian).

Shardakov, V.S., 1953. *The Water Relations of Cotton.* Akad. Nauk Uzbek. S.S.R., Tashkent (in Russian).

Shardakov, V.S., 1956. Determination of irrigation data for cotton plants by the suction tension of their leaves. *Tr. Akad. Nauk Uzbek. S.S.R.*, (in Russian).

Stanyukovich, K.W., 1949. *Plant Cover of Eastern Pamir.* Nauka, Moscow, 159 pp. (in Russian).

Stanyukovich, K.W., 1971. Biological productivity of the main types of vegetation in the Tadzhik S.S.R. In: *Biological Productivity and Mineral Cycling.* Nauka, Leningrad, pp. 164–170 (in Russian).

Stanyukovich, K.W., 1973. *Mountain Vegetation of the U.S.S.R.* Akad. Nauk Tadzhik. S.S.R., Dushanbe, 310 pp. (in Russian).

Stanyukovich, K.W. and Krivonogova-Stanyukovich, M.B., 1973. Thorn-cushion species of the genus *Acantholimon* in Pamir and their geobotanical characterization and altitudinal distribution. In: *Problemy Biogeocoen. Geobot. i Geogr. Rast.* Akad. Nauk S.S.S.R., Leningrad, pp. 280–287 (in Russian).

Steshenko, A.P., 1956. Formation of the semi-shrub structure in the high mountains of Pamir. *Tr. Akad. Nauk Tadzhik. S.S.R.*, 50: 162 (in Russian).

Steshenko, A.P., 1973. Determination of the age of the cushion-plant *Sibbaldia tetrandra* Bunge in the high mountains of Pamir. *Bot. Zh.*, 58: 1004–1011 (in Russian).

Steshenko, A.P., et al., 1969. Studies of edificators in Pamir alpine deserts. *Probl. Osvoyeniye Pustyn*, 1: 32–42 (in Russian).

Sveshnikova, V.M., 1962. *The Water Relations of Plants and Soils in the High Mountain Deserts of Pamir.* Akad. Nauk Tadzhik. S.S.R., Stalinabad, 247 pp. (in Russian).

Sveshnikova, V.M., 1971. On biological productivity of some major formations in the Pamir. In: *Biological Productivity and Mineral Cycling.* Nauka, Leningrad, pp 156–164 (in Russian).

Sveshnikova, V.M., 1973. Water regimes of plants under the extreme conditions of high-mountain deserts of Pamir. In: *UNESCO, Proc. Uppsala Symp. 1970*, pp. 555–561.

Tyurina, M.M., 1957. The significance of negative turgor pressure for the water relations of leaves. *Bot. Zh.*, 42: 1035–1043 (in Russian).

Walter, H., 1968. *Die Vegetation der Erde, II. Die gemässigten und arktischen Zonen.* Gustav Fischer-Verlag, Jena, Stuttgart, 1001 pp.

Walter, H., 1974. *Die Vegetation Osteuropas, Nord- und Zentralasiens. Vegetationsmonographien der einzelnen Großräume, Band VII.* Gustav Fischer-Verlag, Jena, Stuttgart, 452 pp.

Zalenski, O.V., 1963. Maximum photosynthetic intensity of Pamir plants and others. *Tr. Pamir. Biol. Stantsii*, 1 (Dushanbe) (in Russian).

Zalenski, O.V., 1973. Peculiarities of the photosynthesis of desert plants. *Tezisy Delegatsk. S'ezda Vses. Bot. Obshch.*, pp. 105–107 (in Russian).

Chapter 10

TEMPERATE DESERTS AND SEMI-DESERTS OF AFGHANISTAN AND IRAN

S.-W. BRECKLE

MAIN TYPES AND GEOGRAPHICAL DISTRIBUTION

The deserts and semi-deserts south of the Amu-Dar'ya River can be likened to those in the north, which were described in the chapter on the Karakum Desert (Ch. 6 of this volume; Walter, 1974, 1976). On the other hand, they also have affinities to the subtropical deserts to the southwest; that is, there are large zono-ecotones existing there. The deserts of southeastern and eastern Afghanistan are surrounded by high mountains; thus, the zono-ecotones lead to orobiomes. Deserts in Iran are mostly located in vast basins, which, again, are surrounded by mountains.

The region of concern here is delimited by the Zagros Mountains (Tehran–Arak–Isfahan–Abadeh–Neyriz) in the west and southwest and by Zohary's (1963) boundary in the south (approximately the line Neyriz–Kerman–Zahedan). In the latter area hot and cold deserts intergrade (see Vol. 12 of this Series: Evenari and Noy-Meir, in press).[1]

The mountain regions in central Afghanistan (Ghorat, Paropamisus and Hazarajat; see Fig. 10.1, *4E*, *4F*) are mostly semi-desertic and thus included here, but the more easterly high mountains (Hindu Kush, Nuristan, Safed Koh; Fig. 10.1, *4G*) are mostly excluded. The northern slopes of the Afghan mountains (Fig. 10.1, *3E*) are covered by open woodlands and shrub semi-deserts, but some of the broad interior valleys are desertic (Fig. 10.1, *2G*).

The southeastern limit is drawn where the first slight influences of the monsoon climate correlate with open woodland vegetation. The boundary in the northeast is arbitrary. The Oxus Basin (Fig. 10.1, *1F*) is desertic; it comprises the middle part of the Amu-Dar'ya. In the north and northwest limits are the southern slopes of the Elburz Mountains (Fig. 10.1, *4A*) and their extensions to the east, the Kopet Dag Mountains (Fig. 10.1, *4D*).

The whole region can be subdivided into several parts (Fig. 10.1) which might be tentatively classified as subzonobiomes. The central Persian desert basins (Fig. 10.1, *1A*, *1B*), the basins of Masileh, the Great Kavir, Bidschistan, Tabas and Lut, are connected with smaller basins (Fig. 10.1, *2A*, *2B*), those of Qom, Ardestan, Na'in, Bafq and Kerman near the Zagros Range; those in the northeast in Khorasan; and in the east the basins of Esfarayen, the lowlands of Kal-Shur and Chusf-Shur.

Separated by the lower east Persian Mountains, the basin of Sistan leads to the south Afghan deserts (Sistan, Dasht-i-Margo and Registan). The Dar'ya–Namakzar region leads to the northwestern Afghan deserts and semi-deserts in the vicinity of the Murgab River and its tributaries. East of Selsela Turkestan, the lowlands of the Amu-Dar'ya Basin (Oxus Basin) are located.

PHYSICO-GEOGRAPHICAL FEATURES

Geology

The geological history of the northern parts of the area were touched on in the chapter on the Karakum (Ch. 6). Derivation of the more recent relief pattern found within the basins and lowlands is discussed there also. Some main features on the geology of Afghanistan are given in several papers

[1] The northernmost date palms in Iran are growing in Torud, but never bear ripe fruit.

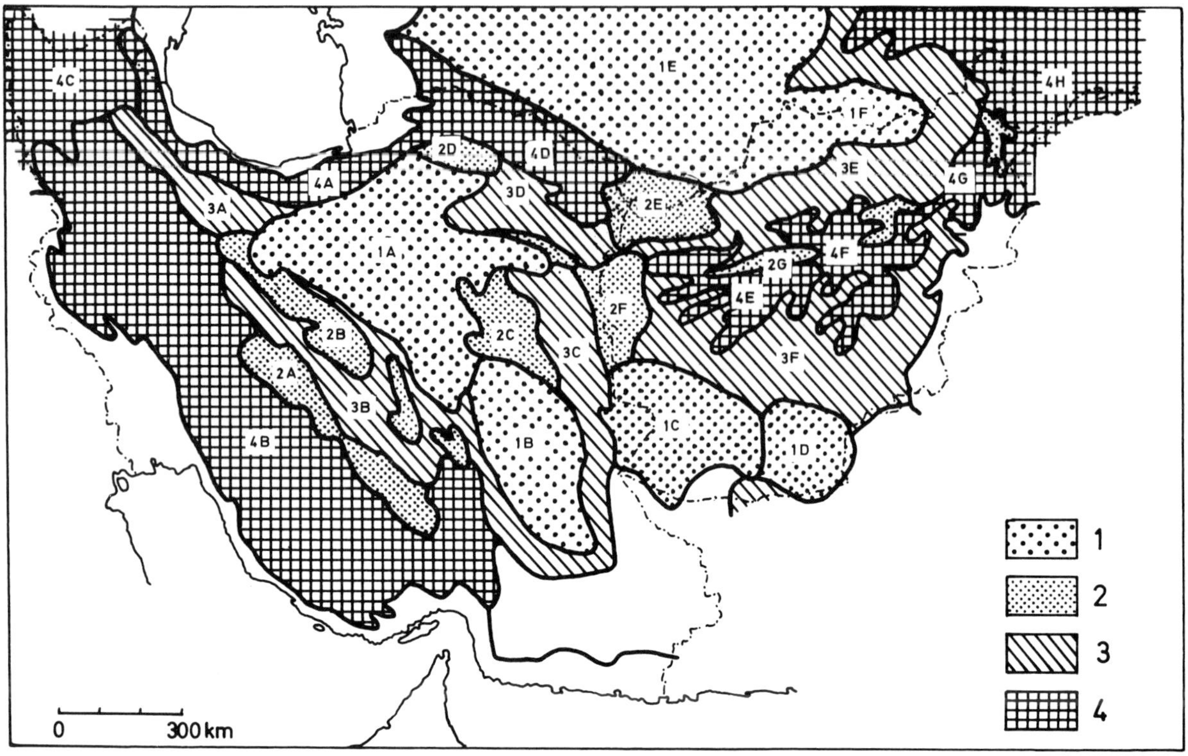

Fig. 10.1. The main regions of the Iran–Afghanistan area (in part from Scharlau, 1969; Schweizer, 1975). Desert regions: *1A* = the Great Kavir and other central Persian desert basins; *1B* = the Dasht-e-Lut Desert basin; *1C* = Sistan and Dasht-i-Margo, south and southwest Afghan deserts; *1D* = Registan, southeast Afghan sandy desert; *1E* = Karakum and adjacent desert regions; *1F* = Oxus (Amu-Dar'ya) Basin.
Smaller marginal deserts and semi-deserts: *2A* = basins within eastern Zagros Ranges; *2B* = marginal deserts in the basins of Ardestan, Na'in, etc.; *2C* = marginal desert in the Chusf-Shur Region; *2D* = marginal desert in the Esfarayen Basin; *2E* = semi-desert in western Afghanistan (Khorasan); *2G* = desertic valleys and basins within Afghan mountains.
Transitional hill regions, semi-deserts and lower mountains: *3A* = plains and hills between the Zagros and Elburz Mountains; *3B* = Kuh-Rud Mountains in central Iran; *3C* = Kuhistan and Zabolistan hill region in eastern Iran; *3D* = Khorasan mountain region; *3E* = northern slopes of the Afghan mountains and circum-Oxian hill region; *3F* = southern slopes and hills of Afghan mountains.
Mountain regions: *4A* = Elburz Range; *4B* = Zagros Ranges; *4C* = eastern Anatolian mountains; *4D* = Turkmenian and Khorasan ranges; *4E* = Ghorat mountain region in western central Afghanistan; *4F* = Hazarajat mountain region in central Afghanistan; *4G* = Hindu Kush high ranges; *4H* = the Pamir high ranges.

of Kaever (1967a, b, c, 1972). The geological mosaic and the stratigraphy of the area is very complex and cannot be easily summarized in the space available here. The reader is referred to Furon (1941), Tromp (1954), Menessier (1961), Hinze (1964), Wirtz (1964), Stöcklin (1968), Wittekindt (1973), Blaise et al. (1977), Boulin and Bouyx (1977), Conrad and Girod (1977), Lang (1977) and Ricou et al. (1977) for details. Some of the main factors will be given here for background information.

The great importance of the Hindu Kush mountain system, which was folded during the Tertiary, is emphasized. The same orogeny applies to the Elburz in Iran and some other ranges. Thus, the present orographical picture of the area is chiefly determined by high-mountain ranges of Tertiary age. In general, they are dominated by crystalline slates, gneisses and intrusives, along with some Paleozoic and Mesozoic metamorphic series. Paleozoic and Mesozoic clay-slates and quartzites are found in some other parts, as, for example, in the Afghan central mountains, in Ghorat and Hazarajat. Limestones are also found in Ghorat. Towards the margins of the mountains, the rocks dip under talus and gravel, sand dunes and loess deposits. Silt and clay deposits characterize the various basins and surrounding deserts. Thick loess deposits occur mainly on the margins of the foothills of the Afghan mountain region and in several parts of northeastern Iran.

Only in the highest regions of the mountains was glaciation prominent in the Quaternary. According to Bobek (1934), only the northern slopes of the Elburz Mountains were glaciated within Iran. The snowline might have dropped to about 1000 m below the present level during glacial periods. Zohary (1973) believed that the rainfall conditions in inner Iran during the Pleistocene were not very different from those of today. There is not the slightest evidence for assuming higher humidity from the very few palynological data available for Iran. The lower basins might, however, have been somewhat less desertic than today (Scharlau, 1958). The differences might have been enough for the existence of large inland lakes where sandy or clay regions often with high percentages of soil salts occur today. Vast areas were filled with gravels, sand, and clay materials during glacial and interglacial periods. Because of the endorheic situation, salts accumulated and are yet accumulating. The lowlands, therefore, are often covered by salt deserts on Quaternary deposits. Alluvial fans, playas and sand dunes with local loess coverings are often mixed together. More details on geological history and the periglacial situation are given by Bobek (1961) and Pias (1977).

The monsoonal influence seems to have occurred rather further to the west of the Kabul area during the Late Tertiary and the beginning Quaternary than today (Breckle, 1967). This can be concluded from findings near Kabul of fossil *Pinus roxburghii* and leaves of other trees in marl sediments. These tree species are no longer growing in Afghanistan but only in Pakistan and eastwards. The limit of the Indian plate on Afghan territory was discussed from a tectonic viewpoint by Carbonnel (1977). Wright and Van Zeist (1963), Wright et al. (1967), and Van Zeist (1967) provided data on the much greater extension of trees in Persia during the Pleistocene, but there are indications that the climate during the pleniglacial was more arid than today. Palynological studies are in progress at the Touran Research Programme (Spooner, 1977), which might give some ideas on the vegetational history of the northeastern part of the Great Kavir.

Hydrology

Rivers of the region are untamed (undammed) mountain streams which discharge highly variable amounts of water. With few exceptions, they are inland rivers, disappearing in the deserts or steppes or flowing into inland lakes. The Zagros Mountains, the Baluchistan and the Makran areas discharge water into the Persian Gulf and the Gulf of Oman. All of central Iran is characterized by several small and some large endorheic systems (vast basins, called *kavirs*; similar to the North African *shotts* with clay and salt swamps, or to the American *playas*) (cf. Krinsley, 1970; Breckle, 1981a). The northern parts of Iran are drained to the Caspian Sea, the greatest inland lake on earth. The northern part of Afghanistan is drained by several rivers which partly join the Amu-Dar'ya and thus discharge to the Aral Sea. The Hari Rud and the Murgab Rud disappear in desert oases and feed large groundwater reservoirs (cf. p. 109) in the Karakum. West, south and central Afghanistan is drained by the Farah Rud, Khash Rud, Helmand and Arghandab Rivers, which discharge into the swampy inland lakes at the Afghan–Iranian border (Hamun-e-Sabori, Hamun-e-Puzak). Only a small section of eastern Afghanistan is drained to the Indian Ocean via the Kabul and the Indus Rivers.

Climate

The entire Irano-Turanian area and almost all Afghanistan has an extremely continental and arid climate. The highly accentuated surface relief, the great variability from year to year and the short period of meteorological records impart an incomplete perspective of the climatic conditions. Some of the data available are used for representative climatic diagrams (Fig. 10.2). Additional sources for climatic data are Bobek (1952), Adle (1960), Ganji (1968), Sabeti (1969), Herman et al. (1971) and Lalande et al. (1974). In general, the climate of the area resembles the arid climate of the adjacent Karakum (Ch. 6).

The monthly mean temperatures show high annual fluctuations varying from 25° to 28°C. The absolute temperature fluctuations in some places reach 70°C. Some stations in arid mountain basins exhibit an absolute minimum of less than −50°C

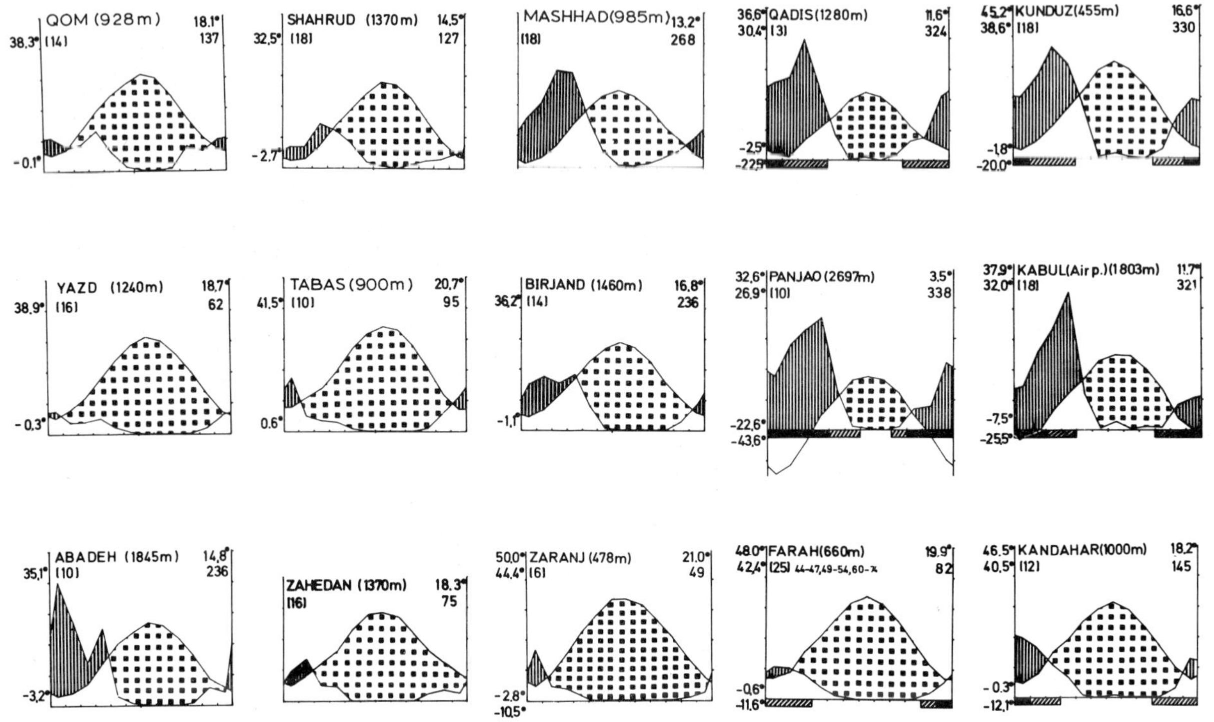

Fig. 10.2. Selected climatic diagrams of the desert and semi-desert areas of Iran and Afghanistan. Data from Afghan Air Authority. (The climatic data are arranged according to the method of Walter and Lieth, 1960–1967; Walter, 1974).

(e.g., Shaharak, in Hazarajat); this type of climate occurs also at Panjao (Fig. 10.2).

The area is characterized by the regular occurrence of frost days (Fig. 10.3). Only a few transitional regions of both countries (e.g., Dasht-e-Lut, or Sistan) exhibit less than thirty days with minimum temperatures below 0°C per year (cf. Vol. 12, Evenari and Noy-Meir, in press). On the other hand, from Dasht-e-Lut and from Sistan, there are reported more than five days per year with an average temperature exceeding 45°C and absolute maxima around 51°C, similar to the conditions shown by the climatic diagram of Zaranj (Fig. 10.2).

Diurnal temperature fluctuations arc likewise remarkable and characteristic of the continental climate of the area. During the dry months they often exceed 20°C. The daily temperature amplitude can reach 50°C or more on the soil surface.

Precipitation varies greatly depending on the elevation and geographic position of the area. Vast areas have total annual precipitation of less than 200 mm yr^{-1} (Fig. 10.4). In the mountains and at their base there are zones with total precipitation (rain and snow) from 200 to 500 mm yr^{-1}. Only at higher elevations and on the exposed mountain ridges (e.g., on north slopes of the Elburz and some parts of the Hindu Kush) does precipitation of more than 1000 mm yr^{-1} occur. There it falls as large amounts of snow during the winter season. The main period of precipitation is during winter and spring in the north, ceasing some weeks later than in the south. The summers are dry and hot; autumn also is an arid season. Exceptions are the non-desertic areas (northern slopes of the Elburz and the monsoon-influenced area of eastern Afghanistan), which are excluded from this chapter.

Of the meteorological stations with similar mean annual precipitation, the more northerly ones report more days with precipitation (Table 10.1) and many more cloudy days than the corresponding stations located in the south. Heavy rainfall with high surface runoff occurs more frequently in the south than in the north.

The variability of the low precipitation in very arid locations is remarkably high (Fig. 10.5). Comparing the different amounts of precipitation

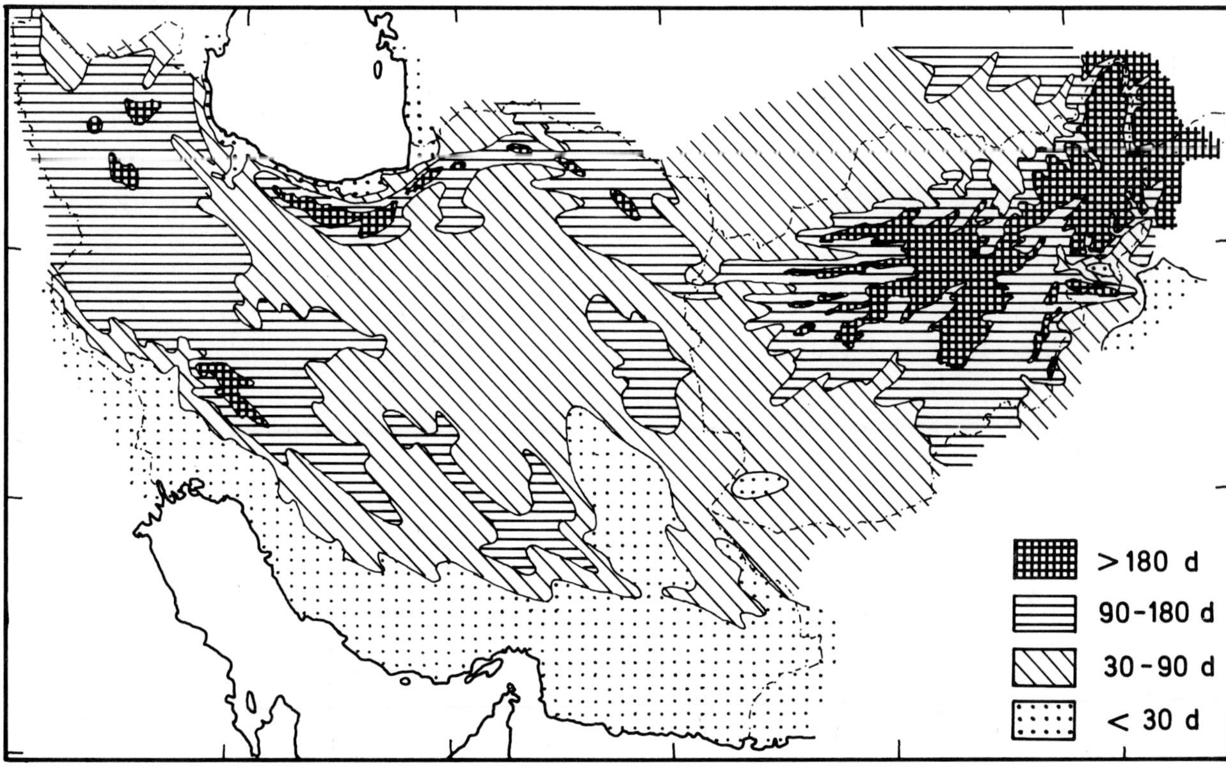

Fig. 10.3. Map of isolines of number of days (d) per year with minimum temperatures equal to or below 0°C (from Breckle, 1981b).

TABLE 10.1

Some characteristics of and data from meteorological stations in Iran (I) and Afghanistan (A) (the location of the stations is shown by abbreviations on the map, Fig. 10.4)

Station	Symbol used in Fig. 10.4	Elevation (m)	Days with rain (yr^{-1})	Days with snow (yr^{-1})	Duration of sunshine ($h\ yr^{-1}$)	Potential evaporation ($mm\ yr^{-1}$)	Annual relative humidity (%)
Abadeh (I)	*A*	1845	*c.*30	*c.*10	*c.*3200	*c.*2300	*c.*41
Birjand (I)	*B*	1460	*c.*35	*c.* 5	*c.*3100	*c.*2500	*c.*40
Farah (A)	*F*	660	21	0	3320	2570	*c.*46
Qom (Ghom) (I)	*G*	928	*c.*45	*c.* 5	*c.*2700	*c.*2500	*c.*43
Kabul (A)	*K*	1800	54	17	3160	1570	55
Kandahar (A)	*H*	1005	31	0	3470	2930	36
Kunduz (A)	*U*	433	50	13	2670	2160	57
Mashhad (Meshed (I)	*M*	985	*c.*45	*c.*10	*c.*2900	*c.*2000	*c.*44
Panjao (A)	*P*	2710	22	50	2230	1030	60
Qadis (A)	*Q*	1280	52	24	2840	1630	54
Shahrud (I)	*S*	1370	*c.*30	*c.* 5	*c.*3000	*c.*1800	*c.*53
Tabas (I)	*T*	900	*c.*15	*c.* 2	*c.*3400	*c.*3000	*c.*38
Yazd (I)	*Y*	1240	*c.*20	*c.* 5	*c.*3000	*c.*2900	*c.*37
Zahedan (I)	*Z*	1370	*c.*15	*c.* 0	*c.*3200	*c.*2600	*c.*40
Zaranj (A)	*R*	478	*c.* 5	*c.* 0	*c.*3500	3040	39

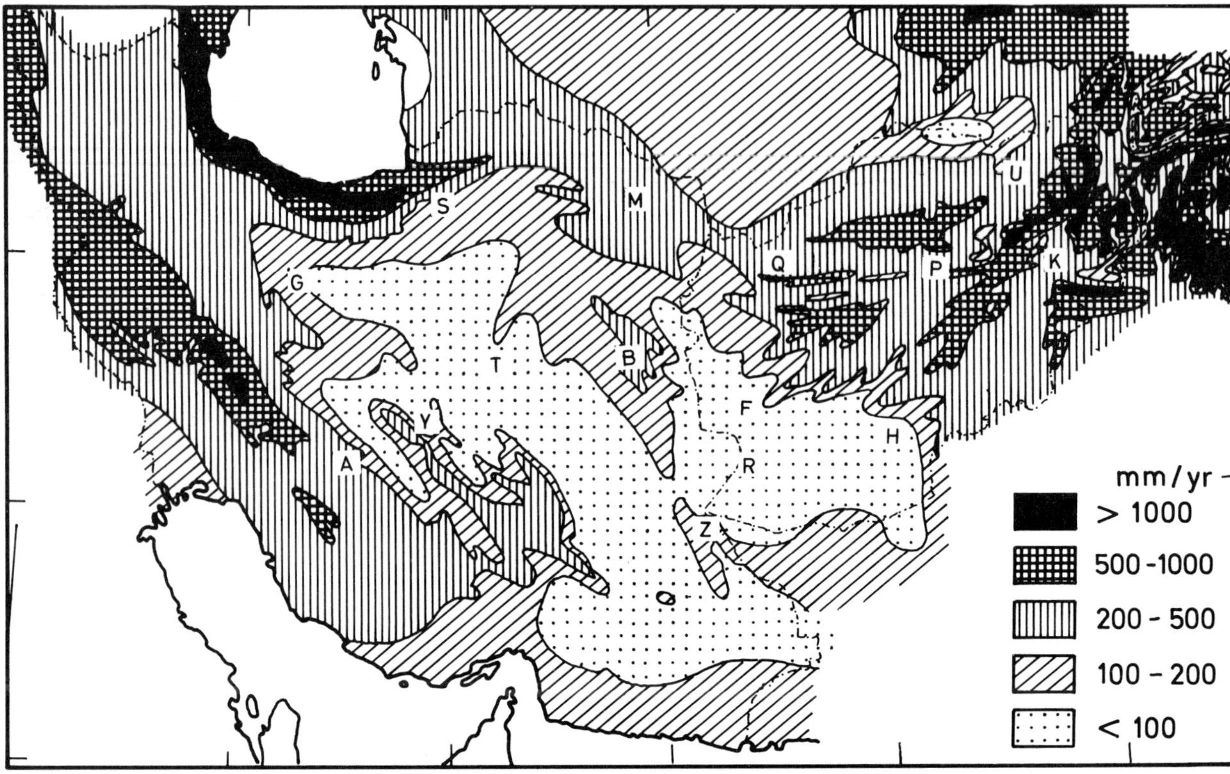

Fig. 10.4. Map of isolines of total annual precipitation (mm). Letters indicate position of meteorological stations for which climatic diagrams are shown in Fig. 10.2; for further data see Table 10.1.

of the rainy season yields a factor greater than 10. The several dry years (from 1970 to 1973) correspond with the arid period experienced in the Sahel region of Africa. These contrast sharply with the exceptionally rainy seasons at the beginning and the end of 1974 (cf. Fig. 10.5).

The global radiation in all the desertic areas is very high. Landsberg (1966) gives values of more than 580 for the north Afghan lowlands, 670 for central Afghan mountains, and 750 kJ cm^{-2} for southern Afghanistan. The number of sunshine hours per year gives a similar impression (Table 10.1; Herman, 1965; Herman et al., 1971).

The relative humidity of the air is low, with annual means around 50% or lower (Table 10.1). Extremely low values (<5%) have been recorded by microclimatological investigations at many stations (Breckle, 1973; Mirwald and Römer, 1967).

It is obvious that evaporation is very high (Tables 10.1 and 10.2). Stenz (1941, 1957) has reported an annual evaporation of 2305 mm for Kabul; Iven (1933) mentions about 3031 mm. Despite the wide divergence, both values confirm that evaporation is many times higher than precipitation and higher than the calculated potential evaporation values.[1]

This might be partly explained by strong katabatic winds dominating in the Kabul Basin during the summer. Dry trade winds, the "120-day-wind," regularly blow during summer in northwestern and western Afghanistan and in eastern Iran. They bear dust and heat and cause rapid migration of the sand dunes in the sandy areas.

[1]These had been calculated by the formula of Ivanov (Herman et al., 1971):

$$E_{pot.} = 0.0018\ (T + 25)^2\ (100 - RH)$$

where $E_{pot.}$ gives the monthly potential evaporation in mm, and T means the monthly mean temperature in °C, and RH the monthly mean relative humidity of the air.

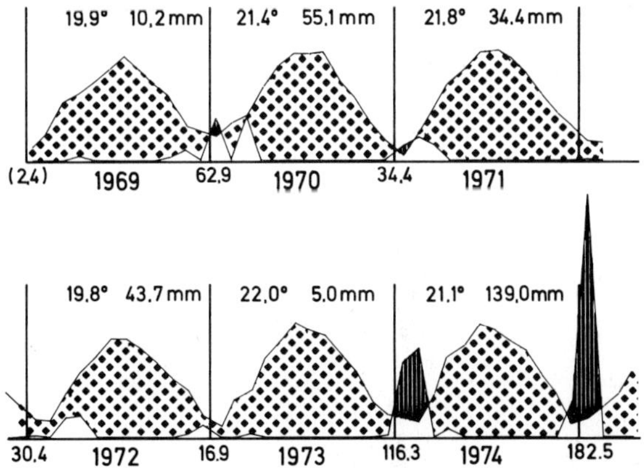

Fig. 10.5. Climatograms for 1969–1974 at the meteorological station of Zaranj (southwestern Afghanistan). Numbers in the upper row indicate annual mean temperature and precipitation, numbers below the diagrams show year and amount of rain per rainy season (from Breckle, 1981b).

Soils

Only generalized descriptions of the various soil types or substrates will be given here. For more details the reader is referred to Petrov (1945, 1955) and Dewan and Famouri (1964). More details will be given when discussing the different vegetation types (see p. 281ff.).

Vast areas are covered by immature soils, derived from unconsolidated materials, such as gravel, rocks or sand. The most common categories (Zohary, 1973) are *hammadas*, *serirs*, *regs*, *ergs*, *nebkas*, *takyry*, *sebkhas* and *salines* — each in a variety of forms.

The term *hammada* in this review will be used as a geomorphological term describing deserts with rough, often strongly dissected rocky substratum. The term *serir* will be used for stony deserts with gravel, pebbles or small, usually rounded stones, forming desert pavement. Sometimes the term *reg* is taken as synonymous to *serir*, especially in describing typical desert pavements with blown-out sand. The *regs* often intergrade with *ergs*. forming mosaics. The terms *erg* describes sandy deserts with either moving or slightly stabilized sand (Fig. 10.6).

Sand stabilization is created by either vegetation growth or by soil humidity or by both. Big dunes break into many, small sand dunes (*nebkas*) as stabilization takes place. The term *takyr* is applied to large, nearly level clay plains (see p. 52), which are normally found at the end of a sedimentation series, e.g., in deltas of wadis and rivers. *Sebkhas* are clay pans in mostly very shallow depressions developed by deflation. Both takyrs and sebkhas are barren deserts incapable of supporting almost any plant life. Salinity of the takyr soils is normally low, whereas sebkha soils usually exhibit high salinity throughout the profile (Fig. 10.7). These are extremes. There are also intermediate and intergrading geomorphological features known.

Almost all upland desert substrates and soils are

TABLE 10.2

Evaporation (24-h values) in some Afghan desert and semi-desert areas (measured by registration of an evaporation balance with open water surface)

Elevation (m)	Locality (province)	Date	Evaporation (mm)
3110	Dasht-i-Nawar (Ghazni)	Aug. 16, 76	9.4
		Aug. 17, 76	8.3
2700	Dragon Valley (Bamiyan)	Sept. 10, 76	10.9
		Sept. 11, 76	11.0
1970	Ab-e-Istada (Paktia)	Aug. 19, 76	19.7
1800	Kabul (Kabul)	Sept. 6, 76	6.9
		Sept. 7, 76	8.8
		Sept. 8, 76	5.1
470	Jehile Puzak (Chakhansur)	Sept. 23, 76	14.6
430	Ali Moghul, Center (Jozjan)	Aug. 30, 76	10.8
330	Sar-e-Namak (Samangan)	Sept. 3, 76	16.3
280	Islam Chomgar (Jozjan)	Aug. 8, 76	12.4

Fig. 10.6. Erg area in northern Afghanistan. Sand dunes partly stabilized with *Stipagrostis karelinii*, *Calligonum* and *Haloxylon persicum*.

calcareous more or less throughout the profile (Table 10.3). This property can be combined with saline and/or sodic properties. Cemented (petrocalcic) horizons often develop in the soil profile (Salem and Hole, 1969). The cementing compounds are usually calcium carbonate, calcium sulfate (Fig. 10.8) and salts of sodium (Figs. 10.9 and 10.10).

Brown soils, sierozems, loess and loess-like soils occupy large portions of the semi-deserts. These soils are rather more influenced by hydration, solution and leaching, than the desert "soils", where mechanical weathering predominates because of deficiency of water. In both cases physical factors predominate in the evolution of the surface cover of the area.

Some soil types from Afghanistan were characterized by Freitag (1971). Normally Aridisols (sierozems) predominate. Halomorphic soils may occur on various substrata, such as loams, loess, and sands, but often they are limited to alluvial lowlands and basins with very fine-textured parent materials. The amount of soluble salts tolerated by plants ranges between 0.3 and 3% in the majority of saline soils (Zohary, 1973). In some cases, higher percentages of salts have been found, but such habitats usually lack plants. Saline soils may, however, harbor quite a number of plant communities and a rather rich flora, even with some endemic species. This depends not only on the amount of salts, but even more on the grain size conditions of the soil and the availability of water. In fine-textured soils evaporation causes losses of water from rather deep in the profile. Thus, deficiency of moisture turns salines into a very adverse habitat for plant life. The plants living in them, the xerohalophytes, have to overcome both drought and excess of salts (Figs. 10.9 and 10.11). Only under more moderate climatic conditions (cf. Breckle, 1976) is plant life possible.

Buringh (1960) divides the solonchaks of Iraq into the following groups (a similar classification can be applied to Iran and Afghanistan):

(a) Flooded solonchaks in shallow depressions, flooded in winter, with a thick salt crust on the surface.

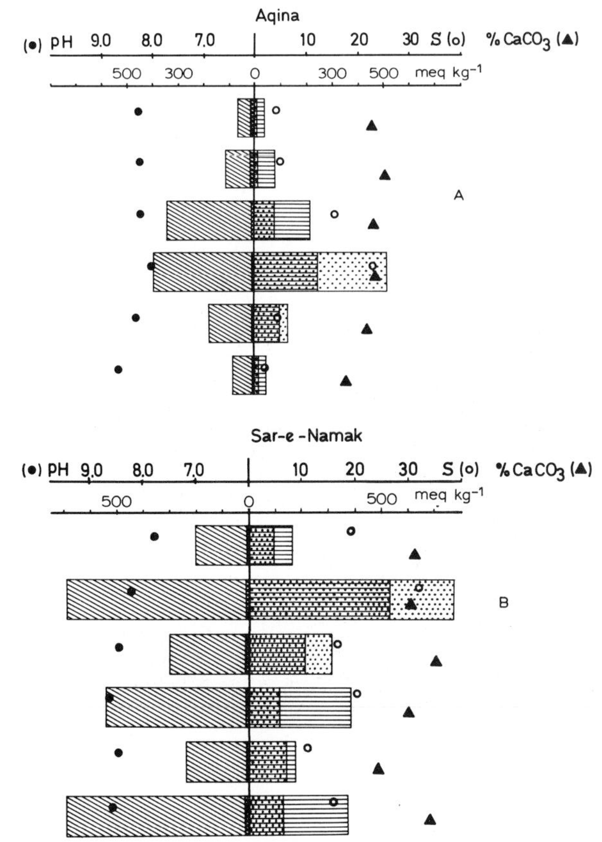

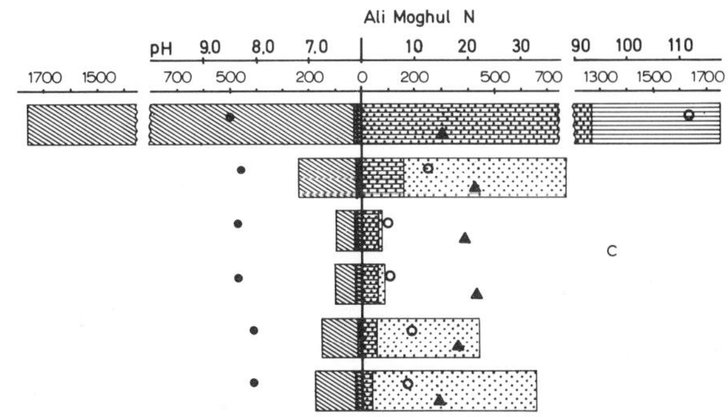

Fig. 10.7. Diagrams of ion concentrations and other features varying with soil depth (S.-W. Breckle, unpubl. data).
Legend: A = soil from takyr plains near Aqina; B = soil from sebkha depressions near Sar-e-Namak; C = solonchak from Ali Moghul. The horizontal columns to the left indicate Na^+ (oblique hatched) and K^+, to the right Cl^- (dotted) and SO_4^{2-}, expressed as meq kg^{-1} dry soil at soil depths 0–1, 1–5, 10–20, 30–40, 60–70, and 100–110 cm. Additionally pH (dots to the left), $CaCO_3$ percentage (triangles to the right), and soil conductivity (open dots to the right, $S = \Omega^{-1}$) are indicated.

Fig. 10.8. Overgrazed *Artemisia* semi-deserts in the hills surrounding Kabul, Afghanistan. Soils here are often protected by thick hard layers of limestone and gypsum crusts.

TABLE 10.3

Characteristics of soils associated with some examples of plant associations (Freitag, 1971; Breckle, 1973, and unpubl. data), from Afghanistan [$(+H_2O)$ = indicates precipitation is less than total water supply]

Association, community or plant-formation	Elevation (m)	Precipi-tation (mm yr^{-1})	Dominant texture (grain size)	Soil pH	Soil-conductivity (mS cm^{-1} = $m\Omega^{-1}$ cm^{-1})	Soil salinity		$CaCO_3$ (%)	Plant-cover (%)
						Cl^-	SO_4^{2-}		
Halocnemum/Seidlitzia	<600	100	salt, gypsum (clay)	7.5–9.5	~30 000	high	high	high	1–10
Halimocnemis/Gamanthus	<600	150	salt (clay, loam)	8–9	~30 000	high	high	high	5–50
Haloxylon salicornicum	<800	150	salt, gypsum (clay, silt)	8–9.5	<30 000	high	high	high	1–20
Seidlitzia/Salsola gemmascens	<600	150	sand, clay	7.5–9.5	<5000	medium	high	high	2–30
Takyr plains (cryptogams, few ephemerals)	<500	200	clay	8–9	<5000	high-medium	high	high	0–1
Zygophyllum/Cousinia arida	<800	150	sand, gravel (clay, silt)	7.5–8.5	<2000	medium	medium	high	5–25
Calligonum/Stipagrostis karelinii/Halox. pers.	<600	150	sand	7.8–8.6	<1000	low	low-medium	high	1–70
Reaumuria stocksii	500–1000	120	hammada	7.8–8.5	<1000	low	low	medium	0–5
Ephemeral semi-desert	<1000	150–300	loess, sand	7.5–8.5	<1000	low	low	high	20–90
Artemisia semi-desert	600–2300	180–350	hammada, serir	7–9	<1500	rather low	low	medium-high	2–40
Amygdalus shrubby semi-desert	900–1200	150–250	hammada, serir	7.5–8	<1000	low	medium	high	10–30
Pistacia vera open woodland	600–1600	250–500	loess, rocks	7–8	<500	very low		high	trees; herbs
Pistacia atlantica open woodland	1000–1800	250	limestone, rocks, scree	6.8–8	<500	very low		high	10–40
Amygdalus open woodlands	1600–1900	>400	rocks, scree	7–8.5	<500	very low		medium-high	10–50
Tragacanthoid cushions (subalpine)	2700–4000	300–800	brown soils, scree, rocks	6.5–7.5	low	very low		medium	20–60
Alpine semi-desert (*Nepeta pamirenis*)	3700–4600	$\gtrsim$600	brown soils, rocks, scree, loam	6–7	very low	none		low-medium	10–30
Subalpine and alpine steppe (*Kobresia*, *Carex*)	2500–4600	>1000 $(+H_2O)$	dark brown soils, loam, sand	5.5–7	low	none		low	100
Alpine meadows (*Primula macrophylla*)	4000–4600	>800 $(+H_2O)$	humus, loam, sand, scree	5–6.8	very low	none		none	20–80

Fig. 10.9. The northeastern margin of the Great Kavir, Iran, marked by the last occurring plants, some scattered *Halocnemum strobilaceum* and *Nitraria schoberi* individuals.

(b) Internal solonchaks with some salt concentrations in deeper parts of the soil (Fig. 10.7); annuals and shallow-rooted dwarf-shrubs may grow successfully.

(c) External solonchaks, more saline in upper layers and with a salt crust on the surface, but saline throughout the soil profile (Fig. 10.7). (These habitats support several kinds of halophytes, varying in degree of salt tolerance.)

(d) Solonchaks with a puffed or fluffy surface layer, almost bare of vegetation in our areas (Fig. 10.9).

(e) *Sabakh* soils with a high salinity and with some magnesium sulfate in the permanently moist surface layer. These soils are often densely vegetated (Fig. 10.10) by meso-halophytes reaching water table, such as *Tamarix* and *Halocnemum*.

Alkali soils or solonetz are rare in the area being considered. They are alkaline, but normally do not exceed a pH of 9.0. The vegetation is very sparse (e.g., *Camphorosma lessingii*). Ecological factors such as aeration, seasonality of flooding, organic matter balance, water table movement and movement of salt within the soil profile are probably important, but need investigation.

THE VEGETATION

Natural vegetation types in general

Vegetation of these deserts can be best classified according to substrate conditions because differentiation of zono- and pedobiomes is next to impossible. Subdivisions can be defined according to degree of aridity, floristic composition and life forms.

The semi-desert vegetation is distinguished by higher canopy cover and density, higher productivity and biomass. It can be classified, like desert vegetation, with respect to substrate conditions or by the dominant life forms.

Fig. 10.10. *Halocnemum strobilaceum* and *Halostachys caspica*, together with some *Seidlitzia rosmarinus*, forming a rather dense mosaic where high water tables occur at the northeastern margin of the local Kavir, Iran. The highly saline soil surface is very soft and rugose.

The following groups can be distinguished:

I. Halophytic vegetation
   1. saline flats (almost bare of vegetation, soil with very high salt content)
   2. euhalophytic vegetation (*halo-hammada*, on gravel, sand, but mainly clayey soils)
   3. mesohalophytic vegetation (less salt accumulation within soil profile)
II. Gypsophilous vegetation (desert and semi-desert, often with dwarf-shrubs, on gypsiferous soils)
III. Psammophytic vegetation (erg; on sand dunes and other sandy areas)
IV. Vegetation on loess (predominantly covered by semi-desert vegetation rich in ephemerals)
V. Vegetation on scree, gravel, rocks (hammada, serir and reg)
   1. desertic vegetation, non-saline (hammada)
   2. semi-desert regions (*Artemisia* semi-desert, very common in the foothills and lower mountain regions)
   3. shrubby semi-desert (only in few, rather inaccessible areas)
   4. open woodlands (semi-desert with single, small trees)
VI. Mountain vegetation (in higher areas, normally above 3000 m)
   1. sub-alpine cushion shrublands and semi-deserts
   2. alpine semi-desert and steppe
VII. Vegetation of oases and river margins (with additional water supply).

Floristics, vegetation and ecology

The central Persian deserts

Vast areas of inner Iran are almost bare of vegetation (Figs. 10.9 and 10.11). These are mostly saline plains with soils of high salt content. The soil is usually fine-textured. Despite the high number of halophytic species, the associations have low diversity. Associations are often monotonous and repeat themselves over vast expanses.

Edaphic conditions in the surrounding regions are characterized by decreasing salt content with increasing soil particle size. Soils of this region that have low amounts of soluble salts are termed "glyco-hammada" by Zohary (1973). A better term

Fig. 10.11. An example of a small, hydrographically isolated plain with moderately saline clay surfaces at Shaq-e-Biar, northeastern Iran. An open mosaic of halophytic annuals, dominated by *Petrosimonia glauca*, *Cressa cretica*, and especially *Halimocnemis pilifera*, occurs depending on flooding in early spring.

would be "non-saline hammada". Zohary (1973) called those regions with high soil salt content "halo-hammada".

The sebkhas are often very saline in the center of the basin. *Halocnemum strobilaceum*, together with *Seidlitzia rosmarinus*, dominate on the most saline sites in these areas as well as in the Kavirs, in eastern Iran, Sistan and in northern Afghanistan (Figs. 10.9 and 10.12). These species grow on the most extreme habitats — for instance, in Dasht-e-Lut and at the margins of numerous salt pans of the Kavirs. The vegetation of these regions is clearly arranged according to grain size and distribution of salt and gypsum layers within the profile. Enormous areas occur in the northern Kavirs where *Halocnemum strobilaceum*, sometimes with *Seidlitzia rosmarinus*, is surrounded by a second belt with *Haloxylon aphyllum* (Figs. 10.13 and 10.14). *Anabasis setifera* also occurs in places on the fringe of the depressions. *Aeluropus lagopodioides*, *Atriplex verrucifera*, *Limonium iranicum*, *Lycium ruthenicum*, *Fortuynia garcinii*, *Alhagi* and *Salsola* spp. form a mosaic of communities on sites of intermediate salinity and intergrade to semi-deserts with *Zygophyllum eurypterum* (Figs. 10.15 and 10.16) or *Artemisia herba-alba* s.l. (Fig. 10.14 and 10.17).

Enormous portions of the saline flats are bare of higher vegetation. *Halostachys caspica* and several *Tamarix* spp. sometimes dominate near water courses or on alluvial fans with lower water tables (Figs. 10.10 and 10.18). *Reaumuria fruticosa* occupies dry, salty ridges (Fig. 10.19). The zonation at the margins of the Kavirs in inner Iran seems to be much more complex and variable than thought earlier (H. Freitag, pers. comm., 1978).

Seidlitzia, *Salsola dendroides*, *Anabasis setifera*, *Halocharis* spp., *Haloxylon salicornicum*, *Reaumuria persica*, *Aellenia glauca*, *Alhagi maurorum* were reported from other parts of the Kavirs on saline

Fig. 10.12. *Halocnemum strobilaceum* stabilizing small sand dunes on very saline soil, northeastern margin of the Great Kavir, Iran.

Fig. 10.13. *Haloxylon aphyllum* community developing in wadis and along runnels. In undisturbed state they reach more than 3 m in height; Touran Biospheric Reserve (northeastern Iran).

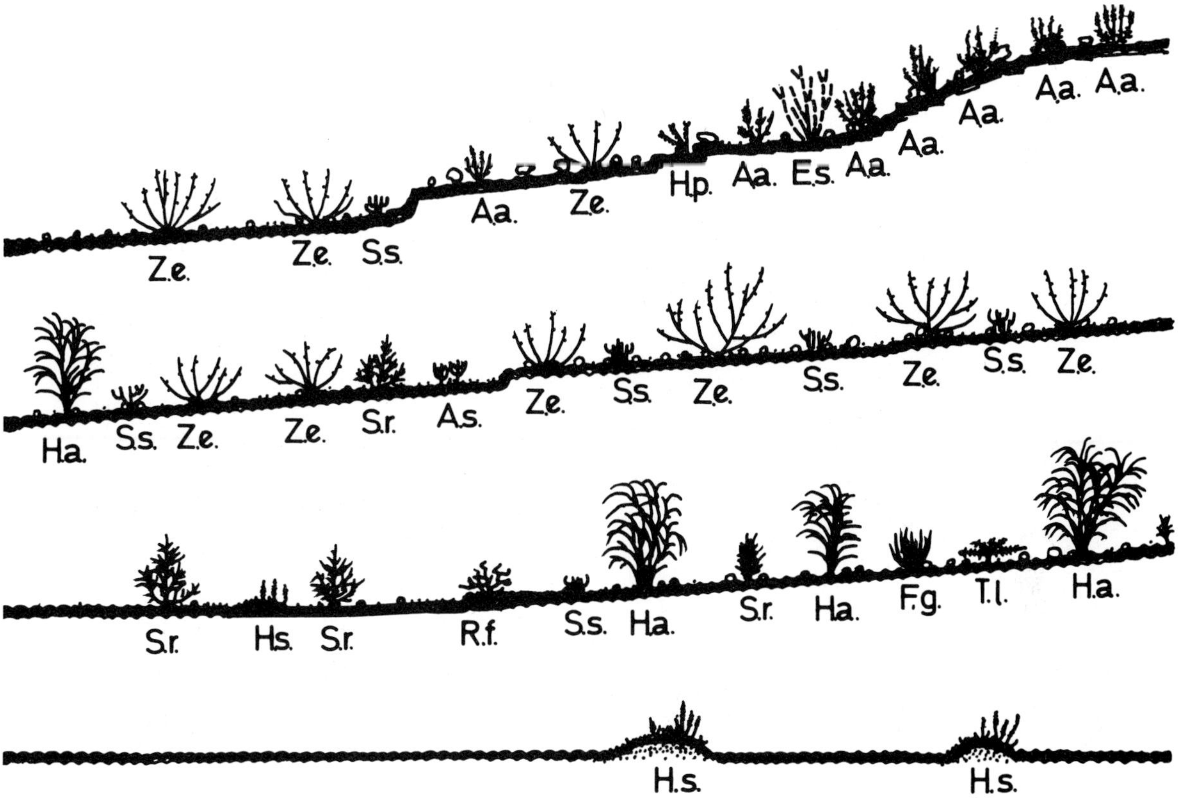

Fig. 10.14. Schematic vegetation profile through the northeastern margin of the Great Kavir, Iran (Touran Biospheric Reserve), showing density and height of the vegetation cover (vertical scale approximately five times enlarged; the highest *Haloxylon aphyllum* reaching about 3 m).
Legend: *A.a.* = *Artemisia herba-alba* (s.l.); *A.s.* = *Anabasis setifera*; *E.s.* = *Ephedra strobilacea*; *F.g.* = *Fortuynia garcini*; *H.a.* = *Haloxylon aphyllum*; *H.p.* = *Hulthemia persica*; *H.s.* = *Halocnemum strobilaceum*; *R.f.* = *Reaumuria fruticosa*; *S.r.* = *Seidlitzia rosmarinus*; *S.s.* = *Salsola* spp. (e.g. *Salsola tomentosa*, etc.); *T.l.* = *Tribulus longipetalus* ssp. *macropterus*; *Z.e.* = *Zygophyllum eurypterum*.

flats (Zohary, 1963). *Tamarix* spp., *Prosopis farcta* and a salt-tolerant subspecies of *Artemisia herba-alba* are found on disturbed sites (Frey and Probst, 1979).

Iranian desert vegetation has a particularly striking number of *Tamarix* species, about forty species for the whole of Iran. Mobayen (1975) mentions *T. aphylla* and *T. macrocarpa*, as occurring with *Prosopis spicigera*, and *Calligonum* sp. on the margin of the more sandy and gravelly parts of the Dasht-e-Lut. The center of Dasht-e-Lut seems to be bare of any vegetation and might be a truly abiotic region (Hallier, 1975, 1978).

Zohary (1963) found vegetal cover of less than 0.1% around Dasht-e-Lut. It consisted of only a few *Anabasis setifera*, *Salsola baryosma*, *S. dendroides*, *Cornulaca leucantha* and *Tribulus longipetalus*. More to the north, he found these plus *Suaeda dendroides*, *Seidlitzia rosmarinus* and *Haloxylon aphyllum*. A more detailed description and a vegetation map of the Dasht-e-Lut can be found in Mobayen (1976).

The Kavir is a vast system of plains. Parts of it have been under protection since 1964. This now constitutes one of Iran's National Parks (Firouz, 1964). The Kavir National Park covers an area of 6094 km^2 and has deserts, semi-deserts, dry steppes and semi-desertic mountains, all within a mosaic of plains and rocky areas (Vahedi, 1976) (Table 10.4). Two hundred eighty-three species of higher plants have been found in that area, more than 50% of which are annuals (Rechinger and Wendelbo, 1976).

Fig. 10.15. Peneplains around the mountains (Touran Biospheric Reserve) in the background are covered by communities dominated by shrubs of *Zygophyllum eurypterum* which reach up to 1.5 m in height. On the slightly inclined pediment plains, *Zygophyllum* intergrades with the *Artemisia* semi-desert.

Fig. 10.16. Community characterized by giant umbellifers (*Dorema*), within the *Zygophyllum* and the *Artemisia* community types. Several *Salsola* species also play a role in this vegetation mosaic around the Great Kavir, Iran.

Fig. 10.17. *Artemisia* semi-desert in the foothills (1300 m elevation) north of the Great Kavir, Iran. *Zygophyllum*, *Salsola* and especially *Pteropyrum* are found near wadis. Some bushes of *Launaea acanthodes* also grow within the wadis. Ungrazed parts of the *Artemisia* semi-desert are characterized by a high percentage of grasses.

Fig. 10.18. A water course where the accumulation of soluble salts is very distinct. In such places several *Tamarix* species are found, along with halophilous *Phragmites australis* ecotypes (Touran Biospheric Reserve).

TABLE 10.4

Principal vegetation types in Kavir Protected Region, Iran (Rechinger and Wendelbo, 1976)

Type	Soil pH	Soil texture (%)				Dominant plant species	Canopy height (m)
		clay	silt	sand	scree		
Salt flats	8.1–8.7	32	41	27	–	*Halostachys caspica*, *Suaeda monoica*	1.5
Sloping grounds next to salt flats	7.9	6	6	88	–	*Seidlitzia rosmarinus* (*Calligonum persicum*)	0.5
Stony-gravelly upland	–	–	–	–	100	*Pteropyrum aucheri*, *Zygophyllum eurypterum*	0.5
Sand dunes	7.9	5	3	92	–	*Haloxylon aphyllum*, *Haloxylon persicum*	2

The family with the greatest representation is the Chenopodiaceae (35 species).

A list of plant species from the ergs of the Kavir National Park is given by Rechinger and Wendelbo (1976). They stressed the remarkably rapid regeneration of the vegetation of the sand dunes as soon as grazing was stopped. The relatively high number of different species in the floristic pool of the area has apparently aided this recovery.

Other Iranian erg areas have been recently studied. *Haloxylon persicum*, *H. recurvum*, *Halimodendron halodendron*, *Calligonum bungei*, *Stipagrostis plumosa* and a few other grasses growing on sand dunes have been reported (Mobayen and Tregubov, 1970; Hallier, 1975). *Haloxylon ammodendron* was also mentioned as being more characteristic of fine-textured, more or less saline soils with an attainable water table (Walter, 1974) (Fig. 10.13). Whereas *Haloxylon aphyllum*[1] seems to be rather widespread in Middle Asia, *H. ammodendron* is restricted to the Zaisan and Ili deserts, reaching Kashgar and Dzungaria, according to the flora of the U.S.S.R. Very likely these two species are synonymous. Some Russian literature applies the name *H. ammodendron* to both taxa. *H. aphyllum* s.l. may form solid stands. The so-called "black saksa'ul forests", which are known from several localities in Iran, have not been found in Afghanistan. The open "white saksa'ul forest" of *Haloxylon persicum* is more widespread, especially in reg and erg areas. These communities are so intensively used for fuel that individuals of saksa'ul occur as a shrubby form.

Parts of the northeastern Kavir area are under protection also. First called the "Touran Protected Area", and later designated "Touran Biosphere Reserve", it serves more or less as an area of wildlife protection and for the scientific study of natural, economic, historical and cultural influences within that biome (cf. Spooner, 1977). The Touran Biosphere Reserve has an area of 14 100 km^2, comprising vast areas of all desert and semi-desert types.

A general example of the zonation at the northern margin of that area is given in Fig. 10.14. There is much variability in this general zonation depending on soil conditions. Sand dune areas are relatively rich in species and vegetation cover. As in similar parts of northern Afghanistan (see p. 297), soils are sandier and have large members of the genera *Ferula* (Fig. 10.20) and *Dorema* (Fig. 10.16). These play a major role in vegetation dominance; however, they do not flower every year. In 1977 only *Dorema aitchisonii* was fruiting; *Ferula foetida*, however, was represented only by many grey, dry stems from 1976 (Fig. 10.20).

Chenopods dominate most communities on all non-sandy areas with precipitation less than 150 mm yr^{-1}. They exhibit the peculiarity of flowering during the arid season and being active during most of the dry summer.

[1] *H. aphyllum* and *H. ammodendron* were often treated as one species. Today Russian taxonomists treat them as separate species (see e.g., Flora U.S.S.R.).

Fig. 10.19. An association of *Reaumuria fruticosa* typically developing on small ridges where there is a high amount of soluble salts in the soil profile. A layer of salt powder is found below the shrubs due to the "pumping" action of salt glands of the leaves (Touran Biospheric Reserve).

A detailed description of the vegetation and lists of plant species occurrences are given by Freitag (in Spooner, 1977) and Rechinger (1977). A few ecology studies have been completed recently. Population studies in *Zygophyllum eurypterum* (Fig. 10.15) show a density of 531 bushes ha^{-1} and cover of 13.2%. Dimensional analysis allowed an estimate of an above-ground biomass of 2.58 t ha^{-1} (Moore and Bhadresa, 1978) (Table 10.5). The density of the smaller subshrub *Artemisia herba-alba* in similar areas in northeast Iran was shown to be 14 544 plants ha^{-1} and its biomass was 1.45 t ha^{-1} (Bhadresa, 1977). The age of *Zygophyllum* trunks ranged between 9 and 88 years. Growth rings are very narrow (Fig. 10.21) but can be used to gauge the past climatic trends in the area (Moore and Bhadresa, 1977/78). They indicate very severe conditions in 1970 and 1971.

Chemical relationships, especially in relation to soil salinity conditions, have been studied in the Chenopodiaceae (Breckle and Bhadresa, unpubl. data). A large set of species with different salt accumulation and growth strategies occur within the different vegetation belts girdling the Kavir areas. Relative ion ratios distinguish classes of response better than absolute salt contents.

Fig. 10.20. *Ferula foetida*, a giant umbellifer, preferring sandy areas within the Touran Biospheric Reserve within a variant of the *Artemisia* community on sand soils. *Salsola sclerantha* and/or *S. leptoclada* is also found in these communities.

TABLE 10.5

Main plant communities in the Touran Biosphere Reserve, with estimated cover and biomass percentages (partly after Freitag, 1977; Moore and Bhadresa, 1977/1978; Breckle, 1979)

No.	Community (ecotope)	Area (km^2)	Area (%)	Plant cover (%)	Community biomass ($t\ ha^{-1}$)
1.	barren Kavir	8400	45.5	0	0
2.	*Zygophyllum–Salsola*	5700	30.9	*c*.15	2.6
	a. *Zygophyllum eurypterum*	4500	27.6	*c*.15	2.6
	b. *Aellenia glauca*	120	0.65	*c*.10	2.5
	c. *Salsola tomentosa*	70	0.35	*c*.10	2.0
3.	*Artemisia–Amygdalus*	1900	10.4	*c*.15	1.6
	a. *Artemisia*	800	4.3	*c*.15	1.5
	b. *Hypocylix kerneri*	600	3.3	*c*.20	2.0
	c. *Ceratoides latens*	350	1.9	*c*.15	1.8
	d. *Gamanthus gamocarpus*	100	0.6	*c*.10	1.0
4.	halophytic	1500	8.1	*c*.30	?
	a. *Seidlitzia rosmarinus*	800	4.3	*c*.30	2.5
	b. *Halocnemum strobilaceum*	250	1.4	*c*.10	1.2
	c. *Halostachys caspica*	50	0.27	*c*.50	4.5
	d. *Petrosimonia, Cressa*	30	0.17	*c*.15	0.8
5.	*Haloxylon aphyllum*	750	4.0	*c*.40	6.0
6.	*Astragalus–Amygdalus*	100	0.6	*c*. 5	0.5
7.	*Calligonum*	70	0.4	*c*.20	3.0

Zarrine-Kafche (1977) reported three soil types and corresponding vegetation types, arranged along a catena in a more disturbed area west of Tehran at Karaj (Fig. 10.22). His table 2, shown here in shortened version (Table 10.6), shows some peculiar and contradictory results. The potassiophilic *Salsola iberica* (= *kali*) contains only 0.7% K^+. On the other hand, the sum of his analyzed ions is about 27.5%. Ash content, however, amounts to only 5.3% in this species. Ionic ratios show wide range of ion accumulation types represented by different plant species.

Deserts and semi-deserts in north and south Afghanistan

Calligonum and *Aristida* (*Stipagrostis*) cover large areas of more or less mobile sand dunes in the sandy parts of the Persian Desert, the north Afghan desert region, and in the south in Sistan and especially in Registan. Cover percentages vary between 1 to 10%, and sometimes increase in larger and stable dune valleys to 25% or more (Fig. 10.6, Table 10.5). Thus, this psammophytic vegetation makes a luxuriant impression when compared to other types. Low salt content and good water

TABLE 10.6

Ash and ionic contents of a few species from a saline area south of Karaj, Iran; values are percentage of dry matter taken from Zarrine-Kafche (1977)

Species	Ash	Cl^-	SO_4^{2-}	P	Ca^{2+}	Mg^{2+}	K^+	Na^+	Cl/S	Na/Ca
Peganum harmala	7.2	0.8	4.5	0.8	2.4	1.2	3.4	0.3	0.18	0.1
Salsola iberica (= *S. kali*)	5.3	9.6	13.2	0.7	0.64	0.7	0.7	2.0	0.7	1.1
Salicornia herbacea	5.7	10.7	5.1	0.5	1.2	1.8	1.5	14.0	1.9	11

Fig. 10.21. *Zygophyllum eurypterum* bush in the Touran Biospheric Reserve, Iran. Degradation separates the small annual rings of the wood.

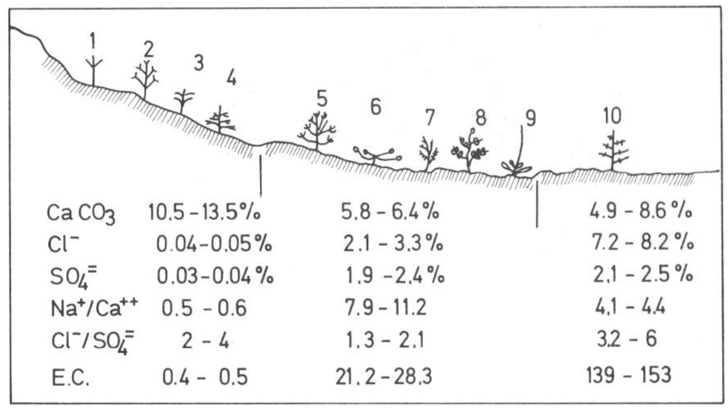

Fig. 10. 22. Profile through a gypsicolous and saline region south of Karaj, Iran. On bottom are given some chemical soil properties (Zarrine-Kafche, 1977), in % of soil, dry weight basis. (E.C. = electric conductivity mS cm^{-1}.)
Calcicolous zone with: *1 = Alhagi camelorum*; *2 = Peganum*; *3 = Chrozophora tinctoria*; *4 = Fagonia* spp.
Gypso-halophytic zone with: *5 = Seidlitzia rosmarinus*; *6 = Aeluropus litoralis*; *7 = Salsola iberica*; *8 = Salsola incanescens*; *9 = Limonium gmelini*.
Halophytic zone with: *10 = Salicornia herbacea*.

conditions in the sand make this high cover possible. Floristic composition in the north differs considerably from the south (Freitag, 1971) and from the Iranian ergs. Besides *Calligonum comosum* and *Stipagrostis pennata* in sandy areas, the following species can be found:

shrubs and subshrubs: *Haloxylon persicum*, *H. salicornicum*, *Aellenia arenaria* (Fig. 10.23), *Calligonum persicum*, *C. turkestanicum*, *C. intertextum*, *C. amoenum*, *C. microcarpum*
perennials: *Stipagrostis karelinii*, *Cyperus conglomeratus*, *Pennisetum divisum*, *Eremostachys regeliana* (Fig. 10.24)
annuals: *Agriophyllum minus*, *A. latifolium*, *Euphorbia cheirolepis*, *Cithareloma lehmanii*, *Cutandia memphitica*, *Horaninowia ulicina*, *Heliotropium arguzioides*, *Sterigmostemum acanthocarpum*, *Diplotaxis harra*, *Isatis minima*, *Alyssum linifolium*, *Sameraria elegans*, *Spirorhynchus sabulosus*
geophytes: *Carex physodes*, *Emimium alberti*, *Merendera robusta*.

The loess-belt in northern Afghanistan and in some local areas of Khorasan with its rather

Fig. 10.23. *Aellenia subaphylla* ssp. *arenaria*, a shrubby species reaching more than 2 m in height, in the semi-deserts of northern Afghanistan.

Fig. 10.24. *Eremostachys regeliana*, a perennial labiate, wilting rapidly in June. This is common in the regs and serir plains in northern Afghanistan.

favorable physical and chemical soil conditions is occupied by peculiar vegetation types. In most cases, however, these areas are under cultivation. The dominant natural life-forms are ephemerals. Hemicryptophytes and geophytes occur too. Small shrubs are rare, bigger shrubs are absent. Freitag (1971) lists among others the following species from loess sites in northern Afghanistan:

perennials: *Poa bulbosa*, *Psoralea drupacea*, and many *Astragalus* species

annuals: *Bromus scoparius*, *B. tectorum*, and some other *Bromus* species, *Vulpia myuros*, *Aegilops tauschii*, *Hordeum leporinum*, *Eremopyrum distans*, *Leptaleum filifolium*, *Medicago radiata*, many *Astragalus* species, *Diarthron vesiculosum*, *Diptychocarpus strictus*, *Arnebia grandiflora*, *A. africana*, *Aphanopleura leptoclada*, *Isatis boissieriana*

geophytes: *Carex stenophylla*, *Merendera robusta*, *Emimium alberti*, some *Gagea* species, several *Tulipa* species

A considerable number of grasses and Cyperaceae are found as part of the loess-related vegetation. Where the precipitation is a little higher and the rainy season lasts until the end of May, the grass vegetation cover on loess is rather dense and forms steppe vegetation. This may be found around Kunduz and Taluqan (east of Kunduz) in the northeastern portion of the Afghan lowlands.

Very few ecophysiological studies have been made on the deserts and semi-deserts of Afghanistan. Phenology, succession, energy flow, nutrient cycling and productivity also have had no detailed study. Thus, only a few more general remarks and observations can be given on ecosystem processes in those deserts.

Because of the distinct seasons associated with a continental climate, the spring flora is very conspicuous. The geophytes rapidly come to flower. The major genera in order of blooming are: *Gagea* and *Crocus*, sometimes *Anemone* and *Corydalis*, then *Tulipa* and *Allium*. In wet years many therophytes develop. In dry years no seedlings appear. The development of vegetation in spring is, of course, highly dependent on the kind of snow cover and melting of the snow during the early spring. In most areas, especially at lower altitudes, the spring vegetation dries up by the end of May. Most dwarf-shrubs and the Chenopodiaceae keep active till mid-summer or longer (Figs. 10.25 and 10.26). During the long, hot and dry summer, vegetation and habitats leave an impression of passivity with grey and brown colors from wilted plants dominating the scene.

The psammophytic, and especially the halophytic vegetation, is green and stays green later, mainly because of some succulent species (*Haloxylon*, *Halimocnemis*, etc.). Mirazai and Breckle (1978) have studied ionic relations in Chenopodiaceae from the northern desert and semi-desert lowlands of Afghanistan. Soil salt content was found to have a great influence on the occurrence of halophytic plant communities, but the salt content of the leaves of individual plants is rather independent of soil salt content. Among the 29 species from 18 genera that were analyzed from northern Afghanistan, several types of halophytes can be distinguished according to accumulation of ions. Ionic balances between sodium and potassium (Table 10.7) have possible systematic significance. Ionic relations are, however, rather independent of localities (Table 10.7 and Fig. 10.7). More data will be available when further plant and soil samples are analyzed.

Fig. 10.25. Semi-desert on sands and clay soils in northern Afghanistan. Various species of *Salsola*, *Aellenia*, *Cousinia* are the dominants.

Fig. 10.26. Serir area in southern Afghanistan, near Delaram, covered with *Halarchon vesiculosum* and a few annuals.

TABLE 10.7

Ionic concentration (in meq l^{-1} of cell sap) in Chenopodiaceae from northern Afghanistan, September 1976; n = number of specimens analyzed (Mirazai and Breckle, 1978)

Species	n	Ionic concentration				Ionic ratio (Na^+/K^+)
		Na^+	K^+	Cl^-	SO_4^{2-}	
Aellenia subaphylla ssp. *arenaria*	4	1280	335	175	60	3.8
Agriophyllum latifolium	1	23	440	155	33	0.052
Anabasis eriopoda	1	1430	210	300	27	6.8
Anabasis turkestanica	1	1910	280	500	26	6.7
Arthrophytum leptocladum	3	170	510	110	55	0.33
Gamanthus commixtus	2	2120	150	950	1160	14.0
Halimocnemis mollissima	5	1660	120	760	590	13.6
Halocharis afghanica	2	1920	270	530	850	7.0
Halocharis hispida	4	2600	310	580	1250	8.2
Halocnemum strobilaceum	1	1870	120	1680	53	15.2
Haloxylon persicum	3	290	450	130	65	0.65
Salsola longistylosa	2	1430	140	750	290	10.2
Salsola orientalis	3	2560	450	740	720	5.7
Salsola paulsenii	1	7	690	175	32	0.01
Salsola richteri	3	660	580	120	36	1.1
Salsola sclerantha	7	2470	450	2200	580	5.5
Seidlitzia rosmarinus	3	1540	180	180	83	8.3

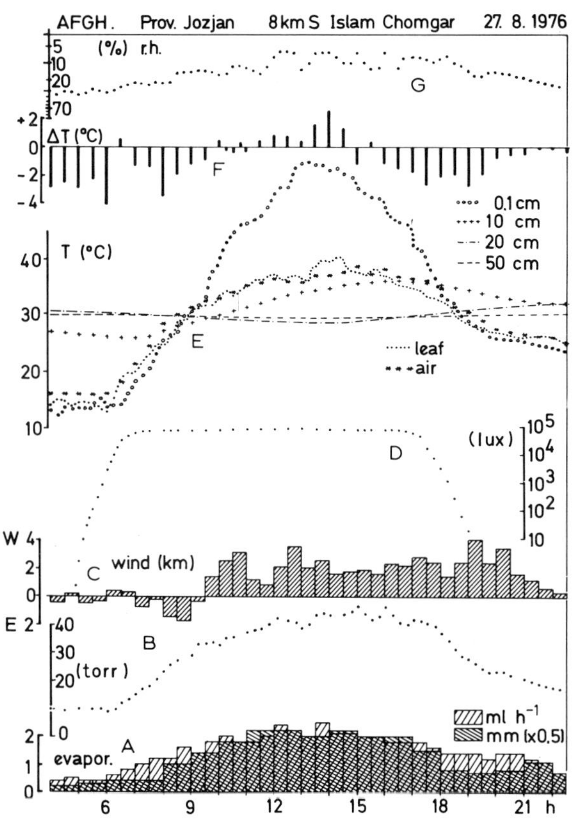

Fig. 10.27 is an example of the microclimatic data taken during late summer in this region. High temperature fluctuations and high evaporation capacity are clear. The sandy soils exhibit a rather high amount of insolation. Temperature fluctuations at the soil surface amount to more than 40°, but are almost absent at a depth of 20 cm in the soil. The leaf temperature allows for a typical two-peaked transpiration curve with stomata closure during mid-day hours.

Fig. 10.27. Daily course of some microclimatic patterns in a *Haloxylon–Calligonum* community, within a mosaic of reg and sebkha (S.-W. Breckle, unpubl. data). The data shown are as follows (upwards):
A. Evaporation, in ml h^{-1}, measured by Piche evaporimeters, with green filter paper, 3 cm ϕ, 1 m above-ground; in mm, measured by an evaporation balance, with open water surface. B. Water vapor saturation deficit (expressed in torr.), calculated from air temperature and relative humidity (r.h.) values. C. Wind: integrated velocity of air, expressed in km per 30 min, measured by a propeller anemometer. D. Light intensity (lux), measured by photo-resistance. E. Temperatures of different soil depths, 50 cm, 20 cm, 10 cm, and sandy soil surface (0.1 cm), registrated by NTC-miniresistances. Temperatures of air (psychrometrically) and of the leaf surface of *Salsola richteri*. F. Temperature difference between air and leaf temperature of *Salsola richteri*. G. Relative humidity (determined psychrometrically in %).

Fig. 10.28. *Salsola brachiata*, a rare species in inner-mountain arid valleys and continental small mountain basins in central Afghanistan. This specimen was near Bamiyan, 2700 m elevation. In the background at this hammada locality is *Anabasis macroptera*.

TABLE 10.8

Above-ground biomass (kg ha^{-1}) of semi-desert species on slightly saline clay in central and east Afghanistan

Dasht-i-Nawar (3110 m) near center of basin	
Polygonum sibiricum ssp. *thomsonii*	72.8
Crypsis aculeata	31.2
Artemisia macrocephala	1.7
Chamaegeron oligocephalum	1.2
Frankenia pulverulenta	1.1
Gentiana minutissima	0.14
Gentiana kauffmanniana ssp. *afghanica*	0.03
Total	*c.* 110
Ab-e-Istada (1970 m) outer vegetation belt with *Artemisia* semi-desert	
Artemisia kurramensis	2690
Atriplex turcomanica	146
Spergularia marina	17.6
Halogeton glomeratus	6.8
Frankenia pulverulenta	4.7
Aeluropus littoralis	1.7
Total	*c.*2860
Ab-e-Istada (1970 m) inner vegetation belt with *Spergularia* on solonchak	
Spergularia marina	795
Artemisia kurramensis	241
Atriplex turcomanica	142
Halogeton glomeratus	16.9
Frankenia pulverulenta	13.7
Halocharis violacea	2.1
Total	*c.*1210

Desertic mountain valleys and small basins

Semi-deserts and deserts develop in some isolated mountain basins because of rain-shadow effects. This phenomenon can be observed in several great mountain regions on earth, such as the Andes, the Himalaya and the Karakorum. It can also be observed within the Hindu Kush. It is called the "Troll Effect" (Troll, 1951; Schweinfurth, 1956; Breckle, 1973). In Hazarajat (near Bamiyan), in Ghorband Valley, in some valleys in the central Afghan mountains, and north of the main range of Hindu Kush, this phenomenon is dominant. The soils of these valleys are often rich in gypsum. Stands of *Salsola maracandica* (Freitag, 1971), some other *Salsola* species (Fig. 10.28), *Anabasis macroptera* (Fig. 10.29), *Arthrophytum lindbergii*, *A. pallidiflorum*, or *Artemisia* and *Acantholimon* species develop.

These localities characteristically show extreme climatic conditions. The daily and annual temperature fluctuations are very great. The temperature at the soil surface on most days reaches freezing early in the morning. The absolute minima in the winter are very low (see Fig. 10.2, climatic diagram Panjao).

In some places transitional to the sub-alpine thorny dwarf-shrub semi-desert and on extreme gypsicolous semi-desert, the *Artemisia–Krascheninnikovia* community is developed. *Ephedra procera* and *E. intermedia* and *Camphorosma lessingii*, may be associates on such sites (Fig. 10.30). Other companion genera are *Acantholimon*, *Acanthophyllum*, *Astragalus* and *Cousinia* (Fig. 10.31), *Arthrophytum lindbergii*, *Polygonum patulum* and *Thevenotia persica* occur on fine-textured soils with some halophytic components (Fig. 10.32) according to salinity and/or alkalinity. Floristic composition

Fig. 10.29. A heavily grazed *Anabasis macroptera*, typical of intermontane basins (Bamiyan, 2700 m).

Fig. 10.30. *Camphorosma lessingii*, a cushion-forming chenopod, found sometimes on sands, but more often on fine-textured and more alkaline soils in mountain regions (Bamiyan, 2700 m).

Fig. 10.31. *Ephedra intermedia*, a major species of a genus widely distributed in deserts, semi-deserts and high up into the mountains. *E. intermedia* is largely found in *Artemisia* semi-deserts, but is in sub-alpine spiny cushion vegetation too (Band-i-Amir, Central Afghanistan).

and the above-ground biomass are exemplified in Table 10.8 (S.-W. Breckle, unpubl. data). The above-ground biomass is very low, but there is probably several times more below ground. Phytomass is much higher at the Ab-i-Istada region. This is a more easterly location at lower elevation. The shallow basin there is mainly surrounded by an association dominated by *Artemisia* (Table 10.8).

Artemisia semi-deserts

Hilly regions have semi-deserts on stony substrata dominated by *Artemisia* (*A.* "*herba-alba*" aggregate) (Figs. 10.7, 10.16). In northern Afghanistan (Breckle and Unger, 1977) and north-eastern Iran they intergrade with communities of giant umbellifers, such as *Ferula foetida* (Fig. 10.20), *F. badrakema*, *F. narthex*, and *Dorema aitchisonii* (Fig. 10.16), or *Zygophyllum eurypterum* (Fig. 10.15) and *Pteropyrum aucheri*. Many annuals (Figs. 10.30, 10.34) and small shrubs like *Salsola arbuscula*, *Krascheninnikovia ewersmanniana*, *Scariola orientalis*, *Gaillonia brugieri*, *Dendrostellera lessertii* and *Pulicaria crispa*, perennials like *Eremostachys regeliana* (Fig. 10.24), graminoids like *Stipa barbata* and *Carex physodes*, and geophytes such as *Iris songarica*, *Allium fibrosum* and *A. caspium* also occur there.

Several *Artemisia* species seem to form separate associations. Despite the fact that *Artemisia* is a very important and often dominant genus, we lack a taxonomic treatment of it for this region. Different *Artemisia* communities can also form mosaics with the open shrub and woodland communities on the mountains.

Many of the non-saline hammadas are dominated by *Artemisia* spp. They occur in so many

Fig. 10.32. The vast basin of Dasht-i-Nawar in central Afghanistan (3110 m elevation) has a mostly saline-alkaline soil and heavily grazed vegetation composed of *Artemisia macrocephala*, *Crypsis aculeata*, *Halocharis violacea* and *Polygonum sibiricum*.

Fig. 10.33. A river oasis within the arid foothills (serir plains) covered by *Artemisia* semi-desert in the Logar Valley south of Kabul.

communities of the central Persian basins and plateaus and surrounding hills and mountains on into Afghanistan (Fig. 10.7) that the vegetation of most areas can be termed *Artemisia* semi-desert (Artemisietea). Perennial grasses must have been once much more common or even dominant in the original composition of most of these communities (Zohary, 1973). Grasses are now almost lacking altogether, except in places inaccessible to grazing.[1]

The original steppe areas must have been much more widespread. Considerable portions of the open woodlands have also been degraded to *Artemisia* semi-deserts.

Hulthemia, *Achillea santolina*, *Acantholimon truncatum*, *Haplophyllum robustum*, *Stipa lagascae* and *Launaea acanthodes* are commonly found within the *Artemisia* semi-desert in western and northern Iran. Some of these species are found occasionally in Afghanistan. Semi-ruderal weeds, such as *Peganum harmala*, *Alhagi*, *Prosopis farcta*, *Goebelia alopecuroides* and *Glycyrrhiza glabra*, are rapidly expanding near cultivated areas.

Despite the fact that the main towns, villages and many cultivated areas are situated within the *Artemisia* semi-desert (Fig. 10.33), very little has been done to investigate the ecology of dominant species in both countries.

Bhadresa (1977) gave a value of 1.45 t ha^{-1} above-ground biomass for *Artemisia herba-alba* communities in northeastern Iran; McArthur et al. (1976) reported that total above-ground dry matter at the beginning of winter at Dasht-e-Hamdamab, northeast of Kohsan (Herat Province) ranged from almost negligible on scalded areas to 4.5 t ha^{-1} on the best sites. Woody stem parts amounted to 2.26 t ha^{-1}, branchlets to 1.74 t ha^{-1}, new leaves 0.43 t ha^{-1} and other plant species 0.062 t ha^{-1}. Total vegetal cover varied from nil to 60%.

[1] In enclosures within the *Zygophyllum* and the *Artemisia* semi-deserts in the Touran Biosphere Reserve in northeastern Iran, after only eighteen months' protection the percentage of grasses had visibly increased.

TABLE 10.9

Some ecophysiologically important characteristics of the three therophytes *Diarthron vesiculosum*, *Scabiosa olivieri* and *Ziziphora tenuior* from the cool semi-desert in eastern Kabul basin (1800 m elevation) during May and June (end of rainy season)

Date	May 8	May 20	June 1	June 10	June 16	June 28
Soil water (% dry basis)	10.0	9.6	4.5	4.3	1.0	1.2
Plant water content (% dry matter)						
Diarthron vesiculosum	332	318	211	181	142	128
Scabiosa olivieri	249	240	181	172	108	91
Ziziphora tenuior	219	200	107	36	6.6	4.1
Osmotic potential of cell sap (bars)						
Diarthron vesiculosum	−11.5	−11.1	−16.9	−22.8	−26.7	−22.0
Scabiosa olivieri	−12.1	−11.0	−17.5	−18.5	−20.9	−27.0
Ziziphora tenuior	−10.4	−11.9	−14.5	−21.1	−*c*.45	
Degree of scleromorphism (ratio: crude fiber/protein)						
Diarthron vesiculosum	0.38	0.39	0.96	1.22	1.25	2.13
Scabiosa olivieri	0.90	1.10	2.0	2.7	3.0	4.8
Ziziphora tenuior	1.92	2.64	3.56	3.84	4.0	4.1
Total content of sugars (% dry matter)						
Diarthron vesiculosum	7.8	7.2	6.1	6.3	6.1	3.7
Scabiosa olivieri	3.3	3.6	3.4	3.8	4.0	4.6
Ziziphora tenuior	2.8	3.8	5.9	2.5	1.6	1.5
Content of slime materials (% dry matter)						
Diarthron vesiculosum	7.0	6.0	4.2	5.0	4.7	3.2
Total content of lipids (% dry matter)						
Diarthron vesiculosum	3.8	4.1	4.3	4.6	6.1	5.3
Scabiosa olivieri	24.1	20.0	18.7	18.4	18.9	19.3
Ziziphora tenuior	2.8	3.2	3.4	4.4	3.6	3.5

A few data on the ecological behavior of three annuals (*Diarthron vesiculosum*, Thymelaeaceae; *Scabiosa olivieri*, Dipsacaceae; and *Ziziphora tenuior*, Lamiaceae) are available from the vicinity of Kabul. The Russian observation that *Diarthron* has an active life throughout the whole arid summer could not be confirmed for Afghanistan (Breckle, 1971b). All of these therophytes wilted by the end of May shortly after rains ceased. The osmotic potential of cell sap and carbohydrate content change rapidly (Table 10.9). It is obvious, however, that these three annuals belong to different subgroups with different physiological behaviour (Breckle and Kull, 1973). *Diarthron* under drought stress yellows, contains appreciable amounts of slime (the other two species none), and the osmotic potential of cell sap increases again (probably by starvation; Table 10.9). *Scabiosa* seems to use lipids as storage material and thus is able to prolong activity and produce relatively more seeds (Kull and Breckle, 1975). Despite its containing essential oils, *Ziziphora* wilts most rapidly in June. Other plants with thin, but long tap-roots stay alive much longer — for instance, *Astragalus tribuloides* or *Ceratocarpus utriculosus*.

To further exemplify the *Artemisia* semi-desert, a transect through a small ravine will be described. It was located on the western slope of Korough Koh, south of Kabul, between hills and ridges of marl and limestone rock. It included broad areas of loess and sand. Some of the areas had been under cultivation by dry-farming and primitive run-off irrigation. Many areas were heavily eroded. One of the smaller ravines was studied and soil conditions analyzed along a profile from south to north (Fig. 10.34 A–E).

The density of vegetation (Fig. 10.34A) was highly variable. The most common weed, occurring

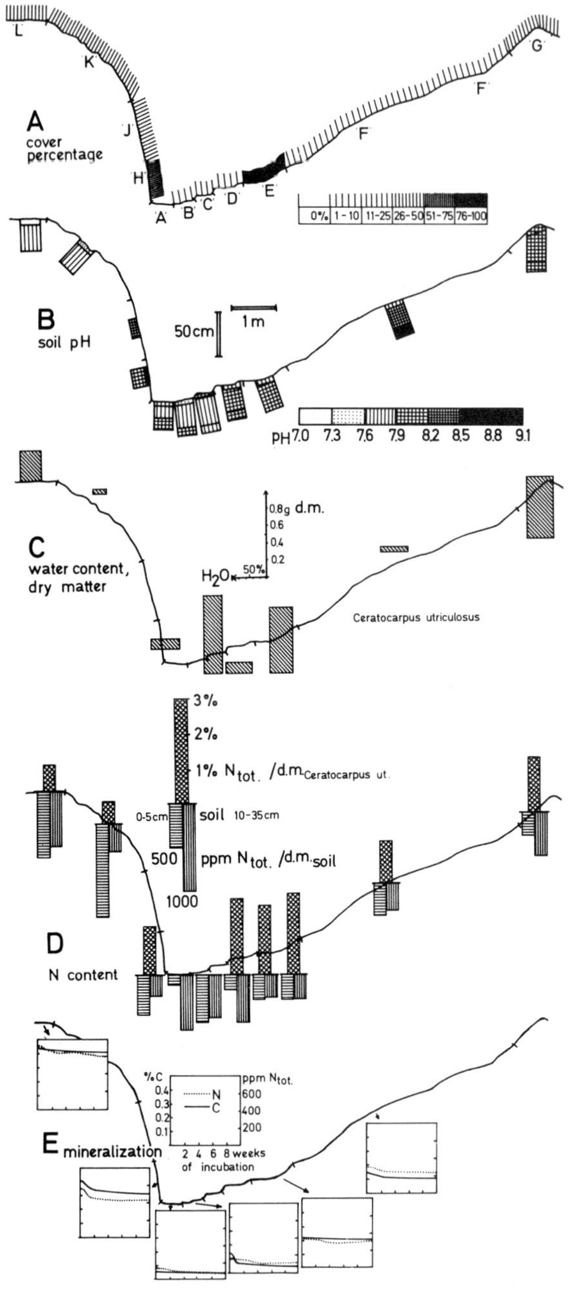

Fig. 10.34. Profile (north–south) of some ecophysiological data through a small ravine south of Kabul, west slope of Korogh Koh (2100 m elevation) (Sept. 1, 1967). A. Cover percentage of vegetation at different spots. B. Soil pH at different spots and depths. C. Water content and dry matter of individuals of the annual *Ceratocarpus utriculosus* (Chenopodiaceae). D. Total nitrogen content of *Ceratocarpus* and of soil in two different depths (dry weight basis). E. Changes in total nitrogen and organic carbon content during a ten week incubation period, incubation of soil (10 cm depth) at 35°C and 15% soil humidity.

almost throughout the transect, was the annual *Ceratocarpus utriculosus*. This was green and active on September 1. In the ravine on Sites *B* and *C*. *Alhagi* was dominant, on Site *D*, *Cousinia decurrens*; on Site *E*, *Perovskia atriplicifolia*, *Alhagi* and *Hertia intermedia* with *Bromus dantoniae* and other grasses formed dense patches of vegetation. Here high dry matter values were found for *Ceratocarpus* (Fig. 10.34C) but not correlated with the plant water content.

On Site *F*, *Scariola orientalis* dominated. On Site *G* and on the Area *L* of the upper plain, *Hertia intermedia*, *Stachys parviflora*, *Boissiera squarrosa*, *Diarthron vesiculosum* and *Artemisia* species occurred. On the shady Site *H* some alkaline salt crusts (carbonates and sulfates) were found with a pH of 9 near the soil surface (Fig. 10.34B), and a rather dense, spotty vegetation of *Alhagi*, *Ephedra*, *Lactuca*, *Astragalus psilacanthus*, *Acantholimon*, etc. On Site *K* a few *Leucopoa karatavica* tussocks and *Trichodesma incanum* were found growing, with *Astragalus psilacanthus*, *Polygonum paronychioides*, *Hertia*, several annual *Astragalus* spp., *Scorzonera*, *Iris*, *Eremurus persicus*, *Gagea*, etc., and a few cushions, like *Gaillonia* (Fig. 10.35).

The total nitrogen content of the soil was rather low (Fig. 10.34D), but that of *Ceratocarpus* was high. The C/N ratio in all soil samples was about 6. The microbial activity in the different soil samples varied along the transect (Fig. 10.34E). Some samples showed distinct decreases in carbon and nitrogen content when incubated at 35°C and constant moisture (15%). In other soil samples, however, almost no activity could be detected — for instance, Sites *A* or *F* (S.-W. Breckle, unpubl. data). Further studies on microbial activity are strongly needed.

There is almost no information on the role of micro-organisms within these desert and semi-desert ecosystems. Soil microbial activity is favored within the neighborhood of ant hills (*Cataglyphis bicolor*) in the *Artemisia* semi-desert south of Kabul (Dasht-e-Khoshi, 2400 m; Breckle, 1971a). Preliminary soil respiration measurements have shown very low values for evolution of carbon dioxide (S.-W. Breckle, unpubl. data). It has to be assumed for the time being that a similar microflora to those studied in Uzbekistan (Ch. 6) occurs on these takyr soils. As for the other regions it must be kept in mind that the relatively short

Fig. 10.35. *Gaillonia macrantha*, a small cushion-plant, common in the eastern parts of Afghanistan within the *Artemisia* semi-deserts; a rubiaceous subshrub normally heavily grazed.

humid season in late winter and spring is the only time when rather good conditions for development of fungi, bacteria, and blue-green and other algae prevail.

Semi-desert open woodlands

Semi-desert shrub communities, with or without small trees, have been described from several parts of the area. A transitional belt of communities dominated by *Amygdalus eburnea* and *A. erioclada* is distinctive of the foothills partly girdling the central Afghan mountains (Freitag, 1971). At higher altitudes this shrubland is replaced by xeromorphic open woodlands, dominated by *Pistacia atlantica*.

Communities dominated by *Pistacia atlantica* form a broad belt from the Hari Rud Valley along the mountains of southern Ghorat and Hazarajat up to the eastern border of Afghanistan (Fig. 10.36) at elevations between 1000 and 1900 m. At lower elevations, *P. khinjuk* may co-dominate. The shrub layer of this semi-desert woodland is formed by some *Amygdalus* spp., *Astragalus mokurensis*, *Convolvulus leiocalycinus*, *Ephedra* spp., and nu-

Fig. 10.36. Some of the last remnants of arboreal vegetation, trees of *Pistacia atlantica* ssp. *cabulica* with strong trunks (mistakenly named *Pistacia vera* in Walter, 1974, fig. 196), in open woodlands on hills in the vicinity of Kabul, 2000 m elevation. This is a characteristic example of "ziarat" vegetation at holy places.

merous dwarf-shrubs like *Cousinia stocksii* (Fig. 10.37), *Astragalus*, *Gaillonia*, *Stachys trinervis*, many *Acantholimon* and *Acanthophyllum* species; some geophytes and many annuals occur in the lower understory.

The *Pistacia* belt in northern Afghanistan is dominated by *P. vera* and lacks the transitional shrub semi-desert with smaller *Amygdalus*. This *Pistacia vera* semi-desert open woodland type has a more moderate climate and extends into northeastern Iran, east and northwest of Mashhad. The tree cover may reach up to 40% on favorable sites. Freitag (1971) reports the following main species from the northern slopes of the Afghan mountains:

trees: *Pistacia vera* (often protected for pistachio nut production), *Amygdalus bucharica*, *Cercis griffithii* (only in the northeast)

shrubs and subshrubs: *Amygdalus spinosissima*, *Cerasus bifrons*, *Rosa* spp., *Colutea gracilis*, *Ephedra ciliata*, *E. intermedia* (Fig. 10.31), *Artemisia* spp., *Acantholimon* spp.

perennials: *Cousinia raddeana*, *C. pseudomollis*, *C. umbrosa*, *Helichrysum plicatum*, *Phlomis bucharica*, *Chaetolimon sogdianum*, *Codonocephalum grande*, *Salvia pterocalyx*, *Solenanthus turkestanicaus*

geophytes: *Carex stenophylla*, *Eremurus bucharicus*, *E. olgae*, *E. regelii*, *Ungernia trisphaera*, *Bellevalia atroviolacea*, *Bongardia chrysogonum*, *Anemone* spp.

annuals: *Aegilops crassa*, many *Bromus* spp., *Taeniatherum crinitum*, *Hypogomphia turkestanica*, *Torularia dentata*, *Thlaspi perfoliatum*, *Malcolmia spryginiodes*.

Dry farming is very common within the *Pistacia vera* woodland zone, mainly on the loess soils. The woodlands are normally overgrazed by livestock.

The upper belt of the open woodlands is represented by *Amygdalus*-dominated communities occurring all around the central Afghan mountains at elevations between 1800 and 2800 m, where precipitation exceeds at least 400 mm yr^{-1}. The original open woodland has been largely replaced by communities of *Artemisia* and other weedy herbs or spiny subshrubs because of overgrazing and other profound anthropic influences. Some soil characteristics in this belt can be seen in Table 10.3.

Similar deciduous woodlands are also found in many mountainous areas of the eastern Zagros, the east Persian mountains, and in some parts of the lower southern slopes of the Elburz (Fig. 10.38). Even on north-facing mountain slopes within the desert regions, such as Kavir, there are remnants of these open semi-desert woodlands. Rechinger and Wendelbo (1976) report a comparatively rich vegetational cover from the Siah Koh, the main mountain in Kavir National Park. Similar conditions can be seen within the Touran Biospheric Reserve with trees such as *Amygdalus scoparia*, *Cerasus microcarpa*, and *Pistacia khinjuk*. *Amygdalus* are scattered here and there and can form rather dense stands in some places. Certain individuals may reach 4 m in height. Zohary (1973) describes an *Amygdalus scoparia–Pistacia khinjuk* association in a very poor and much degraded state. Rechinger and Wendelbo (1976) list, among others, *Acantholimon talagonicum*, *Acanthophyllum sordidum*, *Aegopordon berardioides*, *Atraphaxis spinosa*, *Bunium persicum*, *Cousinia onopordioides*, *Eremurus luteus*, *Galium setaceum*, *Jurinea carduiformis*, *Lepyrodiclis holosteoides*, *Minuartia meyeri*, *Pennisetum orientale*, *Rheum ribes*, *Scandix stellata*, as characteristic species of normal sites in this belt. *Campanula incanescens*, *Parietaria judaica*, *Paronychia kurdica*, *Phagnalon persicum*, *Psychrogeton obovatus* occur in rock crevices.

According to Mobayen and Tregubov (1970) a zone of *Pistacia atlantica* ssp. *mutica* is situated on the lower slopes of the Elburz between 1600 and 1800 m. *Rhus coriaria* and *Celtis australis* also grow sporadically there. This zone is much overgrazed, with degraded sites having *Artemisia*, *Cousinia*, *Stipa*, etc. Increasing *Juniperus* remnants can be found in higher regions. Similar areas in the Zagros have *Cotoneaster nummularia*, *Ephedra procera* and *Scariola orientalis*. Most of this vegetation type has been greatly disturbed and the soils degraded. This is particularly true of watersheds near settlements having irrigation-based agriculture (Fig. 10.38).

A few ecophysiological observations are available for *Pistacia atlantica* ssp. *cabulica* (Fig. 10.36), *Cercis griffithii* in east Afghanistan (Kull and Breckle, 1972), and *Quercus baloot* (Breckle and Kull, 1971). The annual curves of the potential osmotic pressure and that of the refractive index of cell sap of leaves of the deciduous species clearly show highest values during the late summer when the highest drought stress occurs. This is in contrast to the level values of evergreen *Quercus baloot*, even on similar stands. The carbohydrate storage in leaves and in bark show distinct seasonal variations, with maxima in winter in the bark and leaves from *Quercus* and in bark of *Cercis*. *Pistacia*, however, reaches its maximum sugar content during summer. Osmotic conditions show a strong

Fig. 10.37. The cushion-forming *Cousinia stocksii* which is widely distributed on hammadas in the deserts and semi-deserts of southern Afghanistan (north of Farah).

Fig. 10.38. Landscapes on the southern side of the Elburz Mountains, Iran where deep valleys are cut in the old terraces. The terraces (probably from glacial periods) carry fields and orchards. On steep, more inaccessible slopes open *Juniperus* woodlands in small remnants can be found. All other areas are covered by a montane *Artemisia* semi-desert. On the higher parts (above 2800 m) spiny cushions occur. The highest peaks (about 3500 m in this view, to some ranges east and southeast of Damavand) carry very open sub-alpine and alpine semi-deserts rich in plant species.

Fig. 10.39. One of the westernmost naturally occurring *Quercus baloot* stands, at the Kotal-e-Latahband, east of Kabul. The small trees, which had a height of 2 m in 1968, have been heavily grazed by sheep and goats of the nomads.

Fig. 10.40. An individual *Quercus baloot*, in 1976, had a height of only about 50 cm. Natural reproduction is already impossible.

correlation with carbohydrate content. Frost conditions during winter seem to be more important than drought near the western limit of the distribution of *Quercus baloot* (Figs. 10.39, 10.40). Thus, the distribution pattern of *Quercus baloot* gives no good definition of semi-desert boundaries in the eastern parts of Afghanistan.

Subalpine semi-deserts and steppes

Subalpine semi-deserts in the Afghan mountains are found between 3300 and 4000 m in the east, between 2900 and 3700 m in the north, and between 2700 and 3600 m in the center of the mountainous part of Afghanistan. Subalpine semi-deserts further to the east are cushion shrublands which are gradually replaced by subalpine thickets, elfin-woods and dense grass-steppes with increasing amounts of summer precipitation. The more arid cushion shrublands are composed of many thorny subshrubs of tragacanthoid character. Parallel-evolved genera like: *Astragalus*, *Onobrychis*,

Fig. 10.41. *Acantholimon* (cf.) *homophyllum* in the mountainous semi-deserts of Dasht-i-Nawar (at about 3600 m elevation). The south-facing side of the hard cushion is partly damaged by frost desiccation.

Fig. 10.42. A central Afghanistan landscape covered by semi-deserts, with spiny cushions or with *Artemisia* and a few grass tussocks. The foreground and the middle are covered by the same type of vegetation, in the background, the Koh-i-Baba Mountains are partly covered by snow in late autumn.

Euphorbia, *Acantholimon* (Fig. 10.41), *Acanthophyllum* and *Platytaenia* are dominant. Other subshrub species of *Cousinia*, *Artemisia*, *Nepeta*, *Scutellaria*, *Salvia*, *Thymus*, *Hymenocrater*, *Scrophularia*, *Trigonella*, *Cicer*, *Polygonum*, *Fibigia*, *Aethionema* and *Lactuca*, and many annuals and geophytes (*Eremurus*, *Tulipa*, *Gagea*, *Anemone*, *Corydalis*, *Allium*, *Iris*, *Isopyrum*, *Rheum*, *Taraxacum*) are also found in this zone (Fig. 10.42). True shrubs, such as *Astragalus cuneifolius* and a few species of *Rosa*, *Rhamnus*, *Ribes* and *Lonicera* are occasionally found in some portions of the zone.

Ecological studies within the subalpine semi-deserts, particularly the thorny cushion communities, have been rare. Soil temperatures at lower elevations can be almost 60°C at noon at the soil surface (1.2 cm depth) and almost 40°C at 20 cm depth (Neubauer, 1952). The temperatures (at 2400 m, in April) are rather high near the soil surface because of the high insolation (Breckle, 1971c). Some plants exhibit leaf temperatures of 10° or even 15° higher than air temperatures. This phenomenon of heating plays an even more important role in the subalpine and alpine semi-deserts. In the semi-desert above 4000 m, many plants grow only at sites afforded wind protection and favorable water conditions (Fig. 10.43).,The belt of dwarf-shrubs and cushions above 3800 m in central Afghanistan and above 4000 to 4300 m in the Hindu Kush is replaced gradually by open alpine semi-deserts. This alpine hammada rather sharply separates into a few types of plant associations according to substrate particle sizes and water conditions (Breckle, 1973; Walter, 1974). On sites where an additional water supply — for instance, by melting snow (Fig. 10.43) or perched water table (Fig. 10.44) — occurs, short *Kobresia* and *Carex* steppes develop. These are grazed heavily during the summer by the livestock of nomadic peoples.

Some of the main species of the alpine semi-desert are: *Androsace villosa*, *Carex nivalis*, *Cousinia chionophila*, *Draba* spp., *Delphinium brunonianum*, *Leontopodium campestre*, *Nepeta pamirensis* (Fig. 10.45), *Phleum alpinum*, *Potentilla gelida*, *P. multifida*, etc., *Rubia tibetica*, *Saussurea gnaphalodes*, *Valeriana fedtschenkoi*, *Waldheimia tridactylites*, etc. (Breckle, 1973).

The main genera forming alpine steppes are: *Kobresia*, *Carex*, *Poa*, *Oxytropis*, and *Cerastium* (Breckle, 1971c, d, 1973). The annual precipitation is of minor importance because, under the severe continental temperature regime at high elevations, only the moisture available during the short growth period in summer is relevant. Thus, additional water supplied by melting snow, near creeks, furrows, etc., is a main factor in forming the vegetation mosaic (Breckle, 1973; Table 10.3).

The typical formation of tragacanthoid subshrubs in many Eurasian mountain regions (Gams, 1956) must have been caused by special ecological, mainly climatological features. High radiation, precipitation during winter as snow, almost no rain, and extreme temperature fluctuations are some of the main factors. The leaf temperature regime and light combinations occurring during different days with rather different weather conditions are shown in Fig. 10.46A, and Fig. 10.46B for *Primula macrophylla* at 4060 m in the Koh-i-Baba Mountains in Hazarajat, in Fig. 10.46C for the dwarf-shrub *Krascheninnikovia ceratoides* (= *Ceratoides latens* = *Eurotia ceratoides*) at 3110 m in southern Hazarajat (Breckle and Reshad, 1971). In Fig. 10.46 the percentage of time during which the different light and temperature combinations prevail is given as the ordinate.

Fig. 10.46C shows data for a totally cloudless day with high temperature amplitudes. A combination of high radiation and high leaf temperatures predominates. A lower, second maximum combining low light and low temperature is also present. The same pattern affects alpine species such as *Primula macrophylla* (Fig. 10.46B). On July 15 when weather conditions were totally clear, the leaf temperatures ranged between −8°C and +38°C between morning and noon. The previous rather cloudy and partly rainy day with low radiation had moderately low leaf temperatures with a rather narrow range of combinations (Fig. 10.46A). No leaf temperatures fell below 0°C. This latter kind of weather is, however, rare.

Data on osmotic potentials of cell sap are available for a few species of the subalpine and alpine steppes and semi-deserts of Afghanistan (Table 10.10). They indicate rather moderate conditions (Breckle, 1973). In contrast, the osmotic potential of cell sap in desert species of Chenopodiaceae reaches −50 to less than −100 bars.

The subalpine timberless vegetation in Iran is semi-desert in many places also, being rather similar to the Afghan vegetation of this type. In both

Fig. 10.43. An example of the subalpine region with an open *Leucopoa* steppe (4000 m elevation). Only near furrows of melting snow water are dense *Kobresia* steppes developed as narrow stripes (Koh-i-Baba, central Afghanistan).

Fig. 10.44. The *Kobresia* and *Carex* steppe in the basin of Dasht-i-Nawar (3110 m elevation) in central Afghanistan is one of the main grazing areas for the nomadic people in summer.

Fig. 10.45. *Nepeta pamirensis*, one of the main species in the alpine semi-deserts in the Hindu Kush.

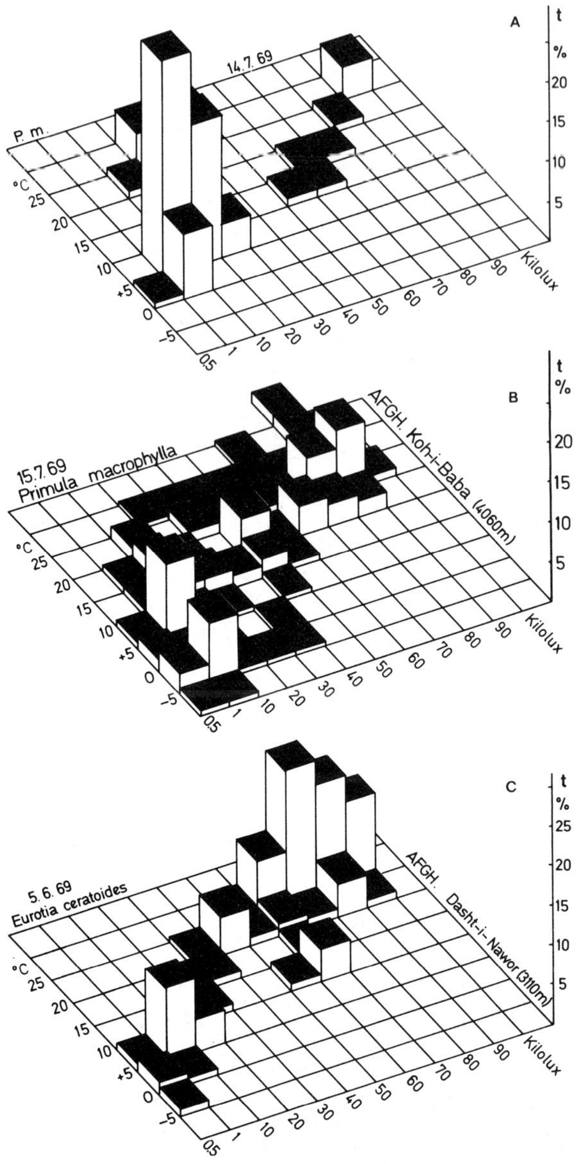

Fig. 10.46. Percentage of day-time with special leaf temperature/light combinations allowing development of *Primula macrophylla*, Koh-i-Baba, main ridge (Fuladi, 4000 m elevation). A. July 14, 1969; a cloudy rainy day. B. July 15, 1969; a bright, very clear day. C. The same with *Krascheninnikovia ceratoides* at Dasht-i-Nawar (3110 m elevation), on June 5, 1969.

cases the extreme species richness must be emphasized. In addition to the scores of *Astragalus* species of spiny cushion and subshrub form, there are over a hundred species of *Acantholimon* and many species of *Acanthophyllum*, *Onobrychis*, *Gypsophila*, *Arenaria*, *Minuartia*, *Tragopogon*,

TABLE 10.10

Maximum and minimum values of the osmotic potential of cell sap of some subalpine and alpine species (Breckle, 1973)

Species	Osmotic potential (bars)	
	max.	min.
Nawar (3110 m)		
Kraschenimnikovia ceratoides	−19.2	−24.7
Nepeta rugosa	−14.1	−15.2
Khrebek (Hindu Kush) (4350 m)		
Carex nivalis	−19.2	−20.8
Phleum alpinum	−19.6	−21.7
Delphinium brunonianum	−16.2	−18.4
Epilobium latifolium	−13.2	−14.6
Oxyria digyna	−10.1	−11.6
Koh-i-Baba (4060 m)		
Rhodiola heterodontum	−6.6	−8.0
Primula macrophylla	−10.8	−11.7
Nepeta pamirensis	−9.2	−13.1
Cousinia polyneura	−14.4	−15.9

Ononis, and others which are quite conspicuous in this formation. Genera such as *Euphorbia*, *Erinacea*, *Genista*, and *Cytisus* also take on a spiny cushion-like form (Zohary, 1973). Despite the uniform appearance of this vegetation type, its species composition varies greatly from place to place. Many endemics are found here (Rechinger, 1963 ff.).

Most species in the spiny cushion semi-desert develop long tap-roots. The hemiglobular shape of many of them must confer adaptive advantages against excessive solar radiation and dry winds. Transpiration is probably reduced within their own small, microclimatically moderated cushion. Detailed ecophysiological studies are strongly needed.

ANIMALS

The importance of the rich fauna of this area to ecosystem function has not yet been investigated. It is, however, known (J. Niethammer, G. Nogge, G. Madel, pers. comm., 1978) that many animal groups have very high seasonal and annual fluctuations in population densities. For instance, near Kandahar (southern Afghanistan) in 1965 many rodents and birds were observed after a wet period.

Fig. 10.47. The fox (*Vulpes vulpes*) is rather common in the Touran Biospheric Reserve in northeastern Iran.

In 1972 during and after an unusually long period of dry months, almost no rodents and only few migrating birds were seen. The population densities fluctuate widely depending on the development of vegetation. Detailed investigations are required if conclusions on the ecological importance of animals can be drawn.

In many regions larger mammals and bird species are extinct or nearly so. Naumann and Nogge (1973) made a survey of the earlier presence of several large mammals and of their recent distribution. *Hyaena hyaena*, *Felis pardus*, *F. caracal*, *Acinonyx jubatus* (eastern Iran) and others were once common but are now rare. *Canis lupus* and *Vulpes vulpes* (Fig. 10.47) are still common in most of the mountains. *Canis aureus* is abundant in lower areas. *Vulpes corsac* is only found in northern Afghanistan. The Touran Biospheric Reserve in northeastern Iran was established mainly to keep some animal populations at selected levels and to protect endangered species (Firouz and Harrington, 1976). The gazelle (*Gazella subgutturosa*), once common in the desert regions in all of Afghanistan, will probably be one of the next species of game animals to disappear from that country. Ungulates play a rather important role in the open woodlands and *Artemisia* semi-desert

Fig. 10.48. The Persian onager (*Equus hemionus*) which has increased to a population of about 800 within the Touran Biospheric Reserve (photo by Rasik Bhadresa).

within the hill and mountain regions. Though the Persian onager (*Equus hemionus* or wild ass) is already extinct from the western Afghan lowlands, *Capra ibex* and *Ovis ammon cycloceros* still exist in most of the Afghan mountain regions (Naumann and Nogge, 1973). The Touran Biospheric Reserve in Iran contains now one of the highest populations (about 800) of the Persian onager (Fig. 10.48). *Gazella subgutturosa* is particularly common in some plains of the protected areas in Iran. In lower areas *G. dorcas* seems to be abundant.

Several rodents are known from the deserts of Iran and Afghanistan. In northern lower areas (below 1500 m) the following are reported from Afghanistan (Janossy and Schmidt, 1970, Hassinger, 1973; Niethammer, 1973): *Spermophilopsis leptodactylus*, *Rhombomys opimus*, *Meriones libycus*, *M. meridianus*, *Allactaga elater* and *Ellobius talpinus*. In the southern deserts and semi-deserts of Afghanistan: *Jaculus blanfordi*, *Allactaga elater*, *Meriones crassus*, *M. libycus*, *Tatera indica*, *Gerbillus nanus* and *G. cheesmanii*. *Calomyscus bailwardii* and *Meriones persicus* occur in rocky areas. In higher regions (above 1500 m), besides some of the species already mentioned, *Citellus fulvus*, *Allactaga euphratica*, *Ellobius fuscocapillus* and *Cricetulus migratorius* occur, and above 3000 m *Marmota caudata* is found, as are hares (*Lepus capensis*) and pikas (*Ochotona*). *Lepus* exhibits a very wide range of altitudinal occurrence. Porcupines (*Hystrix indica*) and the vole (*Blanfordimys afghanus*) are found only in moderately arid regions.

Most of the rodents mentioned for Afghanistan are also common in the Touran Biospheric Reserve in Iran (Harrington, 1976, 1977). Studies on their role in the ecosystem processes are under way (Table 10.11). The most abundant species are *Calomyscus bailwardii* and *Meriones persicus*. Both seem to play an important role as herbivores in rocky biotopes. The same is true of *Cricetulus migratorius* near farmland areas and on well-drained sites. The principal wild herbivores are

TABLE 10.11

Estimates of rodent density, biomass and energy consumption (Brown, 1976; Spooner, 1977), in the Touran Biospheric Reserve, Iran

Site No.	Dominant plant species	Mean number of rodents per km^2	Mean rodent biomass (kg km^{-2})	Rodent energy consumption (kJ km^{-2} yr^{-1} × 10^3)[1]
1	*Artemisia–Amygdalus–Ephedra* on rocky hillside	52	0.9	1386
2	*Artemisia*	175	8.1	12 600
3	*Zygophyllum*	450	28.8	44 100
4	*Haloxylon*	400	22.8	26 860
5	*Halocnemum–Phragmites* on salt flat	455	14.4	22 260
6	*Calligonum–Stipa–Haloxylon* on sand dunes	832	26.0	39 900

[1] 1 kcal = 4.2 kJ.

TABLE 10.12

Estimates of biomass and energy flow of ungulates in the Touran Biospheric Reserve, Iran (Spooner, 1977); mean values from *n* observation areas

Species	Sample numbers (*n*)	Biomass (kg km^{-2})	Food Intake (kJ km^{-2} yr^{-1})	Production (kJ km^{-2} yr^{-1})
Gazella	6	2.25	352 800	46 200
Goat and sheep	3	1300	201 × 10^6	26.8 × 10^6

Fig. 10.49. *Marmota caudata*, one of the main herbivores at higher altitudes, e.g. in the alpine semi-deserts of northeastern Afghanistan and Vakhan.

Ochotona, *Marmota caudata* (Fig. 10.49), *Citellus fulvus*, *Spermophilopsis leptodactylus*, *Lepus*, some *Meriones* species, *Ellobius*, and *Blanfordimys*. *Lepus capensis* seems to feed heavily upon *Zygophyllum*; however, it never occurs in high densities.

Native ungulates play a noteworthy role in grazing only in some protected areas (Table 10.12).

J. Niethammer (pers. comm.) found only *Citellus fulvus* and *Meriones libycus* near Mukur (southeastern Afghanistan). *M. libycus* is active the year around, whereas *C. fulvus* is active only from April to early summer. In the spring the burrows of *M. libycus* and the mounds in front always contained parts of ruptured *Artemisia* leaves. The carnivores *Vulpes* and *Vormela* seem to depend on these two common rodents there. Some birds use the rodent holes as nesting places — for instance, *Oenanthe isabellina* and *Montifringilla theresae*.

Rodents of these arid regions possess several nutritional strategies (J. Niethammer, 1970; pers. comm., 1978):

(1) Seed-consuming species: These occur year-around in vast areas of the open plains. Food available all seasons (*Jaculus*, *Alactaga*, *Gerbillus*).

(2) Species consuming plant parts above-ground: Active year-around but only in semi-deserts or steppes with rather dense vegetation (*Meriones*, *Spermophilopsis*).

(3) Hibernating and drought-passive species: They are only active during those seasons when enough green plants are available (*Citellus fulvus*, *Marmota caudata*).

(4) Species depending on underground plant parts: Only active when the soil is moist, digging extensive burrows near ground surface (*Ellobius*, *Nesoka*). Food available all seasons of the year.

(5) Rock-inhabiting species: In those biotopes the food supply seems to be more balanced (*Calomyscus*, *Meriones persicus*).

(6) Seed-storing species: *Cricetulus*.

One of the best known groups of animals in this region are the birds. Good surveys have been given by Paludan (1959) and Hue and Etchecopar (1970). A list of characteristic breeding birds from the arid areas in Afghanistan (J. Niethammer, pers. comm., 1978) follows:

Alaemon alaudipes
Ammomanes cincturus
A. deserti
Apus melba
Athene noctua
Bubo bubo
Bucanetes githaginea
Calandrella acutirostris
Caprimulgus aegypticus
C. mahrattensis
Chlamydotis undulata (Fig. 10.50)
Columba eversmani
C. livia
Coracias garrulus
Cursorius cursor
Emberiza bruniceps
E. buchanani
E. cia
E. stewarti
Galerida cristata
Hirundo daurica
H. obsoleta
H. rupestris
Leucosticte brandti
L. nemoricola
Merops apiaster
M. superciliosus
Montifringilla theresae
Oenanthe albiniger
O. deserti
O. finschi
O. isabellina
O. picata
O. pleschanka
O. xanthoprymna
Pterocles alchata
P. coronatus
P. orientalis
Rhodopechys obsoleta
R. mongolica
R. sanguinea
Sitta tephronota

Scott in Harrington (1976, 1977) recorded 158 species of birds in the Touran Biospheric Reserve in Iran, 35 of them being all-year residents. Distribution maps of all birds of Iran are given by Scott et al. (1975).

An appreciable number of reptiles live within these desert regions. Monitor lizards (*Varanus*) occur in eastern Afghanistan and in some parts of Iran (Fig. 10.51). *Phrynocephalus* and *Acantho-*

dactylus can be found in Iran and southern Afghanistan. *Eremias*, *Uromastix* and several species of *Agama* (Fig. 10.52), as well as several snakes, appear to be widely spread in the arid areas of these countries. One of the most common reptiles, the tortoise (*Testudo horsfieldi*), is found by the hundreds in the spring (Fig. 10.53), eating ephemeral vegetation of the loess region of the *Artemisia* semi-desert, as well as within the cushion dwarf-shrub vegetation of higher areas in Afghanistan.

It is impossible to discuss the ecological role of the numerous arthropods. Although the faunistic literature is already immense, and it might be possible to elucidate some ecologically relevant data, nobody has done so yet. Thus, only a few examples will be mentioned. Scorpions (*Buthus caucasicus*, *Buthotus alticola* and *Priomurus* spp.) are found practically throughout the whole area (Fischer, 1968). Some arthropod species have been studied — for instance, *Hemilepistus aphganicus* (Schneider, 1971a) (Fig. 10.54), *Cataglyphis bicolor* (Schneider, 1971b; Breckle, 1971a), and the parasitic *Przhevalskiana silenus* (Madel, 1969).

Fig. 10.51. *Varanus*, one of the larger reptiles of the moderately warm deserts of Afghanistan.

Fig. 10.50. *Chlamydotis undulata*, one of the largest birds in the Touran Biospheric Reserve, Iran.

Fig. 10.52. The genus *Agama* is represented by several species in sandy deserts and semi-deserts.

Fig. 10.53. *Testudo horsfieldi*, which exhibits two periods of activity and two periods of dormancy (winter cold, summer drought), is particularly abundant within the loess belt.

Fig. 10.54. *Hemilepistus aphganicus*, a plant-feeding woodlouse widely distributed within the desert areas of the Middle East.

The role of insects for pollination of desert plants in this area is almost unknown. Apart from the high percentage of wind-pollinated plants a high proportion of the zoogamous desert species seem to be pollinated by Diptera. In cultivated, lower areas bees (*Apis*) seem to be important pollinators (Schneider, 1976; Schneider and Djalal, 1970). In higher altitudes, in Afghanistan above 3000 m, bumble bees (*Bombus*) occur (Madel, 1968; Breckle, 1971d).

For more detailed information on fauna, the reader is referred to Smith (1940), Humlum (1959), Paludan (1959), Lay (1967), Niethammer (1969, 1972, 1973, 1975), Kullmann (1972), Hassinger (1973), Prakash and Ghosh (1975).

HUMAN IMPACTS

Although it is difficult to determine the extent of environmental degradation which may have had its origins several millennia ago (Firouz and Harrington, 1976), degradation of the vegetation and soils due to cultivation, grazing, and collecting of fuel has in many regions of Iran reached a stage one would believe was beyond any repair. Over vast stretches one meets a bleak country without trees and shrubs and with hardly any other vegetal cover. Water erosion, windblown sand, lack of fuel and lack of good grazing make life hard for the local people (Wendelbo, 1976). Similar or even more severe degradation occurs in Afghanistan (Freitag, 1971). The need for fuel and charcoal has led to an immense destruction of vegetation. Often all woody plants, even all small subshrubs, have disappeared near villages. A similar effect results from livestock grazing. There are about 2.5 million nomads (Fig. 10.55) in Afghanistan. Almost all of the Afghan semi-deserts, steppes, and mountains have been excessively grazed (Freitag, 1971). Almost every hill is covered with a network of livestock paths. Range management based on grazing capacity is strongly needed but, of course, very difficult to introduce because of the socio-economic system.

The transformation and destruction of ecosystems is a matter of great concern. The alteration of a specific plant community might have taken centuries, but with a burgeoning human population and advanced technology, radical transformations have been wrought on the natural environment in a mere decade (Firouz and Harrington, 1976).

The mechanism of rangeland destruction seems to be that, as the vegetation deteriorates, the

Fig. 10.55. Nomadic pastoralists challenge efforts to plan land utilization.

herdsmen turn to more tolerant species of livestock; sheep are substituted for cattle, and then goats replace sheep. The problem normally is aggravated by the persistent tendency to convert easily accessible sections of such areas to dry-farming, even if they are ill-suited or at best marginal for such purposes (Firouz and Harrington, 1976). Thus, less area is available for grazing resulting in more pressure on the land. Furthermore, range depletion is not only dependent on the grazing of livestock but also on exploitation of fuel sources. Trees, then shrubs, and finally perennial and even annual vegetation are cut and burned (Freitag, 1971; Holloway, 1976). Wind and water erosion (Fig. 10.56) complete the degradation process. Sand storms and dune accumulation threaten more and more areas.

Plowing of natural vegetation on mountain slopes, depletion of mountain forests, and overgrazing have subjected vast areas to the influence of water and wind erosion and have lost potential for productivity. Soil erosion in particular is the greatest concern as it does enormous, almost irreparable, damage to national economics (Mirimanian, 1976).

Social organization based on herding exists today as it has for many centuries (Zohary, 1973). This pattern must have started rather early in postglacial times, particularly in the Near East (Fertile Crescent). Zohary (1973) gave a comprehensive account of the domestication and use of plants through the presegetal and segetal ages with increasing synanthropic flora and vegetation. He stressed that sedentary and nomadic pastoralism has altered the composition of plant communities and their appearance drastically, but it remains uncertain to what extent. The vast areas of *Artemisia* semi-desert in the hammada regions are a result of pastoralism. The area was once mostly a grass-steppe. The grass components have been lost almost entirely through excessive grazing, similar to the pattern Walter (1956) has demonstrated for Turkey.

Large areas, especially those with woodlands

Fig. 10.56. Heavily eroded badlands within serir plains covered by *Artemisia* semi-deserts.

situated on the borders of steppes, semi-deserts and deserts, have lost their arboreal vegetation cover forever (De Planhol, 1969). Soil erosion is impeding regeneration of the degraded vegetation and results in an irreversible destruction and degradation of biological resources (Robertson, 1976). Many processes leading to desertification are touched upon within the Case Study of the Turan Research Project in northeastern Iran (Spooner, 1977). It is to be hoped that this interesting pilot project may be continued on an international, interdisciplinary level.

The changes of vegetation by human interference have resulted in selection for characters of survival value. Zohary (1973) pointed out that "as a result of the grave destructive action by man, selection pressure brought to dominance plants with morphological and biochemical properties entirely or partly disliked by man or his livestock. Such disanthropous and anti-zous characters saved many plant communities from total extinction". Phytochemical properties like bitter principles, toxic alkaloids, essential oils, as, for instance, in species of *Artemisia*, *Euphorbia*, *Ephedra*, *Peganum*, *Eremurus*, *Nepeta*, *Dorema* and *Ferula* (cf. Breckle and Unger, 1977), are noteworthy. Morphological traits that render plants less palatable are lignification and spininess (tragacanthic species like *Astragalus*, *Acantholimon*, *Gaillonia* and *Acanthophyllum*, spiny plants like *Alhagi*, *Amygdalus*, *Fagonia* and *Scariola*), or those with thistle habit as *Cousinia*, *Echinops*, *Eryngium*, etc.). There is a whole series of plant communities dominated by spiny dwarf-shrubs or thorny cushions (e.g., *Hulthemia*, *Noaea*).

NATURE CONSERVATION

A comprehensive program of wildlife research and management has been underway in Iran. A large reserve system covering 59 national parks, protected areas, wildlife refuges and national nature monuments, covering a total area of about 75 090 km^2 or 4.6% of the total area of Iran, has been created (Fig. 10.57). Conspicuous success has been attained within a relatively short time (Harrington, 1976). The Iranian Program for

Fig. 10.57. The Touran Biospheric Reserve is the largest area in Iran under protection. It was established by the Iranian Department of Environment and covers most of the typical desert ecosystems in northeastern and central Iran.

Nature Conservation has been very comprehensive; it insures proper resource protection within established reserves to safeguard viable remnants of ecosystems and biotic communities (Firouz and Harrington, 1976). It will be very important to keep this degree of protection operating in the future.

Only a few steps have been taken in Afghanistan to establish a similar system. Some wildlife protection programs are on the way but should be enlarged and broadened to provide habitat protection for representative disturbed and undisturbed biomes.

Many proposals for range and pasture management have been put forward (Pabot, 1967). Only one example of possible improvement measures against desertification will be mentioned. It demonstrates the possibilities for successful measures in harmony with the ecological capacity of the habitat. Firouz and Harrington (1976) pointed out the low cost of sand dune stabilization through protection from grazing, compared with conventional treatments. Natural stabilization seems to become very efficient due to the great variety of shrubs, subshrubs, and annuals. The latter seem to be especially important. The use of a wide range of native species might lead to safer, quicker, and cheaper results in reclamation of sandy areas.

REFERENCES

Adle, A.M., 1960. *Régions climatiques et végétations en Iran.* Publ. Univ. Tehran, No. 626, Tehran.

Bhadresa, R., 1977. *Population Structure, Pattern, Biomass and Production in a Semidesert Shrub,* Artemisia herba-alba *Asso, in the Touran Biosphere Reserve.* Department of Environment, Tehran, 6 pp.

Blaise, J., Bordet, P., Desparmet, R., Marin, Ph. and Montenat, C., 1977. Géologie dans les Montagnes Centrale de l'Afghanistan (Hazarajat et bordure orientale). *Mém. Soc. Géol. Fr.*, No. 8: 117–143.

Bobek, H., 1934. Reise im nordwestlichen Persien. *Z. Ges. Erdkd. (Berlin)*, 1934: 359–369.

Bobek, H., 1952. Beiträge zur klimaökologischen Gliederung Irans. *Erdkunde*, 6: 65–84.

Bobek, H., 1961. Die Salzwüsten Irans als Klimazeugen. *Anz. Österr. Akad. Wiss., Philos-Hist. Kl.*, 1961 (3): 7–19.

Boulin, J. and Bouyx, E., 1977. Introduction à la géologie de l'Indou-Kouch occidental. *Mém. Soc. Géol. Fr.*, No. 8: 87–106.

Breckle, S.-W., 1967. Fossile Pflanzenreste am Latahbandpass in Afghanistan. *Science (Kabul)*, 4: 16–21.

Breckle, S.-W., 1971a. Die Beeinflussung der Vegetation durch hügelbauende Ameisen (*Cataglyphis bicolor* Fabricius) auf der Dasht-i-Khoshi (Ost-Afghanistan). *Ber. Dtsch. Bot. Ges.*, 84: 1–18.

Breckle, S.-W., 1971b. Ist *Diarthron vesiculosum* (Thymelaeaceae) ein ökologisches Rätsel? Studien an Therophyten in der Steppe von Kabul/Afghanistan. I. Allgemeines, Morphologie und osmotische Verhältnisse. *Bot. Jahrb. System.*, 90: 550–561.

Breckle, S.-W., 1971c. Ökologie und Mikroklima in der alpinen Stufe des afghanischen Hindukusch. *Ber. Dtsch. Bot. Ges.*, 84: 721–730.

Breckle, S.-W., 1971d. Vegetation in alpine regions of Afghanistan. In: P.H. Davis et al. (Editors), *Plant Life of South West Asia.* Botanical Society of Edinburgh. pp. 107–116.

Breckle, S.-W., 1973. Mikroklimatische Messungen und ökologische Beobachtungen in der alpinen Stufe des afghanischen Hindukusch. *Bot. Jahrb. System.*, 93: 25–55.

Breckle, S.-W., 1976. Zur Ökologie und den Mineralstoffverhältnissen absalzender und nicht absalzender Xerohalophyten (in Utah/USA). *Diss. Bot.*, 35: 1–169.

Breckle, S.-W., 1979. Salinity as an ecological factor promoting desertification. In: *Postplenary Session of the Xth Int.*

Congress of Anthropological and Ethnological Sciences. Jodhpur, 1978.

Breckle, S.-W., 1981a. The time-scale of salt-accumulation in a desertic hydrotope in north-central Iran. *Beih. Tübinger Atlas des Vorderen Orients, Reihe A*, 8: 51–60.

Breckle, S.-W., 1981b. Zum Stand der Erforschung von Flora und Vegetation Afghanistans. In: C. Rathjens (Editor), *Neue Forschungen in Afghanistan.* Deutsches Orient-Institut, Hamburg, pp. 87–104.

Breckle, S.-W. and Kull, U., 1971. Osmotische Verhältnisse und Zuckergehalte in Jahresgang bei Bäumen Ostafghanistans. I. *Quercus baloot* Griffith. *Flora*, 160: 43–59.

Breckle, S.-W. and Kull, U., 1973. Ist *Diarthron vesiculosum* ein ökologisches Rätsel? II. Die Wirkung der Dürre auf Mineralstoff-verhältnisse und Kohlenhydrathaushalt. *Bot. Jahrb. System.*, 93: 539–561.

Breckle, S.-W. and Reshad, M., 1971. Ökophysiologischer Tagesgang des Halophyten *Krascheninnikovia* in der Dasht-e-Nawor (Zentral/Afghanistan) an einem Strahlungstag. *Biol. Plant.*, 13: 402–404.

Breckle, S.-W. and Unger, W., 1977. Afghanische Drogen und ihre Stammpflanzen. I. Gummiharze von Umbelliferen. *Afgh. J. (Graz)*, 4: 86–95.

Brown, R.E., 1976. *Rodent Ecology in the Toran Protected Area, Iran.* Unpublished Report submitted to the Department of Environment, Tehran, 18 pp.

Buringh, P., 1960. *Soils and soil conditions in Iraq.* Rep. Iraq Ministry of Agriculture, Baghdad, 322 pp.

Carbonnel, J.P., 1977. La limite de la plaque indienne en Afghanistan. Nouvelle données géologiques et seismologiques. *Mém. Soc. Géol. Fr.*, No. 8. 145–152.

Conrad, G. and Girod, M., 1977. Les formations continentales tertiaires et quaternaires du Bloc du Lou (Iran): importance du plutonisme et du volcanisme. *Mém. Soc. Géol. Fr.*, No. 8: 53–76.

De Planhol, X., 1969. Le déboisement de l'Iran. *Ann. Géogr.*, 78: 625–635.

Dewan, M.L. and Famouri, J., 1964. *The Soils of Iran.* FAO, Rome, 319 pp.

Evenari, M. and Noy-Meir, I. (Editors), in press. *Hot Deserts and Arid Shrublands. Ecosystems of the World*, 12. Elsevier, Amsterdam.

Firouz, E., 1964. Environmental and nature conservation in Iran. *Environ. Conserv.*, 3: 33–42.

Firouz, E. and Harrington, F.A., Jr., 1976. Iran: Concepts of biotic community reservation. *IUCN Publ., N.S.*, No. 34: 147–169.

Fischer, L., 1968. *Afghanistan. Medizinische Länderkunde, 2.* Springer, Berlin, 180 pp.

Freitag, H., 1971. Die natürliche Vegetation Afghanistans. *Vegetatio*, 22: 285–344.

Freitag, H., 1977. Touran biosphere reserve, preliminary vegetation map. In: B. Spooner, *Case Study on Desertification* — Iran, Touran. Department of Environment, Tehran, p. 86.

Frey, W. and Probst, W., 1979. *Flora und Vegetation von Iran.* Encyclopedia Persica (in press).

Furon, R., 1941. Géologie du Plateau Iranien. *Mém. Mus. Hist. Nat., N.S.*, 7: 177–414.

Gams, H., 1956. Tragacantha-Igelheiden der Gebirge um das Kaspische, Schwarze und Mittelländische Meer. *Veröff. Geobot. Inst. Rübel*, 31: 217–243.

Ganji, H.M., 1968. Climate. In: W.B. Fisher (Editor), The land of Iran. University of Cambridge Press, Cambridge, pp. 212–249.

Hallier, U.W., 1975. *Die abiotische Zone der Lut/Zentraliran.* Unpublished manuscript, 15 pp.

Hallier, U.W., 1978. Die Wüste Lut. *Bild Wiss.*, pp. 46–56.

Harrington, F.A., Jr. (Editor), 1976. Iran: Wildlife research as a base for management. *IUCN Publ., N.S.*, No. 34: 114–132.

base for management. *Iucn Publ., N.S.*, No. 34: 114–132.

Harrington, F.A., Jr., 1977. *A Guide to the Mammals of Iran.* Department of Environment, Tehran, 89 pp.

Hassinger, J.D., 1973. A survey of the mammals of Afghanistan. *Fieldiana, Zool.*, 55: 1–195.

Herman, N.M., 1965. *Le climat de l'Afghanistan.* Météorologie Nationale, Paris, 55 pp.

Herman, N.M., Zillhardt, J. and Lalande, P., 1971. Recueil de données des stations météorologiques de l'Afghanistan. *Publ. Inst. Météorol., Kaboul*, No. 2: 58 pp.

Hinze, C., 1964. Die geologische Entwicklung der östlichen Hindukusch-Nordflanke. *Beih. Geol. Jahrb.*, 70: 19–75.

Holloway, C.W., 1976. Conservation of threatened vertebrates and plant communities in the Middle East and southwest Asia. *IUCN Publ., N.S.*, No. 34: 179–188.

Hue, F. and Etchecopar, R.D., 1970. *Les oiseaux du proche et du moyen Orient.* Boubée, Paris, 948 pp.

Humlum, J., 1959. *La géographie de l'Afghanistan.* Scandinavian University Books, Kopenhagen, 421 pp.

Iven, H.E., 1933. Das Klima von Kabul. *Geogr. Wochenschr., Breslau, Beih.*, 5.

Janossy, D. and Schmidt, E., 1970. Die Nahrung des Uhu (*Bubo bubo*); regionale und erdzeitliche Änderungen. *Bonn. Zool. Beitr.*, 21: 25–51.

Kaever, M., 1967a. Historische Entwicklung und derzeitiger Stand der geologisch-paläontologischen Erforschung Afghanistans. *Zentralbl. Geol. Paläontol.*, I: 174–181.

Kaever, M., 1967b. Das Tertiär Afghanistans. *Zentralbl. Geol. Paläontol.*, I: 351–368.

Kaever, M., 1967c. Die Kreide Afghanistans. *Zentralbl. Geol. Paläontol.*, 1: 1853–1880.

Kaever, M., 1972. Geologie und Lagerstätten. In: W. Kraus, *Afghanistan. Erdmann Ländermonogr.*, 3: 68–78.

Krinsley, D.B., 1970. *A Geomorphological and Palaeoclimatological Study of the Playas of Iran.* U.S. Government Printing Office, Washington, D.C., 486 pp.

Kull, U. and Breckle, S.W., 1972. Osmotische Verhältnisse und Zuckergehalte im Jahresgang bei Bäumen Ost-Afghanistans. II. *Cercis griffithii* und *Pistacia cabulica. Flora*, 161: 586–603.

Kull, U. and Breckle, S.W., 1975. Fettsäuren in einigen Therophyten der Steppe von Kabul/Afghanistan. *Z. Pflanzenphysiol.*, 75: 332–338.

Kullmann, E., 1972. Die Tierwelt. In: W. Kraus, *Afghanistan. Erdmann Ländermonogr.*, 3: 68–78.

Lalande, P., Hermann, N.M. and Zillhardt, J., 1974. Cartes climatiques de l'Afghanistan. *Publ. Inst. Météorol., Kaboul*, No. 4: 47 pp.

Landsberg, H.E., 1966. Verteilung der Sonnen- und

Himmelsstrahlung auf der Erde. In: R. Rodenwald and H.J. Jusatz (Editors), *Weltkarten zur Klimakunde*, Springer-Verlag, Heidelberg, 3rd ed.

Lang, J., 1977. Évolution géodynamique de bassins intramontagneux cénozoiques d'Afghanistan Central. *Mém. Soc. Géol. Fr.*, No. 8: 107–116.

Lay, D.M., 1967. A study of the mammals of Iran, resulting from the Street Expedition of 1962–63. *Fieldiana, Zool.*, 54: 282 pp.

Madel, G., 1968. Hummelbeobachtungen in Wakhan. *Science (Kabul)*, 5: 21–28.

Madel, G., 1969. Die Ziegendasselfliege *Crivellia silenus* in Afghanistan. *Angew. Parasitol.*, 10: 204–211.

McArthur, I.D. et al., 1976. *A Field Guide to Some Important Grazing Plants of Herat Province in Western Afghanistan and Near Chakcharan in Ghor Province, Central Afghanistan.* Herat Livestock Development Corp., Exper. Range Improvement Center, Herat, 35 pp.

Menessier, G., 1960. *Lexique stratigraphique international; Afghanistan, 3.* C.N.R.S., Paris, 172 pp.

Mirazai, N.A. and Breckle, S.W., 1978. Untersuchungen an afghanischen Halophyten. I. Salzverhältnisse in Chenopodiaceen Nord-Afghanistans. *Bot. Jahrb. System.*, 99: 565–578.

Mirimanian, K.P., 1976. Measures for protection and rational land utilization in arid mountain regions. *IUCN Publ. N.S.*, No. 34: 224–226.

Mirwald, P. and Römer, H., 1967. Beobachtungen in Wakhan. *Erdkunde*, 21: 48–57.

Mobayen, S., 1975. Aspects, phytogéographiques de l'Iran et particularités de végétation désertique du Lut. *Envir. Stud. (Tehran)*, 4: 181–190.

Mobayen, S., 1976. Structure géobotanique du Loute. *Acta Ecol. Iran.*, 1: 73–86.

Mobayen, S. and Tregubov, V., 1970. Guide pour la carte de la végétation naturelle de l'Iran. *UNDP/FAO IRA 7, Bull.*, No. 14: 89 pp.

Moore, P.D. and Bhadresa, R., 1977/78. *The Study of Growth Rings of* Zygophyllum eurypterum *Bss. & Buhse in the Touran Biospheric Reserve, Northeast Iran.* Unpublished manuscript, 7 pp.

Moore, P.D. and Bhadresa, R., 1978. Population structure, biomass and pattern in a semi-desert shrub, *Zygophyllum eurypterum* Bss. & Buhse, in the Touran Protected Area of Eastern Iran. *J. Appl. Ecol.*, 5: 719–727.

Naumann, C. and Nogge, G., 1973. Die Großsäuger Afghanistans. *Z. Kölner Zoo*, 16: 79–93.

Neubauer, H.F., 1952. Notizen über die Temperatur der Bodenoberfläche in Afghanistan. *Wetter Leben*, 4: 165–168.

Niethammer, J., 1969. Zur Taxonomie der Ohrenigel in Afghanistan (Gattung *Hemiechinus*). *Z. Säugetierkd.*, 34: 257–274.

Niethammer, J., 1970. Die Wühlmäuse (*Microtinae*) Afghanistans. *Bonn. Zool. Beitr.*, 21: 1–24.

Niethammer, J., 1972. Störche über Afghanistan. *Z. Kölner Zoo*, 15: 47–54.

Niethammer, J., 1973. Zur Kenntnis der Igel (*Erinaceidae*) Afghanistans. *Z. Säugetierkd.*, 38: 271–276.

Niethammer, J., 1975. Zur Taxonomie und Ausbreitungsgeschichte der Hausratte (*Rattus rattus*). *Zool. Anz.*, 194: 405–415.

Pabot, H., 1967. *Pasture Development and Range Improvement Through Botanical and Ecological Studies. Report to the Government of Iran.* FAO No. TA 2311, Rome, 129 pp.

Paludan, K., 1959. *On the Birds of Afghanistan.* Vid. Medd. Dan. Naturhist. For., 22. Copenhagen, 322 pp.

Petrov, M.P., 1945. The conditions of soil formation and types of soils of northern Iran. *Pochvovedeniye*, 3–4: 175–181 (in Russian).

Petrov, M.P., 1955. *Iran.* Gos. Izd. Geogr. Lit., Moscow, 184 pp. (in Russian).

Pias, J., 1977. Différentes aspects de l'évolution périglaciale en Afghanistan. *Mém. Soc. Géol. Fr.*, No. 8: 169–178.

Prakash, I. and Ghosh, P.K., 1975. *Rodents in Desert Environments.* Junk, The Hague, 624 pp.

Rechinger, K.H. (Editor), 1963–1981. *Flora Iranica*, Akad. Verl. Druck-Anstalt Graz, Lief. 1–46.

Rechinger, K.H., 1977. Plants of the Touran Protected Area, Iran. *Iranian J. Bot.*, 1: 23–56.

Rechinger, K.H. and Wendelbo, P., 1976. Plants of the Kavir Protected Region. *Iran. J. Bot.*, 1: 23–56.

Ricou, L.E., Braud, J. and Brunn, J.H., 1977. Le Zagros. *Mém. Soc. Géol. Fr.*, No. 8: 33–52.

Robertson, V.C., 1976. Land-use planning: mineral and fuel exploration, agriculture and forestry development in arid areas. *IUCN Publ., N.S.*, No. 34: 61–70.

Sabeti, H., 1969. *Les études bioclimatiques de l'Iran.* Tehran, No. 1231: 266 pp.

Sabeti, H., 1976. *Forests, Trees and Shrubs of Iran.* Ministry of Agriculture and Natural Resources, Tehran, 874 pp.

Salem, M.Z. and Hole, F.D., 1969. Soil geography and factors of soil formation in Afghanistan. *Soil Sci.*, 107: 289–295.

Scharlau, K., 1958. Zum Problem der Pluvialzeiten in Nordost-Iran. *Z. Geomorphol.*, 2: 258–277.

Scharlau, K., 1969. Iran. In: *Westermann Lexikon der Geographie, 2.* Westermann, Braunschweig, 936 pp.

Schneider, P., 1971a. Lebensweise und soziales Verhalten der Wüstenassel *Hemilepistus aphganicus* Borutsky 1958. *Z. Tierpsychol.*, 29: 121–133.

Schneider, P., 1971b. Vorkommen und Bau von Erdhügelnestern bei der Afghanischen Wüstenameise *Cataglyphis bicolor* Fabricius. *Zool. Anz.*, 187: 202–213.

Schneider, P., 1976. Honigbienen und ihre Zucht in Afghanistan. *Afghanistan J.*, 3: 101–104.

Schneider, P. and Djalal, A.S., 1970. Vorkommen und Haltung der östlichen Honigbiene (*Apis cerana* Fabr.) in Afghanistan. *Apidologie*, 1: 329–341.

Schweinfurth, U., 1956. Über klimatische Trockentäler im Himalaya. *Erdkunde*, 10: 297–302.

Schweizer, G., 1975. Der Naturraum. In: U. Gehrke and H. Mehner (Editors), *Iran. Erdmann Ländermonogr.*, 5: 13–39.

Scott, D.A., Hamadani, H.M. and Mirhozzeyni, A.A., 1975. *The Birds of Iran.* Department of Environment, Tehran, 410 pp.

Smith, M.A., 1940. Contribution to the herpetology of Afghanistan. *Annu. Mag. Nat. Hist., N. Ser.*, 5: 382.

Spooner, B. (Editor), 1977. *Case Study on Desertification-Iran: Touran.* Department of Environment, Tehran, 97 pp.

Stenz, E., 1941. On the evaporation capacity in Kabul. *Min. Publ. Works, Afghan Meteorol. Serv. Publ.*, No. 2: 8 pp.

Stenz, E., 1957. Precipitation, evaporation and aridity in Afghanistan. *Acta Geophys. Polon.*, pp. 244–263.

Stöcklin, J., 1968. Structural history and tectonics: a review. *Am. Assoc. Pet. Geol. Bull.*, 52: 1229–1258.

Troll, C., 1951. Die Lokalwinde der Tropengebirge und ihr Einfluß auf Niederschlag und Vegetation. *Bonn. Geogr. Abh.*, 9: 124–182.

Tromp, S.W., 1954. The stratigraphy and main structural features of Afghanistan, part 1, *Proc. K. Ned. Akad. Wiss.*, Ser. B., 57: 370–394.

Vahedi, M., 1976. *Ecological Data of the Kavir Protected Region.* Department of Environment, Tehran, 23 pp.

Van Zeist, W., 1967. Late Quaternary vegetation history of western Iran. *Rev. Palaeobot. Palynol.*, 2: 301–311.

Walter, H., 1956. Vegetationsgliederung Anatoliens. *Flora*, 143: 295–326.

Walter, H., 1974. *Die Vegetation Osteuropas, Nord- und Zentralasiens. Vegetationsmonographien, VII.* Fischer-Verlag, Stuttgart, 452 pp.

Walter, H., 1976. *Die ökologischen Systeme der Kontinente (Biogeosphäre). Prinzipien ihrer Gliederung mit Beispielen.* Fischer-Verlag, Stuttgart, 131 pp.

Walter, H. and Lieth, H., 1960–1967. *Klimadiagramm-Weltatlas.* Fischer-Verlag., Jena, Stuttgart.

Wendelbo, P., 1976. Endangered flora and vegetation, with notes on some results of protection. *IUCN Publ. N.S.*, No. 34: 189–195.

Wirtz, D., 1964. Zur regionalgeologischen Stellung der afghanischen Gebirge. *Beih. Geol. Jahrb.*, 70: 5–18.

Wittekindt, H.P., 1973. *Erläuterungen zur geologischen Karte von Zentral und Südafghanistan 1: 500 000.* Hannover, 109 pp.

Wright, A.E. and Van Zeist, W., 1963. Preliminary pollen studies at Lake Zeribar, Zagros mountains, SW-Iran. *Science*, 140: 65–67.

Wright, A.E., McAndrews, J.H. and Van Zeist, W., 1967. Modern pollen rain in Western Iran, and its relation to plant geography and Quaternary vegetation history. *J. Ecol.*, 55: 415–443.

Zarrine-Kafche, M., 1977. Etude de la relation existant entre la composition chimique des plantes et celle des sols dans le Sud de Karadj (près de Teheran). *Acta Ecol. Iran.*, 1: 60–69.

Zohary, M., 1963. On the geobotanical structure of Iran. *Bull. Res. Counc. Isr., Sect. D, Bot., Suppl.*, 1: 113 pp.

Zohary, M., 1973. *Geobotanical Foundations of the Middle East.* Fischer-Verlag, Stuttgart, 739 pp.

Chapter 11

OVERVIEW OF NORTH AMERICAN TEMPERATE DESERTS AND SEMI-DESERTS[1]

NEIL E. WEST

DEFINITIONS

I will be concerned here with the ecosystems occurring where arid temperate climates are expressed (Zonobiome VIIa of Walter and Box, 1976; map inside covers). In North America all such areas are in the interior of the continent (Fig. 11.1). This region is known as the Intermountain West because it lies between the Cascade Mountains–Sierra Nevada chain on the west and the Rocky Mountains to the east. The northern extent is truncated by the wetter forested latitudes beginning slightly above the Canadian border. The southern boundary corresponds largely to a rapid decrease in altitudes and commencement of deserts of subtropical affinities (Walter, 1971, 1979; Daubenmire, 1978; MacMahon, 1979; MacMahon and Wagner, in press).

The distinctiveness of this area has long been recognized by biogeographers. Merriam (1898) divided the continent on the basis of the average annual sum of temperatures above 0°C. The bulk of the region in Fig. 11.1 fell into his Upper Sonoran Zone. Considerable criticism was subsequently aimed at this single-factor approach (Daubenmire, 1946). Nevertheless, other, multi-dimensioned approaches have yielded map units with similar boundaries. These include the Great Basin Desert of Shreve (1942) and McGinnies et al. (1968), the Artemisian Province of Dice (1943), the Cold Desert and Semi-Desert of Shelford (1963), the Great Basin Province of Gleason and Cronquist (1964), and the Great Basin Biogeographic Province of Udvardy (1975). All of these labels derive from the close correspondence of vegetal physiognomy and floristics with macroclimatic patterns. Daubenmire (1978) included all this area within his Steppe Region and subdivided the more arid portion into *Agropyron spicatum* and *Atriplex confertifolia* Provinces. The hottest and driest areas of the former province were said to be *Artemisia tridentata* shrub-steppe. Although this terminology conflicts with Eurasian usage, it is retained here because it is widely used in North America and is descriptive of areas where mixtures of shrubs and grasses prevail, yet change with successional influences.

MacMahon's (1979) Great Basin Desert excludes southwestern Wyoming and all of the Colorado Plateau because of the absence of animals he considered indicative of deserts. These areas are included here because vegetational, climatic, and edaphic characteristics preclude them from being considered in other volumes. Subdivisions of the entire North American temperate desert and semi-desert area rely on more detailed floristic and faunistic analyses, and topographic and edaphic differences.

PRESENT CLIMATE

The major reason why this distinctive region is easy to delineate, at least in coarse focus, is the dominance of the aridity factor. Very sharp decreases in total annual precipitation are encountered to the east of the Cascades and Sierra (Fig. 11.2). This is due to the predominance of orographic "rain shadow" effects developing in a region where winter precipitation comes from air masses moving from the west (Bryson and Hare, 1974). The Rocky Mountains also separate the region from the weather fronts developing in the

[1] Manuscript completed June, 1980.

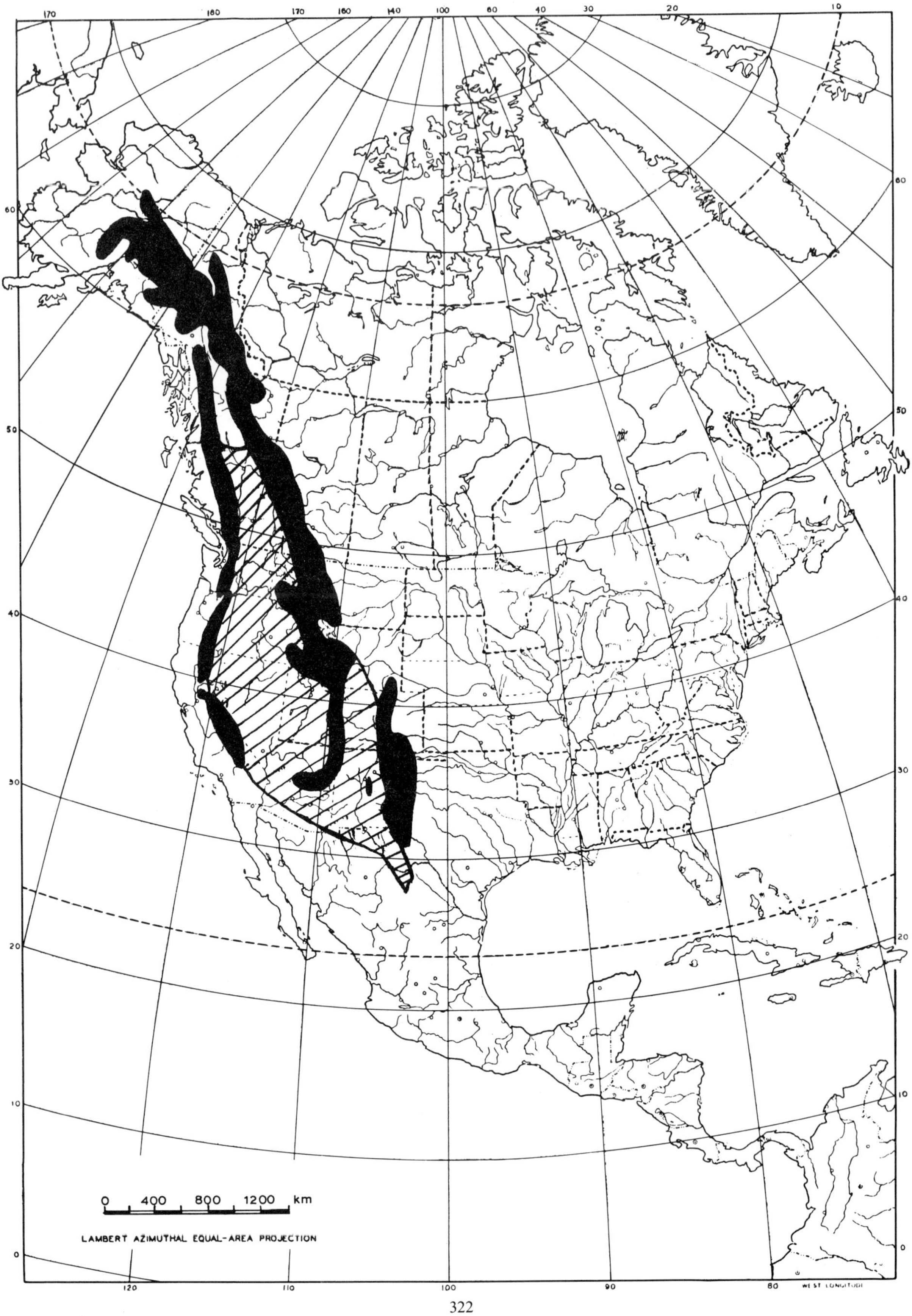
0 400 800 1200 km
LAMBERT AZIMUTHAL EQUAL-AREA PROJECTION
WEST LONGITUDE

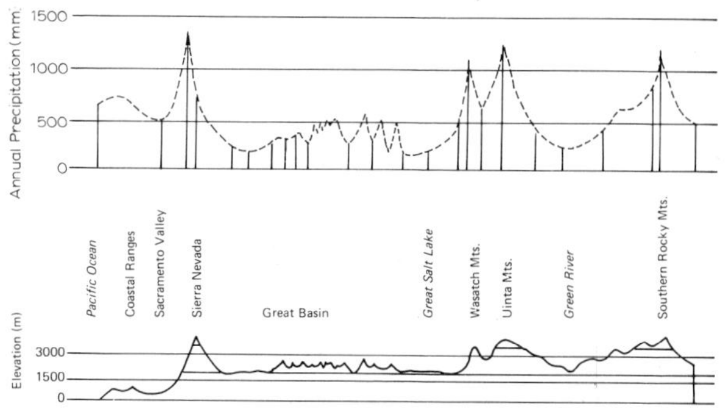

Fig. 11.2. Differences in annual precipitation in the mountains, plateaus and basins of the western United States near 40°N lat. The curve is very similar to the topographic profile because the precipitation varies according to altitude (from U.S.D.A. Yearbook, 1941).

Great Plains to the east. The proportion of total annual precipitation coming from other sources increases in the southwestern portion of the region (Houghton, 1969; Sellers and Hill, 1974), but nearly all such storms are high-intensity, short-duration events that come during the warmer part of the year. Infiltration of water into the soil under these conditions is minimal and of marginal value to most of the plants that live there. Winter precipitation and soil moisture recharge deep in the profile then, and much lesser replenishment during the growing season, are the main reasons for shrub dominance in the Intermountain Region. Plains reaching similar elevations, and thus having similar temperatures and total annual precipitation, occur east and south of the Rocky Mountains; but because much more of the total rainfall comes from Gulf Coastal air masses, rain is concentrated during the growing season. Accordingly, shallow-rooted grasses are favored and dominate there. These semi-arid areas will be included in Volume 8 in this series.

The relatively high elevations of the Intermountain regions are largely responsible for the relatively cool average temperatures and cold winters with abundant snowfall. Even though less precipitation falls on this desert region than those to the south, it largely comes at times when evaporation and transpiration are minimal. Thus, about half of the precipitation enters the soil profile in the Intermountain lowlands. The low-elevation deserts to the south have much higher average temperatures and evaporation rates and thus much less soil moisture storage per unit precipitation (MacMahon and Wagner, in press). Furthermore, hardly any of the precipitation in the southern deserts comes as slowly melting snow.

PALEOECOLOGICAL INFLUENCES

The present is partially a reflection of the past. Accordingly, we need to consider the kinds of ecosystems that have prevailed before. Abundant fossil evidence of changes during the Cretaceous and Tertiary have given American paleoecologists unparalleled opportunities to describe how ecosystems evolved in western North America (Axelrod, 1950, 1958, 1967, 1972, 1976, 1979; Wolfe, 1975, 1978).

The origin of the region can be traced to a series of geosynclines beginning in the Precambrian and extending into the Mesozoic. During the Cretaceous, a sequence of folding, faulting and uplifting of these sediments and erosional deposition toward the east was begun. There were considerable marine embayments in the region during the Cretaceous. The land surfaces available were occupied by primitive tropical forests of tree ferns, cycads, and other primitive gymnosperms. In the Early Tertiary, no large areas with arid or semi-arid environments apparently existed in North America. The western half of the continent was largely a level plain occupied by forests. A mixed temperate conifer–deciduous tree-dominated type (Arcto-Tertiary Geoflora) occurred in the north and the subtropical and tropical forest (Neotropical Tertiary Geoflora) increased in importance to the south.

It has apparently been characteristic for high-pressure systems to develop in lower middle latitudes on the western sides of continents. Thus, in southwestern North America, rainfall was probably more seasonal than either to the north or south, and it was thus possible for local sub-humid areas to exist on the lee slopes of low ranges beginning in Late Mesozoic times. Annual rainfall in such situations may have been 900 to 1000 mm and dry deciduous tropical and subtropical forest

Fig. 11.1. Map of the regions of North America characterized by temperate arid to semi-arid climates (hatched area). Black area is cordillera (from Walter and Box, 1976).

and scrub could have developed. The continued expansion and intensification of dry climate probably resulted in a rapid evolution and wide radiation of the sub-humid floras beginning in Cretaceous times. When first recorded in Eocene and Oligocene deposits from the then elevated central Rocky Mountain region, they contain many genera and even species closely related to modern forms. This distinctly sub-humid element has been called the Madro-Tertiary Geoflora. While many of its plants were probably derived from the Neotropical Geoflora, a few, especially those active in the cooler seasons, apparently came from Arcto-Tertiary plants that had adapted to the drier climates.

During Miocene time, the Madro-Tertiary Geoflora was more generalized in composition than its presently related communities. The Great Basin and Colorado Plateau had plains with rolling hills and minor mountains. The basal elevation was near sea level. The principal communities included oak (*Quercus*) woodland, conifer woodlands, chaparral, subtropical scrub, coastal sage scrub, and plains grassland. A large number of fossil representatives of these communities now have their nearest living relatives on opposite sides of the present subtropical deserts to the immediate south. This clearly indicates that the Madro-Tertiary Geoflora also extended across the present southern desert areas. At the same time, the northern Great Basin was dominated by the Arcto-Tertiary Geoflora. An ecotone in which species of both floras formed a mosaic of overlapping vegetation types characterized the central Great Basin region in Nevada.

These three broad vegetation provinces persisted into the Early Pliocene, with dominant coniferous forest at the north gradually giving way to woodland and scrub at the south. The trend toward a more arid climate culminated in the Middle Pliocene and was reflected in the widespread restriction of forest and woodland from the lowlands. Semi-arid grasslands and shrubland are inferred to have dominated the lowlands throughout the present desert regions. Late Pliocene climate is believed to have been cooler. This permitted some expansion of forest and woodland with a consequent restriction of grassland and shrubland. The Late Pliocene was also characterized by a great deal of mountain-building, and epeirogenic uplift, a process which actively continued well into the Pleistocene. The Sierra Nevada were raised from 1500 to 2800 m above the 1000 m Miocene elevation. The Cascades underwent an uplift to only a slightly less degree. If desert-like environments had previously existed, they must have been quite restricted. But, with the elevation of such high ranges, the reduction of precipitation in their lee was sufficient to permit the widespread development of deserts from these nuclei beginning in the Late Pliocene.

The Pleistocene changes within the Great Basin have been recently reviewed by Mehringer (1977) and Wells (1980). The Quaternary paleoecology of the Colorado and Columbia–Snake River Plateaus and Wyoming Basin is less well understood but similar in as far as is known (Baumhoff and Heizer, 1965; Kottlowski et al., 1965; Martin and Mehringer, 1965).

Although Pleistocene glacial advances did not directly affect much of the land area under consideration in this volume, there were numerous indirect effects. The decrease in average temperatures was probably more important than any increase in precipitation. Less evapotranspiration meant more runoff in the streams feeding the basins. Lower summer temperatures meant less evaporation from the approximately ninety lake surfaces in the Great Basin. Temperature and its impact on effective soil moisture, as well as growing season thresholds, probably pushed conifer forests or woodlands into parts of the valleys not covered by lakes. A rapid upslope retreat of trees and expansion of grasslands and shrublands during the warmer interglacials probably took place. Although data are ambivalent, the prevailing view is that there was a relatively drier (xerothermic interval) period from 4000 to 7500 years B.P. How much this may have caused biota to shift in distribution is debatable.

Post-xerothermic interval cooling probably caused another elevational and latitudinal depression of vegetation. More recent climatic changes have also caused minor movement of lake levels and vegetation boundaries (LaMarche, 1974; Spaulding and Van Devender, 1977).

The major direct effect of Quaternary climatic changes has been the direct submergence and consequent drying of the lowland areas of the Great Basin under pluvial lakes (Fig. 11.3). The bottom of each lake basin became salinized and/or alkalinized

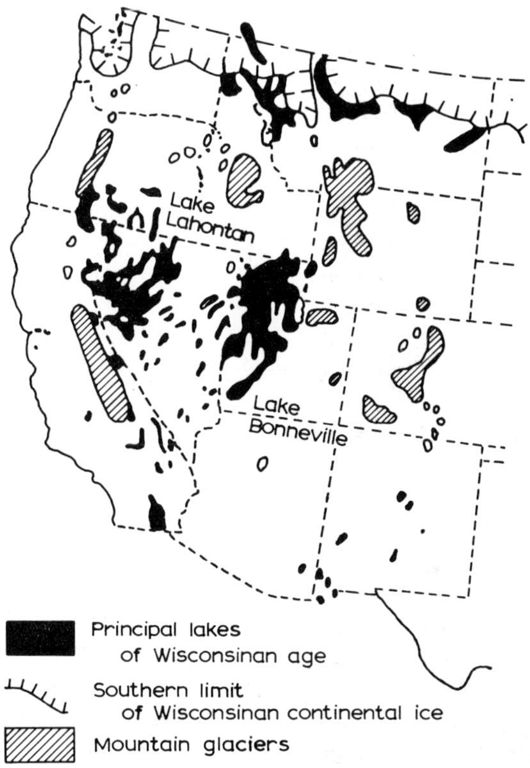

Fig. 11.3. Distribution of glacial ice and principal lakes in the western United States at their maxima during Late Wisconsinan time (from Hunt, 1974; © 1974 W.H. Freeman and Company).

as the water retreated. The moderately salty areas are where we today find the ecosystems of salt-desert shrub type. The saltiest areas are nearly bare of life.

There were great shifts in the distribution of major biota in this area during the Quaternary (Mehringer, 1977; Wells, 1980). There was also rapid evolution, particularly apparent in the dominant plants (McArthur, 1978; Stutz, 1978). Stebbins (1975) believed, however, that the comparatively low floristic diversity of this region was more a result of rapid extinction than rates of evolution. There was an apparently accelerated rate of extinction of the megafauna (Mehringer, 1977) near the end of the Pleistocene. Whether the entrance of man on the North American continent is directly related to these losses is still being debated. Regardless of the causation, the loss of large browsers has probably influenced the subsequent development of the vegetation. Many other aspects of ecosystem organization are also still rebounding from the dislocations of the Quaternary.

Although the issue of timing is by no means settled, man's entry into the New World over the land bridge (Beringia) between Siberia and Alaska is believed to have taken place sometime before 10 000 B.C., perhaps as early as 30 000 B.C. But, although man may virtually have been dispersed (however thinly) over the New World by 20 000 years ago, only after about 13 000 B.P. are his remains numerous and unchallenged as to date of occurrence (Jennings, 1978).

REGIONAL SUBDIVISIONS

The intermountain deserts and semi-deserts may seem to be highly uniform and, therefore, monotonous and unappealing to the non-ecologist, especially to those with little familiarity with the region. With some short experience in the region, however, one can begin to note considerable differences across the landscape. First of all, the prevalent landforms differ considerably. These differences are commonly recognized as physiographic provinces. Although there are some minor differences in boundaries drawn (Fenneman, 1928; Atwood, 1940; Thornbury, 1965; Hammond, 1970; Hunt, 1974; Bailey, 1976, 1978), because of differing methodology with its varying stresses on different features, all agree that four different provinces exist where temperate deserts and semi-deserts exist in North America. These are (Fig. 11.4) the Great Basin portion of the Basin and Range Province, the Columbia–Snake River Plateau, the Colorado Plateau and the Wyoming Basin.

Great Basin

Most of the Great Basin has no drainage to the sea. The elevation of the valley floors exceeds 1200 m. There are over 200 fault-block mountain ranges, mostly north–south trending, some of which exceed 4000 m in elevation. The mountain masses above 1850 m [the approximate average lower boundary of present pinyon–juniper (*Pinus–Juniperus*) woodland] however, occupy about 40% of the area (Billings, 1978). About 60% of the landscape has been formed from gently sloping

Fig. 11.4. Physiographic provinces of the western United States (from Hunt, 1974; © 1974 W.H. Freeman and Company). Heavily dotted lines correspond to cross-sections given in Fig. 11.5.

loessial, alluvial, or lacustrine fill where the pluvial lakes existed.

Colorado Plateau

The Colorado Plateau is a vast epeirogenic upwarp of largely horizontal Mesozoic sedimentary strata that has been deeply dissected by the Colorado River and its tributaries. The elevation of the plateau varies from about 1200 to 4600 m with about 45% below 1850 m (West, 1969). The soils vary from badlands where marine shales outcrop to skeletal residual profiles on mesa tops. Small areas of colluvium next to cliffs, sand dunes on the tablelands, and alluvial fill along the rivers and washes also occur. Either current or fossil lakebeds are rare in this province.

Columbia–Snake River Plateau

The Columbia–Snake River Plateau was built from Miocene and Pliocene outpourings of lava. The relief is nearly flat to rolling. The depth of soils depends mostly on how long the materials have weathered and how much loess has been deposited or removed. Most of the land surface here is lower than in the other regions that is, about 70% is less than 1500 m (Hunt, 1974).

Wyoming Basin

The Wyoming Basin is the smallest province in the Intermountain Region. Its predominant land forms are valleys and plains between the northern and central Rocky Mountain Provinces. The majority of this province varies between 1500 and 2200 m elevation. Eocene lakes that occurred here formed shales and marlstones. These loosely consolidated rocks have eroded into badlands or have been covered by plains built from alluvium brought down from the high mountains nearby (Hunt, 1974).

ECOLOGICAL SUBDIVISIONS

Although we could find distinctive differences in the biological communities occupying the different physiographic provinces, there are more overall similarities than dissimilarities. Much greater separation is seen by viewing all of these areas along vertical gradients. The complex gradient of elevation involves changes in climate and soils that largely control vegetation. However, since climates and soils are harder to measure directly, and vegetation reflects these differences and makes up 99% of the standing crop biomass in terrestrial ecosystems, it seems logical to use vegetation as the primary discriminator of ecosystem boundaries.

The subdivisions of temperate deserts and semi-deserts that we will use here were derived from Küchler's (1964, 1970) map of the potential vegetation of the United States. Küchler's approach was actually quite similar to that used by Walter and co-authors for Eurasia (Walter, 1979). The map units stress the vegetation expected to occur on upland sites of moderate relief when disturbance is minimal. This is essentially a zonal approach. There is, however, provision for recognizing influences other than climate. If soils on level sites modify the effective climate, and thus the pristine vegetation, a separate map unit (pedobiome) is recognized. Similarly, the sharp climatic gradients associated with mountains strongly sort out vegetation and create further map types (orobiomes).

The single exception to the use of the Küchler map units involves the combining of his "wheatgrass–needlegrass (*Agropyron–Stipa*) shrub steppe" type with "sagebrush steppe". The former type involves a comparatively small area in the Big Horn Basin of central northern Wyoming. Not enough ecological research has been conducted there for one to be able to write about more than the vegetation and its relationship to soils and climate. Since even these patterns are not greatly different from those of the sagebrush steppe, it was judged illogical to attempt a separate chapter about that area (H.G. Fisser, pers. comm., 1979).

The adoption of the Küchler scheme does not mean that we will discuss only pristine vegetation. We use it only as an organizing mechanism and point of reference to try and understand how and why changes have resulted. We will discuss all aspects of ecosystems for which data exist. The entire range of landscapes found within each of the Küchler map units will also be considered.

Fig. 11.5 shows examples of the elevational and topographic positions of ecosystem types in some representative cross-sections of the four physiographic provinces. This should help the reader to better visualize the relationship of the following chapters to each other.

REFERENCES

Atwood, W.W., 1940. *The Physiographic Provinces of North America*. Ginn and Co., Boston, Mass., 536 pp.

Axelrod, D.I., 1950. Evolution of desert vegetation in western North America. *Carnegie Inst. Wash. Publ.*, 590: 215–306.

Axelrod, D.I., 1958. Evolution of the Madro-Tertiary geoflora. *Bot. Rev.* (*New York*), 24: 435–509.

Axelrod, D.I., 1967. Drought, diastrophism and quantum evolution. *Evolution*, 21: 201–209.

Axelrod, D.I., 1972. Edaphic aridity as a factor in angiosperm evolution. *Am. Nat.*, 106: 311–320.

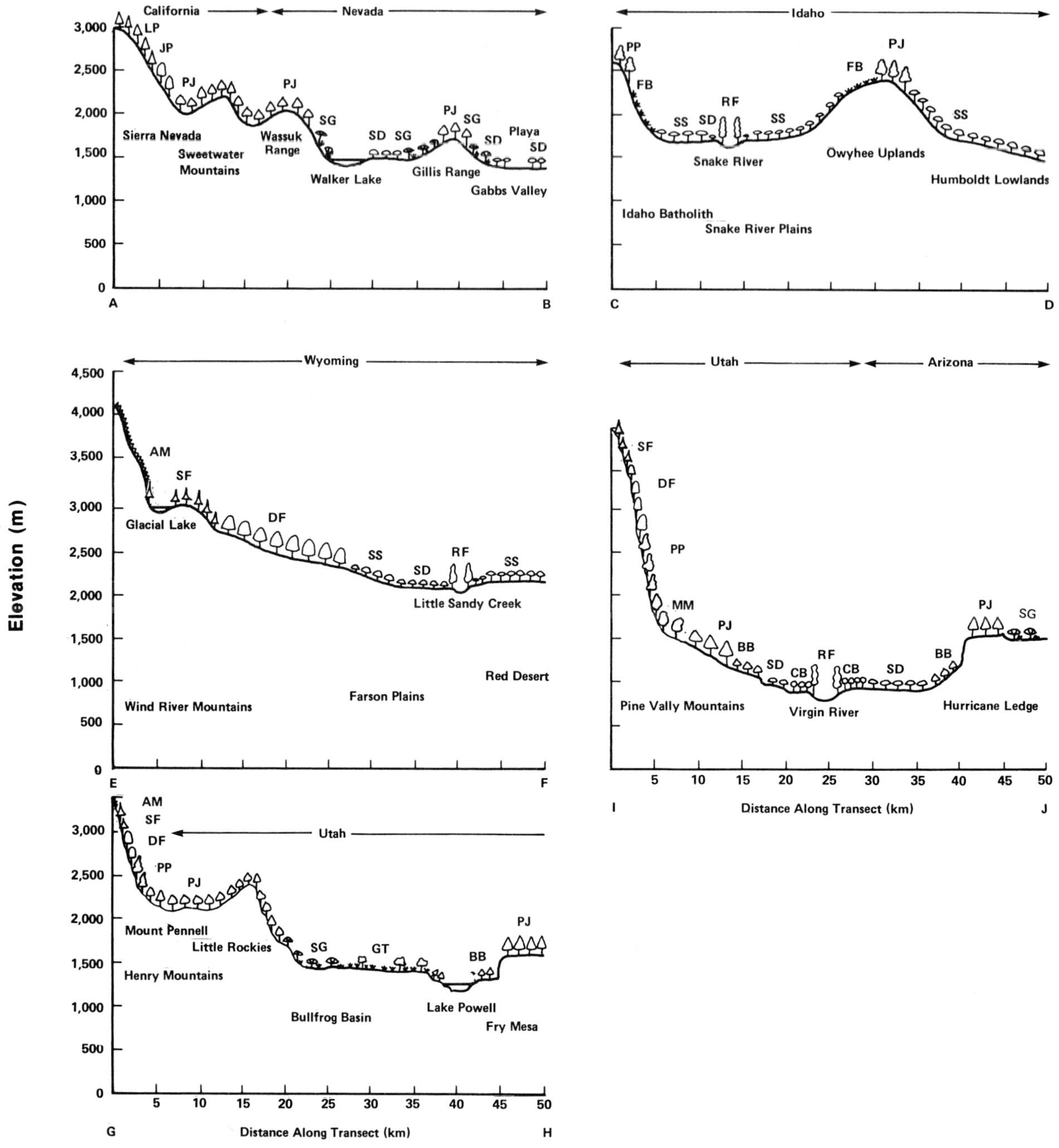

Fig. 11.5. Cross-sections of the four physiographic provinces showing elevational and topographic positioning of ecosystem types. Geographic locations of the cross-sections indicated with dotted lines in Fig. 11.4.

Legend: *AM*=alpine meadow and fellfield; *BB*=blackbrush (*Coleogyne ramosissima*) semi-desert; *CB*=creosotebush (*Larrea tridentata*) desert; *DF*=douglas fir (*Pseudotsuga menziesii*) forest; *FB*=wheatgrass (*Agropyron*)–bluegrass (*Poa*) grassland; *GT*=galleta (*Hilaria jamesii*)–threeawn (*Aristida* spp.) shrub steppe; *JP*=Jeffrey pine (*Pinus jeffreyi*) forest; *LP*=lodgepole pine (*Pinus contorta*) forest; *MM*=mountain mahogany (*Cercocarpus* spp.)–oak (*Quercus* spp.) scrub; *PJ*=pinyon (*Pinus monophylla–P. edulis*)–juniper (*Juniperus osteosperma, J. occidentalis*) woodlands; *PP*=Ponderosa pine (*Pinus ponderosa*) forest; *RF*=riparian forest; *SD*=salt-desert shrubland; *SG*=sagebrush (*Artemisia*) semi-desert; *SF*=spruce (*Picea engelmanii*)–fir (*Abies concolor, A. lasiocarpa*) forest; *SS*=sagebrush (*Artemisia*) steppe.

Axelrod, D.I., 1976. History of the coniferous forests, California and Nevada. *Univ. Calif. Publ. Bot.*, 70: 62 pp.

Axelrod, D.I., 1979. Desert vegetation, its age and origin. In: J.R. Goodin and D.K. Northington (Editors), *Arid Land Plant Resources*, Int. Center for Arid and Semi-arid Land Studies, Texas Technical University, Lubbock, Texas, pp. 1–72.

Bailey, R.G., 1976. *Ecoregions of the United States (map). Scale 1:750 000.* U.S. Dep. Agric., For. Serv., Intermountain Region, Ogden, Utah.

Bailey, R.G., 1978. *Description of the Ecoregions of the United States.* U.S. Dep. Agric. For. Serv., Intermountain Region, Ogden, Utah, 79 pp.

Baumhoff, M.A. and Heizer, R.F., 1965. Postglacial climate and archeology in the Desert West. In: H.E. Wright, Jr. and D.G. Frey (Editors), *The Quaternary of the United States.* Princeton University Press, Princeton, N.J., pp. 697–708.

Billings, W.D., 1978. Alpine phytogeography across the Great Basin. *Great Basin Nat. Mem.*, No. 2: pp. 105–118.

Bryson, R.A. and Hare, F.K. 1974. Climates of North America. Elsevier, New York, N.Y., 420 pp.

Daubenmire, R.F., 1946. The life zone problem in the northern Intermountain Region. *Northwest Sci.*, 20(2): 28–38.

Daubenmire, R.F., 1978. *Plant Geography with Special Reference to North America.* Academic Press, New York, N.Y., 338 pp.

Dice, L.R., 1943. *The Biotic Provinces of North America.* University of Michigan Press, Ann Arbor, Mich., 78 pp.

Fenneman, N.M., 1928. Physiographic subdivisions of the United States. *Ann. Assoc. Am. Geogr.*, 18: 261–353.

Gleason, H.A. and Cronquist, A., 1964. *The Natural Geography of Plants.* Columbia University Press, New York, N.Y., 420 pp.

Hammond, A.E., 1970. Physical subdivisions (map at scale of 1:7 500 000). In: U.S. Geol. Surv., *The National Atlas of the U.S.A.* U.S. Government Printing Office, Washington, D.C., p. 61.

Houghton, J.G., 1969. *Characteristics of Rainfall in the Great Basin.* Desert Research Inst., Univ. Nevada System, Reno, Nev., 205 pp.

Hunt, C.B., 1974. *Natural Regions of the United States and Canada.* W.H. Freeman and Co., San Francisco, Calif., 725 pp.

Jennings, J.D., 1978. *Prehistory of Utah and the eastern Great Basin. Univ. Utah Anthropol. Pap.*, 98: 263 pp.

Kottlowski, F.E., Cooley, M.E. and Ruke, R.V., 1965. Quaternary geology of the Southwest. In: H.E. Wright, Jr. and D.G. Frey (Editors), *The Quaternary of the United States.* Princeton Univ. Press, Princeton, N.J., pp. 287–298.

Küchler, A.W., 1964. Potential, natural vegetation of the coterminous United States (map and manual). *Am. Geogr. Soc., Spec. Publ.*, 36: 116 pp.

Küchler, A.W., 1970. Potential natural vegetation (map at scale of 1:7 500 00). In: *U.S. Geol. Survey, The National Atlas of the U.S.A.* U.S. Government Printing Office, Washington, D.C., pp. 90–91 (revision of 1964 map at scale of 1: 3 168 000).

LaMarche, V.C., Jr., 1974. Paleoclimatic inferences from long tree-ring records. *Science*, 183: 1043–1048.

MacMahon, J.A., 1979. North American deserts: their floral and faunal components. In: D.W. Goodall and R.A. Perry (Editors), *Arid-Land Ecosystems: Structure, Functioning and Management, 1.* Int. Biol. Progr. 16. Cambridge University Press, Cambridge, pp. 21–82.

MacMahon, J.A. and Wagner, F.H., in press. The Mojave. Sonoran and Chihuahuan Deserts of North America. In: M. Evenari and I. Noy-Meir (Editors), *Hot Deserts and Arid Shrublands. Ecosystems of the World, 12.* Elsevier, Amsterdam, in press.

McArthur, E.D., 1978. Sagebrush systematics and evolution. In: Anonymous, *The Sagebrush Ecosystem: A Symposium.* Utah State University, College of Natural Resources, Logan, Utah, pp. 14–22.

McGinnies, W.G., Golman, B.J. and Paylore, P., (Editors), 1968. *Deserts of the World, An Appraisal of Research into Their Physical and Biological Environments.* University of Arizona Press, Tucson, Ariz., 788 pp.

Martin, P.S. and Mehringer, P.J., Jr., 1965. Pleistocene pollen analysis and biogeography of the Southwest. In: H.E. Wright, Jr. and D.G. Frey (Editors), *The Quaternary of the United States.* Princeton University Press, Princeton, N.J., pp. 433–452.

Mehringer, P.J., Jr., 1977. Great Basin Late Quaternary environments and chronology. In: D.D. Fowler (Editor), *Models and Great Basin Prehistory. Desert Res. Inst. Publ. Soc. Sci.*, 12: 113–167.

Merriam, C.H., 1898. Life zones and crop zones of the United States. *U.S. Dep. Agric. Div. Biol. Surv. Bull.*, 10: 9–79.

Sellers, W.D. and Hill, R.H., 1974. *Arizona Climate 1931–1972.* University of Arizona Press, Tucson, Ariz., 616 pp.

Shelford, V.E., 1963. *The Ecology of North America.* University of Illinois Press, Urbana, Ill., 610 pp.

Shreve, F., 1942. The desert vegetation of North America. *Bot. Rev.*, 8: 195–246.

Spaulding, W.G. and Van Devender, T.R., 1977. Late Pleistocene montane conifers in southeastern Utah. *Southwest Nat.*, 22: 269–270.

Stebbins, G.L., Jr., 1975. Shrubs as centers of adaptive radiation and evolution. In: H.C. Stutz (Editor), *Proc. Symp. and Workshop on Wildland Shrubs.* U.S. Forest Service and Brigham Young University, Provo, Utah, pp. 120–140.

Stutz, H.C., 1978. Explosive evolution of perennial *Atriplex* in western America. *Great Basin Nat. Mem.*, No. 2: 161–168.

Thornbury, W.B., 1965. *Regional Geomorphology of the United States.* John Wiley and Sons, New York, N.Y., 609 pp.

Udvardy, M.D.F., 1975. A classification of the biogeographical provinces of the world. *IUCN Occas. Pap.*, No. 18: 48 pp.

U.S. Department of Agriculture, 1941. *Climate and Man.* Agric. Yearbook, U.S.D.A., Washington, D.C. 1248 pp.

Walter, H., 1971. *Ecology of Tropical and Subtropical Vegetation.* Translation from German by D. Mueller-Dumbois. Oliver and Boyd, Edinburgh, 539 pp.

Walter, H., 1979. *Vegetation of the Earth and Ecological Systems of the Geo-Biosphere.* Translated from the third, revised German edition by Joy Wieser. Springer-Verlag, New York, N.Y., 2nd ed., 274 pp.

Walter, H. and Box, E.O., 1976. Global classification or natural terrestrial exosystems. *Vegetatio*, 32: 75–81.

Wells, P.V., 1980. Quaternary vegetational history of the Great Basin. *Bull. Ecol. Soc. Am.*, 61: 106 (abstract).

West, N.E., 1969. *Soil–Vegetation Relationships in Arid South-eastern Utah. Int. Conf. on Arid Lands in a Changing World.* University of Arizona, Tucson, Ariz., 9 pp.
Wolfe, J.A., 1975. Some aspects of plant geography of the northern hemisphere during the late Cretaceous and Tertiary. *Ann. Mo. Bot. Gard.*, 62: 264–279.
Wolfe, J.A., 1978. A paleobotanical interpretation of Tertiary climates in the northern hemisphere. *Am. Sci.*, 66: 694–703.

Chapter 12

GREAT BASIN–COLORADO PLATEAU SAGEBRUSH SEMI-DESERT[1]

N.E. WEST

DEFINITION

This ecosystem type occurs where dominance by woody sagebrushes (*Artemisia*) prevails even when pristine or near-pristine conditions exist (Fig. 12.1). The component of herbaceous species has apparently never been as great in this type as the more mesic but related Intermountain Sagebrush Steppe (Ch. 13). These two types have, because of some structural similarities, often been combined in other reviews (Anonymous, 1979). This author, however, considers this to be a mistake since the responses of the two ecosystem types to perturbation are distinctly different. The more arid ecosystem type that will be discussed below is more akin to deserts, whereas the sagebrush steppe reacts similarly to a semi-arid grassland. For instance, the recovery of vegetation following grazing or fire is much more rapid in the sagebrush steppes. Damage to the sagebrush semi-desert is much less reversible. Manipulations are less successful (Shown et al., 1969).

Fig. 12.1. Relic area with pristine (unaltered) vegetation, Bridger Jack Mesa, Grand County, Utah (Van Pelt, 1978). Photo by N. Van Pelt.

Division into types reflects more our need to communicate than the existence of discrete boundaries in nature. In actuality, there is a continuum from almost pure, stunted sagebrush-dominated sites of low productivity under the driest conditions at the lowest elevations, particularly in the south, to increasing amounts of perennial herbaceous species on gradients up slope and/or to the north where more effective precipitation is received and greater soil development usually ensues. The boundaries used on the maps presented here are arbitrary, but the real differences remain and the concept is useful to those trying to understand and manage these wildlands.

GEOGRAPHICAL LOCATION

The type is largely restricted to the Great Basin and Colorado Plateau Physiographic Provinces (Fig. 12.2). There is about four times as much of the type in the Great Basin as in the Colorado Plateau (West, 1979). Nevada has the greatest area, followed by Utah, Arizona, Colorado, California, and New Mexico (Table 12.1). This ecosystem type is located south of the sagebrush steppes and generally north of the creosotebush and blackbrush types. It is elevationally positioned between the salt-desert shrublands found on halomorphic soils at the lowest elevations in the region and more xerophytic conifer woodlands and forests on the lower flanks of the mountains (Fig. 11.5). Sage-

[1] Manuscript completed June, 1980.

TABLE 12.1

Area occupied by sagebrush semi-desert (derived from Küchler, 1970)

State	Area ($\times 10^6$ ha)	Percentage of total
Nevada	10.6	59.2
Utah	2.7	15.1
Arizona	2.4	13.5
Colorado	0.9	5.1
California	0.8	4.4
New Mexico	0.5	2.7
Total	17.9	100.0

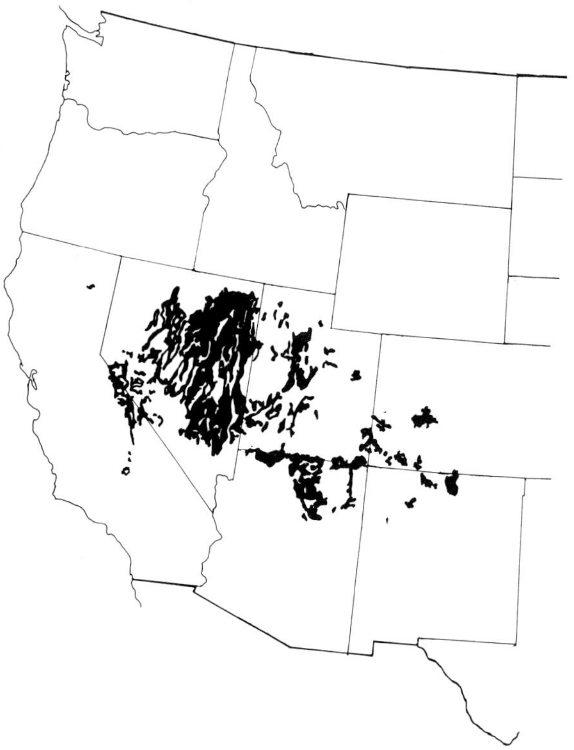

Fig. 12.2. Map of the Great Basin–Colorado Plateau sagebrush semi-desert ecosystem type. Derived from Küchler (1970).

brushes are also a major understory component of the adjacent juniper–pinyon woodlands (West et al., 1975, 1978).

EXTENT

About 17.9×10^6 ha has been indicated for this type in the U.S. National Atlas (Küchler, 1970). There are, of course, inclusions of smaller extent, unmappable at the national scale, that are found embedded in the associated juniper–pinyon woodland, saltbush–greasewood (Ch. 14), blackbrush (Ch. 15), and galleta–threeawn shrub steppe (Ch. 16) map types.

OWNERSHIP AND DOMINANT USES

The lands occupied by this ecosystem type are currently nearly all in public ownership. The Bureau of Land Management (B.L.M.) is the major federal agency concerned with the resources in this ecosystem type. The dominant current use is as rangeland for domestic livestock production.

ECOSYSTEM STRUCTURE AND FUNCTION

Climate

This ecosystem type occurs in a semi-arid temperate continental climate. The type is a major component of the cold-winter Great Basin Regional Desert (Shreve, 1942). Both seasonal and diurnal temperature ranges are extreme (Fig. 12.3). Average annual precipitation ranges from 158 to 419 mm, but the coefficient of variation is approximately 29%. The seasonality of precipitation varies from almost no summer precipitation in central Nevada to about 40% of the total coming from monsoonal influences in southern Utah, Arizona, and New Mexico. Snow cover develops during most winters and may last for several weeks.

Topography and geology

This ecosystem type occurs nearly exclusively on high, wind-swept plateaus. In the Great Basin these take the form of broad alluvial valleys at elevations of 1200 to 1800 m between fault block mountain ranges trending north–south. In the Colorado Plateau, the type occurs at intermediate (1300–1800 m) elevations. Pediments, bajadas and foothills are the predominant land forms under this ecosystem type in the Great Basin. Mesa tops, benches or pediments with a sandy-gravelly covering are predominantly associated with the type in the Colorado Plateau Physiographic Province.

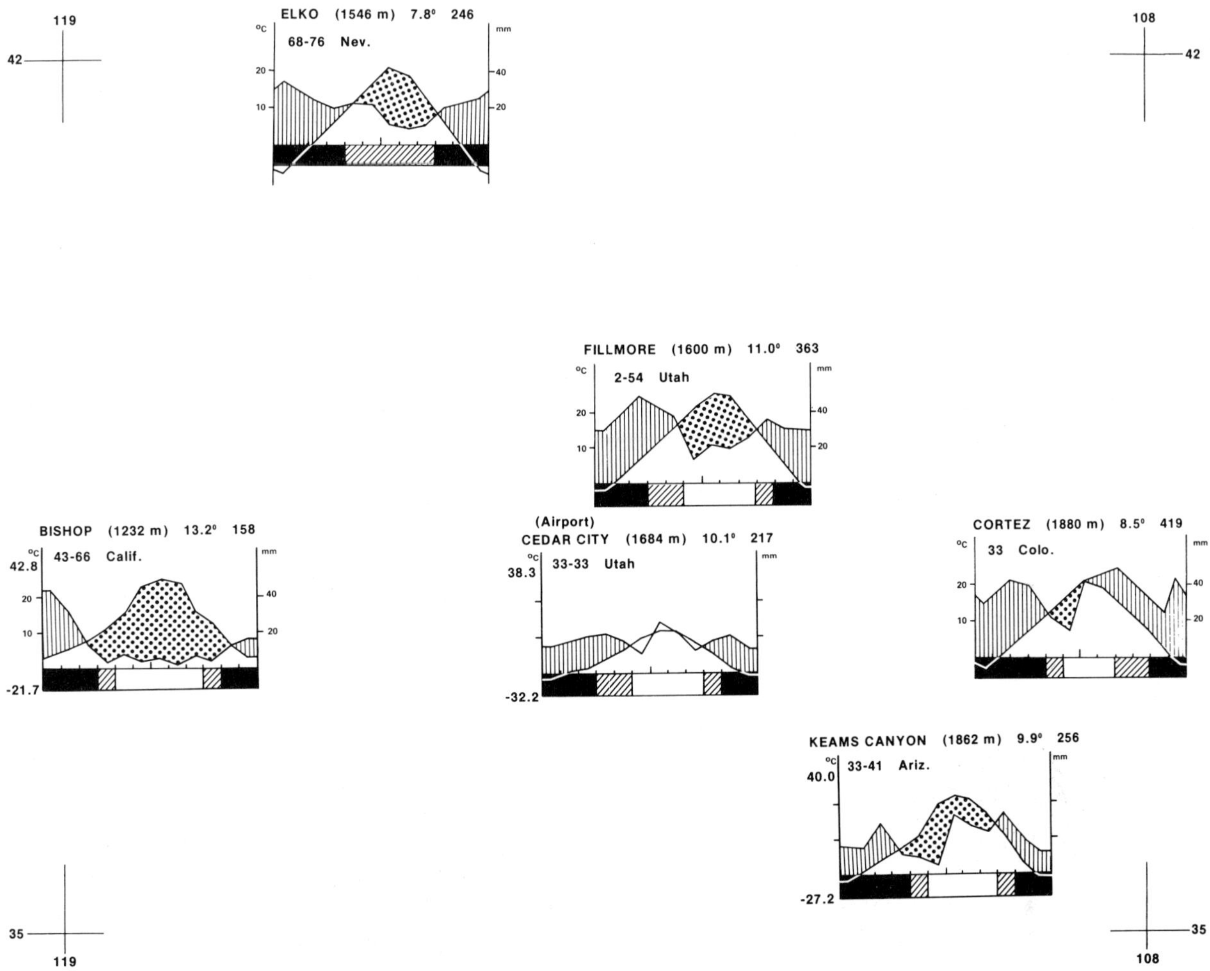

Fig. 12.3. Climatic diagrams for representative stations within the Great Basin–Colorado Plateau sagebrush semi-desert ecosystem type. Data from National Weather Service network.

Microtopography is usually pronounced, with clusters of sagebrushes on slight mounds and the interspaces being bare most of the year (Eckert et al., 1978).

Soils

The soils where this type occurs grade from Aridisols toward Mollisols and Alfisols. It is only on sites immediately above the halomorphic soils, mostly in the bottom of the valleys where the Salt Desert Shrub type (Ch. 14) occurs, that sagebrushes may be considered climax dominants (Weaver and Clements, 1938) if fire is excluded as a factor of climate. The major *Artemisia* species, particularly the *Artemisia tridentata* complex, will not tolerate salinities corresponding with conductivities greater than 18 mS cm^{-1} (Gates et al., 1956; Caldwell, 1979). Salinity in soil profiles where sagebrush dominates usually averages less (Branson et al., 1976). The predominance of halomorphic soils in the lowlands is a major reason why the sagebrush semi-desert type is found in the upland portion of the Great Basin valleys and on the lower mountain foothills. Much more of the Colorado Plateau receives adequate precipitation to foster sagebrush, but the outcropping of marine shales leads to halomorphic soils and salt desert shrub rather than sagebrush-dominated communities (Fig. 14.8). Elsewhere sandstones have weathered and been redeposited by wind. In such sites the galleta–threeawn shrub steppe (Ch. 16)

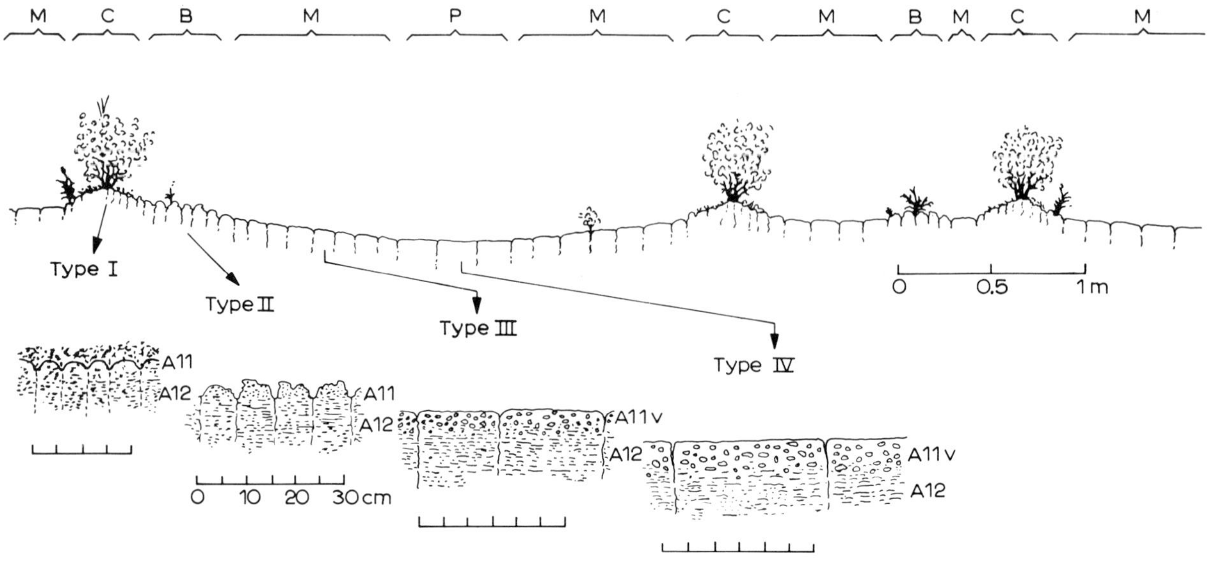

Fig. 12.4. Schematic cross-sectional diagrams of the microtopographic positions and associated surface-soil morphological types of gently sloping, shallowly loess-mantled Argids of the Humboldt loess belt with Wyoming big sagebrush (*Artemisia tridentata* ssp. *wyomingensis*) plant communities. Microtopographic positions are: C = coppice; B = coppice bench; M = intercoppice microplains; P = playette. Vertical scale somewhat exaggerated; intercoppice microplains and playettes can be much wider than shown here; several coppices can be linked together. Vertical lines under soil indicate sides of crust polygons (*A11v*) which continue downward as sides of prisms in the compoundly weak prismatic and moderately platy *A12* horizon. Type I is litter-covered. Circles indicate vesicles in crust (*A11v*). Only types III and IV are significantly crusted. Types V–VII do not occur commonly in the Humboldt loess belt of Nevada. (From Eckert et al., 1978.)

with a sand sagebrush (*Artemisia filifolia*) component is favored.

Soil varies greatly with microtopography and the clustering of shrubs. The differences (Fig. 12.4) are both chemical (Charley and West, 1975, 1977) and physical (Eckert et al., 1978). Whether these features were part of the pristine situation or due to livestock effects during the past hundred years is unknown (Young et al., 1979). The surface soil in the interspaces is highly vesiculated, especially on the east side of playas in the Great Basin where accumulation of silt leads to slower, less effective moisture infiltration and poorer seedling survival (Wood et al., 1978). The soil in the mounded portion of the micro-relief is not vesiculated and has a much higher content of organic matter and nutrients. It is on these portions of the landscape that perennial plants usually persist.

Soil depth and stoniness have a considerable effect on soil moisture budgets. The species composition, both of plants and small animals, responds dramatically to these variables. For instance, *Artemisia arbuscula* usually occurs on old landscapes with clayish soils. *A. nova* usually occurs on younger calcic soils (Young et al., 1977; Fig. 12.5). Small burrowing animals prefer stone-free profiles. A moderate degree of stoniness creates greater plant cover because moisture is forced deeper into the profile and evapo-transpirative losses are less (Walter, 1973).

Primary producers

The vegetation of this type seldom reaches over one meter above the soil surface and is monoto-

TABLE 12.2

Major taxa of woody *Artemisia* found in the Great Basin and Colorado Plateau (from West, 1979)

Great Basin	Colorado Plateau
A. tridentata wyomingensis	*A. tridentata wyomingensis*
A. tridentata vaseyana	*A. tridentata vaseyana*
A. tridentata tridentata	*A. tridentata tridentata*
A. nova	*A. nova*
A. arbuscula	
A. longiloba	
A. spinescens	
	A. bigelovii

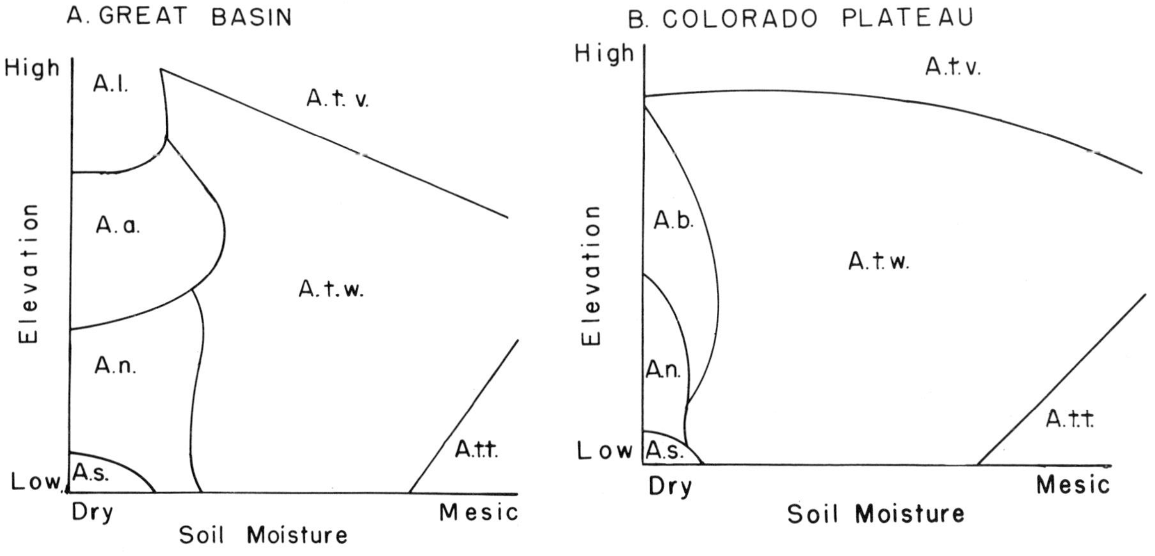

Fig. 12.5. Ordinations of major sagebrush taxa found in the Great Basin (A) and Colorado Plateau (B) against gradients of elevation and effective soil moisture. *A.l.* = *Artemisia longiloba*, etc., as derived from Table 12.2, using the first letters of the genus, species, and subspecies names. (From West, 1979.)

nously simple in floristic composition (Fig. 12.6). Various sagebrushes (Table 12.2) overwhelmingly dominate the community in both near-pristine and degraded conditions. Sagebrush usually makes up over 70% of the relative cover and 90% of the biomass. The densities of these shrubs varies between about 1000 to 7000 plants per hectare. This high degree of dominance by one life form lends a characteristic grayish-green cast to the vegetal cover on these kinds of landscapes. Absolute cover of higher plants varies from about 10 to 40% (West, 1979).

Major associated plant species (Table 12.3) on undisturbed sites are most commonly perennial bunch grasses and perennial forbs. There is a trend from cool season bunch grasses of Arcto-Tertiary origin in the Great Basin to an increasing component of sod grasses of Madro-Tertiary origin that grow with the greater precipitation of late summer in Arizona and New Mexico.

A high fraction of the phytomass in these systems is commonly found in the soil (West, 1972). This is probably an adaptation to the cold, dry atmospheric climate that makes the soil a relatively more favor-

Fig. 12.6. The big sagebrush/needle-and-thread (*Stipa* spp.) habitat type in (A) excellent, (B) good, and (C) poor condition. Sites B and C were in the same drainage bottom with C being closer to a source of livestock water, southern Elko County, Nevada. (From Tueller and Blackburn, 1974.) Photo by P.T. Tueller.

TABLE 12.3

Major perennial plant species at representative lowland locations within sagebrush-dominated areas of the Great Basin and Colorado Plateau; sites ordered from northwest on right-hand side of the table to southeast on left-hand side (location of sites described in footnote; from West, 1979)

Species	1	2	3	4	5	6	7	8
SHRUBS								
Artemisia tridentata	×	×	×	×	×	×	×	×
Chrysothamnus viscidiflorus	×	×	×		×			×
Eriogonum microthecum	×				×			
Xanthocephalum sarothrae	×			×			×	×
Prunus andersonii	×							
Purshia tridentata	×							
Tetradymia canescens	×							
Tetradymia glabrata	×							
Leptodactylon pungens		×						
Chrysothamnus nauseosus				×		×		
Ephedra viridis			×					×
Ephedra nevadensis			×					
Grayia spinosa		×			×			
Ceratoides lanata					×			
Atriplex canescens					×			
SUCCULENTS								
Opuntia polyacantha		×		×	×	×	×	×
GRASSES								
Agropyron spicatum	×	×						×
Bromus tectorum	×	×						
Oryzopsis hymenoides	×	×	×		×		×	
Poa canbyi	×							
Poa secunda	×	×					×	
Sitanion hystrix	×	×	×			×	×	×
Stipa thurberiana	×	×						
Elymus cinereus		×						
Stipa columbiana				×				
Bouteloua gracilis				×	×		×	×
Agropyron smithii				×				
Poa longiligula				×			×	
Hilaria jamesii					×	×		×
Sporobolus airoides					×			×
Sporobolus cryptandrus					×		×	×
Stipa speciosa					×		×	
Stipa comata								
FORBS								
Astragalus spp.	×	×						×
Calochortus spp.	×							
Castelleja spp.	×						×	
Caenactis douglasii	×	×						
Crepis acuminata	×			×				
Epilobium paniculatum	×	×						
Lomatium spp.	×						×	
Mentzelia albicaulis	×			×				
Penstemon spp.	×						×	
Phlox gracilis	×						×	
Viola beckwithii	×			×				
Zigadenus paniculatus	×							
Phlox diffusa		×	×					×

TABLE 12.3 (*continued*)

Species	1	2	3	4	5	6	7	8
Lupinus caudatus		×		×				
Erigeron pumilis		×						
Cryptantha jamesii		×						
Polygonum nuttallii				×				
Aster spp.			×	×				×
Gilia spp.							×	
Erigeron spp.							×	
Townsendia spp.							×	
Sphaeralcea spp.			×		×		×	×

Key to site locations: 1 = Sites No. 1 and 2, Carson Valley near Reno, Nev. (Billings et al., 1954); 2 = *Artemisia tridentata/Poa secunda/Phlox diffusa* community, Coils Creek, Eureka County, Nev. (Blackburn et al., 1969); 3 = *Artemisia tridentata/Hilaria jamesii* association, Hot Creek Valley, Nye County, Nev. (Tueller et al., 1972); 4 = exclosures near Duchesne, Duchesne County, Utah (Dastrup, 1963); 5 = Cedar Mesa relict area, Canyonlands National Park, Utah (Van Pelt, 1978); 6 = experimental watershed paddocks, Badger Wash, near Fruita, Mesa County, Colo. (Branson et al., 1976); 7 = *Artemisia* site, Fishtail Mesa, Grand Canyon National Park, Ariz. (Jameson et al., 1962); 8 = inside fenced highway (N.M. 197) near Rio Puerco, Sandoval County, New Mexico (Dahl et al., 1976).

able micro-environment for biological activity. The roots of the perennials grow into the soil profile under the interspaces, but are not usually found in the surface 5 cm of soil. This is because of the extreme fluctuations in temperature and moisture found there. These patterns are even more pronounced than those illustrated for the Intermountain Sagebrush-Steppe Type (Fig. 13.7).

Although one is most likely to see this ecosystem type in dry weather, muddy conditions, impeded drainage and poor soil aeration can occur and cause sorting of some of the major species. *Artemisia tridentata* cannot tolerate conditions of poor soil aeration (Caldwell, 1979); other species such as *A. arbuscula, A. nova*, and *A. cana* can. Rapid snowmelt or rain does occasionally saturate the surface soil over a deeper frozen layer in fall or spring, especially where concave micro-relief occurs. This is believed to be a major cause of almost pure, nearly symmetric stands of the secondary sagebrushes (Passey et al., in press). The sagebrushes other than *A. tridentata* are also usually more successful on the rockiest and/or shallowest soils.

A sometimes overlooked but important component of the plant communities of such systems is the microphytic (moss–lichen–alga–fungi) crusts occurring on the soil surfaces. The crust, if unbroken by animal trampling or vehicles, provides considerable protection against wind and water erosion. The blue-green algal component fixes nitrogen. Nitrogen is second only to water as a limiting factor to primary production in these systems (West and Skujins, 1978a).

Total above-ground standing-crop phytomass in this type may vary between 2000 and 12 000 kg ha^{-1} (Shepard, 1977). Part of this variation is due to site differences, part due to successional status and/or stand age of the brush. These various contributions to the total variation have not been separated in most instances. The degraded sites will have more phytomass in general, because the proportion of brush will be higher.

Below-ground phytomass is similar in magnitude to that above ground in the few instances when it has been measured (Wallace and Romney, 1972; Caldwell et al., 1975). Litter standing crops are about half the total above-ground phytomass. An unknown percentage of the below-ground phytomass could be considered litter.

Annual net above-ground primary production usually varies between about 500 and 1500 kg ha^{-1}, depending on site, successional status, stand age, and the preceding climatic conditions. Most of this production is by sagebrush whose structure is mostly woody. Its foliage has high concentrations of secondary metabolites, such as terpenoids and coumarins, that create digestive problems for ungu-

TABLE 12.4

Listing of vertebrates, with abundance estimates of some, at a sagebrush semi-desert site, Fernow Valley, Utah (Fautin, 1946) (between parentheses: estimated density ha^{-1})

Scientific name	Common name	Scientific name	Common name
MAMMALS		*Falco mexicanus*	prairie falcon
		Haliaeetus leucocephalus leucocephalus	southern bald eagle
Antilocapra americana americana	pronghorn	*Lanius ludovicianus nevadensis*	Great Basin shrike
Canis latrans lestes	coyote	*Mimus polyglottos leucopterus*	western mockingbird
Citellus leucurus leucurus	antelope ground squirrel	*Oberholseria chlorura*	green-tailed towhee
Citellus townsendi mollis	piute ground squirrel	*Oreoscoptes montanus*	sage thrasher
Dipodomys microps bonnevillei	kangaroo rat (8.5)	*Otocoris alpestris utahensis*	Great Salt Lake horned lark
Dipodomys ordii celeripes	kangaroo rat (11.3)	*Passerella iliaca schistacea*	slate-colored fox sparrow
Eptesicus fuscus pallidus	pallid big brown bat	*Phalaenoptilus nuttalii nuttallii*	Nuttall's poor-will
Eutamias minimus consobrinus	wasatch chipmunk (3.0)	*Pica pica hudsonia*	American magpie
Lepus californicus deserticola	black-tailed jackrabbit	*Polioptila caerulea amoenissima*	western gnatcatcher
Perognathus parvus olivaceus	Great Basin pocketmouse (0.3)	*Pooecetes gramineus confinis*	western vesper sparrow
Peromyscus maniculatus sonoriensis	white-footed mouse (20.3)	*Salpinctes obsoletus obsoletus*	rock wren
		Sayornis saya saya	say phoebe
Reithrodontomys megalotis megalotis	harvest mouse (4.3)	*Selasphorus platycercus platycercus*	broad-tailed hummingbird
Sylvilagus nuttalii grangeri	cottontail rabbit	*Sialia currucoides*	mountain bluebird
Taxidea taxus	badger	*Spizella breweri breweri*	Brewer's sparrow
		Spizella passerina arizonae	western chipping sparrow
BIRDS		*Tyrannus verticalis*	Arkansas kingbird
Accipiter cooperii	Cooper's hawk	*Zenaidura macroura marginella*	western mourning dove
Amphispiza belli nevadensis	northern sage sparrow	*Zonotrichia leucophrys*	white-crowned sparrow
Amphispiza bilineata deserticola	desert black-throated sparrow		
Aquila chrysaetos canadensis	golden eagle	REPTILES	
Asio otus wilsonianus	long-eared owl	*Cnemidophorus tessellatus tessellatus*	whip-tail lizard
Buteo jamaicensis calurus	western red-tailed hawk		
Buteo swainsoni	Swainson's hawk	*Coluber taeniatus taeniatus*	striped racer
Cathartes aura teter	western turkey vulture	*Crotalus viridus lutosus*	Great Basin rattle snake
Chordeiles minor	nighthawk	*Crotaphytus collaris baileyi*	collared lizard
Circus cyaneus hudsonius	marsh hawk	*Phrynosoma platyrhinos*	desert horned toad
Corvus corax sinulatus	American raven	*Pituophis catenifer deserticola*	gopher snake
Cyanocephalus cyanocephalus	piñon jay	*Sceloporus graciosus graciosus*	sagebrush lizard
Euphagus cyanocephalus cyanocephalus	Brewer's blackbird	*Uta stansburiana stansburiana*	brown-shouldered uta

lates. These substances also permeate the air with a distinctive smell, especially after summer rains. Grazing capacities for domestic animals are much lower than would be expected on the basis of phytomass present because of the large fraction that is undigestible to these animals. The original livestock grazing capacity[1] of this type of range was estimated at 0.83 AUM ha^{-1} (McArdle and Costello, 1936). Largely because of losses in the perennial grass component, this declined to an average of 0.27 AUM ha^{-1} in the 1930's (McArdle et al., 1936) and had increased slightly to an estimated 0.31 AUM ha^{-1} by 1970 (Forest-Range Task Force, 1972).

Consumers

A listing of native vertebrates (Table 12.4) that are common in this ecosystem type is not nearly as

[1] Originally expressed as 1.2 ha AUM^{-1}, where AUM means animal unit month. An animal unit is one mature cow and calf or their equivalent grazing for one month.

extensive as in more mesic, adjacent systems. Only the smaller ground-dwelling mammals and the invertebrates spend the entire year in this type of habitat. None of the animals listed are restricted exclusively to this ecosystem type.

The relatively low animal diversity and productivity is related to the dominance of the type by sagebrush (McAdoo and Klebenow, 1979). Few species thrive by eating only sagebrush. Some of the vertebrates, such as pronghorned antelope and sagegrouse (*Centrocercus urophasianus*), eat large quantities of the shrub during the winter, but both adults and juveniles require other forages for good nutrition. In the geologic past perhaps browsers now extinct, such as camels and the ground sloths (*Nothrotheriops*), made better use of this type (Martin, 1970). Reintroduction of browsing animals such as the saiga antelope from Eurasia would perhaps lead to higher meat production than is currently obtained from livestock and game animals on these kinds of ranges.

Black-tailed jackrabbits (*Lepus californicus*) are by far the most important mammalian herbivore. Small mammal trapping in Curlew Valley (Balph et al., 1973) has shown the biomass of all rodents combined to be lower than jackrabbit biomass at a population low of the latter. Rabbit populations oscillate from about 60 km^{-2} to 600 km^{-2} (Gross et al., 1974), and have considerable impact on the vegetation (Rice and Westoby, 1978).

Sagebrush voles (*Lagurus curtatus*) are small herbivores found on the wetter sites. Their diet is the grasses, but have been noted to girdle sagebrushes during winters of heavy snowpack (Frischknecht and Baker, 1972). They have probably greatly declined in abundance with pronounced vegetation changes over the past 100 years. Seed-eaters and broad-spectrum feeders such as *Peromyscus*, on the other hand, have probably increased in abundance (Larrison and Johnson, 1973). They and the heteromyids play a role in dispersing seeds of the weedier plants (LaTourrette et al., 1971).

MacMahon (1976) has grouped the non-flying mammals in this area, of jackrabbit size or smaller, into guilds based on food habits and adult size. He found that there are usually seven guilds present: (1) granivores, (2) a "carnivorous" mouse, (3) large browsers (*Lepus californicus*), (4) small browsers (*Sylvilagus*), (5) a micro-omnivore (*Peromyscus*),

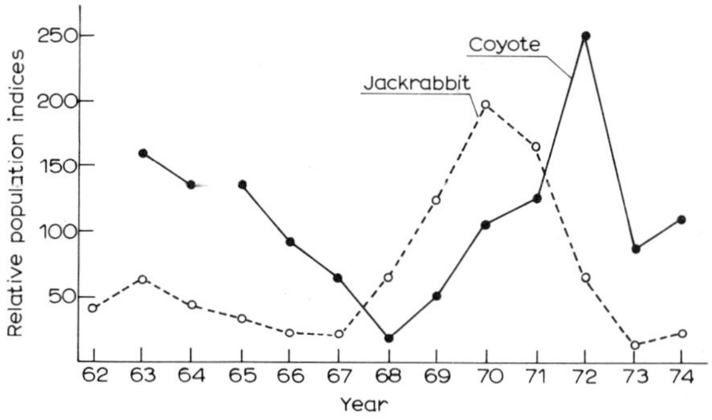

Fig. 12.7. Annual fluctuations in black-tailed jackrabbit and coyote populations in Curlew Valley, northwestern Utah and southern Idaho, 1962–1974. Coyote data and indices are from Clark (1972) and Knudsen (1976), jackrabbit data and indices are from Gross et al. (1974) and Stoddart (1976).

(6) a "pack-rat", and (7) a diurnal medium-sized omnivore (*Spermophilus*). This concept recognizes that a number of species of small mammals may be functionally similar, in particular examples of sagebrush semi-desert. Ecological equivalents may or may not be genetically close.

The predominant secondary consumer is the coyote (*Canis latrans*). The coyote is a very effective predator on jackrabbits. The two populations are linked in an unstable equilibrium approaching a classical predator–prey oscillation with the predator population buildup lagging behind that of the hares (Fig. 12.7).

Badgers (*Taxidea*) have about one-half the biomass of coyotes and are largely rodentivores. Bobcat (*Lynx rufus*) populations constitute only about 1/20th the biomass of coyotes (Wagner, 1971). The significant raptorial feeders on hares — golden eagle, ferruginous hawk (*Buteo regalis*), and western red-tailed hawk — each have a biomass only 1 to 3% of that of the coyotes (Wagner, 1971).

Although a considerable diversity of invertebrates is present in such ecosystems (Table 12.5), the only insect that has been noted to distinctly reduce populations of sagebrushes is the webworm (*Aroga websteri*). The near-monoculture of sagebrush probably facilitates the population eruptions of these species (Tempte, 1977). This moth has destroyed vast areas of sagebrush, particularly during dry periods. Psyllids (Balph et al., 1973) and thrips (Tingey et al., 1972) have also been noted to defoliate sagebrush in some areas. In the spring of some years there are vast populations of adult

TABLE 12.5

Taxonomic summarization of invertebrates collected at a sagebrush semi-desert site, Fernow Valley, Utah (Fautin, 1946)

Orders	Families	Genera	Species	Specimens
Araneida	5	9	11	91
Acarina	1	1	1	3
Collembola	0	0	0	0
Orthoptera	1	2	5	43
Thysanoptera	0	0	0	0
Hemiptera	7	13	15	193
Homoptera	5	14	16	766
Coleoptera	7	15	18	193
Lepidoptera	3	× [1]	× [1]	6
Diptera	5	6	7	81
Hymenoptera	5	9	10	17
Totals	39	69	83	1393

[1] Larvae and pupae not determinable to species.

cicadas present. The larvae must consume considerable amounts of shrub roots, but little more is known than that they influence the development of soil profiles (Hugie and Passey, 1963). Harvester ants build up distinctive mounds. They appear to be more abundant on rangelands degraded by livestock grazing, where annual plants are more frequent, but more investigation of their role in vegetation dynamics and nutrient cycling is needed. The occasional eruptions of populations of Mormon crickets (*Anabrus simplex*) (Young, 1978) and grass-hoppers can cause considerable destruction of herbaceous vegetation. There is little more than acknowledgement of the presence of at least one thousand other species of insects and other invertebrates in such ecosystem types (Fautin, 1946; Knowlton, 1974). Further study will undoubtedly lead to a better understanding of the role of more invertebrates in these systems.

Decomposers

Very little is known about the decomposers in such ecosystems. Since litter is not piling up to distinctive levels, apparently all the kinds of microorganisms needed to effect cycling of minerals are present. Additional amounts of soil carbon would help immobilize nitrogen, the most critical element of these systems (West and Skujins, 1978b).

Decomposition proceeds in surges during the cooler, wetter periods of the year, particularly as snow melts (Comanor and Staffeldt, 1978). Decomposition and subsequent nutrient mineralization is also highly patterned in relation to microrelief (Charley and West, 1977). Litter accumulates at the soil surface and near shrubs. This is where the soil microbial activities are centered. Anything that destroys this patterning diminishes the overall activity on a landscape basis. This is because these desert environments have too little moisture and critical nutrients to carry on high levels of activities across the entire landscape.

ECOSYSTEM INTERACTIONS

This ecosystem type appears deceptively simple upon first observation. The expanses dominated by sagebrush and low populations of large animals lull the casual observer into assuming more homogeneity than there really is. The low floristic diversity and concentration of dominance in so few species (Tables 12.2 and 12.3) result in loose species packing. This appears to be compensated for by high intra-specific diversity (Van Valen, 1965; Van Valen and Grant, 1970). The vast degree of infraspecific variation in major taxa such as *Artemisia* (McArthur, 1979), *Chrysothamnus* (Anderson, 1975), and *Sitanion hystrix* (Clary, 1975) means that the morphologically determined taxa possess considerable genetic and ecological variation and thus functionally different communities can result. Whether grazing animals have been causing selection for more grazing-repellent or grazing-tolerant ecotypes is not yet known.

Sagebrushes possess both a deep tap-root and fibrous root system (Sturges, 1977). They are long-lived (up to 150 years) and may produce allelochemics controlling seed germination of herbaceous plants (Schlatterer and Tisdale, 1969) and nitrogen-fixing and nitrification processes (Skujins, 1980). The full occupancy of many soil profiles in this type by sagebrush may explain why so few other species occur with it. Even evolution of a rich annual flora may have been inhibited by the intense and pervasive competition and interference this brush provides (Robertson, 1972; Young et al., 1972). The drastic changes in the total community when it is removed (Vale, 1974) demonstrate its function as a keystone species.

The predominant difference in vegetation after

sagebrush is destroyed is a switch to herbaceous species, usually alien annuals (Young et al., 1972; Young and Evans, 1973). The annuals favor different animals, have much more variability in abundance and productivity, and greatly increase the frequency of fire.

Fire was apparently not especially frequent or pervasive as a factor in this ecosystem type under pristine conditions. Otherwise, there would have been much more abundance of the root-sprouting shrubs such as *Chrysothamnus* when the early observers first saw it.

A more likely perturbation was probably periodic population explosions of webworms (*Aroga websteri*). Whether drought had a role in promoting these defoliators is not known (Kirkland, 1972; Tempte, 1977).

Total food webs or energy flow diagrams have not yet been worked out for examples of this ecosystem type. The closest approach is that of Fautin (1946).

Successional patterns are brought out in the discussion of land-use history to follow. Suffice it to say here that the pristine ecosystem was probably minimally stable and poorly resilient to the impacts of the white man's livestock. Soil erosion has been severe as vegetation has changed. Nutrient cycles have definitely been accelerated and pools reduced and/or rearranged spatially. The changes may well be irreversible on some sites. That is, productivities may be permanently lowered until soil development proceeds again. Further comparison of degraded sites with relict areas are needed to define how great the dystrophic changes might be.

LAND-USE HISTORY

Aboriginal use

The most important part of the landscape to aboriginal peoples, wildlife, and the earlier livestock users were the areas of either sandier soil or alluviation where greater effective soil moisture and thus greater amounts of grass and forbs occurred. The Indians harvested the seeds of grasses such as Great Basin wildrye (*Elymus cinereus*) (Lesperance et al., 1978) and Indian ricegrass (*Oryzopsis hymenoides*). Areas where these grasses were dominant probably constituted less than 10% of the pre-Columbian landscape. Even less abundant were the naturally wet or irrigable meadows which still provide almost 50% of the livestock forage base of the region (McCormick et al., 1979). Sagebrush dominance in the uplands may once have been less because of probable high populations of now extinct browsers such as camels and ground sloths (Martin, 1970). All these had been lost by the close of the Pleistocene. Even though the sagebrush evolved under high browsing pressures, the native grasses and forbs show poor competitive abilities with sagebrush in this environment. Other shrubs prospering under greater disturbance (*Xanthocephalum*, *Chrysothamnus*, *Tetradymia*) are highly competitive also.

Horses were reintroduced to the southwest by the Spanish in the 1500s. The Paiute Amerindian people, who occupied this area, did not acquire the horse. They were too busy trying to hunt and gather enough food to survive and escape the slave trading of other tribes. The difficulties of Jedediah Smith (1826–27) in trying to find forage for his horses in this area in the 1820s and 1830s makes it doubtful the native peoples could have developed a viable horse-based culture anyway. The aboriginal hunter–gatherer population in this ecosystem type was low. Farb (1978) estimates populations of one to two persons per 250 km^2. Limited availability of pinyon pine nuts probably allowed this area to sustain even this low population density. No more than a few families could remain together for any length of time because there was simply not enough food to go around. A few families might come together for cooperative rabbit drives when concentrations of these animals were found. They had to disperse again when that resource was depleted and rely more heavily on plant materials.

Nineteenth century livestock impacts

Livestock other than horses were first brought into this area in the late 1840s by Mormon pioneers and Californians raising cattle for gold miners. The horses, cattle, and sheep preferred the grasses and other herbaceous components of the brush-dominated communities. Although there was probably somewhat more herbaceous cover within the more northerly or upland sagebrush-occupied sites (Young et al., 1976), many areas apparently never had much grass (Vale, 1975). The greatest amount of grass was found on the sandy soils and alluvial sites along the streams. These latter areas were the pathways the immigrants followed and chose to develop first into farms and ranches. These azonal and intrazonal grasslands in a zonal sagebrush matrix were quickly grazed out (Lang and Young, 1978) and converted to barrens or became more completely dominated by shrubs. Hull (1976) has determined from historical records that these dramatic alterations in both the lowlands and uplands generally took less than fifteen years under conditions of unrestricted livestock use. The Indians of this area, who had been the least wealthy on the entire continent, were reduced to abject poverty. The Mormons survived by cooperative construction and maintenance of irrigation works and its attendant agriculture. Their rapidly developing range-livestock herds were also exploiting the wealth in the rangeland vegetation.

The winter of 1889–90 was the harshest in 500 years. The near-total loss of herds that had been free-roaming in the Spanish style caused a change to fencing of the bottoms to save the grass for standing winter feed, or conversion of these sites to the production of hay aided by irrigation (Young et al., 1979).

As the grass component of the uplands began to become depleted in the late 1800s, sheep were increased to utilize the browse. The sagebrush-type was, however, only one link in the seasonal migration between the desert in winter and the mountains in the summer. The deterioration in watershed conditions, most noticeable in mountains above the irrigation dams, was arrested around the turn of the twentieth century by the establishment of forest reserves under the management of the U.S. Forest Service.

The twentieth century

Livestock use

The sagebrush ranges, being largely in the public domain, continued to carry large numbers of sheep in the first half of this century. Many of these animals were owned by people who had no land and thus no Forest Service permits. They grazed their flocks year-round in the deserts and mountains that were not under Forest Service control. Many of these people were uneducated, non-English-speaking Basques and Greeks who came to America to build a new life for themselves and their families. These entrepreneurs were using a free common resource. Who got there first used it. There was no reward for using the land lightly. On the contrary, if one did not use it, another would. The grasses and forbs used all their food storage in the spring for regrowth. When the new, highly nutritious herbaceous forage is utilized in the spring, it lacks the reserves to recover (Cook, 1971). Sagebrush and associated shrubs have the protection of distasteful foliage, perhaps evolved when the Pleistocene megafauna provided many more browsers.

Furthermore, shrubs of this ecosystem type are partially evergreen. The whole above-ground portion of the plant does not have to be reconstructed each year. They draw water and nutrients from deep in the soil profile. They carry on photosynthesis year-round, although spring is the most productive period for them also (Caldwell, 1979). Sagebrush achenes are small and have a pappus which facilitates wind dispersal. Production of the heavier seeds of the native herbaceous species is sporadic and seedling establishment even rarer.

With this knowledge of plant characteristics and land-use history, it is not hard to see why the degradation of this type has been severe. Grazing pressure has led to vegetation almost exclusively dominated by sagebrush where fire has not been a factor (Fig. 12.8). Perennial grasses are much more susceptible to fire than annual vegetation. As the proportion of annuals increases, the frequency of fire increases, organic matter, and nutrients are lost, and a declining spiral of productivity ensues (Evans and Young, 1978). An associated decline in watershed and wildlife values can easily result. These are the reasons why wildfires should be controlled and prescribed timing used very carefully as a management tool in this ecosystem type.

The pitiful conditions of rangelands of this and other types was finally publicly acknowledged during the drought and economic depression of the 1930s. The Taylor Grazing Act of 1934 brought most of these lands under federal control for the purpose of eliminating the itinerant grazers and stabilizing the livestock industry. The Grazing Service, eventually to become the Bureau of Land Management, was created to determine livestock grazing capacities. The number and kinds of animals and seasons of use were adjusted to bring

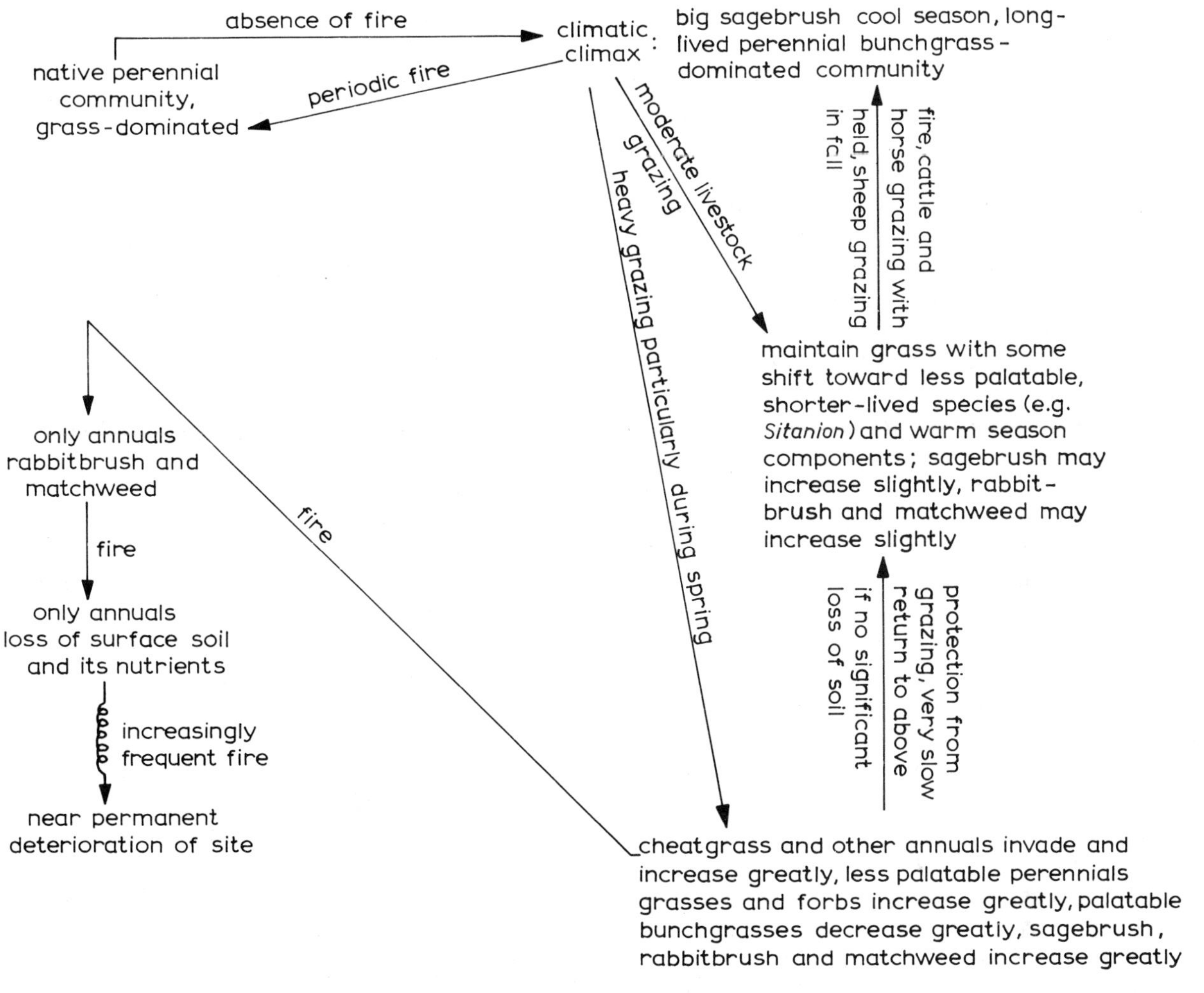

Fig. 12.8. Major pathways of succession in plant communities on important lowland sagebrush semi-desert sites in the Great Basin and Colorado Plateau (from West, 1979). Big sagebrush: *Artemisia tridentata*; rabbitbrush: *Chrysothamnus* spp.; matchweed: *Xanthocephalum* spp.; cheatgrass: *Bromus tectorum*.

about improvement in the vegetation so that a self-sustaining community would be attained. In the case of the Great Basin–Colorado Plateau sagebrush ecosystem type, this process has been slow and painful. The B.L.M. lacked the legislative and administrative powers of the Forest Service. Administration of the law did eliminate the tramp herds and flocks, but most of those who held commensurate, private property[1] continued business as usual up until the 1950s and 1960s when adjudication (formal determination of grazing capacity) was determined. Some ranchers who intuitively knew the public lands were less productive than they had earlier hoped reduced their use voluntarily. The presence of a bacterial disease of cattle, redwater (Lang and Young, 1978), on the largely private low-lying grasslands, had forced deferment of summer use. These areas thus became important for either hay or exclusive winter grazing (McCormick et al., 1979).

After a knowledge of the carbohydrate reserve cycle in grasses became known, a logical management strategy was spring deferment of livestock grazing. The first problem was what to do with the livestock during the spring. These rangelands had

[1] Legally sufficient to ensure year-round availability of forage.

traditionally been used during spring and fall. The ranchers had to use their private lands for hay production, cash crops, or winter pasture in order to sustain a viable year-round operation. Thus, spring deferment could not be instituted until alternate pastures were created. Beginning in the 1930s the Eurasian wheat and rye grasses were introduced to try to fill this need. The exotic grasses had evolved under heavy grazing pressure and could withstand spring use much better than the native grasses. In this ecosystem type, it was desert wheatgrass (*Agropyron desertorum*) that was the best adapted to the aridity of the area. Vast areas with the gentlest slopes and deepest soils were plowed and planted to this grass in the 1940s, 1950s, and 1960s (Vale, 1974). The livestock grazing capacities on these pastures became at least ten times better than on the adjacent grass-depleted native rangelands, and rates of livestock weight gain doubled (Frischknecht and Harris, 1968; Frischknecht, 1978). These pastures could not be grazed unmercifully, however. Heavy, spring grazing opened them to shrub invasion. Judicious winter use of sheep, herbicidal spraying or mechanical treatment was needed to keep the shrub component down and livestock grazing capacity high.

The Russian wheatgrass pastures have allowed the land managers to defer spring use of the native range. The results have not, however, been as desirable as in the related Intermountain Sagebrush Steppe (Chapter 13). This is because the climatic regime of this type naturally favors shrub growth. The grass component has probably always been low because it is at a competitive disadvantage where winter recharge of soil moisture is the most important. Growing-season precipitation is less in amount and dependability here. Native grass seed sources and opportunities for seed production and seedling re-establishment are much less abundant here than in other types where sagebrushes occur. Even desert wheatgrass does not invade outside the pastures where it has been planted in this type.

Wildlife managers have objected to wide-scale vegetation conversions of this type because reductions in total animal diversity and population levels have been noted (Vale, 1974; Wagner, 1978; Reynolds and Trost, 1980). Population explosions of the forage-destroying blackgrass bug (*Labops hesperius*) have also been observed on these near-monocultures (Haws, 1978). Reinvasion of sagebrush causes reductions in these insects, apparently because of its repellent properties (Ostlie, 1979) or facilitation of predators.

Wood production

Wood production has not usually been a value attributed to this ecosystem type, but the settlers and miners used sagebrush (Young et al., 1979) when they found themselves distant from larger fuels. Sagebrush provides a hot fire. The content of essential oil probably contributes to the high caloric value. Sagebrush does grow into thick stands that can reach 1 to 3 m high, particularly along ephemeral water courses. There is a considerable amount of biomass that has a potential for conversion into energy or bacterial protein. Whether this potential is ever used will depend on how desperate the resource problems become elsewhere.

Watershed values

With potential evapotranspiration exceeding precipitation by a factor of over 5 (Major, 1963), this ecosystem type does not produce perennial streamflow or recharge of groundwater tables. Most of the groundwater is a remnant of Pleistocene conditions. Most of the surface water comes from flows generated at higher, wetter elevations. Contribution to streamflow within the sagebrush semi-desert comes only from rapid melting of snow over frozen surface soil or intense rainfall, usually from intense convectional storms during the summer. The relatively low vegetal cover combined with occasional high-intensity storms makes it one of the most erodible ecosystem types in the western United States (Branson, 1975). Average sediment yields frequently exceed 3 t ha^{-1} yr^{-1} on the Colorado Plateau (Langbein and Schumm, 1958) and 1 t ha^{-1} yr^{-1} in the Great Basin (Blackburn and Skau, 1974). Since water erosion is exponentially related to storm intensity, most of the damage is done from the exceptional storms. Consequently, rilling and arroyo formation is particularly pronounced in the Colorado Plateau portions. The prevalent winds can also cause erosional movements of the sandier soils. Dunes and desert pavements are, however, quite minor in extent. The only places where they dominate the landscapes are where rivers ran into the pluvial lakes of Pleistocene time, and these areas are very small in proportion to the total area of the type.

How much of the now common depression and hummock micro-relief (Eckert et al., 1978) is of historical origin is debatable, since very few relict areas in the type exist, especially in the Great Basin.

Wildlife

Huntable wildlife has probably never been particularly diverse or composed of abundant populations, but was more prevalent when more perennial herbaceous plants were found here (Leopold, 1959). As with livestock, the most important areas to wildlife have also been the stringers of more mesic habitat along the stream courses. It is here the young pronghorned antelope and sagegrouse could find the forbs (and insects in the case of the grouse) essential to their well-being. The destruction of the grass and forbs and the thickening of sagebrush caused more transpirative demands and probably increased interception and evaporation. The amount of water flowing in streams thus declined. Riparian shrubs such as willows (*Salix* spp.) and plums (*Prunus* spp.) were either browsed out or died due to the drouth stress. Arroyo cutting lowered water tables and caused xerification of meadows (Cottam and Stewart, 1940).

In addition to declines in the populations of large vertebrates, this probably caused a parallel decline in the sagebrush vole (*Lagurus curtatus*), songbirds, and insects associated with this more open habitat. The increased abundance of annuals on the uplands likely caused an expansion of seed-eating, small mammals such as the deer mice (*Peromyscus*), pocket mice (*Perognathus*), and kangaroo rats (*Dipodomys*). Although no data exist to describe the pristine condition, we would expect a corresponding change in raptors to have occurred. The nocturnal owls may have replaced the diurnal hawks as the predominant avian predators. As sagebrush declined in response to vegetation change, another Middle Eastern import, the chukar partridge (*Alectoris chukar*) (Fig. 12.9) expanded with the increased importance of cheatgrass (*Bromus tectorum*). This species now constitutes the major game bird in this ecosystem type. The prospects for its continued success are bright (Christensen, 1970). Other animals that have increased in this ecosystem type are the black-tailed jackrabbits (Gross et al., 1974) and the coyote (Wagner, in press). How vegetation changes may be related is not yet known, but they are of considerable interest to the declining sheep industry which is affected by both vegetation and coyotes.

Fig. 12.9. Close-up of a chukar partridge in degraded sagebrush semi-desert, central Nevada.

Recent studies by Clark (1979) indicate that forage availability does not control coyote population size, since even hares at maximum observed population densities consume less than 1% of the available forage.

An emerging problem in part of the type is the burgeoning population of feral horses (Cook, 1975; Hyde, 1978). Being year-round residents that prefer grasses, and under recent protection of federal law, they are exerting considerable competition for forage with domestic livestock, and even big game. The ranges they occupy are rapidly becoming the most deteriorated of any found in this type.

Recreational use

Until the past two decades, people other than ranchers, miners, and hunters were rarely visitors to this ecosystem type. Even though these areas are usually far from cities and have among the lowest densities of permanent human residents, an increasing presence of non-consumptive visitors has been noted. "Rock hounds" (seekers of rock specimens) and other users of off-road vehicles roam these deserts during the climatically more pleasant months. Sand dune areas and reservoirs attract virtual crowds at certain times. With increased human populations and greater recreational pressure in more mesic systems, this can only be expected to increase. Even retirement or second-home developments are now known in several different places within this ecosystem type.

Land development

Relatively few mineral resources have been discovered under the Great Basin portion of this ecosystem type, but exploration and its attendant disturbance is very active as our known supplies decline elsewhere. The deep mining for metals and drilling for oil, geothermal sources, and geologic water in Nevada is less disruptive than the strip-mining for coal that lies under some of this ecosystem type in Arizona, New Mexico, and Utah. Given the harsh climate and lack of irrigation water, it will be difficult to rehabilitate these areas.

THE FUTURE

An environmentally conscious generation has forced passage of legislation that is having a pronounced impact on the management of this ecosystem type. Whereas what was formerly done was largely between the rancher or miner and the B.L.M. manager, every action is now open to public scrutiny. The requirement of environmental impact statements is showing the interrelatedness of past, current, and proposed uses. The deteriorated livestock grazing capacities and wildlife habitat are becoming public knowledge (Anonymous, 1975).

The opportunities for improving the situation, however, are fading as fencing, water development, spraying, seeding, and other means of livestock control and vegetation alteration become economically untenable in this kind of ecosystem. Many of the treatments have only short-term value to livestock before brush re-invasion. When coupled with some side effects on wildlife, watersheds, and aesthetics (Vale, 1974) that a good many voters do not desire on public lands, there seems little room for compromise. Cost–benefit ratios should not merely have cows and grass in the equation.

Since most of the private land in this ecosystem type is the more productive fraction that has already been fenced, deferred, and/or had its brush controlled and grasses better managed or reseeded, further subtantive progress in increasing economic outputs lie on the public land. Some estimate that these lands could grow three to ten times more livestock forage with more intensive management (Cook and Stoddart, 1959). The problem becomes how to effect it in economically reasonable ways that do not excessively impair the other values the public now sees in these lands.

Part of what happens in the future is dependent on the demand for beef, lamb, and wool, and how much public concern for wildlife (including feral horses) and aesthetics is voiced. The technology exists to convert more of these lands, even the cheatgrass areas on gentler slopes (Evans and Young, 1977, 1978), to productive pastures. A policy decision that subsidized domestic animal production in the national interest would be needed to allow these lands to become more productive of consumer goods. Many sites occupying vast areas are so steep and rocky that contemplation of manipulative action is non-sensical. Mother Nature heals our past abuses at very slow rates in this ecosystem type. Mere removal of livestock does not result in dramatic change (Potter and Krenetsky, 1967; Rice and Westoby, 1978). No economically viable alternative use of this ecosystem type currently seems to be ready for implementation. Although logical, introduction of a semi-domesticated browser such as the saiga antelope (*Saiga tartorica*) would face many political stumbling blocks. Regardless of what is done, it is difficult to measure recovery of vegetation and soils, even after decades. Whether the currently touted panacea of rest rotation grazing will be effective in this ecosystem type is highly doubtful, because it works only where the densities of perennial grasses do not need to be increased. In this environment, even occasional heavy grazing of grasses during the spring will result in accelerated plant mortality (Cook, 1971). Seed production is low because of depleted native grass populations. Seed production and seedling establishment are very rare even on lands that have been lightly grazed.

Although the current outlook for higher economic output from this land is pessimistic, perhaps some of the readers will be challenged to provide new insight, allowing us to more rapidly improve the productivity of these highly interesting ecosystems.

REFERENCES

Anderson, L.C., 1975. Modes of adaptation to desert conditions in *Chrysothamnus*. In: H.C. Stutz (Editor), *Proc. Symp. and Workshop on Wildland Shrubs*. U.S. Forest Service, Shrub Sciences Lab., Provo, Utah, p. 141.

Anonymous, 1975. *Effects of Livestock Grazing on Wildlife, Watershed, Recreation, and Other Resource Values in Nevada.* U.S. Dep. of Interior, Bureau of Land Management, Washington, D.C., 47 pp.

Anonymous, 1979. *The Sagebrush Ecosystem: A Symposium.* College of Natural Resources, Utah State University, Logan, Utah, 251 pp.

Balph, D.F. (coordinator) et al., 1973. Curlew Valley Validation Site report. *US/IBP Desert Biome Res. Memo.*, 73–1: 336 pp.

Billings, W.D., Humphreys, M.H.K. and Darling, J.B., 1954. *Environmental Studies in the Cold Deserts and Semi-Deserts of the Western Great Basin of North America.* Project termination report to Quartermaster Research and Development Command, U.S. Army, Natick, Mass., 231 pp.

Blackburn, W.H. and Skau, C.M., 1974. Infiltration rates and sediment production of selected plant communities in Nevada. *J. Range Manage.*, 27: 476–479.

Blackburn, W.H., Eckert, Jr., R.E. and Tueller, P.T., 1969. Vegetation and soils of the Coils Creek watershed. *Nev. Agric. Exper. Sta. Bull.*, R48: 80 pp.

Branson, F.A., 1975. Natural and modified plant communities as related to runoff and sediment yields. In: A.D. Hasler (Editor), *Coupling of Land and Water Systems.* Springer-Verlag, New York, N.Y., pp. 157–172.

Branson, F.A., Miller, R.F. and McQueen, I.S., 1976. Moisture relationships in twelve northern desert shrub communities near Grand Junction, Colorado. *Ecology*, 57: 1104–1124.

Caldwell, M.M., 1979. Physiology of sagebrush. In: Anonymous, *The Sagebrush Ecosystem: A symposium.* College of Natural Resources, Utah State University, Logan, pp. 74–83.

Caldwell, M.M., et al., 1975. Gas exchange, translocation, root growth, and soil respiration of Great Basin plants. *US/IBP Desert Biome Res. Memo.*, 75–8: 44 pp.

Charley, J.L. and West, N.E., 1975. Plant-induced soil chemical patterns in some shrub-dominated semi-desert ecosystems of Utah. *J. Ecol.*, 63: 945–964.

Charley, J.L. and West, N.E., 1977. Micro-patterns of nitrogen mineralization activity in soils of some shrub-dominated semi-desert ecosystems of Utah. *Soil Biol. Biochem.*, 9: 357–365.

Christensen, G.C., 1970. The chukar partridge: Its introduction, life history, and management. *Nev. Dep. Fish Game Biol. Bull.*, 4: 82 pp.

Clark, F.J., 1972. Influence of jackrabbit density on coyote population change. *J. Wildl. Manage.*, 36: 343–356.

Clark, W.R., 1979. *Population Limitations of Jackrabbits: An Examination of the Food Hypothesis.* Thesis, Utah State University, Logan, Utah, 242 pp.

Clary, W.P., 1975. Ecotypic adaptation in *Sitanion hystrix. Ecology* 56: 1407–1415.

Comanor, P.L. and Staffeldt, E.E., 1978. Decomposition of plant litter in two western North American deserts. In: N.E. West and J. Skujins (Editors), *Nitrogen in Desert Ecosystems.* US/IBP Synthesis Series 9. Dowden, Hutchinson and Ross, Stroudsburg, Pa., pp. 31–49.

Cook, C.W., 1971. Effects of season and intensity of use on desert vegetation. *Utah Agric. Exper. Sta. Bull.*, 483: 48 pp.

Cook, C.W., 1975. Wild horses and burros: A new management problem. *Rangeman's J.*, 2: 19–21.

Cook, C.W. and Stoddart, L.A., 1959. For increased livestock production improve foothill ranges by seeding to wheatgrass. *Utah Farm Home Sci.*, 20: 5–6, 8.

Cottam, W.P. and Stewart, G., 1940. Plant succession as a result of grazing and meadow desiccation by erosion since settlement in 1862. *J. For.*, 38: 613–626.

Dahl, B.E., Sosebee, R.E., Fish, E.B. and Goen, J.P., 1976. *Range Plant and Range Ecosystem Studies Applicable to the Rio Puerco Area, New Mexico.* Final Report to the Bureau of Land Management from Range Management personnel, Texas Tech University, Lubbock, Texas, 174 pp.

Dastrup, D.B., 1963. *Vegetational Changes of the Uinta Basin Since Settlement.* Thesis, Brigham Young University, Provo, Utah, 117 pp.

Eckert, Jr., R.E., Wood, M.K., Blackburn, W.H., Peterson, F.F., Stephens, J.L. and Meurisse, M.S., 1978. Effects of surface-soil morphology on improvement and management of some arid and semi-arid rangelands. In: D.N. Hyder (Editor), *Proc. First Int. Rangeland Congress.* Society for Range Management, Denver, Colo., pp. 299–302.

Evans, R.A. and Young, J.A., 1977. Weed control–revegetation systems for big sagebrush–downy brome rangelands. *J. Range Manage.*, 30: 331–336.

Evans, R.A. and Young, J.A., 1978. Effectiveness of rehabilitation practices following wildfire in a degraded big sagebrush–downy brome community. *J. Range Manage.*, 31: 185–188.

Farb, P., 1978. *Man's Rise to Civilization From Primeval Times to the Coming of the Industrial State.* E.P. Dutton and Co., New York, N.Y., 2nd ed., 253 pp.

Fautin, R.W., 1946. Biotic communities of the northern desert shrub biome in western Utah. *Ecol. Monogr.*, 16: 251–310.

Forest-Range Task Force, 1972. *The Nation's Range Resources — A Forest-Range Environmental Study.* Forest Resource Report 19. Forest Service, U.S. Dep. Agric., Washington, D.C., 147 pp.

Frischknecht, N.C., 1978. Effects of grazing, climate, fire, and other disturbances on long-term productivity of sagebrush-grass ranges. In: D.N. Hyder (Editor), *Proc. First Int. Rangeland Congress.* Society for Range Management, Denver, Colo., pp. 633–635.

Frischknecht, N.C. and Harris, L.E., 1968. Grazing intensities and systems on crested wheatgrass in central Utah: response of vegetation and cattle. *U.S. Dep. Agric. Tech. Bull.*, 1388: 47 pp.

Frischknecht, N.C. and Baker, M.F., 1972. Voles can improve sagebrush rangelands, *J. Range Manage.*, 25: 466–468.

Gates, D.H., Stoddart, L.A. and Cook, C.W., 1956. Soil as a factor influencing plant distribution on salt deserts of Utah. *Ecol. Monogr.*, 26: 155–175.

Gross, J.E., Stoddart, L.C. and Wagner, F.H., 1974. Demographic analysis of a northern Utah jackrabbit population. *Wildl. Monogr.*, 40: 68 pp.

Haws, B.A. (coordinator), 1978. *Economic Impacts of* Labops hesperius *on Production of High Quality Range Grasses.* Final Report to Four Corners Regional Commission, Utah Agric. Exper. Sta., Logan, Utah, 21 pp.

Hugie, V.K. and Passey, H.B., 1963. Cicadas and their effect upon soil genesis in certain soils in southern Idaho, northern Utah, and northeastern Nevada. *Soil Sci. Soc. Am. Proc.*, 27: 78–82.

Hull, Jr., A.C., 1976. Rangeland use and management in the Mormon West. In: Anonymous, *Symposium on Agriculture, Food and Man — A Century of Progress.* Brigham Young University, Provo, Utah, 21 pp.

Hyde, W.F., 1978. Wild horses and allocation of public resources. *Rangeman's J.*, 5: 75–78.

Jameson, D.A., Williams, J.A. and Wilton, E.W., 1962. Vegetation and soils of Fishtail Mesa, Arizona. *Ecology*, 43: 403–410.

Kirkland, R.L., 1972. *Ecology of* Aroga websteri *Clarke in Curlew Valley, Utah–Idaho.* Thesis, Utah State University, Logan, Utah, 77 pp.

Knowlton, G.F., 1974. Insect studies in Curlew Valley of Utah and Idaho. *Utah Acad. Sci., Arts Lett., Proc.*, 51: 42–45.

Knudsen, J.J., 1976. *Demographic Analysis of a Utah–Idaho Coyote Population.* Thesis, Utah State University, Logan, Utah, 195 pp.

Küchler, A.W., 1970. Potential Natural Vegetation (map at scale of 1:7 500 000). In: U.S. Geol. Survey, *The National Atlas of the U.S.A.* U.S. Government Printing Office, Washington, D.C., pp. 90–91 (revision of 1964 map at scale of 1:3 168 000).

Lang, B.Z. and Young, J.A., 1978. Red water. *Rangeman's J.*, 5: 107–109.

Langbein, W.B. and Schumm, S.A., 1958. Yield of sediment in relation to mean annual precipitation. *Am. Geophys. Union. Trans.*, 39: 1076–1084.

Larrison, E.J. and Johnson, D.R., 1973. Density changes and habitat affinities of rodents of shadscale and sagebrush associations. *Great Basin Nat.*, 33: 255–264.

LaTourrette, J.E., Young, J.A. and Evans, R.A., 1971. Seed dispersal in relation to rodent activities in seral big sagebrush communities. *J. Range Manage.*, 24: 118–120.

Leopold, A., 1959. Big game management. In: *Survey of Fish and Game Problems in Nevada.* Nevada Legislative Counsel Bureau Bull. 36. Nevada State Printing Office, Reno, Nev., pp. 85–99.

Lesperance, A.L., Young, J.A., Eckert, Jr., R.E. and Evans, R.A., 1978. Great Basin wildrye. *Rangeman's J.*, 5: 125–127.

McAdoo, J.K. and Klebenow, D.A., 1979. Native faunal relationships in sagebrush ecosystems. In: Anonymous, *The Sagebrush Ecosystem: A Symposium.* College of Natural Resources, Utah State University, Logan, Utah, pp. 50–61.

McArdle, R.E., et al., 1936. The whiteman's toll. In: H.A. Wallace, *Letter From the Secretary of Agriculture to the U.S. Senate — A Report on the Western Range — A Great but Neglected Resource.* Senate Document 199, pp. 81–116.

McArdle, R.E. and Costello, D.F., 1936. The virgin range. In: H.A. Wallace, *Letter From the Secretary of Agriculture to the U.S. Senate — A Report on the Western Range — A Great but Neglected Resource.* Senate Document 199, pp. 71–80.

McArthur, E.D., 1979. Sagebrush systematics and evolution. In: Anonymous, *The Sagebrush Ecosystem: A Symposium.* College of Natural Resources, Utah State University, Logan, Utah, pp. 14–22.

McCormick, J., Young, J.A. and Burkhardt, W., 1979. Making hay. *Rangelands*, 1: 203–206.

MacMahon, J.A., 1976. Species and guild similarity of North American desert mammal faunas: a functional analysis of communities. In: D.W. Goodall (Editor), *Evolution of Desert Biota.* University of Texas Press, Austin, Texas, pp. 133–148.

Major, J., 1963. A climatic index to vascular plant activity. *Ecology*, 44: 485–497.

Martin, P.S., 1970. Pleistocene niches for alien animals. *BioScience*, 20: 218–221.

Ostlie, K.R., 1979. *Abundance and Dispersal of* Labops hesperius *Uhler in Relation to Vegetation.* Thesis, Utah State University, Logan, Utah, 184 pp.

Passey, H.B., Hugie, V.K. and Williams, E.W., in press. *Soil–Plant Community–Climate Relationships on Intermountain West Rangelands.* Soil Conservation Service, U.S. Dep. Agric., Washington, D.C.

Potter, L.D. and Krenetsky, J.C., 1967. Plant succession with released grazing on New Mexico rangelands. *J. Range Manage.*, 20: 145–151.

Reynolds, T.D. and Trost, C.H., 1980. The response of native vertebrate populations to crested wheatgrass planting and grazing by sheep. *J. Range Manage.*, 33: 122–125.

Rice, B. and Westoby, M., 1978. Vegetative responses of some Great Basin shrub communities protected against jackrabbits or domestic stock. *J. Range Manage.*, 31: 28–33.

Robertson, J.H., 1972. Competition between big sagebrush and crested wheatgrass. *J. Range Manage.*, 25: 156–157.

Schlatterer, E.F. and Tisdale, E.W., 1969. Effects of litter of *Artemisia, Chrysothamnus*, and *Tortula* on germination and growth of three perennial grasses. *Ecology*, 50: 869–873.

Shepard, D.F., 1977. *Prediction of Sagebrush–Grass Site Biomass From Growth Ring Analysis.* Thesis, University of Nevada, Reno, Nev., 55 pp.

Shown, L.M., Miller, R.F. and Branson, F.A., 1969. Sagebrush conversion to grassland as affected by precipitation, soil, and cultural practices. *J. Range Manage.*, 22: 303–311.

Shreve, F., 1942. The desert vegetation of North America. *Bot. Rev.*, 8: 195–246.

Skujins, J., 1980. The nitrogen cycle in desert ecosystems. *Ecol. Bull.*, 33: 477–492.

Smith, Jedediah S., 1826–1827. *The Southwest Expedition of Jedediah S. Smith. His Personal Account of the Journey to California. 1826–1827.* Edited with an introduction by George R. Brooks. Arthur H. Clark Co., Boston, Mass., 259 pp.

Stoddart, L.C., 1976. *Population Dynamics, Movement, and Home Range of Black-Tailed Jackrabbits* (Lepus californicus) *in Curlew Valley, Utah–Idaho.* Progr. Rep. for Contract AT(11-1)-1329, U.S. Energy Research and Development Administration, 89 pp.

Sturges, D.L., 1977. Soil water withdrawal and root characteristics of big sagebrush. *Am. Midl. Nat.*, 98: 257–274.

Tempte, A.K., 1977. *The Demography of the Sagebrush Defoliator* Aroga websteri *Clarke* (Lepidoptera: Gelechiidae) *and Its Impact on* Artemisia tridentata. Thesis, Utah State University, Logan, Utah, 52 pp.

Tingey, W.M., Jorgensen, C.D. and Frischknecht, N.C., 1972.

Thrips of the sagebrush–grass community of west central Utah. *J. Range Manage.*, 25: 304–308.

Tueller, P.T. and Blackburn, W.H., 1974. Condition and trend of the big sagebrush/needle and thread habitat type in Nevada. *J. Range Manage.*, 27: 36–41.

Tueller, P.T., Bruner, A.D. and Davis, J.B., 1972. Ecology of Hot Creek Valley: Vegetation and soil response to underground detonation. *Univ. Nev. Agric. Exp. Sta. Rep.*, NVO-409-1: 131 pp.

Vale, T.R., 1974. Sagebrush conversion projects: An element of contemporary environmental change in the western United States. *Biol. Conserv.*, 6: 274–284.

Vale, T.R., 1975. Presettlement vegetation in the sagebrush–grass area of the Intermountain West. *J. Range Manage.*, 28: 32–36.

Van Pelt, N., 1978. *Woodland Parks in Southeastern Utah.* Thesis, University of Utah, Salt Lake City, Utah, 83 pp.

Van Valen, L., 1965. Morphological variation and width of ecological niche. *Am. Nat.*, 99: 377–390.

Van Valen, L. and Grant, P.R., 1970. Variation and niche width reexamined. *Am. Nat.*, 940: 589–590.

Wagner, F.H., 1971. *Predator–Prey Instability and Diversity of the Curlew Valley Ecosystem.* Paper presented at the meetings of the Southwestern and Rocky Mountain Division of the American Association for the Advancement of Science, Tucson, Ariz.

Wagner, F.H., 1978. Livestock grazing and the livestock industry, In: H.P. Brokaw (Editor), *Wildlife and America.* Council on Environmental Quality, Wash., D.C., pp. 121–145.

Wagner, F.H., in press. Livestock and wildlife — fisheries ecology. In: *Proc. of Conf. on Livestock and Wildlife Fisheries Relationships in the Great Basin, Sparks, Nevada.*

Wallace, A. and Romney, E.M., 1972. *Radioecology and Ecophysiology of Desert Plants at the Nevada Test Site.* U.S. Atomic Energy Comm. TID-26954. Office of Information Services, Springfield, Va., 439 pp.

Walter, H., 1973. *Vegetation of the Earth in Relation to Climate and the Ecophysiological Conditions.* Springer-Verlag, New York, N.Y., 237 pp., (translated from the second German edition by Joy Wieser).

Weaver, J.E. and Clements, F.E., 1938. *Plant Ecology.* McGraw-Hill, New York, N.Y., 2nd ed., 520 pp.

West, N.E., 1972. Biomass and nutrient dynamics of some major cold desert shrubs. *US/IBP Desert Biome Res. Memo.*, 72-15: 24 pp.

West, N.E., 1979. Basic synecological relationships of sagebrush-dominated lands in the Great Basin and the Colorado Plateau. In: Anonymous, *The Sagebrush Ecosystem: A Symposium.* College of Natural Resources, Utah State University, Logan, Utah, pp. 33–41.

West, N.E. and Skujins, J. (Editors), 1978a. *Nitrogen in Desert Ecosystems.* Dowden, Hutchinson and Ross, Stroudsburg, Pa., 307 pp.

West, N.E. and Skujins, J., 1978b. Summary, conclusions, and suggestions for further research. In: N.E. West and J. Skujins (Editors), *Nitrogen in Desert Ecosystems.* US/IBP Synthesis Series 9. Dowden, Hutchinson, and Ross, Stroudsburg, Pa., pp. 244–253.

West, N.E., Rea, K.H. and Tausch, R.J., 1975. Basic synecological relationships in juniper–pinyon woodlands. In: G.F. Gifford and F.E. Busby, Jr. (Editors), *The Pinyon–Juniper Ecosystem: A Symposium.* College of Natural Resources, Utah State University, Logan, Utah, pp. 41–53.

West, N.E., Tausch, R.J., Rea, K.H. and Tueller, P.T., 1978. Taxonomic determination, distribution, and ecological indicator values of sagebrush within the pinyon–juniper woodlands of the Great Basin. *J. Range Manage.*, 31: 87–92.

Wood, M.K., Blackburn, W.H., Eckert, Jr., R.E. and Peterson, F.F., 1978. Interrelations of the physical properties of coppice dune and vesicular dune interspace soils with grass seedling emergence. *J. Range Manage.*, 31: 189–191.

Young, J.A., 1978. Mormon crickets. *Rangeman's J.*, 5: 193–195.

Young, J.A. and Evans, R.A., 1973. Downy brome — intruder in the plant succession of big sagebrush communities in the Great Basin. *J. Range Manage.*, 26: 410–415.

Young, J.A., Evans, R.A. and Major, J., 1972. Alien plants in the Great Basin. *J. Range Manage.*, 25: 194–201.

Young, J.A., Evans, R.A. and Tueller, P.T., 1976. Great Basin plant communities — pristine and grazed. In: R. Elston (Editor), *Holocene Environmental Change in the Great Basin. Nev. Archaeol. Surv. Res. Pap.*, 6: 186–214.

Young, J.A., Evans, R.A. and Major, J., 1977. Sagebrush steppe. In: M.G. Barbour and J. Major (Editors), *Terrestrial Vegetation of California.* Wiley and Sons, New York, N.Y., pp. 763–796.

Young, J.A., Eckert, Jr., R.E. and Evans, R.A., 1979. Historical perspectives regarding the sagebrush ecosystem. In: Anonymous, *The Sagebrush Ecosystem: A Symposium.* College of Natural Resources, Utah State University, Logan, Utah, pp. 1–13.

Chapter 13

WESTERN INTERMOUNTAIN SAGEBRUSH STEPPE[1]

N.E. WEST

DEFINITION

This ecosystem type occurs wherever there is or was a more or less equal sharing of dominance by sagebrush (woody *Artemisia*) and bunchgrasses (caespitose forms) under semi-arid, pristine conditions (Fig. 13.1). The question of what the relative amounts of brush and grass were or could be has stirred a considerable debate (Marquiss and Lang, 1959; Vale, 1975; Cramer 1976; Beetle, 1979; Hironaka, 1979; West, 1979). Actually a whole gradient of situations probably occurred and continue to be found. This type is intermediate between grassland or woodlands on more mesic sites and shrublands on effectively drier sites. Topographic and edaphic conditions that promote aridity favor the dominance of *Artemisia* and companion shrubs. Heavy livestock grazing during the growing season for grass and direct or indirect prevention of fires also favor the brush (Pickford, 1932). Therefore, the question of how much grass and how much brush can be expected must be judged on a site-by-site basis. We will necessarily generalize here on all the sites that currently grow or have potential for more or less equal mixtures of these two dominant forms. We are considering an ecosystem type that is more diverse, more productive, and more resilient than the areas discussed in Chapter 12 (Shown et al., 1969).

Fig. 13.1. Eastern edge of Carey kipuka relic area, Blaine Co., Idaho (Tisdale et al., 1965). Major species in view are *Artemisia tridentata*, *Festuca idahoensis* and *Balsamorhiza sagittata*. Photo provided by M. Hironaka, taken by M. Fosberg, 20 June 1957.

GEOGRAPHIC LOCATION

Large fractions of the Columbia–Snake River Plateau and Wyoming Basin Physiographic Provinces, as well as the northern edge of the Great Basin are covered by this ecosystem type (Fig. 13.2). Several small outliers also occur in the topo-edaphically driest portions of the Rocky Mountain and Great Plains Physiographic Provinces.

Sagebrush steppes border the Palouse Prairie in eastern Washington and northeastern Oregon, but generally occur at slightly lower and drier contours than the fescue–wheatgrass (*Festuca–Agropyron*) and wheatgrass–bluegrass (*Agropyron–Poa*) types (Daubenmire, 1970). Sagebrush steppes have boundaries on the west with various kinds of conifer forests along the east flank of the Cascade Mountains and Sierra Nevada. Western juniper (*Juniperus occidentalis*), ponderosa (*Pinus ponderosa*), and Jeffrey (*P. jeffreyi*) pines are the major tree species there (Franklin and Dryness, 1973). Ponderosa pine also occupies the next higher and wetter zone above

[1] Manuscript completed June, 1980.

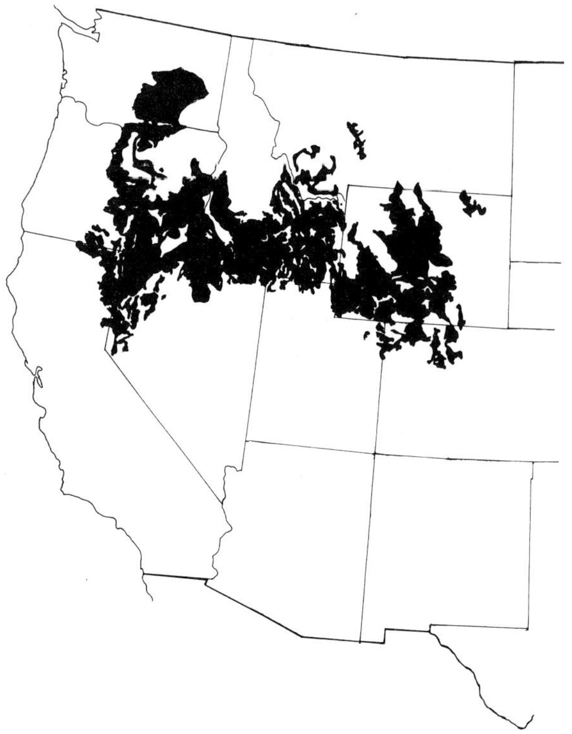

Fig. 13.2. Map of the Western Intermountain Sagebrush Steppe ecosystem type. Derived from Küchler (1970).

sagebrush steppe in northeastern Oregon and central Idaho. In the northeastern and eastern portions, the adjacent conifer forest is dominated by Douglas fir (*Pseudotsuga*) or spruce–fir (*Picea–Abies*). Sagebrush steppe intermingles with Utah juniper–single-needled pinyon (*Juniperus osteosperma–Pinus monophylla*) woodlands or mountain mahogany–oak scrub (*Cercocarpus–Quercus*) at places along its southern boundary. The salt-desert shrub type (*Atriplex–Sarcobatus*) (Ch. 14) occurs at lower elevations where salty soils have developed. Either the salt-desert shrub or Great Basin–Colorado Plateau Sagebrush Semi-Desert (Ch. 12) ecosystem types, which both have much less grass, provide most of the contact at southern boundaries. There is a transition to Great Plains grasslands where the type occurs in central Wyoming to southern Montana (Johnson, 1979).

Some mountains in north-central Nevada lack conifers altogether (West et al., 1978), and sagebrush goes from valley floor to mountain peak. Southern exposures and wind-swept slopes of the higher mountains of the Great Basin have a montane version of sagebrush–grass steppe occurring up to 3000 m (Billings, 1951; Zamora and Tueller, 1973) (Fig. 13.3).

EXTENT

This is the largest of the North American temperate semi-desert ecosystem types. The area delineated in Fig. 13.2 comprises 44.8×10^6 ha. Table 13.1 details the state-by-state contributions to this total.

Fig. 13.3. Sagebrush steppe near crestline of the Ruby Mountains (3100 m elevation), Elko Co., Nevada. The sagebrush is *A. tridentata vaseyana*.

TABLE 13.1

Area occupied by sagebrush steppes (derived from Küchler, 1970)

State	Area ($\times 10^6$ ha)	Percentage of total area
Wyoming	10.9	24.3
Idaho	10.3	22.9
Oregon	9.2	20.5
Nevada	4.7	10.6
Washington	3.8	8.5
California	1.8	4.1
Colorado	1.7	3.8
Montana	1.3	2.9
Utah	1.1	2.4
Total	44.8	100.0

CURRENT OWNERSHIP AND DOMINANT USES

Most of this type is in public ownership. However, about one-third of the area has passed into private control. This is a higher percentage of private control than for any of the other ecosystem types we will consider in this volume. These conversions were largely undertaken in the present century. Some additional opportunities for expansion of irrigated land in this ecosystem type occur in Idaho, Oregon, Wyoming and Colorado.

Dry farming is usually unsuccessful, since too many years have too little rainfall to produce grain crops where sagebrush steppe occurred before. Irrigation, however, allows all small grains, alfalfa (*Medicago*), tame grass pastures, potatoes, onions, corn (*Zea*), sugar beets and a few other temperate-zone crops to be grown. The bulk of the landscape occupied by this ecosystem type is too far from or above irrigation water and/or too rocky for cultivation. The dominant use on these wildlands has historically been and continues to be livestock grazing.

ECOSYSTEM STRUCTURE AND FUNCTION

Climate

The climate for this ecosystem type is semi-arid, temperate and continental. The variations in the basic elements of climate over the type are depicted in Fig. 13.4.

The predominant air masses moving over this region, especially in the winter, originate over the Pacific Ocean. The lack of water vapor and cloudi-

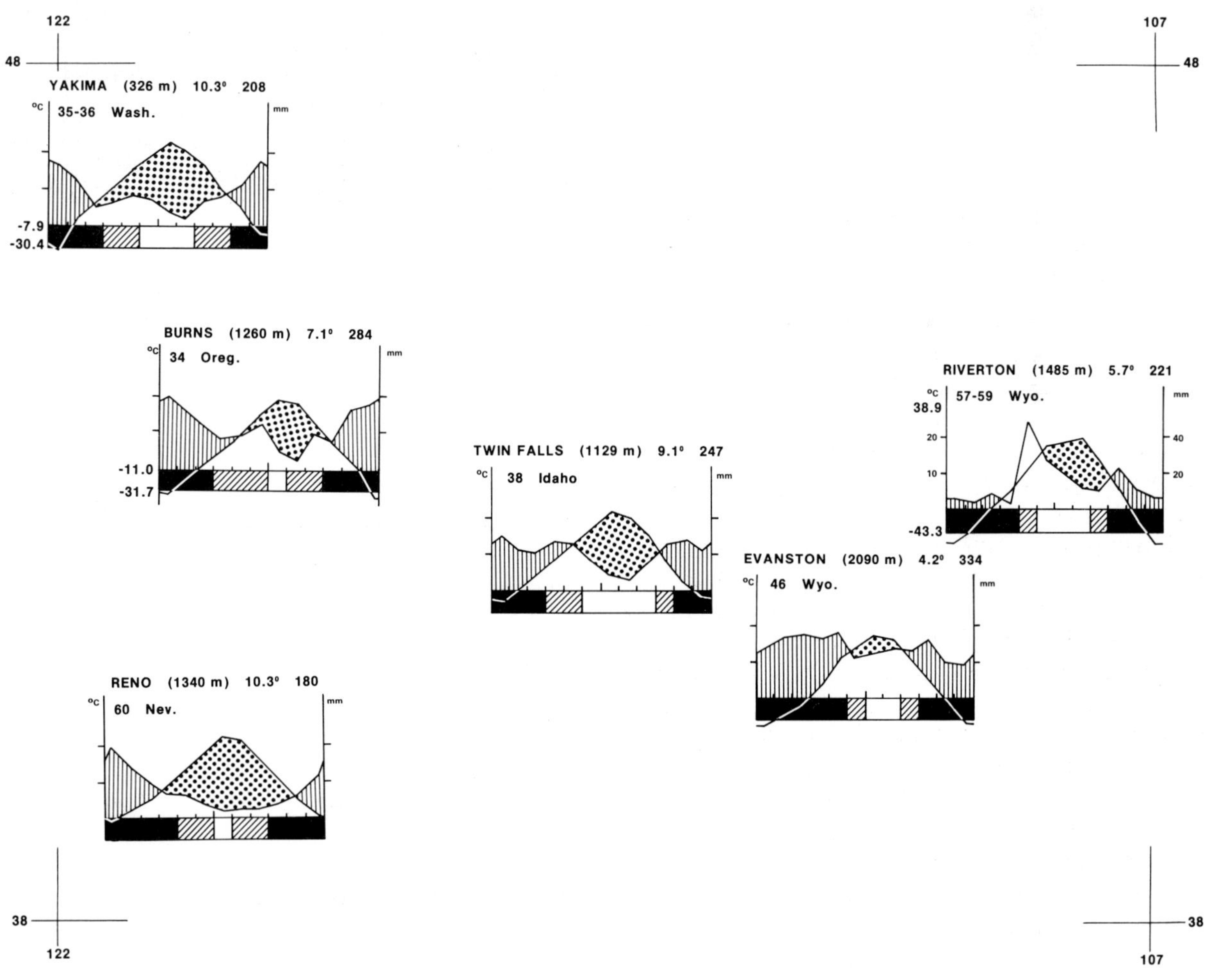

Fig. 13.4. Climatic diagrams for representative stations within the Western Intermountain Sagebrush Steppe ecosystem type. Data from National Weather Service network.

ness during the summer permits intense heating during the day and rapid radiation cooling at night. These conditions result in large daily ranges of air temperature. Except for the portions in Wyoming and Montana, mountain barriers to the north prevent most cold, polar air masses from reaching the area in winter. The primary source of water vapor most of the year is the Pacific Ocean, with low-pressure centers moving across the area and causing rain and snow. Convergence of moisture from air masses derived from the Gulf of Mexico contributes to a late spring–early summer peak of precipitation in the eastern part of the region. Summers are very dry in the western extremities of the type. Occasional hail or rain from summer convectional storms occurs, but is rarely effective in recharging soil moisture (Rawls et al., 1973). The total annual precipitation for the selected stations averages 246 mm. The average coefficient of variation is 20%.

The usual ratio is 10:1 snow to rain. Much of the moisture in snow is lost as vapor from the system via sublimation and thus the environment is effectively more arid than would be expected from precipitation data (May et al., 1971).

The average mean temperature of the stations as depicted in Fig. 13.4 is 7.8°C. Mean of the warmest month is 20.4°C. Mean of the coldest month is −2°C. Extreme temperatures recorded have been −46 and 39°C. Diurnal temperature variations can approach 35°, particularly in the autumn. The higher elevation stations have recorded freezing temperatures in all months.

The average length of the frost-free season is 120 days, but is reduced to 80 days at certain mountain sites.

Although wind data are rarely taken, wind is known to be frequent and often of high velocity in this ecosystem type. For instance, at the Idaho National Engineering Lab in southern Idaho, an average wind velocity of 12 km h^{-1} has been recorded. The spring season is particularly windy. A 15-year average of 15 km h^{-1} rate was recorded in May–June at the above station (Yansky, 1966). Wind greatly adds to potential evaporation which exceeds precipitation by a factor of five or more during the frost-free season. Actual annual evapotranspiration usually equals precipitation (Campbell and Harris, 1977), thus there is little water yield from this ecosystem type except with convectional storms, rain on snow, or rapid snowmelt over frozen soils.

Geology and topography

The bedrock under most of this ecosystem type is layer upon layer of Middle and Late Cenozoic basalts and related volcanics. The only significant exceptions are the sagebrush-steppe areas of the Wyoming Basin and Rocky Mountain Provinces which are largely underlain with Tertiary to Recent sedimentary deposits of alluvial or lacustrine origin (Hunt, 1974).

The strata in both cases are nearly horizontal, giving rise to plains topography. In the case of the lava-underlain topography, it can be rolling (Fig. 13.1) with volcanic cones protruding. The landscapes developing from fluvial deposition are flatter (Fig. 13.5) with pediments often at several levels. Shallower soils developing on low hills and ridges also support sagebrush-steppe ecosystems. Even wind-swept high ridges and peaks in the region occur as sagebrush-steppe islands within a woodland or forest matrix.

Prevailing elevations in the type vary from less than 150 m along the Columbia River between Oregon and Washington to about 2200 m in the Wyoming Basin. The Great Basin and Snake River Plains portions occur mostly at elevations of 500 to 2000 m.

Fig. 13.5. A variant of the Western Intermountain Sagebrush Steppe ecosystem type in the Shirley Basin, Carbon Co., Wyoming. Photo by K.L. Johnson.

Soils

The soils associated with this ecosystem type vary from deep loessal and alluvial pockets in aggrading micro-topography to skeletal profiles on ridgetops. The taxonomic placement varies from Ustolls and Xerolls to Orthids and Argids. Inceptisols (Andepts) derived from volcanic ash are conspicuous on the Columbia Plateau.

A considerable amount of work has been done on soil–vegetation–productivity relationships (Eckert, 1957; Thatcher, 1959; Passey and Hugie, 1962a, b, 1963a, b; Hironaka, 1963; Fosberg and Hironaka, 1964; Hugie and Passey, 1964; Hugie et al., 1964; Passey et al., 1964, in press; Tisdale et al., 1969; Rumsey, 1971; Sabinske and Knight, 1978). All of these studies have pointed to the vast differences in plant community composition and productivity that can be found on different soils, all within the same macro-climate. Soils are very important in interpreting potential vegetation where the original vegetation has been disturbed (Hironaka, 1979).

Polygonal surface patterns are common in the shrub interspaces. Coarse particles may be sorted to the edge of the polygons, and plant seedlings concentrated in the cracks. Miniature terraces may develop on sloping sites. These patterns are apparently produced by cycles of alternate freezing and thawing of the surface soil followed by desiccation (Hugie and Passey, 1964). Prominent vesicular pores often develop in dry surface horizons.

Primary producers

The floristic diversity of this vegetation type is moderate (Hironaka, 1979; Johnson, 1979; West, 1979). Daubenmire (1975a) found an average of twenty species in 1000 m^2 on relic sites in central Washington. Tisdale et al. (1965) found from 13 to 24 species in examples of each of three community types on an ungrazed area in southern Idaho. Zamora and Tueller (1973) found 54 vascular plant species occurring in a selection of 39 high-condition stands in northern Nevada.

The vertical and horizontal structure of vegetation on all these sites is remarkably similar (Fig. 13.6). There is a shrub layer reaching 0.5 to 1.0 m in height. The shrubs commonly have a cover of about 10 to 80%, depending on site and successional status. The grass and forb layer reaches about 30 to 40 cm high during the growing season. The herbaceous cover varies more widely, between 0 to 100% depending on site and successional status. The sum of the cover figures, species by species, on relictual sites usually exceeds 80% and can approach 200% on the most productive sites (Daubenmire, 1970).

The herbs on relic areas are largely hemicryptophytes (Daubenmire, 1975a). The proportion of therophytes (annuals) has increased considerably with disturbance (Tueller, 1962). The proportion of geophytes approaches 20%. A microphytic cover in the interspaces between the vascular perennials is often present.

There is considerable sorting of species against environmental gradients. Since the sagebrushes so dominate the biomass, they have been extensively studied (Beetle, 1960, 1979; McArthur, 1979). This is a rapidly evolving group that shows radiation into particular niches. Fig. 13.7 attempts to summarize our current understanding of how the major sagebrush taxa usually segregate along major environmental gradients in this ecosystem type.

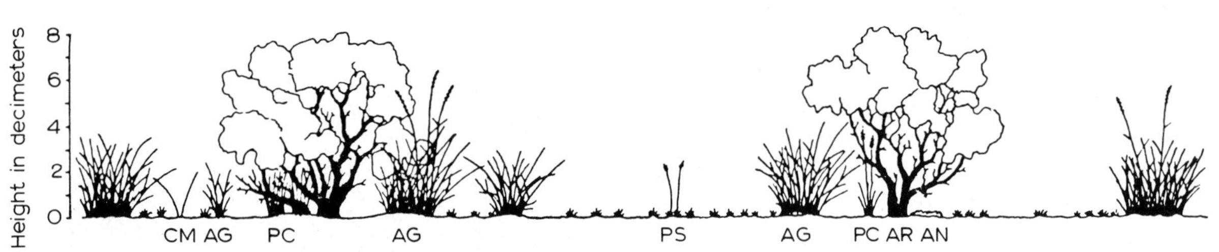

Fig. 13.6. A bisect of *Artemisia tridentata–Agropyron spicatum* relic stand, Yakima Co., Washington (from Daubenmire, 1970). Drawings to scale including all vascular plants with any basal area impinging on a transect 2 cm wide by 400 cm long. Key to species symbols: *AG* = *Agropyron spicatum*; *AN* = *Antennaria dimorpha*; *AR* = *Artemisia tridentata*; *CM* = *Calochortus macrocarpus*; *PC* = *Poa cusickii*; *PS* = *Poa secunda*.

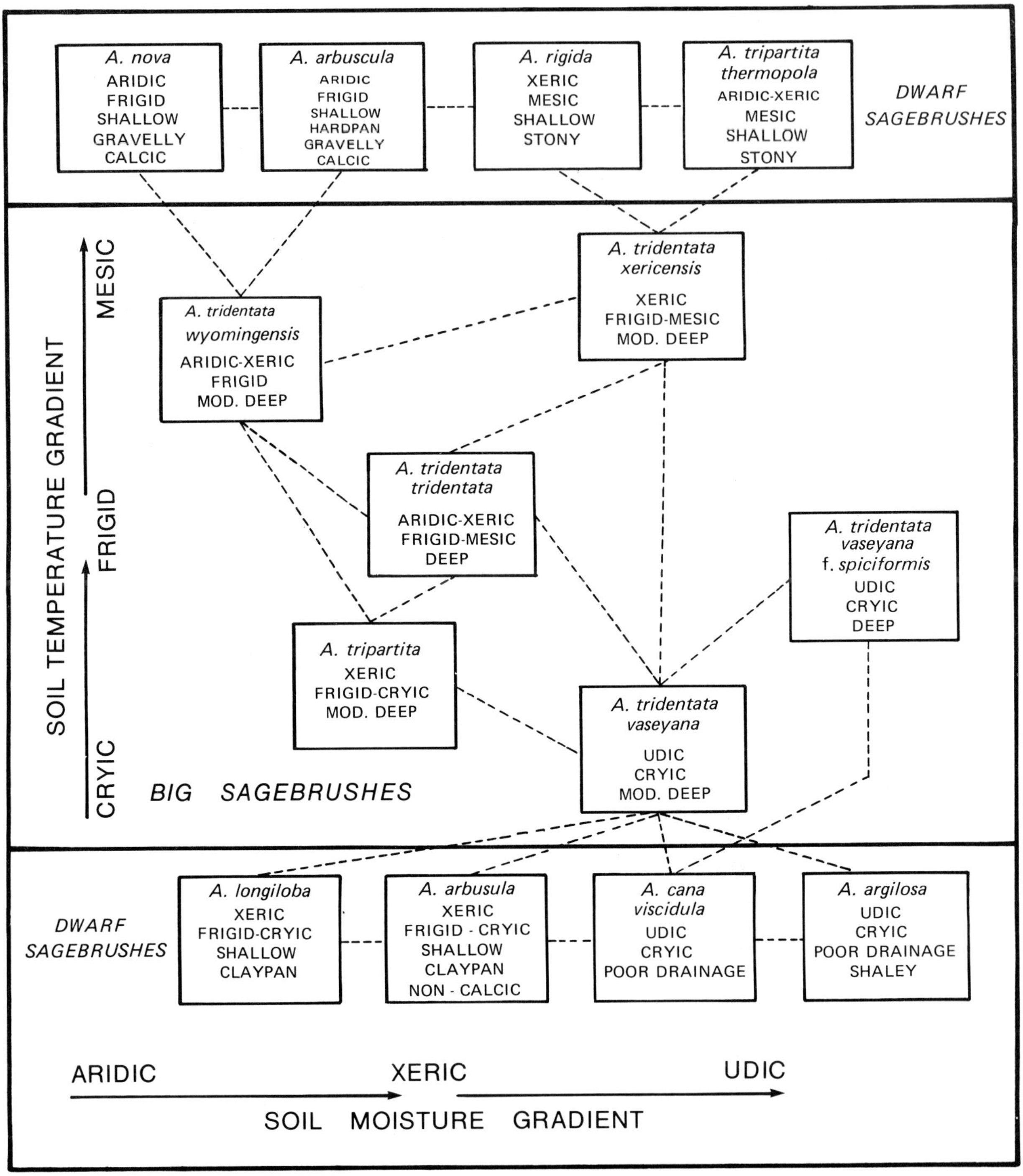

Fig. 13.7. Ordination of major sagebrush taxa against gradients of temperature and aridity. Adapted from Hironaka (1979). *A. argilosa* added from Robertson et al. (1966). For explanation of terms, see Soil Survey Staff (1975).

The ecology of this and related ecosystem types cannot be understood in any detail until the various species of sagebrush are recognized and related to their environmental preferences. Subspecies of *Artemisia tridentata* also have decided environmental preferences and utility. They should be recognized if a full interpretation of conditions is to be accomplished.

The success of sagebrush is due to many factors, not the least of which is its seasonal leaf dimorphism. That is, it has two distinct kinds of leaves adapted to different seasons (Daubenmire, 1975b; Goodwin, 1956; McDonough et al., 1975). There are large, ephemeral leaves that develop in the spring and remain on the plant until soil moisture stress develops in the summer. Smaller, persistent, over-wintering leaves develop in late spring but last through the winter, carrying on photosynthesis through the winter months (Caldwell, 1979).

Big sagebrush (*A. tridentata*) has both a fibrous root system that can feed near the surface and a taproot that can draw moisture and nutrients from deep in the profile (Sturges, 1977). However, it does not survive if flooding creates anaerobic conditions (Caldwell, 1979). Some other species of sagebrush, such as *A. arbuscula* and *A. cana*, may owe their dominance more to occasional anaerobic conditions due to flooding than any other factor (Passey et al., in press).

The perennial grasses found associated with sagebrushes vary considerably. *Festuca idahoensis* and other species in this genus are common in the northwestern portion of the type and at the highest altitudes and latitudes elsewhere. *Agropyron spicatum* is probably the most widespread and important herbaceous component. Sod-forming *Agropyron* species (*A. dasystachyum*, *A. smithii* and others) become more important in the eastern portion of the type, where greater growing-season precipitation occurs. *Stipa* species are important along the southwestern boundary of the type, with *S. thurberiana*, *S. arida* and *S. speciosa* sorting out along a gradient of decreasing elevation (Young et al., 1977).

The perennial grasses have only fibrous roots and limited storage of carbohydrate reserves. They must rebuild all of their above-ground tissues during the short period in late spring and early summer when favorable temperatures and adequate soil moisture exist. Perennial forbs and the geophytes have the same constraints, but can store more reserves in their fleshy underground organs.

Soil moisture (Fig. 13.8) becomes limiting to and truncates the phenology of the shallow-rooted herbs far earlier than for the deeply rooted shrubs (Fig. 13.9). Summer precipitation is usually ineffective at recharging the deeper part of the soil profile, thus perennial grasses are only slightly more favored in this ecosystem type than they are in the Great Basin–Colorado Plateau sagebrush semi-desert (c.f. Ch. 12).

The total phytomass of some representative sites varies from about 2 to 10 t ha^{-1}. About half of the total occurs below ground. Above-ground phytomass in older stands frequently shows only about 15% current annual growth.

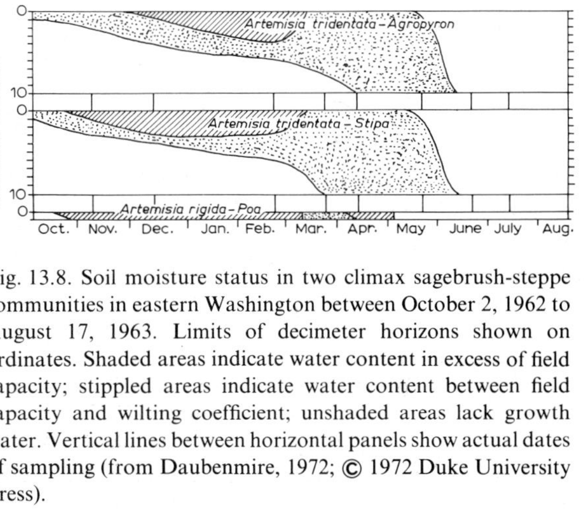

Fig. 13.8. Soil moisture status in two climax sagebrush-steppe communities in eastern Washington between October 2, 1962 to August 17, 1963. Limits of decimeter horizons shown on ordinates. Shaded areas indicate water content in excess of field capacity; stippled areas indicate water content between field capacity and wilting coefficient; unshaded areas lack growth water. Vertical lines between horizontal panels show actual dates of sampling (from Daubenmire, 1972; © 1972 Duke University Press).

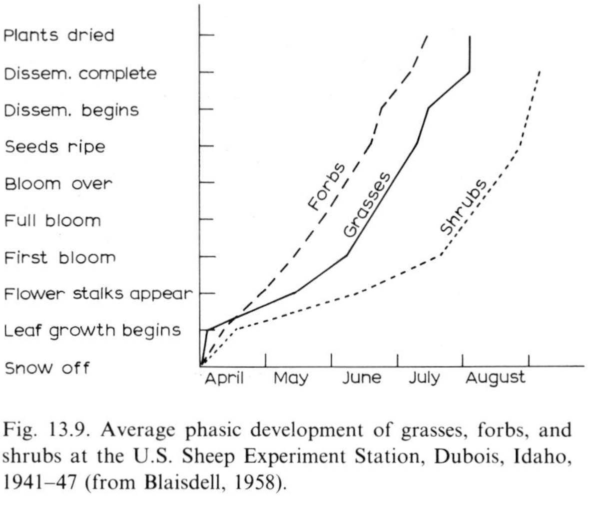

Fig. 13.9. Average phasic development of grasses, forbs, and shrubs at the U.S. Sheep Experiment Station, Dubois, Idaho, 1941–47 (from Blaisdell, 1958).

Sagebrush may contribute up to 70% of the total phytomass, even on sites where livestock grazing has been light or absent. The dominance is even greater where livestock grazing has reduced the perennial grass component.

Sagebrushes are long-lived once they are beyond the seedling stage. West et al. (1979) have calculated mean life spans of about four years for seedlings of *Artemisia tripartita* that have survived their first year on the Snake River Plains of southeastern Idaho. The maximum longevity exceeds forty years. *A. tridentata* lives even longer (Ferguson, 1964).

The accumulation of standing dead material and litter can reach about 1700 kg ha^{-1} (Murray, 1975). Above-ground litter production is of the order of 150 to 800 kg ha^{-1} yr^{-1}, but highly variable, largely due to the great developmental plasticity of sagebrush.

Above-ground net annual primary productivity on representative sites varies between 800 and 2500 kg ha^{-1} yr^{-1} (Pearson, 1965, 1966; Daubenmire, 1975a). The proportion of this productivity that is herbaceous is highly variable depending on site, successional status, shrub age distribution, grazing history and preceding climatic conditions. Sites that have had periodic burns which reduced the brush show much more herbaceous productivity. Light grazing has also been found to stimulate productivity (Pearson, 1965).

Productivity of the herbaceous components, even on relict areas, varies strongly with climatic conditions (Passey and Hugie, 1963a; Passey et al., 1964). A 300% variation in end-of-season peak above-ground standing crop has been noted (Passey and Hugie, 1963b). The species contribution to total standing crop also varies considerably depending on the climatic pattern of the preceding year or so (Blaisdell, 1958; Passey et al., in press).

Efforts to quantitatively relate total net annual above-ground primary productivity to meteorological conditions have met with mixed success (Isaac, 1974; Passey et al., in press). This is likely due to differing effectiveness of precipitation infiltrating the soil profile. In addition to the sublimation losses from snow, snowmelt and/or rainfall on an already frozen soil results in a high percentage of runoff. Summer convectional rains rarely penetrate deep enough into the profile for evaporation losses to be reduced.

Therefore, it is spring precipitation as rain or light, gently melting snow that often shows the best correlation with plant growth, even for shrubs (Fetcher and Trlica, 1980). Attempts to predict productivity of grasses of this ecosystem type on the basis of climate data of preceding months have been quite successful (Sneva and Hyder, 1962; Sneva, 1977). Cool, wet springs are more productive for grass than warm, dry ones.

Class of livestock, season of use and palatabilities of the various species have to be taken into account if phytomass data are interpreted for livestock grazing capacity. Largely because of high content of secondary plant compounds (phenolics, coumarins, etc.), foliage of sagebrush is usually unutilized if livestock have anything herbaceous to eat. The best sites for total productivity are also the sites with the highest terpenoid contents of sagebrush (Powell, 1970). Sheep will utilize some sagebrush in the fall when herbs are dry and low in protein. Cattle will eat some brush during extremely cold, snow-covered conditions.

Original livestock grazing capacity was estimated to average 0.83 AUM ha^{-1} (McArdle and Costello, 1936). Because of the loss of palatable herbaceous species and an increase in shrubs and annuals, the grazing capacity was judged to have declined between 60 and 90% by the early 1930s (McArdle et al., 1936). Thus, 37 to 62 cattle, or five times as many sheep, could have been maintained on one square kilometer fifty to sixty years ago (Aldous and Shantz, 1924). The average grazing capacity was judged to be 0.28 AUM ha^{-1} in 1970 (Forest-Range Task Force, 1972).

To estimate livestock grazing capacities of specific sites under good management, it is best to use a long-term series of end-of-growing-season values for herbaceous standing crops on examples of range in excellent condition. Passey et al. (in press) provide some valuable numbers in this regard. By relating these values to soils and climatic information elsewhere, one can judge how much the current productivity diverges from what could be.

Consumers

The native animals using this ecosystem type are a mixture of grassland and desert species. Table 13.2 lists the vertebrates occurring on a rather dry variant of the type in southern Idaho.

TABLE 13.2

List of vertebrates occurring on a sagebrush-steppe site in southeastern Idaho

Scientific name	Common name
MAMMALS (MacMahon, 1976a)	
Antilocapra americana americana	pronghorn
Canis latrans lestes	coyote
Dipodomys ordii columbianus	Columbian kangaroo rat
Eutamias minimus pictus	least chipmunk
Lepus californicus deserticola	black-tailed desert jackrabbit
Myotis californicus californicus	California little brown bat
Odocoileus hemionus hemionus	mule deer
Perognathus parvus idahoensis	Uinta pocket mouse
Peromyscus maniculatus sonoriensis	Great Basin white-footed mouse
Reithrodontomys megalotis megalotis	Great Basin harvest mouse
Spermophilus townsendii artemisiae	sagebrush ground squirrel
Sylvilagus idahoensis	pygmy cottontail
Sylvilagus nuttallii grangeri	Black Hills cottontail
Taxidea taxus taxus	badger
BIRDS (Trost, 1976)	
Amphispiza belli	sage sparrow
Asio flammeus	short-eared owl
Bubo virginianus	great horned owl
Buteo regalis	ferruginous hawk
Chordeiles minor	nighthawk
Circus cyaneus	marsh hawk
Eremophila alpestris	horned lark
Falco mexicanus	prairie falcon
Falco sparverius	kestrel
Molothrus ater	brown-headed cowbird
Oreoscoptes montanus	sage thrasher
Salpinctes obsoletus	rock wren
Spizella breweri	Brewer's sparrow
Sturnella neglecta	western meadowlark
Zenaidura macroura	mourning dove
REPTILES (MacMahon, 1976b)	
Crotalus viridis lutosus	Great Basin rattlesnake
Eumeces skilitonianus utahensis	Great Basin skink
Phrynosoma d. *douglassi*	pygmy horned lizard
Pituophis melanoleucas deserticola	Great Basin gopher snake
Sceloporus g. *graciosus*	northern sagebrush lizard
Thamnophis elegans vagrans	wandering garter snake

Many more species would be expected on more mesic sites, particularly at higher elevations (McAdoo and Klebenow, 1979). Many of the same species found in the Great Basin–Colorado Plateau Sagebrush Semi-Desert ecosystem type (Ch. 12) are also found here, but in greater abundance. This is probably due to the higher levels of primary productivity and greater availability of surface water. There are additional animal species, probably because of the greater diversity of shrubs, and an increased proportion of herbaceous plants.

The kinds and numbers of large animals were probably greater at the end of the Pleistocene, but many species became extinct over the succeeding years. P.S. Martin (1970) considered that it was more than coincidence that elephants, mammoths, tapirs, camels, horses, various species of bison, and ground sloths occurred in this area, but became extinct when evidence of man's occupation increased. He believed that the new predator was too effective for all but the fleeter and more wary species.

Only small populations of bison (*Bison bison*) and pronghorn (*Antilocapra americana*) were found when the early trappers and explorers first traversed the area in the 1820s. They also found that the Bannock and Shoshoni peoples had obtained the horse reintroduced by the Spanish explorers (Ogden, 1950 and 1971). The abundance of bison, pronghorn and the horse was, however, much lower than those found on the plains to the east of the Rocky Mountains (Buechner, 1953).

The ecosystem type still serves as habitat for mule deer (Fig. 13.10), wapiti (*Cervus canadensis*), and Rocky Mountain bighorn sheep (*Ovis canadensis*). Although a few year-round residents can be found, these animals have traditionally moved down from summer ranges in the surrounding mountains to the lower elevations to utilize forage where less snowpack accumulates.

Feral horses compete with big game animals and livestock at some places in this type. The dietary overlap with cattle and bighorn sheep is greatest because of their strong preferences for grasses.

The most abundant vertebrates in this ecosystem type are the rodents and lagomorphs. Among these, the black-tailed jackrabbit (*Lepus californicus*) and pigmy cottontail (*Sylvilagus idahoensis*) are the most conspicuous. Coyote (*Canis latrans*), the chief predator, has population trends which lag behind the swings in rabbit populations (Fig. 12.7).

Rodents such as Townsend's ground squirrel (*Spermophilus townsendii*) can also be a major herbivore on some sites within this ecosystem type. They and species such as the Great Basin pocket mouse (*Perognathus parvus*), sagebrush vole (*Lagurus curtatus*), and the western harvest mouse (*Reithodontomys megalotis*) may have declined in abundance as the perennial grass component declined with livestock grazing and fire control (Larrison and Johnson, 1973; O'Farrell et al., 1975; Becker, 1978). Voles (particularly *Microtus montanus*) may have population eruptions which cause temporary reductions in sagebrush (Mueggler, 1967).

Fig. 13.10. Large male mule deer using sagebrush steppe as a winter range.

Soil textures and structures directly influence the composition and abundance of the rodent populations living in these ecosystems by affecting the ease in which they can construct and maintain burrows. An indirect effect of plant community composition and architecture can also be noted (Montgomery, 1976).

The various small mammals also divide up the resources on a seasonal basis (O'Farrell, 1975). *Perognathus* is most active in the period between spring and fall, showing torpor in the winter. *Peromyscus* favors the fall-to-spring period, using hypothermia during the driest part of the summer and mid-winter.

The avian predators, particularly raptors such as owls and small hawks, may have declined as their food-base shrank (Howard and Wolfe, 1976). Larger hawks and golden eagles (*Aquila chrysaetos canadensis*) have probably been favored when more jackrabbits became available.

The sagegrouse (*Centrocercus urophasianus*) (Fig. 13.11) is a large game bird which reaches its greatest abundance in this type. Adults have a unique gastro-intestinal tract that allows them to feed on sagebrush leaves and buds. The young require insects and forb buds and thus must spend the early part of their lives in moist meadow areas. In addition, open "booming" grounds (leks) are required for sexual display and mating.

Smaller birds occur in this type with surprising abundance. Table 13.2 includes a list of the species present at a sagebrush–grass site in southeastern Idaho. The species composition and relative abundance of this part of the biotic community may also have changed as the vegetation has been altered by man and his livestock (McAdoo and Klebenow, 1979).

Fig. 13.11. Male sagegrouse on spring strutting grounds (leks).

Amphibians are absent from the uplands. Reptiles are less diverse here than in the southern deserts (Table 13.2). None of these are restricted to just this ecosystem type.

Over 1000 species of insects have been observed on a sagebrush–grass site in southern Idaho (Bohart and Knowlton, 1976). If known, spiders mites, ticks (Allred, 1968) and scorpions would probably add several hundred additional species to this list.

Relatively little is known about the habitat preferences, trophic relationships, and other aspects of the roles of invertebrates in this ecosystem type. Only a few have been noted to make drastic changes in the vegetation. These include tent caterpillars (*Malacosoma*) on *Purshia tridentata* and *Prunus* and thrips on sagebrush (Tingey et al., 1972). The sagebrush webworm (*Aroga websteri*) can make dramatic reductions in sagebrush, especially in dry years (Gates, 1964; Hall, 1965; Tempte, 1977). Psyllids can also sometimes reduce sagebrush (Balph, 1973). Chrysomelid beetles also occasionally cause defoliation of big sagebrush (Halford et al., 1973). Sagebrush can show abundant effects of galls from flies and wasps (Fronk et al., 1964). Kamm et al. (1978) found that root aphids and root coccids were the most abundant phytophagous insects on native sagebrush rangelands in eastern Oregon. They also found that more phytophagous insects were present on range under a managed livestock grazing system than where livestock had been excluded for 37 years.

Grasshoppers have been shown to have some impact on both the shrub and herbaceous components of sagebrush–grass vegetation (Allred, 1941; Sheldon and Rogers, 1978). Cicadas (Cicadidae) are abundant and have a role in soil formation on some sagebrush-steppe sites (Hugie and Passey, 1963). The vast populations of these larvae are bound to take a sizeable fraction of the root production.

Harvester ants are abundant, especially on degraded sites. They collect and move below-ground vast quantities of seeds and leaves. Clark and Comanor (1975) estimated that annual plants were cut at the rate of 157–226 million ha^{-1} yr^{-1} by *Pogonomyrmex occidentalis* on sagebrush-steppe sites near Reno, Nevada. Sneva (1979) has shown how colony density decreased, but the denuded disc area increased as big sagebrush crown cover increased and as herbaceous productivity decreased. Termites harvest the decaying wood of sagebrush and other shrubs and move the nutrients below ground. The conspicuous darkling beetle populations (Rickard et al., 1974; Rogers and Rickard, 1975) also have a role in comminuting litter and accelerating nutrient cycling.

Soil-inhabiting nematodes have been enumerated in an example of this ecosystem type of southeastern Washington (Smolik and Rogers, 1976). Phytophagous forms prevailed, and there were no consistent differences in density or biomass values between grazed, ungrazed or burned areas.

Decomposers

Very little study of the decomposition process has been done in this ecosystem type. Comanor and Staffeldt (1978) summarize what little is known. They noted that surface litter decomposition was rapid during the cool, wet months before above-ground higher plant growth becomes active. Turnover times of material are not much slower than those observed in more mesic woodlands and forests (Mack, 1971; Murray, 1975).

Root litter decomposition occurs mainly during the late spring and early summer when higher temperatures are convergent with adequate soil moisture (Wildung et al., 1975).

The secondary compounds in *Artemisia* litter can limit the processes of nitrogen fixation and nitrification under laboratory conditions (Skujins, 1980). Since nitrogen availability is a limiting factor to primary production when two years wetter than average occur in succession, this could possibly be another indirect means by which *Artemisia* strengthens its dominance in the plant community.

Ash-forming elements are thought to be seldom limiting to plant growth in these environments, but lag times are accentuated by the longevity of sagebrush and slow decomposition of its woody litter (Mack, 1971). Grazing and fire serve to accelerate the recycling of ash elements, but result in gaseous losses of nitrogen and sulfur (Murray, 1975). Whether nutrient losses due to livestock weight gains and conversion of native stands to monocultures of exotic grasses are of near-term concern will require more investigation (Mayland and Murray, 1979).

ECOSYSTEM INTERACTIONS

This ecosystem type has decided vertical and horizontal structural organization. The soil is where over half of the biomass occurs, and thus where most of the total metabolic activity takes place. Plant competition is probably keenest for soil moisture, although *Artemisia* is also known to produce allelochemic substances that reduce seed germination (Hoffman and Hazlett, 1977), seedling growth (Schlatterer and Tisdale, 1969) and root respiration of its competitors (Weaver and Klarich, 1977) and activity of soil microbes (Weaver and Klarich, 1976; West and Skujins, 1978). The most important struggle is between shrub and herbaceous forms. The herbs have to rebuild all their above-ground tissues each growing season. This requires a lot of food reserves and extraction of water and nutrients. Temperatures and soil moisture are most favorable for this during the time following shortly after the period of snowmelt. Growing season precipitation is less dependable for soil moisture recharge than snowmelt, temperatures contribute to greater evapotranspirative losses. The shrubs, to be successful, only have to rebuild, on the average, enough new tissue to replace that which was shed as litter in the previous year. Their photosynthetic tissue is active even through the winter, albeit at marginal rates (Caldwell, 1979). Sagebrush puts out its spring leaves for the favorable spring period. The plants are thus set to put on a flush of growth with comparatively less expenditure of metabolic effort. They can extend growth longer (Fig. 13.9) because they draw moisture from deep within the soil profile. Furthermore, content of secondary compounds in sagebrush is high, probably discouraging herbivory.

The grass species have tissues of high succulence and nutrient content, but little, if any, mechanical or chemical protection. Most of the forbs, however, have some chemical protection against herbivory, at least against some insects. Many have fleshy roots that provide more food reserves should drought or herbivory affect their growth during a particular year. The constrained period of growth for grasses, their lack of protection against and desirability to herbivores, have thus set up a situation where brush has nearly all the advantages when livestock are imposed on the system. Only when fire occurs does grass have temporary advantage over sagebrush. Sagebrush is killed by fire, does not resprout, and must cast seed to reestablish itself. The perennial herbs are not usually killed if fire, as is usual, comes during the dormant season.

The plants partition out the environment in space and time. The herbs use the surface soil of the interspaces between the shrubs when resources are abundant in the spring. Sagebrush can draw its moisture and nutrients from deep in the profile or through fibrous roots near the surface (Robertson, 1943; Daubenmire, 1975b; Sturges, 1977), depending on the occupancy of the community. The shrubs exhibit extreme flexibility and resistance against all environmental extremes, except fire and occasional insect attack. By their extreme abundance, long lives, and consistent seed set, they exhibit great reproductive capacity. In sum, they are extremely competitive in this environment. Any environmental perturbation that diminishes growth or reproduction of forbs or grasses allows sagebrush and companion shrubs to gain a further degree of dominance in this environment. With periodic fire, dominance concentration is relatively well distributed (Daubenmire, 1975a); without any fire, excessive fire and/or livestock grazing, diversity falls and phytomass becomes more concentrated in fewer plant species. The food web may also become truncated.

The pristine system was apparently only weakly stable. The perturbations triggered by the introduction of livestock and some European weeds have caused it to seek new equilibria (discontinuously stable, according to Hanley, 1979). The major pathways of successional change are similar to those of the sagebrush semi-deserts illustrated in Fig. 12.8, although resilience is greater in this type. The rates of change are only relatively faster. Primary succession is slow on absolute scales since rock must be weathered to soil or aeolian materials deposited (Eggler, 1941). Secondary succession on highly disturbed areas is much slower than in old fields in more mesic environments (Hironaka and Tisdale, 1963).

Cheatgrass (*Bromus tectorum*), and in some places with heavier soils medusa-head (*Taeniatherum asperum*), produce vast numbers of seedlings after the first fall rain. Growth, particularly that of their root systems, goes on throughout most of the winter. This is because the roots of annual grasses can grow as long as soil temperatures are above 0°C, whereas bluebunch wheatgrass (*Agropyron spicatum*) roots only grow above a threshold of 4°C (Harris, 1967, 1977). The annuals have built an extensive root system by the time spring comes, and are extracting soil moisture earlier and at higher rates than the native grasses. The annuals have a compressed phenology. They generally dry out and cast their seeds by mid-June. The dry swards then become very susceptible to wild fire. The fires kill the sagebrush. If fires are not too frequent, rabbitbrush (*Chrysothamnus* spp.) and/or horsebrush (*Tetradymia* spp.), which are resprouters, may occupy the site in a few years. If fire is quite frequent, all perennials may give way to a community dominated by annuals. These annual communities have much greater variation in annual primary production than the perennials. The period of green, nutritious feed is also much shorter and the herbage cures less well on the stem. The result is a much less desirable vegetation for livestock production and watershed protection. Cattle may be forced to utilize this forage and even make gains (Murray and Klemmedson, 1965; Murray, 1971); however, perennials would be preferred because of their greater dependability in herbage production. If the more dependable productivity of perennial grasses is desired, then rather drastic management manipulations are required.

LAND-USE HISTORY

Aboriginal use

Aboriginal populations in this ecosystem type were low (Farb, 1978). The Shoshoni peoples who occupied this area in pre-Columbian times were hunter–gatherers of simple cultural patterns. Their impact on the land was probably almost imperceptible. Acquisition of the horse in the late 1700s led to adoption of some of the Plains Indian style of life, but the bison populations were low and exterminated by the 1850s (Farb, 1978).

Impacts of European-derived cultures

Livestock use

Intensive use of these rangelands only began in earnest with white man's agricultural colonizations after 1847 (Young et al., 1979). Archeological (Butler, 1976) and palynological (Davis et al., 1977) evidence indicates that the introduction of domestic livestock had more effect on the environment than any event in the previous 1000 years. Livestock populations built up rapidly in the late 19th century until grazing capacities of these ranges were exceeded around the turn of the century (Griffiths, 1902). The loss of the native perennial grass and expansion of the native shrubs took only ten to fifteen years (Hull, 1976). Loss of soils took longer (Robertson and Kennedy, 1954). The expansion of cheatgrass followed around the turn of the century, but it did not become dominant at many sites until about 1950 (Young et al., 1976). Medusa-head grass invasion on heavy soils has been even more recent (Young and Evans, 1970). The rapidity of vegetation change was due to the ecophysiological and reproductive disadvantages of the native perennial grasses and the great competitive advantages of native shrubs and introduced annual grasses in this kind of environment, as detailed above.

The high nutritive quality of the bunchgrass forage when it is green, and the usual availability of ranges at both higher and lower altitudes at other seasons, made spring and fall use a prevailing pattern (Pechanec, 1949; Rittenhouse and Vavra, 1979). Open water or snow was also most likely to be available for livestock during these seasons. Consistent spring use could not be sustained because the native grasses disappeared with even moderate use (Stoddart, 1946). Those ranges far from water escaped earlier heavy use and thus had some grass left by the time scientific understanding of range management suggested some ways of improving the balance of grass to brush.

One approach was to alternate between fall and spring grazing and give areas some rest in the spring (Craddock and Forsling, 1938). Sheep do make some use of sagebrush in the fall and thus open up the community to grasses and forbs, if plants or seed sources are available. Some striking improvements in the diversity and productivity of such ranges have been experimentally induced (Fig. 13.12; Laycock, 1967, 1970) by using sheep in the fall. This grazing strategy was applied by ranchers and land management agencies where it was feasible. Abundant problems, however, remained where the vegetation was so depleted of grasses that no good seed source remained. Even when a substantial perennial herbaceous component is present, it should be remembered that these native species do not produce much seed or establish many seedlings in most years, and sagebrush does not easily give up its dominance. Thus, reduction of livestock numbers and changes in class of livestock or season of use does not generally bring rapid results in a semi-arid environment such as this (Robertson, 1947, 1971; Lommason, 1948; Harniss and West, 1973).

Fig. 13.12. Fenceline contrast at the U.S. Sheep Experiment Station near Dubois, Idaho. Area to left of fence has been grazed by sheep in the spring and fall for more than 25 years. The area to the right of the fence was separated off in 1923 and grazed only in the fall for the succeeding 25 years (Mueggler, 1950).

Hydrologic changes

Vegetation changes have also had profound effects on watershed characteristics. The increased brush increased the transpirative demands of these systems. Sagebrush and companion shrubs carry on transpiration year-around. They also intercept more precipitation than grasses or forbs, especially snow. Rooting of sagebrush is much deeper and laterally extensive. Thus, as the brush component increased, the amount of water going into deep seepage and stream flow decreased. Sediment production increased because total vegetal cover and litter declined as the interspaces between shrubs had less cover (Sturges, 1979). Annuals, though effective in preventing erosion, are also highly susceptible to fire. Burns usually occur in the summer and ranges remain bare until the next spring. They are very susceptible to erosional losses during this period (Hinds and Sauer, 1974).

Semi-arid areas such as those occupied by this ecosystem type have naturally higher average rates of sediment production than deserts or grasslands (Fig. 13.13). This balanced against very low water yields (0–1 cm per year) has made justification of land management for improved watersheds more difficult than for areas yielding dependable runoff (Sturges, 1975a, b; Johnson and Hanson, 1976).

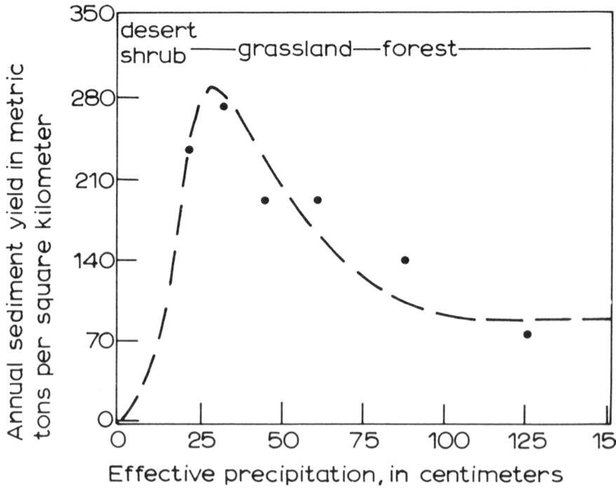

Fig. 13.13. Relationship of sediment yields to effective precipitation and vegetation type. From Branson (1975).

Influence of fire

Grasses were favored under pristine conditions due to periodic fire. Wright et al. (1979) have estimated fire return frequencies of between 32 and

70 years in examples of this type. It was logical for early range managers and later watershed managers to propose fire as a means of reducing the brush and encouraging the grass. There was, however, great public and administrative resistance to this. It took research to show under what conditions fire brought desirable results (Blaisdell, 1953; Pechanec, 1954). It was found that fire was undesirable on sites where rabbitbrushes (*Chrysothamnus* spp.) and horsebrushes (*Tetradymia* spp.) are successful. These shrubs generally have low palatability to livestock. They resprout after fire, but have relatively short lifespans. If burning is repeated after 20 to 25 years, they may resprout vigorously and continue to dominate the vegetation (Wright et al., 1979).

Fire in the spring or summer will often kill the native perennial grasses which the manager is trying to encourage. Wildfires which occur in the summer can burn very hotly. Prescribed burns either very early in the spring or after fall rains when the soil is moist and the bunchgrasses are dormant generally bring the best results (Britton and Ralphs, 1979; Wright et al., 1979).

If fires occur too frequently, annuals are encouraged. The annuals create much more readily flammable fuel. An accelerating frequency of wildfire can be created, leading to a spiral of deterioration in the site because of depletion of volatile nutrients and accelerated soil erosion (Fig. 12.8).

High-technology alterations

Because use of fire requires careful planning, creates special legal liabilities, and has a decided element of chance, managers turned to other ways of reducing brush in the 1940s to 1960s. On relatively level sites without excessive amounts of rocks, mechanical treatment was possible. In a few cases, there were enough native grasses left under the protection of shrubs that a surface treatment with a chain, cable, or disk or pipe harrow was sufficient to clear away old, brittle sagebrush. Where the brush was younger and more supple, it rode under the implement, regrew and produced abundant new seedlings. Rabbitbrush, if present, responded vigorously to such treatment (Parker, 1979).

The densities of native grasses were so depleted on most sites, and the chance of rabbitbrush and annual plant dominance so high, that it was best to till the soil thoroughly and seed perennial grasses. The native grasses are so susceptible to grazing during the growing season that exotics which can stand heavy spring use were used. These have been largely the Russian wheat and ryegrasses (*Agropyron* and *Elymus*) (Gates, 1974). Other grasses commonly used for revegetating sagebrush-steppe ranges are compared to crested wheatgrass (*Agropyron cristatum*) in Table 13.3 (Keller, 1979). Crested wheatgrass has been regarded as a godsend by those interested in improving livestock carrying capacities. Within a few years annual herbage yields

TABLE 13.3

How crested wheatgrass (*Agropyron cristatum*) compared with other species, in stands and yield, by number of studies (from Keller, 1979)

Other species	Crested wheatgrass was, by number of studies					
	superior		not different		inferior	
	yield	stand	yield	stand	yield	stand
Intermediate wheatgrass (*A. intermedium*)	5	5	3	6	5	6
Pubescent wheatgrass (*A. pubescens*)	6	2	5	4	3	2
Crested wheatgrass (*A. cristatum*)	3	–	3	–	–	–
Siberian wheatgrass (*A. sibericum*)	1	–	2	1	2	–
Russian wildrye (*Elymus junceus*)	4	2	–	–	2	1
Bluebunch wheatgrass (*A. spicatum*)	1	2	1	1	1	–
Beardless bluebunch wheatgrass (*A. inerme*)	–	2	2	2	1	–
Big bluegrass (*Poa ampla*)	–	2	1	–	3	1
Totals for yield	20		17		17	
Totals for stand		15		17		10

in these plantings exceed by two to three times that of high-condition range and up to fifteen times that of degraded native rangelands on the same sites (Rumsey, 1971).

Vast areas of this ecosystem type were converted to near monocultures of the Eurasian grass species in the 1940s, 50s, and early 1960s. Only a small fraction of the type was suitable for tillage, even with "stump-jump" plows. Thus, brush on steeper and/or rockier sites had to be controlled in other ways.

Herbicides came to be extensively employed for sagebrush control in the 1940s to 1960s. The availability of war surplus equipment and the cheap costs of fuel and herbicide meant that millions of hectares of the type could be easily treated to improve livestock forage production (Sneva, 1972). Herbaceous broad-leaved plants were killed by materials such as 2,4,5-trichlorophenoxyacetic acid (2,4,5-T) (Schumaker and Hanson, 1977; Evans et al., 1979). As wildlife biologists began to protest against the indiscriminate use of herbicides here, applications became more sophisticated and reseeding with mixtures of desirable forb and browse species with grasses increased. Critical wildlife habitats were by-passed as use of helicopters allowed more precise control and less spray drift. However, as the public became aware of the medical and environmental side-effects of herbicides, they clamored for restrictions on their use. Because of such regulations and much higher energy and material costs, very little spraying of sagebrush is done at the present time.

The effects of spraying are often short-lived (Thilenius and Brown, 1974). Careful management of treated areas, whether tilled or not, is necessary to retard the re-invasion of shrubs (Robertson, 1947; Johnson and Payne, 1968; Hull and Klomp, 1974). Repetitive treatments are necessary if herbaceous species are to maintain dominance. For instance, Johnson (1969) noted that some treated areas in central Wyoming had more sagebrush on them than adjacent untreated areas fourteen years after the manipulations were made.

The great plowing, seeding and spraying epoch of the 1940–60s has almost come to a halt (Vale, 1974). The best sites have already been treated, but sagebrush continues to be a serious competitor for grasses both in the native stands, and as a re-invader of seeded areas (Rittenhouse and Sneva, 1976). Little action can now be taken. The first impediments to this action were public arousal against single-use management of wildlands. The earlier treatments had been done solely to increase livestock production, and thus economic returns to the grazier. Little concern was shown to the wildlife, whether huntable or not. Even soil, and thus watershed stability, was a secondary concern. The dangers of creating a monoculture were overlooked. As the public became environmentally aware, they began asking embarrassing questions of the land management and regulatory agencies. When the answers showed little concern for other than direct economic interests, political pressures were brought to bear to pass tighter regulatory controls on pesticides and demand public input into the planning process for federal lands, of which the sagebrush-steppe forms a large fraction. Environmental impact statements showed that we only understood the effects of the treatments on the above-ground portion of vegetation, not the rest of the ecosystem. With lack of predictability of the total ramifications of treatments and possible negative impacts on aesthetic qualities valued by many voters, the land management agencies were forced to reduce or forgo mechanical and chemical treatments in this and other ecosystem types. This change is regrettable in some ways because the technology now exists for herbicidal control of cheatgrass so that perennial grasses can be established and close the community (Evans and Young, 1977).

The huge infusions of public capital to improve the grazing capacity of deteriorated rangelands of this type have proven biologically successful (Heady and Bartolome, 1977). The economics were also once acceptable, but the great increase in costs of fuel, herbicides, and all materials and labor in recent years has made treatment now often economically untenable. There has been a return to prescribed burning as a cheaper means of brush control. The public is now more knowledgeable about fire as a natural force in ecosystems and thus does not object as much as earlier. However, increased restrictions on working hours and other personnel constraints make prescribed burning difficult on public lands.

Recent efforts to improve forage production on these native rangelands through fertilization have been to no avail (Sneva, 1978). Nitrogen and sulfur amendments only seem to make the native herbaceous perennials more susceptible to drought damage. Fertilization encourages sagebrush and annuals more than it does the native bunchgrasses. The introduced Eurasian grasses do respond favorably to fertilization (McCormick and Workman, 1975; Sneva, 1978; James and Jurinak, 1978), but the cost of getting earlier and greater plant growth is rarely economical any longer.

The sagebrush steppe could still produce much more forage for livestock and wildlife and be in better condition as a watershed. The only currently economically viable and publicly acceptable way to accomplish this on a wide scale is through management of livestock. By varying class, seasons, and intensity of animal use, the vegetation could be manipulated toward a more productive state. We do know that sheep use in the fall definitely reduces the brush and increases the perennial grasses (Laycock, 1967, 1970), but sheep, because of external economic problems, lack of herders and predator losses, are much fewer here now than cattle.

Whether the type can be changed as rapidly with careful livestock management in other seasons is debatable. Gibbens and Fisser (1975) were unable to detect any statistically meaningful differences between rest rotation, deferred and season-long grazing after four years in an example of this ecosystem type in south-central Wyoming. The land management agencies have embraced rest rotation as the panacea for improving these kinds of rangelands. However, no long-term controlled and replicated experiments with this grazing system on this ecosystem type have been done. The reports that support rest-rotation grazing have used observations where livestock numbers were simultaneously reduced; thus, changes in numbers were confounded with changes in timing and other improvements in management. Consequently it remains to be seen whether implementation of grazing systems will bring the desired results without undesirable side-effects.

Unless land allocations to individual graziers already in place could be ignored (Bullock, 1978), there will be a need to build more fences and create more water supplies to employ sophisticated grazing systems. Therefore, it is unlikely that the transition to more intensive management will be rapid. The most likely immediate action will be the general reduction in livestock use in order to bring numbers in harmony with the total carrying capacities of these ecosystems for multiple uses. This will occur because environmental impact statements will show how much usable forage remains and how it must be shared with wildlife and feral horses if range conditions are ever to improve.

Responses of wildlife to other land uses

Wildlife constitutes an important recreational and aesthetic resource of this ecosystem type. The major huntable species are pronghorn, mule deer, bighorn sheep (*Ovis canadensis*), sagegrouse, and chukar partridge (*Alectoris chukar*).

The major federal refuges for pronghorn (Hart Mountain and Charles Sheldon) are largely covered by this ecosystem type. Pronghorns are found ranging over almost all of the rest of this ecosystem type that is not excessively fenced. Wagner (1977) has noted that pronghorn populations have not returned in as great abundance here as in the Great Plains. He speculated that early livestock abuse has permanently reduced the nutritional outputs of this system, and that pronghorn populations will never be high again.

Mule deer were rare in this type under pristine conditions. As livestock reduced the herbaceous species, they opened the communities to brush expansion. Deer, being primarily browsers in all seasons except spring, thrived on a habitat with an expanded food supply. Deer populations in this region reached a peak in the 1950s. They have since declined. Much debate has ensued on the reasons for this general reduction of herd size (Workman and Low, 1976). Although hunting policies, predators and weather may affect shorter-term trends, the ultimate cause is most probably a change in vegetation due to reduction of livestock on public ranges in the sagebrush steppe (Urness, 1979). As livestock numbers have been curtailed, the herbaceous component of the plant communities on critical winter ranges has increased at the expense of browse (Wagner, 1977; Willns et al., 1979). The grazing capacity for deer has been accordingly reduced. Any subsequent improvement of range condition for livestock, particularly cattle production will mean further declines in deer numbers (Urness, 1979).

The life cycle of the sagegrouse requires a variety of habitats. The loss of seeps and small meadows that produce the combination of buds of forbs and insects for adequate nutrition of the chicks appears to be the current bottleneck for increasing their populations. Livestock, by encouraging sagebrush, have caused springs and streams to dry up. Spots of meadow have been obliterated in much of this ecosystem type. Open areas for lek formation have also diminished as the stature of the brush increased. It is also possible that as brush has matured it has less nutritional value for adult populations (Wagner, 1977) and allows greater predation by coyotes. Any manipulation of livestock use or range improvement practices will have an influence on these birds (Klebenow, 1972).

Sagebrush steppes occurring next to the more rugged mountain and canyon areas serve as winter range for Rocky Mountain or California bighorn sheep. These animals prefer bunchgrasses. Morgan (1971) has argued that livestock use in other seasons has reduced the bunchgrass component and thus created wintering areas with poor nutrition for the wild sheep. With poor nutrition they become susceptible to diseases and parasites. These are the phenomena that led to public demand for more comprehensive environmental impact statements from the Bureau of Land Management on renewal of livestock grazing leases.

Chukar partridge do well only on the rockiest, most degraded kinds of sagebrush-steppe ranges (Christensen, 1970). They thrive where winter annual grasses dominate, the blades of which provide them with nutritious feed throughout the winter. They switch to seed and insects during the rest of the year. High-condition range for chukars is thus low-condition range for the ungulates.

The conversion of much of the degraded sites with deeper soils and gentler topography to stands of the Russian wheat and ryegrasses, especially *Agropyron cristatum*, has been cause for concern by wildlife managers. The near-monocultures unquestionably can improve grazing capacities for livestock, especially cattle (Robertson et al., 1970). What effects these treatments have on wildlife has yet to be thoroughly investigated, but most indications point to detrimental reductions in diversity and populations (Vale, 1974; Wagner, 1978; Reynolds and Trost, 1980). To begin with, sagebrush, even if over-mature and impoverished nutritionally, provides some browse for mule deer, pronghorn and sagegrouse. To convert this entirely to grass and fail to leave an interspersion of remnant native stands, runs the risk of eliminating winter browse for these species entirely. Secondly, wheatgrass plantings are rather sterile monocultures lacking the perennial forbs, suffrutescent half-shrubs, and perhaps native perennial grasses which pronghorn and sagegrouse appear to need during the warm seasons. While conclusive evidence is lacking, most wildlife biologists who have observed the situation are skeptical of the value of sagebrush removal and wheatgrass seedings to wildlife (cf. Patterson, 1952; Deming, 1963; Schneegas, 1967; N.S. Martin, 1970).

Studies in northwestern Utah have shown more than twice the rodent biomass in sagebrush stands as in adjacent wheatgrass seedings (Balph, 1973). Differences in jackrabbit numbers are even more pronounced (Westoby and Wagner, 1973). The question whether these animals may have also increased in abundance and developed more cyclic population trends due to changes in the native vegetation and predator control cannot be addressed since the effects are confounded and large unaltered relict areas do not remain to serve as controls. Circumstantial evidence indicates that jackrabbit populations may vary ten-fold in areas where the vegetation has been seriously degraded by livestock grazing (Stoddart, 1976).

Population explosions of grass bugs (*Labops* spp.: Hemiptera) in introduced grass seedlings may be related to simplification of the system (Todd and Kamm, 1974).

Herbicidal treatment and prescribed burning of mature sagebrush steppe vegetation may simultaneously improve watershed stability and livestock pasturage by favoring perennial herbaceous plants, but the treatments should ordinarily be done in small patches so as to minimize the disruption to wildlife species (Laycock, 1979). Wildlife considerations are so important because the major source of recreation on these lands is hunting. Although the density of such use is relatively high, it is dispersed over a huge area.

Some damage to the vegetal cover and thus erosion has been produced in the past by use of off-road vehicles for hunting or sight-seeing; the increased cost of motor fuel is however likely to curtail this activity except nearer to the cities.

THE FUTURE

Land development

Some potentially arable soils occur within this ecosystem type near the Columbia and Snake Rivers. Whereas legal constraints to their tillage (federal ownership) were the main impediment in the past, the cost of energy to pump water from rivers or wells now looms as the principal controlling factor. Not many new farms are expected since most of the landscape occupied by this ecosystem type is too distant from sources of water suitable for irrigation and/or too steep or rocky for tillage.

Hardly any valuable minerals are found under the prevailing basalt. The only economically positive aspect of this orogeny is the potential for geothermal sources of energy. These are rapidly being explored with little surface disturbance.

The Wyoming Basin, Rocky Mountain Valley and Great Plains portions of this type have some underlying deposits of phosphates, and vast seams of coal and oil shale. Prospects for oil and gas production also exist in that portion of the type. Considerable influx of people and disturbance of ecosystems is anticipated there.

The technology exists to restore and even enhance the productivity of these wildlands for livestock production. Current economic conditions and public resistance because of negative impacts on wildlife and aesthetics make many of these manipulations unfeasible. Therefore, it is likely that these wildlands will continue to have high proportions of brush- and annual-dominated communities which have low species diversities and considerable fluctuations in productivity. Animal production by domestic ruminants will continue to be the major way society gains wealth from these lands since they are largely too dry, too rocky, and too cold for most crops.

Graziers will employ increasingly sophisticated grazing systems only if America's priorities for range livestock production increase. Increased demand for livestock could be satisfied with the encouragement of higher prices, increased trade protection or production subsidies. Livestock grazing strategies will have to show sensitivity for all wild animals, fisheries and watershed values. Pronghorn will continue to thrive in portions of this type as long as hunting is regulated and fences are built so as to permit them to crawl underneath. The semi-domesticatable saiga antelope would likely be able to consume the forage here, but such introductions would face many political obstacles. Feral horses will continue to degrade some areas until populations are brought into balance with long-term grazing capacities.

Recreation will increase because the human population of the region is growing rapidly. The main impacts will be through dispersed camping, rock-hounding, hunting, bird-watching, etc. These uses will increase at a lower rate in this ecosystem type than in the adjacent mountains which most people find more aesthetically rewarding and cheaper to travel to. Comparatively little of this landscape will be torn up for mineral extraction since lava bedrock predominates. Restoration of those portions that are disturbed by miners will present problems for those concerned about re-vegetation and ground-water hydrology.

REFERENCES

Aldous, A.E. and Shantz, H.L., 1924. Types of vegetation in the semiarid portion of the United States and their economic significance. *J. Agric. Res.*, 28: 99–127.

Allred, B.W., 1941. Grasshoppers and their effects on sagebrush on the Little Powder River in Wyoming and Montana. *Ecology*, 22: 387–392.

Allred, D.M., 1968. Ticks of the National Reactor Testing Station. *Brigham Young Univ., Sci. Bull.*, 10(1): 1–29.

Balph, D.F. (Coordinator), et al., 1973. Curlew Valley Validation Site Report. *US/IBP Desert Biome Res. Memo.*, 73-1: 336 pp.

Becker, E.L., II., 1978. *Plant Food Preferences of Two Sympatric Rodents and Their Estimated Impact on a Great Basin Shrub–Bunch–Grass Community.* Thesis, Utah State University, Logan, Utah, 37 pp.

Beetle, A.A., 1960. A study of sagebrush — The section Tridentata of *Artemisia. Wyo. Agric. Exper. Stn. Bull.*, 368: 83 pp.

Beetle, A.A., 1979. Autecology of selected woody sagebrush species. In: Anonymous, *The Sagebrush Ecosystem: A Symposium.* Utah State University, College of Natural Resources, Logan, Utah, pp. 23–25.

Billings, W.D., 1951. Vegetation zonation in the Great Basin of western North America. In: *Les bases écologiques de la régéneration de la végétation des zones arides. Union Int. Sci. Biol., Ser. B*, No. 9: 101–122.

Blaisdell, J.P., 1953. Ecological effects of planned burning of sagebrush–grass range in the Upper Snake River Plains. *U.S. Dep. Agric. Tech. Bull.*, 1075: 39 pp.

Blaisdell, J.P., 1958. Seasonal development and yield of native plants in the upper Snake River Plains and their relation to certain climatic factors. *U.S. Dep. Agric. Tech. Bull.*, 1190: 68 pp.

Bohart, G.N. and Knowlton, G.E., 1976. Invertebrates. Appendix D–IV. In: Argonne Natl. Lab., *Final Environmental Impact Statement for the Sodium Cooled Class III Design Safety Research Experiment Faculty*, Idaho National Engineering Lab., pp. D-47 to D-58.

Branson, F.A., 1975. Natural and modified plant communities as related to runoff and sediment yields. In: A.D. Hasler (Editor), *Coupling of Land and Water Systems*. Springer Verlag, New York, N.Y., pp. 157–172.

Britton, C.M. and Ralphs, M.H., 1979. Use of fire as a management tool in sagebrush ecosystems. In: Anonymous, *The Sagebrush Ecosystem: A Symposium*. Utah State University, College of Natural Resources, Logan, Utah, pp. 101–109.

Buechner, H.K., 1953. Some biotic changes in the State of Washington particularly during the century 1853–1953. *Wash. State Univ. Res. Stud.*, 21: 154–192.

Bullock, A.E., 1978. Managing sagebrush–grass ranges of the north-western Great Basin. *Rangeman's J.*, 5: 89–91.

Butler, B.R., 1976. Evolution of the modern sagebrush–grass steppe biome on the eastern Snake River Plain. In: R. Elston (Editor), *Holocene Climatic and Ecological Change in the Great Basin. Nev. Archeol. Surv. Res. Pap.*, No. 6: 4–39.

Caldwell, M.M., 1979. Physiology of sagebrush. In: Anonymous, *The Sagebrush Ecosystem: A Symposium*. Utah State University, College of Natural Resources, Logan, Utah, pp. 74–85.

Campbell, G.S. and Harris, G.A., 1977. Water relations and water use patterns for *Artemisia tridentata* in wet and dry years. *Ecology*, 58: 652–569.

Christensen, G.C., 1970. The chukar partridge: Its introduction, life history, and management. *Nev. Dep. Fish Game Biol. Bull.*, 4: 82 pp.

Clark, W.H. and Comanor, P.L., 1975. Removal of annual plants from the desert ecosystem by western harvester ants, *Pogonomyrmex occidentalis. Environ. Entomol.*, 4: 52–56.

Comanor, P.L. and Staffeldt, E.E., 1978. Decomposition of plant litter in two western North American deserts. In: N.E. West and J. Skujins (Editors), *Nitrogen in Desert Ecosystems*. US/IBP Synthesis Series 9. Dowden, Hutchinson and Ross, Stroudsburg, Pa., pp. 31–49.

Craddock, C.W. and Forsling, C.L., 1938. The influence of climate and grazing on spring–fall sheep range in southern Idaho. *U.S. Dep. Agric. Tech. Bull.*, 600: 43 pp.

Cramer, H.R., 1976. The Range and Hudspeth's Cutoff — Southeastern Idaho. *Rangeman's J.*, 3: 156–159.

Daubenmire, R., 1970. Steppe vegetation of Washington. *Wash. Agric. Exper. Sta. Tech. Bull.*, 62: 131 pp.

Daubenmire, R.F., 1972. Annual cycles of soil moisture and temperature as related to grass development in the steppe of eastern Washington. *Ecology*, 53: 419–424.

Daubenmire, R., 1975a. An analysis of structural and functional characters along a steppe–forest catena. *Northwest Sci.*, 49: 120–140.

Daubenmire, R., 1975b. Ecology of *Artemisia tridentata* ssp. *tridentata* in the state of Washington. *Northwest Sci.*, 49: 24–35.

Davis, O.K., Kolva, D.A. and Mehringer, Jr., P.J., 1977. Pollen analysis of Wildcat Lake, Whitman Co., Washington. *Northwest Sci.*, 51: 13–30.

Deming, O.V., 1963. Antelope and sagebrush. *Trans. Annu. Meet. Interstate Antelope Conf.*, 14: 55–60.

Eckert, Jr. R.E., 1957. *Vegetation–Soil Relationships in Some* Artemisia *Types in Northern Harney and Lake Counties, Oregon*. Thesis, Oregon State University, Corvallis, Ore., 179 pp.

Eggler, W.A., 1941. Primary succession on volcanic deposits in southern Idaho. *Ecol. Monogr.*, 11: 277–298.

Evans, R.A. and Young, J.A., 1977. Weed control-revegetation systems for big sagebrush–downy brome rangelands. *J. Range Manage.*, 30: 331–336.

Evans, R.A., Young, J.A. and Eckert, Jr., R.E., 1979. Herbicides as a management tool. In: Anonymous, *The Sagebrush Ecosystem: A Symposium*. Utah State University, College of Natural Resources, Logan, Utah, pp. 110–116.

Farb, P., 1978. *Man's Rise to Civilization From Primeval times to the Coming of the Industrial State*. E.P. Dutton and Co., New York, N.Y., 2nd ed., 253 pp.

Ferguson, C.W., 1964. *Annual Rings in Big Sagebrush*. Pap. Lab. Tree-Ring Res., University of Arizona Press, Tucson, Ariz., 95 pp.

Fetcher, N. and Trlica, M.J., 1980. Influence of climate on annual production of seven cold desert forage species. *J. Range Manage.*, 33: 35–37.

Forest-Range Task Force, 1972. *The Nation's Range Resources — A Forest-Range Environmental Study*. Forest Resource Report 19. Forest Service, U.S. Dep. Agric., Washington, D.C., 147 pp.

Fosberg, M.A. and Hironaka, M., 1964. Soil properties affecting the sagebrush communities in southern Idaho. *Am. Soc. Agron. Spec. Publ.*, No. 5: 230–236.

Franklin, J.F. and Dyrness, C.T., 1973. Natural vegetation of Oregon and Washington. *U.S.D.A. For. Serv. Gen. Tech. Rep.*, PNW-8: 417 pp.

Fronk, W.D., Beetle, A.A. and Fullerton, D.G., 1964. Dipterous galls on the *Artemisia tridentata* complex and insects associated with them. *Ann. Entomol. Soc. Am.*, 57: 575–577.

Gates, D.H., 1964. Sagebrush infested by leaf defoliating moths. *J. Range Manage.*, 17: 209–210.

Gates, D.H., 1974. Restoration of grasses in the sagebrush–grass steppes of the northwestern USA. In: *Proc. XII Int. Grassland Congr.*, pp. 144–148.

Gibbens, R.P. and Fisser, H.G., 1975. *Influence of Grazing Management Systems on Vegetation in the Red Desert Region of Wyoming*. Scientific Monograph, 29. Agric. Exper. Stn., University of Wyoming, Laramie, Wyo., 32 pp.

Goodwin, D.L., 1956. *Autecological Studies of* Artemisia tridentata. Thesis, Washington State University, Pullman, Wash., 79 pp.

Griffiths, D., 1902. Forage conditions on the northern border of the Great Basin. *U.S. Dep. Agric. Bur. Plant Ind. Bull.*, 15: 59 pp.

Halford, S.A., Rich, G.B. and Bergis, I., 1973. A chrysomelid beetle defoliating big sagebrush in South-Central British Columbia. *Can. J. Plant Sci.*, 53: 383–384.

Hall, R.C., 1965. Sagebrush defoliator outbreak in northern

California *U.S.D.A. For. Serv.*, PSW-75: 12 pp.

Hanley, T.A., 1979. Application of an herbivore-plant model to rest-rotation grazing management on shrub-steppe ranges. *J. Range Manage.*, 32: 115–118.

Harniss, R.O. and West, N.E., 1973. Changes in *Artemisia tridentata/Sitanion hystrix* vegetation of the National Reactor Testing Station, southeastern Idaho, 1950–1965. *Utah Acad. Sci., Arts Lett. Proc.*, 50: 10–16.

Harris, G.A., 1967. Some competitive relationships between *Agropyron spicatum* and *Bromus tectorum*. *Ecol. Monogr.*, 37: 89–111.

Harris, G.A., 1977. Root phenology as a factor of competition among grass seedlings. *J. Range Manage.*, 30: 172–176.

Heady, H.F. and Bartolome, J., 1977. The Vale rangeland rehabilitation program: The desert repaired in southeastern Oregon, *U.S.D.A. For. Serv. Resour. Bull.*, PNW-70; 139 pp.

Hinds, W.T. and Sauer, R.H., 1974. Soil erodibility, soil erosion, and revegetation following wildfire in a shrub-steppe community. In: Anonymous, *Proc. Symp. on Atmosphere–Surface Exchange of Particulate and Gaseous Pollutants*. U.S. Energy Res. Develop. Admin., Richland, Wash., pp. 571–590.

Hironaka, M., 1963. *Plant–Environment Relations of Major Species in Sagebrush–Grass Vegetation of Southern Idaho*. Thesis, University of Wisconsin, Madison, Wisc., 124 pp. (*Diss. Abst.*, No. 63: 5747).

Hironaka, M., 1979. Basic synecological relationships of the Columbia River sagebrush type. In: Anonymous, *The Sagebrush Ecosystem: A Symposium*. Utah State University, College of Natural Resources, Logan, Utah, pp. 27–30.

Hironaka, M. and Tisdale, E.W., 1963. Secondary succession in annual vegetation in southern Idaho. *Ecology*, 44: 810–812.

Hoffman, G.R. and Hazlett, D.L., 1977. Effects of aqueous *Artemisia* extracts and volatile substances on germination of selected species. *J. Range Manage.*, 30: 134–137.

Howard, R.P. and Wolfe, M.L., 1976. Range improvement practices and ferruginous hawks. *J. Range Manage.*, 29: 33–37.

Hugie, V.K. and Passey, H.B., 1963. Cicadas and their effect upon soil genesis in certain soils in southern Idaho, northern Utah, and northeastern Nevada. *Soil Sci. Soc. Am. Proc.*, 27: 78–82.

Hugie, V.K. and Passey, H.B., 1964. Soil surface patterns of some semi-arid soils in northern Utah, southern Idaho, and southwestern Nevada. *Soil Sci. Soc. Am. Proc.*, 28: 786–792.

Hugie, V.K., Passey, H.B. and Williams, E.W., 1964. Soil taxonomic units and potential plant community relationships in a pristine area of southern Idaho. *Am. Soc. Agron. Spec. Publ.*, No. 5: 190–205.

Hull, Jr. A.C., 1976. Rangeland use and management in the Mormon West. In: Anonymous, *Symposium on Agriculture, Food and Man — A Century of Progress*. Brigham Young University, Provo, Utah, 21 pp.

Hull, Jr., A.C. and Klomp, G.J., 1974. Yield of crested wheatgrass under four densities of big sagebrush in southern Idaho. *U.S. Dep. Agric. Tech. Bull.*, 1483: 38 pp.

Hunt, C.B., 1974. *Natural Regions of the United States and Canada*. W.H. Freeman and Co., San Francisco, Calif., 725 pp.

Isaac, R., 1974. *A Climatically Determined Simulation Model For the Prediction of Forage Quantity and Quality of Bluebunch Wheatgrass At Dubois, Idaho*. Dissertation, University of Southern Illinois, Carbondale, Ill., 94 pp.

James, D.W. and Jurinak, J.J., 1978. Nitrogen fertilization of dominant plants in the northeastern Great Basin desert. In: N.E. West and J. Skujins (Editors), *Nitrogen in Desert Ecosystems*. Dowden, Hutchinson and Ross, Stroudsburg, Pa., pp. 219–231.

Johnson, C.W. and Hanson, C.L., 1976. Sediment sources and yields from sagebrush rangeland watersheds. In: *Proc. Third Inter-Agency Sedimentation Conf., Denver, Colo.*, pp. 1–71 to 1–80.

Johnson, J.R. and Payne, G.F., 1968. Sagebrush reinvasion as affected by some environmental influences. *J. Range Manage.*, 21: 209–215.

Johnson, K.L., 1979. Basic synecological relationships of the sagebrush types on the high plains of Montana, Wyoming, and the Dakotas. In: Anonymous, *The Sagebrush Ecosystem: A Symposium*. Utah State University, College of Natural Resources, Logan, Utah, pp. 42–48.

Johnson, W., 1969. Life expectancy of a sagebrush control in central Wyoming. *J. Range Manage.*, 22: 177–182.

Kamm, J.A., Sneva, F.A. and Rittenhouse, L.R., 1978. Insect grazers on the cold desert biome. In: D.N. Hyder (Editor), *Proc. First Int. Range Congr., Denver, Colo.*, pp. 479–484.

Keller, W., 1979. Species and methods for seeding in the sagebrush ecosystem. In: Anonymous, *The Sagebrush Ecosystem: A Symposium*. Utah State University, College of Natural Resources, Logan, Utah, pp. 129–136.

Klebenow, D.A., 1972. The habitat requirements of sagegrouse and the role of fire in management. *Proc. Tall Timbers Fire Ecol. Conf.*, 11: 305–315.

Küchler, A.W., 1970. Potential natural vegetation (map at scale of 1:7 500 000). In: U.S. Geol. Survey, *The National Atlas of the U.S.A.* U.S. Government Printing Office, Washington, D.C., pp. 90–91. (revision of 1964 map at scale of 1:3 168 000).

Larrison, E.J. and Johnson, D.R., 1973. Density changes and habitat affinities of rodents of shadscale and sagebrush associations. *Great Basin Nat.*, 33: 255–264.

Laycock, W.A., 1967. How heavy grazing and protection affect sagebrush–grass ranges. *J. Range Manage.*, 20: 206–213.

Laycock, W.A., 1970. The effects of spring and fall grazing on sagebrush–grass ranges in eastern Idaho. *Proc. Tenth Int. Grassland Congr.*, pp. 51–54.

Laycock, W.A., 1979. Factors affecting choice of management strategies within the sagebrush ecosystem. In: Anonymous. *The Sagebrush Ecosystem: A Symposium*. Utah State University, College Natural Resources, Logan, Utah, pp. 230–236.

Lommason, T., 1948. Succession in sagebrush. *J. Range Manage.*, 1: 19–21.

McAdoo, J.K. and Klebenow, D.A., 1979. Native faunal relationships in sagebrush ecosystems. In: Anonymous, *The Sagebrush Ecosystem: A Symposium*. Utah State University, College of Natural Resources, Logan, Utah, pp. 50–61.

McArdle, R.E. and Costello, D.F., 1936. The virgin range. In: H.A. Wallace, *Letter from the Secretary of Agriculture to the U.S. Senate — A Report on the Western range — A Great but Neglected Resource*. Senate Document 199, pp. 71–80.

McArdle, R.E., et al., 1936. The whiteman's toll. In: H.A. Wallace, *Letter from the Secretary of Agriculture to the U.S. Senate — A Report on the Western Range — A Great But Neglected Resource*. Senate Document 199, pp. 81–116.

McArthur, E.D., 1979. Sagebrush systematics and evolution. In: Anonymous, *The Sagebrush Ecosystem: A Symposium*. Utah State University, College of Natural Resources, Logan, Utah, pp. 14–22.

McCormick, P.W. and Workman, J.P., 1975. Early range readiness with nitrogen fertilizer: an economic analysis. *J. Range Manage.*, 28: 181–184.

McDonough, W.T., Harniss, R.O. and Campbell, R.B., 1975. Morphology of ephemeral and persistent leaves of three subspecies of big sagebrush grown in a uniform environment. *Great Basin Nat.*, 35: 325–326.

Mack, R.N., 1971. *Mineral Cycling in* Artemisia tridentata. Thesis, Washington State University, Pullman, Wash., 117 pp.

MacMahon, J.A., 1976a. Mammals. Appendix D–VII. In: Argonne Natl. Lab., *Final Environmental Impact Statement for the Sodium Cooled Class III Design Safety Research Experimental Facility*, Idaho National Engineering Lab., pp. D-98 to D-115.

MacMahon, J.A., 1976b. Reptiles and amphibians. Appendix D–V. In: Argonne Natl. Lab., *Final Environmental Impact Statement for the Sodium Cooled Class III Design Safety Research Experiment Facility*, Idaho National Engineering Lab., pp. D-59 to D-68.

Marquiss, R. and Lang, R.L., 1959. Vegetational composition and ground cover of two natural relict areas and their associated grazed areas in the Red Desert of Wyoming. *J. Range Manage.*, 12: 104–109.

Martin, N.S., 1970. Sagebrush control related to habitat and sagegrouse occurrence. *J. Wildlife Manage.*, 34: 313–320.

Martin, P.S., 1970. Pleistocene niches for alien animals. *BioScience*, 20: 218–221.

May, M., Lang, R., Lugan, J., Jacoby, P. and Thompson, W., 1971. Reclamation of strip mine spoil banks in Wyoming. *Res. J. 51 Agric. Exper. Sta., Univ. Wyo., Laramie*, 32 pp.

Mayland, H.F. and Murray, R.B., 1979. Mineral-cycling aspects within the sagebrush ecosystem. In: Anonymous, *The Sagebrush Ecosystem: A Symposium*. Utah State University, College of Natural Resources, Logan, Utah, pp. 62–73.

Montgomery, S.J., 1976. *Rodent–Habitat Relationships in Great Basin Desert Shrub Communities*. Thesis, Utah State University, Logan, Utah, 73 pp.

Morgan, J.K., 1971. *Ecology of the Morgan Creek and East Fork of the Salmon River Bighorn Sheep Herds and Management of Bighorn Sheep in Idaho*. Thesis, Utah State University, Logan, Utah, 156 pp.

Mueggler, W.F., 1950. Effects of spring and fall grazing by sheep on vegetation of the upper Snake River Plains. *J. Range Manage.*, 3: 308–315.

Mueggler, W.F., 1967. Voles damage big sagebrush in southwestern Montana. *J. Range Manage.*, 20: 88–91.

Murray, R.B., 1971. Grazing capacity, sheep gains: Cheatgrass, bunchgrass ranges in southern Idaho. *J. Range Manage.*, 24: 407–409.

Murray, R.B., 1975. *Effect of* Artemisia tridentata *Removal on Mineral Cycling*. Thesis, Washington State University, Pullman, Wash., 190 pp. (Univ. Microfilm Order, No. 76-322).

Murray, R.B. and Klemmedson, J.O., 1965. Cheatgrass range in southern Idaho: Seasonal cattle gains and grazing capacities. *J. Range Manage*, 21: 308–313.

O'Farrell, T.P., 1975. Seasonal variations in populations of small mammals on Rattlesnake Mountain, Washington. *Am. Midl. Nat.*, 94: 190–204.

O'Farrell, T.P., Olson, R.J., Gilbert, R.O. and Hedlund, J.D., 1975. A population of Great Basin pocket mice, *Perognathus parvus*, in the shrub-steppe of south-central Washington. *Ecol. Monogr.*, 45: 1–28.

Ogden, P.S., 1950. *Peter Skene Ogden's Snake Country Journals, 1824–25 and 1825–26*. E.E. Rich and A.M. Johnson (Editors). Publication XIII. Hudson's Bay Record Soc., London.

Ogden, P.S., 1971. *Peter Skene Ogden's Snake Country Journals, 1827–1829*. Glyndwr Williams (Editor), Publication XXVIII. Hudson's Bay Record Soc., London, 201 pp.

Parker, K.G., 1979. Use of mechanical methods as a management tool. In: Anonymous, *The Sagebrush Ecosystem: A Symposium*. Utah State University, College of Natural Resources, Logan, Utah, pp. 117–120.

Passey, H.B. and Hugie, V.K., 1962a. Sagebrush on relict ranges in the Snake River Plains and northern Great Basin. *J. Range Manage.*, 15: 273–278.

Passey, H.B. and Hugie, V.K., 1962b. Application of soil–climate–vegetation relations to soil survey interpretations for rangelands. *J. Range Manage.*, 15: 162–166.

Passey, H.B. and Hugie, V.K., 1963a. Fluctuating herbage production on an ungrazed Sierozem soil in Idaho. *J. Soil Water Conserv.*, 18: 8–11.

Passey, H.B. and Hugie, V.K., 1963b. Some plant–soil relationships on an ungrazed range area of southeastern Idaho. *J. Range Manage.*, 16: 113–118.

Passey, H.B., Hugie, V.K. and Williams, E.W., 1964. Herbage production and composition fluctuations of natural plant communities as related to climate and soil taxonomic units. In: *Forage Plant Physiology and Soil–Range Relationships. Am. Soc. Agron. Spec. Publ.*, 5: 206–221.

Passey, H.B., Hugie, V.K. and Williams, E.W., in press. *Soil–Plant Community–Climate Relationships on Intermountain West Rangelands*. Soil Conservation Service, U.S. Dep. Agric., Washington, D.C.

Patterson, R., 1952. *The Sagegrouse in Wyoming*. Sage Books, Denver, Colo., 234 pp.

Pearson, L.C., 1965. Primary production in grazed and ungrazed desert communities of eastern Idaho. *Ecology*, 46: 279–284.

Pearson, L.C., 1966. Primary productivity in a northern desert area. *Oikos*, 15: 211–228.

Pechanec, J.F., 1944. Sagebrush burning: good and bad. *U.S. Dep. Agric. Farmer's Bull.*, No. 1948: 1–34.

Pechanec, J.F., 1949. Grazing spring–fall sheep ranges of southern Idaho. *U.S.D.A. Circ.* No. 808: 1–34.

Pickford, G.D., 1932. The influence of continued heavy grazing and of promiscuous burning on spring–fall ranges in Utah. *Ecology*, 13: 159–171.

Powell, J., 1970. Site factor relationships with volatile oils in big sagebrush. *J. Range Manage.*, 23: 42–46.

Rawls, W.J., Zuzel, J.F. and Schumaker, G.A., 1973. Soil moisture trends on sagebrush rangelands. *J. Soil Water Conserv.*, 28: 270–272.

Reynolds, T.D. and Trost, C.H., 1980. The response of native vertebrate populations to crested wheatgrass planting and grazing by sheep. *J. Range Manage.*, 33: 122–125.

Rickard, W.H., Cline, J.F. and Gilbert, R.O., 1974. Pitfall trapping and direct counts of darkling beetles in cheatgrass communities. *Northwest Sci.*, 48: 96–101.

Rittenhouse, L.R. and Sneva, F.A., 1976. Expressing the competitive relationship between Wyoming big sagebrush and crested wheatgrass. *J. Range Manage.*, 29: 326–327.

Rittenhouse, L.R. and Vavra, M., 1979. Nutritional aspects of native and seeded sagebrush range for domestic livestock. In: Anonymous, *The Sagebrush Ecosystem: A Symposium.* Utah State University, College of Natural Resources, Logan, Utah, pp. 147–191.

Robertson, D.R., Nielsen, J.L. and Bare, N.H., 1966. Vegetation and soils of alkali sagebrush and adjacent big sagebrush ranges in North Park, Colorado. *J. Range Manage.*, 19: 17–20.

Robertson, J.H., 1943. Seasonal root development of sagebrush (*Artemisia tridentata* Nutt.) in relation to range reseeding. *Ecology*, 24: 125–126.

Robertson, J.H., 1947. Responses of range grasses to different intensities of competition with sagebrush (*Artemisia tridentata* Nutt.). *Ecology*, 28: 1–16.

Robertson, J.H., 1971. Changes on a sagebrush–grass range in Nevada ungrazed for 36 years. *J. Range Manage.*, 24(5): 397–400.

Robertson, J.H. and Kennedy, P.B., 1954. Half-century changes on northern Nevada ranges. *J. Range Manage.*, 7: 117–121.

Robertson, J.H., McAdams, K.R., Neal, D.L. and Tueller, P.T., 1970. Spring grazing of crested wheatgrasses in Nevada: A case history. *Nev. Agric. Exper. Stn. Bull.*, 23: 22 pp.

Rogers, L.E. and Rickard, W.H., 1975. A survey of darkling beetles (*Philolithus stenomorpha*) in desert steppe vegetation after a decade. *Ann. Entomol. Soc. Am.*, 68: 1069–1070.

Rumsey, W.B., 1971. Range seedings versus climax vegetation on three sites in Idaho. *J. Range Manage.*, 24: 447–450.

Sabinske, D.W. and Knight, D.H., 1978. Variation within the sagebrush vegetation of Grand Teton National Park, Wyoming. *Northwest Sci.* 52: 195–204.

Schlatterer, E.F. and Tisdale, E.W., 1969. Effects of litter of *Artemisia*, *Chrysothamnus*, and *Tortula* on germination and growth of three perennial grasses. *Ecology*, 50: 869–873.

Schneegas, E.R., 1967. Sage grouse and sagebrush control. *Trans. North Am. Wildl. Nat. Res. Conf.*, 32: 270–274.

Schumaker, G.A. and Hanson, C.L., 1977. Herbage response after mechanical and herbicide treatment of big sagebrush (*Artemisia tridentata*) in southwestern Idaho. *U.S. Dep. Agric., Agric. Res. Serv. West. Reg.*, 15 pp.

Sheldon, J.K. and Rogers, L.E., 1978. Grasshopper food habits within a shrub-steppe community. *Oecologia*, 32: 85–92.

Shown, L.M., Miller, R.F. and Branson, F.A., 1969. Sagebrush conversion to grassland as affected by precipitation, soil, and cultural practices. *J. Range Manage.*, 22: 303–311.

Skujins, J., 1980. The nitrogen cycle in desert ecosystems. *Ecol. Bull.*, 33: 477–492.

Smolik, J.D. and Rogers, L.E., 1976. Effects of cattle grazing and wildfire on soil-dwelling nematodes of the shrub-steppe ecosystem. *J. Range Manage.*, 29: 304–305.

Sneva, F.A., 1972. Grazing returns following sagebrush control in eastern Oregon. *J. Range Manage.*, 25: 174–178.

Sneva, F.A., 1977. Correlations of precipitation and temperature with spring, regrowth, and mature crested wheatgrass yields. *J. Range Manage.*, 30: 270–275.

Sneva, F.A., 1978. Nitrogen and sulfur impacts on the Cold Desert Biome. In: D.N. Hyder (Editor), *Proc. First Int. Range Congr., Denver, Colo.*, pp. 768–780.

Sneva, F.A., 1979. The western harvester ants: their density and hill size in relation to herbaceous productivity and big sagebrush cover. *J. Range Manage.*, 32: 46–47.

Sneva, F.A. and Hyder, D.N., 1962. Forecasting range herbage production in eastern Oregon. *Agric. Exper. Sta. Ore. State Univ., Corvallis, Ore., Sta. Bull.*, 588: 22 pp.

Soil Survey Staff, 1975. *Soil Taxonomy on a Basic System of Soil Classification for Making and Interpreting Soil Surveys.* U.S. Dep. Agric. Handbook 436, 754 pp.

Stoddart, L.A., 1946. Some physical and chemical responses of *Agropyron spicatum* to herbage removal at various seasons. *Utah Agric. Exper. Sta. Bull.*, 324: 28 pp,

Stoddart, L.C., 1976. *Population Dynamics, Movement, and Home Range of Black-Tailed Jackrabbits* (Lepus californicus) *in Curlew Valley, Utah–Idaho.* Progr. Rep. for Contract AT(11-1)-1329, U.S. Energy Res. and Develop. Admin., 89 pp.

Sturges, D.L., 1975a. Sediment transport from big sagebrush watersheds. In: *Watershed Meet. Symp. Held by the ASCE Irrigation and Drainage Division, Logan, Utah, 1975*, pp. 728–738.

Sturges, D.L., 1975b. Hydrologic relations on undisturbed and converted big sagebrush lands: The status of our knowledge. *U.S.D.A. For. Serv. Res. Pap.*, RM-140, 23 pp. (Rocky Mt. For. and Range Exper. Stn., Fort Collins, Colo.)

Sturges, D.L., 1977. Soil water withdrawal and root characteristics of big sagebrush. *Am. Midl. Nat.*, 98: 257–274.

Sturges, D.L., 1979. Hydrologic relations of sagebrush lands. In: Anonymous, *The Sagebrush Ecosystem: A Symposium.* Utah State University, College of Natural Resources, Logan, Utah, pp. 86–100.

Tempte, A.K., 1977. *The Demography of the Sagebrush Defoliator,* Aroga websteri *Clarke (Lepidoptera: Gelechiidae), and Its Impact on* Artemisia tridentata. Thesis, Utah State University, Logan, Utah, 52 pp.

Thatcher, A.P., 1959. Distribution of sagebrush as related to site differences in Albany County, Wyoming. *J. Range Manage.*, 12: 55–61.

Thilenius, J. and Brown, G.R., 1974. Long-term effects of chemical control of big sagebrush (*Artemisia tridentata*). *J. Range Manage.*, 27: 223–224.

Tingey, W.M., Jorgensen, C.D. and Frischknecht, N.C., 1972. Thrips of the sagebrush–grass community of west central Utah. *J. Range Manage.*, 25: 304–308.

Tisdale, E.W., Hironaka, M. and Fosberg, M.A., 1965. An area of pristine vegetation in Craters of the Moon National

Monument, Idaho. *Ecology*, 46: 349–352.

Tisdale, E.W., Hironaka, M. and Fosberg, M.A., 1969. The sagebrush region in Idaho — a problem in range resource management. *Ida. Agric. Exper. Stn. Bull.*, 512: 15 pp.

Todd, J.G. and Kamm, J.A., 1974. Biology and impact of a grassbug (*Labops hesperius* Uhler) in Oregon rangelands. *J. Range Manage.*, 27: 453–458.

Trost, C.H., 1976. Birds. Appendix D–VI. In: *Argonne Natl. Lab. Final Environmental Impact Statement for the Sodium Cooled Class III Design Safety Research Experimental Facility*, Idaho National Engineering Lab., pp. D-69 to D-96.

Tueller, P.T., 1962. *Plant Succession on Two* Artemisia *Habitat Types in Southern Oregon.* Dissertation, Oregon State University, Corvallis, Ore., 249 pp.

Urness, P.J., 1979. Wildlife habitat manipulation in sagebrush ecosystems. In: Anonymous, *The Sagebrush Ecosystem: A Symposium.* Utah State University, College of Natural Resources, Logan, Utah, pp. 169–178.

Vale, T.R., 1974. Sagebrush conversion projects: An element of contemporary environmental change in the western United States. *Biol. Conserv.*, 6: 274–284.

Vale, T.R., 1975. Presettlement vegetation in the sagebrush–grass area of the Intermountain West. *J. Range Manage.*, 28: 32–36.

Wagner, F.H., 1977. Livestock and wildlife–fisheries ecology. In: *Proc. Conf. on Livestock and Wildlife — Fisheries Relationships in the Great Basin*, 28 pp.

Wagner, F.H., 1978. Livestock grazing and the livestock industry. in: H.P. Brokaw (Editor), *Wildlife and America.* Council on Environmental Quality, Wash., D.C., pp. 121–145.

Weaver, T. and Klarich, D., 1976. Toxic effects of volatile exudates from *Artemisia tridentata* Nutt. on soil microbes. *Proc. Montana Acad. Sci.*, 36: 80–85.

Weaver, T.W. and Klarich, D., 1977. Allelopathic effects of volatile substances from *Artemisia tridentata* var. *vaseyana. Am. Midl. Nat.*, 97: 508–512.

West, N.E., 1979. Basic synecological relationships of sagebrush-dominated lands in the Great Basin and the Colorado Plateau. In: Anonymous, *The Sagebrush Ecosystem: A Symposium.* Utah State University, College of Natural Resources, Logan, Utah, pp. 33–41.

West, N.E. and Skujins, J. (Editors), 1978. *Nitrogen in Desert Ecosystems.* Dowden, Hutchinson and Ross, Stroudsburg, Pa., 307 pp.

West, N.E., Tausch, R.J., Rea, K.H. and Tueller, P.T., 1978. Phytogeographical variation within juniper–pinyon woodlands of the Great Basin. *Great Basin Nat. Mem.*, 2: 119–136.

West, N.E., Rea, K.H. and Harniss, R.O., 1979. Plant demographic studies in sagebrush–grass communities of southeastern Idaho. *Ecology*, 60: 376–388.

Westoby, M. and Wagner, F.H., 1973. Use of a crested wheatgrass seeding by black-tailed jackrabbits. *J. Range Manage.*, 26: 349–351.

Wildung, R.E., Garland, T.R. and Buschbom, R.L., 1975. The interdependent effects of soil temperature and water content on soil respiration rate and plant root decomposition in arid grassland soils. *Soil Biol. Biochem.*, 7: 373–378.

Willns, W., McLean, A., Tucker, R. and Ritchey, R., 1979. Interactions between mule deer and cattle on big sagebrush range in British Columbia. *J. Range Manage.*, 32: 299–304.

Workman, G.W. and Low, J.B. (Editors), 1976. *Mule Deer Decline in the West: A Symposium.* Utah State University, College of Natural Resources, Logan, Utah, 134 pp.

Wright, H.A. Neuenschwander, L.F. and Britton, C.M., 1978. The role and use of fire in sagebrush grass and pinyon-juniper communities: A state-of-the-art review. *U.S. For. Serv. Gen. Tech. Rep.*, INT-58: 48 pp.

Yansky, G.R., 1966. *Climatography of the National Reactor Testing Station.* Report IDO-12048. Air Resour. Field Res. Stn., N.R.T.S., Idaho Falls, Idaho, 112 pp.

Young, J.A. and Evans, R.A., 1970. Invasion of medusahead into the Great Basin. *Weed Sci.*, 18: 89–97.

Young, J.A., Evans, R.A. and Tueller, P.T., 1976. Great Basin plant communities — pristine and grazed. In: R. Elston (Editor), *Holocene Environmental Change in the Great Basin. Nev. Archeol. Surv. Res. Pap.*, No. 6: 186–215.

Young, J.A., Evans, R.A. and Major, J., 1977. Sagebrush steppe. In: M.G. Barbour and J. Major (Editors), *Terrestrial Vegetation of California.* Wiley Interscience, New York, N.Y., pp. 763–796.

Young, J.A., Eckert, Jr., R.E. and Evans, R.A., 1979. Historical perspectives regarding the sagebrush ecosystem. In: Anonymous, *The Sagebrush Ecosystem: A Symposium.* College of Natural Resources, Utah State University, Logan, Utah, pp. 1–13.

Zamora, B. and Tueller, P.T., 1973. *Artemisia arbuscula, A. longiloba,* and *A. nova* habitat types in northern Nevada. *Great Basin Nat.*, 33: 225–242.

Chapter 14

INTERMOUNTAIN SALT-DESERT SHRUBLAND[1]

N.E. WEST

DEFINITION OF TYPE

A major portion of the temperate desert region in North America is occupied by shrubs and half-shrubs of the Chenopodiaceae. Since their dominance is usually associated with halomorphic soils, the descriptor "salt desert shrub" has gained widespread usage (Branson et al., 1967; West, 1968). Although the correspondence of chenopod dominance with salty soils is strong, it is not universal. We will also include here the chenopod-dominated areas on dry, but non-saline soils in the extreme rain shadow of the Sierra Nevada in western Nevada and adjacent California (Billings, 1949, 1951). We will not include discussion here of the rather small but much more diverse and productive inclusions of marshlands and meadows where permanent high free water tables exist at or near the soil surface. These have been reviewed elsewhere by Chapman (1974) and Bolen (1964).

Fig. 14.1. Salt flat with sparse cover of vegetation dominated by *Salicornia* and *Allenrolfea*, Lander Co., Nevada.

Where extreme aridity and/or saltiness is present, the type approaches desert, in the international sense — it is virtually devoid of vegetation (Fig. 14.1). Where aridity, salinity and/or alkalinity is more moderate, the landscape has a greater plant cover, largely composed of shrubs in the Chenopodiaceae (Fig. 14.2). Because this type extends over a wide variety of climates found in the arid and semi-arid portions of the continent, including subtropical latitudes, it can be thought of as a pedobiome (Walter and Box, 1976). The other ecosystem types we are considering here have less climatic variation within their boundaries. That is, they are controlled more by a characteristic climate than soils.

GEOGRAPHICAL LOCATION

Mappable portions of this type occur in eight western states (Fig. 14.3). These, plus small unmapped areas elsewhere, place it in all four regional deserts (Fig. 14.4), parts of the Great Plains, Great Central Valley of California, and valleys within the Rocky Mountains. We will consider only the portions of the Great Basin Regional Desert and Rocky Mountains here, since the variants of this ecosystem type in the southern, subtropical deserts of North America are considered elsewhere in this series (MacMahon and Wagner, in press). The types where extensive upper-elevation boundaries are shared are the Great Basin–Colorado Plateau Sagebrush Semi-Desert (Ch. 12), Intermountain Sagebrush Steppe (Ch. 13), and Colorado Plateau–Mohavian Blackbrush Semi-Desert (Ch. 15).

[1] Manuscript completed June, 1980.

Fig. 14.2. Upland salt-desert shrub site dominated by *Ceratoides lanata* and *Atriplex confertifolia*. Extreme western part of Millard Co., Utah. (Photo courtesy of the U.S. Forest Service.)

EXTENT

This ecosystem type covers 16.9×10^6 ha (Table 14.1), with the most extensive portion in the Great Basin and secondarily the Colorado Plateau Physiographic Provinces.

CURRENT OWNERSHIP AND DOMINANT USES

Because of low natural productivity, little open water, and limited potential for intensive agriculture, most of this area has been traditionally regarded as wasteland. Very few hectares of this type were homesteaded or otherwise claimed for transfer into private ownership. The only major uses have been for range livestock grazing, mining of the accumulated minerals, and military testing and maneuvers. The major landlords are thus the Bureau of Land Management, Department of Defense and Department of Energy — all federal agencies.

ECOSYSTEM STRUCTURE AND FUNCTION

Climate

Because different portions of the type are so widely distributed and more related to edaphic than climatic factors, the climate is harder to characterize than that of other ecosystem types (Fig. 14.5). The major similarity is aridity and temperature extremes greater than those of the sagebrush steppes or sagebrush semi-deserts. Since this ecosystem type is usually located in valley bottoms, the elevations are relatively low and rain-shadow effects are well developed. Much of the area being considered lacks drainage to the ocean. Salty soils form in the lower portions of stream drainages where waters have evaporated and left the salts that had been originally dissolved during weathering of rocks in upland positions. There has not been enough leaching of the profiles to carry the salts away.

Total average annual precipitation for most of

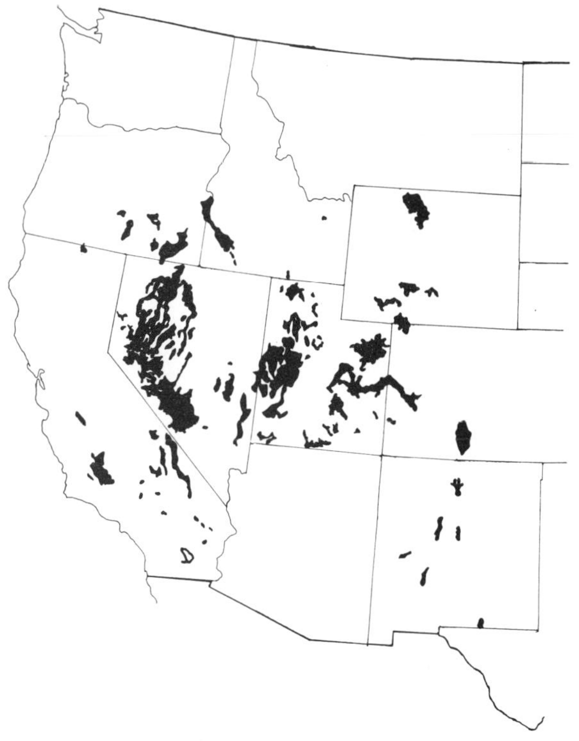

Fig. 14.3. Map of the Intermountain Salt Desert Shrubland Ecosystem Type. Derived from Küchler (1970).

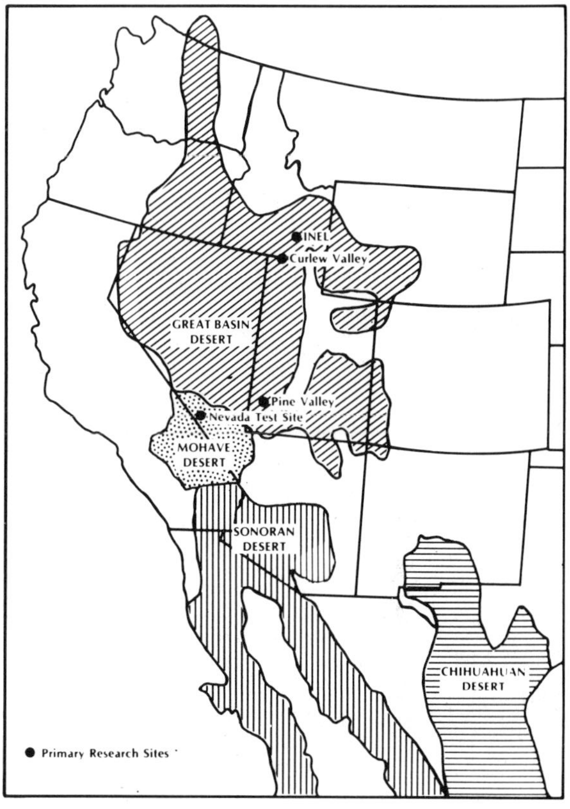

Fig. 14.4. Regional deserts of North America, according to Shreve (1942), with location of major research sites mentioned in this and other chapters.

TABLE 14.1

Area occupied by salt-desert shrublands (derived from Küchler, 1970)

State	Area ($\times 10^6$ ha)	Percent of total
Nevada	6.2	37
Utah	5.0	30
California	1.5	8
Oregon	1.1	7
Wyoming	1.1	7
Colorado	1.1	7
New Mexico	0.5	2
Idaho	0.4	2
Total	16.9	100

this ecosystem type is less than 200 mm. Seasonality of precipitation varies from winter concentrations in the western Great Basin to late summer peaks on the Colorado Plateau and Rio Grande River drainages.

The effectiveness of precipitation from particular storms varies greatly depending on the size and intensity of the storms and the temperature when it is deposited. Winter precipitation is generally more effective, but sublimation and frozen layers may cause reduced infiltration. Soil moisture must generally infiltrate below the surface centimeter to be effective in plant growth. Prewetting may aid deeper recharge in summer (Brewster, 1968).

Diurnal and seasonal variations in temperature are among the highest on the continent (Wein and West, 1972). This is promoted by the extremely clear air, very low average relative humidities, and low elevational position of the basins where this type occurs. Cold air often drains into these lowlands during winter months and accumulates during periods of high atmospheric pressure. This results in the lower average temperatures and colder winter extremes than for nearby stations at sagebrush-dominated sites. Summer temperatures are often greater than in the types immediately above.

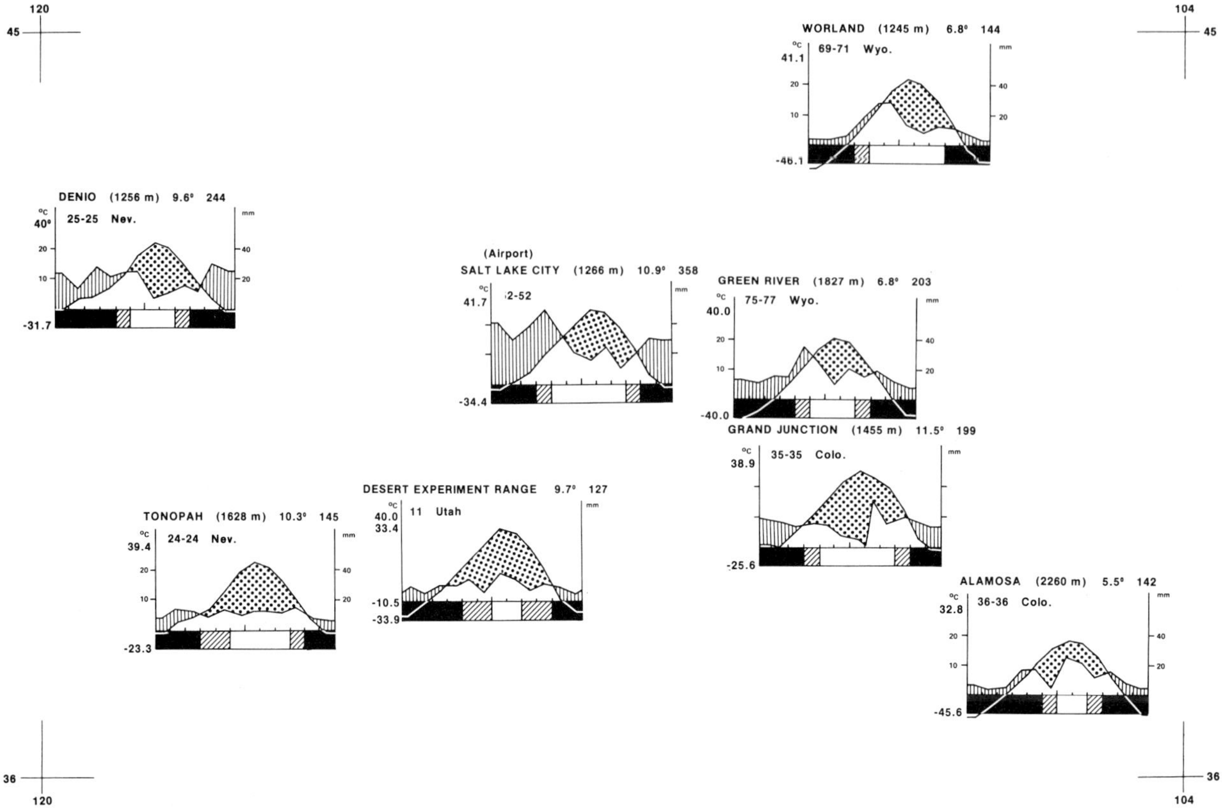

Fig. 14.5. Climatic diagrams for representative stations within the Intermountain Salt Desert Shrubland Ecosystem type. Data from National Weather Service network.

The differential heating and cooling of bare and vegetated areas causes much air turbulence. Small vortices called "dust devils" commonly sweep across the landscapes in summer afternoons and tornadoes are occasionally observed. Average wind velocities have been rarely recorded. Salt Lake City Airport, Utah, has recorded an average of 14.2 km hr^{-1} over the last twenty years. High wind periods lead to dust storms that can greatly reduce visibility. When combined with either frontal or conventional rain, "mud storms" can result. These storms can also move salt in windblown sediments back upstream.

Geology

This ecosystem type occurs mostly in two kinds of situations which promote extreme soil salinity and/or alkalinity. These are either at the bottom of drainages in enclosed basins or where marine shales outcrop. The endorheic (enclosed) basins or bolsons of the Basin and Range Province (Hunt, 1974) where salts and fine-textured fluvial or lacustrine materials have not had a chance to escape to the ocean constitute the major area occupied by this type in the Great Basin.

The sediments from former seas outcropping in dry climates have also come to be occupied by halomorphic soils. Part of the Snake River Plains in southwestern Idaho and the Big Horn Basin and Red Desert in Wyoming, and the "Painted Desert" along the Upper and Little Colorado River drainages in the Colorado Plateau are examples of this situation.

These sediments are largely shales of Cretaceous age (Mancos, Bear Paw, Pierre, Green River Formations) that were deposited in shallow seas. Soils derived from these formations are so salty and fine-textured that they can create shrub habitat in a macro-climate that would normally produce grassland. The Badlands of South Dakota and the breaks along the Missouri River are such situations in the Great Plains. Where shales outcrop in regions of truly arid climates, little or no widely dispersed vegetation develops (Fig. 14.6) because infiltration of soil moisture is almost nil under such circumstances. Plants become restricted to run-on, concave micro-topography.

Fig. 14.6. Example of shale-derived "badlands" near Capital Reef National Park, Utah.

Topography

Two types of topography are predominantly associated with salt-desert shrub vegetation — playas (salt flats or salines within bolsons; Fig. 14.1) and badlands (Fig. 14.6). Playas form at the lowest points of endorheic (enclosed) basins. If annual evaporation exceeds inflow, temporary lakes quickly dry away. Vast areas of the Basin and Range Physiographic Province were covered by lakes during the Pleistocene when conditions were wetter and/or cooler (Fig. 11.3). As conditions subsequently became drier and hotter, the lakes evaporated leaving salty lakes or dry salt pans. The latter are occasionally refilled during wet periods. Playas are the flattest areas of natural topography existing on the earth.

Badlands develop where shales outcrop in desert to semi-desert climates. Soil moisture inputs are so limited that there is not enough to leach out salts or encourage much vegetation. Erosion rates are very great there because of steep slopes and low vegetal cover. High sodium content causes the fine particles of disintegrating shale to disperse (slake), preventing further infiltration. The result is a very high loss as runoff and evaporation from the rain that does fall on such soils.

In the very dry Sierra Nevada rain-shadow area of western Nevada and California, chenopods may be found on a much wider variety of landforms and soils than elsewhere (Young et al., 1977). In that region climatic aridity dominates over edaphic controls.

Soils

Soils usually either derive from flat, deep lacustrine deposits or badlands. The development of soils on the lacustrine deposits depends largely on their salt content, texture and water table. Gradients from Histosols in marshes to Aridisols on the uplands develop (Fig. 14.7). The Great Groups — Camborthids, Calciorthids, Natrargids, and Salorthids — prevail (Naphan, 1966). Entisols may be found where sand dunes and shale badlands occur.

The age and rate of soil change is related to elevation since both water and salts accumulate in the lowest portion of the basins. The particular complement of salt present can determine whether saline and/or alkaline soils are present. This in turn is related to the bedrock geology of the surrounding mountains and the paleoclimatology of the area.

Badland soils are excessively drained in the extreme. The surface of some is eroding away almost as fast as the underlying rock weathers. Because so few plants exist on such sites (Fig. 14.6), some would say that soil does not exist there. Those who are more flexible usually describe the profiles as Entisols. Further classification largely depends on topographic placement and temperature regime (Naphan, 1966).

Soils vary greatly on a micro-scale. Organic matter and nutrients are concentrated on the surface and around shrubs (Charley and West, 1975). Mounded micro-relief is common around the clumps of shrubs. The interspaces have few perennial vascular plants. The surface crust is commonly vesicularized (Springer, 1958) and colonized by a microphytic crust made up of intermingled lichens, mosses, fungi, algae, and bacteria if fine textured. A "desert pavement" of small stones can occur where gravels are in the parent material.

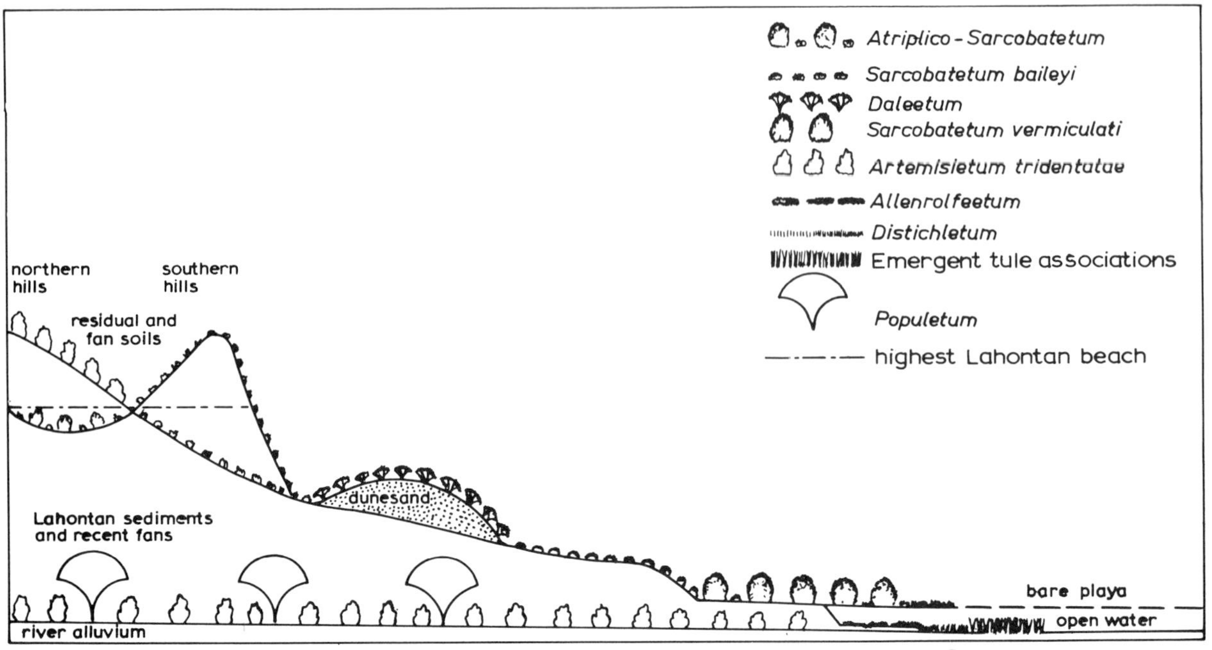

Fig. 14.7. Diagrammatic representation of the topographic and geologic positions of the principal associations in the Carson Desert region of western Nevada. From Billings (1945).

Vegetation composition and productivity are so intimately related to differences in soil characteristics in this ecosystem type that considerable study of these interrelationships has been undertaken. We will return to this topic after discussion of the general characteristics of the vegetation.

Primary producers

Floristics

The overwhelming majority of the vascular plants found in this ecosystem type are members of the Chenopodiaceae. There are occasional members of the Asteraceae, Brassicaceae, Fabaceae, and Poaceae, but they are hardly dominants. A total species list for any given plot would be very brief. Perhaps the total list for the entire type would be less than that of a hectare of eastern deciduous forest or midwestern grassland. There are very few plants that can withstand the rigors of high salinity and/or extreme aridity, and both high and low temperatures. Table 14.2 lists the major plants of this ecosystem type. The shrub and half-shrub (suffrutescent) life forms prevail. The few annuals are largely exotic. Geophytes are largely absent.

Vegetation

Total cover of higher plants in this ecosystem type varies from zero on the salt pans upwards to 20% on the upland sites. The amount of vegetation on level, upland sites largely depends on the average annual precipitation, but can be modified a great deal by soil texture and salt content as it affects soil moisture infiltration and storage. Soils with low amounts of salt and coarser textures (up to about 50% rock) have better infiltration and thus much more effective soil moisture storage and moisture release characteristics. Variation in cover on sites with a seasonal or permanent water table depends on the chemical characteristics of the soils and waters where the plants are rooted (Vest, 1962; Flowers and Evans, 1966; West and Ibrahim, 1968; Steger, 1970; Tueller et al., 1972; Brotherson, 1975). The various species apparently sort out in relation to seasonal water stress patterns developing from both matric and osmotic potential (Detling and Klikoff, 1973).

Height of upland vegetation is usually less than 50 cm. The shrubs are widely spaced and usually occur in clusters (West and Goodall, 1980). The average size of the plants that occur here is much

TABLE 14.2

Major vascular plant species of the salt-desert shrub type, their growth form and preference in regard to subsurface moisture

Habitat	Growth forms		
	shrubs	half-shrubs	herbs
Upland water table well below 1 m xerohalophytes	*Artemisia spinescens* *Atriplex confertifolia* *A.* (*Grayia*) *spinosa*	*Atriplex corrugata* *A. cuneata* *A. falcata* *A. gardneri* *A. tridentata* *Ceratoides* (*Eurotia*) *lanata* *Kochia americana*	*Bromus tectorum* *Elymus cinereus* *Halogeton glomeratus* *Lepdium perfoliatum* *Oryzopsis hymenoides* *Salsola kali* *Sitanion hystrix* *Sporobolus airoides*
Lowland free water table at least occasionally present at the surface and remaining within 1 m hygrohalophytes	*Sarcobatus vermiculatus*	*Allenrolfea occidentalis* *Salicornia utahensis* *Suaeda torreyana*	*Distichlis stricta*

smaller than those of the sagebrush semi-desert. Densities vary from near zero on playas and badlands to approximately two perennials per square meter on sites with relatively salt-free soils of coarse texture (Fautin, 1946).

The interspaces between the higher plant clusters are commonly covered by a microphytic crust. The surface will be rugose and soft when stepped upon, if the site has not been compacted by the feet of animals or wheels of vehicles. Wagner (1980) speculates that the microphytic and vesicular crusts common in this ecosystem type may functionally substitute for the organic mulch layer found in more mesic systems.

There are often distinct boundaries within this and adjacent ecosystem types. This is probably due to the sharp changes in solubility of salts at certain concentrations in the former immersing lake water or the striking differences in soils where marine shales outcrop (Fig. 14.8). The glycophytes do not venture into salty terrain, and the chenopods do not compete well with dominants on adjacent non-halomorphic soils (Barbour, 1970).

Within the chenopods there are considerable differences in tolerance to salinity and aridity. The few species that can survive in these environments tend to sort themselves out along a moisture/salinity gradient. Branson et al. (1967, 1976) stressed the importance of total soil moisture stress

Fig. 14.8. Boundary between salt-desert shrub areas (below) and sagebrush semi-desert and pinyon–juniper woodland (above). Colorado National Monument, Colorado.

(TSMS) in explaining plant distribution. Salts contribute to the osmotic portion of the stress. Physiological drought is not a major problem, however, since the native plants either have means to exclude uptake of salts or take them up and anatomically isolate and/or excrete them. The major shrubs may also carry on some photosynthesis in mid-winter (Caldwell, 1974). Roots of *Atriplex confertifolia* may continue growing at deeper depths into the middle of the summer (Fernandez and Caldwell, 1975). Carbohydrate reserve cycles are less distinctly patterned than those of more mesic environments (Coyne and Cooke, 1970), but not less critical in adjusting season and intensity of livestock use (Trlica and Cook, 1971).

Where topographic gradients are very gradual, monospecific stands of perennial chenopods may develop (Fig. 14.9). The causation of these alterns has attracted considerable research interest, mostly involving edaphic relationships which were not definitive (Gates et al., 1956; Mitchell et al., 1966). How biological competition and other factors enter into sharpening of the boundaries is unknown. About 20% of the area of the type in the Basin and Range Province is found as these mosaics of pure stands (Fig. 14.10).

The sorting of species may follow different sequences in different valleys. At least some of this lack of predictability in soil–vegetation relationships may be due to ecotypic variation (Workman and West, 1969; Goodman and Caldwell, 1971; Goodman, 1973). Chenopodiaceae are known to evolve very rapidly and divergent populations of the same morphological species may be forming incipient species in each isolated valley (Stutz, 1978).

Budbreak of upland xerohalophyte species starts in late winter, flowering comes in the spring and seed development continues throughout the summer (West and Gasto, 1978). The upland plants draw only on vadose water and turn a grayish-green only in the spring. Several of the major species (e.g., *Atriplex confertifolia*) have a set of larger spring leaves that are lost as soil drouth develops. They develop a second set of much smaller, overwintering leaves that can carry on some photosynthesis over the winter. The upland vegetation looks dead during the summer, but turns reddish-yellow in the fall.

Fig. 14.10. *Ceratoides lanata–Atriplex confertifolia* alterne in Curlew Valley, Utah. Between *A* and *C* in Figure 14.9.

Lowland vegetation, being dominated by hygro-

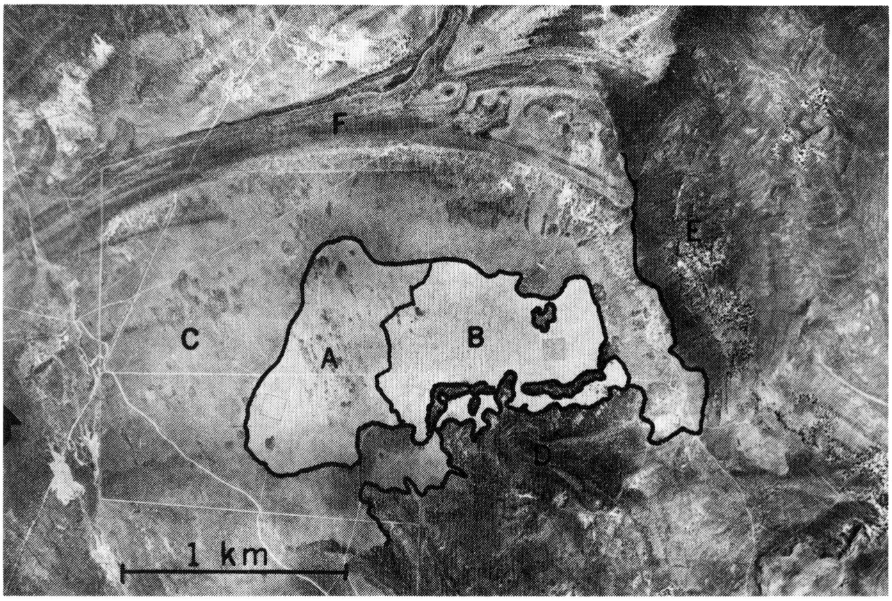

Fig. 14.9. Aerial view of plant community patterns in Curlew Valley, Utah. *A* = *Ceratoides lanata*; *B* = *Atriplex falcata*; *C* = *Atriplex confertifolia*; *D* = *Artemisia tridentata*; *E* = *Artemisia nova*; *F* = Stansbury Water Plain. From Mitchell et al. (1966).

halophytes, does not get leaves and take on a greenish cast until summer. This is when water tables are highest because of snow melt in the mountains and runoff into the basins. The lowland species are mostly phreatophytes that draw on a temporary or permanent water table. Unlike the upland species, they can delay their major development until the warmest period of the year when they are essentially sub-irrigated.

Above-ground biomass in the upland plant communities varies between nearly zero on dry salt pans to around 6 t ha^{-1} on the upland sites with the greatest input of vadose moisture (Bjerregaard, 1971; Holmgren and Brewster, 1972). The one known attempt to measure the biomass of the microphytic crust yielded a weight of up to 200 kg ha^{-1}, even after correction for ash content (Lynn and Vogelsberg, 1974).

Over half of the vascular plant biomass is below ground in these systems and approaches 80% in the case of communities dominated by the suffrutescent species. This makes the root-to-shoot ratios among the highest in the world (Rodin and Bazilevich, 1967). Bjerregaard (1971) measured a three-year average of about 8 t ha^{-1} of below-ground biomass in a pure stand of *Ceratoides lanata* in northwestern Utah. Holmgren and Brewster (1972) present data of similar magnitude for a mixed-species stand in Pine Valley, Utah.

Caldwell et al. (1977) measured a rate of total annual net primary productivity of 192 g C m^{-2} yr^{-1} at a site in Curlew Valley, Utah, dominated by *Atriplex confertifolia*. Since these plants are about 50% carbon we can double this value for an estimate of about 4 t ha^{-1} yr^{-1} as the total rate of net primary production. Only about 25% of this productivity was in the above-ground fraction (Caldwell and Camp, 1974). This is one of the wettest portions of the upland salt-desert shrub-type and should not be considered representative. The estimates of Holmgren and Brewster (1972) of 280 kg ha^{-1} above-ground and about 1 t ha^{-1} below-ground net primary production were derived from a more representative upland site.

No estimates of biomass or productivity are apparently available for lowland sites with high water tables. The poisonous nature of the forage for livestock (high oxalate content) has made these variants of the ecosystem type subject to much less investigation. The supplemental moisture allows these lowland areas to produce much more phytomass. Since the species resprout there, it is conceivable that these communities could be periodically harvested to yield phytomass for energy and food production via chemurgic processes.

Fall estimates of standing crops of available forage made at the Desert Experimental Range in southwestern Utah (Fig. 14.11) show that there is around an 800% variation in above-ground production between the driest and wettest years (Fig. 14.11A). When production is plotted against precipitation of the preceding 12 to 15 months, a closely linear relationship is demonstrated (Fig. 14.11B). Fig. 14.12 shows how the various life

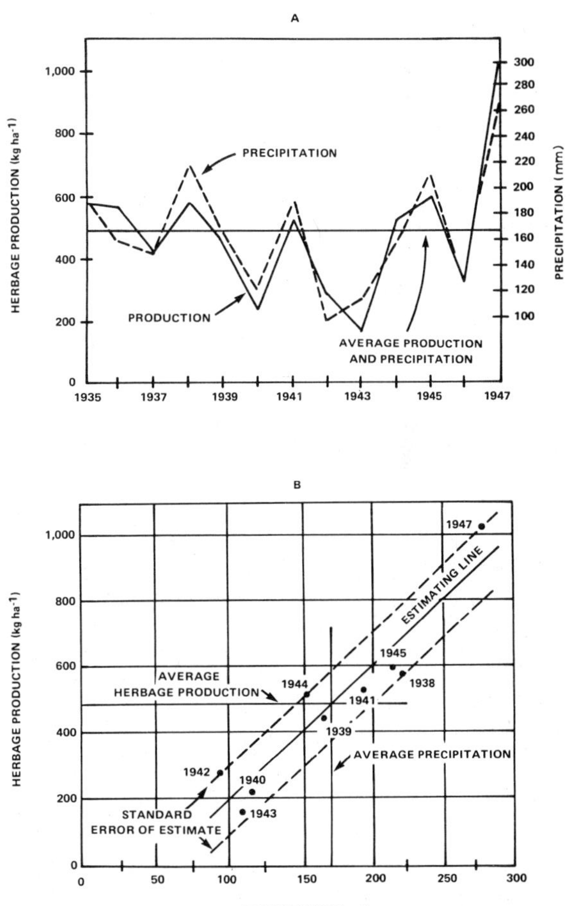

Fig. 14.11. October herbage standing crop in relation to precipitation, 1935–47. A. Year-to-year variation in chronosequence. B. Linear regression of herbage on precipitation of the preceding twelve months. Desert Experimental Range, Millard Co., Utah. From Hutchings and Stewart (1953).

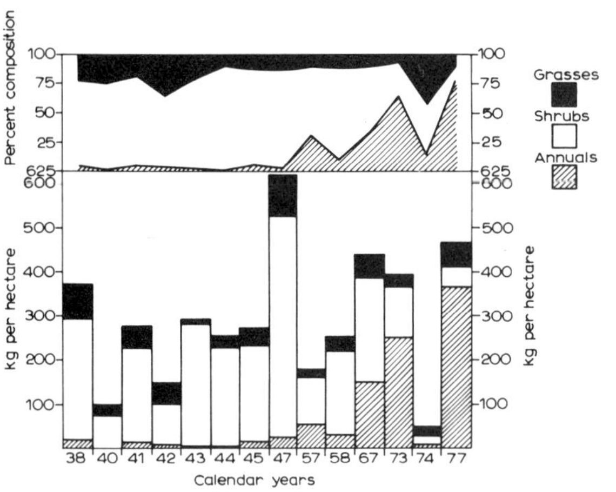

Fig. 14.12. Contribution of three growth forms to October standing crops of herbage on Pasture 18 over fourteen years of record. Desert Experimental Range, Millard Co., Utah. From Gutierrez-Garza (1978).

forms contribute to this productivity. Fetcher and Trlica (1980) have recently shown that the spring precipitation component is most highly correlated with annual growth. The above-ground biomass available in the fall has been of greatest interest to the livestock grazier since winter use by sheep prevailed. A high proportion of the standing crop can be safely consumed at that season without risk to the survival of the shrubs. Sheep grazing requirements of 0.4 to 1.4 ha per month on sites in good condition were possible at that season (Table 14.3).

If the graziers prefer to use cattle or switch sheep use to the growing season, much lower grazing capacities should be established (Gutierrez-Garza, 1978). This is partially because the plants are much less tolerant to foliage removal during the growing season (Cook, 1971). Grazing of shrubs during the winter of drought periods may actually make the plants less susceptible to drouth because the remaining part of the plant has apparently less moisture demands (Chambers, 1979). Grasses sustain much more damage from combined drouth and grazing stresses than do shrubs in these environments (Chambers, 1979).

Cattle prefer the herbaceous component which is usually a small part of the plant production in this ecosystem type. Thus, those graziers who decided to switch from sheep to cattle have often done so at conversion rates of 10–20 sheep to 1 cow rather than the 5 to 1 rate allowed on grassland ranges.

Consumers

The native mammalian fauna is dominated by rodents and lagomorphs (Table 14.4). The only native ungulate present is the pronghorned antelope (Fig. 14.13), for which this is presently marginal range (Hancock, 1966). Wagner (in press) considered that this animal was once more abundant but declined as the quantity and quality of forage was altered by heavy livestock use. Populations of pronghorned antelope are now low because fawn

TABLE 14.3

Grazing capacities of some salt-desert shrub ranges in Pine Valley, Utah (from Hutchings and Stewart, 1953)

Vegetation type	Hectares per sheep month	
	good condition	fair to poor condition
Shadscale–grass	1.4	2.3
Shadscale–winterfat–grass	0.9	1.5
Blacksage–shadscale–grass	0.4	0.9
Winterfat–small rabbitbrush–grass	0.5	1.1
Winterfat	0.4	1.0
Gray molly–Gardner's saltbush–winterfat	0.8	1.1

Shadscale — *Atriplex confertifolia*; winterfat — *Ceratoides (Eurotia) lanata*; blacksage — *Artemisia nova*; small rabbitbrush — *Chrysothamnus stenophyllus*; gray molly *Kochia americana*; Gardner's saltbush — *Atriplex gardneri*; grasses — *Oryzopsis hymenoides*, *Sporobolus cryptandrus*, *Hilaria jamesii*, *Sitanion hystrix*, and others.

TABLE 14.4

Listing of vertebrates and abundance estimates (ha^{-1}) of rodents in sites dominated by shadscale (*Atriplex confertifolia*) and greasewood (*Sarcobatus vermiculatus*) in White Valley, Millard Co., Utah (Fautin, 1946)

Latin name	Common name	Shadscale	Greasewood
MAMMALS			
Antilocapra americana americana	pronghorn	×	×
Antrozous pallidius pallidus	desert pallid bat	×	
Canis latrans lestes	coyote	×	×
Ammospermophilus (*Citellus*) *leucurus leucurus*	antelope ground squirrel	×	×
Spermophilus (*Citellus*) *townsendii mollis*	Paiute ground squirrel	×	×
Dipodomys microps bonnevillei	kangaroo rat	34.0	10.2
Dipodomys ordii celeripes	kangaroo rat	2.8	13.5
Eptesicus fescus pallidus	pallid big brown bat		×
Lepus californicus deserticola	black-tailed desert jackrabbit	×	×
Lynx rufus	bobcat	×	
Microdipodops megacephalus paululus	kangaroo mouse	1.0	0.2
Mustela frenata nevadensis	Nevada long-tailed weasel	×	×
Myotis subulatus melanorhinus	black-nosed bat	×	×
Onychomys leucogaster brevicaudus	grasshopper mouse	0.3	0.7
Perognathus longimembris nevadensis	Nevada pocket mouse	1.3	×
Peromyscus maniculatus sonoriensis	white-footed mouse	2.8	11.7
Reithrodontomys megalotis megalotis	harvest mouse		9.5
Spilogale gracilis saxatilis	Great Basin spotted skunk	×	
Sylvilagus nuttallii grangeri	cottontail rabbit	×	
Tadarida mexicana	Mexican free-tailed bat	×	
Taxidea taxus	badger	×	×
Thomomys bottae centralis	pocket gopher	×	×
Vulpes macrotis	kit fox	×	
BIRDS			
Accipiter cooperii	Cooper's hawk		×
Accipiter striatus velox	sharp-shinned hawk	×	×
Aeronautes saxatilis saxatilis	western white-throated swift	×	
Agelaius phoeniceus	red-winged blackbird	×	×
Amphispiza belli nevadensis	northern sage sparrow	×	×
Amphispiza bilineata deserticola	desert black-throated sparrow	×	×
Anas platyrhynchos platyrhynchos	common mallard		×
Aquila chrysaetos canadensis	golden eagle	×	×
Ardea herodias treganzia	Treganza's great blue heron		×
Buteo swainsoni	Swainson's hawk	×	×
Calamospiza melanocorys	lark bunting	×	×
Carpodacus mexicanus	house finch	×	
Cathartes aura teter	western turkey vulture	×	×
Charadrius vociferus vociferus	killdeer		×
Chondestes grammacus strigatus	western lark sparrow	×	
Chordeiles minor	nighthawk	×	×
Circus cyaneus hudsonius	marsh hawk	×	×
Corvus corax sinuatus	American raven	×	×
Dendroica aestiva	yellow warbler		×
Dendroica aubodoni	northern audobon warbler		×
Ereunetes mauri	western sandpiper		×
Euphyagus cyanocephalus cyanocephalus	Brewer's blackbird	×	×
Falco mexicanus	prairie falcon	×	×
Falco sparverius sparverius	eastern sparrow hawk	×	×
Geothylpis trichas occidentalis	western yellow-throat		×
Haliaeetus leucocephalus leucocephalus	southern bald eagle	×	
Hirundo rustica erythrogaster	barn swallow		×

TABLE 14.4 (*continued*)

Latin name	Common name	Shadscale	Greasewood
Icteria virens auricollis	long-tailed chat		×
Lanius ludovicianus nevadensis	Great Basin shrike	×	×
Melospiza melodia	song sparrow	×	
Mimus polyglottus leucopterus	western mockingbird		×
Molothrus ater artemisiae	Nevada cowbird	×	×
Myiarchus cinerascens cinerascens	ashy-throated flycatcher	×	
Oberholseria chlorura	green-tailed towhee		×
Oreoscoptes montanus	sage thrasher	×	×
Otocoris alpestris utahensis	Great Salt Lake horned lark	×	×
Passer domesticus domesticus	English sparrow	×	
Passerculus sandwichensis nevadensis	Nevada savannah sparrow		×
Phalaenoptilus nuttallii nuttallii	Nuttall's poor-will	×	
Pooecetes gramineus confinus	western vesper sparrow	×	×
Prozana carolina	sora rail		×
Sayornis saya saya	Say's phoebe	×	
Selasphorus platycercus platycercus	broad-tailed hummingbird	×	
Selasphorus rufus	rufous hummingbird	×	
Speotyto cunicularia hypogaea	western burrowing owl	×	×
Spinus tristis pallidus	pale goldfinch	×	
Spizella breweri breweri	Brewer's sparrow	×	×
Spizella passerina arizonae	western chipping sparrow	×	
Tachycineta thalassina lepida	violet-green swallow	×	
Tyrannus verticalus	Arkansas kingbird	×	×
Xanthocephalus xanthocephalus	yellow-headed blackbird	×	×
Zenaidura macroura marginella	western mourning dove	×	×
Zonotrichia leucophrys	white-crowned sparrow	×	×
REPTILES			
Cnemidophorus tessellatus tessellatus	whip-tail lizard	×	×
Crotalus viridus lutosus	Great Basin rattlesnake	×	×
Crotaphytus wislizenii	leopard lizard	×	×
Masticophis (Coluber) taeniatus taeniatus	striped racer	×	×
Phrynosoma platyrhinos	desert horned-toad	×	×
Pituophis catenifer deserticola	Great Basin gopher snake	×	×
Rhinocheilus lecontei	long-nosed snake	×	
Sceloporus graciosus graciosus	sagebrush lizard		×
Uta stansburiana stansburiana	northern side-blotched lizard	×	×

survival is as undependable as the summer precipitation that produces the necessary forbs for lactation and direct consumption (Beale and Smith, 1970). These animals exist here only because they have been pushed off the better ranges by livestock. Mule deer (*Odocoileus hemionus*), bighorn sheep (*Ovis canadensis*) and wapiti (*Cervus canadensis*) use these areas only when extreme snowfalls cover up sagebrush ranges.

The major vertebrate consumer in this ecosystem type is the black-tailed desert jackrabbit. Currie and Goodwin (1966) have demonstrated that about seven of these animals are equivalent to one sheep. A large part of the impact is due to the clipping of foliage, much of which is left on the ground unconsumed (McKeever and Hubbard, 1960). These animals particularly seek out *Kochia americana* (Westoby, 1973). Their population densities may approach two per hectare during the spring of years with cyclical population highs (Olsen, 1971; Gross et al., 1974).

The present abundance of seed-eating rodents, especially the geomyids (Fig. 14.14), may be related to an increase in annual plants with livestock grazing disturbance (Larrison and Johnson, 1973).

The chisel-tooth kangaroo rat has the unique

Fig. 14.13. Young pronghorn (*Antilocapra americana*). (Photo by J. Yoakum.)

adaptation of being able to separate out the hypersaline tissues when it eats foliage of the perennials (Kenagy, 1972).

The only introduced vertebrate that thrives in this ecosystem type is the chukar partridge (*Alectoris chukar*) (Fig. 12.9). This bird occurs, however, only where rocky escape cover is adjacent to rangelands dominated by winter annuals.

Quite a few birds are occasional visitors to this ecosystem type, but few (Table 14.4) spend much of their time there. Only the horned lark is ever seen in any abundance.

Both mammalian and avian predator populations are similar to the Great Basin–Colorado Plateau Sagebrush Semi-Desert but occur in lower numbers because of the lower herbivore food base.

The herpetofauna (Table 14.4) is both relatively low in diversity and low in absolute abundance compared to the southern deserts (Pianka, 1967).

Invertebrates have been largely ignored in this ecosystem type. Only faunistic lists exist for a few areas. Table 14.5 is such an example.

The most conspicuous insects of the uplands are harvester ants (King, 1963). The area occupied by the nests and foraging may take up to 5 to 10% of a salt-desert shrub range (Sharp and Barr, 1960) (Fig. 14.15). Whether this fraction has increased with an increase of annuals after livestock disturbance can be debated (Kirkham and Fisser, 1972). Because there is some increase in production of plants around the edge of their feeding areas (Wight and Nichols, 1966), not all bare area means a linear loss of shrub production.

Occasional loss of browse plants has been attri-

TABLE 14.5

Taxonomic summarization of invertebrates collected at sites dominated by shadscale (S) and greasewood (G) in White Valley, Millard Co., Utah (Fautin, 1946)

Orders	Families		Genera		Species		Specimens	
	S	G	S	G	S	G	S	G
Araneida	4	6	8	14	10	19	32	138
Acarina	2	3	3	3	4	3	92	20
Collembola	1	–	1	–	1	–	3	–
Orthoptera	1	1	3	2	3	2	6	16
Thysanoptera	–	1	–	1	–	1	–	2
Hemiptera	4	7	6	14	6	17	64	98
Homoptera	4	6	11	14	13	15	96	832
Coleoptera	7	7	10	12	11	13	19	132
Lepidoptera	1	4	*	*	*	*	*	*
Diptera	7	14	9	16	9	19	39	229
Hymenoptera	3	7	6	13	7	18	33	146

* Larvae and pupae which were not determinable to species.

Fig. 14.14. Ord's kangaroo rat (*Dipodomys ordii*), a major granivore in the salt-desert shrubland ecosystem type.

buted to round-headed borers (Cerambycidae) (Hutchings, 1952) or cut worms (larval Noctuidae) (L. Sharp, pers. comm., 1967). The interaction of these outbreaks with drouth or wet periods is suspected, but cannot yet be verified.

Since upwards of 100 species of invertebrates may occur in the more highly vegetated portions of this ecosystem type (Fautin, 1946), much more study needs to be done of the ecological roles of insects in this system.

Decomposers

Because there is relatively little above-ground primary or secondary productivity, decomposer activity on the soil surface is low. Most of the activity takes place below ground and in cells centered on the clumps of shrubs (Charley and West, 1977). It is on this small fraction of the landscape where moisture infiltration and nutrients are concentrated that a level of production and decomposer activity approaches that of more mesic systems. There simply is not enough moisture in the upland desert phases for a continuous sward of perennial vegetal cover and thus associated biological activity to be maintained. It is too salty on or near the playas.

Fig. 14.15. Closeup of a western harvester ant (*Pogonomyrmex* sp.) nest on an *Atriplex falcata* site in northwestern Utah. Note regularly dispersed bare spots in Fig. 14.9 due to nest construction and foraging by ants.

Apparently only microbial processes associated with the nitrogen cycle have been so far examined in this ecosystem type (West and Skujins, 1977). There are abundant opportunities for structural and functional aspects of the microbial communities to be studied further.

ECOSYSTEM INTERACTIONS

Because of the floristic and faunistic simplicity of the communities in this ecosystem type, food webs are short and comparatively simple. Energy flow and nutrient cycling processes have large physical components. Although one might intuitively expect the biological modification of micro-climate to be minor, the *edifikator* effect of the scattered shrubs is extremely important in focusing water infiltration, litter inputs and thus animal and microbial activities into small patches (Gasto, 1969; Charley and West, 1975, 1977; West and Goodall, 1980).

The major biota of the interspaces are connected with the microphytic soil crusts. The blue-green algal component, both free-living and as a symbiont in lichens, is the major fixer of nitrogen in the system (Rychert and Skujins, 1974; West and Skujins, 1977). Nitrogen is a second-order limiting factor whenever soil moisture is above normal.

Dominance–diversity relationships for higher plants and animals verge on monocultures in these systems because so few organisms can really be successful in this harsh environment. These same environmental constraints make ecosystem development (succession) appear either very slow or truncated. Since the same species or species similar in appearance and stature often succeed each other after disturbance, autosuccession probably best describes what occurs here. Range condition and trend methodology, being based on linear-deterministic models of stages as hypothecated in Clements's day, is thus hard to employ here.

The probable halosere involved in primary succession is probably related to the gradient of decreasing salinity and deepening water table. Fig. 14.16 depicts the usual sequence of native higher plant species located along this gradient. Species positions change largely in response to changes in salinity. For instance, Rickard (1964) has documented how the salt-pumping action of *Sarcobatus* has caused *Artemisia* to decline as

Fig. 14.16. Probable halosere around retreating saline lakes of the Great Basin (after Flowers and Evans, 1966).

micro-topographic differences were eroded downward.

We do know that the pristine system was somewhat unstable and that considerable compositional change did occur under grazing perturbation (Young et al., 1976), particularly on the best sites within this type. The shrubs that were most palatable to livestock, were the most sensitive to spring grazing, and had the least reproductive capacity have greatly declined — for instance, *Artemisia spinescens*, *Ceratoides lanata*, and *Kochia americana*. There is some evidence that grasses such as *Hilaria jamesii*, *Oryzopsis hymenoides*, *Sitanion hystrix*, and *Sporobolus cryptandrus* were also more abundant once. The less palatable species with high reproductive capacities have come back the most rapidly after control of grazing. These trends are, however, hard to distinguish from annual fluctuations and longer-term responses to weather conditions (Norton, 1978).

Holmgren and Hutchings (1972) have noted that *Atriplex confertifolia* (shadscale) is more sensitive to drought conditions than *Ceratoides lanata* (winterfat) and *Artemisia spinescens* (budsage). Since drought years occur periodically, shadscale populations are expected to decline with each drought and over a period of years be replaced by winterfat and budsage, if lightly grazed or ungrazed.

Other researchers (Harper, 1959; Norton, 1978; Rice and Westoby, 1978) studying permanent quadrats during a rebound period have also noted low resilience or slow general recovery of the palatable species. It is important to note that the only microsites where new seedlings of perennials survive long are where shrubs currently are or once occurred (Gasto, 1969; Norton, 1978).

Because higher plant cover is generally so sparse, this is one of the few ecosystem types in which we do not usually have to consider the perturbations due to fire. Other disturbing factors are also minimal except in localized areas where recreational and military activities occur.

LAND-USE HISTORY

Use by aboriginal peoples

This ecosystem type was little used by native peoples in pre-industrial times. The Indians who were pushed into these habitats had extremely low material standards of living. They were forced to try and live largely by gathering seeds of *Suaeda* and hunting waterfowl in the marshes (Aikens, 1970; Harper and Alder, 1972). They spent so much of their effort in surviving that they developed the lowest cultural complexities of any people in North America (Farb, 1978). Ironically, it was simple cultures like those of the Paiutes, Goshutes, and Shoshoni peoples who lived here that were least altered by the incursion of European-based culture. This is not to say that there were not major disruptions in social and economic patterns — only that they were comparatively less disastrous.

Use by European-derived cultures

Very little human alteration of the ecosystem occurred until minerals were discovered and the surrounding sagebrush semi-deserts and more mesic systems were beginning to be depleted by livestock grazing. It was then that graziers began looking further afield for free winter forage. This places the first livestock usage of most of this ecosystem type after 1860. Livestock numbers were augmented after 1869 when the first railroad across the continent gave impetus to growing and exporting livestock products.

Much of this ecosystem type has never been suitable to any livestock grazing. The salt flats and some badlands had no palatable forage or potable water. The high water table fringing the salt flats produced salty water and plant materials with very high oxalate and other salt content. The salts, particularly the oxalates, are physiologically harmful to domestic livestock. The oxalates precipitate calcium out of the blood and causes death if the diet contains a high proportion of this material. Nevertheless, the areas where grasses such as *Distichlis*, *Elymus cinereus*, and *Puccinellia* spp., and the grass-like *Juncus* spp., exist are heavily grazed. These areas tend to be near the sources of potable water and thus usually in private ownership.

The upland sites where only vadose water supplies the plants have mostly browse plants with extremely high nutrient content (Table 14.6). For instance, *Ceratoides lanata* (winterfat) has winter protein content that rivals alfalfa (*Medicago*) hay (Harris, 1968; Cook and Harris, 1968). It and other suffrutescent forms have very little woody tissue. The aboveground portions can be consumed almost entirely. Cattle can easily survive on salt-desert ranges where grasses exist for a high-energy source and potable water is available. This latter condition is, however, rare in the salt-desert shrub type. Sheep, however, need less water and can use snow for their water needs. Therefore, sheep in the winter became the usual means by which man exploited this eco-

TABLE 14.6

Average nutrient content of salt desert browse and grass species compared to good sun-cured alfalfa (lucerne, *Medicago*) hay (from Cook, 1966)

	Digestible protein (%)	Metabolizable energy (J kg^{-1})	Phosphorus (%)	Carotene (mg kg^{-1})
Native grasses	0.2	1324	0.07	0.10
Browse	4.7	1171	0.12	3.27
Alfalfa	10.5	1713	0.21	3.59

system type. The area was largely open range (public domain) subject to no governmental control or fee collecting until 1934. The areas without potable water supplies received heavy use during the winter but were largely vacated when there was no snow. Use into the growing season was not possible except around waterholes. As other types deteriorated and Forest Service control came to most of the mountain areas around the turn of the century, tramp flocks (those owned by people with no private lands and thus no Forest Service permits) were forced to use the desert rangelands longer. They had to develop springs, dig wells, construct pipelines or haul water in order to be able to use the sparsely scattered vegetation over vast landscapes.

The great demand for wool during World War I led to some flocks of wethers being kept year-around where water was available. Because the major forage plants can stand much less herbivory during the growing season (Cook, 1971), it was not long before the vegetation was depleted and sand dunes and/or arroyos started to form in such areas.

The devastation on this and other public domain lands reached a crisis point during the drought and economic depression of the 1930s. A survey of that period (McArdle et al., 1936; McArdle and Costello, 1936) estimated that there had been 70% depletion of the original forage supply on these lands. After passage of the Taylor Grazing Act in 1934 and creation of the Grazing Service, later to become the Bureau of Land Management (B.L.M.), tramp herds and flocks of livestock were immediately eliminated. Those who did not have private land to carry their livestock for part of the year were denied permits to graze public ranges. Those who remained slowly underwent a process of reduction of number and season of use to bring livestock use within the grazing capacity of the land. Because of lack of strong legal force until recently (1976), few personnel, lack of agency resolve, over-optimism and lack of basic ecological knowledge, this adjustment has continued to the present (Murdock and Welsh, 1971).

The process of adjustment in livestock numbers was accelerated during the 1950s when the Central Asian annual *Halogeton glomeratus* became spread over these ranges (Cook and Stoddart, 1953; Frischknecht, 1967; Williams, 1973). This plant has up to 30% sodium oxalate content. Livestock who consume much of this plant die by internal bleeding. The oxalates cause precipitation of calcium from the blood and prevent clotting. Fortunately, this summer annual is not aggressive. It may, however, permanently change the soil surface via its salt-pumping action (Eckert and Kinsinger, 1960). This impedes moisture infiltration and enhances evaporation. Thus, it may trigger xerification of these ranges. Closed communities in good condition are fortunately not readily invaded by this weed.

Other exotic annuals such as *Bromus tectorum, Lepidium perfoliatum*, and *Salsola kali* had invaded earlier where disturbance was severe, but have rarely reached the fire-carrying densities so common in the sagebrush-dominated ecosystem types. Further expansion of annuals is generally retarded by rational grazing levels that prevent reductions in the densities of native perennials. However, slow expansion of annuals has been noted even under moderate livestock grazing (Fig. 14.11).

Halogeton-infested areas can be avoided by fencing or herding the livestock away from them (James and Cronin, 1974). Livestock do not usually seek out *Halogeton*, if native perennial forage is available in abundance (Cook et al., 1952).

Considerable losses of livestock to *Halogeton* in the 1950s precipitated research programs on the ecology of this desert vegetation, physiology of the plants, and range animal nutrition. When the importance of season and intensity of use on vegetation dynamics was discovered (Hutchings, 1966; Cook, 1971), the B.L.M. began to force the removal of livestock from these ranges in the spring. This, combined with the understanding of the nutritional needs of livestock on these ranges (Harris, 1968), showed why winter-only sheep use of these areas was to be preferred. Recent computer-assisted modelling of season and intensity of defoliation has largely substantiated these earlier conclusions (Wilkin, 1973; Williams, 1979). High-condition ranges have forage with very well-balanced nutrient content. Animals on degraded ranges may need supplementation depending on the forage composition (Cook et al., 1954; Harris et al., 1956; Pieper et al., 1959; Harrison and Thatcher, 1970). Unfortunately, the decline of the range sheep industry has forced many former sheep men into trying to graze cattle on these browse-dominated ranges with only moderate success (Malechek and Smith, 1975).

The plant species that are favored by disturbance are less nutritious for livestock (Stewart et al., 1940; Hutchings and Stewart, 1953; Hutchings, 1954; Holmgren and Hutchings, 1972; Holmgren, 1973). It is not yet clear whether the reductions in livestock following passage of the Taylor Grazing Act were great enough to have resulted in overall improvement of this ecosystem type. The nutritional levels of these rangelands may have been nearly permanently altered by the early abuses, and the present low grazing capacities for livestock and pronghorn may be very difficult to improve even with more intensive management (Wagner, in press).

Recent developments

The Federal Land Management and Policy Act of 1976 finally acknowledged that the public intention was to retain these lands and manage them for the long-term good of all United States citizens. The National Environmental Policy Act requires that the public have a chance to comment on all federal actions that can have a significant effect on the environment. The Natural Resources Defense Council, a private "watchdog" group, has made sure this was the case for grazing leases on such lands. The preparation of plans considering livestock forage, along with all other present, often competing uses, has brought everything out in the open and usually demonstrates that fewer livestock can be grazed if we wish to stop "mining" the forages, have the vegetation return to higher productivity, and have less soil erosion.

It is doubtful if the rest-rotation grazing systems often proposed for such lands can be successful. Such plans often assume availability of funds for investment in fences and water development. Heavy use during the growing season, even if followed by rest for several years, causes heavy mortality of the most desirable plants (Cook and Child, 1971). Rest periods rarely coincide with years of high seed production followed by favorable conditions for seedling establishment (West and Gasto, 1978).

Technology for widespread economical artificial seeding of desirable forage species in this ecosystem type is not yet available (Hull, 1963; Bleak et al., 1965). Only very intensive work with seedlings and stem cuttings in containers is beginning to show promise on mined areas within this ecosystem type (Crofts and Van Epps, 1975; Richardson et al., 1979). Somewhat easier success has been obtained in re-vegetating moderately saline sites with high water tables (Stuart et al., 1973; Eckert et al., 1973). These sites, however, compose only a small fraction of the area being considered.

Fertilization is an ill-advised practice on these ranges. The native perennials have low nutrient requirements and are adapted mainly to moisture stress. Extra tissues produced during wet years must be lost in dry years. Fertilization makes them only more susceptible to drouth (Goodman, 1973). Fertilization also favors the annuals, usually the undesirable *Halogeton*.

The wisest grazing management strategy remains moderate use during the non-growing season only. Hutchings (1966) has estimated that a two- to nine-fold increase in the livestock grazing capacities could be reached by following these recommendations. Herded sheep use would be preferred because the flocks can be taken over a wider array of landscape than cattle, they are well adapted to the browse and to lack of water, and rarely graze the same areas year after year. It is doubtful, however, whether this logic can be followed due to external economic pressures and concern for other values. Among these are watershed and wildlife concerns.

Erosion

Wind erosion is often severe around the salt pans of the Basin and Range Province. Delta areas remaining where rivers entered Pleistocene lakes are also highly susceptible to wind erosion. The interspaces between shrubs are deflated and material is redeposited around the shrubs. Windy periods may move vast amounts of particulate matter into the air and blow them generally eastward.

The soil surface may have once been held more in place by microphytic soil crusts. The action of animal feet and vehicles disrupts this cover and accelerated erosion may have begun during the past 100 years when increased activity by man and his livestock began in this region.

Wind is generally the major cause of erosion in flat, dry places. Water becomes more important in less dry and steeper terrain. Water erosion prevails as an erosional force forming the shale badland topography of the Wyoming Basins, Colorado

Plateau, and parts of the Great Plains (Fig. 14.6). The erosion rates are so pronounced there that considerable effort has been expended in trying to control it (Coltharp and West, 1966; Coltharp, 1969).

Contour furrowing, gully plugging, ripping and pitting have all been tried on these kinds of lands. All of these treatments attempt to increase infiltration and reduce overland flow and thus enhance vegetation (Gifford et al., 1978a). Although successful in some other, largely grassland, ecosystem types, these treatments are often misplaced here. The shales are so impenetrable and vegetation cover so sparse (generally less than 5%) that almost more vegetation is disturbed than encouraged. The treatments are quickly breached by erosion and/or filled with sediment (Gifford et al., 1978b; Hessary and Gifford, 1979).

The concern is not as much with the negligible water production from these lands as it is with sediment and thus salt input to the Colorado River. Frail lands in the Upper Colorado River Basin, largely shale badlands in the salt-desert shrub type, yield about 85% of the sediment, but only 1% of the water in the Colorado River (Lusby et al., 1963; Lusby, 1965, 1970). Sediment input which destroys dam storage capacity is enormous. Salinity of the water is also of considerable concern to irrigation and other water uses downstream in California, Arizona, and Sonora. A treaty with Mexico guarantees delivery of water that will allow crop growth. This can presently only be done by augmenting the flow with artificially desalinized water. This is very expensive. Since Thompson (1968), Turner (1971), and Lusby et al. (1971) have implicated disturbance of salt-desert ranges by livestock hooves as a major source of sediment, the Bureau of Land Management is currently considering the removal of livestock from most of the type on the Colorado Plateau. Others (Hawkins et al., 1977) have shown that most of the salts come from the sides of arroyos. Resolution of the matter will take further research.

OTHER PROBLEMS

An additional problem is that the shale soils are so unique that numerous endemic plants have evolved in these special habitats. Some of them are placed on the listing of endangered and threatened flora. Legal requirements for protection of these rare plants will probably cause conflicts for any users of these lands that threaten these species.

Wild animal species are so few and populations so low that very little wildlife management has been attempted in this ecosystem type. There is, however, growing concern for the recreational values of wildlife here (Hancock, 1966; Wagner, in press).

The major forms of recreation pursued on these areas employ offroad vehicles to reach scenic, hunting, and rock-hounding areas. The wheels of motorcycles and four-wheel drive vehicles can destroy vegetation, including the microphytic crusts. Wind and water erosion can be accelerated if traffic is concentrated. The B.L.M. is now charged with regulating these activities.

Since this ecosystem type has such low biological productivity and is distant from concentrations of human populations, it has often been used for "nuisance" type activities such as military weapons testing and nuclear reactor development. Large areas of this ecosystem type are now off-limits to unauthorized persons. More of such areas will be lost to multiple-use management if the proposed MX missile system is deployed. There have been relatively few conflicts caused by these activities in the past, since only a few livestock graziers and miners were directly displaced by such reserves. This single military use will probably continue and greatly increase if it becomes necessary to develop weapons that operate over large areas of terrain.

Mineral development in the past was mostly restricted to extraction of salts from non-vegetated salt pans. The particular salts obtained depend on the kinds of minerals in the mountains surrounding the basin. For instance, magnesium is beginning to be electrolytically extracted from the waters of the Great Salt Lake. Many other salts and their associated minerals and metals could be extracted from this and other salt-desert locales, e.g., lithium, zeolites, sodium sulfate, etc. Petroleum and gas exploration is proceeding over most of this region. Discoveries have already been made and fields exist within this ecosystem type, particularly in Utah.

A few of the most spectacular salt pans and badland areas have been reserved in national and state parks, but the majority of the type is not aesthetically pleasing to most viewers. Second-home or retirement developments are not likely there. The cheapness of the non-government land has, however, attracted some unethical promoters to buy and resell small tracts to people who have never seen them. Taxation and management of such tracts creates terrific problems for local governments.

THE FUTURE

The salt-desert shrub type will probably largely remain as wildland with extensive management based on ecological restraints and economic constraints. Because of very low grazing capacities, slow rates of recovery and erosional problems, some livestock will probably be removed from part of these salt desert areas where they have historically grazed.

Even though average productivities are low, there is such a huge area involved that the total is significant. Unfortunately, these ranges improve slowly with progressive livestock management, and little can be done otherwise to accelerate improvement. Lack of potential for sensible agricultural and urban use will prevent large relocations of populations from elsewhere. The land has such low potential for natural primary production that more intensive management of livestock cannot be economically justified. Thus, this ecosystem is likely to remain a vast area where air pollutants can be diluted and the government can carry on the nuisance activities of military weapons testing and deployment. Particular graziers and miners will undergo economic displacement and the nation will be forced to obtain that portion of its basic food and mineral resources elsewhere. Much area will, however, remain unfenced and accessible to all who care to travel there. For particular kinds of devotees of wilderness, it may even serve as one place to escape the congestion of the more commercially aggrandized ecosystems.

REFERENCES

Aikens, C.M. (Editor), 1970. *Hogup Cave. Univ. Utah Anthropol. Pap.*, No. 93: 286 pp.

Barbour, M.G., 1970. Is any angiosperm an obligate halophyte? *Am. Midl. Nat.*, 84: 105–120.

Beale, D.M. and Smith, A.D., 1970. Forage use, water consumption, and production of pronghorn antelope in western Utah. *J. Wildl. Manage.*, 34: 570–582.

Billings, W.D., 1945. The plant associations of the Carson Desert region, western Nevada. *Butler Univ. Bot. Stud.*, 8: 89–123.

Billings, W.D., 1949. The shadscale vegetation zone of Nevada and eastern California in relation to climate and soils. *Am. Midl. Nat.*, 42: 87–109.

Billings, W.D., 1951. Vegetational zonation in the Great Basin of western North America. In: *Les bases ecologiques de la régéneration des Zones Arides*. U.I.S.B., Paris, pp. 101–122.

Bjerregaard, R.S., 1971. *The Nitrogen Budget of Two Salt Desert Shrub Plant Communities of Western Utah*. Thesis, Utah State University, Logan, Utah, 123 pp.

Bleak, A.T., Frischknecht, N.C., Plummer, A.P. and Eckert, R.E., Jr., 1965. Problems in artificial and natural revegetation of the arid shadscale vegetation zone of Utah and Nevada. *J. Range Manage.*, 18: 59–65.

Bolen, E.G., 1964. Plant ecology of spring-fed salt marshes in western Utah. *Ecol. Monogr.*, 34: 143–166.

Branson, F.A., Miller, R.F. and McQueen, I.S., 1967. Geographic distribution and factors affecting the distribution of salt desert shrubs in the United States. *J. Range Manage.*, 20: 287–298.

Branson, F.A., Miller, R.F. and McQueen, I.S., 1976. Moisture relationships in twelve northern desert shrub communities near Grand Junction, Colorado. *Ecology*, 57: 1104–1124.

Brewster, Jr., S.F., 1968. *A Study of the Effectiveness of Precipitation in the Salt Desert Shrub Type*. Thesis, Brigham Young University, Provo, Utah, 66 pp.

Brotherson, J.D., 1975. Environmental gradient analysis and its applications to the distribution patterns of salt desert shrub species. In: H.C. Stutz (Editor), *Proc. Symp. and Workshop Wildland Shrub*. U.S. For. Ser. Shrub Sciences Lab., Provo, Utah, pp. 159–160.

Caldwell, M.M., 1974. Physiology of desert halophytes. In: R.J. Reinhold and W.H. Queens (Editors), *Ecology of Halophytes*. Academic Press, New York, N.Y. pp. 355–378.

Caldwell, M.M. and Camp, L.B. 1974. Belowground productivity of two cool desert communities. *Oecologia*, 17: 123–130.

Caldwell, M.M., White, R.S., Moore, R.T. and Camp, L.B., 1977. Carbon balance, productivity, and water use of cold-winter desert shrub communities dominated by C_3 and C_4 species. *Oecologia*, 29: 275–300.

Chambers, J.C., 1979. *The Effects of Grazing on Salt Desert Shrub Species Survival During a Period of Below Average Precipitation*. Thesis, Utah State University, Logan, Utah, 101 pp.

Chapman, V.J., 1974. Halophytes: An overview of salt marshes and salt deserts of the world. In: J. Reimold and W.H. Queen (Editors), *Ecology of Halophytes*. Academic Press, New York, N.Y., pp. 3–19.

Charley, J.L. and West, N.E., 1975. Plant-induced soil chemical patterns in some shrub-dominated semi-desert ecosystems in Utah. *J. Ecol.*, 63: 945–963.

Charley, J.L. and West, N.E., 1977. Micro-patterns of nitrogen mineralization activity in soils of some shrub-dominated semi-desert ecosystems of Utah. *Soil Biol. Biochem.*, 9: 357–365.

Coltharp, G.B., 1969. Use of contour furrows on soil moisture regimes of frail rangelands in Utah. In: *Abstr. Conf. on Arid*

Lands in a Changing World. AAAS Univ. Arizona, Tucson, Ariz.

Coltharp, G.B. and West, N.E., 1966. Effects of surface soil treatments on soil, water, and vegetation in Utah's East Desert Area. In: *Proc. Salt Desert Shrub Symposium.* Bureau of Land Management, U.S. Dep. Inter., Cedar City, Utah, pp. 88–97.

Cook, C.W., 1966. Nutritional value of salt-desert shrub ranges. In: *Proc. Salt Desert Shrub Symposium.* Bureau of Land Management, U.S. Dep. Inter., Cedar City, Utah, pp. 157–159.

Cook, C.W., 1971. The effect of intensity and season of use on desert vegetation. *Utah Agric. Exper. Stn. Bull.*, 483: 68 pp.

Cook, C.W. and Child, R.D., 1971. Recovery of desert plants in various states of vigor. *J. Range Manage.*, 24: 339–343.

Cook, C.W. and Harris, L.E., 1968. Nutritive value of seasonal ranges. *Utah State Univ. Agric. Exper. Stn. Bull.*, 472: 55 pp.

Cook, C.W. and Stoddart, L.A., 1953. The *Halogeton* problem in Utah. *Utah State Univ. Agric. Exper. Stn. Bull.*, 364: 44 pp.

Cook, C.W., Stoddart, L.A. and Hart, V.L., 1952. Good range management prevents losses from *Halogeton* poisoning. *Utah Farm Home Sci.*, 13(2): 25–26, 45.

Cook, C.W., Stoddart, L.A. and Harris, L.E., 1954. The nutritive value of winter range plants in the Great Basin. *Utah State Univ. Agric. Exper. Stn. Bull.*, 372: 56 pp.

Coyne, P.O. and Cook, C.W., 1970. Seasonal carbohydrate reserve cycles in 8 desert range species. *J. Range Manage.*, 23: 438–444.

Crofts, K. and Van Epps, G., 1975. Use of shadscale in revegetation of arid disturbed sites. In: H.C. Stutz, (Editor), *Proc. Symp. and Workshop Wildland Shrub.* U.S. For. Ser. Shrub Sciences Lab., Provo, Utah, pp. 151–152.

Currie, P.O. and Goodwin, D.L., 1966. Consumption of forage by black-tailed jackrabbits on salt-desert ranges of Utah. *J. Wildl. Manage.*, 30: 304–311.

Detling, J.K. and Klikoff, L.G., 1973. Physiological response to moisture stress, as a factor in halophyte distribution. *Am. Midl. Nat.*, 90: 307–318.

Eckert, Jr., R.E. and Kinsinger, F.E., 1960. Effects of *Halogeton glomeratus* leachate on chemical and physical characteristics of soils. *Ecology*, 41: 764–772.

Eckert, R.E., Jr., Bruner, A.D. and Klomp, G.J., 1973. Productivity of tall wheatgrass and great basin wildrye under irrigation on a greasewood–rabbitbrush range site. *J. Range Manage.*, 26: 286–288.

Farb, P., 1978. *Man's Rise to Civilization: The Cultural Ascent of the Indians of North America.* E.P. Dutton, New York, N.Y., 2nd ed., 314 pp.

Fautin, R.W., 1946. Biotic communities of the northern desert shrub biome in western Utah. *Ecol. Monogr.*, 16: 251–310.

Fernandez, O.A. and Caldwell, M.M., 1975. Phenology and dynamics of root growth of three cool semi-desert shrubs under field conditions. *J. Ecol.*, 63: 703–714.

Fetcher, N. and Trlica, M.J., 1980. Influence of climate on annual production of seven cold desert forage species. *J. Range Manage.*, 33: 35–37.

Flowers, S. and Evans, F.R., 1966. The flora and fauna of the Great Salt Lake region, Utah. In: H. Boyko (Editor), *Salinity and Aridity.* Monographica Biologicae, 16. W. Junk, The Hague, pp. 367–393.

Frischknecht, N.C., 1967. How far will *Halogeton* spread? *J. Soil Water Conserv.*, 22: 135–139.

Gasto, J., 1969. *Comparative Autecological Studies of* Eurotia lanata *and* Atriplex confertifolia. Dissertation, Utah State University, Logan, Utah, 293 pp.

Gates, D.H., Stoddart, L.A. and Cook, C.W., 1956. Soil as a factor influencing plant distribution on salt-deserts of Utah. *Ecol. Monogr.*, 26: 155–175.

Gifford, G.F., Hancock, V.B. and Coltharp, G.B., 1978a. Effects of gully plugs and contour furrows on the soil moisture regime in the Cisco Basin, Utah. *J. Range Manage.*, 31: 293–295.

Gifford, G.F., Hawkins, R.H. and Jurinak, J.J., 1978b. Impacting diffuse salt production from rangelands within the Upper Colorado River Basin. In: D.N. Hyder (Editor), *Proc. First Int. Range Congress, Denver, Colo.*, pp. 335–338.

Goodman, P.J., 1973. Physiological and ecotypic adaptations of plants to salt desert conditions in Utah. *J. Ecol.*, 61: 473–494.

Goodman, P.J. and Caldwell, M.M., 1971. Shrub ecotypes in a salt desert. *Nature*, 232: 571–572.

Gross, J.E., Stoddart, L.C. and Wagner, F.H., 1974. Demographic analysis of a northern Utah jackrabbit population. *Wildl. Monogr.*, 40: 68 pp.

Gutierrez-Garza, J.S., 1978. *Potential for Cattle Grazing on Sheep Range in Southwest Utah.* Thesis, Utah State University, Logan, Utah, 79 pp.

Hancock, N.V., 1966. Wildlife use of the salt desert shrub areas of the Great Basin. In: Anonymous, *Salt Desert Shrub Symposium.* Bureau of Land Management, U.S. Dep. Inter., Cedar City, Utah, pp. 101–112.

Harper, K.T., 1959. *Vegetational Changes in a Shadscale–Winterfat Plant Association During Twenty-Three Years of Controlled Grazing.* Thesis, Brigham Young University, Provo, Utah, 68 pp.

Harper, K.T. and Alder, G.M., 1970. The macroscopic plant remains of the deposits of Hogup Cave, Utah, and their paleoclimatic implications. In: *Hogup Cave. Univ. Utah Anthropol. Pap.*, No. 93: 215–240.

Harris, L.E., 1968. *Range Nutrition in an Arid Region.* Utah State University, Faculty Honor Lecture, Logan, Utah, 99 pp.

Harris, L.E., Cook, C.W. and Stoddart, L.A., 1956. Feeding phosphorous, protein, and energy supplements to ewes on winter ranges of Utah. *Utah State Univ. Agric. Exper. Stn. Bull.*, 398: 28 pp.

Harrison, B.J. and Thatcher, A.P., 1970. Winter sheep grazing and forage preference in southwestern Wyoming. *J. Range Manage.*, 23: 109–111.

Hawkins, R.H., Gifford, G.F. and Jurinak, J.J., 1977. Effects of land processes on the salinity of the Upper Colorado River Basin. *Utah State Univ. Agric. Exper. Stn.*, 196 pp.

Hessary, I.K. and Gifford, G.F., 1979. Probable impacts of various range improvement practices on diffuse salt production. *J. Range Manage.*, 32: 189–193.

Holmgren, R.C., 1973. The desert experimental range: description, history, and program. In: *Arid Shrublands — Proc. of the Third Workshop of the United States/Australia Rangelands Panel, Tucson, Ariz.* pp. 18–22.

Holmgren, R.C. and Brewster, Jr., S.F., 1972. Distribution of organic matter reserve in a desert shrub community. *U.S.D.A. For. Serv. Res. Pap.*, INT-130: 15 pp.

Holmgren, R.C. and Hutchings, S.S., 1972. Salt desert shrub response to grazing use. *U.S.D.A. For. Serv. Gen. Tech. Rep.*, INT-1: 153–164.

Hull, A.C., Jr., 1963. Seeding salt desert shrub ranges in western Wyoming. *J. Range Manage.*, 16: 253–258.

Hunt, C.B., 1974. *Natural Regions of the United States and Canada.* W.H. Freeman and Co., San Francisco, Calif., 725 pp.

Hutchings, S.S., 1952. Snout moth damage to shadscale in southern Idaho. *Natl. Woolgrowers*, 42: 24–25, 38.

Hutchings, S.S., 1954. Managing winter sheep range for greater profit. *U.S.D.A. Farmer's Bull.*, No. 2067: 46 pp.

Hutchings, S.S., 1966. Grazing management of salt desert shrub ranges in the western United States. In: *Tenth Int. Grassland Congr. Proc.*, pp. 1619–1625.

Hutchings, S.S. and Stewart, G., 1953. Increasing forage yields and sheep production on intermountain winter ranges. *U.S.D.A. Circ.*, 925: 63 pp.

James, L.G. and Cronin, E.H., 1974. Management practices to minimize death losses of sheep grazing halogeton infested ranges. *J. Range Manage.*, 27: 424–426.

Kenagy, G.J., 1972. Saltbush leaves: excision of hypersaline tissue by a kangaroo rat. *Science*, 178: 1094–1096.

King, C.L., 1963. *Harvester Ant Forage Utilization Studies in the Big Horn Basin of Wyoming.* Western Association of State Game and Fish Commissioners, Tucson, Ariz., 29 pp.

Kirkham, D.R. and Fisser, H.G., 1972. Rangeland relations and harvester ants in northcentral Wyoming. *J. Range Manage.*, 25: 55–60.

Küchler, A.W., 1970. Potential natural vegetation (map at scale 1:750 000.) In: U.S. Geol. Survey, *The National Atlas of the U.S.A.* U.S. Government Printing Office, Washington, D.C., pp. 90–91 (revision of 1964 map at scale of 1:3 168 000).

Larrison, E.J. and Johnson, D.R., 1973. Density changes and habitat affinities of rodents of shadscale and sagebrush associations. *Great Basin Nat.*, 33: 255–264.

Lusby, G.C., 1965. Causes of variation in runoff and sediment yield from small drainage basins in western Colorado. *U.S. Dep. Agric. Misc. Publ.*, 970 (Pap. 14): 94–98.

Lusby, G.C., 1970. Hydrologic and biotic effects of grazing vs. non-grazing near Grand Junction, Colorado. *J. Range Manage.*, 23: 256–260.

Lusby, G.C., Turner, G.T., Thompson, J.R. and Reid, B.H., 1963. Hydrologic and biotic characteristics of grazed and ungrazed watersheds of the Badger Wash Basin in western Colorado, 1953–58. *U.S. Geol. Surv. Water Supply Pap.*, 1532-B: 73 pp.

Lusby, G.C., Reid, V.H. and Knipe, O.D., 1971. Effects of grazing on the hydrology and biology of the Badger Wash Basin in western Colorado, 1953–1966. *U.S. Geol. Surv. Water Supply Pap.*, 1532-D: 90 pp.

Lynn, R.I. and Vogelsberg, M.C., 1974. The role of algae in crust formation and nitrogen cycling in desert soils. *US/IBP Des. Biome Res. Memo.*, 74–41: 83–92.

McArdle, R.E. and Costello, D.F., 1936. The virgin range. In: H.A. Wallace, *Letter from the Secretary of Agriculture to the U.S. Senate — A Report on the Western Range — A Great but Neglected Resource.* Senate Document 199, pp. 71–80.

McArdle, R.E., et al., 1936. The whiteman's toll. In: H.A. Wallace, *Letter from the Secretary of Agriculture to the U.S. Senate — A Report on the Western Range — A Great but Neglected Resource.* Senate Document 199, pp. 81–116.

McKeever, S. and Hubbard, R.L., 1960. Use of desert shrubs by jackrabbits in northeastern Calif. *Calif. Fish Game*, 46: 271–277.

MacMahon, J.A. and Wagner, F.H., in press. The Mojave, Sonoran, and Chihuahuan Deserts of North America. In: M. Evenari and I. Noy-Meir (Editors), *Hot Deserts and Arid Shrublands. Ecosystems of the World*, 12. Elsevier, Amsterdam, in press.

Malechek, J.C. and Smith, A.D., 1975. Square pegs in round holes? Cattle on traditional sheep range. *Utah Sci.*, 5–7.

Mitchell, J.E., West, N.E. and Miller, R.W., 1966. Soil physical properties in relation to plant community patterns in the shadscale zone of northwestern Utah. *Ecology*, 47: 627–630.

Murdock, J.R. and Welsh, S.L., 1971. Land use in Wah Wah and Pine Valleys, western Utah. *Brigham Young Univ. Sci. Bull., Biol. Ser.*, XII(4): 25 pp.

Naphan, E.A., 1966. Soils of the salt desert shrub area and their productive capabilities. In: Anonymous, *Proc. Salt Desert Shrub Symposium.* U.S. Dep. Inter., Bureau of Land Management, Cedar City, Utah, pp. 44–68.

Norton, B.E., 1978. The impact of sheep grazing in long-term successional trends in salt desert shrub vegetation of southwestern Utah. In: D.N. Hyder (Editor), *Proc. First Int. Range Congress, Denver, Colo.*, pp. 610–613.

Olsen, P.F., 1971. *Ecology Studies in Western Utah.* Ecodynamics Inc., Salt Lake City, Utah, Annual Report.

Pianka, E.R., 1967. On lizard species diversity: North American flatland deserts. *Ecology*, 48: 333–351.

Pieper, R., Cook, C.W. and Harris, L.E., 1959. Effect of intensity of grazing upon nutritive content of the diet. *J. Anim. Sci.*, 18: 1031–1037.

Rice, B. and Westoby, M., 1978. Vegetation responses of some Great Basin shrub communities protected against jackrabbits or domestic stock. *J. Range Manage.*, 31: 28–33.

Richardson, S.G., Barker, J.R., Crofts, K.A. and Van Epps, G.A., 1979. Factors affecting rooting of stem cuttings of salt desert shrubs. *J. Range Manage.*, 32: 280–283.

Rickard, W.H., 1964. Demise of sagebrush through soil changes *BioScience*, 14: 43–44.

Rodin, L.E. and Bazilevich, N.I., 1967. *Production and Mineral Cycling in Terrestrial Vegetation.* Oliver and Boyd, Edinburgh, 288 pp. (Translated from Russian by G.E. Fogg).

Rychert, R.C. and Skujins, J., 1974. Nitrogen fixation by blue-green algae-lichen crusts in the Great Basin Desert. *Soil Sci. Soc. Am. Proc.*, 38: 768–771.

Sharp, L.A. and Barr, W.F., 1960. Preliminary investigations of harvester ants on southern Idaho rangelands. *J. Range Manage.*, 13: 131–134.

Shreve, F., 1942. The desert vegetation of North America. *Bot. Rev.*, 8: 195–246.

Springer, M.E., 1958. Desert pavement and vesicular layer of some soils of the desert and of the Lahontan Basin, Nevada. *Soil. Sci. Soc. Am. Proc.*, 22: 63–66.

Steger, R.E., 1970. *Soil Moisture and Temperature Relationships of Six Salt Desert Shrub Communities in North-Central Wyoming.* Thesis, University of Wyoming, Laramie, Wyo., 109 pp.

Stewart, G., Cottam, W.P. and Hutchings, S.S., 1940. Influence of unrestricted grazing on northern salt desert plant associations in western Utah. *J. Agric. Res.*, 60: 289–316.

Stuart, D.M., Eckert, Jr., R.E., Dylla, A.S. and Schuman, G.E., 1973. Seeding coppice dune areas in northern Nevada. *Univ. Nev. Agric. Exper. Stn. Bull.*, 11: 17 pp.

Stutz, H.C., 1978. Explosive evolution of perennial *Atriplex* in western America. *Great Basin Nat. Mem.*, 2: 161–168.

Thompson, J.R., 1968. Effect of grazing on infiltration in a western watershed. *J. Soil Water Conserv.*, 23: 63–65.

Trlica, Jr., M.J. and Cook, C.W., 1971. Defoliation effects on carbohydrate reserves of desert species. *J. Range Manage.*, 24: 418–425.

Tueller, P.T., Bruner, A.D. and Davis, J.B., 1972. Ecology of Hot Creek Valley — Vegetation and soil response to underground detonation. *Univ. Nev. Agric. Exper. Stn. Rep.*, NVO-409-1: 131 pp.

Turner, G.T., 1971. Soil and grazing influences on a salt-desert shrub range in west Colorado. *J. Range Manage.*, 24: 31–37.

Vest, E.D., 1962. Biotic communities in the Great Salt Lake Desert. *Inst. Environ. Biol. Res. Ecol. Epizool. Ser.*, No. 73: 122 pp. (Div. Biol. Sci. Univ. Utah, Salt Lake City, Utah).

Wagner, F.H., in press. Livestock and wildlife-fisheries ecology. In: *Proceedings of Conf. on Livestock and Wildlife-Fisheries Relationship in the Great Basin*, 28 pp.

Wagner, F.H., 1980. Integrating and control mechanisms in arid and semi-arid ecosystems — considerations for impact assessment. In: *Proc. Symp. on Biol. Evaluation of Environmental Impact.* Council on Environ. Quality, Fish and Wildl. Serv., U.S. Dep. Int., Washington, D.C., FWS/OBS-80/26, pp. 145–158.

Walter, H. and Box, E.O., 1976. Global classification of natural terrestrial ecosystems. *Vegetatio*, 32: 75–81.

Wein, R.W. and West, N.E., 1972. Physical microclimates of erosion-control structures in salt desert areas. *J. Appl. Ecol.*, 9: 703–720.

West, N.E., 1968. Ecology and management of salt desert shrub ranges: A bibliography. *Utah State Univ. Agric. Exper. Stn. Mimeo Ser.*, 505: 39 pp.

West, N.E. and Gasto, J., 1978. Phenology of the aerial portions of shadscale and winterfat in Curlew Valley, Utah. *J. Range Manage.*, 31: 43–45.

West, N.E. and Goodall, D.W., 1980. Dispersion patterns in salt desert shrub vegetation of western Utah. *Bull. Ecol. Soc. Am.*, 61: 131 (abstract).

West, N.E. and Ibrahim, K.L., 1968. Soil–vegetation relationships in the shadscale zone of southeastern Utah. *Ecology*, 49: 445–456.

West, N.E. and Skujins, J., 1977. The nitrogen cycle in North American cold-winter semi-desert ecosystems. *Oecol. Plant.*, 12: 45–53.

Westoby, M., 1973. *The Impact of Black-Tailed Jackrabbits* (Lepus californicus) *on Vegetation in Curlew Valley, Northern Utah.* Thesis, Utah State University, Logan, Utah, 93 pp.

Wight, J.R. and Nichols, J.T., 1966. Effects of harvester ants on production of a saltbush community. *J. Range Manage.*, 19: 68–71.

Wilkin, D.C., 1973. Special purpose models: Model of response of a desert shrub community to sheep grazing. *US/IBP Desert Biome Rep. Memo.*, 73-59: 19 pp.

Williams, B.K., 1979. *Optimal Stochastic Control of Salt Desert Shrubs by Season and Intensity of Defoliation.* Thesis, Colorado State University, Ft. Collins, Colo., 261 pp.

Williams, M.C., 1973. *Halogeton* — a sheep killer in the west. *Weeds Today*, 4: 10–11, 22.

Workman, J.P. and West, N.E., 1969. Eco-typic variation of *Eurotia lanata* populations in Utah. *Bot. Gaz.*, 130: 26–35.

Young, J.A., Evans, R.A. and Tueller, P.T., 1976. Great Basin plant communities — pristine and grazed. In: R. Elston (Editor), *Holocene Environmental Change in the Great Basin. Nev. Archeol. Surv. Res. Pap.*, 8: 186–215.

Young, J.A., Evans, R.A. and Major, J., 1977. Sagebrush steppe. In: M.G. Barbour and J. Major (Editors), *Terrestrial Vegetation of California.* Wiley–Interscience, New York, N.Y., pp. 763–796.

Chapter 15

COLORADO PLATEAU–MOHAVIAN BLACKBRUSH SEMI-DESERT[1]

N.E. WEST

DEFINITION OF TYPE

The shrublands that characterize this ecosystem type are dominated by blackbrush (*Coleogyne ramosissima*). This shrub dominates ecosystems occurring in the lowest, driest portions of the Colorado Plateau portion of the Great Basin Regional Desert (Fig. 15.1) and through the Grand Canyon into the boundary area between the Mohave and Great Basin portions of the Basin and Range Physiographic Province.

Fig. 15.1. Blackbrush-dominated vegetation on thin regolith over the Moenkopi Formation, Pete's Mesa, Canyonlands National Park, Utah. Photo by Kay Forsythe.

Fig. 15.2. Map of the Colorado Plateau–Mohavian Blackbrush Semi-Desert ecosystem type. Derived from Küchler (1970).

GEOGRAPHICAL LOCATION

The small-scale map of the potential vegetation of the U.S. (Küchler, 1970) shows blackbrush to dominate terrain only along the Colorado, Green, and San Juan Rivers in Utah and Arizona (Fig. 15.2). The type adjoins salt-desert shrub vegetation on saltier soils, and riparian forests of cottonwood (*Populus*), willow (*Salix*) and tamarisk (*Tamarix*) where water tables are high and permanent. At slightly higher elevations, and thus under higher precipitation, one encounters either sagebrush, galleta–threeawn shrub-steppe or juniper–pinyon woodlands. Blackbrush inclusions in the grasslands or as woodland understory may be found wherever soil depth is restricted (Fig. 15.3). We do not consider here the areas that were

[1] Manuscript completed June, 1980.

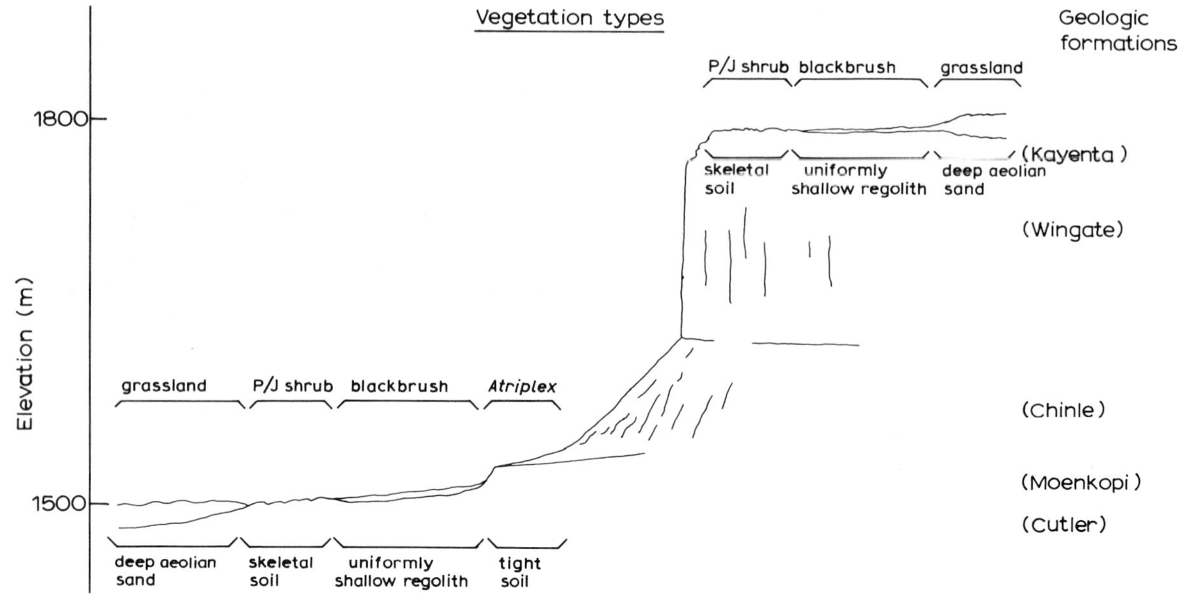

Fig. 15.3. Relationship of vegetation types to elevation and regolith in Canyonlands National Park, Utah (from Loope and West, 1979).

originally semi-desert grasslands, but have been invaded by blackbrush.

As with all ecosystem types, smaller areas within a matrix of other vegetation types had to be ignored for the sake of standardization. This even-handed treatment is necessary to prevent us from accumulating more total area than we know exists for temperate deserts to semi-deserts within the continent. We thus have circumspectly avoided double counting of areas as being simultaneously occupied by two or more types.

In the case of blackbrush, there was a great temptation to make an exception. This is because we know that blackbrush dominates large but slightly studied areas in the Mohave Desert. In fact, it is a dominant over most of its range of distribution (Fig. 15.4). We must draw on data collected in these other areas in order to relate the maximum amount of information we can in the space given here.

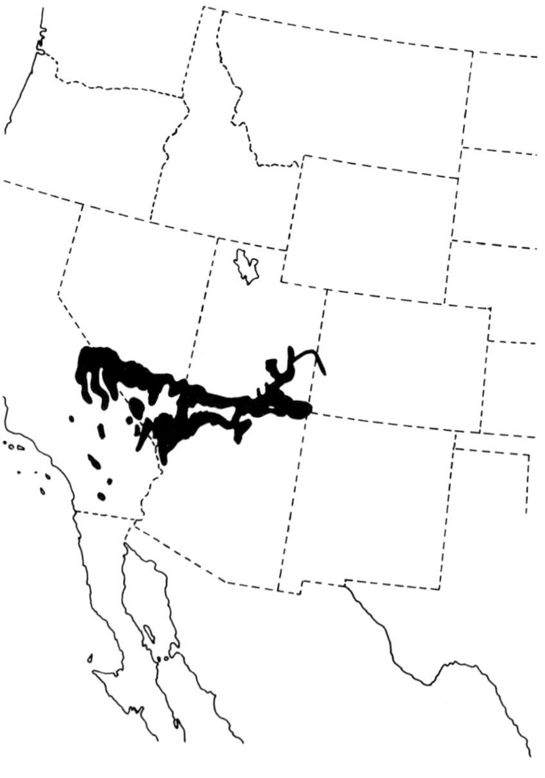

Fig. 15.4. Geographic distribution of blackbrush (from Bowns and West, 1976).

EXTENT

The mappable extent of the ecosystem type totals 1.3×10^6 ha (Table 15.1) with 92% of the total occurring in Utah. The small remaining portion is found in Arizona.

TABLE 15.1

Area occupied by the blackbrush semi-desert (derived from Küchler, 1970)

State	Area ($\times 10^6$ ha)	Percentage of total
Utah	1.2	92
Arizona	0.1	8
Total	1.3	100

CURRENT OWNERSHIP

This ecosystem type occurs in some of the wildest, least inhabited landscape that remains in the United States. Because hardly any of it has been suitable for intensive agriculture, the major landlords are agencies of the U.S. Department of Interior, particularly the National Park Service and the Bureau of Land Management. A large portion of this ecosystem type is also found on the Navajo Indian Reservation.

DOMINANT USES

The major economic use of this ecosystem type was formerly grazing by domestic livestock. This use continues even on the National Park Service lands. The enabling legislation for most of the park units involved (Canyonlands N.P., Capitol Reef N.P., and Glen Canyon National Recreation Area) provides for livestock grazing for many more decades — usually until the death of the current grazier.

The capacity for domestic livestock is low. Although Aldous and Shantz (1924) and Humphrey (1955, 1964) noted this type as largely wasteland from a grazier's point of view, it is rated as fair forage for sheep and cattle during the winter in southwestern Utah (Bowns and West, 1976). Grazing capacities on the order of 0.04 to 0.08 AUM ha^{-1} are used by the Bureau of Land Management in that area.

It is one of the few North American temperate semi-desert ranges used extensively by goats. Most of these animals are owned by the Navajo Indians (Fig. 15.5).

Some of the less accessible areas serve as winter range for small numbers of mule deer (*Odocoileus hemionus*), desert bighorn sheep (*Ovis canadensis*) and feral burros (*Equus asinus*).

Fig. 15.5. Mixed flock of goats and sheep browsing blackbrush on the Navajo Indian Reservation, Arizona.

Most human non-residents of the area probably associate highly scenic vistas with this ecosystem type (Fig. 15.6). A large area of blackbrush was flooded by Lake Powell when Glen Canyon Dam was closed in 1963. Much of the ecosystem type is now more affected by recreational users than livestock.

Fig. 15.6. Blackbrush under Rainbow Bridge. Rainbow Bridge National Monument, Utah.

ECOSYSTEM STRUCTURE AND FUNCTION

Climate

The climate of this area is arid, temperate and continental (Fig. 15.7). The average annual precipitation is less than 160 mm but has a coefficient

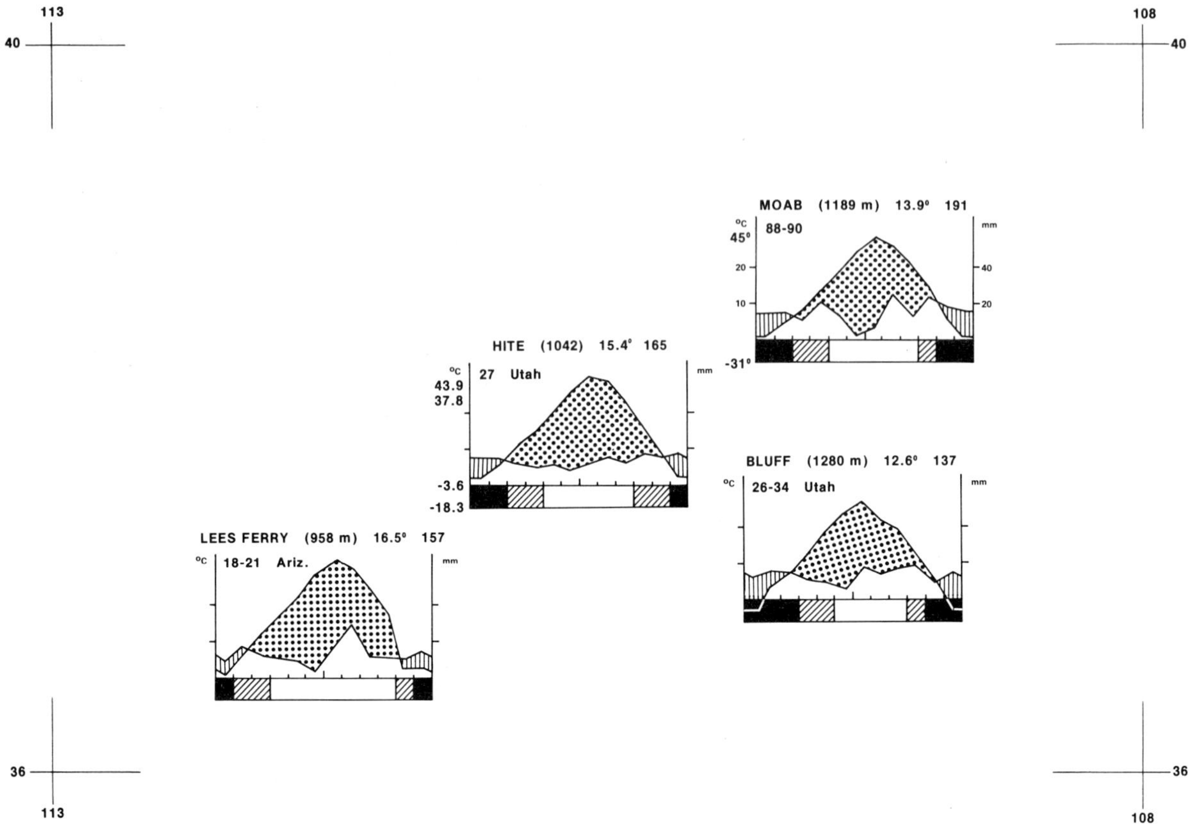

Fig. 15.7. Climatic diagrams for stations located within the blackbrush ecosystem type.

of variation of 33%. Most of the precipitation comes in late summer from convectional storms associated with Gulf Coastal air masses. Winter precipitation comes from the westerly cyclonic events and is largely as snow. The snow rarely lasts for more than a day or two before melting or sublimating.

From the few stations occurring in this ecosystem type, we know that both average and maximum temperatures are higher than in the nearby salt-desert shrub, sagebrush, pinyon–juniper woodlands or galleta–threeawn shrub-steppes. The coldest temperatures are also lower, on the average. Because of the greater heat load, the evaporative demands are also probably higher in this type. Thus, in terms of climate, the blackbrush type is intermediate between the truly temperate and subtropical deserts (Beatley, 1975).

Geology

The level sedimentary strata of the Colorado Plateau are the major parent materials for the soils supporting the blackbrush type. The shales are usually too salty and droughty for blackbrush. Sandstones provide the major substrata for blackbrush in the area indicated on the map. Limestone and limestone-derived alluvium seem to prevail elsewhere (Thatcher et al., 1976; Beatley, 1975; Bowns and West, 1976; Thatcher, 1976).

Topography

Blackbrush mainly occurs on well-drained benchlands, colluvial slopes, pediments or bajadas between 800 and 1600 m in elevation (West, 1969) (Fig. 15.8). It is rarely found in concave microtopography, indicating a probable low tolerance to salinity, excessive soil moisture and impeded soil aeration (Table 15.2). Further, circumstantial evi-

TABLE 15.2

General characteristics of environment and vegetation associated with the six vegetation units[1] in the Kaiparowits Basin, Kane County, Utah recognized by Jaynes and Harper (1980); standard deviations follow the mean for each value

	Mat saltbush	Gray molly–shadscale	Shadscale–Grass	Grassland–shrub	Blackbrush	Sagebrush
Abiotic factors						
Elevation (m)	1308 ± 5	1279 ± 3	1335 ± 26	1522 ± 47	1583 ± 65	1659 ± 5
Texture						
% sand	13.8 ± 4.2	47.2 ± 7.0	64.4 ± 11.4	74.4 ± 8.8	72.7 ± 6.1	70.5 ± 3.0
% silt	22.0 ± 3.0	17.2 ± 3.1	13.0 ± 5.9	8.1 ± 4.9	6.7 ± 3.3	11.0 ± 3.5
% clay	64.2 ± 4.2	34.3 ± 4.9	24.4 ± 8.4	17.3 ± 5.1	20.7 ± 3.4	18.3 ± 2.7
Mean % rock	2.0 ± 0.9	36.2 ± 4.1	16.9 ± 8.6	5.6 ± 9.2	6.4 ± 6.5	11.8 ± 4.8
Mean soil depth (cm)	49.5 ± 16.4	17.7 ± 4.4	31.7 ± 14.7	46.7 ± 21.6	22.2 ± 14.9	19.2 ± 4.9
Soil pH (1:1)	8.1 ± 0.1	7.6 ± 0.4	7.9 ± 0.4	8.0 ± 0.3	8.1 ± 0.2	8.1 ± 0.1
Infiltration rate (cm min^{-1})	1.8 ± 0.6	0.7 ± 0.3	2.6 ± 1.2	3.1 ± 1.5	3.3 ± 3.3	1.1 ± 0.5
Salinity (ppm)	11200 ± 10500	353 ± 120	490 ± 450	281 ± 470	147 ± 74	133 ± 10
Biotic factors						
% Living cover	12.0 ± 6.8	14.6 ± 4.0	19.4 ± 6.4	27.0 ± 6.5	26.2 ± 6.8	21.8 ± 4.2
Composition of cover						
% Shrubs	46 ± 15	36 ± 7	34 ± 15	34 ± 11	43 ± 18	40 ± 11
% Perennial forbs	2 ± 3	14 ± 15	6 ± 4	16 ± 8	11 ± 4	20 ± 11
% Perennial grasses	7 ± 18	18 ± 15	33 ± 18	40 ± 17	26 ± 23	25 ± 17
% Annuals	45 ± 12	32 ± 15	27 ± 14	10 ± 7	20 ± 12	15 ± 13
Species diversity (number of species in 0.25 m^{-2} plots)	2.0 ± 0.9	2.8 ± 0.7	3.2 ± 0.9	3.6 ± 0.8	3.3 ± 0.9	4.1 ± 0.4

[1] Plant species composition for each community type is detailed in Table 15.3.

dence is its failure to act as a riparian element. It does especially well on long bajada slopes with shallow soils.

This preferred topographic position may also be related to temperature tolerances. Beatley (1975) has shown that blackbrush is less tolerant of cold air drainage than chenopods and *Larrea*.

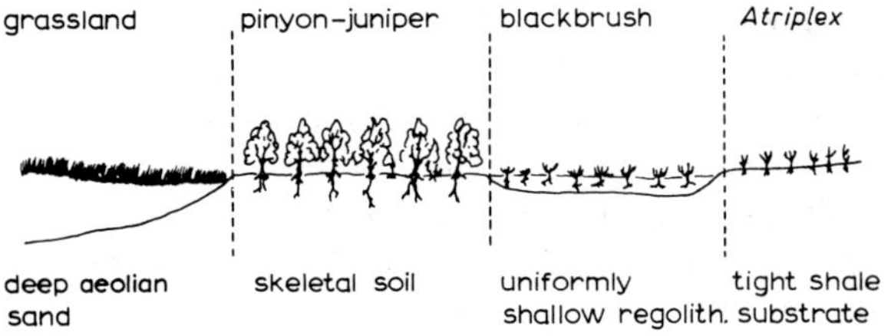

Fig. 15.8. Regolithic control of vegetation types in Canyonlands National Park, Utah (from Loope, 1975).

Soils

Blackbrush is a shrub that tends to dominate where shallow soils are underlain with a petrocalcic (caliche) layer. It can also occur on deeper loessial soils, but these are thought to be invasions into what once was galleta–threeawn shrub-steppe (Loope, 1975). Thatcher et al. (1976) have noted a gradient of increasing amounts of grasses and declining brush with increasing depth to caliche or other restrictive layers for a section of the Hualapai Indian Reservation, Arizona, which has never had livestock grazing because of lack of water.

Most of the soils supporting blackbrush are classified as Aridisols. Paleorthids would likely be the most common because the upland bench land or bajada positions where the type prevails are old stable landforms that have had the maximum development that can be attained in desert environments.

Erosion pavements, microphytic and vesicular crusts are common on the surface of these soils, especially on the limestone-derived bajadas.

Primary producers

There are not many species of plants that can endure the rigors of this environment. Only in the spring is there any showing of greenness, mainly on the part of the thirty or so annual plant species that occur here. Table 15.3 gives a listing of the most frequent species to be found in this and adjacent types of shrubland and grassland in a representative area of southeastern Utah.

Blackbrush dominates both in terms of height and cover (Fig. 15.1). Blackbrush is a very round-shaped shrub reaching about 0.5 m high. It is round because its terminal twigs die back leaving spines. Subsequent growth is always sub-terminal (Bowns and West, 1976). Other shrubs are rare but may include, in addition to those in Table 15.3, desert peachbrush (*Prunus fasciculata*), spiny hopsage [*Atriplex* (*Grayia*) *spinosa*], turpentine bush (*Thamnosma montana*), various *Ephedra* spp., wolfberries (*Lycium* spp.), and, if disturbed, snakeweed [*Xanthocephalum* (*Gutierrezia*) spp.].

Some occasional yucca and cacti may be found interspersed between the gray-green matrix of shrubs. Joshua tree (*Yucca brevifolia*) is frequently associated with the ecosystem type in the Mojave Desert.

TABLE 15.3

The average percent frequency (in plots of 0.05 acre = 202 m^2) of the most abundant species in each of the six vegetation types recognized by Jaynes and Harper (1980) in the Kaiparowits Basin, Kane County, Utah; the frequency values of the commonest in each type are in bold type

Species[1]	Lifeform[2]	Mat saltbush	Gray molly–shadscale	Shadscale–grass	Grassland–shrub	Blackbrush	Sagebrush
Atriplex corrugata	S	**47**			1		
Cleomella palmerana	A	**43**					
Oenothera eastwoodiae	F	**23**		1			
Androstephium breviflorum	F	**4**	1	4	6		
Eriogonum inflatum	A	**3**	1	**21**			
Salsola kali	A	**22**	6	**23**	9		
Phacelia demissa	A	**40**	**15**	**20**			
Atriplex confertifolia	S		**16**	**26**	6	3	
Oryzopsis hymenoides	G	**11**	**21**	**17**	**25**	9	**25**
Hilaria jamesii	G	1	**14**	**40**	**55**	**36**	**34**
Festuca octoflora	A		9	**16**	**14**	**18**	**27**
Gilia leptomeria	A	1		**23**	**28**	**20**	**25**
Gutierrezia sarothrae	S		1	**12**	**34**		**51**
Chaenactis stevioides	A			**14**	4	**15**	**16**
Lepidium montanum	F			2	**14**	**17**	**39**
Ephedra viridis	S			2	**17**	**19**	5
Mentzelia albicaulis	A		1	6	**24**	**25**	14
Eriogonum deflexum	A			7	**11**	1	1
Sporobolus cryptandrus	G			3	10		
Grayia spinosa	S				4	**12**	3
Coleogyne ramosissima	S				2	**32**	
Gilia scopulorum	A				8	**40**	**57**
Bouteloua gracilis	G				6	9	**15**
Aster arenosus	F			2	9	5	**15**
Artemisia tridentata	S						**40**
Cymopterus purpurascens	F		**19**	1	3		
Astragalus lentiginosus	F		**17**		3	1	1
Kochia americana	S		**32**		1		
Lappula occidentalis	A		**14**	2	1		1
Langloisia setosissima	A		**21**	**10**	1		
Chaenactis macrantha	A		**27**				

[1] Nomenclature follows Welsh and Moore (1973).

[2] Letter following plant name indicates life form: S = shrub; G = perennial grass; F = perennial forb; A = annual.

The perennial herbaceous component is limited to a few bunch grasses such as galleta (*Hilaria jamesii*) and big galleta (*H. rigida*), Indian ricegrass (*Oryzopsis hymenoides*), threeawn (*Aristida* spp.), needle grasses (*Stipa arida*, *S. speciosa*) and gramas (*Bouteloua eriopoda* and *B. gracilis*).

Annuals can be abundant for about six weeks in the spring during wet years, but composition and productivity vary greatly. The value of winter annuals as forage and soil cover is, however, undependable. Only opportunistic use can be made of this portion of the community. This cover is believed, however, to be important in stabilizing soil against erosion and fixing nitrogen. For instance, Beatley (1969a) recorded a five-year maximum of 7.8% cover and a density of 976 plants per square meter for winter annuals in a blackbrush community on the Nevada Test Site.

The shrubs are usually clustered on mounded micro-topography. The total soil depth under such clusters is often greater than in the interspaces because the shrubs apparently trap aeolian materials, and greater root activity and weathering cause a depression of the petrocalcic layer directly under the shrubs. This plus nitrogen-fixing activities associated with *Coleogyne* (Wallace and Romney, 1972) make for highly patterned accumulations of organic matter and key nutrients such as nitrogen (Charley and West, 1975).

The total cover of blackbrush-dominated vegetation is consistently the highest (37–51%) of that in the regional vegetation (Beatley, 1975). This comparatively high amount of cover for semi-desert environment is probably due to the great age that these shrubs can attain (Christensen and Brown, 1963) and the accumulation of much standing dead material in the centers of the shrubs.

This type also has abundant soil microphytic crusts developed on fine-textured soils in the shrub interspaces where livestock grazing and human traffic has not been excessive.

Blackbrush, most of the cool-season grasses, and winter annuals seem to depend on winter soil-moisture recharge. They begin growth in the early spring (mid-March), and cease when that moisture is depleted, usually by early June. Summer rainfall may activate the microphytic crust and warm-season grasses such as *Hilaria*, but it has little or no effect on the rest of the community since summer annuals are rare except on disturbed sites.

No phytomass data have apparently been collected in the mapped area of this type. The only known data on stand phytomass and organic matter were collected in San Bernardino County, California (Wells, 1967). Since this area is known to have been heavily grazed, it may not be representative. The high proportion of *Hilaria rigida* at the site more distant from livestock water indicates that grasses may have been more important in the pristine system here (Table 15.4).

One aspect of the phytomass distribution to be noted is that there is much less below-ground phytomass here than in other types in the Great Basin Regional Desert. This low root:shoot ratio is more common in the southern deserts (MacMahon and Wagner, in press), where it is related to the rather shallow petrocalcic horizons.

Although net annual primary production has not been precisely measured, we know that growth rates of the shrubs are extremely low (Bowns and West, 1976). A low growth rate and limited root development may be due to the often shallow soil profiles where these ecosystems occur. The depth to caliche which impedes root growth and soil moisture movement is usually 30 to 50 cm below the surface. Average net primary productivity is apparently low. Thatcher et al. (1976) noted a range in herbage production from 302 kg ha^{-1} in nearly pure blackbrush stands on shallow soils to 1688 kg ha^{-1} on sites with mostly *Bouteloua eriopoda* and *Muhlenbergia porteri* and minor amounts of *Coleogyne* on relatively deep gravelly loams. All their sites on the Hualapai Indian Reservation of northwestern Arizona were pristine (had never been grazed by livestock). Lower net primary productivities may be expected to prevail elsewhere where livestock grazing has taken place.

While considerable shrub growth can be made following a wet winter and spring, the shrubs frequently die back during drought by forming intercalary cork and isolating portions of the stem into separate vascular traces. This is the main cause of the multi-stemmed nature of this shrub.

Blackbrush is protected against heavy ungulate use by a combination of spines and woodiness, low nutrient quality and secondary plant compounds.

Production of annuals varies widely, probably in the order of several thousand percent (Beatley, 1969a). The major winter annuals, *Erodium botrys*, *Bromus tectorum* and *B. rubens*, are comparatively

TABLE 15.4

Cover, phytomass and total organic matter (kg ha^{-1}) on two sites in Gold Valley, San Bernardino County, California (from Wells, 1967)

Species	Cover (%)	Leaves	Woody stems	Root crowns and major roots (to 1 m depth)	Total	Surface litter	Soil organic matter (including fine roots) (to 1 m depth)		Total organic matter
							>2 mm	<2 mm	
Site A (near water)									
Coleogyne ramosissima	32.2								
Ephedra californica	0.5								
Yucca baccata	1.9								
Dead	1.3								
Total	35.9	3450	14 610	5190	22 450	970	13 900	80 540	117 860
Bare area						340	460	52 570	53 370
Site B (farther from water)									
Coleogyne ramosissima	1.6								
Ephedra californica	3.7								
Other live shrubs	1.4								
Dead shrubs	0.6								
Hilaria rigida	13.8								
Total	21.0	10 800	23 930	4730	40 030	3110	14 050	100 780	157 870
Bare area					29	340	150	37 750	38 530

more dependable than spring annuals which can form a nearly continuous carpet in the interspaces during wet springs or almost nothing in dry years. The predictable food base for ungulates is thus limited in this type, causing it not to be highly regarded as rangeland for livestock production. Desert rodent populations fluctuate greatly in response to the wide variation in abundance of seeds and herbage from the annuals (Beatley, 1969b, 1976a).

Consumers

The fauna of this ecosystem type is a mix of Great Basin and Mojave Desert elements. A few of the northern desert vertebrates wander into the type, particularly in the winter, whereas southern desert elements are marginally present year-round or in the summer (Table 15.5). The northernmost distributions of numerous herpetofauna are found in this type (Tanner, 1978). This is probably related to the comparatively warmer temperatures that extend up the Colorado River drainage into Utah (MacMahon and Wieboldt, 1978).

The only native ungulate that seems to be closely tied to this type is the desert bighorn sheep (*Ovis canadensis nelsoni*) (Fig. 15.9). Blackbrush-dominated communities form an important part of its winter range (Bradley, 1965; Wilson, 1967). Mule deer (*Odocoileus hemionus*) only occasionally move into the area if snow packs at higher elevations are particularly heavy. Feral burros (*Equus asinus*) occur in the Grand Canyon and Death Valley portions of this ecosystem type, where they are apparently displacing the native sheep (Wells and Wells, 1961; Russo, 1956).

The major animals from the point of view of biomass and influence are the heteromyid rodents. These animals, largely seed-eaters, may even control the plant composition by their differential consumption. Beatley (1976b) has documented how man-caused disturbances of this ecosystem type will engender changes in rodent species composition and population levels.

No obviously adverse effects of small mammals or insects on blackbrush itself have been reported. Caching of seeds by rodents may aid shrub regeneration (Bowns and West, 1976), but the animal's welfare seems to be tied more to the soil seed reserves, largely of annuals (Beatley, 1969b).

Tent caterpillars (*Malacosoma:* Lepidoptera) have been observed to keep *Prunus fasciculata* defoliated much of the time. This is a possible reason for its minor role in this ecosystem type.

TABLE 15.5

Checklist of vertebrates found in the blackbrush type of southern Nevada (from Bradley and Deacon, 1967)

A = abundant	R = rare	*For birds*:
C = common	× = record of occurrence	PR = permanent resident
FC = fairly common	* = introduced	SR = summer resident
U = uncommon		WR = winter resident
		V = visitant
		AC = accidental
		× = record of occurrence

Species	Abundance
MAMMALS	
Antilocapridae	
Antilocapra americana	×
Bovidae	
Ovis canadensis	A
Canidae	
Canis latrans	C
Urocyon cinereoargenteus	U
Vulpes macrotis	C
Cervidae	
Odocoileus hemionus	U
Cricetidae	
Neotoma lepida	C
Onychomys torridus	U
Peromyscus crinitus	A
Peromyscus eremicus	C
Peromyscus maniculatus	C
Reithrodontomys megalotis	U
Equidae	
Equus asinus	*
Felidae	
Lynx rufus	U
Geomyidae (gophers)	
Thomomys umbrinus	C
Heteromyidae	
Dipodomys merriami	A
Dipodomys microps	U
Dipodomys panamintinus	R
Perognathus formosus	A
Perognathus longimembris	A
Leporidae	
Lepus californicus	C
Sylvilagus audubonii	C
Molossidae	
Tadarida brasiliensis	C
Mustelidae	
Spilogale gracilis	U
Taxidea taxus	C
Sciuridae	
Citellus leucurus	C
Citellus variegatus	U

TABLE 15.5 (*continued*)

Species	Abundance
Tayassuidae	
Tayassu tajacu	×
Vespertilionidae	
Antrozous pallidus	U
Corynorhinus townsendii	C
Eptesicus fuscus	U
Myotis californicus	A
Pipistrellus hesperus	A
BIRDS	
Accipitridae	
Buteo jamaicensis (red-tailed hawk)	U, SR, U, WR
Apodidae	
Aeronautes saxatilis (white-throated swift)	C, SR
Caprimulgidae	
Chordeiles acutipennis (lesser nighthawk)	C, SR
Chordeiles minor (common nighthawk)	FC, SR
Phalaenoptilus nuttallii (poor-will)	C, SR
Cathartidae	
Cathartes aura (turkey vulture)	FC, SR
Corvidae	
Corvus corax (common raven)	C, PR
Cuculidae	
Geococcyx californiana (roadrunner)	U, PR
Falconidae	
Falco sparverius (sparrow hawk)	C, WR
Fringillidae	
Amphispiza belli (sage sparrow)	C, PR
Carpodacus mexicanus (house finch)	U, PR
Junco hyemalis (slate-colored junco)	U, WV
Icteridae	
Icterus parisorum (Scott's oriole)	C, SR
Mimidae	
Oreoscoptes montanus (sage thrasher)	U, WR
Paridae	
Psaltriparus minimus (common bushtit)	C, V
Phasianidae	
Lophortyx gambelii (Gambel's quail)	×
Picidae	
Dendrocopus scalaris (ladder-backed woodpecker)	FC, PR
Scolopacidae	
Numenius phaeopus (whimbrel)	AC
Troglodytidae	
Campylorhynchus brunneicapillus (cactus wren)	FC, PR
Catherpes mexicanus (canon wren)	FC, WR
Salpinctes obsoletus (rock wren)	FC, PR
Turdidae	
Myadestes townsendii (Townsend's solitaire)	×
Sialia currucoides (mountain bluebird)	FC, WR

TABLE 15.5 (*continued*)

Species	Abundance
Sialia mexicana (western bluebird)	C, WR
Tyrannidae	
Myiarchus cinerascens (ashy-throated flycatcher)	FC, SR
Sayornis saya (Say's phoebe)	C, WR
REPTILES	
Tortoises	
Testudinidae	
Gopherus agassizi	×
Lizards	
Gekkonidae	
Coleonyx variegatus	×
Helodermatidae	
Heloderma suspectum	R
Iguanidae	
Callisaurus draconoides	R
Crotaphytus collaris	C
Crotaphytus wislizenii	×
Phrynosoma platyrhinos	R
Sauromalus obesus	×
Sceloporus magister	C
Uta graciosa	R
Uta ornata	R
Uta stansburiana	C
Teiidae	
Cnemidophorus tigris	C
Xantusidae	
Xantusia vigilis	A
Snakes	
Colubridae	
Masticophus flagellum	C
Masticophus taeniatus	×
Pituophis catenifer	×
Crotalidae	
Crotalus mitchelli	R

Much more study of the native fauna and their role here could be undertaken, especially the scarcely studied insects associated with this ecosystem type.

Decomposers

Except for the *in vitro* studies of Charley and West (1977), there is a complete lack of investigation of soil micro-biological processes in this ecosystem type. As with other desert shrublands, biomass and available soil nutrients are highly patterned. The materials and activities are centered in the surface soils around the typical clusters of shrubs. The interspaces usually have vesiculated soil surface and are nearly devoid of higher plant life, except during wet springs.

Fig. 15.9. Mature desert bighorn ram (*Ovis canadensis nelsoni*).

ECOSYSTEM INTERACTIONS

Very little attention has been given to phenomena at ecosystem level in this type. Food webs are presumably simple because of the steep dominance/diversity curves that prevail. Blackbrush seems to be a keystone species that stabilizes the community because, once removed, it is hard to predict which plants will succeed (Bowns and West, 1976). Whether the high levels of secondary plant compounds in the foliage of blackbrush have any effect on other plants, browsers or micro-organisms is unknown.

Blackbrush has difficulty reseeding itself into large disturbed areas after fire (Bowns and West, 1976), and a variety of seres develop in a seemingly unpredictable fashion.

It is anomalous that enough fuel builds up in these dry climates to carry fire. The relatively close spacing and cover for this desert-shrub type (Beatley, 1966), plus the resinous nature of the blackbrush foliage, appear to make it easy for fire to start and spread.

The earliest community appearing after fire is usually dominated by annual grasses, such as *Bromus tectorum* and *Bromus rubens*. The seed of the short-lived half-shrub *Xanthocephalum* spp. may be blown into these areas because it is light and borne by pappus. The annual-dominated communities are subject to recurring fire following wet years and buildup of fuels. Blackbrush and other long-lived perennials reappear very slowly. The large seeds and lack of root-sprouting may be the cause of such poor resilience of blackbrush to fire.

Because of the mechanical and chemical defenses of blackbrush and most of its companion shrubs, it takes a great deal of livestock use to damage the shrub component. The scattered perennial grasses are of most importance to the desert tortoise (*Gopherus agassizi*), bighorn sheep and livestock. Blackbrush is not utilized heavily until these are nearly depleted. The nutritional value of winter browse from this shrub is below maintenance levels for livestock (Bowns and West, 1976); thus, it is a forage of last resort.

LAND-USE HISTORY

This land was apparently not much used by aboriginal people. The biological communities did not have much of value to them, and the lands were not conducive to early agriculture. European-derived culture was largely absent until the late nineteenth century. The only feasible use was extensive winter livestock grazing. Much of this ecosystem type has never been used by people because rugged topography has made access unfeasible. Roads were lacking in most of the region until the uranium exploration boom beginning in the 1950s.

Only in winter was open water or snow likely to be available for livestock. What little grass was available was probably quickly scavenged by horses and cattle. In addition to harmful levels of defoliation, the method of grazing by cattle may have pulled the bunch grasses directly out of the shallow soils. The numerous relic areas (Jameson et al., 1962; Loope, 1975; Thatcher, 1976) that remain in this rugged topography have relatively little grass where shallower soils prevail. As cattle were replaced by sheep from about 1900 to 1920, more of the browse was taken. It did not please the sheepmen to have to use these ranges, because the forage is not of high nutritional quality and the spines of the blackbrush tend to pull wool from the animals.

Except for the Navajo Indian Reservation where no better forage remains and the type is used year-around (Dwyer, 1977), the livestock industry no longer uses as large a portion of this type as it once did. In part, this is due to the limited life of grazing leases on areas that have come under control of the National Park Service.

The Navajo Indian Reservation lands are so depleted that goats have become the animal that can best survive on the harsh browse and the prickly culms of *Aristida* spp. and *Muhlenbergia torreyana*.

Watershed management becomes a concern only for sediment, not water production. The potential evaporation is too high for any but the most extreme events to carry highly silt-laden water from these ecosystems. If the vegetation, including the microphytic crust, is broken, sheet and gully erosion may ensue, especially where natural erosion pavement has not already developed.

The Bureau of Land Management burned vast areas of these lands in Nevada in the 1940s and 1950s. The burned areas became occupied by annuals and the poisonous half-shrub *Xanthocephalum*, and were subject to recurring fires. Herbage production, although eight to ten times higher than the original range for short periods in wet years, was almost nil in dry years. Subsequent appraisal of erosion from burned and unburned areas (Brunner, 1972) showed that much more erosion was taking place from the altered sites. Because organic matter and nutrients are concentrated in the soil surface around the shrubs (Charley and West, 1975), and more stable, perennial-dominated plant communities are so slow in returning, further intentional burning is not recommended in this ecosystem type.

Although sophisticated grazing management systems are now being proposed, there is little evidence that great changes will be rapidly effected. For instance, Halliday (1957) found no change in the relative abundance of grass and blackbrush when livestock grazing was reduced or excluded for nearly twenty years from an area in southeastern Nevada. Hughes (1979) reported no desirable improvements in the abundance of desirable grasses in northwestern Arizona following a switch from season-long to rest-rotation grazing.

Wildlife management is not a major activity in this type because so few animals occur there. If the small herds of desert bighorn sheep and groups of desert tortoise are to be maintained, more care will have to be given to their preferred forage species and other needs. Both of these animals prefer the perennial bunch grasses and solitude. This puts them in direct conflict with both domestic and feral livestock. Domestic livestock have depleted the grasses on relatively gentle topography. In areas of more rugged topography, primarily Grand Canyon, most livestock have never penetrated. Nevertheless, feral burros, released by miners decades ago, have now exploded into populations that greatly exceed the long-term grazing capacity of this ecosystem type. This unfortunate situation developed because of the well-intentioned, but misguided, species protection demanded by people who do not understand that limitations have to operate in any ecosystem context. Because burros are so flexible in their diet and highly competitive with bighorn sheep, it is ironic that an introduced domestic animal with essentially no economic value can displace the one large native ungulate and symbol of wilderness that evolved with this ecosystem.

Nearly all of this ecosystem type is under federal control. Much has been placed in reserves such as national recreational areas, monuments and parks and given formal wilderness status. These controls mean that very little will be commercially developed. Even further mining is prohibited in much of the type. More minerals and energy sources, such as potash, salt, uranium, vanadium, tar sand, oil and gas, probably occur here; but the deposits that were more easily reached have already been tapped and excluded from the reserves. The road building that accompanied the great search for uranium and recreational opportunities during the past several decades has only recently made this area accessible to many more people through vehicular traffic.

Since secondary succession is so slow on most sites within this type, some research is needed on ways to re-vegetate areas disturbed by man (Fig. 15.10). Otherwise, wind and water erosion will make the eventual repair of the damage even more difficult.

Fig. 15.10. Route of El Paso Natural Gas Company pipeline through blackbrush area in Arches National Park, Utah. The pipeline was laid 21 years previous to the photo. Photo taken by W. L. Loope.

THE FUTURE

An attempt to use goats to reduce biologically the brush dominance and encourage more forage for other classes of livestock (Provenza, 1978) was unsuccessful because the regrowth of blackbrush was unpalatable to both goats and cattle. This is apparently an evolved mechanism to prevent excessive browsing. These results were discouraging since we currently lack other ways of enhancing the productivity of this ecosystem type. Most attempts to kill the shrubs and reseed desirable forage grasses have failed. Costs of such manipulations are so high and productivity potentials are too low to continue much more than extensive, low-level livestock management in this ecosystem type. Biological success is more probable and, thus, economic rates of return are much greater if vegetation alteration and intensive livestock management is concentrated on more mesic ranges.

Even though some potential forage is present in this ecosystem type, most of it is effectively unavailable for more livestock use because of rugged topography, lack of water, inclusion in various kinds of scenic reserves, or combinations of these constraints. Continued public pressure to preserve remnant populations of wildlife, such as the desert bighorn sheep and desert tortoise, will further reduce the area available for livestock use. Feasibility of shrub control and perennial grass reseeding in such dry, remote and rocky areas is likely to remain economically untenable, even if we had well-proven technology to do so. Thus, this is one ecosystem type that will probably continue to serve society mainly as a place to cleanse the air and the spirit.

REFERENCES

Aldous, A.E. and Shantz, H.L., 1924. Types of vegetation in the semi-arid portion of the United States and their economic significance. *J. Agric. Res.*, 28: 99–129.

Beatley, J.C., 1966. Ecological status of introduced brome grasses (*Bromus* spp.) in desert vegetation of southern Nevada. *Ecology*, 47: 548–555.

Beatley, J.C., 1969a. Biomass of desert winter annual plant populations in southern Nevada. *Oikos*, 20: 261–273.

Beatley, J.C., 1969b. Dependence of desert rodents on winter annuals and precipitation. *Ecology*, 50: 721–724.

Beatley, J.C., 1975. Climates and vegetation patterns across the Mojave/Great Basin transition of southern Nevada. *Am. Midl. Nat.*, 93: 53–70.

Beatley, J.C., 1976a. Rainfall and fluctuating plant populations in relation to distributions and numbers of desert rodents in southern Nevada. *Oecologia*, 24: 21–42.

Beatley, J.C., 1976b. Environments of kangaroo rats (*Dipodomys*) and effects of environmental change on populations in southern Nevada. *J. Mammol.*, 57: 67–93.

Bowns, J.E. and West, N.E., 1976. Blackbrush (*Coleogyne ramosissima* Torr.) on southwestern Utah rangelands. *Utah State Univ. Agric. Exper. Stn. Res. Rep.*, 27: 27 pp.

Bradley, W.G., 1965. A study of the blackbrush plant community of the Desert Game Range. *Trans. Desert Bighorn Council*, 11: 56–61.

Bradley, W.G. and Deacon, J.E., 1967. The biotic communities of southern Nevada. *Nev. State Mus. Anthropol. Pap.*, 13: 202–295.

Brunner, J.R., 1972. A 20-year trend record of woody vegetation in southern Nevada. *Abstr. Annu. Meet. Soc. Range Manage.*, 23: 33–34.

Charley, J.L. and West, N.E., 1975. Plant-induced soil chemical patterns in some shrub-dominated semi-desert ecosystems of Utah. *J. Ecol.*, 63: 945–964.

Charley, J.L. and West, N.E., 1977. Micro-patterns of nitrogen-mineralization activity in soils of some shrub-dominated semi-desert ecosystems of Utah. *Soil Biol. Biochem.*, 9: 357–365.

Christensen, E.M. and Brown, R.C., 1963. A blackbrush over 400 years old. *J. Range Manage.*, 16: 118.

Dwyer, D.D., 1977. *History of Use and Abuse of the Navajo Indian Reservation.* A report prepared for Rodey, Dickason, Sloan, Akin and Robb, P.A., 62 pp.

Halliday, B.O., 1957. Nineteen years of range improvement on the Crystal Springs Range Demonstration Area in Nevada. *J. Range Manage.*, 10: 203–207.

Hughes, L.E., 1979. Rest-rotation grazing on the Arizona Strip: An observation. *Rangelands* 1: 106–107.

Humphrey, R.R., 1955. Forage production on Arizona Ranges IV. Coconino, Navajo and Apache Counties. *Ariz. Agric. Exper. Stn. Bull.*, 266: 84 pp.

Humphrey, R.R., 1964. Forage production on Arizona Ranges III. Mohave County. *Ariz. Agric. Exper. Stn. Bull.*, A-34: 77 pp.

Jameson, D.A., Williams, J.A. and Wilton, E.W., 1962. Vegetation and soils of Fishtail Mesa, Arizona. *Ecology*, 43: 403–410.

Jaynes, R.A. and Harper, K.T., 1980. *Composition and Environment of Major Plant Communities in the Kaiparowits Basin of Utah.* Unpublished manuscript.

Küchler, A.W., 1970. Potential natural vegetation (map at scale of 1:7 500 000). In: U.S. Geol. Survey, *The National Atlas of the U.S.A.* U.S. Government Printing Office, Washington, D.C., pp. 90–91 (revision of 1964 map at scale of 1:3 168 000).

Loope, W.L., 1975. *Vegetation in Relation To Environment of Canyonlands National Park, Utah.* Dissertation, Utah State University, Logan, Utah, 157 pp.

Loope, W.L. and West, N.E., 1979. Vegetation in relation to environments of Canyonlands National Park. In: R.M. Linn (Editor), *Proc. First Conf. on Scientific Research in the National Parks, 1. U.S. Dep. Inter., NPS Trans. Proc. Ser.*, No. 5: 195–199.

MacMahon, J.A. and Wagner, F.H., in press. The Mojave Sonoran and Chihuahuan Deserts of North America. In: M. Evenari and I. Noy-Meir (Editors), *Hot Deserts and Arid Shrublands. Ecosystems of the World, 12.* Elsevier, Amsterdam, in press.

MacMahon, J.A. and Wieboldt, T.F., 1978. Applying biogeographic principles to resource management: A case study evaluating Holdridge's Life Zone model. *Great Basin Nat. Mem.*, 2: 245–258.

Provenza, F., 1978. *Biological Manipulation of Blackbrush* (Coleogyne ramosissima Torr.) *by Browsing With Goats.* Thesis, Utah State University, Logan, Utah, 68 pp.

Russo, J.P., 1956. *The Desert Bighorn Sheep in Arizona.* Arizona Game and Fish Dep., Phoenix, Ariz., 153 pp.

Tanner, W.W., 1978. Zoogeography of reptiles and amphibians in the Intermountain Region. *Great Basin Nat. Mem.*, 2: 43–54.

Thatcher, A.P., 1976. The amount of blackbrush in the natural plant community is largely controlled by edaphic conditions. In: H.C. Stutz (Editor), *Proc. Symp. and Workshop Wildland Shrub.* U.S. For. Ser. Shrub Sci. Lab. Provo, Utah, pp. 155–156.

Thatcher, A.P., Doughty, J.W. and Richmond, D.L., 1976. Amount of blackbrush in the pristine plant communities controlled by edaphic conditions. *Abstr. Annu. Meet. Soc. Range Manage., Omaha, Nebr.*, 27: 12.

Wallace, A. and Romney, E.M., 1972. Radioecology and ecophysiology of desert plants at the Nevada Test site. *U.S. Atomic Energy Comm.*, TID-25954: 439 pp.

Wells, K.F., 1967. *Aspects of Shrub Herb Productivity in an Arid Environment.* Thesis, University of California, Berkeley, Calif., 49 pp.

Wells, R.E. and Wells, F.B., 1961. *The Bighorn of Death Valley. Fauna of the National parks. Series 6.* U.S. Government Printing Office, Washington, D.C., 242 pp.

Welsh, S.L. and Moore, G., 1973. *Utah Plants: Tracheophyta.* Brigham Young University, Provo, Utah, 3rd ed., 474 pp.

West, N.E., 1969. Soil–vegetation relationships in arid southeastern Utah. In: *Int. Conf. on Arid Lands in a Changing World. University of Arizona, Tucson, Ariz.*, 9 pp.

Wilson, L.O., 1967. *The Distribution, Ecology and Factors Affecting the Desert Bighorn Sheep of Southeastern Utah.* Thesis, Utah State University, Logan, Utah, 217 pp.

Chapter 16

SOUTHEASTERN UTAH GALLETA–THREEAWN SHRUB STEPPE[1]

N.E. WEST

DEFINITION OF THE TYPE AND ITS LOCATION

This chapter concerns a little known type of desert grassland with occasional shrubs (Fig. 16.1). The type occurs in the vicinity of the confluence of the Green and Colorado Rivers (Fig. 16.2) on relatively deep sandy soils of the Colorado Plateau. In pre-Columbian times it probably had more abundant perennial grass cover and less shrub and weedy annual cover than it now possesses (Fig. 16.3). The grasses are a mix of bunch and sod-formers; the woody vegetation is largely half-shrubs or low shrubs breaking into clusters with many stems.

The salt-desert shrub type occurs on shale-derived soils nearby, whereas blackbrush occurs on shallower soils at generally lower elevations. Juniper–pinyon woodlands occur at slightly higher and wetter elevations directly above it. These grasslands have considerable affinities to the grama–galleta steppes to the south discussed in Volume 8 of this Series.

Fig. 16.1. *Ephedra viridis* growing in sand mounds in Gray's Pasture, Island in the Sky, Canyonlands National Park, Utah (photo by W.L. Loope).

Fig. 16.2. Map of the southeastern Utah Galleta–Threeawn Shrub Steppe ecosystem type. Derived from Küchler (1970).

EXTENT

The mappable area of this ecosystem makes it the smallest type we will deal with. About 0.5×10^6 ha of this kind of ecosystem occur, entirely within Utah.

[1] Manuscript completed June, 1980.

Fig. 16.3. Relic area (never been grazed by livestock) of galleta–threeawn shrub steppe at the southern end of Chesler Park, Canyonlands National Park, Utah (photo by W.L. Loope).

CURRENT OWNERSHIP AND DOMINANT USE

Nearly all of this area is publicly owned rangeland used by cattle. The major controlling agencies are the Bureau of Land Management and the National Park Service. Livestock grazing on the National Park Service land is being phased out by legislation, usually not allowing renewal of permits after the present grazier dies.

ECOSYSTEM STRUCTURE AND FUNCTION

Climate

Since there is very little permanent human occupation of this area, very limited climatic information is available. Data from the most relevant long-term station is presented in Fig. 16.4. The climate of this area is arid, temperate continental. It is as dry and cold as the adjacent salt-desert shrub type but somewhat colder on the average than the blackbrush type and drier than the juniper–pinyon woodlands developing at higher elevations nearby. The sandy texture of the soils has a great deal to do with increasing the effectiveness of the precipitation.

The distribution of precipitation is a mixture of winter snow and growing-season rain. Late summer convectional storms are brief but violent, often unleashing considerable downpours.

Winds, both frontal and convectional with thermal heating in summer, are a major evaporative and erosional feature.

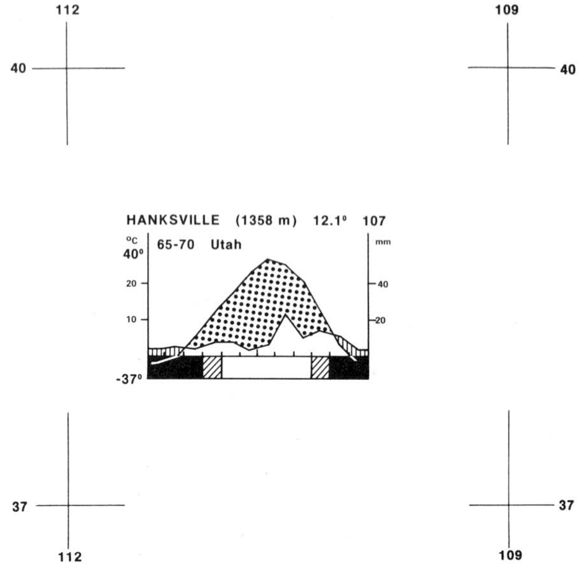

Fig. 16.4. Climatic diagram for Hanskville, Utah.

Geology

This shrub steppe only develops well on silt and sandy deposits from weathered sandstone. Depth of these deposits has a great bearing on the kind of vegetation that develops (Fig. 15.8). Where sand is less than a meter deep, either shrubs or trees can prevail. This is probably related to the blockage of the wetting front and creation of some soil moisture storage (Loope, 1975). On deep sands only vadose moisture can remain. The fibrous root structure of the grasses is better adapted to absorb the moisture as it quickly moves through the profile. Overwinter storage is not as important as it apparently is for the less extensive and more slowly absorbing root systems of shrubs or trees.

Topography

Plains on the nearly level stratigraphy of the Colorado Plateau are where most of the silts and sands have come to rest. The particles were, however, probably produced by wind and water erosion in the exposed sandstone cliffs, hogbacks, or pinnacles that ring the area. The San Rafael Swell is a greatly folded and highly dissected monocline in the center of the area. Its sparse cover and highly erosive strata are probably the source of most of the wind-blown material. The particles in the Cedar

Mesa, Organ Pipe, Kayenta, and Navajo Sandstone Formations were blown about much more in the Permian Period when they were formed. Patches of this grassland type may be found throughout the Colorado Plateau, sequestered between the rocks where lenses of deep sand have accumulated.

Soils

Because of considerable movement by wind, the soils of this ecosystem type do not have well-developed profiles (Fig. 16.5). Most of them would be classified as Psamments. Petrocalcic, argillic or other horizons are lacking because silts or clays are minor. Because of the texture and low salinity, infiltration of moisture is rapid (see Table 15.2). The vertical and horizontal gradients of nutrients that occur are largely related to accumulations of organic matter (Kleiner and Harper, 1977a).

Fig. 16.5. Soil profile in thick aeolian deposits near Murphy Point, Canyonlands National Park, Utah. Depth of wetting is 50 cm after a day with heavy summer rain. (Photo by W.L. Loope.)

Primary producers

The flora is composed of a mix of long-lived bunch and sod-forming grasses and palatable half-shrubs (Tables 15.3 and 16.1). The name of the type derives from galleta (*Hilaria jamesii*) a major sod-forming grass of the southwestern United States (West, 1972) and threeawns (*Aristida* spp.) which are short-lived bunchgrasses. Both genera usually become more abundant with grazing disturbance; but many more major taxa are usually found on examples of the type (Table 16.1), thus the type is locally known as "semi-desert grassland". The total amount and specific shrub components probably depend on surrounding vegetation types for seed sources. As in other shrub steppes, the proportion of shrubs and annuals has increased greatly with grazing disturbances. These successional changes will be detailed later.

The bunchgrasses usually put on their growth in the cool season, whereas the sod-formers develop later and can grow and reflower whenever the soil is remoistened in the summer and early fall (West, 1972).

Although the sward is low-growing (usually less than 30 cm high), the total vascular plant cover on such sites is considerable (25–60%) for a semi-arid environment (Allan, 1977). To this must be added the soil microphytic crusts in the interspaces between the higher plants (Fig. 16.6). This cover is particularly well-developed in ungrazed relics, where it occupies nearly all the bare ground (Kleiner and Harper, 1972, 1977b; Loope, 1975).

On ungrazed areas, there is a richer mix of species, nearly all of which are perennials and well segregated into a mosaic in which each occupies the micro-environment in which it apparently best competes (Kleiner and Harper, 1972). With livestock grazing, the microphytic crust is broken, the diversity decreases, and the grasses become more mixed across the landscape. The palatable half-shrubs decline and shorter-lived grasses, especially *Aristida* spp. and *Muhlenbergia torreyana*, and forbs invade. Tables 16.2 and 16.3 compare a pristine area (Virginia Park) with a moderately grazed area (Chesler Park) immediately adjacent but protected by a natural rock wall. These areas are nearly identical in physical characteristics and differ mostly in regard to grazing history.

Grazing heavier than that found in Chesler Park

TABLE 16.1

Summary of vegetation data for transects in deep aeolian material, Canyonlands National Park, Utah (from Loope, 1975)

Plant name[1]	Average frequency (%)	Coefficient of variation for frequency	Average density individuals per 250 m²	Average cover (%)	Constancy (%)	Constancy × frequency index
SHRUBS						
Ephedra viridis	55	29	5	3	36	1980
Ceratoides lanata (Pursh), J.T. Howell	18	53	9	3	44	792
Opuntia spp.[2]	14	78	6	1	42	588
Gutierrezia spp.[3]	16	68	8	2	26	416
Atriplex canescens	8	55	4	1	20	160
Coleogyne ramosissima	11	40	4	t	6	66
Echinocactus whipplei	5	40	1	t	6	30
Sarcobatus vermiculatus	4		4	t	2	8
PERENNIAL FORBS						
Sphaeralcea coccinea	35	57	18	1	62	2170
Hymenopappus filifolius	17	47	9	5	26	442
Arenaria fendleri	16	19	4	t	22	352
Eriogonum microthecum	31	54	1	5	8	248
Cryptantha confertiflora	11	68	15	t	22	242
Gilia longiflora	21	33	3	t	8	168
Abronia fragrans	8	57	5	t	20	160
Lepidium montanum	7	57	1	t	16	112
Sphaeralcea parvifolia	6	50	1	t	18	108
Gaillardia pinnatifida	18	11	5	t	6	108
Linum aristatum	10	20	4	t	8	80
Gilia gunnisonii	10	70	2	t	4	40
Astragalus mollissimus	5	20	1	t	8	40
Townsendia incana	6	20	6	t	6	36
Eriogonum ovalifolium	4		1	t	4	16
Penstemon cyanocaulis	4		1	t	2	8
Eriogonum inflatum	4		1	t	2	8
GRASSES						
Hilaria jamesii	67	34		5	78	5226
Stipa comata	61	45		5	80	4880
Oryzopsis hymenoides	32	45		2	80	2560
Bouteloua gracilis	71	26		9	34	2414
Vulpia octoflora (Walt.) Rydberg	33	50		1	42	1386
Bromus tectorum	27	44		2	30	810
Sporobolus cryptandrus	22	50		2	32	704
Aristida fendleriana	18	52		1	30	540
Sporobolus flexuosus	12	50		t	20	240
Sporobolus airoides	31	83		1	6	186
Sitanion hystrix	8	50		1	12	96
ANNUAL HERBS						
Malacothrix sonchoides	44	49		1	56	2464
Plantago patagonica	57	44		3	42	2394
Eriogonum cernuum	59	27		1	24	1416
Gilia inconspicua	32	68		3	30	960
Stephanomeria exigua	22	68		1	32	704
Cymopterus fendleri	25	69		t	26	650
Helianthella uniflora	22	72		1	26	572

TABLE 16.1 (*continued*)

Plant name[1]	Average frequency (%)	Coefficient of variation for frequency	Average density individuals per 250 m^2	Average cover (%)	Constancy (%)	Constancy × frequency index
Lappula redowskii	16	37		t	34	544
Senecio multilobatus	14	53		t	26	364
Oenothera trichocalyx	6	50		t	32	192
Salsola kali	23	87		1	8	184
Oenothera caespitosa	8	74		t	18	144
Lupinus pusillus	35	77		t	4	140
Machaeranthera bigelovii	17	98		1	8	136
Chaenactis stevioides	11	72		t	12	132
Aster leucelene	29	13		1	4	116
Streptanthella longirostris	42			1	2	84
Descurainia sophia	37			t	2	74
Cleome lutea	11	18		t	6	66
Astragalus ceramicus	7	28		t	6	42
Chenopodium album	4			t	6	24
Calochortus nuttallii	4			t	4	16
Microphytic soil crusts				10 (0–40)		

[1] All nomenclature follows Holmgren and Reveal (1966) except where authorities are given. Plants are listed by growth form and arranged by decreasing constancy × frequency index.
[2] Includes *Opuntia polyacantha, Opuntia erinacea* and others.
[3] Includes *Xanthocephalum sarothrae* and *Xanthocephalum microcephala.*
Coefficients of variation not calculated when frequency values were below 5%.

TABLE 16.2

Prevalent species and their relative abundance (C × F index) in Virginia and Chesler Parks (test criterion for the relative abundance data was χ^2-analysis based on presence and absence of each species in quadrats of the two parks; Kleiner and Harper, 1972; © 1972 Duke University Press)

Species	C × F index		Significance of χ^2-value	Species	C × F index		Significance of χ^2-value
	Virginia	Chesler			Virginia	Chesler	
Collema tenax	9710	5548	**	*Machaeranthera pulverulenta*	465	296	NS
Tortula ruralis	9130	660	**	*Eriogonum cernuum*	312	138	*
Buellia papillata[a]	6450	99	**	*Bromus tectorum*	183	9	**
Hilaria jamesii	5206	3553	**	*Lepidium densiflorum*	163	64	NS[b]
Fulgensia bracteata	4533	0	**	*Aster leucelene*	36	230	**
Stipa comata	3626	4902	**	*Stephanomeria exigua*	36	4	NS
Plantago purshii	3304	516	**	*Cryptantha crassisepala*	27	105	*
Lecidea decipiens	2040	0	**	*Calochortus nuttalii*	18	0	NS
Festuca octoflora	1876	2890	**	*Chaenactis steviodes*	17	0	*
Toninia caeruleonigricans	1544	0	**	*Bouteloua gracilis*	16	522	**
Opuntia aurea	1368	9	**	*Opuntia littoralis*	9	30	NS
Oryzopsis hymenoides	1141	1343	*	*Ephedra viridis*	3	360	**
Lecidea rubiformis[a]	806	0	**	*Ceratoides (Eurotia) lanata*	3	90	**
Sporobolus cryptandrus	663	1530	**	*Oenothera trichocalyx*	1	15	*
Descurainia obtusa	588	48	**	*Salsola kali*	0	24	**
Gilia sinuata	538	0	**	*Sphaeralcea leptophylla*	0	1463	**
Lappula redowskii	476	6	**				

[a] Identification tentative.
[b] $\chi^2=0$; species distributed at random.
** = highly significant; * = significant; NS = not significant.

TABLE 16.3

Floristic and vegetational characteristics of Virginia and Chesler Parks (Kleiner and Harper, 1972; © 1972, Duke University Press) (Sampling by twenty quadrats of 50 m^2 in each stand)

Characteristic	Virginia	Chesler
Floral		
Average number of species per stand (species density)	22	13
Average number of vascular species per stand	15	11
Average number of cryptogam species per stand	6	2
Average number of species per quadrat	7	3
Vegetational		
Average total living cover (%)	55	22
Vascular plant living cover (%)	17	17
Cryptogamic cover (%) (1/2 moss–1/2 lichen)	38	5
Average number of hits per stand[1] on living vascular plant tissue	23	26
Litter cover (%)	12	10
Contribution of annuals to the total vegetation (%)	12	16

[1] Out of 100 point quadrats.

Fig. 16.6. Well-developed microphytic crusts in an ungrazed relic area. Virginia Park. Canyonlands National Park, Utah.

reduces the vegetal cover further, and wind begins to cause micro-dunes. The less palatable shrubs such as *Coleogyne*, *Ephedra*, *Quercus* and *Xanthocephalum* are least grazed and also catch moving sand. A low-relief dune and slack microtopography apparently prevails as long as the grasses are selectively grazed.

Phytomass data have apparently never been recorded in an example of this ecosystem type. Net primary productivity estimates are also lacking. The best appraisals of annual above-ground herbage production are available from the Soil Conservation Service (1971). They indicate that "southern semi-desert sand sites" in good condition can produce up to 1500 kg ha^{-1} of forage in a single favorable year and average 900 kg ha^{-1} yr^{-1} in a set of favorable years. They provide no data for other range conditions or unfavorable years. If the management "rule of thumb" of using half and leaving half of the herbage is applied, these grasslands would be grazed at levels of less than 1 AUM ha^{-1}.

Consumers

Very little is known of the fauna in this ecosystem type. Large native ungulates have rarely used this ecosystem type. Large populations could not have been common because of the lack of dependable surface water on much of this area. Pronghorn (*Antilocapra americana*) and bison (*Bison bison*) have recently been re-introduced into these areas. Huntable herds of bison and bighorn sheep (*Ovis canadensis*) have been built up on such areas in both Utah and Arizona.

The Gunnison's prairie dog (*Cynomys gunnisoni*) and Utah prairie dog (*C. parvidens*) and their predator, the black-footed ferret (*Mustela nigripes*), were probably once widespread on this type (Collier and Spillett, 1975). Vegetation deterioration and poisoning have endangered the survival of the Utah prairie dog. Whether the black-footed ferret is extinct here is not known.

The sandy soils make for poor burrowing for most small mammals except kangaroo rats. These and other seed-eating animals may have increased in abundance because annuals produce more seed.

Virtually nothing is known of the ecological relationships of invertebrate fauna in this ecosystem type. Dung beetles (Scarabaeidae) and other detritivores are conspicuous in the spring, but their role in energy flow and nutrient cycling remains to be studied.

Decomposers

No published information on decomposer activities in this ecosystem type could be located.

ECOSYSTEM INTERACTION

Because information is lacking on so many important components in this ecosystem type, it is difficult to discuss food webs, guilds, keystone species, and other elements of community organization and feedback mechanisms. The greatest amount of extant information deals with the interrelationships of vegetation to environments and livestock grazing impacts. These plus the interesting observation of Kleiner and Harper (1972) on patterning have already been related.

We can surmise that this ecosystem was somewhat unstable and poorly resilient to heavy livestock grazing. Whether it will move back toward ecosystems similar to those we see in relic areas is unknown. The rates of secondary succession are slow in such semi-arid environments (Loope, 1975). We also lack research on how to accelerate the recovery of the perennial plants. The periodic droughts undo a lot of vegetal recovery, particularly if livestock grazing reductions are not adjusted rapidly enough.

Fortunately, not much besides livestock grazing has perturbed these systems. The fuel loads are usually not great or continuous enough to carry fire.

LAND-USE HISTORY

Aboriginal peoples utilized this type a great deal, particularly those areas near cliffs and ephemeral streams. The earliest pictorial record of North American agriculture can be found in the Maze Section of Canyonlands National Park. Pre-Columbian people harvested native grass seeds and planted corn (*Zea mays*), beans (*Phaseolus*) and squash (*Cucurbita*) in the sandy soils of canyon bottoms. They kept turkeys (*Meleagris gallopavo merriami*) and hunted bighorn sheep. Rock overhangs were used for protection from harsh weather. Into these alcoves were later built elaborate masonry apartment-type dwellings and granaries. These cliff dwellers flourished from about 1100 to 1300 A.D. They departed suddenly following an apparent period of declining summer precipitation, moving to the Rio Grande Valley where irrigation diversions were more feasible (Jennings, 1978).

Few, if any, Indians occupied this ecosystem type when European-based culture colonized the area in the 1860s and 1870s. This was the last area of the Colorado Plateau to be developed by the livestock industry. Although the grasses were plentiful, water was hard to develop (Baker, 1976).

The rugged topography nearby made this a favorite area for outlaws to keep their horses. The area thus became more famous for its colorful desperados than beef production. Sheep grazing was never common in this type, partly because of tradition and partly because of the low proportion of good browse. Grasses are very low in protein during the winter months when sheep use is most feasible.

The unrestrained use of lands near natural and developed water sources, plus the drought and economic depression of the 1930s, made this a small dust bowl. When the federal Grazing Service finally assumed control of the public domain in 1934, much of these lands had been greatly degraded (Fig. 16.1). The process of bringing the livestock numbers and seasons of use into congruence with the ecological carrying capacity of the land has been a long and painful process that is not yet finished. Those ranchers who were fortunate enough to have irrigated pasture and hay lands and summer range in the mountains have been able to use those other ranges in the spring and fall.

As the cattle were moved to the mountains, the semi-desert grasses had a chance to recover. The late summer rains provided green feed or standing dead material when they returned in the fall. Irrigated lands and mountain ranges are, however, in short supply in this region. Those without these advantages have attempted to use these ranges at all seasons. Seasonal rotation helps, but during drought there is little or no forage anywhere in the lowland ranges. Even light use during such periods causes considerable damage to the vegetation. It is difficult to hold off the livestock and let the vegetation recover in the critical period when drought finally breaks. The conclusion is, that those who do not have access to the more dependable forages on irrigated lands and mountain pastures to carry their animals in drought periods should not be allowed on these ranges. Although no grazing management research has been done in this type, it is observable and logical that spring or summer use is tolerable on these ranges, if some seasonal deferment takes place to favor a balance of species and forms.

It is not surprising that ranchers using these ranges have not been happy to see the re-introduction of pronghorns and bison. These animals go where they please, when they please, and compete with cattle for the largely grassy forage. In addition, bison can make a shamble of light fences. Pronghorn require fences that can be crawled under. Resolving these conflicts will be a major job of the land manager.

This type is not preferred by many recreationists. It is too open, bleak, windswept and lacking in water for campers and second-home development. Because it is nearly all federal land, many permits are required for industrial development.

THE FUTURE

The portion managed by the Bureau of Land Management is likely to remain as livestock range. The portions under Park Service control eventually will revert to natural areas managed for their wilderness character, if they have not already done so. Some is *de facto* wilderness and has been proposed for wilderness status, particularly in the unroaded sections. Wilderness status does not in itself prevent livestock use, only the degree to which management can be pursued (fences, jeep trails for salting, water development, herbicide usage, reseeding, etc., are all prohibited). These considerations influence our estimates of potential increases in the livestock carrying capacity of this range. Any increase in animal production will largely have to be from better adjustment of season and intensity of use, as no chemical or mechanical technology is known or even proposed to control brush and seed grasses in such enviroments. The economic returns from the extensive management that is used now is marginal.

The biggest clouds on the horizon are due to potential energy development, revegetation of mined areas and pollution control mandates. There is a lot of coal in the San Rafael Swell and the Henry Mountains, and uranium and tar sands elsewhere in the region. Some would like to see large mine-mouth coal-fired power plants developed in the area. Geologic water sources for this apparently exist. Air quality would, however, deteriorate; and it is questionable if strip-mined lands can be revegetated under such dry conditions. The presence of National Park lands nearby and their legal and administrative protection against polluted air has thus far delayed this kind of development. Whether large mining scars and smoke stacks ever appear in this unique kind of semi-desert steppe depends on trade-offs society is willing to make.

Until miners and engineers begin to be active in this area, both on and off the job, it will likely remain a sparsely populated region contributing a small amount of beef and many scenic vistas for the few who may be able to travel in this area as gasoline becomes too precious for most to visit it.

REFERENCES

Allan, J.S., 1977. *The Plant Communities of Arches National Park*. Thesis, Brigham Young University, Provo, Utah, 98 pp.

Baker, P.B., 1976. *Robber's Roost Recollections*. Utah State University Press, Logan, 194 pp.

Collier, G.D. and Spillett, J.J., 1975. Factors influencing the distribution of the Utah prairie dog (*Cynomys parvidens*/Sciuridae). *Southwest Nat.*, 20: 151–158.

Holmgren, A.H. and Reveal, J.L., 1966. Checklist of the vascular plants of the intermountain region. *U.S. Dep. Agric., For. Serv., Res. Pap.*, INT-32: 160 pp.

Jennings, J.D., 1978. Prehistory of Utah and the eastern Great Basin. *Univ. Utah (Salt Lake City), Anthropol. Pap.*, 98: 263 pp.

Kleiner, E.F. and Harper, K.T., 1972. Environment and community organization in grasslands of Canyonlands National Park. *Ecology*, 53: 299–309.

Kleiner, E.F. and Harper, K.T., 1977a. Occurrence of four major perennial grasses in relation to edaphic factors in a pristine community. *J. Range Manage.*, 30: 286–289.

Kleiner, E.F. and Harper, K.T., 1977b. Soil properties in relation to cryptogamic groundcover in Canyonlands National Park. *J. Range Manage.*, 30: 202–205.

Küchler, A.W., 1970. Potential natural vegetation (map at scale 1:7 500 000). In: U.S. Geol. Survey, *The National Atlas of the U.S.A.* U.S. Government Printing Office, Washington, D.C., pp. 90–91.

Loope, W.L., 1975. *Vegetation in Relation to Environment of Canyonlands National Park, Utah.* Dissertation, Utah State University, Logan, Utah, 151 pp.

Soil Conservation Service, 1971. *Yield and Composition of Utah's Range Sites.* U.S. Dep. Agric., Soil Conservation Service, Salt Lake City, Utah, 153 pp.

West, N.E. (Editor), 1972. Galleta: Taxonomy, Ecology and Management of *Hilaria jamesii* on Western Rangelands. *Utah State Univ. Agric. Exper. Stn. Bull.*, 487: 31 pp.

Chapter 17

DESERTS AND SEMI-DESERTS OF PATAGONIA

ALBERTO SORIANO

With sections by W. VOLKHEIMER, H. WALTER, E.O. BOX, A.A. MARCOLIN, J.A. VALLERINI, C.P. MOVIA, R.J.C. LEÓN, J.M. GALLARDO, M. RUMBOLL, M. CANEVARI, P. CANEVARI and W.G. VASINA

INTRODUCTION (A. Soriano)

The vast territory (Fig. 17.1) in southern Argentina generally known as Patagonia extends from approximately 40° to 55°S. From a broad, ecological point of view it can be considered a cool semi-desert. According to Meigs (1952), most of this territory belongs to the climatic class **Aa12** (A = arid, a = rainy season lacking, 12 = cool winters, mild summers). Walter and Box (1976) also place this region in the arid temperate desert category. The only other "cool" desert in South America occurs at high elevations in the Puna Region of northwestern Argentina and the southwestern Bolivian high plateau (Altiplano). The latter region has much less seasonal variation in temperature because of low latitudes.

The landscape is mainly molded on sedimentary and volcanic rocks of Mesozoic and Cenozoic age, forming a system of plateaus and hills. In some areas, Upper Paleozoic sedimentation and even the crystalline basement are forming hilly landscapes.

Far from being uniform, the region exhibits quite a rich spectrum of vegetation types, from the real desert to shrub or grass steppe. No estimates have been made of the area actually covered by each one of the different physiognomic units present, but a rough approximation could be: 45% shrub desert, 30% shrub-grass semi-desert, 20% grass steppe and 5% water surface and minor types like meadows, locally named "*mallines*", "*cañadones*" and "*vegas*".

In referring to ecosystems present in Patagonia today, immediate mention has to be made of the sheep industry in the region. Since the beginning of this century, sheep covered the whole territory of Patagonia. The European hare (*Lepus europaeus*) spread all over too, having been supposedly introduced in 1911 (Jones, 1961). Coincidentally, pumas (*Felis concolor*), guanacos (*Lama guanicoe*) and lesser rhea (*Pterocnemia pennata pennata*) were much reduced in number.

By now, it is not possible to find areas where sheep and hare grazing has not taken place. Because no direct documentation of the degree of change produced by sheep flocks is available and no relic areas remain to infer changes, it is difficult to describe the pre-Columbian biotic communities of this region. Descriptions of explorers made during the Indian period[1] only contain broad descriptions of the landscape and, sometimes, commentaries on special resources, but do not allow useful comparisons with present ecosystems. Even from the work of enlightened botanists, like Lucien Hauman (1920), it is not possible to draw precise ideas about what vegetation deterioration has occurred. The reports of early naturalists (Soriano, 1948) contain valuable and broad descriptions of the landscape and the flora. Among them Charles Darwin (1840) deserves special mention because of the momentous results of his discoveries on the voyage of H.M.S. *Beagle* from 1831 to 1836.

According to Morrison (1917), sheep were introduced around 1885 to Rio Gallegos, and ten years later there were 369 264 sheep in the Province of Santa Cruz (243 943 km^2). By 1908 the number reached 2 387 566 and in 1914, there were 3 940 616 sheep in the same territory. Thus, it is conceivable

[1] General Julio A. Roca, in his Desert Campaign took his troops to the Rio Negro in 1879. The Campaign put an end to the dominance of Indians in those territories and to their incursions farther north.

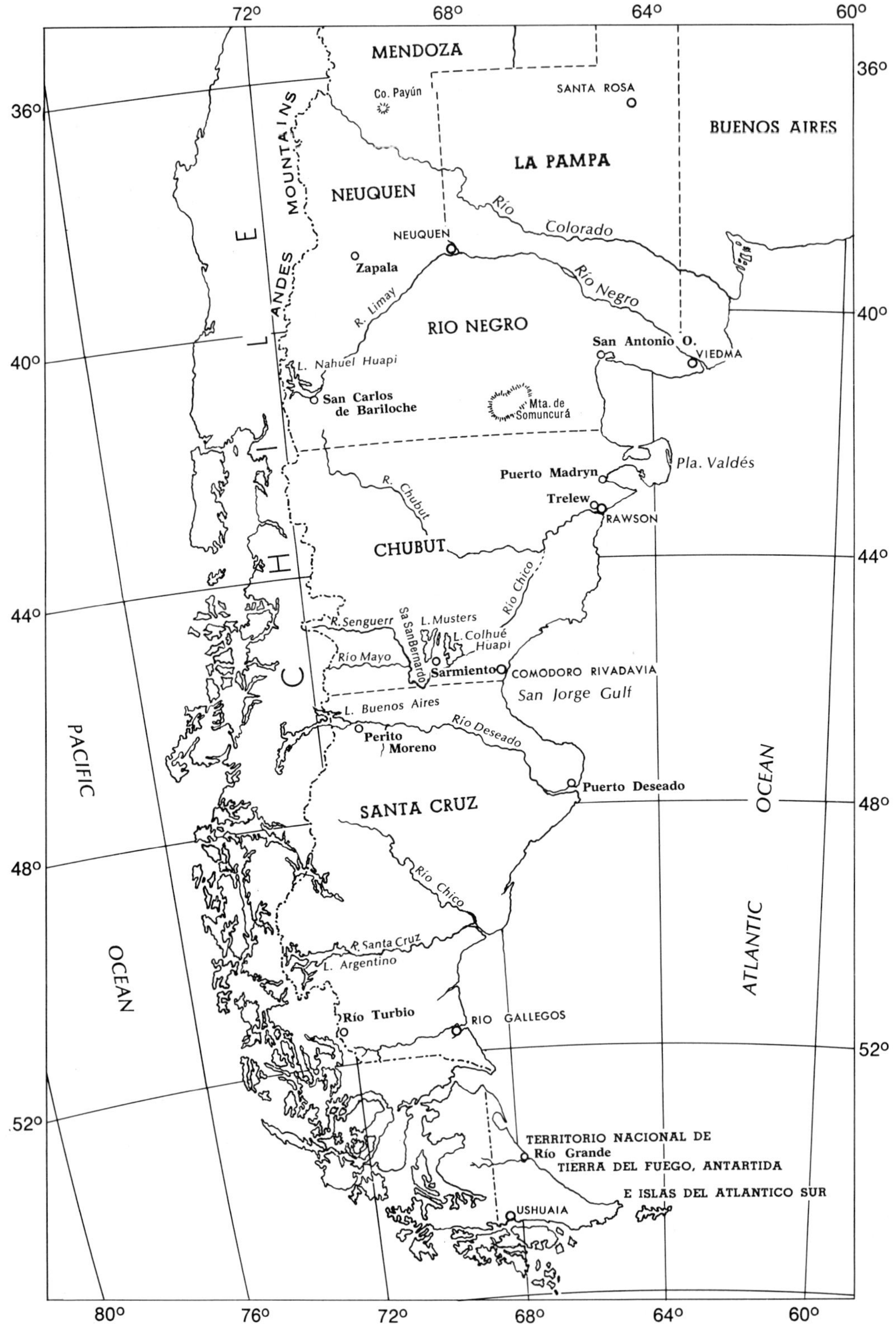

Fig. 17.1. Map of the southern tip of South America showing longitude, latitude, national and provincial boundaries and extent of Patagonia.

that the most important change in vegetation cover occurred during this early period.

Although the current human population density in Patagonia is very low (0.6 people km^{-2}), man's impact has become evident not only through the sheep industry already mentioned, but also through wood cutting (some shrubs are said to have been much more abundant early in the century), the oil industry, dam construction, and town and road building.

GEOLOGY OF EXTRA-ANDEAN PATAGONIA

(W. Volkheimer)

Introduction

Extra-Andean Patagonia extends approximately from 39° to 55°S. It is limited to the west by the Andean Cordillera with elevations mainly between 2000 and 2500 m, and exceptionally higher peaks, as (from north to south) the Volcan Lanin (3774 m), Mt. Tronador (3554 m), Cerro Tres Picos (2600 m), Cerro San Valentin (4058 m), Cerro Fitzroy (3375 m), Cerro Murallón (3600 m) and others.

To the north, the plateaus (*mesetas*) of extra-Andean Patagonia merge into the Pampa plains, and to the east they are limited by the South Atlantic Ocean.

In northern Patagonia (Rio Negro Province) the high plateaus begin at the sub-Andean foothills at levels between 800 and 1200 m and descend towards the east to the Atlantic. The plateaus are frequently crowned by young volcanic forms (e.g. Sierra de Somuncurá: level of the high meseta at the foot of Cerro Corona, 1200 m; Cerro Corona, 1974 m).

Extra-Andean Patagonia from 42° to 46°S (Chubut Province) is characterized by plateaus which commence at the sub-Andean foothills at 700 to 1000 m. In Chubut there are two parallel extra-Andean orographic units: (1) to the east of meridian 71°, the Precordillera Patagonica (Sierra de Tecka and continuations to the north and south, with elevations of 1300 to 1500 m, and some higher elevations, e.g. Cerro Putrachoique, 1700 m); (2) the Central Hills of Patagonia (Sierras Centrales Patagonicas), located between the meridians 69° and 70° and limited by the Rio Chubut to the north and the Rio Senguerr to the south.

In southern extra-Andean Patagonia from 46° to 51°S the high plateaus begin at 1200 to 1600 m, and in the south at up to 2000 m. Glacial forms, corresponding to the last glaciation, advance upon the plateaus far to the east, and south of the Rio Gallegos they reach the Atlantic coast.

The plateaus from 41° to 55°S, in southern Santa Cruz and Tierra del Fuego, commence north of the Rio Gallegos at 900 to 1200 m; south of this river at 200 m or less (Garleff, 1977).

For general aspects of Patagonia geology and complete literature up to 1949 see Feruglio (1949–50).

Geologic history

During Late Paleozoic (Anthracolithic) times a marine basin extended from NNW to SSE in extra-Andean Patagonia. This Central Patagonian Basin was limited by continental areas at the ENE and WSW. Some of these areas were glaciated to some extent during Carboniferous times. Several intercalations of diamictites observed in the Anthracolithic sequence of the basin have been interpreted as possible glaciomarine sediments. The glacial phenomena of Carboniferous age and the presence of *Glossopteris* and *Gangamopteris* in Lower Permian indicate continental relations with other Gondwanic regions like Antarctica, South Africa, India and Australia during Late Paleozoic times.

In some areas the Anthracolithic sequence was intruded by "Permotriassic" granites and granodiorites.

The Triassic sediments known from extra-Andean Patagonia are of continental origin and are characterized by a *Dicroidium* flora and prosauropod dinosaurs (El Tranquilo Formation: Bonaparte, 1978).

Marine strata bearing *Dactylioceras* and other ammonites were deposited in Chubut during the Early Jurassic approximately within the same boundaries as the marine Carboniferous beds (Central Patagonian Basin). Marine conditions continued up to the Aalenian or Early Bajocian. Lower Jurassic, *Thaumatopteris*, *Otozamites*–anuran and supposed lacertilian-bearing sediments of Santa Cruz province are of continental origin (Herbst, 1965).

It is possible that the predominantly marine beds of Chubut and the continental strata of Santa Cruz mentioned correspond to one narrow Liassic basin which extended from NNW to SSE over extra-Andean Patagonia.

During the lower Middle Jurassic, the ocean withdrew from this basin, and continental strata with *Amygdalodon patagonicus* were deposited. But the most conspicuous parts of the Middle Jurassic sequence are the mesosilicic and acidic effusive rocks of the Complejo Porfirico and Porfiritico (La Matilde Formation), which extend from south-eastern Chubut to Santa Cruz and Tierra del Fuego. They may have covered most of extra-Andean Patagonia and seem to have been related to important epeirogenic movements (Stipanicic and Reig, 1956).

Slight tectonic movements near the close of the Middle Jurassic took place in Chubut. They are responsible for the unconformity between the effusiva mentioned and the mainly lacustrine, *Cyzicus* (euestheriids and lioestheriids)-bearing Cañadón Asfalto Formation with saurischian dinosaurs. This formation is of probably Callovian to Oxfordian age, having filled many small basins within a pronounced relief in central northern Patagonia.

Due to strong tectonic movements during the Late Jurassic, fundamental paleogeographic changes took place in Patagonia. The basement was fractured and three large basins, the Colorado, San Jorge, and Magallanes Basin evolved. They were separated by highland areas: the North Patagonian Nesocraton between the Colorado Basin and the San Jorge Basin, and the Deseado Nesocraton and Rio Chico high between the San Jorge Basin and Magallanes Basin (Fig. 17.2).

More than 2500 m of sediments were deposited during the Late Jurassic and earlier Cretaceous (Pozo D 129 Formation) in the central part of the continental San Jorge Basin. In Early to Middle Cretaceous times relief increased and high-energy sediments (conglomerates, sandstones), pyroclastic rocks and tuffites filled the basin and made up the conspicuous, and in some places petroliferous, Chubut Group.

In Senonian times slight orogenic ("inter-senonian") movements took place. They are responsible for the unconformity between the Chubut Group and the overlying Trebol Formation.

Marine *Eubaculites*-bearing sediments were deposited on an unstable platform during the Maastrichtian, in northwestern Chubut and south-western Rio Negro. In Danian (Early Paleocene) times an Atlantic marine ingression covered north-eastern Santa Cruz, eastern Chubut and south-eastern Rio Negro.

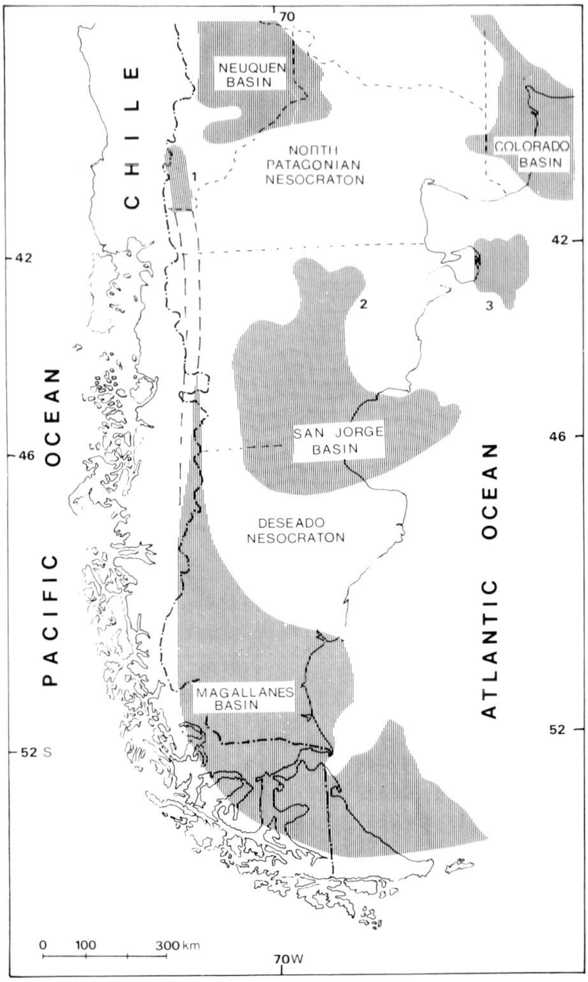

Fig. 17.2. Geological provinces of extra-Andean Patagonia. Outlines of sedimentary basins from different sources (Zambrano and Urien, 1970; Russo and Flores, 1972; and others).

In Eocene times strong volcanic activity (*Serie Andesitica, Tobas de Sarmiento*, etc.) took place. Oligocene marine (Patagonia Formation, etc.) and continental sediments were deposited in several areas. The Miocene tuffs, tuffites, and continental sediments of the Collon Cura Formation (Santacrucense–Friasense) filled north–south oriented intermontane troughs in northwestern extra-Andean (and Andean) Patagonia.

Strong effusive activity (*Basalto I*) took place during Late Miocene times in many areas of Patagonia. This activity was recurrent during Pliocene, Pleistocene and Holocene times (Basalts II to VII). The Meseta basalts resulted and the Plio-Pleistocene Patagonian Gravel, observed 140 years ago by Charles Darwin (1851), are striking features of the recent landscape of extra-Andean Patagonia.

Several mammal stages have been recognized in the Tertiary of Patagonia. These fossils permitted the establishment of provincial ages for South America.

Geological provinces in extra-Andean Patagonia

The following geotectonic units will be briefly outlined:

Pre-Cretaceous basement

1. North Patagonian "Nesocraton"
2. Deseado "Nesocraton"

Middle to Upper Mesozoic and Cenozoic sedimentary basins

3. Neuquén Basin
4. Colorado Basin
5. Ñirihuau Basin
6. Valdés Basin
7. Golfo San Jorge Basin
8. Magallanes Basin (=Austral Basin)

The only partially extra-Andean (and mainly Andean) Neuquén Basin is an exception, which occurs at the southern end of the larger Chilean Basin (Harrington, 1962). Its sedimentary history began during the Early Jurassic. Most basins in extra-Andean Patagonia originated during the Late Jurassic or Early Cretaceous (Zambrano and Urien, 1970). All units mentioned, except 3 and 5, continue onto the continental shelf.

(1) North Patagonian Nesocraton. This area contains exposures of highly metamorphic rocks considered classically as Precambrian (Harrington, 1962; Stipanicic, 1967). Slightly metamorphic rocks of the Sierra Grande and Gastre area are of doubtful Precambrian or Early Paleozoic age (Stipanicic, 1967). Sedimentites of Silurian to Early Devonian (?) age (iron formation) outcrop in Sierra Grande (Klammer, 1964). A part of the acid intrusives within the North Patagonian Massif is now considered to have a Late Permian age; some of them are affecting, in the southwestern margin of the massif, the Upper Paleozoic (Gondwanic) sedimentites of the area of Tecka, Nueva Lubecka and Languineo (Stipanicic, 1967). Continental sediments bearing *Dicroidium* were deposited (Los Menucos Formation) during the Late Triassic and overlain by Late Triassic and/or Early Triassic acid volcanics, which are followed in some places by marine sediments of Early Jurassic age. The Middle Jurassic is represented by volcanics, and the Upper Jurassic in central and northern Chubut by the essentially lacustrine Cañadon Asfalto Formation.

(2) Deseado Nesocraton. This area, like the North Patagonian Massif, is a sub-positive, sub-stable area, the oldest rocks being the slightly metamorphic La Modesta Schists of the central Santa Cruz province, which may be of Early Paleozoic age (Lesta and Ferello, 1972). The metamorphic basement lies unconformably on the continental Lower Permian La Golondrina Group with *Glossopteris*, *Ganygamopteris*, etc. This is overlain by Triassic, Jurassic and Cretaceous continental beds with volcanic intercalations.

(3) Neuquén Basin. This region contains marine and continental Jurassic and Cretaceous sediments. It is located north of the North Patagonian Nesocraton.

(4) Colorado Basin. This basin is situated immediately at the NNE end of the North Patagonian Nesocraton. It strikes approximately WNW–ESE. Most of its areal extension and its largest sedimentary thickness have been inferred to exist on the continental shelf.

On a basement of slightly metamorphic rocks of possibly Early Paleozoic age, acid intrusives (249 ± 10 million years: Stipanicic, 1967) are known from a few bore holes. They lie unconformably clastic and are presumably continental sediments of a probably Early Cretaceous (Fortin Formation) and Late Cretaceous (Colorado Formation) age. The Lower Paleocene Pedro Luro Formation is marine (dated by planktonic foraminifers) (Kaasschieter, 1965).

The Upper Tertiary and Quaternary are represented by up to more than 1000 m of marine sediments (in the area of the Colorado Delta). Much of the Cenozoic sedimentary filling of the basin was contributed by the Rio Negro and Rio Colorado. The base of the Neogene was found at the Colorado-1 well at 1500 m below sea level (Kaasschieter, 1965). At the present, fluviomarine (Colorado Delta) and marine sedimentation are continuing in the Colorado Basin.

(5) Ñirihuau Basin. This depression extends from the eastern Andean border to the area of Ñorquincó–Cushamen. It was filled during the Tertiary by more than 2500 m of marine and continental deposits.

Above a Precambrian or Lower Paleozoic metamorphic basement (Cushamen Formation: Volkheimer, 1964) which has been intruded by tonalites and granites, and in a few areas on marine or continental Lower Jurassic beds, lie unconformably up to 2000 m of the Eocene Vulcanitas Nahuel Huapi (Cazau, 1972). Its lava flows and tuffs belong to a continental environment. Above an erosion surface composed of the volcanics mentioned, rest, at some places, marine lutites and tobaceous sediments with *Venericardia planicosta* of Late Eocene age (Camacho, 1967). At Cholila and Ñirihuau, the marine (Lower Miocene) sandstones and scarce clays of the Rincón de Cholila Formation correspond mainly to a littoral to sublittoral high-energy environment (Cazau, 1972). These deposits are covered by up to 1900 m of continental sandstones, clays and tuffs, bearing a fossil flora of *Nothofagus* and fossils of mammals (Megatheridae and cf. *Eutypotherium*). This Ñorquincó Formation is overlain by the Upper Miocene continental Collon Cura Formation. The latter is characterized by widespread outcrops of tuffs, tuffites, frequently cross-bedded sandstones, clays and silts. Intercalations of paleosols with scarabeid nests and fossils of Edentata can be found in this continental formation.

(6) Valdés Basin. This is a small, Late Cretaceous basin whose presence is inferred from seismic studies (Zambrano and Urien, 1970). It is situated south of the Colorado Basin. Its sediments are covered by Cenozoic deposits.

(7) Golfo San Jorge Basin. The most complete study based on subsurface and surface data of the Golfo San Jorge Basin, is that of Lesta (1968). The pre-Upper Jurassic basement of this sedimentary basin is built up by: (1) Early Paleozoic to Precambrian metamorphic rocks, intruded by (2) granites and granodiorites of mostly Permian to Triassic age (Ugarte, 1966); (3) the Middle Jurassic, mainly volcanic Chon Aike Formation (rhyolites, volcanic breccias, sandstones and conglomerates).

The sedimentary history of the basin begins in the Early Jurassic with the essentially lacustrine Pozo D-129 Formation, characterized by oolitic limestones, tuffs, tuffites, clays, siltstones and some sandy layers. The Cretaceous Chubut Group lies unconformably on the Pozo D-129 Formation.

The Upper Cretaceous part of these continental deposits (sandstones, siltstones, claystones, tuffites, tuffs) contains most of the petroleum found in the basin. The emerged western part of the Golfo San Jorge Basin is represented by Upper Cretaceous angiosperm-bearing beds in Sierra de San Bernardo. The Paleocene marine Salamanca Formation is overlain by the Lower Tertiary Rio Chico and Sarmiento Formations with abundant mammal faunas.

(8) Magallanes Basin. The area is also known as the Austral Basin. Metamorphic rocks of a possible Precambrian or Early Paleozoic age, the Chon Aike Formation (Middle Jurassic) and the "Tobifera Formation" (Middle to lower Upper Jurassic) compose the basement of the Magallanes Basin. The formations mentioned have been recognized in some wells.

The stratigraphy of the Cretaceous and Tertiary basin was synthesized by Russo and Flores (1972). The Lower Cretaceous, partly marine, sandstones of the Springhill Formations are overlain by the gray marine shales and occasional sandstone intercalations of the Lago San Martin Formation, which may range from the Aptian to the Cenomanian (Leanza, 1973). The mainly marine standstones of the Piedra Clavada Formation (with *Actaeonella patagonica*) and the very fossiliferous siltites and claystones of the Mata Amarilla Formation (with the ammonite *Peroniceras santacrucense*) are of Coniacian age. The lower Tertiary Man Aike Formation (glauconitic sandstones) is overlain by the Rio Leona Formation (plant-bearing claystones with sandstones and some conglomerates), and this by the marine Patagonia Formation, characterized by more or less calcareous sandstones, alternating with fossiliferous claystones and siltstones of Miocene age (Camacho, 1967). The limit with the tuffs, tuffites, claystones and sandstones of the Upper Miocene Santa Cruz Formation is transitional. This latter formation bears the richest Tertiary mammal fossil faunas (Edentata, Rodentia, Notoungulata, Marsupialia, etc.) of South America.

ASPECTS OF VEGETATIONAL HISTORY

(W. Volkheimer)

Composition and distribution of Patagonian fossil floras have recently been summarized by Menéndez (1972). The oldest plant remains from Patagonia are of Pennsylvanian age. A few specimens of Lycopodiales (*Lepidodendron*) and Articulatae (*Calamites, Annularia*) are the only finds from this period.

Permian

The Permian register comprises a larger number of Lycopodiales (*Lepidodendron*, *Lepidophloios*, *Lycopodiopsis*, *Sigillaria*), Articulatae (*Annularia*, *Phyllotheca*, *Sphenophyllum*), Filices (*Asterotheca*, *Dizeugotheca*, *Pecopteris*), possible Pteridospermae (*Eremopteris*, *Merianopteris*), Glossopteridales (*Glossopteris*, *Gangamopteris*, *Palaeovittaria*), Cordaitales (*Cordaites*, *Cordaicarpus*, *Cordaianthus*) and a few Coniferae (*Paranocladus*, *Genoites*). One of the best preserved floras is that of Bajo de La Leona, Santa Cruz province, studied by Archangelsky (1958) and Archangelsky and De la Sota (1960). It is characterized by the presence of both Gondwanic and Holarctic elements. Another flora from the Permian of eastern Chubut province (Rio Genoa, Nueva Lubecka) is of prevailing Gondwanic affinity.

Triassic

Floras from the Triassic of Patagonia are known from Neuquén, Rio Negro and Santa Cruz provinces. Amongst the ferns, *Cladophlebis*, *Dictyophyllum* and trunks of *Osmundites* are the most frequent elements; the guide form *Dicroidium*, which characterizes the Gondwanic Triassic, belongs to the Pteridospermae.

The Araucariaceae are represented by *Araucarites* and the Ginkgoaceae by *Ginkgoites*, *Sphenobaiera* and other forms.

Famous localities for Triassic floras are Paso Flores (Neuquén) and El Tranquilo (Santa Cruz). The Triassic floras from these localities are similar to those from the Triassic of western and northwestern Argentina (La Rioja and San Juan).

Jurassic

Floras of Jurassic age are known from Neuquén, Chubut and Santa Cruz. The Araucariaceae have left the most conspicuous fossil register (petrified forests of Santa Cruz). Ferns are still well developed, as testified by the micro- and megafloristic register. Macrofossils of Gleicheniaceae (*Gleichenites*) appear first in the Middle Jurassic, but *Gleicheniidites* spores already in the Lias.

The Dipteridaceae (*Dictyophyllum*, *Thaumatopteris*) are frequent. Caytoniales (*Sagenopteris*, *Vitreisporites*) are most frequent in the Lower and Middle Jurassic and the Podocarpaceae are represented, for instance, by *Pagiophyllum*, *Elatocladus*, *Podocarpidites* (bisaccate pollen) and *Microcachryidites* (tri- and polysaccate pollen).

Athrotaxis (Taxodiaceae) is frequent in the Lower and Middle Jurassic. The Cheirolepidiaceae (conifers), as shown by the palynological evidence (*Classopollis* spp.), were one of the most important components of the coastal vegetation during the Jurassic and Early Cretaceous times. In some areas of northern Patagonia the pollen genus *Classopollis* covers up to 90% of the spectra in the Lias, and up to more than 50% in the Dogger and Malm Formations.

Lower Cretaceous

The macropaleobotanical evidence of the Lower Cretaceous is in many aspects similar to that of the Jurassic. Thus it is difficult, in some cases, to distinguish Upper Jurassic and Lower Cretaceous megafloras. Fortunately, the microfloristic associations show clear differences between both periods.

Among the Lower Cretaceous ferns, the Schizaeaceae (*Ruffordia* and the spore genus *Cicatricosisporites*), Osmundaceae (some species of *Cladophlebis* ?) and Gleicheniaceae (*Gleichenites*) may be mentioned. The Cycadophytes are represented, for instance, by *Dictyozamites*, *Otozamites*, *Williamsonia*, *Zamites*, *Ptilophyllum* and *Nilssonia*; the Ginkgoales by *Baiera* and *Ginkgoites*.

Amongst the conifers the Cheirolepidaceae (e.g. *Tomaxellia* with *Classopollis* pollen grains) prevail.

The Podocarpaceae (*Podocarpus*, *Apterocladus* and *Podocarpidites* pollen) are present in most Lower Cretaceous floras. Pre-Albian Lower Cretaceous microfloras of southern Neuquén are characterized by *Cyclusphaera psilata*, a conspicuous coniferalean pollen type (Volkheimer and Sepulveda, 1976). The best known Lower Cretaceous plant fossil locality is Tico, Santa Cruz province, with well-preserved specimens belonging to more than fifty species (Archangelsky, 1967 and earlier papers; Menéndez, 1966). This locality provided a *Clavatipollenites* microflora (Shell Cia. Argentina de Petroleo S.A., 1965; Archangelsky and Gamerro, 1966).

The oldest angiosperm palynoflora in Argentina with tricolpate to stephanocolpate pollen grains was found at the type locality of the Aptian–Albian Huitrin Formation, Neuquén Basin (Volkheimer and Salas, 1975).

Upper Cretaceous

Upper Cretaceous floras have been described from Sierra de San Bernardo, Chubut (Menéndez, 1961) and from the Senonian of Mata Amarilla (Berry, 1928, 1937; Frenguelli, 1953).

In addition to a few ferns (*Gleichenites*), Articulatae (*Equisetites*), and Cycadophytae (*Nilssonia*), the main bulk of the plant remains are angiospermous leaves which correspond probably to the following families: Polygonaceae, Annonaceae, Monimiaceae, Lauraceae, Menispermaceae, Nymphaeaceae, Saxifragaceae, Anacardiaceae, Vitaceae, Sterculiaceae, Myrtaceae, Bignoniaceae, Potamogetonaceae, Poaceae, Cyperaceae (Menéndez, 1972).

Tertiary

The best known Tertiary plant fossil localities of Patagonia are Laguna de Hunco, Chubut province (Berry, 1925), with plant remains of a probable Late Paleocene age (Archangelsky, 1974); Rio Pichileufu, Rio Negro province (Berry, 1938) of Late Paleocene to Eocene age, and Rio Turbio, Santa Cruz province, with paleofloras of probable Eocene age (Hünicken, 1966).

A few ferns (*Dicksonia*, *Alsophila*, *Coniopteris*, *Blechnum*, *Dryopteris*, *Asplenium*), Cycadales (*Zamia*), Ginkgoales (*Ginkgo*), and Coniferopsidae (*Fitzroya*, *Podocarpus*, *Araucaria*, etc.) have been described from these floras, but leaves of dicotyledoneous angiosperms are the most frequent components and total nearly 250 species belonging to 55 families. (For a complete list of fossils see Menéndez, 1972.) In addition there occur some Monocotyledoneae (Poaceae and Cyperaceae).

PALEOCLIMATES (W. Volkheimer)

Mesozoic

Mesozoic climates of Patagonia have been outlined elsewhere (Volkheimer, 1967, 1969a, b, 1970). There is evidence for strong climatic changes during the Cenozoic.

Tertiary

The frequency of palm fruits and palm trunks in the Paleocene Roca Formation of northern Patagonia and their presence in the Salamanca Formation (Paleocene) of southern Chubut indicates that a warm climate prevailed. Forms characterizing cool-temperate humid climates (flora with *Nothofagus*) are encountered in the lower part of the Eocene in the Santa Cruz (Rio Turbio Formation). In Middle Eocene strata, forms associated with warm climates, like *Annona*, *Bignonia*, and *Nectandra*, prevail (Hünicken, 1955). At the end of the Eocene the latter migrated to the north, leaving the habitat to the *Nothofagus* flora. The association of "temperate" and "warm" floral elements has been explained according to their different topographic positions (Menéndez, 1969). Romero (1978) suggests the existence of a stable, composite vegetation unit which he calls "Paleoflora Mixta".

Large amounts of coal accumulated in the Rio Turbio Basin[1] during the Early Tertiary. In addition to the more than 1000 million tons of Tertiary coals of southern Chile (Arauco and Concepción Provinces), the total of Lower Tertiary coal accumulated in southern Patagonia reaches 1500 million tons — a clear evidence of extremely moist conditions.

Warm climates are also indicated by the existence

[1] Estimated reserves: more than 375 million t.

of large boid snakes (*Madtsoia bai*) in the Eocene *Notostylops* Beds of the Chubut province (Simpson, 1933); Boidae were distributed through all the tropical and subtropical areas of the world at that time (Werner, 1921).

Crocodiles are also useful climatic indicators. Crocodiles are presently distributed up to the 10° (to 15°) isothermal line for the coldest month of the year (Berg, 1964). Experiments with living alligators have shown that at 4°C there is a danger of perishing with cold (Colbert et al., 1946). The finding of *Necrosuchus ionensis* from the Paleocene Salamanca Formation of Chubut province (Simpson, 1937) and *Eocaiman cavernensis* from the Middle Eocene Casamayor Formation (*Notostylops* Beds) of Chubut, could indicate that minimum temperatures of not less than 10° to 15°C during the coldest month prevailed during that period.

In the Miocene, arborescent ferns (*Alsophilocaulis calveloi*, a Cyatheaceae stipit) grew in the Neuquén province. The current distribution of Cyatheaceae indicates that they inhabit humid zones of more than 1000 mm of annual rainfall; it may be inferred that during Miocene times the climate of Neuquén was moist, with much more rainfall than now occurs (Menéndez, 1961). That speculation agrees with the results of Rutland et al. (1965) who showed that the major part of the Andean uplift occurred since the Late Miocene and continued during the Pliocene and Pleistocene. Thus, the moist Pacific western winds were able to reach the extra-Andean region of Argentina until the Late Miocene.

The presence of Megalonychidae in the Pliocene Rio Negro Formation of northern Patagonia indicates the presence of trees and somewhat warmer and moister conditions than those prevailing at the present time within this region (Scillato Yane et al., 1975).

As a result of the study of climatic indicators from several Tertiary basins of Patagonia, it can be deduced (Volkheimer, 1971) that:

"(1) For the Paleocene and Eocene, the 10°C isotherm of the coldest month can be traced at more than 1500 km from its present position in southern Patagonia. Main indicators: crocodiles and palms.

(2) During the Eocene, moist-temperate climate (**Cfb**-Climate of Köppen) prevailed in southernmost Patagonia. The mean temperature of the warmest month was below 22°C. Main indicators: fossil floras and coals.

(3) Climate was subtropical and moist during most of the year (**Cfa**, Virginian Climate) in northern Patagonia (Rio Negro and Chubut) during the Eocene (and Upper Paleocene). The mean temperature of the warmest month was higher than 22°C, and that of the coolest month between 18° and 10°C. (The lower limit of 10°C is given by the distribution of crocodiles. The value was probably around 12–13°C). Precipitation was distributed throughout the year and was (probably much) higher than 1000 mm per year. Main indicators: fossil floras of Laguna de Hunco (Chubut), Rio Pichi Leufu (Rio Negro) and Lotena–Coronel (Chile) and their statistical analysis.

(4) The climate was semi-arid during the Eocene in central West Argentina (lithologic indicators).

(5) The Miocene climate was cold on the Antarctic Peninsula and moist-temperate in Patagonia. At the latitude of the Nahuel Huapi Lake (Rio Negro Province) climate was transitional between **Cfa** (subtropical and moist during most of the year) and **Cfb** (moist-temperate)."

Quaternary

Interesting contributions to the Pleistocene climatic history of Patagonia can be found in the papers of Caldenius (1932), Auer (1956, 1958, 1965), Flint and Fidalgo (1963), Groot and Groot (1964, 1966), Polanski (1965), and Mercer and Sander (1975). Caldenius (1932) concluded that extra-Andean Patagonia was not covered by inland ice during the Pleistocene. He established that the Würm (=Wisconsin) glaciation was the most extensive one and described final moraine arcs in many areas of the eastern slope and eastern base of the Patagonian Cordillera. He compared the four Wisconsin stages of Patagonia with Scandinavian stages, calling them Inicioglacial, Daniglacial, Gotiglacial and Finiglacial. Caldenius, and more recently Flint and Fidalgo and Polanski, emphasized that the southeastern winds (*sudestada*) which prevailed in the Patagonia region during summer, did not provide rainfall to the arid Patagonian region. "Rainfall comes only from fronts and instability lines that result in rare summer storms. There is no evidence whatsoever that winds from the Atlantic provided moisture sufficient for glaciation" (Polanski, 1965, p. 458). Misinterpretation of arid environment deposits as till have greatly retarded the understanding of morphological and sedimentological features in extra-Andean Patagonia.

The hypothesis of arid conditions in extra-Andean Patagonia during the Pleistocene was verified by Groot and Groot (1966) who studied the pollen spectra from deep-sea sediment cores taken in the southwest Atlantic and in the southeast Pacific Ocean. They obtained evidence "that central and southern Chile and the Andean region of Argentina have experienced alternating 'warm-dry' and 'cool-moist' periods throughout the Quaternary and probably during the latest Tertiary as well". The Pliocene pollen spectra indicate milder climate. Extra-Andean Argentina remained, during Quaternary time, a semi-arid steppe area which probably enlarged greatly during the glaciations, due to glacial eustatism.

The advantage of Groot and Groot's work lies in the fact that palynological study of deep-sea sediments provides a very complete record of pollen and spores from the neighboring continent, thus avoiding a limited approach involving local environments, as results in the case of palynological studies of continental beds, like peats (Auer, 1958). In the Pacific core (34°30′S and 74°19′W, at a 4078 m depth of water, at 175 km off the Chilean coast) the Postglacial, four glaciations and three interglacials are represented in its upper 550 cm. The Atlantic core was obtained from 43°58′S and 52°09′W, about 875 km offshore in the southwestern part of the Argentine Rise (depth of water: 5785 m).

Recently Mercer and Sander (1975) concluded by means of radiometric dating and magnetic polarity measurements of lava flows interbedded with glacial drift, that the major glaciations in southern Argentina began during the Pliocene, about 3.5 million years ago. The major glaciations were broadly contemporary with cycles of basaltic vulcanism. The greatest glaciation occurred after 1.2 million years ago (Mercer and Sander, 1975). Its till covers basalt 1.2 million years old in southernmost Santa Cruz Province and a till possibly belonging to this glaciation is covered by basalt 1.0 million years old, south of Lago Argentino. Mercer and Sander (1975) conclude that the Patagonian Gravel (Rodados Patagonicos) in the area of Lago Viedma–Lago Argentino–Rio Santa Cruz "is outwash gravel of the same age as the Pliocene–Early Pleistocene tills that occur farther to the west".

CLIMATE OF PATAGONIA (H. Walter and E.O. Box)

All of Patagonia falls within the realm of strong, prevailing west winds known as the Westerlies, which blow throughout the year in a generally eastward direction about the South Pole and present the most obvious feature of the Patagonian climate. These west winds are very moist over the Pacific but lose their moisture over the Andes, especially on the western slopes, in the form of heavy rain or snow. On the eastern side these air masses descend into the warmer foothills and become very dry, much like the well-known *Föhn* winds of Europe. As a result, precipitation decreases very rapidly east of the Andes, as one can see in a rainfall profile (Fig. 17.3) taken along about 41° south latitude. Rainfall on the higher western slopes of the Andes can often be as high as 6000 mm annually. At the eastern foot of the Andes near Puerto Blest (western shore of Lake Nahuel Huapi) rainfall can still reach 4000 mm per year. At Bariloche, however, 45 km further east, it has dropped to only 1080 mm, and 20 km further, near the airport, rainfall is only about 700 mm annually. Still 15 km further east, at the east end of Lake Nahuel Huapi annual rainfall has dropped to about 500 mm, and at Pilcaniyeu, 40 km further east, it is only 300 mm.

In the adjacent, more level landscape east to the Atlantic, rainfall decreases more slowly: 173 mm annually at Maquinchao (150 km east of Pilcaniyeu), with similar values (down to a low of about 160 mm) for the remaining 300 km of the

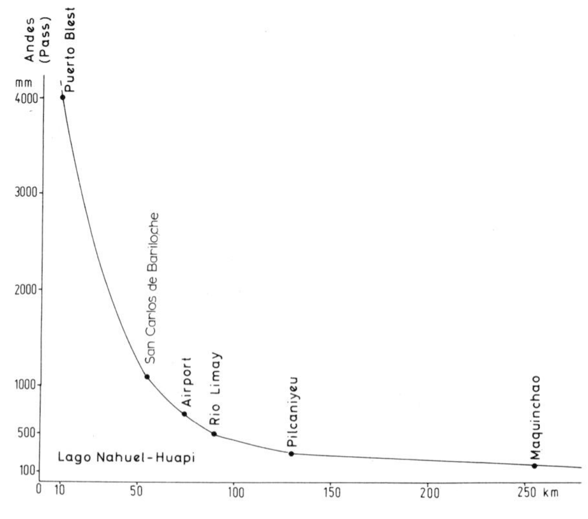

Fig. 17.3. Rapid decrease of annual amount of precipitation east of the Andes along the 41°S latitude. Ordinate = precipitation (mm); abscissa = distance (km) from the Andes to the East Coast.

Patagonian semi-desert. Only immediately on the east coast, as at San Antonio, does rainfall increase again to slightly over 200 mm.

The east–west precipitation profile further south in Patagonia must show a similar pattern, but the corresponding meteorological stations are lacking there.

The increasing dryness eastward from the crest of the Andes is reflected quite clearly in the corresponding eastward change in vegetation. At Puerto Blest one finds the Valdivian rain forest characteristic of the Andes west slope. This vegetation is rich in species and has various dominating conifers together with considerable *Nothofagus.* Just a few kilometers to the east the forest becomes poorer in species, and soon only *Nothofagus dombeyi* remains. At Bariloche this is replaced by a dry conifer forest composed of *Austrocedrus* (*Libocedrus*) *chilensis*, reminding one of a dry pine forest. The dryness limit of the forest region is reached near the Bariloche airport, with only scrub remaining. Especially striking here is the proteacean *Embothrium coccineum*, with its brilliant red flowers.

Treeless Patagonia begins immediately to the east, first with a narrow steppe zone composed of tussock-grass species (*Festuca pallens*, *Stipa speciosa*, etc.). On moist sites in this climatic zone one also finds introduced Central European species, such as *Taraxacum officinale*, *Trifolium repens*, and *Poa pratensis*.

With increased dryness, the vegetation approaches that of a semi-desert, dominated by the cushion-plant growth forms which are well adapted to a windy climate. First come the large, loose but spiny cushions of *Mulinum spinosum* (Apiaceae). Then, in the true semi-desert, the cushions become ever lower and denser, with the total vegetation cover falling to about 30 to 40% or less.

The diagrams (Fig. 17.4) of Bariloche (41°44′S, 71°10′W, near the eastern border of the forest zone) and Maquinchao (41°15′S and 48°44′W in the semi-desert) show the humid climate of Bariloche and the dry climate of Maquinchao quite clearly. The third climate diagram, from Rio Grande on Tierra del Fuego, shows a humid climate further to the south, outside the Patagonian semi-desert and near the boundary between grass steppe and boggy subantarctic forest. The temperature patterns of these three stations vary only slightly. The mean annual temperature is around 8 to 9°C at the first two stations, in northern Patagonia, and is only slightly lower, about 5°C, at the southern tip of South America. The annual variation is small at all three stations, but no month is absolutely frost-free. The cold season lasts four months in the semi-desert and five months elsewhere, but the mean minimum daily temperature of the coldest month is only 1 to 3° below freezing. The climate is definitely maritime.

The dryness of the Patagonian semi-desert is caused not so much by high temperatures during the summer as by the strong wind, which causes high evaporation rates. This has been studied at a great many stations in Argentina by means of measurements with the Class A Pan.[1] This made it possible to include the curves of average monthly evaporation on the climatic diagrams in Fig. 17.4. The evaporation curve runs almost parallel to the temperature curve, as shown by comparing it with an exaggerated temperature curve, which is also included. The temperature curve is shown on all climatic diagrams according to the relationship 10°C = 20 mm precipitation (i.e., double-scaled, t × 2) in order to distinguish humid and dry seasons. In order to make the temperature curve approach the evaporation curve, however, the exaggeration factor must be increased: to 10°C = 120 mm evaporation (t × 12) at Bariloche, to 10°C = 100 mm (t × 10) at Maquinchao, and to 10°C = 150 mm (t × 15) at Rio Grande. This extremely high evaporation is due to the effect of the wind. In the less windy, humid but hot climate of Misiones (northeastern Argentina), one must increase the temperature curve by a factor of only five in order to approximate evaporation, and even in the semi-arid to arid pampa by a factor of only seven to nine. The cool, humid climate of Tierra del Fuego is extremely windy, and an exaggeration factor of t × 15 is necessary.

Of course the evaporation curve and the amplified temperature curve do not match exactly (Fig. 17.4). The evaporation curve often rises earlier in the spring than does the temperature curve (as at Bariloche), which is the result of the rapidly increasing dryness of the air in the spring. The

[1] The evaporation values and many other data were very obligingly made available to us by the directors of the Servicio Meteorologico Nacional in Buenos Aires. Refinements of the evaporation values (such as for oasis effects) were not made.

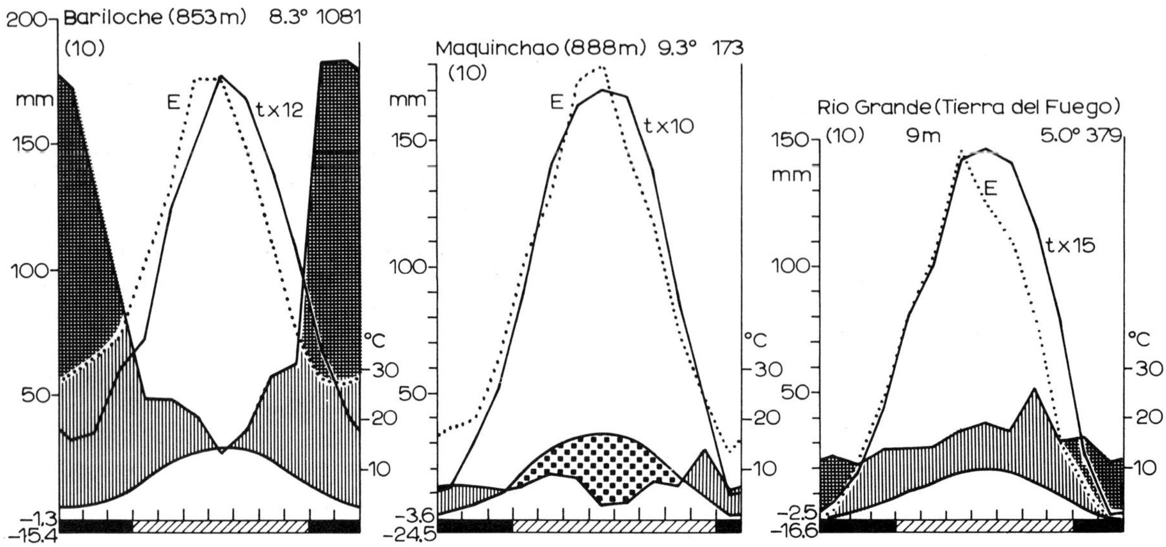

Fig. 17.4. Climatic diagrams of northern Patagonia (Bariloche in the forest region, Maquinchao in the Patagonian semi-desert) and of Tierra del Fuego (Rio Grande in the Antarctic forest–grassland ecotone). Positions of the stations are indicated in Fig. 17.6. E=evaporation curves and t × 12, t × 10, t × 15 = increased temperature curves (detailed explanation in the text). After Walter (1968).

evaporation curve usually falls earlier in the autumn, as a result of increasing air moisture (see Rio Grande).

A comparison of the climatic water balance of Patagonia with those of other parts of Argentina is especially interesting. The climatic water balance is understood here to refer to the difference between annual precipitation and annual (potential) evaporation. This water balance is shown in Fig. 17.5 for all those Argentine stations at which evaporation is measured. One sees that the water balance is positive — that is, annual precipitation exceeds annual potential evaporation, on the average — only in the warm-humid region of northeastern Argentina. The difference is zero along the La Plata River and in Tierra del Fuego, that is, the annual budget is balanced. The balance is negative in the pampa region, increasing to −700 mm to the west. The most strongly negative balance is found in windy Patagonia, with values of over −1000 mm, and in the hot desert of northwestern Argentina (over −1800 mm).

The average wind speed at Bariloche, at the foot of the mountains, is about 6.5 to 7 m s^{-1} in the summer months. Somewhat to the south in Esquel it is almost 8 m s^{-1}. Wind speed decreases in the plain to the east. A value of 4 to 4.5 m s^{-1} is given as an average for Maquinchao, but it is often much higher. When one drives along with a tailwind of 60 km h^{-1} over unpaved roads one is constantly blinded temporarily by his own dust clouds.

In this dry region, as is always the case, the variation in annual precipitation from year to year is quite large. Extremes at Pilcaniyeu are given as 157 mm and 519 mm.

The climate of the northern part of the province Rio Negro diverges from that of Patagonia. It is still arid, but the constant west wind is lacking and the temperature is much higher. Here occur woody species such as *Prosopis denudans* and frost-sensitive *Larrea* species, including *Larrea divaricata*. This is referred to in Argentina as the "monte" formation, which extends far to the north along the eastern base of the Andes. This formation is actually a part of the hot or subtropical desert to semi-desert region referred to in Volume 12 of this series.

Fig. 17.6 shows the climatic divisions on a climate-diagram map of Patagonia, along with the marginal areas not belonging to the Patagonian semi-desert. In the north, in the Rio Negro Valley, are stations with mean annual temperatures around 15°C: Cipoletti, Choele Choel, and Patagones. In the west are the humid stations at the eastern base of the Andes: Bariloche, Esquel, and Colonia 16 de Octobre. In the south are Rio Grande on Tierra del

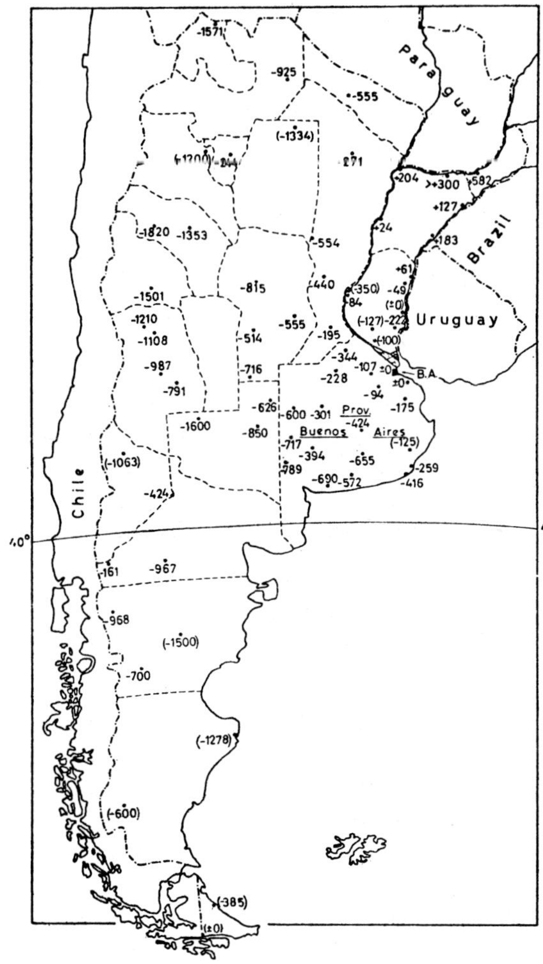

Fig. 17.5. Hydrological water balance (annual precipitation minus annual potential evaporation, in mm) of some meteorological stations in Argentina (after Walter, 1968).

Fuego, Ushuaia in the moor and forest region, and Dungeness in the steppe zone. From the Patagonian semi-desert only the stations Maquinchao, Sarmiento and Santa Cruz are known, plus Puerto Madryn and Trelew on the Atlantic coast.

SOILS (A.A. Marcolin and J.A. Vallerini)

Introduction

From a geographic point of view, Patagonia is considered to be the whole region extending south of Rio Colorado in Argentina. It comprises the provinces of Neuquén, Rio Negro, Chubut, Santa Cruz and the Argentine part of the island of Tierra del Fuego. This is a larger territory than the deserts and semi-deserts considered here; these are enclosed within the Patagonian Tablelands, which constitute a discrete physiographic unit.

The arid and semi-arid part of Patagonia begins at the Precordillera (Pre-Andean Range) foothills in the west, and its altitude decreases gradually towards the Atlantic coast in the east. Besides transverse valleys running from west to east, the tableland is cut by longitudinal steps. These latter features interrupt the continuity of the gradual slope towards the Atlantic. Within the extended *mesa* which constitutes the dominant physiographic feature a variety of land forms, such as isolated extended plateaus and terraces with very gentle slopes and smooth surfaces grading up to the west, and other forms shaped by water and wind erosion, can be found within the region.

Only dominant soils and their associates in this area are considered. Soil units to be mapped were chosen to remain as separate entities (Fig. 17.7).

History

Knowledge of Patagonian soils is far from being complete. Some of the explorers sent by different governments arrived in Patagonia during the second half of the nineteenth century. Darwin (1840) visited different points in Patagonia during 1833–34 in his trip around the world, leaving a detailed description of the natural resources found in some areas, including soils.

Musters (1873), in 1869, travelled with the aboriginal people through the Andes foothills from the Santa Cruz Province to Patagones, in the south of Buenos Aires Province. He made a general description of the region including references to soils. Patagonia started to be considered a potentially rich region worthy of development at the start of the twentieth century. Consequently, several surveys were made, some of which included descriptions of soils and vegetation (Willis, 1912, 1914).

Vessel (1946) reviewed these early studies. Cappannini and Dominguez (1959) also wrote a broad description of the Argentinian soils. The Instituto de Suelos y Agrotecnia produced a map in 1960 which compiled a great part of the knowledge published or otherwise available about soils of Argentina at that time. Etchevehere (1971) pub-

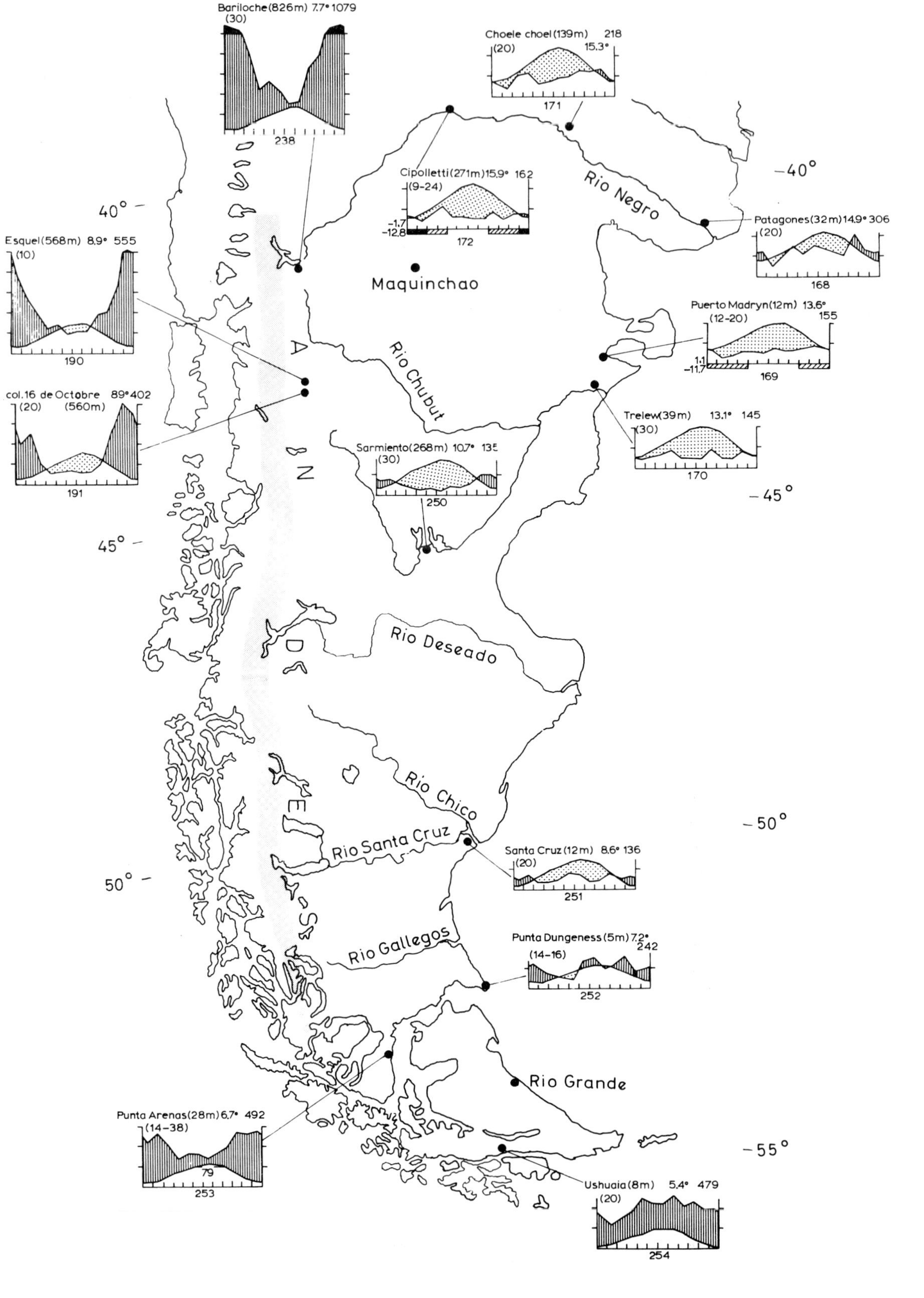

Bariloche(826m) 7.7° 1079
(30)
238
Choele choel(139m) 218
(20) 15.3°
171
Cipolletti(271m)15.9° 162
(9-24)
-1.7
-12.8
172
Rio Negro
Patagones(32m)14.9° 306
(20)
168
Maquinchao
Esquel(568m) 8.9° 555
(10)
190
Puerto Madryn(12m) 13.6°
(12-20) 155
1.1
-11.7
169
Rio Chubut
col.16 de Octobre 8.9° 402
(20) (560m)
191
Trelew(39m) 13.1° 145
(30)
170
Sarmiento(268m) 10.7° 135
(30)
250
40°
45°
50°
-40°
-45°
-50°
-55°
Rio Deseado
Rio Chico
Rio Santa Cruz
Santa Cruz (12m) 8.6° 136
(20)
251
Rio Gallegos
Punta Dungeness(5m) 7.2° 242
(14-16)
252
Rio Grande
Punta Arenas(28m) 6.7° 492
(14-38)
79
253
Ushuaia(8m) 5.4° 479
(20)
254

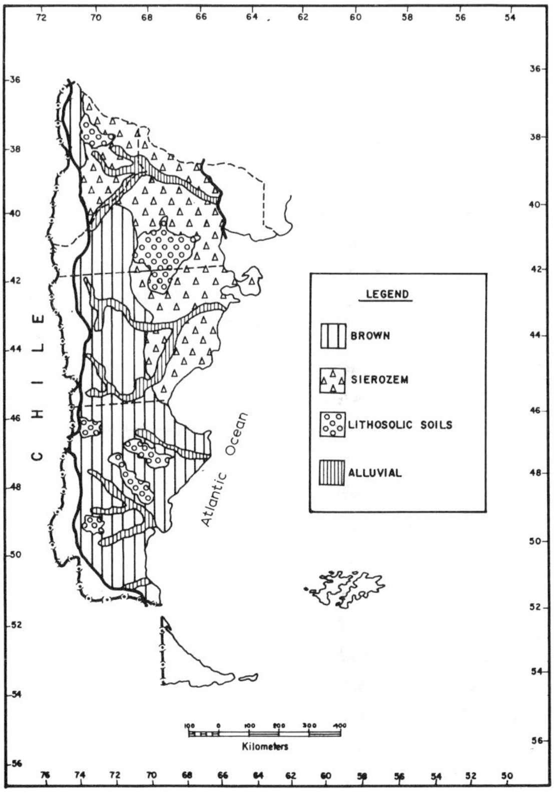

Fig. 17.7. Map of soils of Patagonia.

lished a map of the Argentinian soils, using the U.S.D.A. 7th Approximation System of soil classification.

Many studies of restricted areas were made to serve different purposes (Fantini, 1954; Bandura, 1960; Italconsult, 1960; Etchevehere and Miaczynski, 1963; Monteith and Laya, 1970; Story et al., 1970; Marcolin, 1973; Vallerini and Marcolin, 1975; etc.). Unfortunately, not all these studies used the same methodology and classification system. In 1973, the Instituto Nacional de Tecnologia Agropecuaria, Bariloche, produced a map of the Patagonian soils, based on earlier works and studies of soil carried out by the experts of the Experimental Station of Bariloche.

The work presented here is a revision of the knowledge of the soils of the arid and semi-arid parts of Patagonia, based on a bibliographical survey and on many observations and descriptions made by the authors. The nomenclature used here corresponds to that given by Thorp and Smith (1949), to the level of the Great Soil Group. Terms have been checked in the Multilingual Vocabulary of Soil Science (Jacks et al., 1960). The equivalent names given by the 7th Approximation to soil classification (U.S.D.A., 1960 and 1967) are included (Table 17.1).

TABLE 17.1

Equivalence between the natural (Thorp and Smith, 1949) and the 7th Approximation (U.S.D.A., 1960 and 1967) classifications of soils

Thorp and Smith (1949)	U.S.D.A. (1960, 1967)
1. Brown soils[1]	Haplustols
Regosols	Psamments
Solonchak	Calciustols
Low Humic Gley	Haplaquents
2. Sierozem	Calciorthids
Solonchak	Salorthids
Regosols	Lithosols
3. Alluvial	Ustifluvents
Solonchak	Salorthids
Low Humic Gley	Ochraquepts

[1] Names preceded by numbers correspond to dominant or principal soils, whereas those without numbers indicate the associated soils.

Soil descriptions

Brown and associated soils. These soils cover most areas with normal relief, and are mainly derived from glacial detritus and volcanic material.

Textures are dominantly coarse with different sizes of gravels and stones (Patagonian pebbles) commonly found throughout the solum.

Associated regosols are located westward of the area covered by brown soils and are related to a stronger glacial influence. Solonchak, and Low Humic Gley Soils are present as intrazonal associated soils. As lagoons formed in small depressions produced by glacial activity in the southern part of the Santa Cruz Province dry up during summer, strong winds blow out fine materials rich in salts. This causes the death of vegetation and the

Fig. 17.6. Climate-diagram map of Patagonia with thirteen diagrams. The diagrams Sarmiento, Santa Cruz, Puerto Madryn, Trelew and Maquinchao (see Fig. 17.4) of the Patagonian semi-desert, the others of adjacent regions.

formation of dunes leeward of the lagoons. This natural feature, together with the overgrazing that has occurred there, has caused extensive areas to deteriorate.

Sierozem and associated soils. These soils predominate in the central and eastern part of Chubut and Rio Negro Provinces and the northeastern part of Neuquén Province. They occupy a vast area of desert and semi-desert environmental conditions, with annual rainfall under 200 mm. Stones and gravels are commonly found throughout the solum. The profiles commonly have shallow depths and sandy texture. Associated soils are Solonchak occurring in shallow depressions, and narrow strips of Regosols appearing mainly adjacent to the Atlantic sea coast. Some Lithosols are associated with outcrops of crystalline rocks.

Alluvial soils of the Patagonian Tablelands. Although small in total area, the valleys of large rivers crossing the Patagonian Tablelands from west to east are of paramount importance for the regional economy. A wide range of soil particle sizes are found in these soils, and are related to the position of a particular site within the valley. Where imperfect drainage conditions appear, saline soils (Solonchak) and Low Humic Gley Soils occur as associates. Some of the soils of the Rio Negro and Chubut Province river valleys are irrigated, and fruit, vegetable, and forage crops are produced under intensive agriculture. Other river valleys (Senguerr, Chico, Chalia, Deseado, Santa Cruz and Coyle) are used mainly for grazing, although some limited forage crops are produced.

Lithosolic soils of the Patagonian Tablelands. These soils are mapped as discrete units in areas where lava beds (basalt) and crystalline outcrops occupy extensive areas. Some aeolian accumulations occur overlying the unweathered rock in places where vegetation dominated by small shrubs and grasses appears. Areas covered by these soils provide a poor grazing resource for goats (in the north of the Patagonian Tablelands) and sheep.

Erosion forms and processes (C.P. Movia)

The whole of extra-Andean Patagonia shows manifold features of wind and water erosion. Wind is more important, by far, as an erosional agent.

The high intensity of winds, their prevailing west–east direction and their seasonality, which coincides with the high evaporation period, as well as the precipitation characteristics and regime, are only some of the factors which determine the lack of soil stability in the region. We also have to consider the occurrence of canyons and large lakes in the Andean foothills, oriented in the wind direction, and enhancing the length of wind fetch.

Several sources of abrasive materials can be found in the region and can be considered as potential starting points of wind erosion processes. These include: the outcrops of soft materials (continental and marine sediments, especially from the Tertiary), the broad and shallow depressions which periodically dry up, temporarily filled ponds, wide and humid valley bottoms, salinized areas and other sites subject to deterioration by overgrazing, the alluvial plains of temporary rivers, and *arroyos*.

The low and sparse vegetation, frequently overgrazed and sometimes accidentally burned, contributes to the unstable equilibrium which can easily be upset, resulting in often spectacular erosion phenomena.

Eroded areas in Patagonia and the seriousness of the process were first mentioned by Auer (1951). Subsequently, Cappannini and Auer (1957), Tallarico and Ferreiro (1957), and Castro (1965) described some particularly serious situations, like those in Lake Tar (Santa Cruz), Chenque Hill (Comodoro Rivadavia, Chubut), Ricardo Rojas (Chubut) and the Lake Belgrano moving dunes (Santa Cruz). Ipucha Aguerre (1965) reviewed the situation at the regional level and made an estimation of the area affected by wind erosion. More recently, Montieth (1972) performed an erosion survey in Patagonia and prepared a map (unpublished; scale 1:1 000 000) with descriptions of dominant erosion forms and evaluation of their importance.

According to this survey, the highest percentage of eroded areas is found in Santa Cruz Province (38%); followed by Chubut Province with 25.7% and the southern portion of the Rio Negro (16.7%). The total eroded area represents 123 230 km^2. The area of concern is considerably larger if incipiently eroded sites are included. Movia (1973) started survey and mapping of areas for which a recent aerial photographic coverage was available.

Wind erosion forms (Figs. 17.8 and 17.9)

Accumulation forms existing in Patagonia are represented by macro- and micro-accumulations, barkans and fixed sand dunes. Gravel pavements and blow-outs of denuded and compacted soil are principal forms of wind erosion. The most active and impressive of all are, however, *lenguas* ("sand tongues") (Monteith, 1972; Movia, 1972). Sand is transported and deposited in narrow strips, parallel to the wind direction. The strip widths do not change markedly, either in time or because of obstacles on the ground. Progression down wind, however, is quite rapid, reaching 2 km yr^{-1}. The "tongue" consists of an active frontal dune, commonly of coarse material, a central area where sand transportation is maximum with only temporary accumulations, and the area of origin where blowing prevails (Movia, 1972). The latter becomes an erosion pavement or a denuded area with compacted soil. The predominant direction of progression is west to east in most of Patagonia, except in the coastal region and in areas where the topography locally changes wind direction.

Fig. 17.8. Barkhan on a valley slope, with gravelly pavement upward (left) and sand cover downwind (right). Rio Mayo, Chubut Province. (Photo by C. Movia, 1975).

Comparing old and recent aerial photographs, as well as LANDSAT II satellite images, the conclusion has been reached that the broadest erosion forms are about 100 years old. This is an age which matches with the arrival of sheep flocks in the region. The area covered by sand tongues has been estimated (Monteith, 1972) to be 52 400 km^2 in Santa Cruz, 26 790 km^2 in Chubut and 5800 km^2 in southern Rio Negro Province.

Sand macro-accumulations leeward of basalt cliffs in Central Patagonia (Somuncura), fixed sand dunes such as the coastal dunes, and barkhans in transverse valleys in Santa Cruz (Shehuen River) and Chubut (Mayo River, Chalia River) are comparatively less active, but nevertheless impressive. The area covered by these latter forms does not exceed 10 000 km^2 (Monteith, 1972).

Fig. 17.9. Active *lengua* ("sand tongue"), arising from the marginal moraines along Argentino Lake in Santa Cruz Province. (Photo, looking toward the west, by C. Movia, 1976.)

Vast pavements bordering the edges of plateaus, overgrazed fields, and stony terraces along the Patagonian rivers are the prevailing deflation forms. The same forms are also found in areas where salinity or low cohesion of surface material allows its denudation. This usually occurs in flat, humid valleys where precipitation is between 100 to 200 mm, in the large endorheic saline basins, and in some areas where glacial deposits have resulted in soils of fine to medium texture. In the last mentioned case, blow-outs up to 60 cm deep can occur.

Water erosion

In regions of less than 200 mm of rain, water erosion forms are impressive. They can be considered largely the result of geologic erosion. These serve as sources of abrasive material for wind erosion.

Badlands, sheet-washing, and active gullies are common features, as well as mud flows resulting from mass sliding, particularly in the endorheic basins appearing in central and east Santa Cruz Province.

In humid, flat bottom valleys, gullying and stream-bank erosion are the most serious forms of water erosion. Erosion lowers the water table and dries up these valleys, increasing the possibilities of deflation and excavation by wind action.

The area seriously affected by water erosion has been estimated at 28 590 km^2 (Monteith, 1972). Probably this is an underestimation because of the difficulty of detection in aerial photography at small scales.

Water–wind erosion

Rarely are erosion forms of a single origin (wind or water). Mixed forms appear as a result of combined forces. The occurrence of a gully transverse to the wind direction may become the source of abrasive material and the origin of a "sand tongue" extending leeward. The same can be said about sites suffering sheet washing, fire or overgrazing. On the other hand, areas denuded by wind erosion, lacking a gravel pavement, are prone to rill or gully formation.

VEGETATION (A. Soriano, C.P. Movia and R.J.C. León)

The Patagonian semi-deserts represent in terms of plant-geographical units the Patagonia Province. Cabrera (1953) has given a thorough account of names and limits given to this unit by different authors.

As a botanical province, it includes several endemic genera of plants. These are *Philipiella*, *Neobaclea*, *Xerodraba*, *Benthamiella*, *Eriachaenium*, and *Duseniella*. The eastern limit of this province is the Atlantic Ocean, and to the west and south it is bounded by the Subantarctic Forest and high Andean Desert. The border between Patagonian vegetation and the scrub characteristic of the Monte Botanical Province to the north-east have been described by Soriano (1949) and Ragonese and Piccinini (1969). This boundary approximately follows the annual isotherm of 13°C.

Up to 1956, vegetation maps presented Patagonia as a vast and uniform territory, without structural or floristic subdivisions. The only exception was Kalela's map, published by Auer (1951), in which a *Stipa* range was distinguished from a less xeric *Festuca* range.

Soriano (1956a) identified five different floristic districts within the territory south of 42°S. Cabrera (1971) included a sixth district, north to 42°S, in his synopsis of the Patagonia Province. This same scheme will be kept in this chapter; however, the two districts named Subandean and Fueguian will not be considered here, because their vegetation is typified as grass steppe, and rainfall generally exceeds 300 mm per year. The rest of the Patagonian Botanical Province, can be considered to be desert or semi-desert. The following subdivisions are recognized (Fig. 17.10):

1. Payunia District
2. Occidental District
3. Central District
   a. Chubutian Subdistrict
   b. Santacruzian Subdistrict
4. San Jorge Gulf District

Payunia District

This northwest corner of Patagonia is under volcanic influence (Cabrera, 1971). It occurs in the southern part of Mendoza and northern part of Neuquén Provinces. No thorough description has been given by Cabrera; neither has he traced the limits of the district. Roig (1960) assigned a part of the southern area in Mendoza to this floristic province by including the Payun Mountains, up to

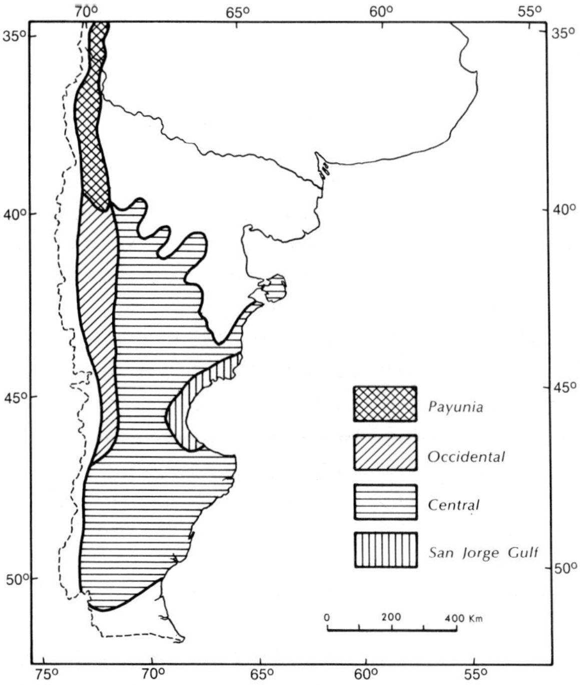

Fig. 17.10. Map of the Patagonian semi-desert and its division into districts (Cabrera, 1978).

a level which has not been determined, and a portion that extends to the west to meet the Andean Desert. According to Roig, the Patagonia Province would extend to the north in a narrow strip along the pre-Andean range, up to the limit between Mendoza and San Juan.

It is obvious that more exploration is needed before we will be able to describe and characterize vegetation units or continua in this complex junction between the scrub formation (Monte Botanical Province) and the Patagonian Botanical Province.

Notwithstanding this lack of information, it is conceivable that the most conspicuous elements of the Patagonian vegetation will be confined to special ecotopes defined by exposure and elevation in the northern part of this district (Ruiz Leal, 1972). These are intermingled with other ecotopes in which scrub vegetation (Monte) prevails.

Occidental District

The Occidental District extends as a strip about 100 to 120 km wide from Neuquén Province to the northwestern corner of Santa Cruz Province. As has already been stated, its relationships with the Payunia District in the north have not yet been clarified. It is situated between the High Andean vegetation to the west and the Monte Scrub to the east. It is conceivable that the district appears as a fragmental territory in the northern part of the Neuquén Province. The border of this district widens steadily to the east from the town of Zapala to the south, leaving outside its realm the town of Piedra del Aguila, and proceeding approximately through Ingeniero Jacobacci, where the Central District begins to appear in the east. The extent of the district south of Ingeniero Jacobacci is approximately the 70° meridian up to the latitude of 46°S. From Mata Magallanes, the boundary goes southwest to Perito Moreno. The northern coast of Buenos Aires Lake completes the southern limit of this district.

To the west the district meets, from north to south, first with the High Andean vegetation, then with the Subantarctic Forest and finally with the fescue steppe of the Subandean District.

The same general vegetation extends over plateaus, mountains, and the non-humid parts of valleys (Fig. 17.11). It is a semi-desert in which the dominant plant form is tussock grass with linear, spiny leaves. Scattered among the grasses, shrubs less than 1 m high, frequently cushion-like, present themselves in a very wide range of densities.

Total cover ranges from 20 to 60%, depending on site and successional status. Tussocks of grasses have considerable standing dead mass. In some of them, the old dead part forms the center of the bunch, encircled by a ring of tillers, bearing live and recently dead leaves. Several features of the shrubs are common to different species, namely hemispheric form; spinosity (either because of true spines present or spring leaves); and a low, hard, cushion-like habit.

Low, perennial herbs grow under the protection of tussocks or spiny shrubs. Most of them are geophytes. Rhizome-bearing plants seem to be more common than bulb or corm-producing ones. A few ephemerals, generally tiny plants, able to complete development with not much more than their cotyledons, complete this list. The cover seldom looks green. Yellow, gray and brown are the dominant colors year-round. A high proportion of the flowers are yellow also.

Fig. 17.11. The semi-desert between Rio Mayo and Mata Magallanes, southwest Chubut Province. *Stipa speciosa*, *S. humilis*, *Mulinum spinosum*, and *Senecio filaginoides* are the dominant species (photo by B.C. Piccinini).

The floristic composition[1] of plant communities living in this part of Patagonia is dominated by the grass family. The genus *Stipa* is heavily represented. *Stipa speciosa* and *S. humilis* are the most frequent. *Stipa ibari*, *S. chrysophylla*, *S. psylantha*, *S. neaei*, *S. subplumosa*, and *S. sorianoi* are conspicuous and locally abundant. *Stipa*, as well as other tussock grasses, goes by the local name *coirón* (Fig. 17.12).

Fig. 17.12. *Poa ligularis*, *Stipa speciosa* and *S. humilis*, the dominant *coirones*. Rio Mayo, Chubut Province. Patagonia pebbles cover the soil surface (photo by A. Soriano).

Poa ligularis, *Bromus setifolius*, *Hordeum comosum*, *Festuca argentina*, and *Poa lanuginosa* are other important components of the flora. *Stipa humilis* and *S. speciosa* usually occur together but alternate in dominance. There are, however, extensive areas where only one of the two is present. The ecological reasons of this difference in pattern are unknown.

Mulinum spinosum, *Adesmia campestris* and *Senecio filaginoides* are three shrubs which, together with the *Stipa* species, are the most constant companions within the communities of this district. Neither the shrubs nor the grasses mentioned are exclusive to this district. Only a very few of these species occur only in Patagonia. Most of them appear in several different districts of Patagonia and elsewhere in Argentina, the continent, or the hemisphere. For instance, *Stipa speciosa* is found all along the Andes in South America and reaches north to California. *Poa ligularis* occurs in much wetter and/or hotter areas such as in the Monte scrub and in the pampean grasslands. *Mulinum*

[1] Plant names are according to Nicora (1978) and Cabrera (1971) regarding Poaceae and Asteraceae, respectively.

spinosum is only sporadically found outside Patagonia. It has not been observed as a dominant outside of the Occidental District.

Two species of *Berberis*, *B. cuneata* and *B. heterophylla*, are the tallest shrubs commonly found in this district. They also attract notice by being a deep green color which contrasts with the rest of the vegetation. Only isolated individuals or small groups are usually found; although along the borders of the terraces and in some flat valleys it is possible to observe them distributed as a continuous strip. The latter situation probably corresponds to a zone of water seepage from geologic strata.

Other commonly found shrubs are *Lycium chilense*, *Verbena ligustrina*, *Acantholippia seriphioides*, *Schinus polygamus* and *Nardophyllum obtusifolium*. Typical hard, cushion-like plants are *Azorella caespitosa* and *Brachyclados caespitosus*. Both of them may look very showy when covered with yellow flowers or fruits. It is not uncommon to find these hard cushions clasping bunches of grasses or tillers of other herbs which have been completely encircled by the growing carpet. Other cushion-like species differ from these in forming not hard crusts but soft covers. Various species of *Adesmia*, *Acaena* and *Verbena* belong in this class.

Several cacti are found in this district and represent the southernmost members of the family. *Maihuenia patagonica*, *Austrocactus* spp. and *Pterocactus* spp. are the two more frequently found. *Maihuenia* forms large patches, no more than 20 cm high but sometimes 1 m in diameter. It exhibits beautiful clusters of large white, pink or tobacco-colored flowers in the summer. The small, nearly spherical stems of *Pterocactus* spp. mimic the brownish-pink color of the sandy soil surface.

A fairly long list of perennial herbs can be given for this district. Some of them are seldom found because of heavy livestock grazing. Among the most frequent ones are the graminoids *Carex argentina* and *C. andina*. Other commonly found herbs are *Calceolaria lanceolata*, *Polygala darwiniana*, *Adesmia lotoides*, *Arjona patagonica*, *Cerastium arvense*, *Perezia recurvata* and several species of *Silene* and *Hypocharis*. Less common are *Loasa argentina*, *Tweedia odonelli*, *Valeriana clarionifolia*, *Oxalis* sp., *Hypochaeris hookerii*, *Huanaca acaulis* and *Leuceria candidissima*. A few annual plants can be mentioned: *Microsteris gracilis*, *Gilia patagonica*, *Polemoniella antarctica*, *Erodium cicutarium*, *Oenothera contorta* var. *divaricata*, *Plantago patagonica* and *Doniophyton patagonicum*.

There are some weeds which infest very degraded sites or dwell in corrals, around houses and along roadsides. The most common are the short-lived grass *Bromus tectorum*, the crucifers *Erysimum repandum* and *Sisymbrium altissimum*, and *Atriplex roseum*.

The district shows considerable vegetational variation over its extent, both in terms of floristics and physiognomy. Many of the species already mentioned occur over the district, generally as members of several community types. Others are limited to certain areas. Among the latter are some shrubs which, when present, exhibit a great degree of abundance. One of these is the columnar composite *Nassauvia axillaris*; another is *Corynabutilon bicolor*, easily recognized by its leafy aspect and grayish color. This latter species also occurs on the east side of the Andes, in Chile, and appears again further south, in Santa Cruz. In the central part of the district another shrub species, occurring in several spots as a dominant, is *Sapium patagonicum*. This is a leafless, poisonous plant, with flexuous branches emerging from the center of the plant.

Wet places, such as springs occurring on slopes, or more frequently in valley bottoms, have grasslands locally called *mallines* (Fig. 17.13). *Juncus* species are often very abundant at such sites. Among the sedges and the grasses, *Carex nebularum*, *C. gayana*, *Hordeum comosum*, *Poa annua*, *Agrostis pyrogea*, *Polypogon australis*, *Deschampsia antarctica* and *Koeleria mendocinensis* are additional common components. *Colobanthus subulatus*, *C. crassifolius*, *Potentilla anserina*, *Vicia megellanica*, *Acaena magellanica*, *Azorella trifurcata*, *Plantago australis* ssp. *cumingiana*, and *Gnaphalium pratense* are all forbs contributing to the dense sward at those rare *mallines* that have not been disturbed by livestock grazing. In places where soil is really wet, at least during part of the year, some of these species are associated with *Eleocharis melanostachys*, *E. albibracteata*, *Arjona pusilla*, *Arenaria serpens*, *Caltha sagittata*, *Ranunculus uniflorus*, *R. cymbalaria*, *R. peduncularis*, *Samolus spathulatus*, *Mimulus glabratus*, *Pratia repens*, etc.

In some valleys, open wells called *menucos* are present. A ring of reeds (*Scirpus californicus*) is

Fig. 17.13. A humid pasture (*mallin*), Tecka, Chubut Province (photo by B.C. Piccinini).

often seen around them. Submerged plants, the Halorrhagaceae *Hippuris vulgaris* and *Myriophyllum elationoides*, appear rooted in the bottom. A plant community dominated by *Festuca pallescens* is commonly found on the margins of some valleys (Fig. 17.14). This grass is also called "white or sweet coiron" and is the main grass of the steppes west and south of the semi-desert. The valleys are inhabited by other communities towards the wetter bottomland. Grasses, sedges and forbs are dominant on such sites. Temporarily flooded saline soils are often found in the valleys. In such sites *Distichlis scoparia*, *D. australis*, and *D. spicata*, together with *Puccinellia skottsbergii*, *Nitrophila australis* var. *kunzei*, *Suaeda patagonica* var. *crassiuscula* and *Lycium repens* are most conspicuous.

The phenology of some of the dominant plants has been studied in this district (Soriano et al., 1976). Phenophases have been recorded with methodology like that used by French and Sauer (1974). The grasses and the most frequent sedges exhibit a high degree of uniformity through the different seasons. They do not show a period of thorough dormancy, having leaves not fully expanded both during the coldest and the driest seasons. Flowering occurs during the last part of spring for all the species mentioned. *Carex argentina* is the earliest, being followed by *Poa ligularis* and *Festuca argentina*. *Stipa psylantha* is the last to flower (Fig. 17.15).

Water economy regulates flowering intensity. This is evident in all the species considered, but seems to be especially clear-cut in *Stipa speciosa* and *S. humilis*, the dominant *coirones*.

Among the dominant shrubs, three of them, namely *Mulinum spinosum*, *Adesmia campestris* and *Acantholippia seriphioides*, have a well-defined dormant period, lasting the whole autumn and winter. *Mulinum spinosum* keeps the dry, spiny leaves of previous years. The other two species are deciduous. *Senecio filaginoides* does not exhibit a period of dormancy; its twigs have expanding leaves in all seasons. Flowering, for all the four species, occurs at the end of spring and beginning of summer (December–January).

Different perennial herbs have, in general, similar phenological patterns; they go through a long dormant period in autumn and winter, have early growth at the end of winter, and flower in spring. The ephemerals germinate at the end of autumn and complete development during winter and early spring.

Fig. 17.14. A flat valley (*cañadón*), Rio Mayo, Chubut Province. *Festuca pallescens* is dominant along the border (photo by B.C. Piccinini).

Growth of individual leaves in marked plants of dominant grasses has been followed during a whole year at one site (Soriano et al., 1976) (Fig. 17.16). From these measurements it can be concluded that in *Stipa speciosa* and *S. humilis*, as well as in the sedge *Carex argentina*, maximum growth occurs in the winter period (June–September). In *Bromus setifolius* and *Poa ligularis* the growth rate of leaves keeps increasing from autumn (March–June) through winter and spring (September–December), reaching its maximum during this latter season.

Measurements of water tension in the xylem have been taken at one site in different seasons through the year for three dominant grasses and three common shrubs (Soriano et al., 1976; Table 17.2). The shrubs (*Mulinum spinosum*, *Senecio filaginoides* and *Adesmia campestris*) always exhibited higher (less negative) water potential and more stable values at different times of the day than the three grasses. Among these, *Stipa humilis* exhibited the highest tension values, repeatedly showing a water potential of less than −40 bar at the time of the day when maximum temperature and vapor deficit occurred. The other two species (*Stipa speciosa* and *Poa ligularis*) responded quite similarly; both showing water potential values of −10 to −40 bar.

The roots of the dominant tussock grasses and those of the frequent shrubs are, apparently, neatly differentiated in terms of location and structure. The roots of the grasses, fibrous and densely covered by hairs, are concentrated in the upper soil. Of the root mass present in the first 30 cm of soil, 76% is found in the upper 15 cm (Soriano and Satorre, 1979). Principal shrubs, on the other hand, generally have tap roots which go to 35 to 45 cm deep and are subdivided into two or more horizontal subdivisions (Fig. 17.17).

Data on primary productivity have been published by Soriano et al. (1976) and by Ares (1978). Periodic harvests during 1975 in a 4 year old exclosure and in a nearby grazed area at Rio Mayo, Chubut, provided data to calculate productivity, senescence and litter formation, in an example of desert grassland. Calculations were accomplished according to Sala et al. (1981), using corrections for senescence and litter fall, for the estimation of primary productivity.

In Fig. 17.18 above-ground net productivity (ANPP) is greater in the grazed than in the ungrazed grassland. Senescence is also greater and in each case is numerically equivalent to production. On the other hand, litter fall is greater in the grazed area than in the exclosure.

Litter fall in the grazed area should probably be

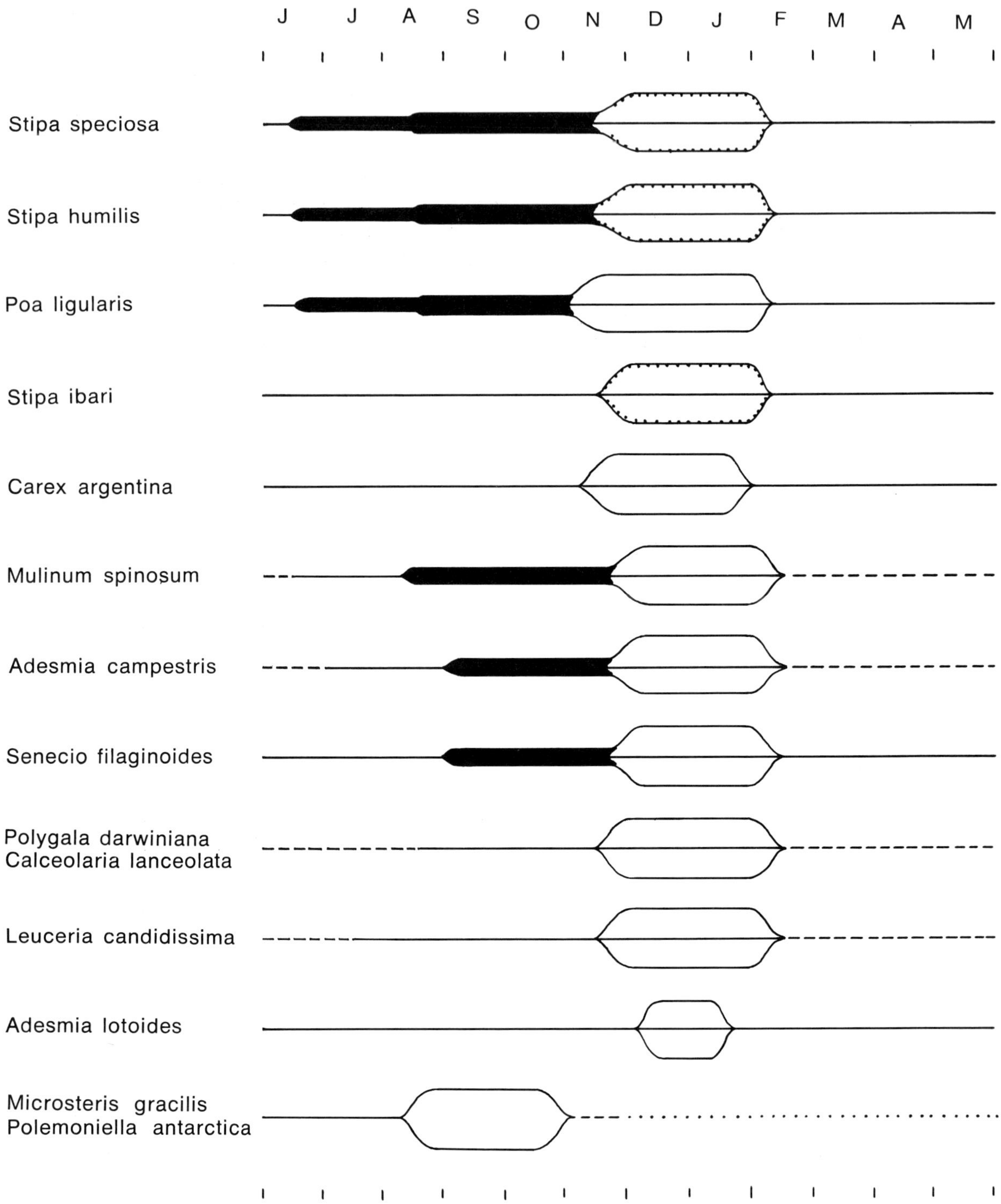

Fig. 17.15. Phenology of some of the dominant species in the Occidental District. Uninterrupted line: vegetative growth (the width of the black line represents intensity); dashed line: dormant period; dotted line: seed; enclosed area: reproductive growth (points on the line means that this phenophase appears only certain years, depending on water).

TABLE 17.2

Water potential (xylem tension), in bars, for three grasses and three shrubs for different seasons and time of day (Soriano et al., 1976)

Date:	Jan. 1975		March 1975			June 1975	Sept. 1975			Dec. 1975		
Time of the day[1]:	M	A	M	A	E	A	M	A	E	M	A	E
Temp. (°C):	15.3	26.8	15.0	18.5	19.0	8.0	3.2	15.9	11.7	9.0	13.5	6.0
Rel. humid. (%):	63	29	65	45	50	48	56	36	48	65	50	70
Grasses												
Stipa speciosa	−31	−40	−28.5	−40	−16.5	–	−25	−33	−12	−22.7	−31.2	−21
Stipa humilis	−28.5	−40	−40	−40	−40	–	−23.5	−40	−22	−29.3	−37.8	−30.7
Poa ligularis	−13	−35	−15.5	−38	−17.5	−28.3	−20.5	−27	−10.5	−21	−29.3	−14.5
Shrubs												
Mulinum spinosum	−12	−26	−18.5	−24	−19	−27	−8.5	−15	−13.5	−18.3	−21	−18.5
Senecio filaginoides	−10	−18	−17.5	−21.5	−18	−24	−8.5	−15.5	−11.5	−13.5	−13.5	−9.5
Adesmia campestris	−21	−27	−19	−24	−22	–	−11.5	−14.5	−14	−13.5	−14.5	−9.5

[1] M = morning; A = afternoon; E = evening.

designated standing-dead disappearance. We do not know how much of this material is consumed by sheep. Anyhow, the slower rate of litter fall in the ungrazed area agrees with the fact that standing dead tends to accumulate steadily when grazing is reduced or excluded. An increment of the total standing-dead material of the order of 145% was found when an area protected since 1954 was compared with the one closed to grazing in 1972. Litter also increased by 48.5%.

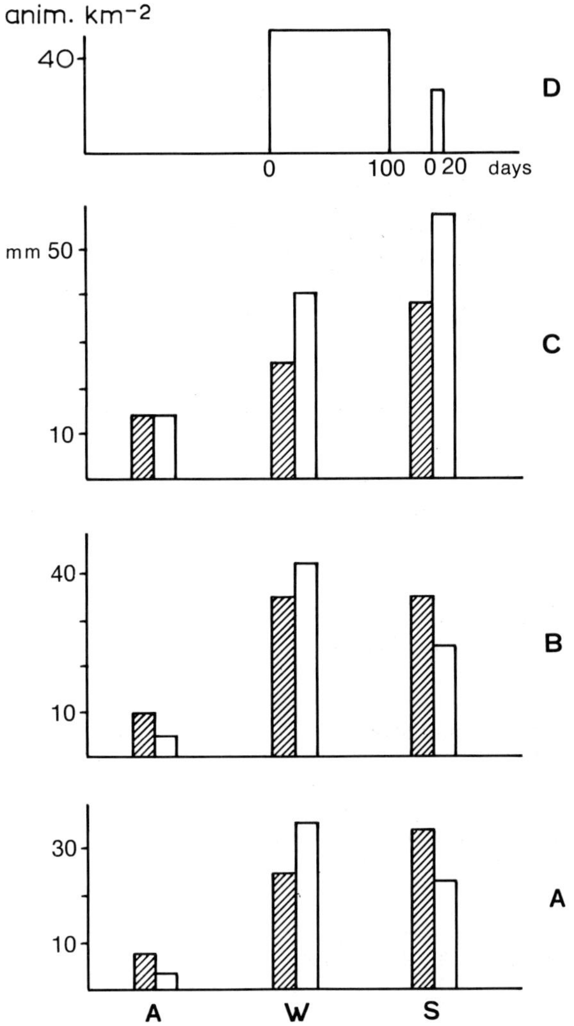

Fig. 17.16. Growth of leaves in the three dominant grasses of the Occidental District. Hatched bars: ungrazed; white bars: grazed. A = autumn (measurement period: 9.III–17.VI); W = winter (measurement period: 17.VI–24.IX); S = spring (measurement period: 24.IX–9.XII) A. *Stipa humilis*. B. *S. speciosa*. C. *Poa ligularis*. D. Animals per square kilometer in the grazed field.

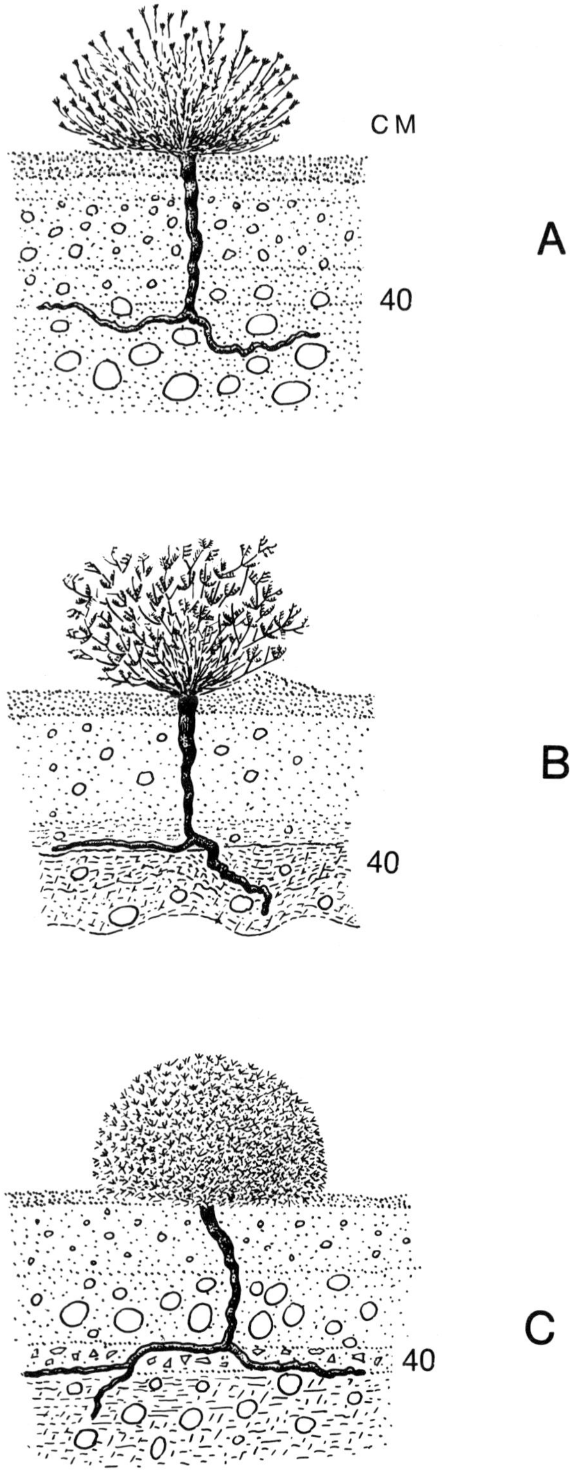

Fig. 17.17. Root systems of A, *Senecio filaginoides*, B, *Adesmia campestris*, and C, *Mulinum spinosum*.

Litter disappearance, calculated as the sum of negative differences in the litter values between sampling dates, shows nearly the same results for both sites, grazed and ungrazed (46 and 36 mg m^{-2} d^{-1}, respectively). These results need further attention, since in the case of this type of grassland disappearance probably involves not only decomposition, but blowing away of dead material, as a result of sheep trampling and wind action.

Data of 1972, from harvests made in the same exclosure, were used by Ares (1978) to fit a model of differential equations to calculate productivity, senescence and death rhythms of above-ground biomass. He used as compiler SIMCOMP 2.1 (Gustafson and Innis, 1973) and obtained an estimate of productivity of 4.4 kJ m^{-2} d^{-1}. Assuming the grazing animals only consume a green diet composed of *Stipa speciosa* and *Poa ligularis* in a proportion of 1:2.5, Ares estimated that, at a carrying capacity of 0.6 animals ha^{-1}, utilization equals productivity, both being 739 kJ m^{-2} yr^{-1} (Fig. 17.19).

Decomposition of organic matter has been studied using filter paper buried in the soil (A.C. Landaburu, unpublished). Results indicate that 57.1% of the material decomposed in winter (June–September) and only 4.2% in spring. When filter paper remained in the soil from December 1975 to October, 1977, only 0.7% of the cellulose was recovered.

Central District

This district includes the greatest part of the Patagonian deserts and semi-deserts. Two subdistricts are recognized: the Chubutian in the north and the Santacruzian in the south. They spread over the most arid part of Patagonia. The plant species which can be considered as the most characteristic and dominant over extensive areas is the composite *Nassauvia glomerulosa*, often replaced by *N. ulicina*. Recent descriptive studies of vegetation in this area include those of Molina Sánchez (1976), Bertiller et al. (1977), and Movia et al. (1977).

Chubutian Subdistrict

This has been described previously as the Central District (Soriano, 1950). Ragonese and Piccinini (1969) have traced the northern boundary between

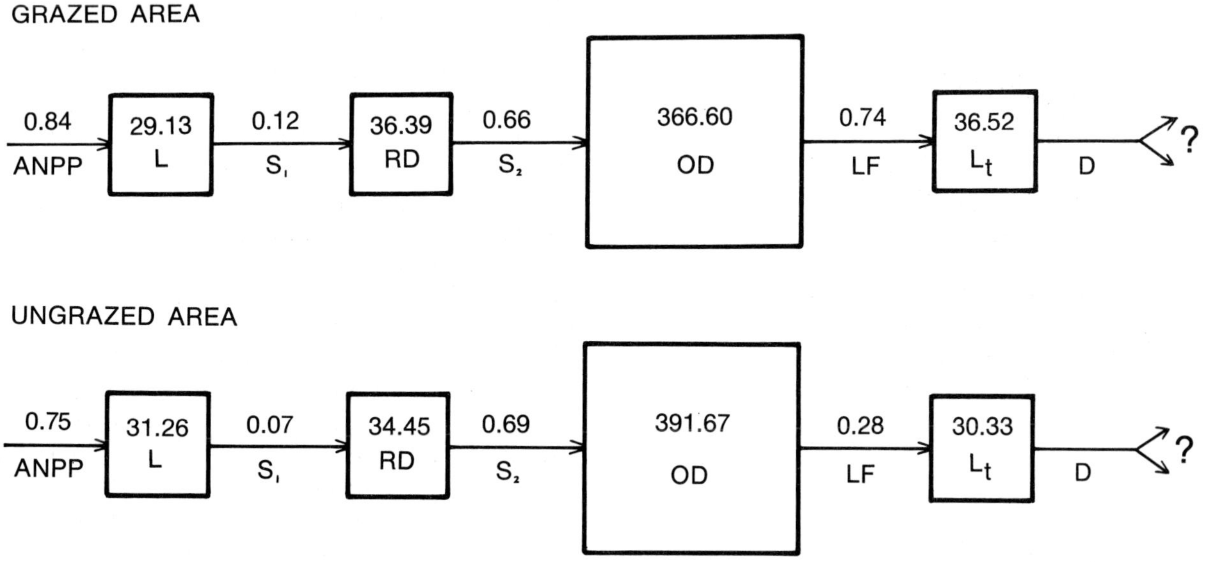

Fig. 17.18. Flow diagram showing above-ground net primary productivity and average biomass in the different compartments of a *Stipa speciosa–S. humilis–Poa ligularis* community. *ANPP* = above-ground net primary productivity; S_1 = senescence to recently dead; S_2 = senescence to old dead; *LF* = litter fall; *D* = disappearance; *L* = live; *RD* = recently dead; *OD* = old dead; L_t = litter. Biomass in g m^{-2}, fluxes in g m^{-2} day^{-1}.

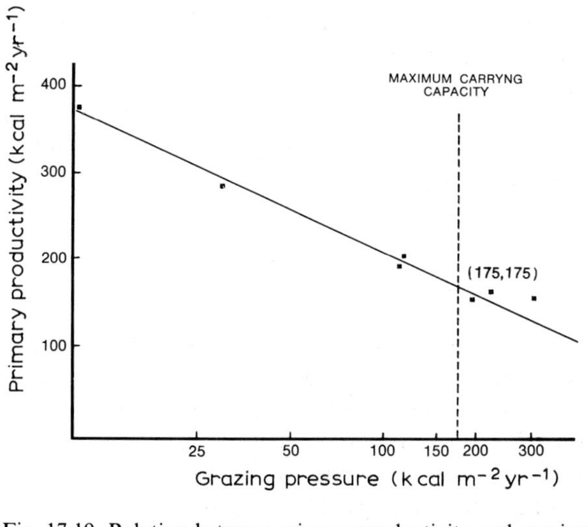

Fig. 17.19. Relation between primary productivity and grazing pressure (Rio Mayo, Chubut Province), according to Ares (1978) (1 kcal is 4.2 kJ).

this subdistrict and the so-called Monte vegetation, in the Provinces of Neuquén and Rio Negro. From the town of Zapala, the limit goes SSE down to 42°S, leaving then the whole northeastern part of Chubut Province within the Monte realm. Valdés Peninsula, Carlos Ameghino Isthmus and the land strip from Rio Chico to the sea coast between Punta Lobos, and Cabo Raso, previously considered as an ecotone (Soriano, 1949), are included in this subdistrict here (Bertiller and Beeskow, 1979).

Two different kinds of vegetation are found on two topographic and physiognomic units. The first is the vegetation of plains, plateaus and hills above 400 m in elevation. The second type occurs in the areas below 200 m. Examples of the latter are the Rio Chico Valley and the Sarmiento Basin.

Vegetation of the plains, plateaus and hills is dominated by *Chuquiraga avellanedae*, *Nassauvia glomerulosa*, *N. ulicina*, *Stipa speciosa*, *S. humilis*, and *Poa ligularis*. A very sparse plant cover occurs here with at least 65% of the soil exposed.

Chuquiraga avellanedae, a hemispheric shrub, generally no more than 60 cm high, with spiny leaves and big yellow heads densely covering the plants in some summers, is consumed in moderate amounts by sheep, but especially by rheas.

Some other shrubs that are larger than *Chuquiraga* sometimes appear as groups or patches or are sparsely distributed. They are *Prosopis denudans*, *Lycium ameghinoi*, *Berberis cuneata*, and *Verbena ligustrina*. However, they are frequently absent and the community is then entirely dominated by the dwarf cushions of *Nassauvia glome-*

rulosa. These plants often mimic the color and shape of pebbles on the soil surface (Fig. 17.20). This contrasts with the straw-colored tussocks of *Stipa* and the rounded bushes of *Chuquiraga*. *Nardophyllum obtusifolium* is another common shrub in some areas of this subdistrict. In very dry seasons it is the only species which remains bright green. *Pleurophora patagonia* is very attractive in summer when covered with pink flowers. It is characteristic of the whole district, its area extending from Neuquén to Rio Deseado.

The lowlands of the Rio Chico Valley and the Sarmiento Basin have vegetation that does not appear to differ fundamentally from the aspects already described. Soil characteristics there can, however, cause the frequent appearance of halophytic communities. *Atriplex lampa*, *A. sagittifolium* and *Frankenia patagonica* are common members of these communities. *Lycium ameghinoi* and *Prosopis denudans* var. *patagonica* are occasionally found there. Vegetal cover is very poor in these places as the shrubs grow very sparsely on the eroded slopes and scarps. This is perhaps the most desert-like landscape in Patagonia.

South of Colhue Huapi and Musters Lakes, and in the narrow strip of land separating them, halophytic communities alternate with others in which *Nassauvia glomerulosa* (or *N. ulicina*) and *Chuquiraga avellandeae* are dominants. Along the lake shores, alkaline meadows and temporarily flooded area a cover of *Juncus* sp. and *Distichlis spicata* can be found.

The vegetation of the badlands around the Sarmiento Basin includes the endemic *Ameghinoa patagonica* and also *Nicotiana ameghinoi*.

Santacruzian Subdistrict

This unit covers nearly all the Argentine territory between 46°S and 51°S, from the seashore to the Andes.

From the physiognomic and floristic points of view, its affinities with the Chubutian Subdistrict are obvious, but the Santacruzian area lacks *Chuquiraga avellanedae*; instead, *Verbena tridens* becomes one of the principal dominants (Fig. 17.20).

To the south and west, this district readily merges into the fescue steppe (*Festuca pallescens* and *F. gracillima*) of the Sub-Andean District.

Two main physiognomic subunits are distinguished here: the semi-desert or dwarf scrub, and the shrub steppe or scrub (Mueller-Dombois and

Fig. 17.20. *Verbena tridens* forming a rim around a circular, flat depression. In foreground, *Nassauvia glomerulosa* is the dominant cushion plant. Piramides, Santa Cruz Province (photo by B.C. Piccinini).

Ellenberg, 1974). The first one is the more widely distributed in the territory. It covers alluvial plains, plateaus, till plains, rounded moraines and gentle slopes, from the sea coast to altitudes around 500 m. Cover is seldom greater than 40%. *Nassauvia glomerulosa*, the dominant species, is associated with *Acantholippia seriphioides*, *Chuquiraga aurea*, *C. kingii*, *Brachyclados caespitosus*, *Azorella* spp., and *Mulinum microphyllum*. Bunchgrasses, represented by *Stipa speciosa*, *S. humilis*, *S. chrysophylla* and the species with feathered spikes (*S. neaei*, *S. psylantha*, *S. subplumosa*) are abundant in some facies, and scarce or completely lacking in others. Whenever a pattern of surface streams (paleo-drainage pattern) or depressions appear, *Verbena tridens*, the characteristic species of the subdistrict, is associated with it.

The shrub steppe or scrub presents itself either as a shrubland or as a thicket. The only species forming real thickets over extended areas is the *Verbena tridens* already mentioned, an evergreen, dark colored shrub, 80–100 cm high.

Scrub is present on quite a broad range of different physiognomic units: alluvial plains, flat or rolling, abrupt rocky slopes and landslides, valley bottoms, fixed dunes and muddy coastal plains.

On the low plateaus (200–300 m elevation) near the Atlantic, *Verbena tridens* is associated with *Nardophyllum obtusifolium*, *Berberis cuneata*, and *Lycium chilense* forming an open shrubland where bunchgrasses, cushion plants and small perennial herbs of the genera *Viola*, *Calceolaria*, *Nassauvia* and *Perezia* intermingle.

On hilly areas, generally covered by a basaltic mantle, a sparse scrub is present. There, several shrubs can be found, without any of them becoming dominant. *Verbena tridens*, *Anarthrophyllum rigidum*, *Schinus polygamus*, *Nardophyllum obtusifolium*, *Verbena ligustrina*, *Lycium chilense* and *Berberis heterophylla* are the most frequent species in those areas.

To the southwest of the district, where altitude determines lower temperatures and better water conditions, scrub living on glacial deposits is dominated by *Nardophyllum obtusifolium* associated with *Festuca pallescens*. On the other hand, in lowlands, like muddy coastal plains and endorheic depressions, *Lepidophyllum cupressiforme* is the dominant shrub and sometimes the only bushy species on these flat sites where a high proportion of the soil is bare and covered by a compacted crust.

San Jorge Gulf District

This district extends from approximately Cabo Raso to Punta Casamayor, forming a narrow arc bordered by the Gulf. It includes the Montemayor, Del Castillo and Espinosa Plateaus. The general geomorphology of this area is represented by these

Fig. 17.21. Wind erosion effects. Pampa del Appeleg, Chubut Province (photo by B.C. Piccinini).

plateaus and numerous narrow valleys dissecting them. Vegetation changes sharply according to these geomorphological characteristics.

A dense scrub extends over the slopes. The dominant shrubs on slopes are *Trevoa patagonica* and *Colliguaya integerrima*. Cover is frequently over 60%.

Trevoa patagonica is a nearly aphyllous shrub, with vigorous stems diverging from the base of the plant. Root nodules have been described in this species (Medan and Tortosa, 1979). It is regularly browsed by sheep. This species exists in Mendoza and has also been found in several points in the fringe between the Monte and Patagonian vegetation. However, in none of these other localities does it appear as a dominant plant cover over a considerable area.

Between the shrubs and underneath the open canopy of *Trevoa* is found a herbaceous stratum dominated by *Stipa* species common to all Patagonia. Other members of this stratum are *Erodium cicutarium*, a cosmopolitan and opportunistic forb, and *Vulpia dertonensis*, an annual grass able to form a continuous sward in wet years and during the spring. Companion herbaceous species of this community type are *Calceolaria polyrrhiza*, *C. lanceolata*, *Silene magellanica*, *Cerastium arvense*, *Sisyrinchium striatum*, *Huanaca acaulis*, *Amsinckia hispida*, *Plagiobotrys calandrinioides*, *Perezia recurvata*, *Microsteris gracilis*, and *Oenothera contorta* var. *divaricata*.

Other shrubs found in the other districts occur here also — *Mulinum spinosum*, *Adesmia campestris*, *Ephedra ochreata*, *Anarthrophyllum rigidum*, *Schinus polygamus*, *Lycium chilense*, etc.

Some of the floristic elements belonging to the Monte vegetation reach at least the northern part of the district. Examples are *Stipa tenuis* and *Prosopis denudans*.

The scrub is sometimes dense enough to form a real thicket. Under the shrubs some tender forbs find protection against grazing animals. *Phacelia magellanica*, *Stellaria media*, and *Galium* spp. can be found there plus a few climbing plants of the district: *Magallana porifolia*, *Loasa argentina* and *Mutisia retrorsa*. South of Pico Salamanca, as Del Castillo Plateau widens and steadily rises up to 900 m in elevation, the vegetation changes rather abruptly. A steppe dominates the flat landscape. *Festuca pallescens*, *Festuca argentina* and *Poa ligularis* are the most abundant grasses here, but *Stipa speciosa* and *S. humilis* are also present. Winds are severe here, and, as in the rest of Patagonia, a very important limiting factor. Both the intensity and constancy of the wind is involved. A few shrubs surpassing the height of the grasses show a dwarfed aspect. In some places, *Adesmia campestris*, *Senecio filaginoides* and *Nardophyllum obtusifolium* form groups or are spread over the steppe. Cushion plants represent a common feature of this steppe structure. *Brachyclados caespitosus*, *Mulinum microphyllum*, *Benthamiella patagonica*, *Cruckshanksia glacialis*, among other species, belong to this morphological group.

It is on these plateaus that the genus *Larrea* reaches its southernmost limit, represented here by *Larrea ameghinoi*, the only species in the genus with a carpet-like habit.

Communities in wet sites, like the valley bottoms, are dominated by *Eleocharis albibracteata*, *Myosurus apetalus*, *Poa annua*, *P. ligularis*, *Plagiobotrys calandrinioides*, *Veronica peregrina*, *Plantago myosurus*, *Taraxacum officinale*, etc. Halophytic vegetation is dominated by species of *Atriplex* and by *Suaeda divaricata*.

Grazing management, desertification and vegetation dynamics

Sheep grazing, introduced nearly 100 years ago, has been the major influence on the present vegetation of arid Patagonia. Around twelve million sheep presently graze in the area. Continuous grazing is generally practiced by the ranchers — that is, animals are kept on the range all year around. The only exception is the division of winter and summer pasture, common for ranches in the western part.

Pastures have border fences, but sheep roam unherded, even within very large paddocks. Breeds most widely used are Merino and Corriedale. Wool production is the main enterprise.

Over about 45 million hectares (the approximate extent of arid Patagonia) about 10 000 ranches are established. Wool production in 1964–69 was 60 000 metric tons. This includes the whole Rio Negro Province and the grass steppes considered outside arid Patagonia (INDEC, 1969).

Grazing pressure has been and is now generally very high. Several authors have expressed the urgent need for reducing numbers of animals by about 30% (Soriano, 1956b, 1958; Laporte and Coccimano, 1970). Sheep numbers per square kilometer range between 20 and 60. Overgrazing has resulted in a broad and persistent process of desertification (Fig. 17.21). Vegetation deterioration and soil erosion are the two main features of this process (Ericksen, 1975). The seriousness of soil erosion has already been detailed in another section of this chapter.

In the *Stipa* desert grasslands of the Occidental District, range condition has been evaluated and a classification has been proposed on the basis of several indicators (Soriano and Brun, 1973). Eight different indicators were chosen beforehand, and points given to each of them. The scale was applied to quadrats sampled in an area of 200 km^2. The samples were then ordinated in a series going from 81 to 32 points. From this ordination, certain floristic indicators not previously considered emerged (Table 17.3).

Applying the same range condition ratings, Schlichter et al. (1978) concluded that the information theory index, H (Lloyd and Ghelardi, 1964)

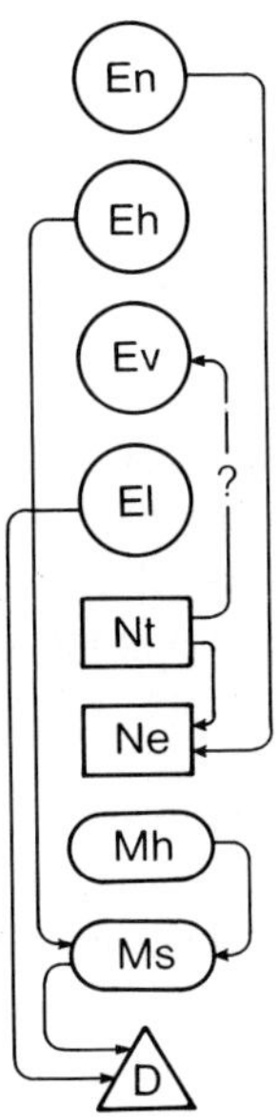

Fig. 17.22. Hypothetical retrogression trends in the Santa Cruz semi-desert vegetation types. E = shrub steppes; N = dwarf scrub; M = humid sites; D = desert; n = *Nardophyllum obtusifolium* dominant; h = mixed shrubs, in flood plains; t = with *Stipa* and other grasses; e = without grass layer; h = humid grasslands; s = saline; v = *Verbena tridens* dominant; l = *Lepidophyllum cupressiforme* dominant.

TABLE 17.3

Ordinated partial table showing species groups restricted or predominantly associated with a range of rating points; species with an asterisk were not included among the floristic indicators originally considered; values of cover-abundance are according to the Braun-Blanquet (1964) scale (Soriano and Brun, 1973)

Relevé	14	16	12	17	2	20	21	15	3	25	19	26	1	5	6	13	27	4	18	10	7	28	41	23	24	9	22	8
Number of species	32	23	28	31	25	23	19	26	25	22	24	23	17	25	25	20	20	22	20	16	23	21	19	17	18	13	18	14
Condition rating points	81	76	71	66	65	65	64	62	59	58	57	57	56	55	54	54	51	51	50	49	47	11	38	37	37	35	34	32
Stipa speciosa	2	2	1	1	2	2	1	1	1	1	1	1	+	1	1	1	1	1	1	1	1	1	1	1	1	2	1	+
Stipa humilis	+	+	+	+	+	+	1		+	+	+	+	+	1	+	+	1	+	+	1	+	1	1	1	1	+	1	1
Huanaca acaulis	+	+	+	+	+			(+)				+											+					
**Arjona patagonica*	(+)	+		+																								
Loasa argentina	+	+	+	+		(+)		(+)		+											(+)							
Hypochaeris sp.	+	+	+	+	(+)	+	+		+					+	+								(+)					+
Acaena platyacantha			+	+				+						(+)														
Polygala darwiniana	+	+	+	+	+	+		+	+	+	+	+	+		+			+	+		+							
Leuceria sp.	+	+	+	+			+	+		+	(+)			+	+	+	(+)			+	(+)							
Calceolaria lanceolata	+	+	+	+	+	+		+	+	(+)	(+)	+	+	(+)	+	(+)	(+)		(+)		+							
**Stipa psylantha*	+		+	(+)	+	+		+	+	+	+			+	+	+	+	+	+				(+)					
**Poa lanuginosa*					+	1					+	+					+	+		1		+		+	+	(+)	+	+
**Euphorbia collina*						+									(+)		(+)				+			+	+	(+)	+	+

reflects very adequately the range condition in the area.

Retrogression produced by overgrazing has been ascertained by using comparison data obtained in exclosures and adjacent pastures (Soriano, 1959; Soriano et al., 1976). Principal changes registered are: reduction in diversity, shrub density decrement, reduction in the relationship of live to dead tissue in the dominant grasses, increments of ephemeroids and reduction of frequency of big denuded patches. It was concluded that the increasing dominance of *Stipa speciosa* and *S. humilis*, with their opportunistic strategies (Soriano et al., 1980), the reduction in ephemeroid abundance and the increasing density of shrubs is related to this grazing history. Compared to the system before sheep introduction, present productivity is lower in periods of average or above-average precipitation (rain plus snow) and equal or perhaps higher during extreme drought periods. This is so because both ephemeroids and shrubs are strictly periodic in their annual cycle. They may contribute very little or no biomass if their growth period coincides with very dry conditions. On the other hand, the opportunist bunchgrasses do not become completely dormant during the annual cycle, even when the driest conditions occur (Soriano, 1978).

Range rehabilitation is, in general, very slow. Different features of this process have, as expected, different rates. Recovery of *Poa ligularis*, one of the best forage grasses in arid Patagonia, is generally accomplished with three or four years of rest. Seed reserves in the soil accumulate very slowly (Alippe and Soriano, 1978). One factor that probably does not favor recovery is seed germination; 17 dominant species out of 20 species tested lack dormancy mechanisms, which would adjust them to periodically favorable environmental conditions (Soriano, 1960).

In the central and southern part of the Santacruzian Subdistrict some coarse indicators of floristic and erosion character allow a tentative scheme of desertification trends. In Fig. 17.22, the principal vegetation types are represented and their relations according to the presumed trends of desertification.

During the last twenty years some seeding trials have been performed in several areas within the semi-desert (Pereyra and Abadie, 1966; Habit, 1973; Molina Sánchez, 1978). Generally speaking, little success has been obtained, although *Agropyron elongatum* and *Poa ampla* are considered to have some future on favorable sites. Sand dunes have been successfully stabilized in several locations with *Elymus arenarius* and *E. sabulosus* (J.M. Castro, pers. comm.).

ANIMALS (J.M. Gallardo, M. Rumboll, M. Canevari, P. Canevari and W.G. Vasina)

The Neotropical Region has been divided (Ringuelet, 1960) into three zoogeographical subregions: Guianan, Araucanian and Andino Patagonian. Patagonia is within the third subregion and has arid portions. Further subdivisions of the district are not justifiable with our present state of knowledge, but there are differences which will be considered later.

Zoogeographical indicators of this district are mammals such as the *mara* (*Dolichotis patagonum*) and *piche* (*Zaedyus pichyi*), birds such as the lesser rhea (*Pterocnemia pennata pennata*) and the Patagonian tinamou (*Tinamotis ingoufi*), and reptiles such as the *yarara ñata* (*Bothrops ammodytoides*), and *chelco* (*Homonota darwini*).

The number of vertebrates within this district is comparatively poor. Pampean and subandean elements diminish in number towards the south.

The mammalian, bird and insect faunas have apparently not evolved as closely with the vegetation as the reptiles have. This is probably due to smaller size and lesser mobility of the reptiles. For example, there are several species of the genus *Liolaemus* which are highly adapted to different combinations of vegetation and landscape (Gallardo, 1971).

Exceptions to the above could be demonstrated in the southern tip of Santa Cruz and the north of Tierra del Fuego where we find several species of mammals, birds and reptiles such as *Akodon lanosus*, *Geositta antarctica*, *Muscisaxicola capistrata*, *Pluvianellus socialis* and *Liolaemus magellanicus* that appear to be limited by vegetal characteristics.

Phytophagous mammals are few in this region. The greatest number are rodents of small size: *Akodon*, *Phyllotis* and *Reithrodon*. Only the Patagonian hare or *mara*, amongst these, must be considered as a species of greater size. The guanaco is the only large herbivore ungulate (Fig. 17.23).

Fig. 17.23. Guanaco (photo by M. Canevari).

The *huemul* (*Hippocamelus bisulcus*) formerly reached the coast in its distribution (Cabrera and Yepes, 1940), but nowadays is quite restricted to some places in the Cordillera.

Predator species are surprisingly numerous though their populations are as a rule low. These include the red fox (*Dusicyon culpaeus*), the grey fox (*Dusicyon griseus*), puma (*Felis concolor*), patagonian colocolo (*Felis colocolo*), Geoffroy's cat (*Felis geoffroyi*), ferret (*Lyncodon patagonicus*), skunk (*Conepatus humboldtii*) and a marsupial (*Lestodelphis halli*).

Due to the great differences in temperature and lack of other types of refuges, most of the animals use cave habitats, digging their own burrows or using those of other species. Also, due to this lack of refuges, the Chiroptera, so typical of the Neotropical Region, do not have many representatives in Patagonia. The most characteristic is *Histiotus montanus*, the long-eared bat, though other genera are found in the northern part of the region (Cabrera, 1957, 1960).

Amongst the birds, we find that the greater part of the species are cursorials, that is, they prefer to run away and not fly from predators. One of them, the lesser rhea or *choique*, is unable to fly. The Tinamiformes are represented here by the Patagonian tinamou (*Tinamotis ingoufi*) and the elegant tinamou (*Eudromia elegans*) (Fig. 17.24) which are fliers, but probably due to the strong winds of the area they prefer to move by walking.

Patagonia has plovers and sand pipers and numerous Passeriformes, Furnaridae (*Geositta*), Tyrannidae (*Neoxolmis*, *Muscisaxicola*) which are also mainly walkers. Other Passeriformes represented are mainly Mimidae, with one species, and several Fringillidae.

Birds of prey are plentiful, but none is exclusively Patagonian. Among the more important raptors are the buzzard eagle (*Geranoetus melanoleucus*), crested caracara (*Polyborus plancus*), great horned owl (*Bubo virginianus*), several falcons, hawks and owls. Two species of vultures and the Andean condor (*Vultur gryphus*) (Cathartidae), plus the crested caracara and the *chimango* (*Milvago chimango*) (Falconidae) are eaters of carrion.

The Patagonian lagoons, either salty or freshwater, have a plentiful avifauna nesting there. Most of the Anseriformes and shore-birds leave during the winter. The more characteristic of these are the

Fig. 17.24. Elegant tinamou (photo by P. Canevari).

TABLE 17.4

Some conspicuous invertebrates of Patagonia

COLEOPTERA

Tenebrionidae

Epipedonata lata
E. tricostata
Emallodera obesa
Nyctelia corrugata
N. fitzrovi
N. lateristriga

Curculionidae

Cylindrorhinus angulatus and another five species

Scarabaeidae

Taurocerastes patagonicus (tumblebug)
Tribostethus villosus
Trox hemisphaericus (in carrion)

Meloidae

Meloe sp.

LEPIDOPTERA

Agrostis saucia
Argopteron aureipennis
Argyrophorus argenteus
Carsia uniformata
Chlamidophora patagiata
Colias lesbia
Gelechia invenustela
Pelamia tehuelcha
Polythysana rubrescens
Satyrus chilensis
Synclerostola pampeana
Xylophasia offuscata

ORTHOPTERA

Acrididae

Bufonacris bruchi
B. terrestris
Clarazella patagonica
Dichroplus elongatus
D. maculipennis
D. vittatus
Eremopachys bergi
Papipappus clarazianus

Oedipodinae

Trimerotropis pallidipennis

ARANEIDA

There are several families of spiders in Patagonia; some of them are:

Theraphosidae

Grammostola cala

Argiopidae

Araneus cinnabarinus

Lycosidae

Alopecosa nigricans

Thomisidae

Petrichus niveus

Amyphaenidae

Gayenna strigosa

flamingo (*Phoenicopteus ruber*), the black-necked swan (*Cygnus melancoryphus*), upland geese (*Cloephaga picta*) and several other Anatidae, coots, grebes, plovers, sea gulls, etc. (Olrog, 1963). There are no aquatic mammals. The amphibian fauna is very poor. There is a species of *Bufo* and one leptodactylid (*Pleurodema bufonina*). Towards the north on the Somuncura Plateau, we find an endemic group, Telmatobiinae, with two species (Cei, 1969; Gallardo, 1970).

Of the few fishes present the genus *Percichthys* prevails in the greater part of the Patagonian rivers.

The native fauna of this region has been greatly affected by the introduction of domestic livestock and the European hare. The reduction in population levels of the large animals is due to a combination of modifications of vegetation, increased soil erosion, and hunting. Foremost among the animals affected is the guanaco (*Lama guanicoe*). Herds of thousands of these animals were seen by early European travelers. The groups of these animals rarely exceed thirty animals today. Since these animals are apparently highly adapted to the coarse grasses of this region, it is a pity that research has not yet been done to compare their diets, productivity and utilization with those of livestock.

Very little is known of insects in this region. All that can be offered at this time is a listing of some of the more conspicuous species (Table 17.4). Opportunities are great for more information and research on ecological controls and function.

REFERENCES

Alippe, H.A. and Soriano, A., 1978. La población de disemínulos en el suelo de un pastizal de *Stipa* en el oeste de Chubut. *Ecología*, 3: 133–137.

Archangelsky, S., 1958. Estudio geológico y paleontológico del Bajo de la Leona (Santa Cruz). *Acta Geol. Lilloana*, 2: 5–133.

Archangelsky, S., 1967. Estudio de la formación Baquero. Cretácico Inferior de Santa Cruz, Argentina. *Museo La Plata, Rev. N.S. Paleontol.*, 5(32): 63–171.

Archangelsky, S., 1974. Sobre la edad de la tafoflora de la Laguna de Hunco, Prov. del Chubut. *Ameghiniana*, 11: 413–417.

Archangelsky, S. and De la Sota, E., 1960. Contribución al conocimiento de las Fílices pérmicas de Patagonia extrandina. *Act. Geol. Lilloana*, 3: 85–126.

Archangelsky, S. and Gamerro, J.C., 1966. Spore and pollen types of the Lower Cretaceous in Patagonia (Argentina). *Rev. Palaebot. Palynol.*, 1: 211–217.

Ares, J., 1978. Un modelo de la dinámica de la biomasa aérea de tres especies dominantes en un pastizal patagońico. *Deserta*, 5: 54–56.

Auer, V., 1933. Verschiebungen der Wald- und Steppengebiete Feuerlands in postglazialer Zeit. *Act. Geogr.*, 5(2): 1–313.

Auer, V., 1951. Consideraciones científicas sobre la conservación de los recursos naturales de Patagonia. *Inf. Invest. Agric.*, No. 40–41.

Auer, V., 1956. The Pleistocene of Fuego-Patagonia, Part I: The ice and interglacial ages. *Ann. Acad. Sci. Fenn., Ser. A III*, 45: 1–226.

Auer, V., 1958. The Pleistocene of Fuego–Patagonia Part II: The History of the Flora and Vegetation. *Ann. Acad. Sci. Fenn., Ser. A III*, 50.

Auer, V., 1959. The Pleistocene of Fuego–Patagonia, Part III: Shoreline displacements. *Ann. Acad. Sci. Fenn., Ser. A III, Geol. Geogr.*, 6: 1–247.

Auer, V., 1965. The Pleistocene of Fuego–Patagonia, Part IV: Bog profiles. *Ann. Acad. Sci. Fenn., Ser. A III*, 80: 1–160.

Bandura, I., 1960. Suelos de la Patagonia y sus aptitudes para el riego. *Inf. Invest. Agric., Supl.*, No. 1: 164–166.

Berg, D.E., 1964. Krokodile als Klimazeugen. *Geol. Rundsch.*, (Stuttgart), 54: 328–333.

Berry, E.W., 1925. A miocene flora from Patagonia. *Johns Hopkins Univ. Stud. Geol.*, 6: 183–252.

Berry, E.W., 1928. Tertiary fossil plants from the Argentine Republic. *Proc. U.S. Natl. Museum*, 73(22): 1–27.

Berry, E.W., 1937. Contributions to paleobotany of South America. An upper Cretaceous flora from Patagonia. *Johns Hopkins Univ. Stud. Geol.*, 12: 11–31.

Berry, E.W., 1938. Tertiary flora from the Rio Pichileufú, Argentina. *Geol. Soc. Am., Spec. Pap.*, 12: 149.

Bertiller, M.B. and Beeskow, A.M., 1979. Las unidades florístico-fisonómicas de la vegetación de la Península Valdes y sus alrededores. *VII Reunión Argentina de Ecología, Resúmenes*, pp. 10–11.

Bertiller, M.B., Beeskow, A.M. and Del P. Irisarri, M., 1977. *Caracteres Fisonómicos de las unidades de vegetación del Chubut*. Puerto Madryn, 101 pp. (mimeogr.).

Bonaparte, J.F., 1978. *El Mesozóico de América del Sur y sus tetrápodos*. Opera Lilloana 26, San Miguel de Tucumán, 596 pp.

Braun-Blanquet, J., 1964. *Pflanzensoziologie*. Springer-Verlag, Berlin, 3rd ed., 875 pp.

Cabrera, A.L., 1953. Esquema fitogeográfico de la Republica Argentina. *Rev. Mus. La Plata (N.S.)*, 8: 87–168.

Cabrera, A., 1957. Catalógo de los mamíferos de América del Sur. *Rev. Mus. Arg. Cienc. Nat.*, 4(1): 1–307.

Cabrera, A., 1960. Catalógo de los mamíferos de América del Sur. *Rev. Mus. Arg. Cienc. Nat.*, 4(2): 308–732.

Cabrera, A.L., 1971. Fitogeografiá de la Republica Argentina. *Bol. Soc. Arg. Bot.*, 14(1–2)1: 42.

Cabrera, A.L., 1978. La vegetación de Patagonia y sus relaciones con la vegetación Altoandina y Puneña. In: C. Troll and W. Lauer (Editors), *Geological Relations Between the Southern Temperate Zone with the Tropical Mountains. Erdwissenschaftliche Forschung, XI*. Franz Seiner Verlag, Wiesbaden, pp. 329–343.

Cabrera, A. and Yepes, J., 1940. *Mamíferos Sudamericanos.* Ediar, Buenos Aires, 370 pp.

Caldenius, C., 1932. *Las glaciaciones cuaternarias de la Patagonia y Tierra del Fuego.* Dir. de Min. Geol. Hidrogeol. (Buenos Aires), 95.

Camacho, H.H., 1967. Las transgresiones del Cretácico Superior y Terciário de la Argentina. *Rev. Assoc. Geol. Arg.*, 22(4): 253–280.

Cappannini, D.A. and Auer, V., 1957. La erosión en la región de los lagos San Martín y Tar. *Inf. Invest. Agric.*, 111.

Cappannini, D.A. and Dominguez, O., 1959. *Suelos. La Argentina, Suma de Geografia.* Peuser, Buenos Aires.

Castro, J.M., 1965. Características, desarrollo e importáncia de la erosión en Santa Cruz. *Inf. Invest. Agric.*, 215: 64–66.

Cazau, L., 1972. Cuenca de Ñirihuau–Ñorquinco–Cushamen. In: A.F. Leanza (Editor), *Geologia Regional Argentina.* Acad. Nacl. Cienc., Córdoba (Argentina), pp. 272–740.

Cei, J., 1969. The Patagonian telmatobiid fauna of the volcanic Somuncura Plateau. *J. Herpetol.*, 3(1–2): 1–18.

Colbert, E.H., Cowles, R.B. and Bogert, C., 1946. Temperate tolerance in the American alligator and their bearing on the habits, evolution and extinction of the dinosaurs. *Bull. Am. Mus. Nat. Hist.*, 86: 327–374.

Crespo, J. and De Carlo, J., 1963. Estudio ecológico de una población de zorros colorados (*Dasycion culpaeus*) en el oeste de la Provincia de Neuquén. *Rev. Mus. Arg. Cienc. Nat., Ecol.*, 1(1): 1–55.

Darwin, C., 1840. *Journal of Researches into the Geology and Natural History of the Various Countries Visited by H.M.S. Beagle, 1832–36.*

Darwin, C., 1851. *Geological Observations on Coral Reefs, Volcanic Islands and on South America.* Smith, Elder & Co., London.

Ericksen, W., 1975. Disruptions in the ecosystems of the steppe and forest regions of Patagonia by climate and man. *Appl. Sci. Devel.*, 6: 127–142.

Etchevehere, P., 1971. *Mapa de Suelos Sistema 7a aproximación.* Centro de Investigaciones de Recursos Naturales, INTA, Buenos Aires.

Etchevehere, P. and Miaczynski, C.R., 1963. *Los suelos de Tierra del Fuego.* I.S.A., La Plata, 21 pp. (unpublished).

Fantini, A., 1954. *Levantamiento agrológico de la zona de influencia del Arroyo Picún Leufú (Neuquén).* I.S.A. (unpublished).

Feruglio, E., 1949–50. *Descripción Geologica de la Patagonia, Vol. I–III.* Dir. Grl. Yacim. Petrol. Fisc., Buenos Aires.

Flint, R.F. and Fidalgo, F., 1963. Geologia glacial de la zona de borde entre los paralelos 39°10′ y 41°20′ de latitud sur en la Cordillera de los Andes (Argentina). *Direc. Nacl. Geol. Min.*, 93.

French, N. and Sauer, R.H., 1974. Phenological studies and modeling in grasslands. In: H. Lieth (Editor), *Phenology and Seasonality Modeling. Ecological Studies, Analysis and Synthesis, 8.* Springer-Verlag, Berlin, pp. 227–236.

Frenguelli, J., 1953. La flora fósil de la región del alto Rio Chalia en Santa Cruz (Patagonia). *Not. Mus. La Plata (Paleontol.)*, 16(98): 237–257.

Gallardo, J., 1970. A propósito de los Telmatobiinae patagónicos. *Neotropica*, 16(50): 73–85.

Gallardo, J., 1971. Algunas ideas sobre la zoogeografica de los saurios patagónicos. *Rev. Mus. Arg. Cienc. Nat. Ecol.*, 1(4): 135–146.

Garleff, K., 1977. Höhenstufen der argentinischen Anden in Cuyo, Patagonien und Feuerland. *Göttingen Geogr. Abh.*, 68: 150 pp.

Groot, J.J. and Groot, C.R., 1964. Quaternary stratigraphy of sediments of the Argentine Basin. *Trans. N.Y. Acad. Sci.*, 26: 881–886.

Groot, J.J. and Groot, C.R., 1966. Pollen spectra from deep-sea sediments as indicators of climatic changes in southern South America. *Mar. Geol.*, 4: 525–536.

Gustafson, J. and Innis, G., 1973. SIMCOMP 2.1, User's Manual and maintenance Document. *U.S./I.B.P. Grassland Biome, Tech. Rep., Fort Collins, Colo.*, 217: 96 pp.

Habit, M.A., 1973. Informe sobre la investigación en la mejora de pasturas. *FAO-SF-ARG 65/514 Inf. Tecn.*, 165 pp. (mimeogr.).

Harrington, H.J., 1962. Paleogeographic development of South America. *Bull. Am. Assoc. Pet. Geol.*, 46(10): 1773–1814.

Hauman, L., 1920. Un viaje botánico al Lago Argentino. *An. Soc. Cient. Arg.*, 89: 179–281.

Herbst, R., 1965. La flóra fosil de la Formación Roca Blanca, Provincia de Santa Cruz, Patagonia. Con consideraciones geológicas y estratigráficas. *Opera Lilloana*, 12: 7–101.

Hünicken, M., 1955. Depósitos neocretácicos y terciarios del extremo sud-sudoeste de Santa Cruz. Cuenca carbonífera de Río Turbio. *Rev. Mus. Arg. Cienc. Nat. B. Rivadavia (Geol.)*, 4: 3–161.

Hünicken, M., 1966. Flora terciaria de los Estratos del Río Turbio, Santa Cruz (Niveles plantiferos del arroyo Santa Flavia). *Rev. Fac. Cienc. Exact., Fis. Nat., Univ. Cordoba, Ser. Cienc. Nat.*, 27(3–4): 139–227.

Hünicken, M., 1967. Flora terciaria de los Estratos de Río Turbio, Santa Cruz. (Niveles plantíferos de arroyo Santa Flavia.) *Rev. Fac. Cienc. Exact., Fis. Nat. Univ., Nacl. Cordoba*, 27: 139–262.

INDEC, 1969. *Censo Nacional Agropecuario.* Buenos Aires, 337 pp.

Ipucha Aguerre, J., 1966. La conservación de los suelos y el futuro patagónico. *Inf. Invest. Agrop.*, 226: 45–58.

Italconsult, 1960. *Plan de desarrollo agrícola de Valle de Viedma.* Italconsult, Rome, 78 pp.

Jacks, G.V., Tavernier, R. and Boalche, D.H., 1960. *Multilingual Vocabulary of Soil Science.* Land and Water Development Division, FAO, Rome.

Jones, T.P., 1961. *Patagonian Panorama.* The Outspoken Press, Bournemouth, 173 pp.

Kaasschieter, J.P.H., 1965. Geologia de la Cuenca del Colorado. *Segundas Jornadas Geol. Arg.* (*Acta Geol. Lilloana*, 7), 3: 251–269.

Klammer, G., 1964. Die Paläzoischen Eisenerze von Sierra Grande, Argentinien. *Z. Erzbergbau Metallhüttenwes*, 17: 534–541.

Laporte, O.J. and Coccimano, M.R., 1970. *Resultado de una experiencia con dos tipos de carga animal en el Desierto Patagónico.* INTA, Estacion Agropecuaria Trelew, 21 pp. (mimeogr.).

Leanza, H.A., 1973. Estudio sobre los cambios faciales de los estratos limítrofes Jurásico–Cretácico entre Loncopue y

Picun Leufú, provincia de Neuquén, República Argentina. *Rev. Assoc. Geol. Agrent.*, 28(2): 97–132.

Lesta, P.J., 1968. Estratigrafia de la Cuenca del Golfo San Jorge. *Terceras Jornadas Geol. Arg.*, 1: 251–289.

Lesta, P.J. and Ferello, R., 1972. Region extrandina de Chubut y norte de Santa Cruz. In: A.F. Leanza (Editor), *Geologia Regional Argentina.* Acad. Nacl. Cienc., Cordoba, pp. 601–653.

Lloyd, M. and Ghelardi, R.J., 1964. A table for calculating the equitability of species diversity. *J. Anim. Ecol.*, 33: 217–225.

Marcolín, A.A., 1973. *Cartografía de los principales grupos de suelos de Cholila y Epuyén (Chubut).* INTA, E.E.R.A. Bariloche. Informe Técnico.

Medan, D. and Tortosa, R.D., 1981. Nódulos actinomicorrícicos en especies argentinas de los géneros *Kentrothamnus, Trevoa* (Rhamnaceae) y *Coriaria* (Coririaceae). *Bol. Soc. Arg. Bot.*, 20 (in press).

Meigs, P., 1952. La répartition mondiale des zones climatiques arides et semi-arides. *C.R. Rech. Hydrol. Zone Aride*, UNESCO, pp. 208–215.

Menéndez, C.A., 1959. Die fossilen Floren Sudamerikas. In: E.J. Fittkau et al. (Editor), *Biogeography and Ecology in South America.* W. Junk, The Hague, pp. 519–561.

Menéndez, C.A., 1961. Flóra Cretácica de la Serie de Castillo, al S. del Cerro Cachetaman, Chubut. *Rev. Assoc. Geol. Arg.*, 14: 219–223.

Menéndez, C.A., 1966. Fossil Bennettitales from the Tico flora, Santa Cruz Province, Argentina. *Bull. Br. Mus (Nat. Hist.) Geol.*, 12: 3–42.

Menéndez, C.A., 1969. Die fossilen Floren Sudamerikas. In: E.J. Fittkau et al. (Editor), *Biogeography and Ecology in South America.* W. Junk, The Hague, pp. 519–561.

Menéndez, C.A., 1972. Paleofloras de la Patagonia. In: M.J. Dimitri (Editor), *La región de los bosques andino-patagónicos. Sinopsis general. Colecc. Cient. INTA*, 10: 129–165.

Mercer, J.H. and Sander, W., 1975. Pliocene and early Pleistocene glaciations in southern Argentina and the origin of the Patagonian gravel (abstr.). *VI Congr. Geol. Arg. Resumenes*, Bahia Blanca, 20 pp.

Molina Sánchez, D., 1976. Aspectos ecológicos pasturiles y crianza lanar. *Cienc. Inv.*, 32: 24–53.

Molina Sánchez, D., 1978. *Pasturas artificiales permanentes en la Provincia de Santa Cruz.* Rio Gallegos AER, 18 pp.

Monteith, N., 1972. Estudios sobre erosión en la Patagonia. *Informe final proyecto INTA–FAO para el Desarrollo Ovino en la Patagonia*, 62 pp. (unpublished).

Monteith, N. and Laya, H., 1970. Estado actual y programa general de la cartografía y clasificación de suelos en la Patagonia. *Rev. Agric. Noroeste Arg.*, 7(1–2): 385–388.

Morrison, J.J., 1917. *La ganadería en la región de las mesetas australes del territorio de Santa Cruz.* Thesis, Facultad de Agronomía y Veterinaria, Buenos Aires, No. 55, 172 pp.

Movia, C.P., 1972. Formas de erosión eólica en la Patagonia. *Photointerprétation*, No. 6/3. (Editions Technip, Paris).

Movia, C.P., 1973. *Relevamiento y cartografía de las formas de erosión eólica de la Patagonia.* Facultad de Agronomia, Buenos Aires (unpublished).

Movia, C.P., Soriano, A. and León, R.J.C., 1977. *Relevamiento de la vegetación de la Cuenca del Río Santa Cruz.* Informe, Inst. Nac. Cienc. Téc. Hidr., Buenos Aires, 72 pp.

Mueller-Dombois, D. and Ellenberg, H., 1974. *Aims and Methods of Vegetation Ecology.* Wiley and Sons, New York, N.Y., 547 pp.

Musters, G.Ch., 1873. *At. Home with the Patagonians.* John Murray, London, 2nd ed., 340 pp.

Nicora, E.G., 1978. Gramineae. In: M.N. Correa (Editor), *Flora Patagónica Vol. III.* Coleccion Cientifica, INTA, 563 pp.

Olrog, C., 1963. Lista y distribución de las aves argentinas. Opera Lilliana, 9. Tucumán, 377 pp.

Pereyra, J.A. and Abadie, C.A., 1966. Resultados preliminares de adaptación de especies forrajeras e implantación de pasturas en el oeste de Chubut. *Bol. Divulg., Centro Reg. Patag.*, No. 5: 36 pp.

Polanski, J., 1965. The maximum glaciation in the Argentine cordillera. *Geol. Soc. Am., Spec. Pap.*, 84: 453–472.

Ragonese, A.E. and Piccinini, B.C., 1969. Límite entre el Monte y el Semi-desierto Patagónico en las provincias de Río Negro y Neuquén. *Bol. Soc. Arg. Bot.*, 11(4): 299–302.

Ringuelet, R., 1960. Rasgos de la zoogeografía de la Argentina. *Sesiones científicas de Zoología, Tucumán, Physis Tomo XII*, No. 63: 151–170.

Roig, F.A., 1960. Bosquejo fitogeográfico de las Provincias de Cuyo. *Fac. Cienc. Agr., Mendoza, Publ.* No. 3: 33 pp. (mimeogr.).

Romero, E.J., 1978. Paleoecología y paleofitogeografía de las tafofloras del Cenofítico de Argentina y áreas vecinas. *Ameghiniana*, 15(1/2): 209–227.

Ruiz Leal, A., 1972. Los confines boreal y austral de las Provincias Patagónia y Central respectivamente. *Bol. Soc. Arg. Bot.*, 13(Suppl.): 89–118.

Russo, A. and Flores, M.A., 1972. Patagonia Austral Extra Andina. In: A.F. Leanza (Editor), *Geología Regional Argentina.* Acad. Nac. Cienc., Córdoba, pp. 707–725.

Rutland, R.W.R., Guest, J.E. and Crastley, R.L., 1965. Isotopic ages and Andean uplift. *Nature*, 208: 677–678.

Sala, O.E., Deregibus, V.A., Schlichter, T. and Alippe, H.A., 1981. Productivity dynamics of a native temperate grassland in Argentina. *J. Range Manage.*, 34(1): 48–51.

Schlichter, T., León, R.J.C. and Soriano, A., 1978. Utilización de índices de diversidad en la evaluación de pastizales naturales en el centro oeste de Chubut. *Ecología*, 3: 125–132.

Scillato Yane, G.J., Uliana, M.A. and Pascual, R., 1975. Un Megalonychidae (Edentata, Pilosa) del Plioceno de la Provincia de Río Negro (República Argentina). Su importancia bioestratigráfica y paleobiogeográfica. *VI Congr. Geol. Arg. Resumenes, Bahía Blanca*, 57 pp.

Shell Cia. Argentina de Petroleo, S.A., 1965. La palinología en la industria petrolera y algunos resultados palinológicos en la Argentina. *Act. II Jornados Geol. Arg.*, 3: 347–353.

Simpson, G.G., 1933. A new fossil snake from the *Notostylops* Beds of Patagonia. *Bull. Am. Mus. Nat. Hist.*, 67: 1–22.

Simpson, G.G., 1937. An ancient eusuchian crocodile from Patagonia. *Am. Mus. Nov.*, 965: 1–20.

Soriano, A., 1948. Las exploraciones botánicas en la Patagonia argentina. *Cienc. Inv.*, 11: 443–453.

Soriano, A., 1949. El límite entre las Provincias botánicas

Patagonica y Central en el territorio del Chubut. *Lilloa*, 20: 193–202.

Soriano, A., 1950. La vegetacion del Chubut. *Rev. Arg. Agron.*, 17: 30–66.

Soriano, A. 1956a. Los distritos florísticos de la Provincia Patagónica. *Rev. Inv. Agric.*, 10: 323–347.

Soriano, A., 1956b. Aspectos ecológicos y pasturiles de la vegetación patagónica, relacionados con su estado y capacidad de recuperación. *Rev. Invest. Agric.*, 10: 349–372.

Soriano, A., 1958. El manejo racional de los campos en Patagonia. *Inv. Invest. Agric.*, 124: 1–7.

Soriano, A., 1959. Síntesis de los resultados obtenidos en las clausuras instaladas en Patagonia en 1954 y 1955. *Rev. Agon. Noroeste Arg.*, 3(1–2): 163–176.

Soriano, A., 1960. Germination of twenty dominant plants in Patagonia in relation to regeneration of the vegetation. *Proc. 8th Int. Grassl. Congr.*, pp. 154–158.

Soriano, A., 1978. Ecological strategies in a semi-desert *Stipa* grassland in Patagonia (Argentina). *2nd Int. Congr. Ecol. INTECOL*, 1: 368.

Soriano, A. and Brun, J., 1973. Valoración de campos en el centro-oeste de la Patagonia: desarrollo de una escala de puntaje. *Rev. Inv. Agropecu., Ser. 2*, 10(5): 173–185.

Soriano, A. and Satorre, E., 1979. Magnitud y dinámica del sistema subterráneo de un pastizal de coirón amargo (*Stipa* spp.) de Río Mayo Chubut. *VII Reun. Arg. Ecol., Resúmenes*, pp. 69–70.

Soriano, A., Alippe, H.A., Sala, O.E., Schlichter, T.M., Movia, C.P., León, R.J.C., Trabucco, R. and Deregibus, V.A., 1976. Ecologia del pastizal de coirón amargo (*Stipa* spp.) del Sudoeste de Chubut. *Acad. Nacl. Agric. Vet.*, 30(11): 1–13.

Soriano, A., Sala, O.E. and León, R.J.C., 1980. Vegetación actual y vegetación potencial en el pastizal de coirón amargo (*Stipa* spp.) del SW. de Chubut. *Bol. Soc. Agric. Bot.*, 19(1–2): 309–314.

Soriano, A., Alippe, H.A., Sala, O.E., Schlichter, T.M., Movia, C.P., León, R.J.C., Trabucco, R. and Deregibus, V.A., 1976. Ecologia del pastizal de coirón amargo (*Stipa* spp.) del sudoeste de Chubut. *Acad. Nacl. Agric. Vet.*, 30(11): 1–13.

Stipanicic, P.N., 1967. Consideraciones sobre las edades de algunas fases magmaticas del Neopaleozoico y Mesozoico. *Rev. Asoc. Geol. Arg.*, 22(2): 101–133.

Stipanicic, P.N. and Reig, A.O., 1956. El Complejo porfirico de la Patagonia Extraandina y su fauna de anuros. *Acta Geol. Lilloana*, I: 185–297.

Story, R., Wijnhoud, S., Sourrouille, A. and Vega, A., 1970. *General Report On the Lands of the Río Gallegos–Río Turbio Area.* FAO–INTA Project, Tech. Rep., Bariloche, 16 pp. (unpublished).

Suero, T., 1961. Paleogeografía del Paleozoico superior en la Patagonia (Republica Argentina). *Rev. Assoc. Geol. Arg.*, 16(1/2): 35–42.

Tallarico, L.A. and Ferreiro, A., 1957. Suelo y erosión del faldeo del Cerro-Chenque (Comod. Rivadavia). *Inf. Invest. Agric.*, 112: 1–9.

Thorp, J. and Smith, G.D., 1949. Higher categories of soil classifications. Order, Suborder, and Great Soil Groups. *Soil Sci.*, 67: 117–126.

Ugarte, F.R., 1966. La cuenca compuesta carbonifero–jurásica de la Patagonia meridional. *Anal. Univ. Patagonia "San Juan Bosco", 2, Cienc. Geol.*, I(1): 37–68.

U.S. Department of Agriculture, 1960. *Soils Classification, A Comprehensive System, 7th Approximation.* Soil Survey Staff, Soil Conservation Service, Washington, D.C., 265 pp.

U.S. Department of Agriculture, 1967. *Supplement to Soil Classification System (7th Approximation).* Soil Survey Staff, Soil Conservation Service, Washington, D.C., 207 pp.

Vallerini, J. and Marcolin, A., 1975. Relevamiento de Suelos de la zona. Cordillerana de Patagonia. *Act. 7° Reun. Arg. Cienc. Suelo, Bahía Blanca. Inf. Invest Agrop. Supl.*, No. 33: 526–530.

Vessel, A.J., 1946. Soil association areas of Argentina and Chile. *Soil Sci. Soc. Am. Proc.*, 11: 464–473.

Volkheimer, W., 1964. Estratigrafía de la zona extrandina del departamento de Cushamen (Chubut) entre los paralelos 42° y 42°30′ y los meridianos 70° y 71°. *Rev. Assoc. Geol. Arg.*, 19(2): 85–107.

Volkheimer, W., 1967. La paleoclimatología y los climas del Mesozoico Argentino. *Soc. Arg. Min. Geol., Rev. Min. Geol. Mineral.*, 28: 41–48.

Volkheimer, W., 1969a. Esporas y granos de polen del Jurásico de Neuquén (República Argentina). II. Asociaciones microflorísticas, aspectos paleoecológicos y paleoclima. *Ameghiniana*, 6: 127–145.

Volkheimer, W., 1969b. Palaeoclimatic evolution in Argentina and relations with other regions of Gondwana. *Gondwana Stratigraphy, Earth Sciences*, 2: 551–587 (UNESCO, Paris).

Volkheimer, W., 1970. Neuere Ergebnisse der Anden-Stratigraphie von Sud-Mendoza (Argentinien) und benachbarten Gebieten und Bemerkungen zur Klimageschichte des südlichen Andenraums. *Geol. Rundsch.*, 59: 1088–1124.

Volkheimer, W., 1971. Aspectos paleoclimatológicos del Terciario Argentino. *Rev. Mus. Arg. Cienc. Nat. Paleontol.*, 1(8): 243–262.

Volkheimer, W. and Salas, A., 1975. Die älteste Angiospermem Palynoflora Argentiniens von der Typus-Lokalität der unterkretazischen Auitrin Folge der Neuquen-Beckens. Ihre mikrofloristische Assoziation und biostratigraphische Bedeutung. *Neues Jahrb. Geol. Paläontol., Monatsh.*, 1975(7): 424–436.

Volkheimer, W. and Sepulveda, E., 1976. Biostratigraphische Bedeutung und mikrofloristische Assoziation von *Cyclusphaera psilata* n. sp., einer Leitform aus der Unterkreide (Agrio-Folge) des Neuquen-Beckens (Argentinien). *Neues Jahrb. Geol. Paläontol. Monatsh.*, 1976(2): 97–108.

Walter, H., 1968. *Die Vegetation der Erde, 2.* Gustav Fischer-Verlag, Jena, 1001 pp.

Walter, H. and Box, E.O., 1976. Global classification of natural terrestrial ecosystems. *Vegetatio*, 32: 75–81.

Werner, F., 1921. Synopsis der Schlangenfamilie der Boiden auf Grundlage des Boulengerschen Schlangenkatalogs (1893–96). *Arch. Naturges.*, 87 A, 7: 230–265.

Willis, B., 1912. Recent surveys in Northern Patagonia. *Geogr. J.*, 40: 607–615.

Willis, B., 1914. Northern Patagonia. Character and resources. *Arg. Min. Publ. Works*, 1: 170 pp.

Zambrano, J.J. and Urien, C.M., 1970. Geological outlines of the basins in southern Argentina and their continuation off the Atlantic shore. *J. Geophys. Res.*, 75: 1363–1396.

Chapter 18

COMPARISONS AND CONTRASTS BETWEEN THE TEMPERATE DESERTS AND SEMI-DESERTS OF THREE CONTINENTS[1]

N.E. WEST

INTRODUCTION

Ecology can be approached in three basic ways — descriptive, functional and evolutionary (Krebs, 1978). The descriptive approach is mainly natural history; that is it deals with the "who, what and where" of ecosystems. An endless tabulation and narration of anecdotal information may ensue. The functional approach seeks out relationships of organisms to one another or organisms to environments. It deals with the "whens and hows". Proximate causes are sought. Because so many possible interrelationships exist, the choice of those to be studied becomes arbitrary. Some relationships are more aesthetically pleasing, fashionable or experimentally tractable than others. Importance of some interrelationships within an ecosystem may be trivial or irrelevant to economic needs to predict responses to perturbations. The evolutionary approach attempts to deal with the "whys" — to find ultimate causes — the historical reasons why natural selection has favored the particular adaptations we now see. Because evolution usually takes much time and is not as experimentally tractable, the evolutionary approach can easily degenerate into undisciplined speculations which can never be tested in the real world. Although each approach has its strengths and pitfalls, we need each to further our understanding of ecosystems.

Descriptive data are most abundant because they are easier to obtain and more people have been engaged in this type of activity for a longer period of time. Descriptive ecology still has to be done on those ecosystems that have still been relatively poorly explored by scientists. Functional studies usually use hypotheses generated by the observations of the descriptive ecologists. Evolutionary understanding becomes firm only after a great deal of descriptive and functional data becomes available.

This volume has dealt mainly with descriptive data because that is the principal kind of information that exists for these ecosystems. We have also reviewed the lesser amount of information available on functional ecology of the main components of these ecosystems. Finally, we have attempted evolutionary explanations where they seem plausible. We have included some speculation in the interest of promoting further research. I will now compare temperate desert and semi-desert ecosystems on all three continents via a mixture of the descriptive, functional and evolutionary approaches. Dissimilarities as well as similarities will be drawn out. Needs for further study will also be mentioned.

DEFINITIONS OF ECOSYSTEM TYPES

The definitions of temperate versus subtropical and desert versus semi-desert differ between ecologists and laymen over the three continents being considered (McGinnies, 1979). We have been concerned with mid-latitude areas that consistently experience winter temperatures below 0°C. Thus, snow commonly accumulates for at least a few days, but rarely lies for more than a few months on the ecosystems we considered.

The vegetation of the areas of our concern was best characterized by scattered shrubs with varying amounts of herbaceous species. The cause of the

[1] Manuscript completed July, 1980.

variation in the herbaceous components is largely due to differences in climatic controls and land-use history. The pristine vegetation of the wetter areas was steppe-like with perennial grasses and forbs declining and shrubs and annuals increasing with livestock grazing pressure. The drier areas apparently always had fewer perennial herbs. Grazing pressures have also accentuated the woody nature of the vegetation there.

We avoided use of the term "cold desert" because it is often used for polar areas. We rarely used the term "desert" unless we were dealing with an area largely devoid of vegetation. This is in keeping with the international use of the term and contrary to how Americans generally use the term. Most of the area we reviewed is termed semi-desert and shrub steppe. Semi-desert implies that there is some vegetation, but not enough to term it a grassland or shrubland. In those cases where total vegetal cover exceeded 50% of the ground surface and shrubs were still conspicuous, along with considerable herbaceous species, we employed the term "shrub-steppe". This will appear contradictory to some since steppes are usually grasslands. Since many of our ecosystems were intermediate between grasslands and deserts, we could find no better alternative.

A way of avoiding the above issues is to stress geographical descriptors. This was the emphasis taken in the Eurasian and Patagonian chapters. The reader can then decide from the descriptions of the climatic, edaphic, floral, and faunal patterns presented how to place particular areas within his preferred classification.

The classificatory approach used for Eurasia is first of all regional and then the multifactor approach developed by Walter (Walter and Box, 1976). For North America, stress was placed on potential natural vegetation (Küchler, 1970). This was facilitated by the greater abundance of relic areas due to a generally shorter, less intense history of land use compared to Eurasia. The subdivisions of Patagonia are geographic and floristic. There appears to have been scarcely enough scientific work there to do otherwise.

The particular approaches to classification chosen were the choices of the authors. Considering the relatively small amount of synecological research available for these ecosystem types, it would be much too premature to attempt more than the most general comparisons. A much more systematic pattern of investigation and publication of much unsummarized data would be necessary to do any more detailed comparisons.

The location of detailed research on all three continents is highly aggregated at a few locations. Co-ordinated ecosystem-level research has only been attempted within the last decade. The great demands for manpower and budgets to do this style of research has limited it to a very few locations. Thus, we have to resist overextrapolation from those few sites since the results may well be peculiar to those specific areas. We scarcely know how representative the sites may be. We do know that political and economic considerations often outweighed biological rationale in the selection of many sites.

Some desert regions have been much more studied than others that are larger, more productive per unit area and even scientifically more interesting. For instance, it appears that an inordinate amount of effort has been expended on sites where high potential for intensive agriculture exists. Governments have financially encouraged such research. This circumstance, however, introduces a possible bias of expecting too much from deserts. If sites with less favorable topography and soils could have been studied in proportion to their occurrence, we would understand better the characteristics of the prevailing types of land.

In all ecological research, the most important decision made is where the researcher stops his car (Ashby, 1935). If a road does not go into an area, its chances of being studied greatly diminish. Road builders favor gentle topography for ease of construction. We therefore often let the topography and the engineer predetermine where we obtain our data. Extraordinary effort has to be expended to overcome this bias. These problems occur in all ecosystem types, they are only more apparent in deserts where cultural features are so few and far between.

GEOGRAPHICAL LOCATION

All of the areas considered (map inside cover) are located between 28 and 50° latitudes. All but Patagonia lack contact with a sea coast. The average elevation of the Patagonian area is also lower than those areas in the Northern Hemisphere. Much of the western portions of the Eurasian deserts and semi-deserts are also found at relatively low elevations. The southern and eastern portions reach very high elevations. The North American representations of this type nearly all exceed 1000 m in elevation.

All of these regions are flanked by grassland areas on the north. Woodlands occur at higher elevations where mountains either fringe the areas or occur within them. Hot deserts occur to the south of the Northern Hemisphere areas and to the northwest of the Patagonian area. Western boundaries of the New World areas are created by large mountain ranges. Mountains fringe the eastern boundary of the North American area, but the sea is east of Patagonia. Patagonia is the only temperate desert in the world occurring on an eastern coast. The western boundary in Eurasia is fringed by grassland, the southern boundary by very high mountains, and the eastern margin by grasslands.

EXTENT

The Eurasian temperate deserts and semi-deserts are by far the largest in area (see Table 1.1). The sheer size of their area contributes to the opportunities for great diversity in all aspects of ecosystem structure and function. The North American temperate desert and semi-desert area is over five times as large as that in Patagonia. The number of pages which have been devoted to these regions roughly reflects these size differentials.

OWNERSHIP AND DOMINANT USES

Productivity of the natural ecosystems discussed in this volume is relatively low compared to forests and grasslands. Surface water is usually missing, making man less covetous of these lands than elsewhere. The earliest use was largely as hunting grounds and then as places to pasture livestock on an extensive basis. In the present century, the drilling of wells and the building of surface water works has allowed intensive, irrigation-based agriculture to develop where tillable soils existed. Such lands have passed from a grazing commons to private ownership in the Western Hemisphere. State ownership prevails in Eurasia.

In the more technologically developed nations much of the temperate deserts and semi-deserts are regarded as wasteland or at least a sacrifice area in terms of placement of "nuisance" activities. Thus, much of the area in both hemispheres is spotted with military development, testing, and deployment sites. Since these areas are sparsely populated by people and relatively few understand their fragility, governments commonly either ignore them or develop policy poorly suited to their management. Advocates of more rational use and development of these areas have difficulties in explaining the characteristics of these lands to legislators and administrators who usually grew up and live in more mesic and urbanized environments. The syntheses in this book will hopefully aid those from such contexts to understand these ecosystems better.

ECOSYSTEM STRUCTURE AND FUNCTION

Climate

In any continent-to-continent comparison consideration of climatic similarities and differences is fundamental. Climatic patterns are also the primary causes of ecosystem differences observed at the within-continent scale. This is especially the case for deserts and semi-deserts where moisture is the primary factor limiting biological productivity.

We have included numerous climatic diagrams to characterize representative stations. We have then discussed how these climates were related to soils, biota, and land management. It is now appropriate to consider the climatic similarities and differences between continents.

Cloudsley-Thompson and Chadwick (1964) list four primary causes of deserts, all of which relate to the climatic complex of aridity. These are (1) latitudinal position, (2) orographic effects, (3) position in continent, and (4) cold ocean currents.

The first cause is related to global air circulation. The majority of the world's deserts are found at latitudes of 25° to 35°. These are the "horse" latitudes where high pressures dominate and warm air usually descends. The dry, stable air masses there resist convective currents. This is the major cause of the "hot" deserts covered in the companion volume edited by Evenari and Noy-Meir (in press).

The deserts discussed here are at higher latitudes

and thus in the belt of the westerly winds that bring moisture from the oceans. In the case of the Western Hemisphere, we have huge mountain ranges in the path of the westerlies. Thus, as moist air is forced up on the windward flanks of these cordilleras, much orographic precipitation is created. The airstream flowing down the lee slope of the mountains expands and warms, taking up moisture rather than yielding much precipitation. In North America there is also a mountain range on the eastern side of the desert region which lessens the monsoonal influence in the late summer and early fall. Similarly an even higher mountain barrier exists on the southern fringe of the Eurasian desert region.

The third cause of climatic aridity is however much more important in Eurasia. The interiors of continents are well known for their climatic extremes. They are drier because the chance of moist oceanic air reaching them is diminished by sheer distance. General lack of cloud cover and low humidity in the air allow temperatures to fluctuate greatly on both a diurnal and a seasonal basis. Absence of large water bodies to store much heat and release it more slowly than earth or rock also contributes to the temperature fluctuations. It is therefore not surprising that indices of continentality are much higher in Eurasia than in North America and Patagonia (Major, 1963). We also see such gradients within continents. For instance, the Wyoming Basin, U.S.A., is both warmer in the summer and colder in the winter on the average than the "High Desert" around Burns, Oregon (Fig. 13.4) which is about 1500 km closer to the Pacific. Burns, however, can experience greater short-term changes than Riverton, Wyoming, as storm fronts pass through the area fresher from the Pacific.

The fourth cause of deserts is operative only for Patagonia. There the presence of the cold Falkland Current accentuates the aridity of the region by inhibiting landward movement of moist air currents. The proximity of the ocean to Patagonia also contributes to flattening of both diurnal and seasonal temperature curves.

Walter (1979) has abundantly pointed out that altitude is not equivalent to latitude in its climatic and thus biotic effects. Eurasia and North America have numerous mountain ranges rising within the perimeters of their arid to semi-arid regions. These mountains frequently have very dry conditions from base to summit. Temperatures drop in accordance with the adiabatic lapse rate. We thus have dry, cold mountain deserts and semi-deserts in the Northern Hemisphere. These "orobiomes" are much more abundant and have been much more studied in the U.S.S.R. (Chs. 7 and 9) than in the U.S.A. (Ch. 13). Any comparison of biota, soils, etc., between continents should be cognizant of these elevational variants within the separate regions.

Another important distinction is the west-to-east decline in concentration of winter precipitation in Eurasia and North America. The westernmost temperate deserts and semi-deserts in that Hemisphere receive nearly all of their precipitation in the winter from cyclonic storms centered around low pressure cells moving out of the west, whereas the southern and eastern margins of their arid to semi-arid regions receive half or more of their precipitation from summer convectional storms developed in the monsoonal pattern and moving up from the south. Those areas in between tend to get precipitation from both sources. This helps increase the plant diversity and biotic community productivity, all other environmental controls being equal.

Wind is a strong environmental feature of all the deserts and semi-deserts discussed. In addition to major storms, there is frequently great differential heating of bare areas and downslope movements to replace the rising air. Convectional cells can develop into spiraling winds that carry mineral and organic particles high into the sky. Such winds develop most prominently in North America where desert valleys and plateaus are commonly separated by mountains. Wind, however, more consistently blows in one direction (usually from the west) on the plains-type topography typical of Eurasia and Patagonia. In the latter instances wind is a major erosional agent, forming vast dune fields. The presence of more dunes in Eurasia and Patagonia however, is probably due at least as much to the differing geological history of these areas as to wind.

Geology

The diversity of bedrock geology of the three continents varies mostly with relative area. The huge area in Eurasia has virtually every type of rock outcropping within its desert region. As with the other continents, the regolith (unconsolidated surface) is of more interest in explaining differences in biota and influencing land use. Large portions of the desert areas of all three continents were covered by seas and lakes in the past. All have areas which have not recently drained to the ocean. These circumstances have led to heavy mineral deposition in their soils. Clays and silt-sized soil particles accentuate aridity by impeding infiltration and retaining salts in exchangeable forms. Sands and gravels are effectively wetter because of the inverse texture principle (Noy-Meir, 1973). Rocky areas are, however, more resistant to erosion, forming mountain massifs, ridges, plateaus, and hogbacks. Sands have the potential to form dunes, if poorly vegetated.

Topography

Most temperate desert and semi-desert areas occur on plains and plateaus. These landforms occur at the lowest elevations. This can be from near sea level to about 400 m in Patagonia, and about 500 m from near the Aral Sea to 3000 to 4000 m on the syrts of the Pamir or in the high valleys of Tibet. The North American desert valley floors occur at about 1000 to 2000 m. Playas (salines, salt flats) are common in basins with internal drainages (bolsons). Dune fields are common where river deltas and beaches occurred along ancient shorelines of inland seas and lakes. Badlands occur where shales outcrop. Where smaller mountains occur within the desert matrix, arid conditions can be found right up to the ridgeline. A wide variety of mountain landforms have desert to semi-desert ecosystems, particularly where rain shadows are accentuated in Central Asia.

Soils

Soils on the three continents could probably be classified into the same taxa of a given system of classification. The system used by the government of the U.S.A. has been used in Patagonia with apparent success. The different terminology commonly used to describe Eurasian soils makes this transferability somewhat speculative there. It is possible that different periods and intensities of livestock grazing on the three continents may have caused degradation of the surface part of profiles enough to cause differing matches of climatic conditions to soil taxonomic characters.

The deserts and semi-deserts of all three continents have extensive saline depressions in the centers of endorheic basins with fine-textured soils and permanent to ephemeral water tables at or near the surface. These variously go by the names: salt flat, salina, sebkha, chott, or kavir. Where these depressions have more moderately salinized soils they are called playa, takyr, pan or clay flats. Above the depressions where salts are better leached, the zonal soils can develop on areas with gentle topography and medium surface texture. Such soils are known as sierozem, brown desert or aridisols.

All three continents also have some stone pavement, vesicular and microphytic crusts developed in their temperate desert regions. Unlike the hot deserts where red prevails, the color of most temperate desert and semi-desert soils is grayish brown. The few cases of reddish to yellowish soil are due to situations where rocks of older age are present. There the iron and manganese of these parent materials were probably oxidized in deserts of Permian times.

Sand dunes are rare in North American temperate deserts, more frequent in Patagonia and quite common in Middle Asia. This relates to the presence of the Tethys Sea in Cretaceous times and its slow retreat to the Aral and Caspian remnants of today. Many deltaic, beach line and other alluvial deposits are the source areas for the sand dunes. Sand dunes in Patagonia are located along the huge rivers flowing eastward from the Andes. These waterways carried even more water and thus coarse material in Pleistocene times. Wind blew the materials from the braided stream channels. Loess forms the bulk of the surface of the plains there and at some semi-arid places in North America and Eurasia. The sands have not moved very far from the rivers in most cases. The large rivers flowing through the North American temperate desert region are usually deeply entrenched in canyons; thus there has not been as much opportunity for sand to blow out onto plains or plateaus.

Primary producers

Floristics

The floras of the Northern Hemisphere temperate deserts and semi-deserts are strikingly similar at the family and generic levels. This is largely due to the long connection that the North American and Eurasian continents have had through geologic time. It was not until the Pleistocene Epoch that North America and Greenland finally separated from Eurasia. Then in the Pleistocene there were several periods during which a land bridge existed between Alaska and Siberia (Beringia). Considering the recency of this connection, it is surprising that floristic differences do exist. If the Asteraceae and Chenopodiaceae were not so dominant and noted for their rapidity of evolution (Stebbins, 1975), we would expect even more similarity at the specific level.

It is easier to point out the major plant genera that are different than those that are similar. The differences seem to be concentrated along the southern boundaries of both desert regions. This is probably related to influx there of groups with southern affinities. These probably had less chance of surviving passage across land bridges at higher latitudes. Examples that come to mind are *Coleogyne*, *Tetradymia*, *Hilaria* and *Aristida* in North America and *Haloxylon*, *Calligonum*, *Reaumuria*, *Anabasis* and *Amygdalus* in Eurasia.

Some groups that have mostly woody species on one continent show a preponderance of herbaceous species elsewhere. For instance, there are numerous woody *Astragalus* and *Salsola* species in Eurasia, whereas most *Artemisia* species there are herbaceous or half-shrubs, some being cut for hay (Petrov, 1979a). In North America, there are few woody *Astragalus* species but the *Artemisia* species are mostly woody. Eurasia also seems to have more cushion-plants represented. These trends may be related to climatic and land-use differences. The climates of the deserts and semi-deserts of Eurasia are more rigorous, especially in regards to temperature fluctuations. Higher-elevation deserts and semi-deserts are also more abundant there. This combined with a much longer period of uncontrolled livestock grazing may account for adaptations favoring survival in more stressful environments.

The present environments, however, do not tell the whole story. For instance, the *Artemisia* species of the two continents have differing secondary compounds. Those in North American taxa seem more repellent to browsers. This may be related to the once higher densities of browsers in North America (Martin, 1973).

There are many weedy species in North America that came from Eurasia. Most of these are winter annuals. This group was poorly represented in pre-Columbian America. The Eurasian species have quickly expanded, apparently because they are more able to utilize resources of water and nutrients at colder temperatures. By constructing root systems over the winter, they can grow much earlier than the native perennial herbaceous species. The same strategy may eventually explain why the Eurasian perennial grasses have been so successfully transported to North America, and why there has been so little transfer of North American plants to Eurasia. Revegetation efforts in Eurasia have utilized native species almost entirely (Petrov, 1979a).

Patagonian deserts and semi-deserts have a quite different non-gramineous flora from those in the Northern Hemisphere. *Stipa* and *Festuca* are common to the semi-deserts of all three continents, but the shrubs and forbs of Patagonia are mostly cushion plants belonging to the Asteraceae, Verbenaceae, and Rubiaceae. The favoring of that life form may be related to the nearly constant, strong winds which create much transpiration stress. The taxonomic uniqueness, however, may be more related to length of separation of the South American continent from the rest of the world's land mass. The earliest splitting of Gondwanaland began nearby (Axelrod, 1979). Additionally, during most of the past 65 million years, South America was an island continent (Simpson, 1980). It has only been between the last 3 to 8 million years that land bridges to North America were reestablished, but much tropical forest intervened between the areas of interest to us here.

None of the three areas is particularly rich in flora, considering the size of the areas involved. This is probably related to the climatic severities experienced in these environments. If this is so, we would expect comparatively more taxa per unit area from Patagonia > North America > Eurasia. Unfortunately, data are unavailable to allow separation of montane and aquatic taxa. Distribution maps are not yet available to decide the affinities of most species. Progression to quantitative comparisons of floristics will require accumulation of much more information on distribution and habitat preferences before such ideas can be adequately addressed.

Vegetation physiognomy

The physiognomy of the temperate desert and semi-desert vegetation in Eurasia and North America is also strikingly similar. Low bunch grasses and small shrubs or half-shrubs prevail in both areas. Since Eurasia has more area and thus climatic and topographic variation, the size ranges of these forms tends to be greater. There are numerous tall shrubs in the lower, southern extent of Eurasian deserts. Many of them are surrounded by phytogenic hillocks of sand.

Microphytic soil crusts are not mentioned much in connection with Eurasian uplands but are abundant on the salinized depressions (takyry) and deltaic areas. The greater abundance of microphytic crusts on upland sites on North America may be due to a much shorter history of livestock grazing.

The prevalence of tall tussock grasses and cushion plants in Patagonia may be related to the strong winds and the comparatively more equable temperatures of that region. Vegetal cover is also apparently higher on the average there, again perhaps most related to comparative climatic equability.

Cover of vascular plants in Eurasian and North American temperate deserts and semi-deserts varies from none on heavily salinized soils to almost 100% where the zono-ecotone with steppe prevails.

Associations or vegetation gradients of strikingly similar plant communities are developed in the Northern Hemisphere versions of the ecosystem types being considered. The patterns of distribution are highly related to the complex gradient of elevation, climate, topography, and soils. Both continents have vegetation-less salt pans, hygro-halophyte-dominated areas around the salt pans where high water tables exist, xerohalophytic zones directly above, and glycophytic shrublands in the uplands. Progressively more perennial grasses and forbs appear as the average annual precipitation increases. Both areas also have *Tamarix* and *Populus* riparian gallery forests along rivers with permanent flow. Open woodlands occur directly above most of the most mesic shrub steppe variants. The only major differences are the more abundant areas of sand dunes in the Karakum and Kyzylkum, and the vast areas of desert pavement in the Gobi. What part the greater climatic extremes and a longer period of livestock use have played in the formation of this difference is impossible to say because of the absence of relic areas in Eurasia.

The more recent role of livestock in changing the character of Patagonia has apparently not been thoroughly examined. Indication of these pressures are considerable areas of sand dunes that have been rapidly spreading outward from the rivers in more recent times.

Phenological patterns

Eurasian and North American desert and semi-desert plants show similar phenological patterns. The western portions of both areas are dominated by winter recharge of soil moisture. Plant growth is concentrated in the short period of spring when temperatures allow plants to utilize that soil moisture. Winter annuals get an advantage over the grasses and forbs by germinating in the fall and growing their root systems throughout the winter. Summers and falls are dry and most plants are quiescent then.

In the eastern portions of the semi-arid to arid parts of both continents the importance of monsoonal rainfall causes a switch to dominance by warm-season plants. This influence is so strong in Central Asia that very few annuals are able to flourish (Petrov, 1979b).

Patagonia lacks the monsoonal influence; thus, phenological patterns are generally like those of the western portions of the Northern Hemisphere areas under consideration.

Phytomass

Phytomass is so highly variable as to virtually defy generalization between continents. Sites with none to 5 t ha^{-1} of phytomass exist within one landscape complex, relating largely to effective soil moisture and disturbance history.

Primary productivity

Extreme variation also exists in primary productivity. The greatest primary productivity within the desert regions of all three continents exists in the riparian forests and fresh-water marshes. There solar radiation is high because of relatively few clouds, and water and nutrients are not limiting. Only the period of cold temperatures constrains the annual period of growth.

Water tables also strongly affect the desert shrublands, but their effects are modified by soil texture and salinity. Coarse-textured upland soils with salinity leached from the surface horizons have the most productive upland communities. Temperature compensates for moisture input at the highest elevations with comparatively less moisture lost to evapotranspiration. If the net water supply is not high enough to allow grasslands or forests to develop, and temperatures are not too cold to prevent desert taxa having at least a short growing season, then montane versions of deserts and semi-deserts may be found, particularly in Eurasia.

Consumers

The faunistic dissimilarities between the three regions are comparatively greater than for the flora. Some of the most conspicuous animals are not even in the same families, let alone genera.

The very long period of heavy use by hunting and domestic grazing animals has probably caused the extinction of some Eurasian forms. However, the advent of man to the New World was coincident with the demise of many forms of animals that continue to survive in Eurasia, albeit in reduced numbers (e.g., horses, asses, camels).

There have been many exchanges of animals between all three continents, especially during the past 65 million years (the age of mammals). Some very interesting patterns have been described to us by paleontologists.

The exchange of mammals between Eurasia and North America has been nearly continuous, with North America being the place of origin of cameloids and equids, and Eurasia originally producing the bovids and cervids. The region of dominance and diversity of each has switched through time. Man could probably do much more to test the present adaptability of forms on the two continents. For instance, America lost most of its desert and semi-desert browsers in the period between 10 000 and 50 000 years ago. Now that livestock have triggered reductions in perennial grasses and increases in brush, it would seem plausible to try the semi-domesticated saiga antelope for meat production on North American temperate deserts and semi-deserts. This seems especially worthwhile since the demise of the sheep industry has caused many graziers to attempt to use cattle on ranges poorly adapted to their use.

The mammals of Patagonia were once more diverse than we presently find them. This is related to absence of continental land connections between South America and other continents during most of the Tertiary Period. Its fauna thus evolved in splendid isolation (Simpson, 1980). As a result, weird and curious animals evolved that were unique to the continent. These included giant sloths, tank-like glyptodonts, and sabertooth marsupials. This isolation was not complete, however, with some arrivals of caviomorphs and marsupials in the Late Eocene. The greatest exchange began taking place, however, about 8 million years ago when the intermingling of the long-separated North and South American land faunas was made possible by the emergence of land in Central America. The Panamanian land bridge was fully in place 3 million years ago and interchange has continued to the present.

It is tempting and almost unavoidable to conjecture that the disappearance of some native South American groups was a consequence of the interchange. For instance, placental carnivores (dog, cats) appeared and marsupial carnivores became extinct. The fossil records are too incomplete to infer competitive exclusion (Simpson, 1980).

Patagonia is left today with one cameloid, the guanaco, a small suite of rodents, dogs, and cats. It appears that birds may fill part of the niches used by mammals in similar habitat on the other two continents. For instance, the large flightless bird, the rhea, is a browser; and ground-dwelling parrots are major granivores. The heteromyids, so abundant in the temperate deserts of the Northern Hemisphere, are lacking in Patagonia but replaced by the birds and a few caviomorphs.

The avifauna of dryland areas in temperate Eurasia and North America is characterized by a large galliform each (sand grouse and sagegrouse), and a large group of fringillids, with the omnivores being the most dense and present for the longest period of the year. The raptor populations can be fairly abundant, largely following the abundance of small mammals. The avifauna of the driest portions of Patagonia also have large galliforms (tinamous). The fringillids and raptors are similar in appearance and diet to those in the Northern Hemisphere.

The herpetofauna of the temperate deserts of each continent is quite unique with few similarities at the generic level. This is probably due to the relatively low mobility of these creatures, particularly their abilities to disperse long distances across seas, mountain ranges, and through colder environments. Nevertheless parallel evolution has produced similar forms such as small insect-eating lizards, larger lizards and snakes that prey on them, small mammals and nestling birds. A herbivorous tortoise occurs in the warmest versions of the temperate deserts of both northern continents. The Patagonian herpetofauna is very limited.

The invertebrate fauna of all these deserts is very poorly known. These creatures provide a focus for future research of a highly interesting nature. Only in North America has it even been observed what a few major insect species may be doing to change vegetation and accelerate decomposition and nutrient cycling. We know virtually nothing about spiders, mites, ticks, solpugids, and other non-insect invertebrates in any of the temperate deserts. We can hopefully begin training and supporting a new generation of ecologically trained entomologists to begin correcting this vast deficiency in our knowledge.

Faunomass inventories have rarely been undertaken in any temperate-desert site. Similarly, productivity of native animal populations is known only for a few nuisance-type species such as hares in North America. Productivity of entire animal communities at temperate-desert and semi-desert sites remains the task of future ecologists.

Decomposers

Because they are individually small and inconspicuous, the decomposer organisms have been poorly studied in most desert ecosystems. The few reviews of these components and their functions (Friedmann and Galun, 1974; Fuller, 1974; Binet, 1978; West, 1980) have concentrated on hot deserts. Virtually all attention has been placed on soil micro-organisms, and hardly any on epiphytic or epizootic organisms and their activity.

Because the organisms and their spores can be wafted across vast distances, there is a high degree of micro-biological taxonomic similarity in the microflora of temperate deserts of the Northern Hemisphere. Probably the same microflora also exists in Patagonia. Studies of such organisms, however, apparently remain to be done there.

Because of the importance of micro-organisms in nutrient cycling, these parts of the ecosystem merit much more study on all continents. Discovery of allelochemic interactions between higher plant litter and certain critical processes like nitrogen fixation and nitrification in North America (West and Skujins, 1978) indicate that explanations as to why shrubs dominate and herbaceous forms only slowly recover when grazing pressure is released may be found when higher plant–microfloral interactions are more thoroughly studied. Overall productivity of these ecosystems, particularly during periods of above-average precipitation, may be limited by the lag in litter decomposition and thus conversion of critical nutrients from organic to inorganic forms.

Microfauna of these ecosystems have been scarcely studied but investigations in more mesic ecosystems indicate that their role in litter comminution may be essential in accelerating nutrient cycling.

ECOSYSTEM INTERACTIONS

Attempts to study entire biotic communities and ecosystems at one location have only barely begun in the temperate deserts and semi-deserts. We have more or less complete structural inventories available at only a few locations. The stress in such descriptive studies has usually been on higher plants and vertebrates. Because of the complexity of such investigations, a teamwork approach has to be mounted. The taxonomic difficulties can easily frustrate work with invertebrate and microfloral aspects of biotic communities. Aggregation by functional groups is suggested as a way around this impasse.

The only places where teams have been mounted to study functional as well as structural aspects of temperate-desert ecosystems are at the Repetek Station in the Karakum Desert of Turkmen S.S.R. (Rodin, 1979) and in salt-desert shrub types at Curlew Valley and the Desert Experimental Range in Utah. Even there the scope of the investigations has not included all possible, or even important parts of the ecosystem, and synthesis of the information is not at all close to a satisfying level of realization. These areas represent only a small fraction of the area which we would like to understand better and eventually predict responses for.

The generally low productivity of these ecosystems, the sparsity of scientists interested in them and thus the modicum of attention they have received compared to more productive, mesic ecosystems makes it presently difficult to generalize about food webs, trophic pyramids, and other topics related to energy flow. Similarly, nutrient cycles have been worked out in detail at only a few sites. More complex feedback mechanisms have been scarcely addressed at all.

Only the environmental controls of climate, soils, fire, and livestock grazing on major plants are known at all well. The effects of these controls on lesser known plants, vertebrates, invertebrates, and microbiota are very poorly known. Therefore, processes such as succession can be described but not well predicted. Stability and resilience are also poorly understood in most temperate deserts and semi-deserts. These are all limitations to improving the scientific basis of land management.

LAND USE

All of the world's temperate deserts and semi-deserts have been extensively modified by man's activities. The oldest and most pervasive impact has been through domestic livestock. Such use still contributes about 70% of the economy of the Mongolian People's Republic (Petrov, 1979b). The Eurasian continent has been grazed for millennia, whereas the New World temperate-desert areas have only seen intensive use for about the past 120 to 150 years. Because heavy use in Eurasia began before written history, it is very difficult to document changes triggered by livestock impacts. In North America there exist written records and relic areas from which pristine conditions can be deciphered, at least for a few ecotopes. It is thus possible to see the enormous changes that can occur within a few decades. Vegetation and larger wild animal populations have changed drastically. The woody, spiny and poisonous plants have been favored. Animals with wider dietary spectra have replaced those with more restricted dietary and behavioral preferences.

Although some rapid and pronounced changes in erosion, vegetation and vertebrates are documented in historical times in Patagonia, the gentle topography of that region has not lent itself to preservation of pristine tracts. Some details of retrogressional sequences are thus not available.

THE FUTURE

Growing human populations across the world and the need to grow more food, fiber, and hides for trade to pay for imports of petroleum, metals or manufactured goods do and will continue to involve these temperate-desert and semi-desert regions in more than regional and national economies. Greater productivity will require more intensive management. The harshness of most of the lands considered will probably limit management procedures to better selection of species, breeds, timing and intensity of grazing use and better herding, fencing, salting, and water development. Nevertheless, the sites with higher potential will see attempts at prescribed burning, brush control, and tillage and planting of more palatable and resilient species. For example, the longer grazing history of Eurasia has apparently resulted in species of *Agropyron* and *Elymus* which are more resistant to livestock use than the native species in the same genera in North America. Americans have tilled degraded rangelands and planted Eurasian grasses, and obtained 10- to 20-fold increases in livestock grazing capacity on some sites. Although large areas of monocultures of these grasses will probably never be established because of concern for pest outbreaks and desirability of native species for wild animals, there are many more Eurasian plant species which the Americans would like to try. Such plantations are likely to allow at least seasonal reductions of livestock grazing on adjacent native rangeland. Grazing systems can then be developed to enhance their productivity for wild animals, water, and other values the public perceives besides livestock.

The erosive action of wind and water are highly recognizable in the temperate deserts and semi-deserts. The reduction of vegetal cover by livestock has led to serious erosion by both wind and water in these areas. The production of water from these lands is not as important as the production of sediment. Particulate matter fills up dams and other man-made systems for water control and conveyance. After the nutrient-rich topsoil has been lost, it may take centuries to restore. Management strategies thus need to be developed to maintain desirable levels of soil protection where ecosystem productivity is presently high and devise ways of improving those lands that have been degraded in the past. There is much evidence on all continents that processes of xerification are recognized and are being addressed. Further efforts to separate the effects of natural geologic erosion from man-induced erosion will aid us in deciding where our efforts should have the highest priorities.

Although livestock grazing is by far the most common economic use of the lands considered here, we can anticipate increasing pressure for alternative prevailing uses. Energy and mineral exploration has been accelerating. The mines and wells, plus the rights of ways they require, will require increasing land disturbance. We have barely begun the necessary research on how to reclaim these areas, especially in the more arid sectors.

Political leaders usually emerge from more populous, mesic areas. Having no experience with deserts and semi-deserts, they regard them as either wastelands or highly durable ecosystems because of the degree of climatic severity. This attitude, plus the generally low populations of desert areas and thus minor political representation, leads to a colonial attitude toward the arid regions (Box, 1978). Many "nuisance"-type developments thus are proposed for such areas. Examples of such activities include military maneuvers, weapons development and testing, and large coal-fired and nuclear power plants. As the world becomes more populous, hungry, materialistic, and possibly bellicose, we can only expect more desert land to be diverted for such intensive, single use. The desert ecologist is thus faced with problems of trying to predict what will change in temperate deserts and semi-deserts with a variety of new and intended uses. Our present degree of understanding of these ecosystems scarcely allows us to do more than speculate on what could happen. Hopefully, some more researchers will be able to find the incentive and support to join the ranks of those who appreciate the vast expanses of temperate deserts and semi-deserts.

REFERENCES

Ashby, E., 1935. The quantitative analysis of vegetation. *Ann. Bot.*, 49: 779–802.

Axelrod, D.I., 1979. Desert vegetation, its age and origin. In: J.R. Goodin and D.K. Northington (Editors), *Arid Land Plant Resources.* ICASALS, Texas Technical University, Lubbock, Texas, pp. 1–72.

Binet, P., 1978. *Short-Term Dynamics of Minerals in Arid Ecosystems.* US/IBP Desert Biome, Utah State University, Logan, Utah, 33 pp.

Box, T.W., 1978. *The Arid Lands Revisited: 100 Years After John Wesley Powell.* 57th Annual Faculty Honor Lecture, Utah State University, Logan, Utah, 30 pp.

Cloudsley-Thompson, J.L. and Chadwick, M.J., 1964. *Life in Deserts.* G.T. Foulis and Co., London, 218 pp.

Evenari, M. and Noy-Meir, I., in press. *Hot Deserts and Arid Shrublands. Ecosystems of the World, 12.* Elsevier, Amsterdam, in press.

Friedmann, E.I. and Galun, M., 1974. Desert algae, lichens, and fungi. In: G.W. Brown, Jr. (Editor), *Desert Biology, II.* Academic Press, New York, N.Y., pp. 166–212.

Fuller, W.H., 1974. Desert soils. In: G.W. Brown, Jr. (Editor), *Desert Biology, II.* Academic Press, New York, N.Y., pp. 32–101.

Krebs, C.J., 1978. *Ecology: The Experimental Analysis of Distribution and Abundance.* Harper and Row, New York, N.Y., 2nd ed., 678 pp.

Küchler, A.W., 1970. Potential natural vegetation (map at scale of 1:7 500 000). In: U.S. Geol. Survey, *The National Atlas of the U.S.A.* U.S. Government Printing Office, Washington, D.C., pp. 90–91 (revision of 1964 map at scale of 1:3 168 000).

Major, J., 1963. A climatic index to vascular plant activity. *Ecology*, 44: 485–497.

Martin, P.W., 1973. The discovery of America. *Science*, 179: 969–974.

McGinnies, W.G., 1979. Arid-land ecosystems — common features throughout the world. In: D.W. Goodall and R.A. Perry (Editors), *Arid-land Ecosystems, 1.* Cambridge University Press, Cambridge, pp. 299–316.

Noy-Meir, I., 1973. Desert ecosystems: Environment and producers. *Ann. Rev. Ecol. Syst.*, 4: 25–51.

Petrov, M.P., 1979a. Land use of semi-deserts in the U.S.S.R. In: B.H. Walker (Editor), *Management of Semi-Arid Ecosystems.* Elsevier, Amsterdam, pp. 301–327.

Petrov, M.P., 1979b. Land use of semi-deserts in central Asia. In: B.H. Walker (Editor), *Management of Semi-Arid Ecosystems.* Elsevier, Amsterdam, pp. 329–349.

Rodin, L.E., 1979. Productivity of desert communities in central Asia. In: D.W. Goodall and R.A. Perry (Editors), *Arid-Land Ecosystems, 1.* Cambridge University Press, Cambridge, pp. 273–298.

Simpson, G.G., 1980. *Splendid Isolation: The Curious History of South American Mammals.* Yale University Press, New Haven, Conn., 266 pp.

Stebbins, G.L., 1975. Shrubs as centers of adaptive radiation and evolution. In: H.C. Stutz (Editor), *Proc. Symposium and Workshop on Wildland Shrubs.* Brigham Young University, Provo, Utah, pp. 120–140.

Walter, H., 1979. *Vegetation of the Earth and Ecological Systems of the Geo-Sphere.* Springer-Verlag, New York, N.Y., 2nd ed., 274 pp.

Walter, H. and Box, E.O., 1976. Global classification of natural terrestrial ecosystems. *Vegetatio*, 32: 75–81.

West, N.E., 1980. Nutrient cycling in desert ecosystems. In: D.W. Goodall and R.A. Perry (Editors), *Arid-Land Ecosystems, 2*. Cambridge University Press, Cambridge, pp. 301–324.

West, N.E. and Skujins, J. (Editors), 1978. *Nitrogen in Desert Ecosystems*. Dowden, Hutchinson and Ross, Stroudsburg, Pa., 307 pp.

AUTHOR INDEX[1]

[1] Page references to text are in roman type, to bibliographical entries in italics.

SYSTEMATIC INDEX

GENERAL INDEX

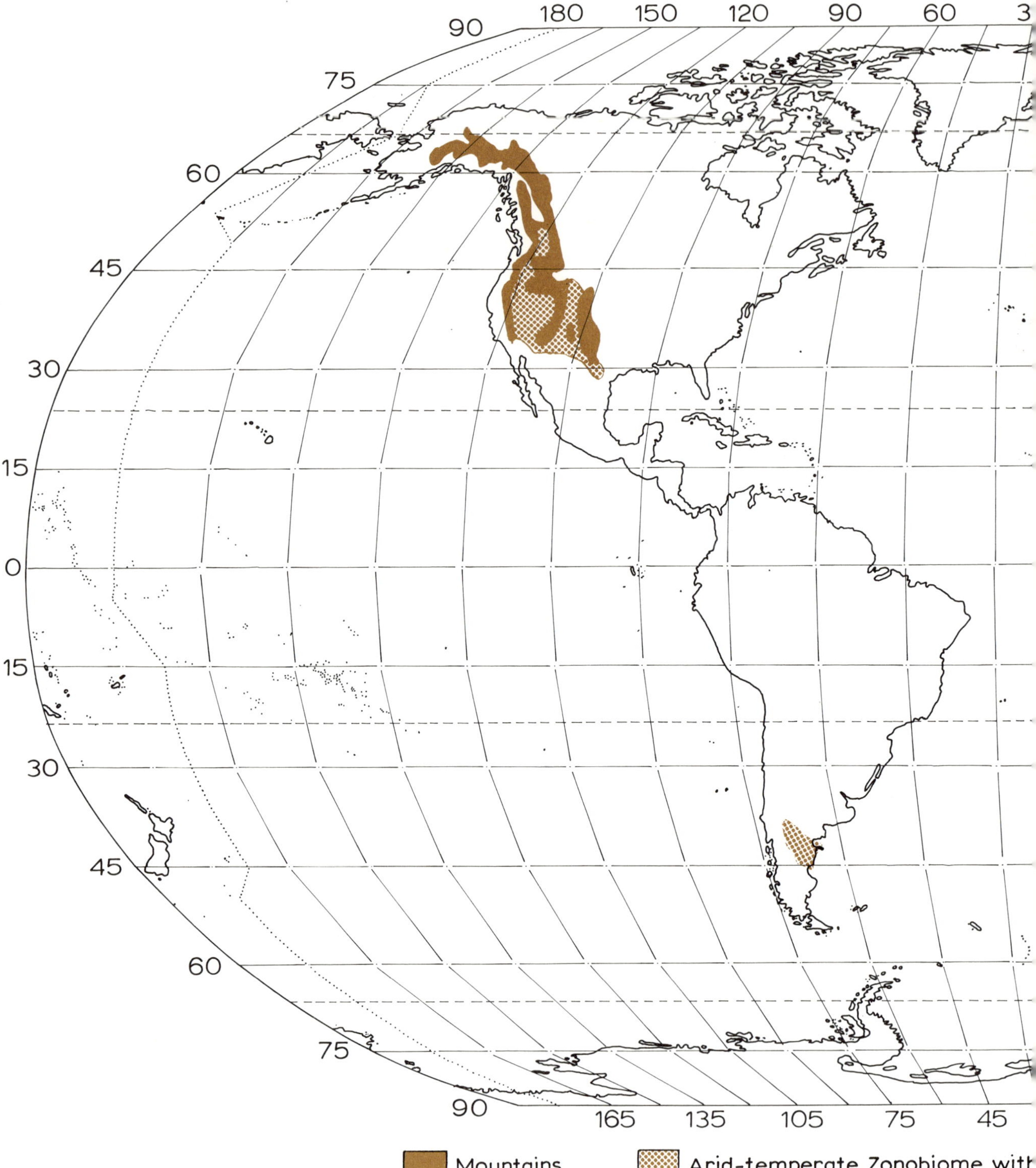

90
180
150
120
90
60
75
60
45
30
15
0
15
30
45
60
75
90
165
135
105
75
45
Mountains
Arid-temperate Zonobiome wit